Iterative and Self-Adaptive Finite-Elements in Electromagnetic Modeling

For a complete listing of the *Artech House Antenna Library*,
turn to the back of this book.

Iterative and Self-Adaptive Finite-Elements in Electromagnetic Modeling

Magdalena Salazar-Palma
Tapan K. Sarkar
Luis-Emilio García-Castillo
Tanmoy Roy
Antonije Djordjević

Artech House
Boston • London

Library of Congress Cataloging-in-Publication Data
Iterative and self-adaptive finite-elements in electromagnetic modeling/ Magdalena Salazar-Palma...
[et al.].
p. cm.
Includes bibliographical references and index.
ISBN 0-89006-895-X (alk. paper)
1. Electromagnetism—Mathematics. 2. Numerical analysis. 3. Finite element method.
4. Iterative methods (Mathematics)
I. Salazar-Palma, Magdalena.
QC760.I88 1998
530.14'1—dc21 98-6553
CIP

British Library Cataloguing in Publication Data
Iterative and self-adaptive finite-elements in electromagnetic modeling
1. Electromagnetic waves 2. Finite element method
I. Salazar-Palma, Magdalena
621.3
ISBN 089006895X

Cover design by Elaine K. Donnelly

685 Canton Street
Norwood, MA 02062

International Standard Book Number: 0-89006-895-X
Library of Congress Catalog Card Number: 98-6553

10 9 8 7 6 5 4 3 2 1

Et veritas liberabit vos...
And the truth shall make you free...
(From the Gospel by John 8:32)

As long as I live so long do I learn
(Ramakrishna)

To our families

Table of Contents

PREFACE

Nowadays it is difficult to analyze electromagnetic properties of the devices utilized by modern technology without any computational tools. Computational electromagnetics deals essentially with the solutions of Maxwell's equations either in the time or in the frequency domain using a differential or an integro-differential equation. Each technique has some advantages and disadvantages.

In this book we focus our attention on the solution of the differential form of Maxwell's equations. The methodology used is the finite element method. In spite of many advantages that will be highlighted throughout the book, the method has primarily two drawbacks.

The first drawback is related to setting up the solution process by subdividing the computational region of interest into finite elements and approximating the unknown over each element by a linear combination of basis functions. The accuracy of the computed solution depends on how the region has been subdivided and how well the approximation has been chosen to expand the unknowns over the finite domains. In short, the mesh (distribution and size of the finite elements and/or the type of approximation) should be well adapted to the behavior of the electromagnetic field. However, its behavior is not known *a priori*. Thus, it is highly desirable to establish some *a posteriori* error estimates or indicators for the solution and then use that information in a self-adaptive procedure that will automatically refine the mesh, or, alternately increase the order of the approximation utilized.

One of the goals of this book is to illustrate the flexibility and the efficiency of self-adaptive procedures. This is implemented in the solution of various electromagnetic field problems. A discussion of the appropriate formulation and the choice of basis functions for each type of problem is presented first. Then the error estimates or the indicators are developed. The theoretical rate of convergence of the self-adaptive mesh algorithm is compared with the computed rate of convergence to illustrate that via appropriate automatic meshing it is possible to achieve the maximum rate of convergence of the finite element method for the problems of interest. This is a unique feature of the book.

A second problem associated with the finite element method is formulating a procedure to solve open region problems where part of the domain can reach up to infinity. Since it is practically impossible for arbitrary shaped structures to generate finite elements up to infinity, a procedure must be set up that terminates the mesh at a finite distance. A radiation condition based on Green's function technique has been used to terminate the mesh very close to the boundary without sacrificing accuracy through an iterative procedure. In this procedure the sparseness of the finite element system matrices is preserved. It is hoped that with these two innovations the cause of the finite element method can be further enhanced.

In this book no attempt is made to compare various techniques since the efficiency of a particular method depends on the nature of the problems. However, results using other methods are utilized to validate the results for the finite element method.

The first chapter provides a summary of the historic development of the finite element method and its use for the solution of electromagnetic field problems.

The second chapter gives a general description of the method in the context of stationary linear elliptic second-order boundary value problems. It describes the mathematical details of how the variational methods of approximation (the weighted residual method and variational principles) are utilized to develop the finite element formulations. The description of various types of basis functions is given in addition to the role of different boundary conditions on the formulation. Special attention is paid to div-conforming and curl-conforming two-dimensional and three-dimensional elements. Results for interpolation errors and *a priori* rates of convergence are summarized. Finally, the chapter closes by listing some commercially available computer programs.

The third chapter illustrates how to apply the finite element method for the analysis of various waveguiding problems utilizing both the quasi-static approximation and the full-wave solution. In both cases several formulations are discussed together with the different finite elements used for their discretization: Lagrange and div-conforming types for the quasi-static approach, and Lagrange and Lagrange/curl-conforming types for the full-wave analysis. The validation of the different finite element procedures is performed by applying them on some benchmark problems. An analysis of the rate of convergence is also carried out.

The fourth chapter presents a self-adaptive mesh procedure which is applied to solve the problems discussed in the previous chapter. It is shown how to use the method to automatically refine the mesh and reach the maximum theoretical rate of convergence. After applying the method to a coarse mesh, the error estimators or indicators are computed for each element, and then the elements in regions with large errors are refined. The method is applied again to the refined mesh. The process continues until some pre-specified error criteria are met.

The fifth chapter provides additional examples for both the quasi-static and full-wave analysis of waveguiding structures utilizing the various types of basis functions, and the self-adaptive mesh procedure.

Chapter 6 develops a procedure for terminating the mesh for two-dimensional open region problems close to the actual physical boundary utilizing a mesh termination scheme without loosing the sparsity of the finite element system of equations. Examples are given for electrostatic problems, and transverse magnetic (TM) and transverse electric (TE) scattering cases.

Finally, Chapter 7 deals with the solution of three-dimensional closed and open region problems. A discussion of what constitutes a spurious solution is made with the idea of distinguishing it from an actual solution. A survey of the different procedures utilized in the literature is given, and the role of vector finite elements (two-dimensional and three-dimensional curl-conforming elements) is stressed. Examples dealing with the analysis of discontinuities in waveguiding structures and responses from waveguide cavities are presented. Then, the methodology of chapter six is extended to the solution of three-dimensional open-region problems. Several examples of electromagnetic radiation and scattering from three-dimensional conducting and dielectric structures have been analyzed to validate the accuracy of this method.

These seven chapters are followed by seven appendixes that provide the derivations of the various mathematical formulas used to compute the parameters of interest illustrated through various examples in this book.

The first two appendixes describe some of the fundamental mathematical definitions related to the concepts of various spaces and convergence.

In Appendix C, the derivation of the basis functions used in the book is carried out. It also gives some mathematical details related to implementing finite element procedures.

Appendix D provides a summary of Maxwell's equations in a nonhomogeneous-anisotropic region under various differential formulations of the waveguiding problems that are being solved in this book.

Appendixes E and F describe the various integral weak formulations for the quasi-static and full-wave analyses, respectively, and their discretization by the finite element method using various types of basis functions.

Finally, Appendix G provides the mathematical equations used for the computation of the various error estimates and error indicators used for the refinement of the mesh in the self-adaptive procedure.

The various mathematical and computational aspects of the finite element method are documented in the bibliography through a large number of papers and textbooks.

It is hoped that this book is going to provide an insight on the various procedures for enhancing the efficiency of the finite element method. In addition, computation of various error criteria are presented that can be utilized in an automatic mesh refinement algorithm, which is an important topic for this method. A large number of references are given to refer interested readers to sources of additional information.

Magdalena Salazar-Palma
Tapan Kumar Sarkar
Luis-Emilio García-Castillo
Tanmoy Roy
Antonije Djordjević

ACKNOWLEDGMENTS

We gratefully acknowledge Professors Félix Pérez-Martínez (Polytechnic University of Madrid, Spain) and Carlos Hartmann (Syracuse University, New York, USA) for their continued support in this endeavor.

We are extremely grateful to Dr. Edmund K. Miller (Santa Fe, New Mexico, USA), and Professors Vladimir Petrovic (University of Belgrade, Yugoslavia) and Luis Ferragut-Canals (University of Salamanca, Spain) for their advice and technical review of the text.

Thanks are due to Dr. José-Félix Hernández-Gil (Madrid), and Professor Carmen Sánchez-Ávila (Polytechnic University of Madrid) for their advice and helpful discussions.

This book would not have been possible without the research works of Mr. José-María Recio-Peláez, Mr. Fernando Blanc-Castillo, Mr. Antonio Pérez-Yuste, and Mr. Luis-Miguel Garrido-Díaz (Polytechnic University of Madrid).

We would like to express our gratitude to Professors Chaowei Su and Jinbiao Zhang (Syracuse University) for a careful reading of the manuscript and for making suggestions for improvements.

Our gratitude goes also to Ms. Dina Galiatsatos (Madrid) for translating some of the chapters of the book into English from the original Spanish version.

Ms. Katia Cervantes-Navarro, Mr. José-Luis Vázquez-Roy, Mr. Mustapha Maphoudi, and Professor Javier Gismero-Menoyo (Polytechnic University of Madrid) have devoted some of their time helping us in many ways. A word of thanks to all of them.

Thanks are also due to Ms. Brenda Flowers and Ms. Maureen Morano (Syracuse University) for their expert typing of the manuscripts.

We would also like to express sincere thanks to Ms. Ellen Magid, Ms. Chamanti Mandadi, Mr. Sheeyun Park, Mr. Wonwoo Lee, Mr. Jinwan Koh, Mr. Srikant Nagraj, and Mr. Kyunjung Kim (all from Syracuse University) for their help with the book.

List of Figures

Chapter 3

Chapter 4

Chapter 5

$h = 6.35$ mm, $W = 3$ mm, $t = 0.3$ mm, $\varepsilon_r = 9.8$. 321
Figure 5.33 Microstrip line, $\theta = 90°$. (a) Initial mesh of one half of the structure. (b) Zoom-in over the last mesh (fifth one, 501 nodes). 321
Figure 5.34 Microstrip line, $\theta = 135°$. (a) Initial mesh of one half of the structure. (b) Zoom-in over the last mesh (sixth one, 537 nodes). 321
Figure 5.35 Dispersion of the normalized propagation constant. Symbols: +, $\theta = 45°$; *, $\theta = 90°$; ×, $\theta = 135°$. Lines: [Alam et al. 1994b] and [Olyslager et al. 1993]. Markers: present method. 322
Figure 5.36 Dispersion of the normalized characteristic impedance. Symbols: +, $\theta = 45°$; *, $\theta = 90°$; ×, $\theta = 135°$. Lines: [Alam et al. 1994b] and [Olyslager et al. 1993]. Markers: present method. 322
Figure 5.37 Results for the single microstrip line. (a) Normalized propagation constant: —, even mode; --, odd mode. (b) Characteristic impedance. Symbols: ×, present method; □, Libra-Super Compact; lines, [Slade and Webb 1992]; +, SDA (same reference). 323
Figure 5.38 Asymmetric microstrip line with $w = h = 0.635$ mm and $\varepsilon_r = 9.7$. 324
Figure 5.39 Effect of the side walls; $d_1/h = d_2/h = (a/h\text{-}w/h)/2$, $b/h = 7$. Symbols: —, [Wu and Chang 1991]; •, present method with nine mesh refinements. 324
Figure 5.40 Effect of the top conductor; $a/h = 10$, $d_1/h = d_2/h = 4.5$. Symbols: —, [Wu and Chang 1991]; •, present method with nine mesh refinements. 324
Figure 5.41 Total attenuation constant. Symbols: ×, present method; •, [Kitazawa 1993]; o, measurements. 325
Figure 5.42 Bilateral circular finline. 326
Figure 5.43 Mesh generated for the analysis of the structure of Figure 5.42. 326
Figure 5.44 Dispersion curve for the structure of Figure 5.42. Symbols: —, Eswarappa et al. [1989]; --, present method. 327
Figure 5.45 Double semi-circular ridge waveguide. Mesh with (a) 561 nodes and (b) 2145 nodes. 327
Figure 5.46 Electromagnetic field of the dominant TE mode: (a) transverse electric field, (b) transverse magnetic field, and (c) longitudinal magnetic field. 328
Figure 5.47 Coplanar line, $c_1/a = 0.4$; $c_2/a = 0.1$; $c_3/a = 0.5$; $d_1 = d_2 = d_3 = d_4 = d_5 = b/5$. . . 329
Figure 5.48 Square of the normalized propagation coefficient for the fundamental quasi-TEM mode. Symbols: *, $\varepsilon_r = (3;5;3)$, $\mu_r = (5;1;1)$; +, $\varepsilon_r = 3$, $\mu_r = 5$; ×, $\varepsilon_r = 3$, $\mu_r = (5;1;1)$; × (lower line): $\varepsilon_r = 3$, $\mu_r = 1$. Lines, [Mazé-Merceur et al. 1993]. Markers: present method. 330
Figure 5.49 Initial mesh for the coplanar waveguide analysis. 330
Figure 5.50 Analysis of the coplanar waveguide for $\varepsilon_r = (3;5;3)$, $\mu_r = (5;1;1)$. $\bar{E}$ formulation: (a) zoom-in over the refined mesh, (b) transverse electric field, and (c) zoom-in over the strip area. $\bar{H}$ formulation: (d) zoom-in over the refined mesh, (e) transverse magnetic field, and (f) zoom-in over the strip area. 331
Figure 5.51 Dispersion of the relative effective dielectric constant of the fundamental mode of the microstrip line with anisotropic substrate. Symbols: •, $\bar{E}$ formulation; ×, $\bar{H}$ formulation; —, [Cano et al. 1989]. 332
Figure 5.52 Geometry of the finline. 333
Figure 5.53 Dispersion of the effective dielectric constant of the fundamental mode of the anisotropic finline. Symbols: —, TLM [Bulutay and Prasad 1993]; •, FEM, $\bar{E}$ formulation; --, SDA [Bulutay and Prasad 1993]; ×, FEM, $\bar{H}$-formulation. . . 333
Figure 5.54 Suspended coplanar waveguide. 334
Figure 5.55 Dispersion of the normalized propagation constant of the first two modes of the structure shown in Figure 5.54. Symbols: —, [Lyons et al. 1993]; •, even mode [Mirshekar-Syahkal 1990]; Δ, odd mode [Mirshekar-Syahkal 1990];

Chapter 6

Chapter 7

Appendix A

Appendix C

List of Tables

Chapter 4

Chapter 5

Chapter 6

Chapter 7

Appendix G

Chapter 1

Introduction

1.1 NUMERICAL METHODS IN ELECTROMAGNETICS

Modern technology greatly depends upon the engineering analysis and synthesis of electromagnetic systems. These, ultimately, are based upon obtaining accurate solutions of Maxwell's equations for the system of interest. On the other hand, the increased research and development in applications as diverse as radiocommunication, microwave and millimeter wave telecommunication, radar, medical diagnosis and therapy, and radiation and scattering, for example, have led to systems that are becoming more complex each day. In many cases, a single device exhibits a very complicated structure that involves a number of conductors, dielectric, and semiconductors of arbitrary shapes and of a complex physical nature. Their expensive fabrication technologies preclude the possibility of modifying a device if its performance is not within the specifications of the designer. Therefore, extremely accurate characterization methods are required to model the structures. The possibility of obtaining such methods is a challenging test for engineers and mathematicians nowadays because, as a consequence of the complex nature of any of the modern electromagnetic devices, their analysis does not lead to closed-form expressions. Thus, numerical methods giving approximate solutions within engineering accuracy are required. It has been said that to develop a numerical method means to apply a small number of general and relatively simple ideas and to combine them with one another in an inventive way. Then, with the help of a computer, one can solve a complicated problem with a large amount of simple numerical calculations.

It has been said by M. N. O. Sadiku [1992; Ch. 1] that the need for numerical solutions of electromagnetic problems is best expressed in the words of D. T. Paris and F. K. Hurd [1969]: "Most problems that can be solved formally (analytically), have been solved." In fact, before electronic computers were available, the objective of scientists and engineers was focused on obtaining analytic solutions through procedures like separation of variables, series expansion, conformal mapping, integral solutions, and variational methods. Many efforts were also devoted to formulating approximate or asymptotic solutions as accurately as possible, making use of static field approximations and variational expressions, for example.

In the field of diffraction problems, analytic solutions for particular structures were given as early as 1881 by Lord Rayleigh (J. M. Strutt). Approximate and asymptotic solutions were devised by G. Kirchoff (1891), A. Sommerfeld (1896), P. Debye (1908), and in the late 1930s and 1940s by a number of researchers like V. A. Fock, W. Pauli, and C. L. Pekeris for example [Umashankar and Taflove 1993; Ch.1].

At microwave and low frequencies, during the last quarter of the nineteenth century, analytic studies regarding wave propagation were performed by H. Hertz, O. Heaviside, Lord Rayleigh, and other researchers on transverse electromagnetic (TEM) transmission lines and metallic waveguides [Bryant 1984], [Packard 1984]. During the first two decades of the twentieth century, surface waves and dielectric rods were analyzed [Oliner 1984], [Packard 1984]. Progress during these years and the next two decades was tremendous and a number of researchers made significant discoveries like G. Marconi, N. Tesla, A. S. Popov, J. C. Bose (wireless communication, radio telegraph, millimeter wave experiments), J. A. Fleming (diode), L. De Forest (triode amplifier), E. H. Armstrong (superheterodyne receiver), J. D. Kraus (directional antennas), R. H. Varian and S. F. Varian (magnetron), and G. C. Southworth (re-discovery of waveguides). Nevertheless, the first large-scale effort to solve practical electromagnetic engineering problems was undertaken at the Massachusetts Institute of Technology (M.I.T.) Radiation Laboratory during World War II. The Nobel prize winner, J. Schwinger, and other researchers applied the variational method to arbitrary structures. The solutions to many of these problems were published in 1951 in a book by N. Marcuwitz entitled *Waveguide Handbook*.

The nature of computation started to change during the 1940s with the invention of computers. The first electromechanical computers were shortly substituted by electronic computers. From then on, numerical analysis was possible and many researchers started to work in related fields like numerical methods and equation solvers. Although the *Finite Difference* (FD) method was used in one form or another almost since the beginning of differential calculus, the work of an engineer, R. V. Southwell [1940, 1946], was the first real application to problem solving. During the 1940s and 1950s, a number of engineers in the field of structural mechanics were setting up the basis for the *Finite Element Method* (FEM). During the 1950s, planar techniques were developed for microwave circuits and the miniaturization of electronic devices was becoming a reality [Howe 1984], [Barret 1984], [Cohn and Levy 1984], [Levy and Cohn 1984]. Different planar and non planar dielectric waveguides were also introduced for millimeter waves [Wiltse 1984]. By the end of this decade the laser was invented [Chang 1984], [Wiltse 1984] and research work in related areas, like optical communications, optic fibers, and optoelectronics was subsequently undertaken. The growing complexity of electromagnetic devices was requiring new tools for their accurate design. In 1954 V. H. Rumsey gave birth to the reaction concept and R. F. Harrington applied a variational method based on such concept to a number of practical electromagnetic problems [Harrington 1961]. A tremendous expansion in electromagnetic engineering due to industrial and defense needs took place in the early 1960s. The first attempts toward monolithic devices also were done during this decade, although true monolithic devices were not available until the 1970s [McQuiddy et al. 1984]. Concurrently, computers with substantial speed and storage capabilities were available for the first time.

These circumstances prompted a number of researchers to investigate the use of numerical methods in solving electromagnetic field problems. However, the first attempt to use numerical solutions can be found in the classical work by J. C. Maxwell (1831-1879) himself in 1864. Also, finite difference approximations for derivatives were already used by L. Euler in 1768. Other early attempts may be traced back to 1908 (when C. Z. Runge applied FD to a simple two-dimensional Poisson's equation), 1910 (when L. F. Richardson published a similar work together with the earliest application of iterative methods to the solution of continuous problems by FD), and 1918 (when H. Liebmann used FD to solve for the electrostatic potential from which the field of a TEM mode is derived and also suggested an improved method of iteration). The first proofs of mathematical convergence of this numerical procedure were presented by J. LeRoux in 1914, and H. B. Phillips and N. Wiener in 1923. The celebrated 1928 paper of R. Courant, K. Friedrichs, and H. Lewy is considered as the birth of modern theory of numerical methods [Ames 1992; Ch. 1]. A pioneering work using the integral equation method can be traced back to 1951 when D. K. Reitan and T. J. Higgins used it to evaluate the static charge of a cube [Morita et al. 1990].

However, it was not until computers were readily available that researchers solved real practical problems. T. W. Edwards, K. K. Mei, and J. Van Bladel developed an integral equation methodology. They used subsectional basis functions and a point-matching method to compute the scattering from rectangular cylinders. Others, like M. G. Andreasen, F. K. Oshiro, and J. H. Richmond, generated similar methods [Umashankar and Taflove 1993; Ch.1]. In 1965 O. C. Zienkiewicz and Y. K. Cheung used FEM to solve two-dimensional (2D) field problems derived from Poisson's equation. This marked the extension of this method to nonstructural problems. A related paper for three-dimensional (3D) electrical problems followed [Zienkiewicz et al. 1967]. The FEM starts from a differential formulation although it discretizes an integral form obtained through any variational method. In 1967 R.F. Harrington worked out a systematic, functional-space description of electromagnetic problems, starting from integral equations and using the reaction concept. He called this technique the *Method of Moments* (MOM) after the name used by the Russian authors L. V. Kantorovich and V. I. Krylov [1964], and Y. U. Vorobev [1965] who developed a similar method several years before. The first papers on the application of FEM to electromagnetic wave problems appear in 1968 [Ahmed 1968], [Arlett et al. 1968]. Since then, FD, MOM, FEM, and many other numerical methods have been developed and refined as they were applied to a wide variety of electromagnetic problems for different frequency bands.

Nowadays one of the main objectives of researchers in the electromagnetic field is to devise *computer aided design* (CAD) tools that make use of numerical methods for the analysis, design and optimization of electromagnetic devices in a fast and reliable way. The use of CAD tools is a common practice nowadays for an electrical engineer. Their development has greatly facilitated an engineer's work, who may simulate, optimize, and correct a given design before its fabrication and experimental characterization. To make that possible, each structure involved in the design must be conveniently characterized by means of some type of circuit parameters (for example, any multiport parameter) or an equivalent circuit. The complexity of any active or passive structure in present day technology precludes the use of simple models, like in circuit simulators, that perform a number of approximations and simplifications. Conversely, modern CAD tools must utilize numerical methods for the electromagnetic analysis of the structure, as derived from the application of Maxwell's equations to it. Thus, they are called electromagnetic simulators. Besides the parameters characterizing a given structure, they usually provide a complete description of the configuration of the electromagnetic field and its visualization. Because of these features nowadays electromagnetic simulators also are used in electromagnetic education [Iskander 1993].

An electromagnetic field problem is defined by means of partial differential equations, integrodifferential equations or integral equations derived from Maxwell's equations and, eventually, a set of boundary and initial conditions, on an infinite-dimensional space. Numerical methods perform a projection of the continuous problem onto a finite-dimensional subspace. The set of partial differential equations (or integrodifferential equations, in general) is converted to a system of algebraic equations that is easily solved on a computer. The parameters of interest of the device and the field configuration, for example, may be obtained, after performing some postprocesses, from the finite number of unknowns of the discrete algebraic system of equations.

Numerical methods differ in the way in which the discretization of the problem from the continuous domain is performed. Important concepts to evaluate the performance of a numerical procedure are its accuracy, rate of convergence, storage and *central processor unit* (CPU) or run-time requirements. By accuracy one understands the error, measured according to a given norm, between the approximate and the exact solutions of the problem. By convergence or rate of convergence it is understood how the accuracy improves when the number of degrees of freedom (or unknowns) of the discrete problem increases. There is no numerical procedure that can solve all electromagnetic problems with high accuracy and in a fast and economical way.

1.1.1 Classification of Methods by Technique

A first classification of the methods utilized for the analysis of electromagnetic problems may be to subdivide them into experimental, analytical and numerical techniques. The first ones are expensive and time consuming. The techniques in the second group do not allow flexibility in the type of structures that may be analyzed. However, it must be mentioned that many numerical techniques make use of analytic methods, in this way improving their efficiency.

1.1.1.1 *Analytical Methods*

In this classification, some of the methods are [Sadiku 1992; Chs. 1 and 2]

- Conformal mapping;
- Separation of variables;

- Series expansion;
- Integral methods (like Laplace and Fourier transforms).

The method of conformal mapping is essentially a static technique. It performs a coordinate transformation in such a way so as to obtain an analytical solution in the mapped domain [Collin 1960; Ch. 4]. This technique is applicable to structures of simple geometry. However, it has also been applied to more complex configurations after performing certain approximations that simplify them. In those cases, semianalytical solutions will be obtained. It has also been utilized in conjunction with other methods. Examples of its use with FEM for the solution of quasi-static and static open problems are [Decreton 1974], [Imhoff et al. 1990], and [Stochniol 1992]. An example for full-wave analysis of open waveguiding structures is [Wu and Chen 1985b]. In these papers open structures are transformed into closed ones. In [Chang et al. 1990, 1991] and [Shih et al. 1988] the same principle has been used to transform open structures with re-entrant corners into closed smooth ones.

The method of separation of variables (also called Fourier method) seeks for a solution that may be given as a product of functions, each of them depending on just one variable. It has a limited flexibility. An example of its use in conjunction with FEM is given in [Gasiorski 1985] where it allows the computation of losses.

The series expansion method writes the solution as an infinite series of functions mutually orthogonal. As the infinite series must be truncated, the method is essentially semianalytical. In many cases it is used together with the previous method or other numerical techniques. Examples are the mode-matching method, and its use with FEM for open-region problems [Cangellaris and Lee 1990, 1991], [Chang and Mei 1976], [Graglia et al. 1993], [Lee and Csendes 1987], [Mei 1974, 1987], [Pelosi et al. 1993].

Integral methods are utilized in conjunction with other numerical techniques. An example is the spectral domain method, in which the Fourier transform is used together with MOM or the Rayleigh-Ritz method (see Section 1.1.2.2).

1.1.1.2 *Numerical Methods*

Table 1.1 shows a classification of the most common numerical methods used to analyze electromagnetic field problems. A number of variations of each numerical procedure was developed during the last few years. A few of them will be mentioned. Some of these variants use two or more of the methods given in Table 1.1.

The classification of Table 1.1 takes into account in the first place the domain in which the solution is sought. Because this book is devoted to techniques that solve the problem in the frequency domain, the time domain techniques will not be described. It may be mentioned that the *Transmission Line Matrix* (TLM) method is the only one that is specific for time domain. It discretizes Huygens's principle, solving a problem analogous to the original electromagnetic one in terms of a distributed parameter equivalent circuit [Hoefer 1985], [Itoh 1989a; Ch. 9], [Sadiku 1992; Ch. 7]. A frequency domain version has also been developed [Jin and Vahldieck 1992].

Table 1.1
Classification of numerical methods

Maxwell's equations			
Frequency domain		**Time domain**	
Differential equations	Integral equations	Differential equations	Integral equations
* Finite Difference Method * Method of Lines * Variational Methods: - Rayleigh-Ritz - Weighted Residuals • Collocation • Subdomain • Galerkin • Least Squares ⇓ **FINITE ELEMENT METHOD**	* Method of Moments[(1)]: - Collocation - Subdomain - Galerkin - Least Squares * Rayleigh-Ritz[(1)] * Boundary Element Method (1) In the spatial domain or the spectral domain	* Finite Difference Method * Method of Lines * Transmission Line Matrix Method * Finite Element Method	* Method of Moments

1.1.2 Classification of Numerical Methods by Type of Formulation

The next criterion for classification is the original type of formulation, that is, partial differential equations or integrodifferential equations.

1.1.2.1 *Methods Based on Partial Differential Equation Formulations*

These methods solve for the electromagnetic field throughout the spatial domain of the problem. They require the discretization of such domain. Thus, they are especially suited for closed structures that may be inhomogeneous and anisotropic. However, specific techniques that allow one to handle open structures have been developed.

The *Finite Difference* method, as mentioned before, was the first numerical method applied to solve an electromagnetic problem [Ames 1992; Ch. 1]. In the 1920s A. Thom and C. J. Apelt developed it under the name of Method of Squares to solve nonlinear hydrodynamic equations [Sadiku 1992; Ch. 3]. A method for its computer implementation was set up by R. V. Southwell [1940, 1946]. FD discretizes both the problem domain and the differential operators, substituting them by the finite differences of the unknown values at the points of the space discretization. The discrete points are placed at the vertices of rectangular or cubic grids, although more flexible versions are available. In any case, FD may not be adequate for the analysis of structures of arbitrarily complex geometry [Booton 1992; Chs. 2 and 3], [Sadiku 1992; Ch. 3]. In the case of open structures, artificial boundaries, absorbing boundary conditions, hybrid methods, or any other technique able to transform the infinite region problem into a finite one is necessary (see Section 6.2.2). The *Measured Equation of Invariance* method (MEI) [Mei et al. 1992, 1994], [Prouty et al. 1993] and iterative procedures [Sandy and Sage 1971], [Djordjević et al. 1995] have been proposed as alternate solutions. The FD method is the least analytical numerical technique [Itoh 1989a; Ch. 1]. The dimensions of the algebraic system of equations may be quite high for problems needing a large number of discretization points. The solution of the algebraic system of equations may be unstable in some cases [Wexler 1969], [Ng 1974].

The *Network Analog Method* is a variation of FD. It discretizes both the domain of the problem and the boundary conditions, at the boundary and media interfaces. Examples of its use may be [Linner 1974] and [Tripathi and Bucolo 1985].

The origin of the *Method of Lines* (MOL) can be traced back to 1965 [Liskovets 1965]. It is also a variation of the FD method. It reduces the dimension of the problem by solving it analytically in one dimension. Hence, structures to be analyzed by this method must be homogeneous in one direction. This condition reduces the range of applicability of the method. On the other hand, it has the advantage of reducing the number of unknowns and, consequently, the CPU time, storage, and memory requirements [Ames 1992; Ch. 1], [Pregla and Pascher 1989]. It has been used in conjunction with FEM in [Wu and Davidowitz 1992], for example.

Variational methods allow one to reduce the problem of solving a differential equation to an equivalent problem in which one seeks a function that gives an extreme of some integral [Mikhlin 1964, 1966], [Mikhlin and Smolitsky 1967], [Rektorys 1977]. Variational methods may be subdivided into two categories: direct and indirect methods.

The direct method is also called the *Rayleigh-Ritz method* or *Ritz method*.[1] It starts from the functional or integral form from which the differential equations defining the problem are derived. The unknowns of the problem are approximated by a linear combination of expansion or basis functions that must be conveniently selected [Sarkar 1983]. Its coefficients are the unknowns of the discrete problem obtained by substituting the approximation of the unknowns in the functional and making it stationary with respect to those coefficients [Sadiku 1992; Ch. 4], [Booton 1992; Ch. 5]. In its original version (whole domain basis functions) it is not adequate for complex geometries.

The indirect methods are collectively referred to as the *Method of Weighted Residuals*. The unknown is approximated as in the Rayleigh-Ritz method. The residue of the differential equation is then weighted by a set of weighting or test functions, again conveniently selected [Sarkar 1983, 1985, 1991], [Sarkar et al. 1985]. The integral of the result over the domain is set equal to zero. In other words, the residue and the weighting functions are forced to be orthogonal in the mean. Depending on the type of weighting functions used, the various methods can be summarized as follows:

1. *Collocation* or *point matching*, in which the weighting functions are Dirac delta functions; hence, the method is equivalent to forcing the residue to zero at a

---

1 Named after Lord Rayleigh [1870, 1876] and W. Ritz [1908].

number of selected points in the domain;
2. *Subdomain*, in which the weighting function is zero in all of the domain except in a given subdomain, where it may be equal to a constant or a linear function, or a sinusoidal function, for example;
3. *Galerkin* or *Bubnov-Galerkin*,[2] in which the test functions are the same as the expansion functions;
4. *Least squares*, in which the weighting functions are the result of applying the differential operator to the test functions; hence, it is equivalent to finding the stationary points of the functional obtained by weighting the residual by itself;
5. *Petrov-Galerkin*,[3] in which the weighting functions can be any other functions.

The solution of a differential equation problem using an indirect method may also be seen as a method consisting of two basic steps [Reddy 1984]:

- Cast the differential equation into an integral form (also called variational form);
- Determine the approximate solution (i.e., the coefficients of the linear combination of basis functions) using one of the methods mentioned previously.

Indirect methods are less analytical than the direct method [Sadiku 1992; Ch. 4]. In any case, it is relatively easy to formulate the solution of many differential and integral equations in variational terms. In general, variational methods give accurate results without making excessive demands on computer storage and time.

The *Finite Element Method* is a variational method in which the unknowns are approximated by subdomains, that is, by means of linear combinations of polynomial functions (or other functions) locally defined in each of the subdomains in which the domain of the problem is subdivided. These subdomains are called finite elements in a geometrical sense. They have simple geometries like rectangular or triangular shapes for 2D problems, or tetrahedrons or hexahedra for 3D problems. This approximation of the unknowns by subdomains is used in conjunction with either the Rayleigh-Ritz method or any weighted residual method, although the most common approach is the Galerkin procedure. If possible, an integration by parts is applied to the corresponding integral expression obtaining an alternate weak integral form. Then, the approximation is applied locally over it resulting in a set of discrete expressions for each element (called local discrete integral form). An important concept is the continuity conditions that the unknowns or certain quantities related to them (given derivatives or components) must fulfill. These conditions allow the assemblage of the local discrete integral forms of all elements obtaining a global discrete form. Finally, the imposition

---

2 Named after the Russian civil engineer B. G. Galerkin who in 1915 introduced this method on analysis of bending and buckling of bars while he was behind bars [Galerkin 1915]. It is also known as Bubnov-Galerkin, after I. G. Bubnov whose work was the basis of Galerkin's developments [Mikhlin 1964].

3 Named after Galerkin and G. I. Petrov. The latter author used different basis and testing functions [Mikhlin 1964].

of the boundary conditions provides the algebraic system of equations [Ames 1992; Ch. 1], [Booton 1992; Ch. 5], [Sadiku 1989, 1992; Ch. 6]. The analytical load of FEM is low [Itoh 1989a; Chs. 1 and 2]. It may be applied to arbitrarily shaped geometries that may be inhomogeneous and anisotropic. The method is very well suited for nonlinear materials (i.e., those whose physical characteristics depend on the unknowns of the problem). The system matrices may have large sizes. However, matrices are banded and highly sparse because the unknowns of an element are connected only to those of contiguous elements. These characteristics allow the use of dedicated solvers leading to moderate storage, memory, and CPU time requirements. For accurate results and a good rate of convergence, the method requires a refined mesh (or, in general, a higher number of unknowns) in regions where the electromagnetic field does not have a smooth behavior, i.e., re-entrant corners, discontinuous boundary conditions, abrupt changes in the physical characteristics of the different media, and proximity and coupling effects, for example. These also are requirements of FD method or its variants. In order to achieve well adapted meshes self-adaptive procedures may be used like the one described in Chapter 4. Open structures may be tackled by means of conformal mapping, infinite elements, artificial boundaries, absorbing boundary conditions, MEI method adaptations, hybrid approaches, or iterative procedures (see Chapter 6 and Section 7.5).

1.1.2.2 *Methods Based on Integral Formulations*

Integral methods start from an integrodifferential formulation of the problem. These methods are best suited for open structures. In fact, open-region problems have traditionally been tackled using integral methods to solve surface or volume integral formulations. Unlike differential equation-based methods, the integral equation formulations solve for the field sources (currents and charges). The sources can be line (e.g., for wire antennas and scatterers), surface (e.g., for metallic plates), or volume sources (e.g., for inhomogeneous dielectric bodies). The integral equations are, generally, formulated starting from the boundary conditions or some local relations between fields or from variational principles. The fields are further expressed in terms of the sources through certain integrals (usually with the aid of potentials), which involve Green's functions. In the final form of the equations, the sources appear as unknowns under integrals. These equations are linear in terms of the field sources.

The integral equation-based methods have numerous attractive strengths. For example, if a surface formulation is used they require only the discretization of the boundary and the media interfaces (i.e., they are more adequate for homogeneous structures or structures with not too many different media). However, they also possess some disadvantages. In particular, integral equations can become impractical when dealing with structures with highly inhomogeneous or nonlinear media. The integral formulation always requires an *a priori* knowledge of Green's functions. However, in electromagnetically complex media, these functions are seldom known. To solve the problem, it is often necessary to formulate the integral equations in terms of equivalent field sources located throughout the whole volume of inhomogeneities [Sarkar et al. 1989b]. In such cases, the unknowns have the same localization and complexity as they have in a differential equation formulation. The analytical load of integral methods is high. In contrast to the integral equation approach, the differential equation formulations are simple to set up regardless of the complexity of the structure analyzed. They amount to describing the whole structure by the basic electromagnetic field equations and suitable boundary conditions. For nonlinear media, the differential equations become nonlinear. Differential equation-based methods can tackle them, while no counterpart exists for the solution of such electromagnetic systems with integral methods.

The integral equation formulations give rise to systems of linear equations with dense coefficient matrices. The evaluation of these coefficients is a hard and time-consuming numerical task. When dealing with a volume formulation, the number of unknowns is usually very high. Consequently, a long CPU time is required to compute the matrix elements. Solving the resulting system of linear equations requires a large amount of fast-access computer memory to efficiently manage all matrix elements. On the other hand, the differential equation formulation results in a matrix of a similar size as for the integral equation formulation. However, the former matrix is sparse and its coefficients are easily evaluated. Hence, the system can be solved much faster using dedicated solvers for sparse matrices. Nevertheless, if a surface integral equation formulation can be applied to a certain problem, then the integral equation approach may be superior to the differential equation approach. The latter method usually involves a significantly larger number of unknowns for the same accuracy, since it must approximate the fields throughout a volume. Hence, the CPU time required to solve the matrix equation resulting from the differential equation formulation usually exceeds the matrix fill-in time for the surface formulation integral equation solution.

The *Method of Moments* is the most well known among integral equation-based methods. It coincides formally with the weighted residual method because the sources (the unknowns) are expanded by a sum of certain basis functions multiplied by unknown coefficients. The residual of the integral equation is weighted using a suitable inner product and a set of weighting functions. This results in a set of linear equations, which can be solved in the usual way. Depending on the definition of the inner product of such a weighting operation, several methods are possible. MOM uses the scalar product (equivalent to the reaction concept) [Harrington 1967, 1968]. The *Conjugate Gradient Method* utilizes the Hilbert product (equivalent to the conservation of power principle) [Sarkar 1991]. The Hilbert product is also the inner product used by the weighted residuals method. MOM has been used together with FEM for the analysis of open-region problems (see Section 6.2.2, where references are given). In [Ramahi and Mittra 1991b], discontinuity problems are solved by a MOM/FEM technique.

The *Rayleigh-Ritz* technique may also be used to obtain integral methods because it may be applied to variational principles involving Green's functions [Kitazawa and Hayashi 1987].

In both MOM and the Rayleigh-Ritz method, the Fourier transform is frequently used to simplify the computation of the Green's function, because the dimension of the original problem is reduced. This procedure gives rise to the *Spectral Domain Approach*

(SDA), introduced by E. Yamashita and R. Mittra [1968]. Conversely, spectral methods are difficult to apply to structures having inhomogeneities along the direction in which the Fourier transform has been used. However, methods have been devised to cope with this difficulty. Spectral methods reduce the dimensions of the system matrices and, consequently, the memory, storage, and CPU time requirements [Itoh 1989a; Chs. 3 and 5], [Booton 1992; Chs. 9 to 11], [Jansen 1985].

The *Boundary Element Method* (BEM) is related to MOM and FEM. It performs a finite element discretization of the boundary using the operator Green's function [Ames 1992; Ch. 3], [Brebbia 1978], [Celia and Gray 1992; Ch. 4]. Some authors consider BEM as just a FEM applied to integral formulations. However, a number of mathematical and computational issues differ from one method to the other. Its use in conjunction with FEM for discontinuity problems may be found in [Hirayama and Koshiba 1989, 1992], [Ise and Koshiba 1988], [Kiyoshi and Masanori 1988], and [Wu et al. 1989], for example.

Boundary-integral (BI) formulations may also be used together with FEM for open-region problems. These techniques are called hybrid FE-BI methods. Procedures named in the literature as FEM/BEM fall under this category. References are given in Section 6.2.2. In Chapter 6 and Section 7.5, an iterative FE-BI method is presented.

There are many methods that are modifications or combinations of the basic numerical and analytical methods mentioned up to this point, like the *singular integral method* [Mittra and Itoh 1971], and the *spectral admittance method* [Itoh 1980].

1.1.2.3 *Comparison between Numerical Methods*

Table 1.2 makes an attempt to summarize in a qualitative fashion the properties of the various numerical methods, following similar criteria as in [Itoh 1989a; Ch. 1]. In short, solutions of a problem in the frequency domain utilizing differential equation-based techniques produce large sparse matrices that often are real, whereas integral equation-based techniques give rise to moderate-sized dense complex matrices. Further comparison between both families of numerical procedures is obtained from the step involving the solution of the matrix equation resulting from the discretization of the problem. The accuracy on its solution depends, for direct solvers, on the condition number of the matrix.[4] For a given accuracy, iterative solvers require a number of iterations that also depend on the condition number. For formulations based on differential equations, the condition number is inversely proportional to the square of the size of the discretization. For an electric field integral equation, it is inversely proportional to the size of discretization. However, for the magnetic field integral equation, it is independent of the size of the discretization, but this formulation cannot be used for infinitely thin (zero thickness) structures [Sarkar et al. 1994].

---

4 The condition number of a matrix, $[A]$, is defined as the ratio of the largest to the smallest singular values, which correspond to the square root of the eigenvalues of $[A]^H[A]$ or $[A][A]^H$ where $[A]^H$ denotes the conjugate transpose of matrix $[A]$.

Table 1.2
Qualitative comparison of numerical methods

Method	Analytical load	Open/ Closed problems	Material generality(1)	Geometric versatility	Dense/ Sparse matrices	Storage	CPU time
FD	Nil	Closed	Good	Good	Sparse	Large	Large
MOL	Large	Closed	Limited	Limited	Sparse	Moderate	Small
FEM	Small	Closed	Very good	Very good	Sparse	Large	Mod./Large
MOM	Moderate	Open	Limited	Good	Dense	Moderate	Moderate
BEM	Moderate	Open	Limited	Good	Dense	Moderate	Moderate
SDA	Large	Open	Limited	Limited	Dense	Small	Small

(1) Homogeneous isotropic linear/ Inhomogeneous anisotropic nonlinear

Books dealing with different aspects of analytical and numerical methods for differential and integral problems are [Courant and Hilbert 1989], [Ghatak et al. 1995], [Hildebrand 1952, 1956], [Pearson 1974], [Stakgold 1979], and [Strang 1986], among others. A general description of analytical and numerical methods may be found in the books by Ames [1992], Celia and Gray [1992], Dahlquist and Björck [1974], Mikhlin [1964, 1966], Mikhlin and Smolitsky [1967], and Rektorys [1977], among others. For electromagnetic problems, the reader is referred to [Binns et al. 1992], [Booton 1992], [Hoole 1989], [Itoh 1989a], [Mittra and Lee 1971], [Sadiku 1992], and [Steele 1987]. Survey papers regarding numerical techniques for these problems are [Ng 1974], [Saad 1985], [Sorrentino 1988], [Tortschonoff 1984], and [Wexler 1969], among others.

1.2 THE FINITE ELEMENT METHOD IN ELECTROMAGNETICS

In 1978, in the preface to *An Introduction to Finite Element Analysis*, D. H. Norrie and G. de Vries stated that in a little over two decades, FEM evolved as a technique of major importance for the solution of a wide range of scientific and engineering problems. In 1998 it may be stated that probably this method is the one with a greater impact in the overall field of engineering. The finite element method indeed has the virtues of simplicity in concept, elegance in development, and potency in application as the previously mentioned authors said twenty years ago.

The success of the finite element method is based largely on the basic finite element procedures used [Bathe 1982], namely,

- The formulation of the original differential equation-based problem in a variational weak integral form, either through the stationarity of a variational principle or through a weighted residual procedure;
- The finite element discretization, through the subdivision of the physical region of the problem into subregions of finite elements and the approximation of the

dependent variables (the unknowns of the problem) over each element and hence over the whole domain (the approximation consists of a linear combination of known simple functions, usually polynomials; the parameters of the linear combination subsequently become the unknowns of the discrete problem; and the substitution of these approximations into the variational formulation that yields a set of discrete integral forms, one for each element);
- The assemblage of the sets of discrete integral forms and the imposition of the boundary conditions that yield the algebraic system of equations in the unknown parameters;
- The effective solution of the resulting finite element equations, which yields the unknown parameters and hence the approximate solution of the problem.

On one hand, FEM overcomes the disadvantages of traditional entire-domain variational methods where the approximation functions for problems with arbitrary domains are difficult to construct. In fact, FEM provides a systematic procedure to construct approximation functions over subregions of the domain [Reddy 1993; Ch.1].

On the other hand, the basic steps mentioned previously are the same whichever problem is considered and provide a general framework. In conjunction with the use of a digital computer it is quite a natural approach in engineering analysis. It is a complete numerical process implemented on a digital computer. The steps are comprised of:

- The formulation of the local finite element matrices;
- The numerical integration to evaluate the coefficients of these matrices;
- The assemblage of the element matrices into global matrices that correspond to the complete finite element problem;
- The enforcement of the necessary boundary conditions;
- The numerical solution of the global system of equations.

Perhaps the main reason for the popularity of FEM is that it is based on a *weak integral* formulation of the problem. This property provides not only the existence of nonregular solutions of differential boundary value or initial value problems but also allows one to formulate the solution by an integral of a given quantity over a domain. The integral of a measurable quantity over a domain may be split into a sum of integrals over an arbitrary set of subdomains, whose union gives the original domain. This principle is of paramount importance. In fact, due to this property, the analysis of a problem may be carried out locally over a typical subdomain. Also, if the subdomains are arbitrarily small it may be justified to select polynomial functions in order to represent the local behavior of the solution in these regions. This possibility of adding up components of the solution is the aspect that is exploited in any FEM [Oden 1990].

At this point, FEM may be defined as a polynomial (in general, any other basis functions are admissible) approximation method by subdomains of a variational formulation for the problem of interest. This simple definition involves two aspects.

The first one is related to the fact that the fundamental equations describing the behavior of the problems analyzed through FEM are of differential type. This implies that they must be manipulated in order to treat them numerically. The mathematical tool that performs such manipulation is the variational calculus.

If the first aspect is purely analytical, the second one is numerical. Approximating the solution of a variational problem over a domain by means of piecewise polynomial functions involves a number of issues such as the appropriate subdivision of the domain, the degree of the polynomials utilized, the error incurred (or to say it in a positive way, the accuracy obtained), and the convergence of the solution, as well as other issues such as the imposition of the boundary conditions of the problem or the initial values. These aspects constitute FEM. The advantages and drawbacks of FEM are quite evident from the description of the method given previously.

Following J. E. Akin [1994; Ch. 1], it may be said that the greatest advantage of FEM is its ability to handle arbitrary geometries. Its next most important features are the ability to deal with general boundary conditions and to include inhomogeneous and anisotropic materials, that is, systems of arbitrary shape made of numerous different material regions may be easily treated. Each material could have constant anisotropic properties or they could vary with spatial location. In addition, the method allows large flexibility in prescribing the excitations for the problem and in the postprocessing of any quantity related to the unknown of the original problem. For elliptic boundary value problems, FEM offers significant computational and storage efficiencies (sparse banded matrices) that further enhance its use. Additionally, FEM gives an important solution technique for nonlinear problems [Akin 1994; Ch. 1], [Johnson 1992].

The drawbacks of FEM are mainly two. First, the accuracy of the method greatly depends on a suitable discretization of the domain and on the order of the polynomials utilized to approximate the solution in these subdomains. Second, the method by itself is not well suited to treat problems with open regions.

The first aspect implies that in order to obtain an accurate solution, the set of subdomains and the order of the polynomials used should be properly adapted to the nature of the exact solution of the problem. Hence, this method is usually implemented in an iterative fashion. The FEM code is first applied to the analysis of a problem over an initial discretization of the domain (in both the spatial subdivision and the order of the polynomials chosen) generated from the experience of the user. Successive refinements of either the subdomains (reducing their size) or the polynomial (increasing their order) or modifying both of them simultaneously results in a repeated application of the code over those refined discretizations. This leads to a more accurate result. In fact, the most cumbersome task for a finite element practitioner is generating an adequate discretization of the domain. This is why J. E. Akin in the preface of *Finite Elements for Analysis and Design* [1994] states that the future of FEM would most probably involve adaptive analysis. Self-adaptive techniques generate automatically a discretization of the domain well adapted to the local behavior of the solution of the problem, ensuring maximum accuracy for minimum computational cost and increased rates of convergence. This book presents a procedure for generating a self adaptive mesh that may overcome the first weakness of the method presented earlier.

The second drawback of FEM, namely, the need to truncate infinite domains, is also well known. A number of techniques have been proposed to ensure that although performing an analysis over a finite domain, a good accuracy may be obtained for open-region problems. In this book, an iterative scheme leading to numerically exact radiation conditions is proposed.

These two techniques, a self-adaptive procedure for refinement of a FEM mesh and an iterative procedure for open-region problems, are the main contributions of this book in the frame of electromagnetic problems.

In what follows, a brief history of FEM will be given. Because the method is based on variational calculus and variational methods of approximation, the historical background of these mathematical tools will be presented first.

The finite element method was born as a tool for the solution of problems in structural mechanics. It has evolved to be a widely utilized and richly varied computational approach for many scientific and technological areas. D. H. Norrie in the preface to the *Finite Element Handbook* edited by H. Kardestuncer and himself [1987], mentions that by the early 1980s there were over 20,000 finite element users worldwide who were estimated to spend about $500 million annually on finite element analysis. Since that time, there has been continuing rapid expansion in both the number of users and the range of applications. A brief summary of the initial history of FEM is given next as well as an account of the development of FEM in electromagnetics.

1.2.1 Variational Calculus and Variational Methods of Approximation

Variational calculus has been demonstrated as one of the most important areas of mathematical analysis during more than two centuries. It is a powerful tool that may be applied to a wide variety of mathematical problems. It may also be used to express the basic principles of mathematical physics in an elegant and unusually simple fashion.

Variational calculus deals with quantities (length of an arc, area of a surface, time of a displacement, and energy, for example) that depend on a curve, as a whole, seeking that which minimizes (or maximizes) the quantity of interest. Variational calculus treats other extrema problems that depend on surfaces, not only curves. For example, if a circular wire is distorted and is dropped into a soap solution, the soap film that will be sustained by the wire will adopt the form of the surface of minimum area among those having the wire as its contour. The mathematical problem consists of finding such surfaces from the wire contour and the property of minimum area.

The oldest extrema problem is the Classical Isoperimetric Problem: *Find the curve that contains the maximum possible area among the closed curves of a given length contained in a plane.* Apparently the Greek philosophers tried to solve this problem early in the fifth century B.C., as Aristotle (384-322 B.C.) already wrote about it. Additionally, the Greek geometricians stated and solved a number of extrema problems. They are mentioned in *Elements* by Euclid and in the works of Archimedes (287-212 B.C.) and Apolonio (260-170 B.C.). A well-known example is the computation of the area of a circle by covering it with an infinite number of triangles.

During the seventeenth century, the renewed scientific activity took care of a number of problems of natural philosophy, among them extrema problems. The starting point was the attempt to explain refraction of light. This topic was also addressed by the Greek philosophers. In particular, Claudius Ptolomey (130 A.D.) of Alexandria was the first to produce a table of very accurate measurements of the incident and refracted angles for different media. He tried unsuccessfully to deduce an experimental law.

Johannes Kepler (1571-1630) nearly arrived at a successful deduction of the laws of refraction in his book, *Supplements to Vitello*, in 1604. Unfortunately, he was misled by some erroneous data previously compiled by Vitello (1270). The correct law of refraction of light was finally established in 1621 by Willebrord Snell (1591-1626). However, the formulation in its present form is due to René Descartes (1596-1650), who deduced the same law using a model in which light was visualized as a pressure transmitted by an elastic medium, as described in his work *La Dioptrique* (1637).

It was in this period when the problem of the physical principles of the previous law was investigated. Pierre de Fermat (1601-1665), without considering the assumptions made by Descartes, deduced the law of refraction from the Extrema Principle of Least Time (1657). Deviating from the postulate of Hero of Alexandria, who asserted that light traverses the shortest allowed path between two points, Fermat stated that the actual path between two points taken by a beam of light is the one that is traversed in the least time, even though it may not follow the shortest route. The Fermat law in its modern formulation is [Hecht and Zajac 1990]: *A light ray going from a point S to a point P, follows an optical path length (OPL) that is stationary with respect to variations of that path.* In other words, the *OPL* for the exact path is equal, through a first approximation, to the *OPL*s of the paths immediately contiguous. Hence, a number of contiguous curves neighboring the exact one will provide nearly equal propagation time for the light following those paths. These simple statements of the problem allow a simple interpretation of the propagation of light. Assume that a light beam is propagating through an isotropic and homogeneous medium in such a way that it goes from point S to point P. The atoms of the medium are excited by the incident perturbation and will reradiate in all directions. In general, these wavelets that originate in the immediate vicinity of the stationary path will arrive at P through routes slightly different between them and, hence, they will reinforce each other. Thus the energy will effectively propagate along the route from S to P fulfilling the Fermat principle.

Fermat's achievement stimulated a big effort in the development of similar variational formulation for Newton's laws of mechanics. From then on, variational principles (the conviction that natural laws may be deduced from stationary principles) took a central place in all natural sciences.

Up to the second half of the seventeenth century there was no general procedure for solving extrema problems. The need to formulate those problems prompted the birth of mathematical analysis. The first procedure for the study of maximum-minimum problems was suggested by Fermat himself in 1629. In modern terms its formulation is: *At the extrema point (of a given function of a variable) the derivative equals zero; thus, the extrema points must be sought among the roots of the derivative.* However, Fermat described this procedure just for polynomials. In its general form the method was given first by Isaac Newton (1642-1727) in 1671 and again rediscovered by Gottfried Wilhelm Leibnitz (1646-1716), who published a famous article that is considered the starting point of mathematical analysis (1684). It is worth noticing the title: "Nova methodus pro maximis et minimis" [Galéev and Tijomírov 1989].

The invention of calculus by Newton and Leibnitz encouraged the study of a wide variety of variational problems, like the famous Brachistochrone problem (from the greek words *brachistos*, shortest, and *chronos*, time) proposed by Johann Bernoulli (1667-1748) in June 1696 in *Acta Eruditorum* (the first and only scientific journal at that time). Bernoulli, a well-known Swiss mathematician (who was a student of Leibnitz and a teacher of Euler), entitled his paper "To the most acute mathematicians of the entire world." Apparently, Newton, who was at that time the director of the Mint (that printed money), recognized it at once as a challenge to himself from the continental mathematicians and in spite of being out of habit of scientific thought, summoned his resources and solved it that evening before going to bed. His solution was published anonymously, but when Bernoulli saw it he wryly remarked, "I recognize the lion by his claws" [Simmons 1993; Ch. 3]. The publication of the Brachistochrone problem is regarded as the starting point of classical variational calculus.

A number of works, mainly those of Pierre de Maupertuis (1698-1759) and Leonhard Euler (1707-1783), led to the principle of Joseph Louis Lagrange (1736-1813) and to the Principle of Least Action of William Rowan Hamilton (1805-1865).

It was Euler who, in 1744, gave to variational calculus the prestige of a coherent branch of mathematical analysis. He discovered the basic differential equations used to minimize a curve. Together with him, another mathematician with a decisive role during that period was Lagrange. The contributions of Lagrange to variational calculus were among his earlier and more important works. In 1755 he informed Euler about his method of multipliers for the solution of isoperimetric problems that for years remained unsolved by Euler because his semigeometric procedures were unable to handle them. Euler was then able to answer a number of questions over which he had thought for a long period of time. However, he treated Lagrange in a generous and friendly way, "not going to press with his own work so as not to detract from him a single bit of the glory he deserved" [Simmons 1993; Ch. 12]. Lagrange continued to work for some years on the analytical version of his calculus of variations. Both Lagrange and Euler applied it to many new types of problems, mainly in the field of mechanics. The development of the theory of variational calculus continued, after that, for more than two centuries. Besides the first-order necessary conditions (Euler-Lagrange equations), other second-order necessary and sufficient conditions for the two types of extremes, strong and weak, were developed by Lagrange, Jacobi, and Weierstrass.

Later on, the Irish physicist and mathematician, W. R. Hamilton, demonstrated that any optical problem may be solved by just one procedure that includes Fermat's Principle of Least Time as a particular case. He extended that technique to mechanical problems and when he was 30 years old, formulated a universal principle (called today the Hamilton Principle) that shows how mechanics and optics are just two different aspects of variational calculus. Thanks to Hamilton's efforts, a new method to handle variational problems was born (Hamilton-Jacobi Theory). It made possible the development of the Kneser and Hilbert theory, in the final years of the nineteenth century .

In modern physics, Albert Einstein (1879-1955) made extensive use of variational calculus in his work and Erwin Schrödinger (1887-1961) used it to discover his famous wave equation which is one of the corner stones of quantum mechanics [Simmons 1993; Ch. 12]. In fact, the astonishing similarity between the Fermat and Hamilton principles played an important role in the development of quantum mechanics which reached its climax when in 1942, Richard Phillips Feynman demonstrated that quantum mechanics may be alternatively formulated from a variational point of view [Feynman 1987].

Variational calculus is also the basis of the methods for the approximate solution of differential and integral equation-based problems which reduce these problems to finite systems of algebraic equations, that is, the variational methods of approximation.

The first attempts to solve variational problems in an approximate way may be found in the work of Lord Rayleigh [1870, 1876]. At the beginning of the twentieth century, W. Ritz [1908] published his method (see Section 1.1.2.1) that was a far-reaching generalization of that of Rayleigh [Mikhlin 1964]. With the Ritz procedure (or Rayleigh-Ritz method), a tool was provided for the approximate solution of problems which were previously quite intractable. Especially good results were obtained for boundary value problems associated with ordinary differential equations. Following the Ritz procedure it was possible to find from the set of admissible approximate functions (those satisfying the boundary conditions) the linear combination giving the minimum value of the variational principle from which the differential equation of the problem is derived. Later on in the 1950s, R. Courant, K. Friedrichs, D. Hilbert, and S. L. Sobolev [Courant and Hilbert 1989], [Sobolev 1963a,b] completed the theoretical aspects of the method with the notions of functional analysis that establish the required space to which the approximate solution should belong and the way in which the different boundary conditions should be fulfilled [Mikhlin 1964].

In 1913, I. G. Bubnov published a report on the work of S. P. Timoshenko who applied the Ritz method to a particular structural problem. Bubnov pointed out that the final equations produced by the Ritz method can also be obtained without resorting to the consideration of the energy of the system (i.e., the variational principle). Bubnov proposed essentially the method that is called nowadays the Bubnov-Galerkin method (or Galerkin method, see Section 1.1.2.1). The contribution of B. G. Galerkin [1915] was that he claimed this method can be applied to any system of differential equations (the equations may also be integrodifferential not just purely differential), provided the basis functions and the test functions are appropriately chosen. Thereafter, a large amount of work was published in which the method was applied to extremely diverse applied problems. The procedure was soon extended to nonlinear problems. In 1940, Y. V. Repman applied it to Fredholm-type integral equations and G. I. Petrov used it for higher-order differential equations [Mikhlin 1964].

With respect to the other indirect variational methods (see Section 1.1.2.1), the least squares method was used in 1922 by M. Piccone to solve the Dirichlet problem. Some of its further applications may be found some years later in the works of V. I. Krylov, and M. F. Kravchuk and coauthors [Mikhlin 1964]. The work of C. B. Biezeno and R. Grammel, in 1933, may be recognized as an application of the subdomain method. Similarly, the work of R. A. Frazer (1937) and coworkers is an early application of the collocation method (or point matching method). In 1940, G. I. Petrov applied the Bubnov-Galerkin method by choosing weighting functions which are different from the basis functions, providing a complete freedom in the choice of both sets of functions. Other early works that may be recognized as indirect variational methods were done by C. F. Gauss (1795), and C. B. Biezeno and J. J. Koch (1923).

In 1956, S. H. Crandall described for the first time the method of weighted residuals as a general variational method of approximation from which all the other indirect methods (collocation, subdomain, Bubnov-Galerkin, least squares, and Petrov-Galerkin) may be recognized as particular cases. In the field of mechanics the method of weighted residuals coincides with the application of the Principle of Virtual Works to equilibrium problems. More recently, B. A. Finlayson has given a complete description of the weighted residual method and its connection to the direct variational method of approximation [Finlayson 1972], [Szabó and Babuška 1991; Chs. 2 and 5], [Zienkiewicz 1977; Chs. 1 and 3], [Zienkiewicz and Taylor 1989; Chs. 1 and 9].

1.2.2 The Origin of the Finite Element Method and Its Development

From the discussions of the previous section, it may appear that FEM emerged from the growing theory on variational calculus, partial differential equations, weak solutions of boundary value problems, Sobolev and Hilbert spaces, variational methods of approximation, and interpolation theory. Instead, FEM grew independently and parallel to the development of such mathematical theory for almost twenty years.

It is not easy to trace the origin of FEM because there are many concepts involved in the definition of what constitutes a finite element method. In any case its origins are related to the work of a group of engineers in structural mechanics. The difficulty of this type of engineering problems is caused not by their mathematical complexity but by the number of individual components present. Over the years such problems have been solved by looking at each particular component and connecting it to the remainder of the structure. Thus, it was natural for an engineer to think of a continuum in terms of a discrete assembly of small components [Zienkiewicz 1973].

Hence, if FEM is viewed as a numerical procedure in which a global approximation of a partial differential equation is obtained from a sequence of local approximations over subdomains, then the invention of FEM should be credited to A. Hrenikoff [1941]. In fact, Hrenikoff was very close to the basic idea of FEM when he introduced a new approach for the study of plane elasticity problems by a set of orthogonal bars, beams and spring elements [Reddy 1993; Ch. 1]. That idea also underlines the work of D. McHenry [1943]. Some claim that such a concept may be recognized even in the work of Leibnitz who employed piecewise linear approximation for the Brachistochrone problem. With this new calculus tool Leibnitz derived the governing differential equation for the problem. However, Leibnitz had no intention of approximating a differential equation; rather, his goal was to derive it. Nevertheless, the fact is that two and a half centuries later, it was realized that differential equations could be approximated by keeping the elements finite in size, instead of taking infinitesimal elements, as in calculus. And this idea is the reason for the term *finite element*.

If FEM is considered as a method of piecewise polynomial approximation over subdomains, then its origin may be attributed to R. Courant. When he prepared his manuscript for his 1942 address to the American Mathematical Society, he added a two-page appendix trying to demonstrate how the Rayleigh-Ritz method could be formulated in a general way for practical use [Silvester and Pelosi 1994]. He described it, one year later, in a paper entitled "Variational Methods for the Solution of Problems of Equilibrium and Vibration." He used piecewise linear approximation for a Dirichlet problem, where the domain was subdivided into a triangular mesh [Courant 1943].

However, not all credits should go to Courant. As early as 1851, Schellbach proposed a solution to the problem of Plateau consisting of the determination of the surface S of minimum area having a given closed curve as a contour. Schellbach used an approximation, S_h, of S by means of a mesh of triangles over which the desired surface was represented by piecewise linear functions. From it he obtained an approximate solution of the Plateau's problem by minimizing S_h with respect to the coordinates of the hexagon formed by six elements [Schellbach 1851]. It was not a proper FEM. However, the way of solving the problem is related to FEM as much as the work by Courant [Oden 1990], [Williamson 1980].

As J. T. Oden has said,there are some differences in the processes of generating a mesh of triangles over a domain, on one hand, and generating the domain of approximation by piecing together triangles, on the other hand [Oden 1990]. Although both processes may be identical, they may differ considerably in the way in which the boundary conditions are enforced. In fact, none of them, Schellbach nor Courant, were careful enough in properly enforcing those conditions or the way in which the boundary of the domain was to be modeled by the method. Nowadays, these are fundamental aspects of any finite element procedure. It may be added that Courant did not gave information related to other important FEM topics like repeatability of computations and the role of nodes, for example [Felippa 1994], [Gupta and Meek 1996].

However, finite elements are much more than a method of piecewise polynomial approximation. Any FEM comprises the subdivision of the domain, the local approximation, the assembly of elements, the application of excitations, the imposition of boundary conditions and the use of other numerical techniques. During the mid-1950s, J. H. Argyris developed a primitive version of such a method using rectangular elements [Argyris 1954, 1955]. He based his work on the studies of G. Kron [1939, 1953] in structural mechanics together with some concepts of system theory and variational approximation methods in an attempt to produce an effective methodology for the analysis of aeronautical structures. Instead, although J. L. Synge [1957] used piecewise linear approximation on triangular meshes, his work is not considered by some researchers to be in the spirit of FEM because of his treatment of boundary conditions. For some other researchers the works of R. J. Duffin [Duffin 1959], [Duffin and Porsching 1959], based on Synge's ideas, are pioneering contributions to FEM [Silvester and Pelosi 1994]. Independently of Argyris, S. Levy developed a matrix method for the analysis of the structural behavior of the wings of an airplane using an assemblage of elements of different shapes. Like Hrenikoff and McHenry he used polynomial approximations [Levy 1953]. Both the methods of Levy and Argyris played an important role in the USA aeronautical industry during the last part of the 1950s.

The formal presentation of FEM is attributed to M. J. Turner, R. W. Clough, H. C. Martin and L. J. Topp in their work "Stiffness and Deflection Analysis of Complex Structures" published in 1956. In this paper both the local approximation of the partial differential equations and the assembly strategies that are essential to FEM are addressed. It is interesting to notice that the local properties of the elements were deduced without using variational principles. In 1960 J. H. Argyris and S. Kelsey published a pioneering book. That same year the name *Finite Element Method* was used for the first time by R. Clough [Clough 1960]. In his 1989 presentation, R. Clough gave all credits for the work developed to M. J. Turner [Clough 1989]. However, his own contributions in topics like convergence and other issues were of major importance [Clough and Wilson 1962]. And, no doubt, the invention of the method name is among the most relevant ones [Gupta and Meek 1996].

During the 1960s the demand for efficient and reliable numerical methods, greatly amplified by the needs of the space program in the United States, was the key motivating factor for the development of FEM. During those years it was performed entirely by engineers. The names of B. Fraeijs de Veubeke, R. H. Gallagher, B. M. Irons, R. J. Melosh, and T. H. H. Pian may be added to those mentioned before. During this time FEM was based on intuitive reasoning, analogies with naturally discrete systems, and numerical experimentation. Discretization errors were handled by uniform or quasi-uniform mesh refinement [Szabó and Babuška 1991]. Once the engineering community realized that new formulations could be obtained from variational principles, this topic dominated the literature. Whenever an operator was not symmetric, it was thought that the problem was unfit for FEM application because it was not leading to a stationary problem in the Rayleigh-Ritz way [Oden 1990]. Of special relevance is the work of O. C. Zienkiewicz, that besides launching FEM into other fields of engineering [Zienkiewicz and Cheung 1965] gave a clear and complete description of FEM and its attributes (e.g., continuity properties of the solution, banded and sparse structure of matrices) [Gupta and Meek 1996]. The FEM conferences, organized in the Air Force Flight Dynamics Laboratory (Dayton, Ohio) led to the gathering of a large number of specialists who contributed to the development of FEM and its expansion worldwide and to other fields like civil, electrical, and electromagnetic engineering [Proceedings 1965, 1968, 1971]. From the mid-1960s FEM rapidly developed in several directions.

- It was reformulated from variational considerations under the form of weighted residuals [Zienkiewicz and Holister 1965], [Greene et al. 1969], [Oliveira 1968].
- High-precision elements [Felippa 1966], [Silvester 1969c], and isoparametric elements [Ergatoudis et al. 1968], [Irons and Zienkiewicz 1968] were developed.
- The first studies of methods that are nowadays recognized as important alternate solutions were done, like spectral finite elements, Hermite and bicubic spline approximations, mixed and hybrid FEMs, and methods for nonlinear systems.
- The mathematical analysis of FEM was addressed and topics like interpolation properties, *a priori* error estimates, and convergence were studied. The paper by M. Zlámal [1968] attracted the attention of mathematicians and the numerical

analysis community. Important contributions in this area were done by P. G. Ciarlet, E. R. D. A. Oliveira, and M. H. Schultz, among others [Oden 1990].

- The FEM was recognized as a general tool for the solution of partial differential equations. It was applied to nonlinear and nonstationary problems in the field of structural analysis and it was extended to other fields: geophysics, fluid mechanics, thermal, and electromagnetic problems, for example.
- Techniques related to the implementation of FEM were also addressed, like frontal solution techniques, and numerical integration schemes [Oden 1990].
- The first books on the method were published, such as [Przemieniecki 1968], and [Zienkiewicz and Cheung 1967], for example.

The decade of the 1970s was of tremendous importance for the mathematical foundations of FEM. The mathematical conferences organized on a regular basis represented landmarks in this area [Aziz 1972], [Whiteman 1973, 1976, 1979, 1982, 1985, 1988, 1991, 1994]. A tremendous progress took place in all the topics addressed previously like, for example, *a priori* error estimates and convergence of linear and nonlinear, elliptic, parabolic, hyperbolic, and eigenvalue problems. The works relevant for the type of problems considered in this book are referenced in Chapter 2. The FEM was derived in the frame of functional analysis and its generalization was completed [Finlayson 1975]. The work of P. G. Ciarlet and P. A. Raviart on interpolation theory deserves special mention [1972a,b]. Several books devoted to the FEM mathematical properties appeared during this decade [Strang and Fix 1973], [Oden and Reddy 1976], [Ciarlet 1978]. During the 1980s the mathematical foundations of FEM for nonlinear problems was an active field of research.

The research activity regarding *a priori* error estimates that took place during the second half of the 1970s continued during the next decade. The distinction between reducing the error by mesh refinement and the alternate approach based on increasing the order of the polynomial basis functions led to the labels h-version and p-version. Usually the symbol h is used to represent the size of the finite elements. Convergence occurs when the size of the largest element ($h_{\max}$) is progressively reduced (hence the name h-version). The polynomial order is usually denoted by the symbol p. Convergence occurs when the lowest polynomial order ($p_{\min}$) is progressively increased (hence the name p-version). The hp-version is the combination of both procedures. An understanding of the different versions and how to combine them was reached by the mid-1980s. Research on *a posteriori* error estimation techniques started in the second half of the 1970s and continues until nowadays. A related area is that of self-adaptive refinement procedures designed to reduce the error with improved efficiency. A historical account is given in Chapter 4. The contribution of I. Babuška and coworkers to both *a priori* and *a posteriori* error estimation must be highlighted.

Formulations employing variables related through constitutive laws, giving rise to mixed, dual, or hybrid FEMs were also studied during the 1970s and 1980s. The vector quantities involved were approximated through new vector elements [Raviart and Thomas 1977a], [Nédélec 1980, 1986], [Brezzi et al. 1985]. Elements of this type play today a crucial role in electromagnetic field problems. Open problems were also addressed during the 1970s. The names of P. Bettess (infinite elements) and C. A. Brebbia (BEM method) may be mentioned. Hybrid FE-BI methods were proposed during the 1970s [Zienkiewicz et al. 1977]. This field is still an active area of research.

The 1970s marked the extension of FEM to all fields of engineering. Thus a large number of books have been published during these thirty years dealing with FEM general theory, implementation, and applications. A number of them are listed in the Bibliography. Research continues in all the topics mentioned previously. Its progress may be followed through the scientific journals listed in the Bibliography. Spectral and hierarchical elements, mesh generation, superconvergent postprocesses, optimization, and intelligent systems are active research areas. A big effort has been initiated to quantify the computational issues in order to reduce the requirements of memory, storage, and overall run-time. A number of general purpose software packages have been developed. They usually require big computers. An excellent summary of the packages available as of 1987 is given in [Kardestuncer and Norrie 1987]. A new generation of computer programs with modular structure, more adapted to small machines, is being developed.

1.2.3 The Finite Element Method in the Field of Electromagnetic Engineering

A pioneering paper applying FEM to field problems derived from Poisson's equation with applications to electrical problems can be traced back to 1967 [Zienkiewicz et al. 1967]. The first paper to be published in a scientific journal (*Electronics Letters*) dealing with the FEM solution of an electromagnetic wave problem is due to S. Ahmed [1968]. The modes propagating along homogeneous metallic waveguides were analyzed employing a scalar variational formulation in terms of the longitudinal component of either the magnetic field for transverse electric (TE) modes, or the electric field for transverse magnetic (TM) modes. First-order scalar Lagrange elements were used. P. P. Silvester presented a similar study at the 1968 URSI International Symposium, but it was published in *Alta Frequenza* only the next year [Silvester 1969a]. Also in 1968, P. L. Arlett, A. K. Bahrani, and O. C. Zienkiewicz, published a paper in *Proceedings of the IEE* applying FEM to the analysis of several homogeneous waveguides, a 3D homogeneous transducer, and a dielectric-loaded waveguide. They did not provide complete information about the formulation used for the two latter devices.

In 1969, higher order Lagrange elements were developed by P. P. Silvester [1969b,c]. They were applied to homogeneous waveguide (eigenvalue) problems and quasi-static (deterministic) problems. The same year S. Ahmed and P. Daly [1969a,b] published two papers on homogeneous and inhomogeneous waveguide analysis, respectively. A vector formulation in terms of both longitudinal components was used in the latter case. Other papers using similar approaches appeared during the 1970s [Csendes and Silvester 1970, 1971a,b], [Daly 1971], including the analysis of anisotropic and optical waveguides [Vandenbulcke and Lagasse 1976], and lossy structures [McAulay 1977]. The appearance of spurious (nonphysical) modes due to the non definite character of the operator forced the researchers to use formulations based on other field components. However, spurious modes also polluted the computed spectrum. Such modes were reported also for 3D eigenvalue problems (cavities) and 2D and 3D full-wave deterministic problems (discontinuities). Procedures were proposed to separate or distinguish them from physical solutions. Spurious modes have been a nightmare for microwave engineers and is one of the reasons for the slow development of FEM in this field. The understanding of this topic has been reached only recently. As an alternate to scalar elements, a 2D vector element (i.e., with vector basis functions) was used for the first time in 1984 [Hano 1984]. It was different from those utilized from 1977 in other fields of engineering. Once the microwave community realized that spurious modes were correctly modeled by these new elements as numerically zero-eigenvalue solutions, researchers proposed different 2D and 3D linear and higher order vector elements. An account is given in Sections 3.3.2 and 7.2.

The relatively reduced use of FEM for high frequency problems up to the late 1970s is due also to the difficulty of applying it to open-region problems, mainly 2D and 3D scattering and radiation problems. Active research in this field started by the 1980s. Methods combining FEM with analytical techniques and integral equation-based procedures have been proposed (see Sections 1.1.1.1, 1.1.2.1, and 1.1.2.2), as well as other approaches. Research continues in this topic. An account is given in Section 6.2.2.

As of this date FEM has been applied to the quasi-static analysis of arbitrary-shaped inhomogeneous anisotropic multiconductor transmission lines, as well as the full-wave analysis of general waveguiding structures, including dielectric waveguides and optical fibers. Problems like discontinuities (posts, irises, junctions, and bends, for example), design of microwave devices, dielectric resonators, 2D and 3D scattering problems, antenna analysis, penetration of electromagnetic waves in biological bodies, and semiconductor devices modeling, have been successfully solved. Losses, nonlinearities, and magnetic materials have been included [Coccioli et al. 1996a]. Nine years were required to publish the second book in FEM electromagnetics [Koshiba 1992] after that of P. P. Silvester and R. L. Ferrari [1983]. However, a third one [Jin 1993], an anthology on the topic [Silvester and Pelosi 1994], and a book reporting recent applications [Itoh et al. 1996] have been published in a short period of time. The number of publications in scientific journals grows year by year. A list of them may be found in the Bibliography. The leading role has been taken by the *IEEE Transactions on Magnetics*. In fact, due to its flexibility and facility in handling nonlinearities, FEM became the most popular numerical tool for the low frequency, magnetics, and power engineering community after the pioneering work of [Silvester and Chari 1970].

In short, nowadays FEM is well introduced in high frequency electromagnetic engineering. New types of elements are being investigated. Cauchy's, Prony's, Padé's, matrix pencil, and similar methods have been used for efficient wideband analysis of electrically large structures. Results on FEM-time domain are available [Coccioli et al. 1996a]. Recent papers on the topic are [Brauer et al. 1997], [Dibben and Metaxas 1997], [Koh et al. 1997], [Kolbehdari et al. 1996], [Lee et al. 1997], [Roy et al. 1998], [Sarkar et al. 1998]. Active research on error analysis and other techniques has been undertaken in order to increase the accuracy and convergence of problems solved by this versatile procedure. This book attempts to address this topic.

Chapter 2

The Finite Element Method

2.1 INTRODUCTION

As mentioned in Section 1.2.2, the origin of the Finite Element Method (FEM) goes back to the early 1940s. It was initially developed in the field of structural mechanics. When the mathematical studies about the method were started in the sixties, it was evident that the method was a general technique for the numerical solution of problems formulated by means of partial differential equations. From then on, the method has been developed by engineers and mathematicians, with applications in many areas of science and engineering. Thus the bibliography on FEM is extremely voluminous. In this book, a partial list of basic texts is given. Additional references that deal with its mathematical and numerical aspects, and applications are available in the Bibliography.

In the electromagnetic field and more specifically in microwave and millimeter waves engineering, FEM was not used until the end of the 1960s (see Section 1.2.3). Its popularity in the solution of problems in this area has increased considerably in the last few years. This chapter is devoted to a general description of the method and the introduction of the concepts that will be utilized in the rest of the book.

The analysis of the electromagnetic structures considered in this book may be reduced to the solution of linear boundary value problems in terms of one or more unknown variables. Thus, the chapter starts with the presentation of these problems, followed by the description of the steps that are necessary for their FEM solution.

After defining the domain of the problem and its classical or strong formulation from a set of partial differential equations, the concept of weak or integral formulations is introduced. An overview is given of the numerical methods based on such weak integral formulations.

Next, the concepts that are more specific for FEM (those related to the way in which the problem is discretized) are explained with more detail. The original domain is split into a set of subdomains having simple geometrical shapes over which the unknowns of the problem will be approximated by means of a linear combination of basis functions (usually, polynomial functions). Typically the coefficients of such linear combination are related to given points of the subdomain called nodes. Each of these subdomains, with its nodes and basis functions, constitutes a finite element. Two-dimensional (2D) and three dimensional (3D) finite elements, referred to as Lagrange elements, are introduced first. Thereafter, infinite 2D elements are described. They are used for the analysis of open-region problems. Two-dimensional div-conforming elements as well as 2D curl-conforming elements are also introduced. Finally, 3D curl-conforming elements are presented. In all the cases, linear and higher order versions as well as straight and curved versions are discussed.

The next section illustrates how to apply the FEM to a weak formulation and how it is used to discretize a continuous problem. The integral formulation over the whole domain is written as the sum of integral forms, one over each element. Then the FEM approximation of the unknowns is introduced into each local integral form. In this way a discretized local integral form (i.e, a set of integral expressions over each element and its boundary) is obtained in terms of computable quantities and the coefficients of the linear combination. The continuity conditions of the unknowns of the continuous problem allow to assemble the local expressions into a global discrete integral form (i.e., a set of integral expressions over the whole domain of the problem and its boundary) in terms of the quantities computed previously and all the coefficients of the local approximations. The assembly process is a typical procedure of FEM. The global set of integral expressions is transformed to a system of algebraic equations after enforcing the boundary conditions. The unknowns (or degrees of freedom) are the coefficients of the linear combination of the FEM approximations. This global system of equations involves sparse matrices. An overview of the available numerical solvers of sparse systems of algebraic equations is given.

Once the discrete problem is solved, there is usually an interest in computing the value of the unknown or other variables at given points. Some considerations about the postprocesses involved are also given. The analysis of the error incurred in the solution of a problem by means of FEM and its rate of convergence are topics of paramount importance. A section is devoted to these concepts.

The chapter ends with an example of a flow chart of a FEM program and a list of some available software packages. Additional material related to this chapter can be found in Appendixes A to C.

2.2 GENERAL PRESENTATION OF THE FINITE ELEMENT METHOD AS APPLIED TO LINEAR BOUNDARY VALUE PROBLEMS

A physical system is characterized by a set of variables that may depend on the space coordinates of a given spatial region (domain) and on time. The system is said to be stationary or time-invariant if there is no time dependence. Some variables of the system (physical properties, dimensions, boundary conditions, or initial conditions, for example) may be known *a priori*. The remaining variables constitute the unknowns of the problem. They also are called dependent variables because they depend on the known variables.

By means of physical laws (such as equilibrium law, energy conservation law and constitutive relations) it is possible to establish a mathematical model that will interrelate the known and the unknown variables. These relations lead, in general, to a system of equations, whose solution will ultimately provide the values for the unknown variables of the problem. The number of parameters that are necessary to determine the unknown variables at any point of the domain at a given instant of time is the number of degrees of freedom of the system of equations. A system is said to be continuous if the number of degrees of freedom is infinite. It is said to be discrete if it is finite.

The behavior of a discrete system may be represented by means of a system of algebraic equations; whereas the behavior of a continuous system is usually described employing a set of differential equations, partial differential equations, or integro-differential equations, with given conditions on the boundary of the domain or at the initial instant of time. In both cases, the system of equations may be linear or nonlinear. For the first case, the known variables that constitute the input data of the problem do not depend on the unknown variables. For the second case, one or more of the input data are not really independent, but dependent upon one or more unknown variables.

The system of algebraic equations for a discrete linear system may be solved, at least theoretically, in an exact way. When the number of equations is large, a computer-based solver must be used. The system of differential equations corresponding to a continuous problem cannot be solved, in general, in an exact analytical way. It is necessary to employ an approximate numerical method in which the first step will consist of the discretization of those equations. That leads to replacing the continuous problem by a system of a finite number of algebraic equations. FEM is one of the numerical procedures that may perform this discretization [Oden 1973], [Wexler 1969], [Zienkiewicz 1973, 1977].

This book deals with time-invariant, linear, and continuous problems formulated by means of a set of differential equations defined in a given domain having various boundary conditions. Such a formulation is called classical or strong formulation. Two categories of problems are considered [Dhatt and Touzot 1981; Ch. 3]:

1. *Equilibrium or deterministic problems.* They are described by a system of partial differential equations, which in matrix notation may be expressed as

$$[\mathfrak{L}]\{u\} - \{f_\Omega\} = \{0\}, \quad \text{in the domain } \Omega \tag{2.1a}$$

with

$$[\zeta]\{u\} - \{f_\Gamma\} = \{0\}, \text{ on the boundary } \Gamma \tag{2.1b}$$

In these equations, $\{u\}$ is a column matrix (a vector) of unknown functions. These functions, in turn, may be scalar or vector quantities, either real or complex. Further, $\{f_\Omega\}$ and $\{f_\Gamma\}$ are column matrices of excitation functions, which are known. Finally, $[\mathfrak{L}]$ and $[\zeta]$ are square/rectangular matrices of differential operators, which are independent of the unknown functions because the system is linear. Figure 2.1(a) illustrates a deterministic problem.

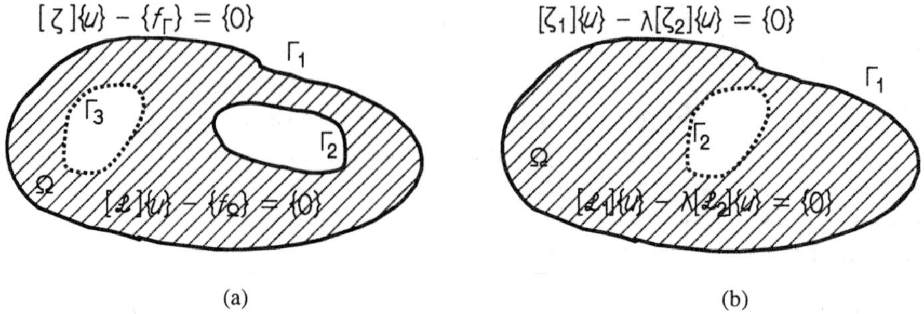

Figure 2.1 Problems analyzed in this book: (a) deterministic problems, (b) eigenvalue problems.

Examples of problems included in this category are the quasi-static analysis of transmission lines (see Section 3.2.3), the electrostatic analysis of 2D open structures (see Section 6.3), the 2D scattering problem (see Sections 6.4 and 6.5), the 3D analysis of discontinuities, electromagnetic scattering, and antenna problems (see Sections 7.4 and 7.5). The column vector of unknown variables may be reduced to a scalar quantity as the electric potential or its dual in the case of the direct or dual formulation of the quasi-static analysis of transmission lines (see Sections 3.2.3.1 and 3.2.3.2, respectively), or the longitudinal component of the electric or magnetic field, as in the 2D scattering problem. It also may consist of a vector variable as the electric or magnetic field in the analysis of discontinuities or the 3D scattering and antenna problems. Or it may consist of a scalar variable and a vector one as, for example, in the mixed formulation of the quasi-static analysis of transmission lines, where the unknowns are the electrostatic potential and the displacement vector (see Section 3.2.3.3). The matrix of differential operators in (2.1a) may be, for example, the generalized harmonic equation, for the case of the direct formulation of the quasi-static analysis of transmission lines, where the coefficients of the excitation vector are zero. In all these examples, the boundary conditions require that certain conditions must be fulfilled at the surface of the conductors involved in the problem or at the symmetry or magnetic walls, if they exist.

2. *Eigenvalue problems*. They constitute an extension of the deterministic problems in which to each critical value (or eigenvalue) of a given parameter (being an unknown of the problem) there corresponds a vector of unknown functions, referred to as an eigenvector. They may be described as

$$[\mathcal{L}_1]\{u\} - \lambda[\mathcal{L}_2]\{u\} = \{0\}, \text{ in the domain } \Omega \tag{2.2a}$$

$$[\zeta_1]\{u\} - \lambda[\zeta_2]\{u\} = \{0\}, \text{ on the boundary } \Gamma \tag{2.2b}$$

Here, λ is the eigenvalue and $[\mathcal{L}_1]$, $[\mathcal{L}_2]$, $[\zeta_1]$, and $[\zeta_2]$ are matrices of differential operators. Figure 2.1(b) illustrates this category of problems. Examples of these classes of problems are the full-wave analysis of waveguiding structures (see Section 3.3) or the analysis of cavities (see Section 7.3). As with the deterministic problems, the unknowns may be scalars, vectors, or mixed scalars and vectors. Examples of the first case are the longitudinal component of the electric or the magnetic field in a full-wave analysis of homogeneous isotropic waveguides or both longitudinal components in a full-wave analysis of inhomogeneous anisotropic waveguides (see Section 3.3.3.1). An example of the second case is the solution of cavity problems, where the unknowns are the electric or the magnetic field vectors (see Section 7.3). Finally, an example of the mixed case is the full-wave analysis of general waveguiding structures where the unknowns are the longitudinal component of the electric or the magnetic field, which is treated as a scalar, and the corresponding transverse component, which is treated as a vector (see Section 3.3.3.2). The matrix of differential operators corresponds to expressions derived from Maxwell's equations for the full-wave analysis of waveguides or cavities.

It must be stressed that the same problem allows several formulations depending on the variables that are selected as unknowns. A judicious choice according to the circumstances should always be done.

In any case, as opposed to other numerical procedures, the FEM does not perform a direct discretization of equations (2.1) or (2.2) in their differential form. In fact, FEM discretizes them once they are transformed into an integral or variational form, called the weak formulation of the problem. Weak integral formulations may be developed by means of the weighted residual method or by making stationary a variational principle from which equations (2.1) or (2.2) are derived. The dependent variables or unknowns of the continuous problem are also called variational variables or variational unknowns.

Once a weak formulation is obtained, the next step consists of the subdivision of the domain into a set of elements having simple geometric shapes. In each of the elements an approximation of the unknown will be made according to

$$\{\tilde{u}^{e}\} = \left[N^{e}\right]\{d^{e}\} \tag{2.3}$$

where $\{\tilde{u}^e\}$ represents the approximate value of the variational unknowns over the eth element; $[N^e]$ is the matrix of approximation functions selected by the user for the eth element; and $\{d^e\}$ is the local column matrix of the coefficients of the linear combination. The term *local* refers to the fact that the unknowns are locally approximated by (2.3) over each element. The coefficients of $[N^e]$ are known basis functions (usually, polynomials functions), appropriately chosen. The coefficients of $\{d^e\}$ constitute the unknowns or *degrees of freedom* for the discrete problem. They are defined as the values at given points of the elements, called nodes, of either the variational unknowns or a quantity related to any of their directional derivatives or any other quantities related to them, like in the case of vector variational unknowns where they are defined as their normal or tangential component with respect to the edges or faces of the elements. There are finite elements in which some of the degrees of freedom, although defined over the element, are not strictly associated with a node. The analytical expressions for the coefficients of $[N^e]$ depend on the type of problem and the formulation used to define it along with the geometry of the element, the number of degrees of freedom associated with such an element, and the location of each node.

Once the domain is divided into subdomains the variational integral form is split into the sum of local integral forms, one over each element. Then the substitution of (2.3) in the eth element leads to a discretized integral form over it and results in one of the following two expressions:

$$\{W^e\} = \left[k^e\right]\{d^e\} - \{f^e\} \tag{2.4a}$$

or

$$\{W^e\} = \left[k^e\right]\{d^e\} - \lambda\left[m^e\right]\{d^e\} \tag{2.4b}$$

where the coefficients of $[k^e]$, $[m^e]$, and $\{f^e\}$ involve easily computable integral expressions in terms of the basis functions, the node coordinates, and the input data.

The continuity conditions that must be fulfilled by the variables of the problem force the continuity of the degrees of freedom shared by two or more elements (i.e, the value of the degrees of freedom shared by two or more elements must be identical). This allows the assemblage of the local expressions (2.4), so that a global expression of the same type is obtained. Imposing on it the boundary conditions a system of equations is derived that may be written in matrix form as

$$[K]\{D\} = \{F\} \tag{2.5a}$$

or

$$[K]\{D\} = \lambda[M]\{D\} \tag{2.5b}$$

where the coefficients of the global matrices $[K]$, $[M]$, and $\{F\}$ are obtained from the coefficients of the matrices $[k^e]$, $[m^e]$, and $\{f^e\}$ for each element, and from the boundary conditions. The column matrix $\{D\}$ represents the global vector of degrees of freedom or unknowns of the discrete problem. In this way, the FEM develops an algebraic system of equations (2.5), that represents a discretization of (2.1) or (2.2). Thus the continuous problem, with an infinite number of degrees of freedom, is transformed into an approximate discrete problem with a finite number of degrees of freedom that is easy to solve with the help of a computer.

For the deterministic problem, the system of equations (2.5a) provides a unique solution for $\{D\}$. For an eigenvalue problem, the solution of (2.5b) is not unique; a set of possible eigenvalues, λ_i, will be obtained. Corresponding to the ith eigenvalue, there will be an eigenvector $\{D^i\}$ of degrees of freedom.

The degrees of freedom of an element are shared only by the neighboring elements. Thus the matrices in (2.5) are highly sparse and banded. When solving the FEM system of equations it is important to use techniques that will take advantage of such a structure of the matrices. This will reduce the memory and storage requirements as well as the overall run-time and central processing unit (CPU) time.

Once $\{D\}$ is computed, the approximate value of any of the variables of interest and the corresponding approximate values of any other quantity related to them for any point of given coordinates belonging to the domain of definition of the problem or over its boundary can be computed. For example, the *a posteriori* estimates or indicators of the errors incurred in the procedure, which are the fundamental tools for the self-adaptive mesh procedures as described in Chapter 4 [Carey and Oden 1984; Ch. 1].

Figure 2.2 provides a summary of the classification of physical systems and illustrates the process of discretization of a continuous system by means of FEM. Figure 2.3 shows a schematic outlining the various steps necessary for the FEM analysis of a linear time-invariant continuous problem. The following sections explain each of them in more detail.

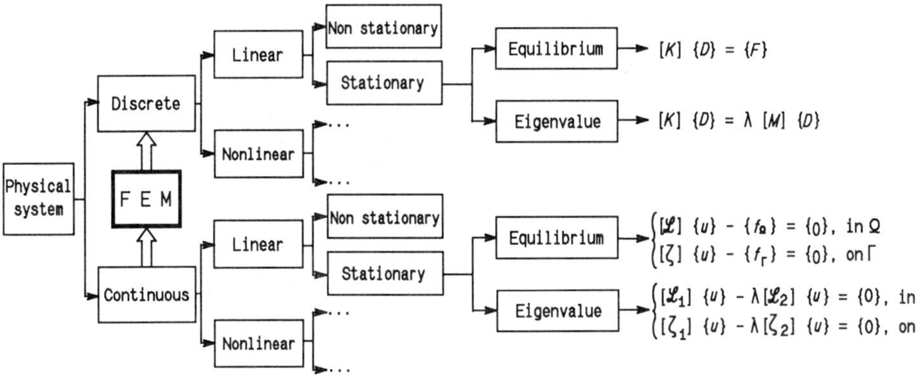

Figure 2.2 Classification of physical systems and discretization process for continuous systems.

2.3 DEFINITION OF THE CONTINUOUS PROBLEM

This section is subdivided into three subsections that treat topics related to the definition of the domain of the problem, its strong or classical formulation, and its variational formulations.

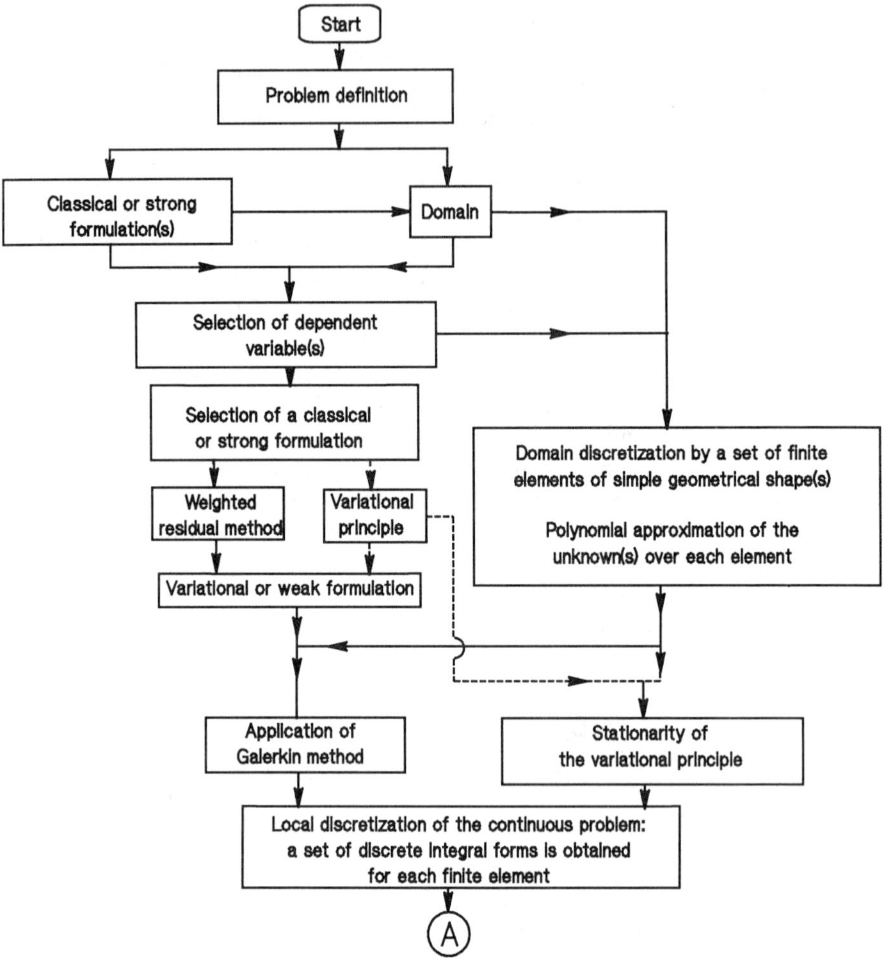

Figure 2.3 Flow chart for the FEM analysis of a continuous time-invariant linear problem.

2.3.1 Definition of the Domain of the Problem

The formulation of a problem implies the geometrical definition of its domain, its boundaries, and its physical characteristics. In this book, 2D and 3D problems are considered. In the 2D problems, the various quantities may be functions of two independent spatial variables that determine the plane in which the domain is located. In 3D problems, the various quantities may be functions of three independent spatial variables.

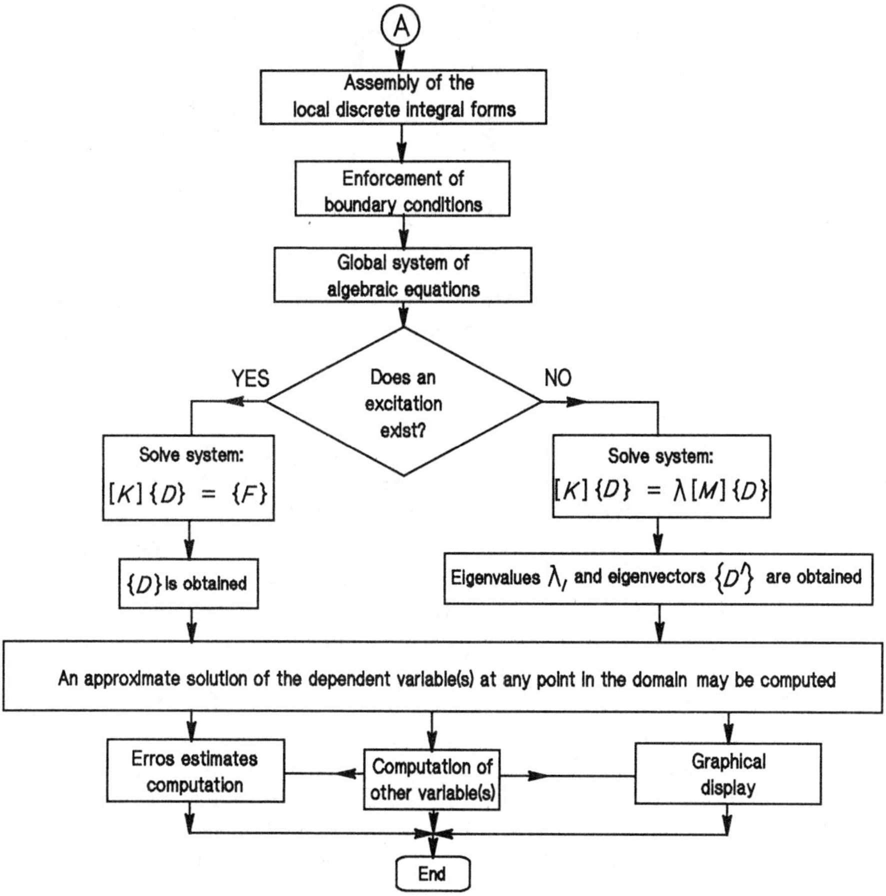

Figure 2.3 (Continued.)

A 2D domain is a set of points in a plane with the property that, given a point in the domain, all points sufficiently close to it belong to the domain. The boundary of the domain is a set of points in the plane such that given a point on the boundary, points belonging to the domain as well as points not belonging to it may be found in its neighborhood. The definitions of a 3D domain and its boundary follow a similar scheme. In this book, domains with only one boundary or with several boundaries are considered, as well as open-region problems (those in which infinity is one of the boundaries) and closed-domain problems [Reddy 1984, 1993; Ch. 2]. Figure 2.4 illustrates these definitions.

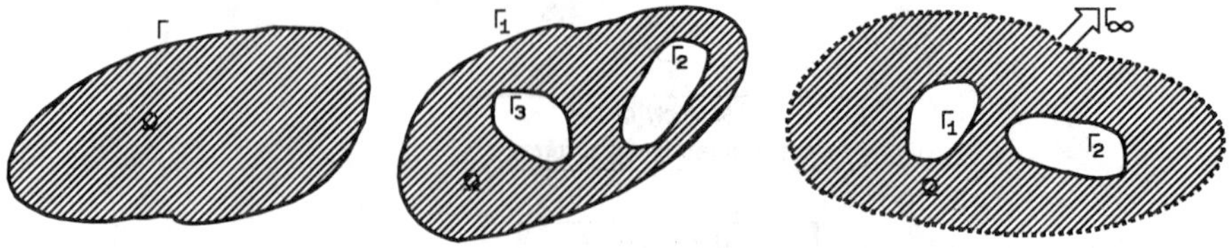

(a) (b) (c)

Figure 2.4 Domain and boundaries of the problem: (a) closed-domain problem with one boundary, (b) closed-domain problem with multiple boundaries, (c) example of an open-region problem.

The geometrical definition of the domain implies the establishment of a global coordinate system that allows the geometrical definitions of the faces, edges, vertices, and nodes of the finite elements that are used to discretize the domain. For the analysis of the problems that are considered in this book, the Cartesian system is utilized.

However, it must be mentioned that in some cases the use of cylindrical or spherical systems of coordinates allows the reduction of the dimension of the problem to be analyzed. This amounts to saving CPU time, and computer memory and storage requirements. Examples are axisymmetric structures where the electromagnetic field solutions also display an axisymmetric configuration [Barton 1982], [Daly 1973b, 1974], [Davies et al. 1982], [English 1971], [English and Young 1971], [Gakũrũ and Ferrari 1992], [Hernández-Gil and Pérez-Martínez 1985], [Hernández-Gil et al. 1987], [Kajfez and Guillon 1986; Ch. 5], [Kooi 1985], [Lee et al. 1993], [Okamoto and Okoshi 1978], [Oldfield and Ide 1985], [Petre and Zombory 1988], [Sakellaris et al. 1992], [Segerlind 1976; Ch. 10], [Wilkins et al. 1991], [Zhu and Landstorfer 1995].

Even if for the sake of generality a Cartesian system of coordinates is preferred, the reduction of the dimension of the problem is feasible in many cases. A typical example is the analysis of waveguiding structures where the translational symmetry allows a 2D analysis (see Section 3.1). Further examples are the analysis of E-plane and H-plane discontinuities, planar networks, and a class of scattering problems [Coccioli et al. 1996b,c], [Hirayama and Koshiba 1989, 1990, 1992], [Ise and Koshiba 1988], [Kanellopoulos and Webb 1990], [Kiyoshi and Masanori 1988], [Koshiba et al. 1982a], [Koshiba 1992; Ch. 7], [Lee and Csendes 1987, 1988], [Ramahi and Mittra 1991a,b], [Silvester 1973], [Webb and Parihar 1986], [Webb 1990], [Wu et al. 1989].

Furthermore, the computational domain may be reduced by considering the characteristics of the domain of the problem, such as the existence of physical or geometrical symmetry, or antisymmetry axes, or planes. This allows in many occasions to reduce the size of the domain by considering those axes or planes as boundaries of the computational domain where special conditions may be imposed. Nevertheless, a word of caution. In some cases, reducing the size of the domain and imposing these special boundary conditions may eliminate some of the possible solutions of the problem. For example, in some eigenvalue problems (like the computation of the modes propagating along a waveguiding structure), the existence of geometrical symmetries does not imply that only symmetric eigenvectors (or at least eigenvectors with a given type of symmetry) may be solutions to the problem [Zienkiewicz 1977; Ch. 20]. In the same way, periodic boundary conditions may be advantageously exploited in a computer program. However, as in the previous case, the reduction of the domain must be done with caution [Gluckstern and Opp 1985].

In the application of FEM to open-region problems, the truncation of the infinite domain is mandatory. In general, FEM cannot tackle directly these type of problems while other numerical procedures are more suited for their solution, such as a surface integral equation formulation in connection, for example, with the method of moments (MOM) or the boundary element method (BEM). Nevertheless, FEM may take into account the case of open boundaries in several ways.

To start with, the boundary at infinity may be simulated just by considering a large enough domain, closed by an artificial boundary where the conditions at infinity are imposed. In order to reduce the size of the computational domain as much as possible, iterative procedures have been proposed [Ikeuchi et al. 1981]. An alternate solution is to divide the infinite domain into two types of subdomains. Those of ordinary type (near-field regions), and those containing the boundaries at infinity or open type (far-field regions). Then, infinite elements may be utilized to discretize the latter subdomains while ordinary finite elements may be used in the former subdomain. In Section 2.5.1.2, references dealing with infinite elements for static problems are given and a 2D infinite element is presented. In Section 3.2.3.1, some results of its application to the quasi-static analysis of open transmission lines are given. Infinite elements for other problems have been presented in [Beer and Meek 1981], [Bettess and Zienkiewicz 1977], [Bettess 1988], [Chiang 1985], [Gratkowski and Ziolkowski 1992], [Gratkowski et al. 1996], [Hayata et al. 1988], [McDougall and Webb 1989], [Nakata et al. 1990], [Petre and Zombory 1988], [Svedin 1991], [Towers et al. 1993], [Zienkiewicz et al. 1985]. Alternate approaches are to recast the far-field region into a closed one through suitable transformations (see Section 1.1.1.1) [Brunotte et al. 1992], [Freeman and Lowther 1988], [Nath and Jamshidi 1980] or to use other nonclassical techniques to treat the far-field region as summarized in Section 6.2.2.

A second group of methods discretizes only the near-field region and imposes suitable boundary conditions at the artificial boundary enclosing it. The topic is summarized in Section 6.2.2 and will not be repeated here. In this book an iterative hybrid finite element-boundary integral (FE-BI) method has been selected that allows a considerable reduction of the computational domain while at the same time provides a numerically exact radiation condition at the artificial boundary (see Chapter 6 and Section 7.5).

Before finishing this section, just some words with respect to the definition of the physical characteristics of the domain. One of the sources of errors in any analysis comes from the approximation utilized when defining them. This book deals with structures in which the physical variables are either scalar (isotropic media) or dyadic (anisotropic media) quantities. Their coefficients are either constant in the whole domain (homogeneous structures) or a function of the spatial coordinates (inhomogeneous structures). The latter is the case whenever the domain is composed of several physical media. In those cases, the original domain must be subdivided into as many subdomains as different physical media are present. The discretization must be done so that each element belongs to only one medium or subdomain. Then, the physical characteristics will be constant in each subdomain, so that the medium, although inhomogeneous, will be homogeneous in each subdomain and element. For a more general case, it may happen that in a given subdomain the physical characteristics may be dependent on the (x,y,z) coordinates. That dependence may be formulated by means of an exact or approximate analytical expression. In these cases, the value of the physical characteristics will not be constant over elements belonging to that subdomain. A way of discretizing the domain is to assign to each element the value of the media parameters at the centroid of the element. If numerical integration is employed to compute the coefficients of the FEM matrices, then their value at each point of integration may be used [Zienkiewicz 1977; Ch. 17]. A structure may be homogeneous with respect to one physical characteristic but not with respect to another. This fact may be relevant when choosing the variable that is going to be the unknown of the problem. In some cases, the same problem may be defined in such a way that the formulation only involves a physical characteristic with respect to which the structure is homogeneous. This fact may be exploited advantageously. In this book, in the analysis of the different electromagnetic problems that are considered, it is always assumed that the coefficients of the dyadics that describe the physical characteristics of the domain are real. Lossy media having complex coefficients are analyzed in an approximate way, by means of perturbational methods.

2.3.2 Classical or Strong Formulation of the Problem

The strong formulations of the continuous problems to be considered are given by (2.1) and (2.2). They are second-order elliptic problems with boundary conditions, involving one or more unknowns. They are also called boundary value second-order elliptic problems. The reader is directed to the bibliography regarding such problems. The mathematical aspects related to FEM may be found in various books, for example, in [Kardestuncer and Norrie 1987; Ch. 3] and [Raviart and Thomas 1983; Chs. 2 and 6].

Although the unknown for some of those problems may be a complex quantity, the formulation will be manipulated to obtain a final result in terms of real operators with real unknowns. In the general presentation that follows, it is assumed that the operators and unknowns are always real. However, most of the definitions, both in this chapter and in the appendixes, may be generalized to the complex case.

The coefficients of the matrices of operators $[\mathfrak{L}]$, $[\mathfrak{L}_1]$, and $[\mathfrak{L}_2]$, of (2.1a) and (2.2a), can have two forms. For scalar problems, as in the quasi-static analysis of transmission lines in terms of the electric scalar-potential or for the full-wave analysis of waveguiding structures in terms of the longitudinal component of the electric and/or the magnetic fields, they are given by

$$\mathcal{L}_{kl} = -\{\nabla_t\}^T [a]\{\nabla_t\} + b \tag{2.6a}$$

where matrix notation has been used. The superscript ^{T} represents the matrix transpose operator; $[a]$ is a matrix of dimensions 2×2; and b is a scalar quantity. Some of the coefficients of $[a]$, as well as b, may be zero. These coefficients are related to the physical characteristics of the medium and, as mentioned before, may be continuous or discontinuous functions of the spatial coordinates. For vector problems, as in the case of the full-wave analysis of waveguiding structures in terms of the electric field or the magnetic field, the coefficients are

$$\{\mathcal{L}_{kl}\} = \{\{\nabla\} \times [a]\{\nabla\} \times + b\} \tag{2.6b}$$

where $[a]$ is now a 3×3 matrix. In (2.6) $\{\nabla_t\}$ and $\{\nabla\}$ are the column matrix operators in two or three dimensions, respectively, defined by

$$\{\nabla_t\} = \begin{Bmatrix} \dfrac{\partial}{\partial x} \\ \dfrac{\partial}{\partial y} \end{Bmatrix} \tag{2.7a}$$

$$\{\nabla\} = \begin{Bmatrix} \dfrac{\partial}{\partial x} \\ \dfrac{\partial}{\partial y} \\ \dfrac{\partial}{\partial z} \end{Bmatrix} \tag{2.7b}$$

and the vector product between two column matrices denoted by × is given by

$$\begin{Bmatrix} u_x \\ u_y \\ u_z \end{Bmatrix} \times \begin{Bmatrix} v_x \\ v_y \\ v_z \end{Bmatrix} = \begin{Bmatrix} u_y v_z - u_z v_y \\ u_z v_x - u_x v_z \\ u_x v_y - u_y v_x \end{Bmatrix} \tag{2.7c}$$

The matrix expressions (2.1a) and (2.2a) show that for problems with several unknowns there may be coupling between the unknowns in the equations [Reddy 1984; Ch. 4].

The independent term in (2.1a) represents the sources of excitation of the deterministic problem. It may be a zero column vector, like in the case of the quasi-static analysis of transmission lines, and then the system of partial differential equations is said to be of the homogeneous type

$$\{f_\Omega\} = \{0\} \tag{2.8}$$

For problems with sources (like scattering, discontinuity, and antenna problems considered in Sections 6.4, 6.5, 7.4, and 7.5) this term is not zero, and the system of partial differential equations is said to be inhomogeneous.

With reference to the conditions on the boundary, Γ, represented by (2.1b) and (2.2b), it is useful to define the conormal derivative of a scalar variable, u_m, with respect to the operator $\mathfrak{L}_{kl}$ of (2.6a) as

$$\frac{\partial u_m}{\partial n_{\mathfrak{L}_{kl}}} = \{a_n\}^T \, [a] \, \{\nabla_t\} \, u_m \tag{2.9a}$$

and for the vector variable $\{u_m\}$ with respect to the vector operator $\{\mathfrak{L}_{kl}\}$ of (2.6b) as

$$\frac{\partial \{u_m\}}{\partial n_{\{\mathfrak{L}_{kl}\}}} = \{a_n\} \times [a] \{\nabla\} \times \{u_m\} \tag{2.9b}$$

where $\{a_n\}$ in matrix notation, or $\bar{a}_n$ in vector notation, is the unit vector normal to Γ.

As indicated previously, the unknowns $\{u\}$ of the problem may be scalar or vector complex quantities. Vectors may have one, two, or three components in the directions of a Cartesian system of coordinates.

Boundary conditions are of the following types:

1. *Dirichlet or first kind.* Given by

$$u_m = u_d, \quad \text{on } \Gamma_D, \text{ for problems where (2.6a) holds} \tag{2.10a}$$

or

$$\{a_n\} \times \{u_m\} = \{u_d\}, \quad \text{on } \Gamma_D, \text{ for problems where (2.6b) holds} \tag{2.10b}$$

The functions, u_m, related to this condition are usually called primary variables. In the FEM context, boundary conditions of this type are called essential boundary conditions.

2. *Neumann or second kind.* Given by

$$\frac{\partial u_m}{\partial n_{\mathcal{L}_{kl}}} = g_n, \quad \text{on } \Gamma_N, \text{ for problems where (2.6a) holds} \tag{2.11a}$$

or

$$\frac{\partial \{u_m\}}{\partial n_{\{\mathcal{L}_{kl}\}}} = \{g_n\}, \quad \text{on } \Gamma_N, \text{ for problems where (2.6b) holds} \tag{2.11b}$$

Please note that the definitions for the scalar and vector conormal derivatives have been used, respectively. The left-hand quantities in these expressions, i.e., the right-hand quantities of (2.9) involve the scalar or vector product of $\overline{a}_n$ with the variables $[a]\{\nabla_t\}u_m$ or $[a]\{\nabla\}\times\{u_m\}$, respectively. These variables are usually called secondary variables. In the FEM context, conditions of this type are called natural conditions.

3. *Third kind.* This case consists of a linear combination of (2.10) and (2.11).
4. *Mixed type.* In this case, part of the boundary has Dirichlet conditions and the rest has Neumann conditions. Thus, (2.10) stands for Γ_D and (2.11) for Γ_N, where $\Gamma_D \cup \Gamma_N = \Gamma$. In this case as well as in the cases for Dirichlet or Neumann conditions, g_n and u_d are constants that might have different values for different portions of the boundaries. For the problems analyzed in this book, these conditions are in many cases homogenous, i.e., zero.
5. *Radiation type.* For open-region problems, a condition at infinity must be specified in order to obtain a unique solution for the problem. In the electromagnetic area this condition is the Sommerfeld radiation condition [Jin 1993; Ch.1], [Morita et al. 1990; Ch. 1].

Once the classical or the strong formulation is described, the next problem is how to select the specific formulation that is going to be used for a particular problem. It is well known that the same continuous problem may admit various formulations, all of which are equivalent.

On one hand, it is possible to formulate the problem by means of different systems of partial differential equations of the same order, in which the number of unknowns may be different. For example, the full-wave analysis of a microwave structure with translational symmetry may be formulated using the three components of the electric field and the three components of the magnetic field as unknowns. It will also admit formulations of the same order in which one or more of those unknowns may be eliminated, thus, reducing their number, which of course will depend on the particular problem and the physical characteristics of the media. In fact, formulations with six, four, three, two, and one unknown are possible. It is possible also to use formulations where the unknowns are not physical quantities, but mathematical ones, like the scalar or vector potentials from which the electric and magnetic fields are derived [Hammond 1982].

On the other hand, starting from a formulation in terms of both the primary and secondary variables (as in a mixed formulation), a different formulation may be obtained by eliminating the secondary variable, giving rise to the standard or direct formulation. It is also possible to formulate the problem only in terms of the secondary variable, obtaining the complementary formulation. In both cases, the formulations will be called irreducible if there is no possibility of reducing the number of unknowns of the same type (primary or secondary) [Zienkiewicz and Taylor 1989; Ch. 12]. Formulations of the complementary type are usually of lower order than the direct ones. The portions of the boundary that have a Neumann condition in the direct problem will have a Dirichlet condition in the complementary formulation for the secondary variable. The selection of one or the other formulation depends, among other considerations, on the variable of interest. Mixed formulations may be preferable in some cases [Arthurs 1980], [Reddy 1984; Ch. 5], [Kardestuncer and Norrie 1987; P. 2, Ch. 9], [Zienkiewicz 1977; Ch. 12], [Zienkiewicz and Taylor 1989; Chs. 12 and 13]. In Section 3.2.3.3, a mixed formulation for the quasi-static analysis of transmission lines is presented.

The number of unknowns or the variable of interest are not the only factors to be taken into account when selecting a formulation of the problem. It is also important to consider the characteristics of the weak integral form related to it. To be more precise, what is important is the numerical behavior of the FEM discretization. Each formulation will have advantages and drawbacks when compared with others. Thus, a compromise must be made. In the following, the factors to be taken into account are briefly highlighted.

The final objective of the analysis should be considered initially. For example, the goal may be to obtain the primary parameters of a transmission line (i.e., the coefficients of the capacitance and inductance matrices: see Section 3.2.2). These parameters must be calculated from the FEM solution obtained for the unknowns of the problem. In general, the accuracy for the line parameters will not be as high as the accuracy of the FEM procedure because a postprocess of the computed results is required. It is possible to perform such a postprocess either from the primary or secondary variables. The use of formulations based on one or the other variable provides, usually, a bound of a different type (a lower or an upper bound) of the exact value of the parameters. Both bounds are usually obtained with similar accuracy. Hence in this case there is no clear criterion for selecting one or the other formulation [Arthurs 1980]. Instead, if the goal is the computation of the electric field propagating along the transmission line it is clear that it will be preferable to choose as unknowns the components of the electric field themselves. Then the field components will be obtained with the highest accuracy allowed by the specific FEM implementation.

If possible, a formulation should be selected that will employ a minimum number of unknown variables in order to reduce the size of the system of equations as well as memory, storage, and CPU time requirements. Nevertheless, the use of a formulation with a higher number of unknowns may be more suitable in some cases because of different reasons, mainly related to the numerical aspects of the problem. To be more specific, whenever possible, formulations leading to selfadjoint positive definite problems should be preferred. Similarly, formulations leading to a discrete system of equations characterized by dense or full matrices should be avoided. The sparsity of the FEM matrices is of paramount importance.

If the exact solution of a given variable is singular at some points of the domain of the problem, such a variable should not be selected as the variational unknown. Otherwise, the errors when applying FEM will be very high and the method will not converge properly. For example, some components of the electric or the magnetic field will have singularities at conductor or dielectric corners. In those cases, formulations in which the singularity will appear in the derivative of the unknown are preferable (hence, the use of potentials is sometimes preferred). However, it is also possible to select as unknown the function that will become singular and alleviate the problem in a different way, like adopting a discretization scheme that will not make use of the values of the unknowns at the singularity points as in the case of vector elements (also called conforming elements). Singular elements may also be used. These latter elements try to mimic the singular behavior of the electromagnetic field [Akin 1982; Ch. 6], [Kardestuncer and Norrie 1987; P. 2, Ch. 6], [Zienkiewicz 1977; Ch. 23]. However, in general, if the exact solution of a problem is not smooth enough these local remedies are not sufficient to obtain accurate solutions and high rates of convergence. An alternate solution that seeks both the global and local improvement of the error incurred is to employ self-adaptive procedures (see Chapter 4).

In the application of the FEM, the continuity across elements of the degrees of freedom (the values of the unknown variables or of quantities related to them at the nodes) is used to assemble the discrete local integral forms of all elements to obtain the global system of equations. Thus, formulations leading to discontinuities of those quantities should be avoided. In other words, the variables chosen as unknowns should be those that, after applying FEM, will produce a system of algebraic equations whose degrees of freedom would be physically continuous over the domain. Otherwise, the use of discontinuous functions will require adding new equations to the original formulation in order to approximate their jump or discontinuity. This, although possible, will complicate the application of FEM and in most cases will lead to nonbanded matrices with lower sparsity. In most of the electromagnetic problems, possible discontinuities of the variables are related to changes of the dielectric permittivity or permeability of the medium of the domain. Thus, in general, those components of the field vectors (the electric field, the magnetic field, the electric displacement vector, or the magnetic flux density vector) that are continuous over the domain should be preferred as unknowns.

Finally, it should be pointed out that if a symmetric system of algebraic equations is obtained, there is the possibility of saving memory, storage, and CPU time requirements because only the upper or the lower triangular part of the matrices need to be dealt with when solving the global system of equations. Solvers specially designed for symmetric sparse matrices are very efficient. This is particularly important in the case of eigenvalue problems. If the problem is formulated from a quadratic variational principle, the symmetry of the global system of equations is ensured. Thus, strong formulations derived from quadratic variational principles are usually preferred.

2.3.3 Weak Formulation of the Problem

Any system of partial differential equations with boundary conditions defining a problem is called a strong formulation of the problem because it requires a solution with strong regularity conditions, i.e., existence and given continuity conditions of the unknown variables and their derivatives up to the order of the partial derivatives involved [Kardestuncer and Norrie 1987; P. 1, Ch. 3].

For this reason, it is convenient to weaken those conditions so that a wider set of approximate solutions may be possible. Thus, weak or generalized formulations of the problem are sought, so that the problem will be easier to solve.

It is obvious that a strong solution of a problem is, at the same time, a weak solution of it, but the reverse is not always true. A problem may or may not have a strong solution. If it exists, the weak solution must be unique and should coincide with the strong solution.

In a strict sense, the denomination *variational formulation* is applied to any weak formulation that has been obtained by utilizing a stationary variational principle. In this book (as it is done in other cases, [Reddy 1984; Ch. 2], [Raviart and Thomas 1983; Ch. 2]), it is also used to refer to a weak integral form obtained by weighting the strong formulation with a test function (or weighting function) and integrating it (eventually, by parts) over the domain. In many problems, that weak formulation is equivalent to making stationary a quadratic functional or variational principle [Oden 1973], [Mikhlin 1964], [Mikhlin and Smolitsky 1967]. Next, the two procedures to obtain weak or variational formulations are presented.

2.3.3.1 *The Weighted Residual Method*

Given a system of differential equations of order s (i.e., involving partial differential equations up to order s) defined by (2.1) or (2.2), the residual over the domain, $\{R(\{u\})\}$, is defined as

$$\{R(\{u\})\} = [\mathfrak{L}]\{u\} - \{f_\Omega\} \tag{2.12a}$$

for deterministic problems of type (2.1) or as

$$\{R(\{u\})\} = [\mathfrak{L}_1]\{u\} - \lambda[\mathfrak{L}_2]\{u\} \tag{2.12b}$$

for eigenvalue problems of type (2.2). The residual will be identically zero when $\{u\}$ is the strong solution of (2.1) or (2.2).

An integral form corresponding to (2.1) or (2.2) may be obtained by weighting (2.12) by a vector of weighting functions, $\{v\}$, and integrating it over the domain resulting in

$$W(\{u\},\{v\}) = \int_{\Omega} \{v\}^T \{R(\{u\})\}\, \mathrm{d}\Omega = \int_{\Omega} \{R(\{u\})\}^T \{v\}\, \mathrm{d}\Omega \tag{2.13}$$

This integral form will consist of a bilinear part and a linear part (see Appendix A.2). The solution of the problem (2.1) or (2.2) is equivalent, under certain conditions, to that of finding the set of functions $\{u\}$ that will make (2.13) zero, according to

$$W(\{u\},\{v\}) = \int_{\Omega} \{v\}^T \{R(\{u\})\}\, \mathrm{d}\Omega = \int_{\Omega} \{R(\{u\})\}^T \{v\}\, \mathrm{d}\Omega = 0 \tag{2.14}$$

for every subset of weighting functions $\{v\}$ belonging to a set $\{V\}$. Here $\{u\}$ is a subset of admissible solutions that satisfy the boundary conditions. The fulfillment of some of the boundary conditions may also be imposed in a weak sense, adding to (2.14) the weighted integral form of the corresponding residual along the boundary. The procedure mentioned previously is called, for obvious reasons, the weighted residuals method. In this general formulation, when $\{u\}$ is real, it coincides with MOM [Harrington 1967]. When $\{u\}$ is complex, $\{v\}$ should be substituted in (2.13) and (2.14) by $\{v^*\}$, where * denotes the complex conjugate (i.e., $\{v^*\}^T$ is the conjugate transpose $\{v\}^H$ of $\{v\}$).

Any solution $\{u\}$ that satisfies (2.1) or (2.2) will also satisfy (2.14), no matter how the set of weighting functions is selected. On the contrary, the solution $\{u\}$ of (2.14) will depend on the selection of $\{V\}$. If the set $\{V\}$ is infinite and consists of the Dirac distribution $\{\delta(x,y,z)\}$ over Ω, the solution $\{u\}$ of (2.14) will also satisfy (2.1) or (2.2). If $\{V\}$ is finite, the solution of (2.14) will be an approximate solution $\{\tilde{u}\}$ of (2.1) and (2.2), that is, a generalized solution of (2.1) or (2.2). The solution $\{\tilde{u}\}$ provided by the weighted residual method makes $\{R(\{\tilde{u}\})\}$ and $\{v\}$ orthogonal in the mean. It is evident that in order to satisfy (2.14) the integral must be defined. Thus, $\{\tilde{u}\}$ and $\{v\}$ must be such that in (2.14) a finite integral is obtained. This implies that there is freedom in selecting $\{\tilde{u}\}$ and/or its derivatives as discontinuous functions as long as the corresponding integral is still finite. In this way, the continuity conditions of $\{u\}$ may be relaxed and the set of admissible functions, $\{\tilde{u}\}$, will be larger. For the moment, the only condition that has been imposed on $\{v\}$ is the integrability of (2.14).

If an even weaker formulation is desired, integration by parts may be used (see Appendix A.2) to yield an expression of the type

$$\begin{aligned} W(\{u\},\{v\}) &= \int_{\Omega} l\left(\{u\},\{v\},\left\{\frac{\partial u}{\partial x}\right\},\left\{\frac{\partial u}{\partial y}\right\},\left\{\frac{\partial u}{\partial z}\right\},\left\{\frac{\partial v}{\partial x}\right\},\left\{\frac{\partial v}{\partial y}\right\},\left\{\frac{\partial v}{\partial z}\right\},\ldots\right) \mathrm{d}\Omega \\ &\quad + \int_{\Gamma} m(\{u\},\{v\},\ldots)\, \mathrm{d}\Gamma = 0 \end{aligned} \tag{2.15}$$

It may be observed that an integral extended to the boundary of the problem has been obtained. That integral may be decomposed into a sum of two other integrals. The first one covers the part of the boundary where the unknown must satisfy Dirichlet type boundary conditions. This term may be eliminated by choosing the weighting functions to be zero on that part of the boundary; i.e., the weighting functions should fulfill homogeneous Dirichlet conditions on that part of the boundary. The second boundary integral extends over the rest of the boundary. It is always possible to write this integrand in terms of the weighting functions and the conormal derivatives of the unknowns. This implies that the unknown variables will not appear in the remaining boundary integral term. In fact, this term involves the product of the weighting functions by the known values of the natural boundary conditions of the problem. Hence, there is no need to force the unknown variables to satisfy these conditions that will be fulfilled in a distributional sense. On the contrary, the unknown variables should be forced to satisfy the Dirichlet conditions over the corresponding part of the boundary. This is why the Dirichlet boundary conditions are called essential.

In (2.15) it may also be observed that in the integrand of the integral over the domain Ω the order of the derivatives of $\{u\}$ has been reduced (e.g., for a second-order problem, from second to first order). Thus, the requirement on the existence and continuity of subsequent derivatives is reduced. The regularity conditions that $\{\tilde{u}\}$ must satisfy are less stringent. However, the first derivatives of the weighting function have appeared. Hence, those derivatives should exist. Therefore when using integration by parts, the regularity conditions that $\{v\}$ must satisfy are more strict.

Examples of the advantages of using such a procedure may be given with reference to the electromagnetic problems solved in this book. The quasi-static analysis of transmission lines and the electrostatic analysis utilize as unknown the electric scalar potential (see Sections 3.2 and 6.3, respectively). Given formulations of the full-wave analysis of waveguiding structures and some 2D scattering problems also employ scalar unknowns that are the longitudinal component of the electric and/or the magnetic field (see Sections 3.3.3.1, 6.4, and 6.5). In all those cases, the problems are defined by second-order differential equations, according to (2.1) or (2.2), with (2.6a), (2.10a), and (2.11a). Starting from that formulation, a weak formulation would be obtained according to which it would be enough, for both the approximate functions $\{\tilde{u}\}$ and the weighting functions $\{v\}$, to belong to the first-order Hilbert space over the domain Ω, $H^1(\Omega)$: continuous functions whose first derivatives may be discontinuous but should be square integrable (see Appendix A.1.3).

In the case of the full-wave analysis of waveguiding structures described in Section 3.3.3.2, the unknowns are a scalar and a vector function, namely, the longitudinal and the transverse components of the electric or the magnetic field. Starting from (2.2), (2.6b), (2.10b), and (2.11b), a weak formulation is obtained, for which it is enough if both the unknown functions and the weighting functions belong to the space $H^1(\Omega)\times H(\text{curl},\Omega)$. This means that the longitudinal component and the scalar weighting function must belong to the aforementioned space, $H^1(\Omega)$, while the transverse component and the vector weighting function must belong to the space $H(\text{curl},\Omega)$. Hence, vectors and their curl may have discontinuous components but they must be square integrable (see Appendixes A.1.5 and A.1.6).

For the case of 3D analysis of cavities, discontinuities, scattering, and antenna problems, the unknown is the electric or the magnetic field (see Chapter 7). The solution obtained through the weighted residual method will belong to the $H(\mathrm{curl},\Omega)$ space.

The application of the weighted residual method to several equivalent strong formulations, with a different number or type of unknowns, will lead to several equivalent integral forms of the complementary or mixed type. The unknown functions and the weighting functions will have different requirements than those of the direct formulation. For example, in the mixed formulation of the quasi-static analysis of transmission lines, the unknown variables are the scalar electrostatic potential and the displacement vector (see Section 3.2.3.3). The first variable must be square integrable. The second one should belong to the $H(\mathrm{div},\Omega)$ space. The later space is the space of vectors having discontinuous but square integrable components and divergence (see Appendix A.1.4) [Raviart and Thomas 1977a], [Thomas and Joly 1981].

It is also possible to generate additional weak integral formulations by modifying the original one. The objective usually is to increase the rate of convergence or to reduce the number of unknowns. Among other techniques, the penalty function method is of interest. It adds to the original formulation an integral form corresponding to a particular condition that must be fulfilled by the variables of the problem multiplied by a penalty factor. This procedure is employed, for example, in the mixed formulation of the quasi-static analysis of transmission line structures in order to eliminate the scalar unknown (see Section 3.2.3.3).

2.3.3.2 *Variational Principles*

Development of a Weak Formulation from a Functional

The weighted residual method provides a weak integral formulation of a problem. In many cases, it will coincide with the final expression obtained from a variational principle. The formulation through a weighted residual method can always be applied. Instead, a weak formulation may be obtained from a variational principle only when the corresponding functional exists.

As mentioned earlier, the set of admissible solutions of a problem defined by (2.1) or (2.2) should belong to a given space and should fulfill certain boundary conditions. For given deterministic and eigenvalue problems, an integral form or functional, $F(\{u\})$, may be constructed. The principle of stationarity may be introduced as follows. Among all those admissible solutions, the one satisfying expressions (2.1) or (2.2) will also produce a stationary point of the functional $F(\{u\})$. This means that its first variation, δF, is zero (see Appendix A.3.1):

$$\delta F(\{u\}) = 0 \tag{2.16}$$

and it is for this reason that the functional $F(\{u\})$ is called a variational principle. Equations (2.1) or (2.2.) are called the Euler equations of that variational principle and they are derived from it. Conversely, the stationary principle may be formulated by stating that the solution $\{u\}$ that makes a variational principle stationary is also a solution of its Euler equations. Thus, when a variational principle exists, a way of solving (2.1) or (2.2) consists of applying to $F(\{u\})$ the stationarity condition (2.16).

The general form of a functional or variational principle is given by

$$\begin{aligned} F(\{u\}) &= F\left(\{u\},\left\{\frac{\partial u}{\partial x}\right\},\left\{\frac{\partial u}{\partial y}\right\},\left\{\frac{\partial u}{\partial z}\right\},\dots\right) \\ &= \int_\Omega f\left(\{u\},\left\{\frac{\partial u}{\partial x}\right\},\left\{\frac{\partial u}{\partial y}\right\},\dots\right)\mathrm{d}\Omega + \int_\Gamma g\left(\{u\},\left\{\frac{\partial u}{\partial x}\right\},\left\{\frac{\partial u}{\partial y}\right\},\dots\right)\mathrm{d}\Gamma \end{aligned} \tag{2.17}$$

After applying (2.16), the following general expression is obtained (see Appendix A.3.1)

$$\begin{aligned} \delta F(\{u\},\{\delta u\}) =& \int_\Omega l\left(\{u\},\{\delta u\},\left\{\frac{\partial u}{\partial x}\right\},\left\{\frac{\partial u}{\partial y}\right\},\dots,\left\{\frac{\partial(\delta u)}{\partial x}\right\},\left\{\frac{\partial(\delta u)}{\partial y}\right\},\dots\right)\mathrm{d}\Omega \\ &+ \int_\Gamma m\left(\{u\},\{\delta u\},\left\{\frac{\partial u}{\partial x}\right\},\left\{\frac{\partial u}{\partial y}\right\},\dots\right)\mathrm{d}\Gamma = 0 \end{aligned} \tag{2.18a}$$

Using integration by parts, the previous expression may be written as

$$\delta F(\{u\},\{\delta u\}) = \int_\Omega \left\{ f_1\left(\{u\},\left\{\frac{\partial u}{\partial x}\right\},\left\{\frac{\partial u}{\partial y}\right\},\left\{\frac{\partial u}{\partial z}\right\},\dots\right)\right\}^T \{\delta u\}\mathrm{d}\Omega = 0 \tag{2.18b}$$

It may be observed that (2.18a) and (2.18b) are formally identical to (2.15) and (2.14), respectively. This is seen by identifying $W(\{u\},\{v\})$ with $\delta F(\{u\},\{\delta u\})$ and $\{v\}$ with $\{\delta u\}$, respectively.

Thus, a variational principle may be used in two different ways:

1. Stationary conditions may be directly applied to it to obtain the solution $\{u\}$ of (2.1) or (2.2).
2. It may be used as a means to obtain a weak formulation of the weighted residual type by identifying the integral form $W(\{u\},\{v\})$ with $\delta F(\{u\},\{\delta u\})$ and the weighting functions $\{v\}$ with $\{\delta u\}$ as indicated by

$$W(\{u\},\{v\}) = \delta F(\{u\},\{\delta u\})_{|\{\delta u\}=\{v\}} = 0 \tag{2.19}$$

If the operators $[\mathfrak{L}]$ of (2.1) and (2.2) are selfadjoint, then it is always possible to find a variational principle from which those equations are derived (see Appendix

A.3.1). Whenever a variational principle exists, the stationarity of that principle will have a clear physical meaning. In these cases, the application of a variational principle is very attractive [Angkaew et al. 1987], [Berk 1956], [Chang et al. 1990], [Chew and Nasir 1989], [Cvetkovic and Davies 1986], [English 1971], [English and Young 1971], [Fernández and Lu 1991], [Hoffmann 1984], [Jeng and Chen 1984], [Kitazawa and Hayashi 1987], [Kitazawa 1989a,b], [Konrad 1976], [Liu and Chen 1981], [McAulay 1977], [Mabaya et al. 1981], [Mishra et al. 1985], [Morishita and Kumagai 1977], [Ohtaka et al. 1976], [Ohtaka and Kobayashi 1990], [Okoshi and Okamoto 1974], [Pichon and Bossavit 1993], [Rozzi et al. 1991], [Wu and Chen 1985a,b], [Yamashita 1968], [Yamashita and Mittra 1968]. Quite frequently, variational principles are derived for problems that do not have them in a strict sense [Lindell 1982, 1992], [Lindell and Sihvola 1983], [Oksanen and Lindell 1989a,b]. Variational principles formulated to include the essential boundary conditions for example, have also been derived [Hazel and Wexler 1972], [Jin 1993; Ch.6].

Generalized, Complementary, and Mixed Variational Principles

The variational principle or the functional corresponding to a classical irreducible formulation in terms of primary variables is usually called a direct variational principle. If a direct variational principle exists, other equivalent variational principles may be obtained.

Consider a problem whose solution makes the functional $F(\{u\},\{q\})$, which is a function of the variables $\{u\}$ and $\{q\}$, stationary. Consider that those variables are interrelated. In other words, it may be said that the problem is derived from a variational principle with conditions. In Appendix A.3.2 it is described how to obtain from F a set of variational principles without conditions. Figure 2.5 summarizes the process. Essentially the alternate variational principles are:

1. A direct variational principle as a function of the primary variables $\{u\}$;
2. Several generalized variational principles as functions of primary variables $\{u\}$, secondary variables $\{q\}$, and one or more Lagrange multipliers;
3. A mixed variational principle as a function of the primary variables $\{u\}$ and secondary variables $\{q\}$;
4. A complementary variational principle as a function of the secondary variables $\{q\}$.

It is interesting to observe that pairs of functionals may exist that are complementary between them. This fact provides a means of computing upper and lower bounds of the exact value of a given quantity. Thus, complementary variational principles allow one to estimate the error of an approximate solution.

Numerical methods employing variational principles seek a good approximation, $\{\tilde{u}\}$, of the exact solution, $\{u\}$, that makes a functional stationary. If the second variation of such a functional is positive, the stationary point is a minimum for the functional. Thus, any approximate solution will give an upper bound $F^u = F(\{\tilde{u}\})$ of that exact minimum value $F(\{u\})$. If the second variation is negative, the exact solution gives a maximum of the functional. Thus, any approximate solution will give a lower bound, $F^l = F(\{\tilde{u}\})$, of that exact maximum value, $F(\{u\})$. The functionals of the first type are called convex functionals, and those of the second type are called concave functionals.

Consider a functional of a mixed type $F_m(\{u\},\{q\})$ that may be expressed in two equivalent forms: a convex functional $F_u(\{u\})$ and a concave functional $F_q(\{q\})$. Formally, with the appropriate relations between $\{u\}$ and $\{q\}$, the following identity holds:

$$F_u(\{u\}) = F_m(\{u\},\{q\}) = F_q(\{q\}) \tag{2.20}$$

The stationarity of any of these two functionals would provide a unique value for F_m. In one case, it will consist of the minimum of the convex functional, and in the other case, of the maximum of the concave functional. If an approximate analysis of the problem is conducted, by means of the double variational formulation, then upper and lower bounds of the exact value would be obtained, according to

$$F_u(\{\tilde{u}\}) \geq F_u(\{u\}) = F_m(\{u\},\{q\}) = F_q(\{q\}) \geq F_q(\{\tilde{q}\}) \tag{2.21}$$

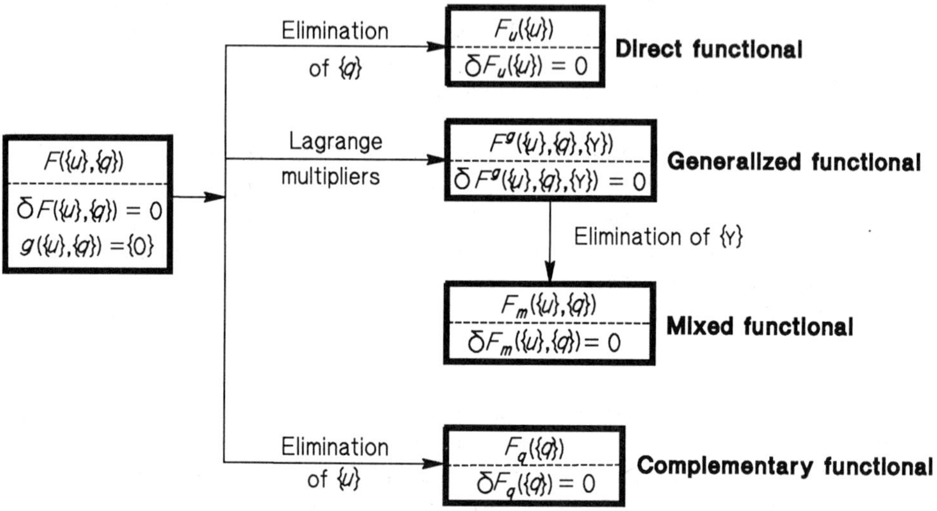

Figure 2.5 Transformation of a variational principle leading to different functionals.

The literature on this topic is quite large [Cheng et al. 1987], [Csendes et al. 1983], [Daly 1984, 1985], [Fremond 1973], [Golias et al. 1994], [Hammond and Penman 1976, 1978], [Hammond and Tsiboukis 1983], [Penman and Fraser 1982, 1983, 1984], [Thatcher 1982]. The work of A. M. Arthurs [1980] contains many references to the classical works in the field and presents a complete methodology on the derivation of complementary functionals for a given set of Euler's equations. The possibility of obtaining upper and lower bounds of the error of an approximate analysis makes it very attractive to use both the direct and the complementary functionals although this implies analyzing twice the same problem. However, formulations based on concave functionals may present problems. Then, it may be preferable to substitute the concave functional by a mixed functional that can also provide the same bound [Arthurs 1980]. However, this double analysis may consume an excessive CPU time. Then, an alternate solution is to develop just the analysis based on the mixed functional that may provide both the primary and the secondary variables, together with adequate error estimates. Hybrid formulations (involving the trace of given variables at the boundaries) have also been developed [Raviart and Thomas 1977a,b].

Modified Functionals

The previous paragraph refers to the so-called natural functionals. These are functionals whose Euler equations coincide with those of the strong formulation of the problem. It is also possible to construct other functionals whose Euler equations will not coincide with the strong formulation of the original problem but with a formulation obtained after some manipulations of the original problem. These manipulations may be interesting for several reasons. For example, when a functional is desired for a problem for which it does not exist in a strict sense. This is the case when the operators of the strong formulation are not strictly selfadjoint [Zienkiewicz 1977; Ch. 3]. Then, modifications may be performed in order to obtain a symmetric functional. In other cases, a modified functional may provide approximate solutions with a higher rate of convergence than those obtained using a weak formulation derived from the original strong formulation. Otherwise, the goal may be to obtain a simple form for the functional.

The best known procedures for obtaining modified functionals are those resulting from the method of least squares:

1. *Least squares.* Given a residual $\{R(\{u\})\}$, a functional may be constructed as

$$F^{ls} = \int_{\Omega} \{R^*(\{u\})\}^T \{R(\{u\})\} \mathrm{d}\Omega \tag{2.22}$$

This functional will always be positive semidefinite, no matter what the differential operators of the original problem are (see Appendix A.3.1). The solution $\{u\}$ that makes F^{ls} stationary will provide the minimum of the integral over the domain of the square of the residuals. It is evident that the original strong formulation will be satisfied in an approximate way [Zienkiewicz 1977; Ch. 3]. For problems where the operator is not selfadjoint it may be more meaningful to apply the method of least squares.

2. *Correction functions.* Consider a problem in which variables $\{u\}$ and $\{q\}$ are related by a set of conditions $\{g\}$ derived from a functional $F_m(\{u\},\{q\})$. A new functional may be constructed as:

$$F' = F_m + \alpha \int_\Omega \{g^*\}^T \{g\} \, d\Omega \tag{2.23}$$

where the value of α may be selected arbitrarily. The stationarity of F' will satisfy the original equations only in an approximate way. This method may also be applied to the boundary conditions on Γ [Zienkiewicz 1977; Ch. 3].

3. *Penalty functions.* Given a problem derived from a functional $F(\{u\})$ and having a residual $\{R(\{u\})\}$, a functional that is a generalization of (2.22) [Reddy 1984; Ch. 2] may be constructed as

$$F' = F + \alpha \int_\Omega \{R^*(\{u\})\}^T \{R(\{u\})\} d\Omega \tag{2.24}$$

4. *Conjugate gradient.* Given a problem in which the residual $\{R(\{u\})\}$ is weighted by a positive definite operator Q, a functional may be defined in matrix form as

$$F'(u) = \int_\Omega \{R^*(\{u\})\}^T [Q^*] \{R(\{u\})\} \, d\Omega \tag{2.25}$$

This is related to the functionals used in iterative procedures for the computation of $\{u\}$ [Sarkar 1991]. When Q is equal to unity, the functional (2.25) coincides with that given by (2.22). Another special case is when Q is given by

$$Q = (\mathcal{L}^* \cdot \mathcal{L})^{-1} \tag{2.26}$$

where $\mathcal{L}$ is the operator of (2.1a) or $\mathcal{L}_1 - \lambda\mathcal{L}_2$ in the case of the problem (2.2). Since vector operators are possible, the scalar product between vectors is used.

2.4 DISCRETIZATION OF AN INTEGRAL FORM

Once an integral or a variational form of the continuous problem is obtained, the next step is to discretize it. The discretization procedure converts the continuous problem, which has an infinite number of degrees of freedom, to a discrete one with a finite number of degrees of freedom. When the discrete problem is solved, it will give an approximate solution of the continuous problem.

The process starts with the selection of an approximation for the continuous unknown variables in a finite-dimensional space. Then, the discretization of the continuous problem is done, either from a weak formulation of the weighted residual type or through the process of making stationary a variational principle.

This section deals first with the topic of the approximation of a function. Next it gives an overview of the methods utilized to discretize weak integral forms. These methods are called, in general, variational methods of approximation. The section finishes with some comments regarding their rate of convergence and other properties.

2.4.1 Approximation of a Function

Given a scalar or a vector function $w(x,y,z)$, its approximation $\tilde{w}$, by means of an interpolation function Πw, is defined as

$$\tilde{w} = \Pi w = \{P(x,y,z)\}^T \{D\} = \sum_{i=1}^{N_d} P_i(x,y,z) d_i \tag{2.27}$$

where $\{D\}$ is a column matrix of N_d coefficients, d_i, which are called the parameters of the approximation or degrees of freedom. $\{P(x,y,z)\}$ is a column matrix of scalar or vector functions; $P_i(x,y,z)$, $i = 1, ..., N_d$, is the ith scalar or vector basis functions or expansion functions of the approximation. These functions are linearly independent. They form a complete set on the N_d-dimensional finite space in which the approximation Πw of w is sought. A component of a vector function (or all of them) may also be approximated treating it as a scalar function, using scalar basis functions according to (2.27). The parameters $\{D\}$ may not have any physical meaning or they may be defined as the value of w (or any of its directional derivatives or any other quantity related to w) at given points of the domain (called nodes). Then the coefficients of $\{D\}$ are called nodal parameters or nodal variables of the approximation. Because the approximation makes use of points, the functions $P_i(x,y,z)$ are usually called interpolation functions. The coordinates of the nodes may be arbitrary. Some of the degrees of freedom may not be related to points but they may be defined as integrals of given quantities related to w over the edges, faces or volume of the domain.

The magnitude of the error involved in the approximation of w by Πw will depend on the type of basis functions selected and the position of the nodes. The magnitude of the error in the approximation of the function w and/or of its derivatives of given order may be measured according to the norms or seminorms defined in Appendices A.1.3 to A.1.5 and depends on the nature of the exact solution w. For the case of a nodal approximation, such an error is called interpolation error. For the nodal variables, the interpolation error will be zero at the nodes. This means that the error in the approximation of the function w or its directional derivatives, for example (depending on the definition of the nodal variable or degree of freedom), will be zero at the nodes.

If the function w is approximated by (2.27) over the entire domain, the functions of the approximation are called complete domain basis functions. In problems where different physical media are present or where the domain has a complex geometry, it seems convenient to split the original domain, Ω, into several subdomains, Ω^e, $e = 1, ..., N_e$, where N_e is the total number of subdomains, and to approximate the unknown variable, according to (2.27), over each of them. This procedure is called approximation by subdomains. If for subdomain Ω^e, (2.27) depends only on the nodes of subdomain Ω^e and its boundary, the approximation is said to be local and the functions of the approximation are called local basis functions. This is the approximation method employed by the FEM. There are other methods for which the approximation also depends on the degrees of freedom belonging to other subdomains.

There are problems that have more than one variational function. Then, it is possible to approximate each of them employing a different number of parameters of the approximation as well as different approximation functions (or interpolation functions). In these cases, the approximation may be expressed in general as

$$\{\Pi w\} = [P]\{D\} \tag{2.28}$$

where $[P]$ is a matrix of basis functions and $\{D\}$ is a vector whose coefficients are all the parameters of the approximation.

In summary, there exists a wide variety of approximation methods. The FEM employs a local approximation method by subdomains, called finite elements. For some classes of elements such an approximation is of nodal type, while for other classes of elements it is not so in a strict sense.

2.4.2 Discretization of a Weak Formulation or Variational Formulation of the Weighted Residual Type

Let us assume an integral form of type (2.14) corresponding to a problem defined by (2.1) or (2.2)

$$W(u,v) = \int_{\Omega} v\, R(u)\, d\Omega = 0 \tag{2.29}$$

The unknown u may be a scalar or a vector quantity. Here the derivation is done with u a scalar quantity, however it may easily be extended to the vector case. If u and v are vectors, the scalar product between two vectors must be used in (2.29). For the sake of clarity, only one unknown is considered. If the strong formulation of the problem involves complex quantities, v should be substituted by v^*. If integration by parts is performed, the integral form in (2.29) will correspond to (2.15). The exact solution for (2.29) consists of finding the function u belonging to an infinite-dimensional space that will satisfy (2.29) for any weighting function v belonging to a given space V of infinite dimension.

The approximate solution of (2.29) consists of replacing the continuous problem by a discrete problem projected into an N_d-dimensional space. The approximate solution will belong to the N_d-dimensional space of functions defined by (2.27). In order to determine the N_d parameters of the approximation $\{D\}$, the weighting functions v must also belong to a N_d-dimensional space. Thus, in order to discretize (2.29), the following steps are necessary:

1. The unknown u must be substituted by the approximation selected according to (2.27). This means that N_d basis functions should be selected and the nature of the degrees of freedom should be defined.
2. The weighting functions (scalar or vector, according to the unknown) must be linearly independent and should be selected from the proper space V. The number of weighting functions must at least be the same as the number of parameters of the approximation, resulting in a square system matrix. If the number of weighting functions is greater than the number of unknowns for the problem, then a numerical least squares solution of a matrix equation needs to be carried out.
3. From (2.29) the following system of equations is obtained for a set of independent weighting functions:

$$\left.\begin{aligned} W_1 &= \int_\Omega v_1\, R(\{P\}^T\{D\})\ \mathrm{d}\Omega = 0 \\ W_2 &= \int_\Omega v_2\, R(\{P\}^T\{D\})\ \mathrm{d}\Omega = 0 \\ &\ \vdots \\ W_{N_d} &= \int_\Omega v_{N_d}\, R(\{P\}^T\{D\})\ \mathrm{d}\Omega = 0 \end{aligned}\right\} \tag{2.30a}$$

4. Taking into account that the operators involved in (2.1) and (2.2) are linear, the expression of the residual can be written in one of the two following forms:

$$R(\{P\}^T\{D\}) = \mathcal{L}(\{P\}^T\{D\}) - f_\Omega = \{\mathcal{L}P\}^T\{D\} - f_\Omega \tag{2.30b}$$

for a deterministic problem or

$$R(\{P\}^T\{D\}) = \{\mathcal{L}_1 P\}^T\{D\} - \lambda\{\mathcal{L}_2 P\}^T\{D\} \tag{2.30c}$$

for an eigenvalue problem. Thus the system of equations may be written as

$$\left.\begin{aligned}
&\left[\int_\Omega v_1 \ \{\mathfrak{L}P\}^T \ d\Omega\right]\{D\} - \int_\Omega v_1 \ f_\Omega \, d\Omega = 0 \\
&\left[\int_\Omega v_2 \ \{\mathfrak{L}P\}^T \ d\Omega\right]\{D\} - \int_\Omega v_2 \ f_\Omega \, d\Omega = 0 \\
&\qquad\qquad\qquad\qquad\vdots \\
&\left[\int_\Omega v_{N_d} \{\mathfrak{L}P\}^T \ d\Omega\right]\{D\} - \int_\Omega v_{N_d} f_\Omega \, d\Omega = 0
\end{aligned}\right\} \tag{2.30d}$$

for a deterministic problem or

$$\left.\begin{aligned}
&\left[\int_\Omega v_1 \ \{\mathfrak{L}_1 P\}^T \ d\Omega\right]\{D\} - \lambda\left[\int_\Omega v_1 \ \{\mathfrak{L}_2 P\}^T \ d\Omega\right]\{D\} = 0 \\
&\left[\int_\Omega v_2 \ \{\mathfrak{L}_1 P\}^T \ d\Omega\right]\{D\} - \lambda\left[\int_\Omega v_2 \ \{\mathfrak{L}_2 P\}^T \ d\Omega\right]\{D\} = 0 \\
&\qquad\qquad\qquad\qquad\vdots \\
&\left[\int_\Omega v_{N_d} \{\mathfrak{L}_1 P\}^T \ d\Omega\right]\{D\} - \lambda\left[\int_\Omega v_{N_d} \{\mathfrak{L}_2 P\}^T \ d\Omega\right]\{D\} = 0
\end{aligned}\right\} \tag{2.30e}$$

for an eigenvalue problem. Equations (2.30d) and (2.30e) lead to algebraic system of equations given by

$$[K]\{D\} = \{F\} \tag{2.31a}$$

for the case of a deterministic problem or by

$$[K]\{D\} - \lambda[M]\{D\} = \{0\} \tag{2.31b}$$

for an eigenvalue problem. $[K]$ and $[M]$ are square $N_d \times N_d$ matrices given by

$$[K] = \int_\Omega \{v\}\{\mathfrak{L}P\}^T \ d\Omega \tag{2.31c}$$

for the deterministic problem or

$$[K] = \int_\Omega \{v\}\{\mathfrak{L}_1 P\}^T \ d\Omega \tag{2.31d}$$

for the eigenvalue problem. In addition,

$$[M] = \int_{\Omega} \{v\} \{\mathcal{L}_2 P\}^T \, d\Omega \tag{2.31e}$$

Here $\{v\}$ is a column matrix of N_d coefficients, given by the N_d weighting functions. $\{F\}$ is a column matrix of N_d coefficients given by

$$\{F\} = \int_{\Omega} \{v\} f_{\Omega} \, d\Omega \tag{2.31f}$$

The coefficients of $[K]$, $[M]$, and $\{F\}$ involve known functions and may be easily computed either by analytical integration or, if necessary, by numerical integration over the domain Ω. If integration by parts was performed over (2.29), then (2.31c) to (2.31e) will involve (for the case of second order problems) the first derivatives of $\{v\}$ and $\{P\}$. An integral extended to the part of the boundary having natural boundary conditions will be added to (2.31f). This integral will involve $\{v\}$ and the values of such boundary conditions. The boundary integral extended to the part of the boundary with Dirichlet conditions is identically zero.

Dirichlet conditions will be imposed by forcing to the prescribed values the related degrees of freedom. The solution of (2.31a) or (2.31b) will provide the unknown parameters of the approximation $\{D\}$.

5. Once $\{D\}$ is computed, an approximate solution $\{\tilde{u}\}$ may be obtained at any point (x,y,z) of the domain according to (2.27). Thus, the coefficients of $\{D\}$ are the finite number of degrees of freedom of the discrete problem. Once $\{\tilde{u}\}$ computed any other quantity related to it may be obtained by means of a postprocess.

This procedure may be immediately extended to the case of more than one variable, $\{u\}$. Figure 2.6 summarizes the aforementioned steps and shows also how the weak formulation resulting from the application of the weighted residual method may be obtained. It is also shown how the weighted residual formulation may be developed from a variational principle as summarized by (2.19).

The possibility of selecting the weighting functions according to different criteria gives rise to alternate methods of discretization. Hence there is an essentially infinite number of ways to discretize the continuous problem. The better known methods are the following:

- *Collocation method (or point matching method)*. The weighting functions are Dirac functions

$$v_i = \delta\,(x - x_i,\ y - y_i,\ z - z_i) \tag{2.32}$$

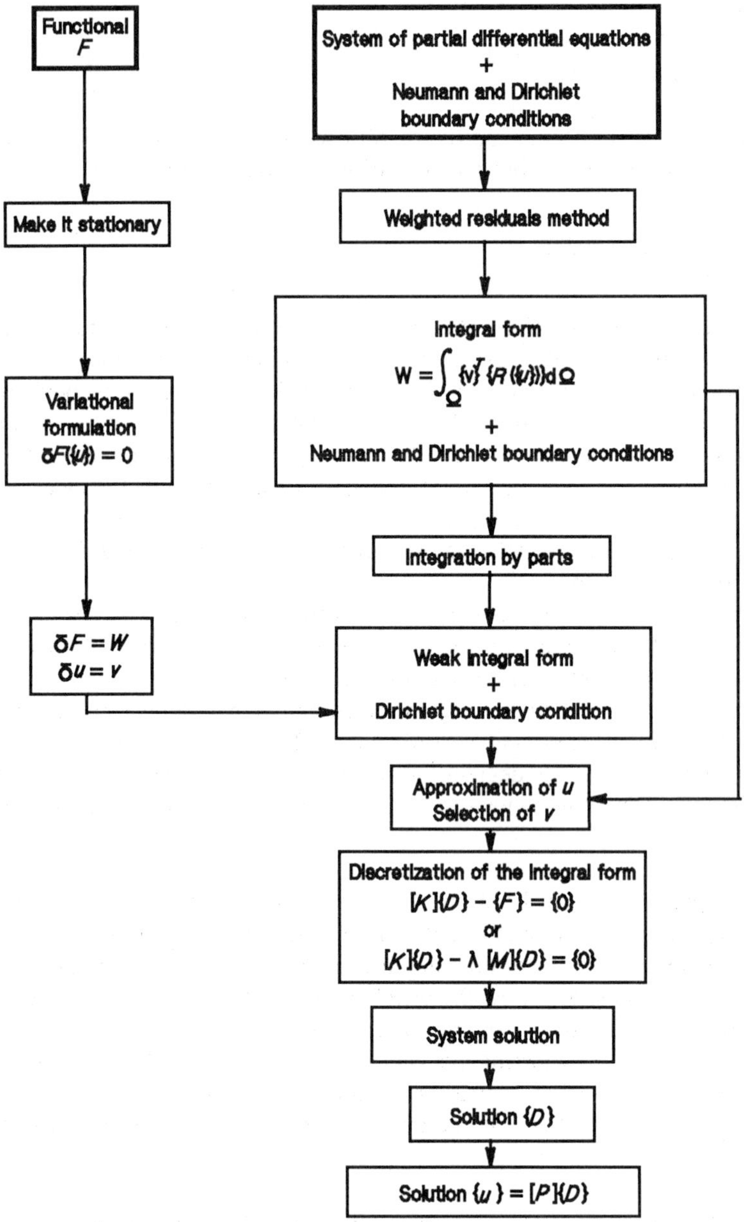

Figure 2.6 Various steps in the discretization process of the continuous problem utilizing a weak formulation derived from the weighted residual method.

at conveniently selected points (x_i,y_i,z_i) for $i = 1, ..., N_d$. This method is equivalent to making zero the residual $R(\{\tilde{u}\})$ at those N_d points, which is equivalent to enforcing the strong formulation to be exactly satisfied at those points. Thus, the integration is not necessary. The correct selection of the points is essential in order to obtain a well-conditioned system of equations and an accurate solution. In numerical electromagnetics this method is better known as the point matching method.

- *Piecewise-constant method* (*or collocation by subdomains*). The weighting functions are equal to unity in given subdomains and zero in the rest. This method is equivalent to enforcing the mean value of the residual to be zero in the related subdomains.
- *Least squares method.* The weighting functions are given by the expression

$$v_i = \frac{\partial R^*(\{P\}^T\{D\})}{\partial d_i}, \qquad 1 \le i \le N_d, \tag{2.33a}$$

which leads to

$$v_i = (\mathcal{L}P_i)^*, \qquad i = 1,\dots,N_d \tag{2.33b}$$

in the case of deterministic problems or to

$$v_i = (\mathcal{L}_1 P_i)^* - \lambda^*(\mathcal{L}_2 P_i)^*, \qquad i = 1,\dots,N_d \tag{2.33c}$$

in the case of eigenvalue problems. This method is equivalent to making stationary the integral expression given by

$$F = \int_\Omega R^*(u)\, R(u)\, \mathrm{d}\Omega \tag{2.34a}$$

once the approximation of u is given by (2.27). The stationarity condition may be written as

$$\delta F(\tilde{u}) = \delta\left(\int_\Omega R^*(\tilde{u})\, R(\tilde{u})\, \mathrm{d}\Omega\right) = 0 \tag{2.34b}$$

Thus, it is equivalent to

$$\delta F(\tilde{u}) = 2\int_\Omega \left(\frac{\partial R^*(\{P\}^T\{D\})}{\partial d_1} R(\{P\}^T\{D\}) \right) \delta d_1 \, d\Omega + \cdots$$
$$+ 2\int_\Omega \left(\frac{\partial R^*(\{P\}^T\{D\})}{\partial d_{N_d}} R(\{P\}^T\{D\}) \right) \delta d_{N_d} \, d\Omega \qquad (2.34c)$$

Expression (2.34c) should be satisfied for every δd_i. Thus, the following system of equations is obtained:

$$\left.\begin{aligned} &\int_\Omega \left(\frac{\partial R^*(\{P\}^T \{D\})}{\partial d_1} R(\{P\}^T\{D\}) \right) d\Omega = 0 \\ &\qquad \vdots \\ &\int_\Omega \left(\frac{\partial R^*(\{P\}^T\{D\})}{\partial d_{N_d}} R(\{P\}^T\{D\}) \right) d\Omega = 0 \end{aligned}\right\} \qquad (2.34d)$$

This method provides a system of equations characterized by positive semidefinite symmetric matrices no matter what the operators of (2.1) or (2.2) are. Nevertheless, it is seldom used because it does not allow integration by parts. This means that the solution u and the basis functions $\{P\}$ should satisfy the regularity conditions in a strict sense.

- *Galerkin method.* As mentioned before, the use of integration by parts in (2.29) will weaken the regularity conditions of the basis functions while making more stringent those of the weighting functions. In fact, for second order problems the conditions for both set of functions will be identical. This is why in the Galerkin method the weighting functions are the same functions as used for the approximation of the unknown. In linear problems in which the operators are of an even order, this choice may lead to algebraic systems with symmetric matrices. This is the procedure that is used more often in FEM. The basis and weighting functions are usually polynomials. In applied electromagnetics, the mode matching techniques may be viewed as Galerkin procedures employing orthogonal functions.
- *Petrov-Galerkin method.* This name is given to the weighted residual method when arbitrary weighting functions are used.

Each of these methods would lead to many others depending on the way in which the approximation of the unknown functions is done as explained in Section 2.4.1.

2.4.3 Discretization of a Variational Principle: Ritz Method

Assume that the continuous problem considered in Section 2.4.2 has a variational principle $F(u)$. Its exact solution would consist of finding the function u (belonging to an infinite-dimensional space) that will make the functional $F(u)$ stationary.

The Ritz method discretizes the continuous problem by seeking an approximate solution, $\tilde{u}$, belonging to a finite-dimensional space, that will make the approximate functional, $F(\tilde{u})$, stationary.

Thus, the following steps are necessary:

1. The approximation for u as given by (2.27) must be substituted in $F(u)$.
2. The stationarity condition must be applied to the approximate functional, $F(\tilde{u})$. This approximate functional may now be considered a function of the degrees of freedom $\{D\}$ by $F(\{P\}^T\{D\})$. Thus, it may be written

$$\delta F(\tilde{u}) = \delta F(\{P\}^T \{D\}) = \left\{\frac{\partial F}{\partial d}\right\}^T \{\delta D\} = 0 \tag{2.35}$$

   where the coefficients of the column vector, $\{\delta D\}$, are the first variations of the degrees of freedom, δd_i, $i = 1, ..., N_d$.
3. Expression (2.35) must be satisfied for every δd_i, resulting in a system of N_d equations, given by

$$\left\{\frac{\partial F}{\partial d}\right\} = \{0\} \tag{2.36}$$

   that has the same form as (2.31). The solution of that system of equations will provide the parameters of the approximation (or degrees of freedom) $\{D\}$.

In Section 2.3.3.2, it was established that the stationarity of a variational principle provides a weak formulation that is identical to that obtained by means of the weighted residual method if $\{\delta u\}$ is considered as the column matrix of weighting functions $\{v\}$. Thus, an alternate expression of (2.35) is

$$\begin{aligned}\delta F(\tilde{u}) &= \int_\Omega \delta\tilde{u}\, R(\tilde{u})\, d\Omega = \int_\Omega \delta(\{P\}^T\{D\})\, R(\{P\}^T\{D\})\, d\Omega \\ &= \int_\Omega \{P\}^T\{\delta D\}\, R(\{P\}^T\{D\})\, d\Omega = 0\end{aligned} \tag{2.37}$$

Taking into account that (2.37) must be satisfied for every δd_i, the following system of equations is obtained

$$\left.\begin{aligned} W_1 &= \int_\Omega P_1 \; R(\{P\}^T\{D\}) \; \mathrm{d}\Omega = 0 \\ &\vdots \\ W_{N_d} &= \int_\Omega P_{N_d} R(\{P\}^T\{D\}) \; \mathrm{d}\Omega = 0 \end{aligned}\right\} \tag{2.38a}$$

Substituting expressions (2.1) and (2.2) for deterministic and eigenvalue problems, respectively, the previous system of equations results in

$$\left.\begin{aligned} &\left[\int_\Omega P_1 \; \{\mathcal{L}P\}^T \; \mathrm{d}\Omega\right] \{D\} - \int_\Omega P_1 \; f_\Omega \; \mathrm{d}\Omega = 0 \\ &\left[\int_\Omega P_2 \; \{\mathcal{L}P\}^T \; \mathrm{d}\Omega\right] \{D\} - \int_\Omega P_2 \; f_\Omega \; \mathrm{d}\Omega = 0 \\ &\qquad\vdots \\ &\left[\int_\Omega P_{N_d} \{\mathcal{L}P\}^T \; \mathrm{d}\Omega\right] \{D\} - \int_\Omega P_{N_d} f_\Omega \; \mathrm{d}\Omega = 0 \end{aligned}\right\} \tag{2.38b}$$

or

$$\left.\begin{aligned} &\left[\int_\Omega P_1 \; \{\mathcal{L}_1 P\}^T \; \mathrm{d}\Omega\right] \{D\} - \lambda \left[\int_\Omega P_1 \; \{\mathcal{L}_2 P\}^T \; \mathrm{d}\Omega\right] \{D\} = 0 \\ &\left[\int_\Omega P_2 \; \{\mathcal{L}_1 P\}^T \; \mathrm{d}\Omega\right] \{D\} - \lambda \left[\int_\Omega P_2 \; \{\mathcal{L}_2 P\}^T \; \mathrm{d}\Omega\right] \{D\} = 0 \\ &\qquad\vdots \\ &\left[\int_\Omega P_{N_d} \{\mathcal{L}_1 P\}^T \; \mathrm{d}\Omega\right] \{D\} - \lambda \left[\int_\Omega P_{N_d} \{\mathcal{L}_2 P\}^T \; \mathrm{d}\Omega\right] \{D\} = 0 \end{aligned}\right\} \tag{2.38c}$$

These systems of equations coincide with (2.30d) and (2.30e), respectively, if the Galerkin method is applied to the Euler system of partial differential equations deriving from the functional. It is also evident that (2.38) coincides with (2.36). In conclusion, when a variational principle exists and the operator is positive definite, the systems of equations obtained through the Ritz and Galerkin methods will coincide. The extension of the Ritz method to the case of more than one variable, $\{u\}$, follows immediately. Figure 2.7 summarizes the relations between the Ritz and Galerkin methods.

When the functional is quadratic (see Appendix A.3.1), the Ritz method provides a system of equations with symmetric matrices. This could result in an efficient solution

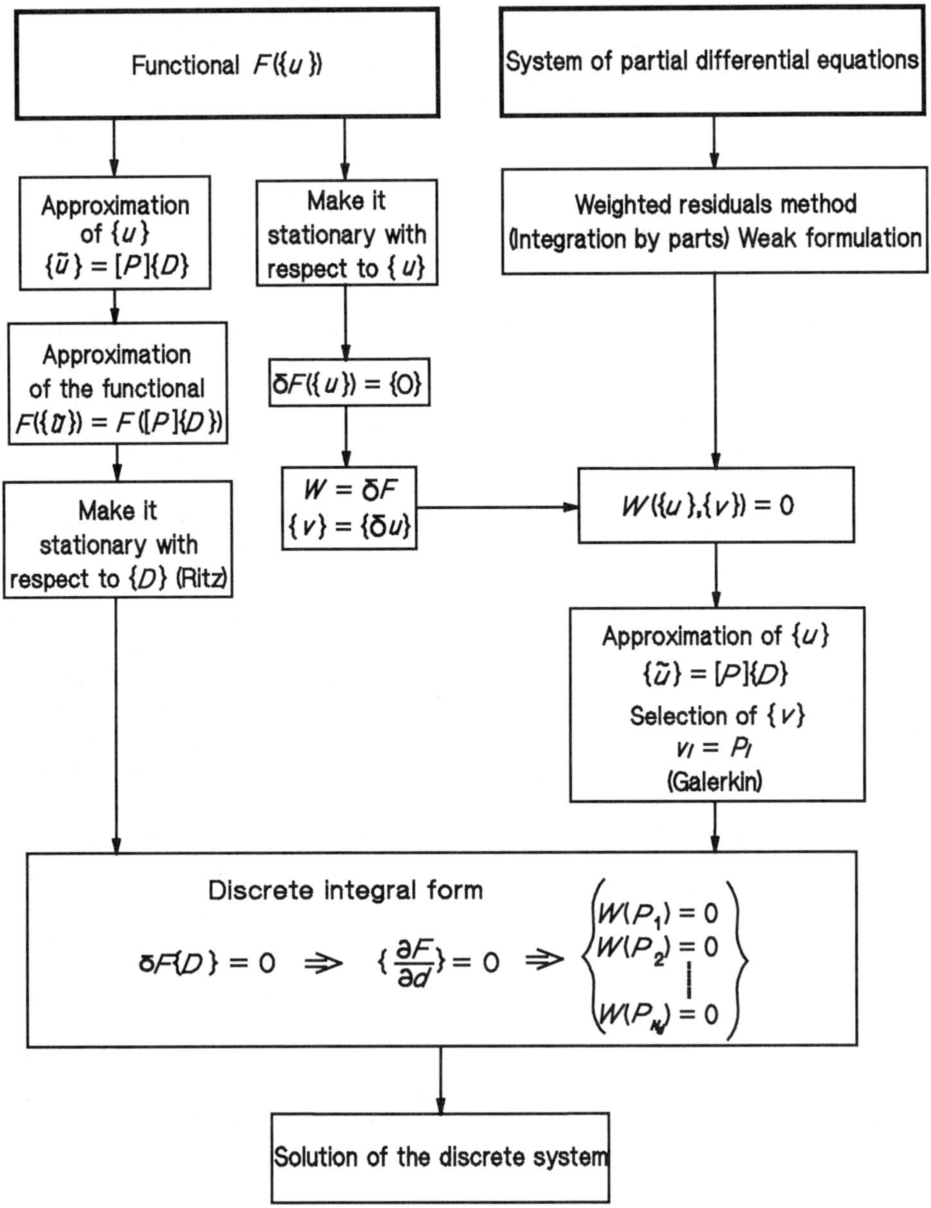

Figure 2.7 Equivalence between the Ritz and the Galerkin methods.

procedure. For that reason, when solving a problem it is always advisable to construct a quadratic functional if possible and apply the Ritz method [Zienkiewicz 1977; Ch. 3].

The Ritz method applied to a modified functional created through the penalty function method, as given by (2.24), is also called the Courant method. This method is very involved, but its convergence is very good [Reddy 1984; Ch. 2].

One possible variant of this procedure is the conjugate gradient method, where the functional (2.25) is minimized at each iteration. The basis functions are selected so that the parameters of the approximation may be obtained in an iterative way without the need for solving an algebraic system of equations at each step [Sarkar 1991].

2.4.4 Convergence and Other Properties of the Variational Methods of Approximation

Every approximate method of solution to a problem must guarantee the existence and uniqueness of the solution and its convergence to the exact solution. In the variational methods of approximation, the solution is sought in a finite-dimensional space. The accuracy of the numerically obtained solution is defined in terms of the error incurred; that is, the difference between the exact solution, u, which can be a scalar or a vector, and the approximate solution, $\tilde{u}$, which is a scalar or a vector, respectively. The accuracy of the solution is directly related, in general, to the number of degrees of freedom of the discrete problem, N_d. This dependence may be expressed, in particular, as a function of one or more parameters, which should either tend to zero (like the size of the subdomains in a subdomain approximation method, as for example in FEM) or increase to infinity (as the order of the polynomials in a FEM procedure) when the number of degrees of freedom are increased [Raviart and Thomas 1983; Ch. 3], [Carey and Oden 1984; Ch. 3].

In any case, by convergence it is understood that the evolution of the accuracy of the approximate solution, $\tilde{u}_{Nd}$, should progress toward zero as the number of degrees of freedom is increased. It is said that the approximate solution, $\tilde{u}_{Nd}$, converges toward the exact solution, u, if given an $\varepsilon > 0$, a value of N_d may be obtained so that the magnitude of the error will be less than ε. The error may be measured according to different norms. Depending on the measure selected, different types of convergence may be defined (see Appendix B.1).

When a variational method is selected for the solution of a given problem, not only should it provide convergent solutions, but also they should be accurate enough without requiring a large number of degrees of freedom. Otherwise, the memory, storage, and CPU time requirements may be too high. In other words, it is very important to have a fast rate of convergence.

It is not possible to mention all the topics related to the existence, uniqueness, and the rate of convergence of the solutions obtained by means of the various variational methods of approximation. In fact, these properties depend not only on the method itself, but also on the type of basis and weighting functions utilized as well as on the regularity of the exact solution of the problem under study [Zienkiewicz 1977;

Ch. 2]. In [Raviart and Thomas 1983; Ch. 3], [Sarkar et al. 1985], and [Sarkar 1983, 1985, 1991] some of these methods have been analyzed. Some general conclusions regarding their rates of convergence are summarized in Appendix B.2.

Any variational method of approximation is employed in an iterative way. The user will apply the chosen method to the analysis of the problem under study with an *a priori* fixed number of degrees of freedom or, for example, with a prespecified size of the subdomains in which eventually the domain is divided. After obtaining a solution, the knowledge of the rate of convergence of the method allows the user to decide how and where the number of degrees of freedom should be increased or the size of the subdomains further reduced.

In most problems the exact solution will not be known *a priori*, so in order to measure the rate of convergence estimates of the norms must be computed. That computation may be done *a priori*. This means that by taking into account the properties of the variational method and the nature of the problem under study, for example, an estimate of the error may be obtained with a specified number of degrees of freedom or with a given value of the parameters related to them.

In problems where the solution is not regular enough, it is not possible to determine reliable *a priori* estimates. Then *a posteriori* error estimates may be computed and used in a similar way to improve the performance of the method in the next iteration step. Once an approximate solution, $\tilde{u}$, is obtained, it is usually possible to compute an *a posteriori* estimate of the error incurred in that computation.

When selecting a variational method of approximation, it should also be desirable that the space of basis and weighting functions be easily accessible (i.e., easy to construct), that the coefficients of the matrices $[K]$, $[M]$, and $[F]$ be easily computable in an accurate way, and that the system of equations be solved by efficient numerical algorithms. The FEM takes into account all those requirements. The rest of the chapter is devoted to describing FEM and its rate of convergence.

2.5 DISCRETIZATION OF THE CONTINUOUS PROBLEM BY MEANS OF THE FINITE ELEMENT METHOD

The variational methods of approximation that are described in Sections 2.4.2 and 2.4.3 are not specific of FEM. What is specific of FEM is the way in which the unknown functions $\{u\}$ are approximated.

FEM employs a local approximation, defined in each different subdomain in which the domain is subdivided. These subdomains are called finite elements. Over each of them the variable is approximated by a linear combination of basis functions. The coefficients of the linear combination (the local degrees of freedom) are defined as the value of the unknown variables (or other quantities related to them) at given points of the element or at it edges, faces, or volume. In the most common approach, the basis functions are polynomials. Some types of singular and infinite elements use exponential or decaying functions. Harmonic and spline functions have also been used, as well as wavelet-like basis functions [García-Castillo et al. 1994, 1995], [Sarkar et al. 1994].

The fact that the unknowns are approximated locally, element by element, implies that the weak integral formulation should be written not over the whole domain but over each element. This is the main difference between FEM and methods using entire-domain basis functions. Thus, the global integral form should coincide with the sum of the local integral forms. Hence, in order to discretize the continuous problem using FEM, a variational method of approximation is applied over each element, resulting in a local algebraic description of the integral form for each one. Under certain conditions, which will be detailed in the following subsections, a discrete global integral form and a subsequent global system of equations may be obtained from the assembly of the local forms and the enforcement of the essential boundary conditions. This global system of equations corresponds to the discretization of the continuous problem over the complete domain. The matrices of this global system of equations are quite sparse (most of the coefficients of the matrices are zero) and have a band structure (nonzero coefficients clustered along the main diagonal of the matrices). Special techniques are employed for storing these matrices (band storage, skyline storage, or nonzero coefficients storage) and specially designed solvers are used (see Section 2.5.2.4).

Next, the FEM approximation of a variable is outlined followed by the various concepts involved in the FEM discretization of weak integral forms. Finally, results for the rate of convergence are given.

2.5.1 Approximation of a Function by Means of the Finite Element Method

FEM performs an approximation of a scalar or a vector variable over each of the different subdomains in which the domain of the definition of the variable has been subdivided. For each subdomain the approximation of the variable will be given by a linear combination of a number of basis functions as expressed by (2.27). Those basis functions must be linearly independent. They must form a complete basis of a linear space of finite dimension defined over each subdomain. Thus, the approximation of the variable over each subdomain will also belong to that space (see Appendix A.1).

If the variable to be approximated is a scalar, the basis functions are also scalar functions, usually, polynomials. If the variable is a vector, several ways of obtaining an approximation are possible. Cartesian components may be expressed as scalar variables multiplied by the unit vector in the corresponding Cartesian direction. Then the scalar variables may be approximated by a linear combination of scalar basis functions. In other cases the vector variable itself may be approximated by a linear combination of vector basis functions. Usually the Cartesian components of these vector basis functions are polynomials. Finally, a vector with components in all three Cartesian directions may be decomposed into a sum of two orthogonal vectors: one along the direction of one of the Cartesian coordinates, and the other one along the transverse direction, i.e., in a plane perpendicular to the given Cartesian direction. Then, the amplitude of the first vector may be approximated by a linear combination of scalar basis functions, while the transverse vector may be approximated by a linear combination of vector basis functions defined in the transverse plane.

In any case, the coefficients of the linear combination are the degrees of freedom (i.e., the unknowns) of the local approximation of the variable within the element. Each degree of freedom may have a different definition. For example, for scalar Lagrange elements the definition for all degrees of freedom is the value of the scalar variable at specified points of the element called nodes. In this case, a node is associated with each degree of freedom. For scalar Hermite elements, the degrees of freedom are defined as the value of the scalar variable and some or all of its partial derivatives of various orders at the nodes. In these elements, one or more degrees of freedom are associated with the same node [Ciarlet and Raviart 1972a]. Div-conforming and curl-conforming elements are used to discretize a vector variable by means of a linear combination of vector basis functions. For such elements, the degrees of freedom are defined as integrals extended to the edges, surfaces, or volume of the element where the integrands are related to given moments involving the vector variable. For practical reasons (among others, in order to deal with the case of curved elements), those integrals may need to be computed by numerical integration. This fact allows one to associate nodes (actually, points for numerical integration) on the edges, or faces of the elements, to some of the degrees of freedom (but eventually not to other degrees of freedom). For the case of div-conforming and curl-conforming elements, within an element there are degrees of freedom that may be associated with the points of integration, while others are associated with the faces or the volume of the element itself, without any specific assignment to a node. However, for practical reasons, nodes placed at the centroids of the faces or the element may be associated with those degrees of freedom [Raviart and Thomas 1977a], [Nédélec 1980], [Monk 1992].

Thus, it may be said that in all cases the FEM approximation is local in nature and is related to the nodes. The number of degrees of freedom (and the location of the nodes) depends on the geometric shape of the element, the actual definition of the degrees of freedom, and the order of the polynomials used. In general, the number of degrees of freedom (and nodes) increases with the order of the polynomials involved. Each basis function is associated with a degree of freedom. Basis functions are also called shape or interpolation functions.

As mentioned before, the values of the degrees of freedom are defined as the value at each node of either the unknown variable itself or a quantity related to the unknown variable to be approximated. The definition of the degrees of freedom depends on the physical meaning of such variable.

Assume two (or more) elements that share a node at the interface between them. It is clear that the value of the degree of freedom at that node (or the degrees of freedom related to common edges or faces) must be the same for those two elements, although the approximation of the variable has been carried out separately over each of them. Thus, the definition of the degree of freedom must be done by a quantity which should be continuous across elements. For example, consider the analysis of a multidielectric multiconductor transmission line. The domain of definition of the problem is the 2D cross-section of the structure. Assume that the variable of interest is the electrostatic potential, $\phi(x,y)$, over the cross section. Then it is reasonable to choose the electrostatic potential itself as the unknown or degree of freedom, because it should be continuous over the domain. Nodes may be placed at any point of the element (vertices, edges, or inner points). Let us assume that the variable of interest is the displacement vector $\bar{D}(x,y)$. The normal component of $\bar{D}(x,y)$ is continuous across dielectric interfaces. It is also well known that dielectric corners may give rise to singularities of the electric field and the displacement vector. Hence, continuity conditions cannot be defined at those dielectric corners. Assume that the decomposition of the domain is such that each subdomain belongs just to one, and only one, dielectric region. Thus, it makes sense, first, to choose the location of the nodes in such a way that they will never be placed at corner points and they will be shared by a maximum of two subdomains (i.e., to place them at the edges, excluding vertices, of the elements); and second, to choose as the degree of freedom the normal component of $\bar{D}(x,y)$ at the element interfaces. This example illustrates how the location of the nodes and the definition of the degree of freedom depends on the type of variable to be approximated.

The term finite element is frequently employed in a geometric sense to refer to each of the subdomains in which the domain of the definition of the variable to be approximated by FEM has been subdivided. However, an n-dimensional finite element is defined, in a strict sense, by

- A compact region, Ω^e, of the space $\mathbb{R}^n$, connected, and nonempty;
- A linear space of finite dimension, composed of scalar or vector functions defined over Ω^e;
- A set of degrees of freedom defined over Ω^e, to which eventually there is associated a set of nodes.

Given the triplet (the three items mentioned previously) defining a finite element, the basis function associated with each degree of freedom may easily be obtained from the definition of the degrees of freedom themselves. Then, the approximation of the unknown variable over Ω^e is easily obtained according to (2.27) [Raviart and Thomas 1977a, 1983; Ch. 4], [Nédélec 1980], [Thomas and Joly 1981; Ch. 2], [Ciarlet and Raviart 1972a], [Monk 1992].

This section is further subdivided into two parts. The first one deals with the geometric decomposition (or mesh generation) of a given domain. In the second part, the finite elements used for the analysis of the electromagnetic problems considered in this book are described.

2.5.1.1 *Discretization of the Domain*

The process of subdivision of the domain into a set of subdomains is called mesh generation or geometric discretization of the domain. The resulting subdomains are usually called finite elements in their geometric sense.

The shape of these finite elements should allow an easy definition of each subdomain and, simultaneously, the discretization of complex geometries. Triangles and tetrahedrons are the 2D and 3D basic element shapes, respectively. They are called simplexes. Any polygon or polyhedron may always be expressed as a union of simplexes [Silvester and Pelosi 1994]. In 2D, the elements generally utilized have triangular or rectangular shapes. In 3D, tetrahedral, prismatic, or cubic shapes are used. All these elements may have curved contours, so the terms triangular and rectangular, for example, should be understood in a broader sense since the shape of elements may be a geometric transformation of triangular and rectangular straight elements, respectively [Babuška and Guo 1988], [Blanc-Castillo et al. 1995], [Brauer et al. 1991], [Ciarlet and Raviart 1972b,c], [Ergatoudis et al. 1968], [Irons and Zienkiewicz 1968], [Richards and Wexler 1972], [Silvester and Rafinejad 1974], [Wang and Ida 1993], [Welt and Webb 1985].

The generation of the mesh of a domain constitutes a more or less complicated process on which the accuracy of the FEM solution greatly depends. Assume a domain Ω with boundary Γ. If the boundary has curved portions the process starts with the definition of a number of points (called geometric points) along the boundary. Connecting them by means of rectilinear segments (in the 2D case) or plane surfaces (in the 3D case), a new boundary $\tilde{\Gamma}$, defining a new domain $\tilde{\Omega}$, will be obtained. $\tilde{\Gamma}$ and $\tilde{\Omega}$ are approximations of the boundary and the domain, respectively, of the original problem (see Figure 2.8).

If the distance between the geometric points is reduced, increasing their number, a better approximation of the boundary Γ and of the domain Ω will be obtained. The approximation will be further improved if in the mesh generation process, elements with curved contours (2D) or faces (3D) are employed to discretize the curved boundaries. The error incurred, because of the fact that $\tilde{\Omega}$ and $\tilde{\Gamma}$ do not exactly coincide with Ω and Γ, respectively, is called the *geometric discretization error*. In some cases (e.g., when discretizing open problems by means of finite domains, closed by artificial boundaries, placed in the far-field region), the original domain will admit a truncated approximate domain and still the approximate solution may be accurate enough. For a closed-domain with noncurved boundaries, $\tilde{\Gamma}$ and Γ will coincide as well as $\tilde{\Omega}$ and Ω, so that the geometric discretization error will be zero.

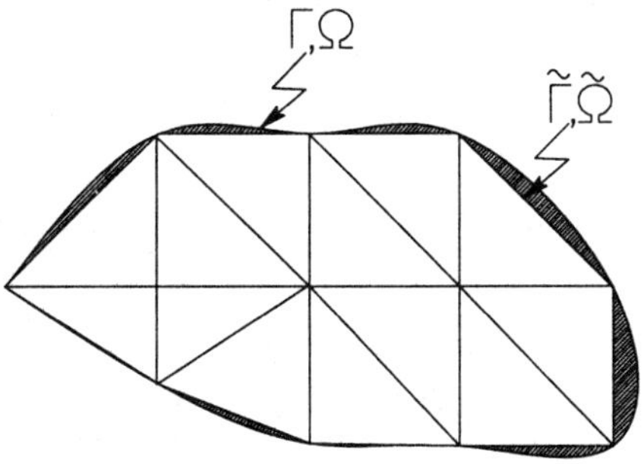

Figure 2.8 Geometric discretization of the domain. The shaded surface shows the geometric discretization error.

The next step consists of subdividing the domain $\tilde{\Omega}$ into several subdomains that should coincide with each different physical medium and its location within the domain. The subdivision will be done following the interfaces between any two media. If those interfaces are curved, it is again important to reduce as much as possible the geometric discretization error using either a large number of points defining the rectilinear or planar surface interface, or curved elements, or both procedures.

Next, each subdomain will be subdivided into a set of finite elements Ω^e, where $1 \le e \le N_e$, $e \in \mathbb{N}$, and N_e is the total number of elements of the domain $\tilde{\Omega}$. The following condition should be fulfilled:

$$\tilde{\Omega} = \bigcup_{e=1}^{N_e} \Omega^e \tag{2.39}$$

together with

- Each element, Ω^e, must have a nonempty interior;
- The interior of two different elements may not have common points; i.e., there must not be any overlapping;
- Any edge or face (3D) of element Ω^1 must be either the edge or the face of another element Ω^2 (in which case Ω^1 and Ω^2 are contiguous) or it must belong to the boundary $\tilde{\Gamma}$ of $\tilde{\Omega}$; i.e., there must not be gaps.

Figure 2.9 shows an example of a 2D mesh in which both triangular and rectangular, straight and curved elements have been used. It is not possible to provide general rules regarding the selection of the shape of the elements to be utilized. In general, triangular (2D) or tetrahedral (3D) elements provide higher accuracy than the rectangular, cubic, or prismatic ones for the same number of nodes. They also allow a better treatment of structures with complex geometry and of regions close to corners, where the solution of the problem may be singular or nonsmooth. In general, given two elements or meshes, the one better fitting the symmetry of the problem provides better results in terms of accuracy of the solution [Reddy 1984; Ch. 4]. The mesh of a structure may include elements of different shapes as long as they are compatible. The

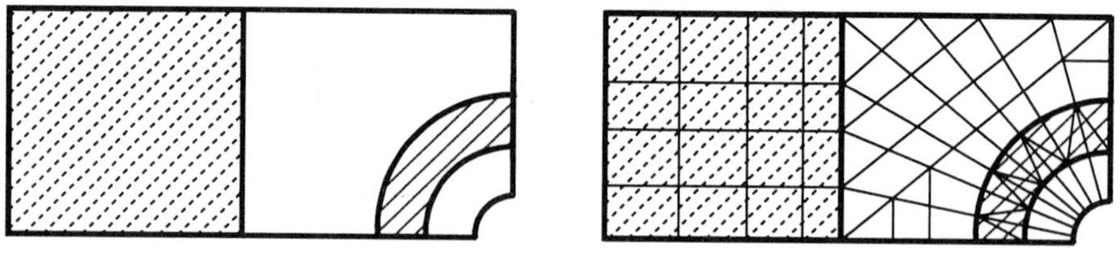

(a) (b)

Figure 2.9 Mesh of a two-dimensional domain: (a) original domain, (b) discretized structure.

concept of compatibility is explained in the next page.

The interpolation error, i.e., the error incurred because of the approximation of the variable according to (2.27), is minimum for equilateral elements (see Section 2.5.1.2). This is the ideal shape for any type of finite element. Thus, every element of a mesh should be as close to its ideal shape as possible. A measure of the distortion of an element, Ω^e, from its ideal shape may be given by the aspect ratio, σ_e, defined by

$$\sigma_e = h_e / \rho_e \tag{2.40}$$

where h_e is the diameter of element Ω^e (or maximum euclidian distance between two vertices of the element) and ρ_e is its roundness (maximum diameter of the circles, in 2D, or the spheres, in 3D, that may be inscribed in element Ω^e). Figure 2.10 illustrates the definition of h_e and ρ_e for triangles and rectangles, including the case of curved elements. For curved elements, the values for h_e and ρ_e are the same values corresponding to those of the straight elements that have the same vertices as the curved one as long as the shape of the curved element does not deviate too much from that of the straight element. A measure of this deviation will be given in Section 2.5.1.2. In

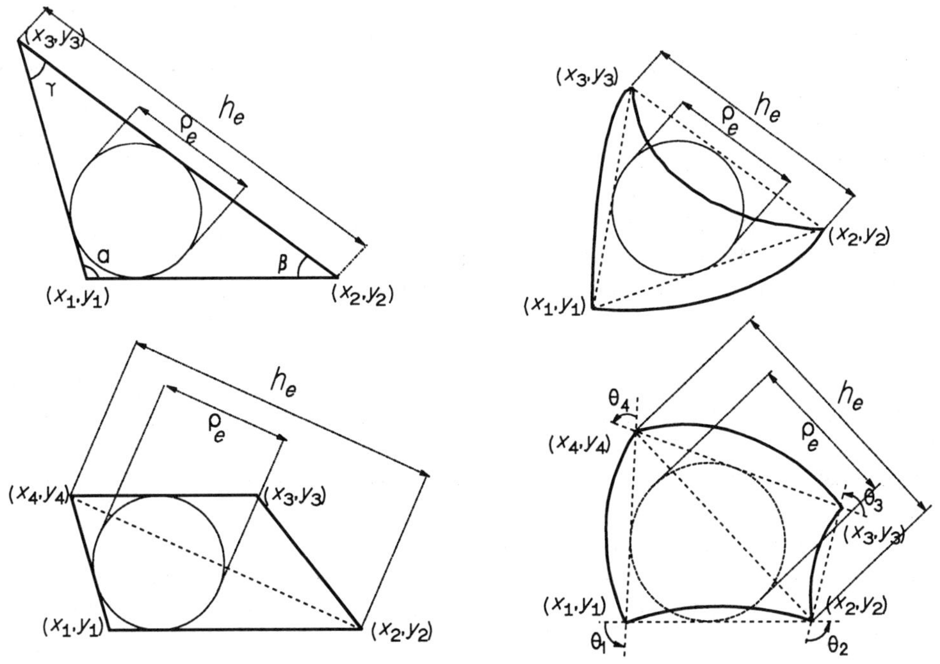

Figure 2.10 Definition of the diameter, h_e, and the roundness, ρ_e, of straight and curved two-dimensional elements.

general, elements with different aspect ratios are required for a mesh to be well adapted to a specific domain. Thus, in practice, nonideal elements or distorted elements are used. Nevertheless, there are limits to the distortion that an element may suffer with respect to its ideal shape. For the elements described in Section 2.5.1.2, the interpolation error of either the variable to be approximated or its first derivatives is proportional to the aspect ratio. For triangular elements, σ_e is inversely proportional to sin θ_e, where θ_e is the smallest internal angle of Ω^e [Zlámal 1968, 1973] or to $\cos(\theta_e'/2)$, where θ_e' is the largest angle of Ω^e [Oden and Reddy 1976; Ch. 6]. Triangular elements with straight or curved contours having very small internal angles or with an internal angle close to 180 degrees are called degenerate triangles [Babuška and Aziz 1974], [Ženíšek 1995]. In a similar way, rectangular elements with straight or curved contours with high aspect ratio or with angles close to 0 or 180 degrees are also called degenerate elements. Similar definitions stand for 3D elements. Meshes with degenerate elements or highly distorted elements should be avoided in order to obtain more accurate results.

Two issues should be considered with respect to the size of the elements of a mesh. The first consideration should be whether the mesh should be uniform (or quasi-uniform) or not [Davies 1994]. For a uniform or quasi-uniform mesh, all elements have the same or similar size. The second issue is the actual size of the elements.

The first issue is related to the rate of convergence of the problem analyzed by FEM. The concept of rate of convergence must be distinguished from that of interpolation error. The latter refers to the measure of the error between the exact value of the variable and its interpolation or approximation over a subdomain, according to (2.27). The rate of convergence refers to the measure of the error between the exact value and the approximate value obtained after the application of the FEM to a weak integral form over the whole domain and its evolution as the number of degrees of freedom is increased.

A mesh is said to be regular if it consists of compatible and nondegenerate elements having a bounded aspect ratio. The concept of compatible elements refers to contiguous elements. Two contiguous elements of a mesh are said to be compatible if they are of the same geometric shape (such as triangles and rectangles), same order (or degree of the polynomial basis functions), and same type (i.e., for scalar or vector approximation and ordinary or infinite type, e.g., see Section 2.5.1.2). Alternately, even if they are of different characteristics, they may still be compatible if at the interface they share the same nodes and values for the degrees of freedom, and the same approximation of the quantity related to the degrees of freedom along the interface [Raviart and Thomas 1983; Ch. 5].

In Section 2.5.3, some results regarding the rate of convergence of the finite element analysis of boundary value problems are summarized. For problems where the exact solution is sufficiently regular in a mathematical sense (a smooth enough function, with no singularities), and if a regular mesh is used, the measure of the error incurred in the approximation of the unknown function or its derivatives is proportional to h^{p+1}, or to h^p, depending on the type of element employed. Here h is the maximum diameter of the elements of the mesh and p is the minimum order of the elements. Thus, for problems with smooth solutions, meshes should not only be regular, but uniform or quasi-uniform, with small enough elements of some high degree in order to obtain the desired good rate of convergence.

In problems having a nonsmooth or nonregular solution (with singularities of the solution or its derivatives, because of domains with sharp edges or corners, e.g., nonconvex domains, complex geometries, abrupt changes of the physical characteristics of the media involved or of the boundary conditions, coupling and proximity effects), the rate of convergence is much lower (see Section 2.5.3) requiring elements of smaller size in order to obtain a good accuracy. In these cases the mesh should not be uniform.

On one hand, a uniform discretization with elements of very small sizes or very high polynomial degree will lead to a large number of degrees of freedom, so the size of the global system of equations will also be large. Thus, when solving the algebraic system of equations by means of any direct or iterative numerical procedures, errors of numerical type (round-off and truncation errors) may become very important [Fried 1973], or a large number of iterations may be required. The computer CPU time, storage, and memory requirements would also be large.

On the other hand, the discretization of the domain of a problem with a uniform mesh would be inadequate. Actually, the results for the measure of the interpolation error summarized in Section 2.5.1.2 demonstrate that for those regions of the domain where the exact solution is smooth (regions far away from any source of nonsmoothness), a rather coarse mesh would provide a good approximation, while in regions close to points where there are singularities of the exact solution or their derivatives, a high density of smaller elements will be adequate. Thus, it is a general practice to employ meshes with elements of different sizes. Then, in order to obtain a good condition number for the matrices of the algebraic global system of equations, the transition between regions with elements of different sizes should be smooth enough [Kardestuncer and Norrie 1987; P. 4, Ch. 3]. These meshes are called graded or adapted. The term graded refers to a mesh in which the sizes of the elements gradually vary from regions of small elements to regions of bigger ones. A mesh is said to be of adaptive type when the sizes of the elements are adjusted to the solution of the problem. These types of meshes are used in Chapters 4, 5 and 7 of the book, generated either in an automatic way by means of a self-adaptive mesh procedure or semiautomatically. Figure 2.11 shows a uniform mesh, a graded mesh, and an adapted mesh for a structure in which the first derivative of the variable is singular at points A and B. Such a structure may be the cross section of a shielded asymmetric stripline. The strip is assumed to have zero thickness and is represented by the line AB. The boundaries of the problem are the strip AB and the enclosing shield.

It has been mentioned already that both the interpolation error (see Section 2.5.1.2) and the rate of convergence (see Section 2.5.3) improve if the order of the elements is increased. However, an increase of the order implies an increment of the number of degrees of freedom. It may be added that, if the order of the basis functions is increased, the width of the band of the matrices of the algebraic system of equations will be also increased, leading to higher computer storage requirements [Zienkiewicz and Craig 1986] (see Section 2.5.2.4). It is admissible to employ elements with a higher degree in certain regions (e.g., near a singularity) and of lower degree in other regions.

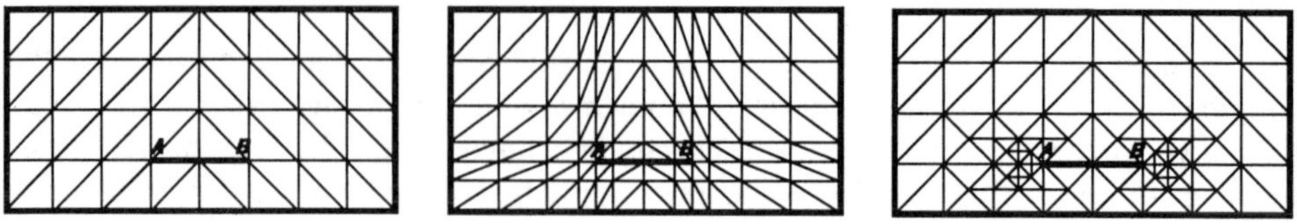

(a) (b) (c)

Figure 2.11 Example of a mesh of (a) uniform type, (b) graded type, (c) adapted type, for a two-dimensional structure having singular derivative of the unknown at points A and B.

For these cases, the elements used must be compatible between them. This requirement gives rise to transition elements and techniques related to them [Zienkiewicz 1977; Ch. 7] as well as hierarchical bases [McFee and Webb 1992], [Rossow and Katz 1978], [Sears 1988], [Wang 1997], [Webb and McFee 1991], [Webb and Forghani 1993, 1994], [Webb and Abuchacra 1995], [Zienkiewicz et al. 1983], [Zienkiewicz and Craig 1986]. The same criterion of compatibility also affects elements with different geometric shapes [Raviart and Thomas 1983; Ch. 5].

It may be observed that within given limits, it is possible to generate quite different meshes to analyze the same problem. In any case, the mesh obtained should be regular in regions of elements of different sizes, with a smooth transition between them [Ciarlet 1973], [Raviart and Thomas 1983; Ch. 4]. The user of any mesh generation algorithm should first choose the algorithm and then choose the input data carefully [Bryant 1985], [Forsman and Kettunen 1994], [Kost and Jänicke 1992], [Lo 1991], [Lowther and Dyck 1993], [McFee and Webb 1993], [Yuan and Fitzsimons 1993], [Ramakrishnan et al. 1992], [Heighway and Biddlecombe 1982], [Heighway 1983], [Shenton and Csendes 1985]. There are a number of mesh generation algorithms available, from the most simple (for which coordinates of the vertices and the sequence of the vertices of each element should be provided) to the most sophisticated (like those generating meshes of quasi-ideal elements from the geometric description of the boundary of the domain).[1] Meshes of the type in Figure 2.11 are called structured meshes. They are generated by subdividing the structure into rectangular elements that are further subdivided into triangular elements. The mesh of Figure 2.12 is an example of a nonstructured mesh obtained by an algorithm based on Voronoi polygons and Delaunay tessellation [Delaunay 1934], [Watson 1981]. It yields quasi-ideal elements. In both cases the MODULEF mesh generator has been used [Bernadou et al. 1985]. The structure of Figure 2.12 may be a shielded microstrip line. In this case a nonzero thickness of the strip has been assumed.

The FEM is used in an iterative way. In fact, only after studying the results of

---

1 Information related to mesh generation algorithms may be found in http://www-users.informatik.rwth-aachen.de/~roberts/software.html.

the analysis of the problem by the application of FEM to the mesh under study, the quality of the analysis may be determined. In some cases, further analysis may be needed of successive meshes from less to more refined (in terms of number of elements and/or degree of the basis functions, i.e., in terms of the number of degrees of freedom) in order to determine which mesh refinement strategy will lead to solutions with a given accuracy [Sussman and Bathe 1985]. Thus, self-adaptive mesh procedures are highly attractive since the meshes are adapted progressively to the electromagnetic field behavior over the domain in an automatic way [Carey and Oden 1984; Ch. 2]. Self-adaptive mesh generation is discussed in Chapter 4 where references are given.

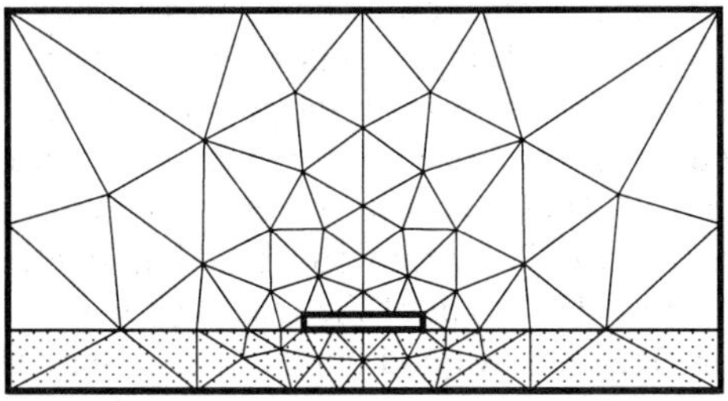

Figure 2.12 Shielded microstrip line: unstructured mesh.

In two dimensions, the geometric definition of the element of a mesh is usually done (manually or, in general, automatically) by means of the coordinates of its vertices, following a specific local order (starting from anyone of them and following always the same direction, usually in a trigonometric sense or counterclockwise). This defines also the edges and their local order. Similar criteria are valid for 3D meshes (see Appendix C.4.3). In the case of elements with curved edges or faces, these are generated by a geometric mapping of the corresponding straight edge or face. This mapping employs geometric points placed at the curved edges or faces. The coordinates of the geometric points defining the curved edges or surfaces are usually computed from the definition of the curved contours or surfaces that are considered. Thus, the analytical description of such curves or surfaces must also be given. Figure 2.13 illustrates the 2D local numbering process. The data regarding the number of degrees of freedom and their corresponding nodes (that may or may not coincide with the vertices and geometric points mentioned above) are given next, following the same order. Other characteristics should be included such as media or physical subdomain to which each element belongs and edges that are located at boundaries, for example.

The process terminates by numbering the elements and globally numbering the degrees of freedom. The method used to solve the system of linear equations may have a great influence in reducing the storage, memory, and CPU time computer requirements. Due to the fact that the FEM approximation is of local type, the global system of equations resulting from the FEM application has the property that only a few unknowns are involved with each equation. Namely, the degrees of freedom of an element and those of contiguous elements. Typically, the number of coefficients that are nonzero is lower than a given constant that depends on the type and order of the element employed, but not on the total number of degrees of freedom or on the refinement of the mesh. Additionally, the nonzero coefficients tend to be clustered in the vicinity of the main diagonal of the matrices, so these matrices have a banded structure. The methods used to solve these systems of equations take advantage of the fact that most of the coefficients are zero (sparse or nearly empty matrices) [Evans 1973]. Furthermore, for most of the problems analyzed in this book, the matrices are symmetric. The width of the band may be controlled by means of a good strategy in global numbering the degrees of freedom. There are several algorithms that optimize this numbering. The MODULEF library includes some of them [Bernadou et al. 1985]. A number of techniques allow one to store the matrices in a banded or skyline form and, furthermore, for the symmetric case, to store just the upper or the lower part of the matrix (see Section 2.5.2.4). Thus, specific solvers, employing these types of storage for sparse systems may be used. They turn out to be highly efficient [George and Liu 1981]. In the frontal method, the length of the front that is stored at each step depends on the numbering of the elements. A good numbering strategy may minimize it. There exist many numbering algorithms as well as a large volume of related bibliography. Some references are [Akin 1982; Ch. 3], [Bathe and Wilson 1976; Chs. 6 and 7], [Carey and Oden 1984; Ch. 3], [Duff 1977], [Dhatt and Touzot 1981; Ch. 4].

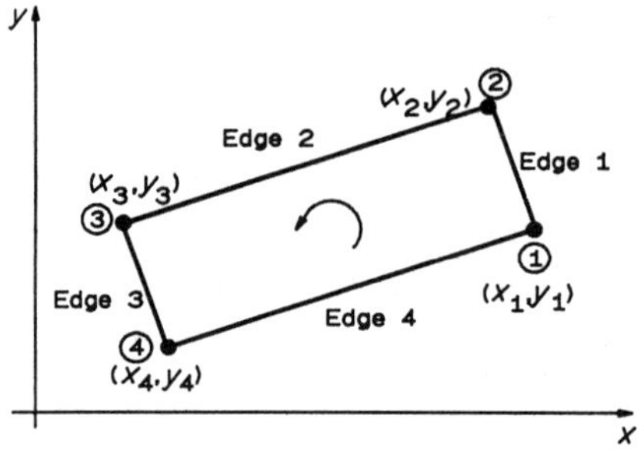

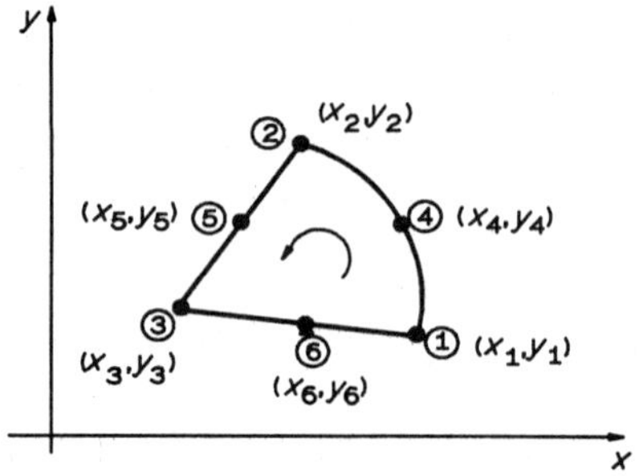

Figure 2.13 Geometric definition of elements. Local numbering of vertices and other geometric points. (a) Element with straight contours and (b) element with curved contour obtained through a second-order geometric transformation: geometric points are required in the curved contours with projections to the midpoints of the straight edges.

2.5.1.2 *Description of the Finite Elements*

Next, the finite elements that are used in this book are described. They are subdivided into three main categories: Lagrange, div-conforming and curl-conforming elements. For all cases, 2D and 3D elements are presented.

Lagrange Elements

Lagrange elements are used in Chapter 3 for the discretization of the direct or standard weak integral formulation for the quasi-static analysis of transmission lines, where the unknown variable is the scalar electrostatic potential. In this case, the approximate solution should be continuous over the domain of the definition of the problem. Its first derivatives may be discontinuous, but they should be square integrable in order to provide solutions of finite energy. Thus, the approximate solution should belong to the Hilbert space $H^1(\Omega)$ (see Appendix A.1.3). Lagrange elements provide solutions fulfilling those conditions.

First, finite elements are presented for closed-domain problems. They have been used for the quasi-static analysis of transmission lines using the direct or standard formulation. They have also been utilized for the dual formulation of the same problem, that is, the one utilizing the magnetostatic vector potential as unknown. Furthermore, they have also been used for the full-wave analysis of waveguiding structures using the longitudinal component of the electric field and/or the magnetic field as unknowns.

Next, infinite elements are presented. They are used in conjunction with the previous ones, for the quasi-static analysis of given open-region transmission line problems. Infinite elements force the approximate function to decay to zero at infinity.

Singular elements may also be used to mimic the singular behavior of the exact solution of problems in nonconvex domains [Akin 1976], [Babuška and Rosenzweig 1972], [Burais et al. 1985], [Gil and Zapata 1994, 1997], [Gil and Webb 1997], [Israel and Miniowitz 1987, 1990], [Mabaya et al. 1981], [Pantic and Mittra 1986, 1988], [Slade and Webb 1992], [Svedin 1989], [Webb 1988b]. However, they are not shown here. In this book the effects of such geometries are automatically taken into account, as well as other sources of nonsmoothness of the solution, using a self-adaptive mesh procedure. Transition elements may also be considered as special elements. They are required as interfaces between noncompatible elements, like when using elements of different order in the same mesh or meshes in which the elements are not connected via their vertices, or for regions between singular and ordinary elements [Gil and Zapata 1995]. However, with the techniques used in this book they are not required. To improve the rate of convergence of a given problem, a self-adaptive mesh algorithm may be applied. Elements of the same order are used at each iteration that may be followed eventually by a uniform increment of the order of the polynomial basis functions.

Lagrange Ordinary Elements

Given a compact region, Ω^e, of $\mathbb{R}^n$, connected and nonempty; a linear space P of finite dimension, n_e, composed of real scalar polynomials functions, $p(\bar{r})$ (where $\bar{r}$, or in matrix notation $\{r\}$, stands for the vector location of a point (x,y) in the 2D case or (x,y,z), in the 3D case), defined over Ω^e; and a finite set, A, of n_e degrees of freedom, to which a finite set, Σ, of n_e points or nodes over Ω^e, $\bar{r}_i$, $1 \le i \le n_e$, may be associated.

Then the triplet (Ω^e, P, A) defines a Lagrange finite element if for any integer i, $1 \le i \le n_e$, there exists a scalar function $N_i(\bar{r})$ belonging to P fulfilling:

$$\left.\begin{array}{l} N_i(\bar{r}_j) = 0\ ,\quad i \neq j \\ N_i(\bar{r}_j) = 1\ ,\quad i = j \end{array}\right\} \quad 1 \le j \le n_e \tag{2.41}$$

Functions N_i (the Lagrange interpolation functions) have the property that for any point $\bar{r}$ of Ω^e they satisfy

$$\sum_{i=1}^{n_e} N_i(\bar{r}) = 1 \tag{2.42}$$

It is easy to see that any scalar variable $w(\bar{r})$ defined over Ω^e may have an approximation, $\tilde{w}(\bar{r})$, by means of an interpolating function or interpolant, $\Pi w(\bar{r})$, employing the set of n_e nodes and n_e degrees of freedom, defined as $d_i = w(\bar{r}_i)$ according to

$$\tilde{w}(\bar{r}) = \Pi w(\bar{r}) = p(\bar{r}) = \sum_{i=1}^{n_e} N_i(\bar{r})\, w(\bar{r}_i) = \sum_{i=1}^{n_e} N_i(\bar{r})\, d_i \tag{2.43a}$$

so that $p(\bar{r})$ belongs to the space P. In other words, the function $w(\bar{r})$ has been approximated by a function, $p(\bar{r})$, of P, so that the value of the approximate function at the nodes, $\Pi w(\bar{r}_i) = p(\bar{r}_i)$, coincides with the value of the exact function at those points

$$\Pi w(\bar{r}_i) = p(\bar{r}_i) = w(\bar{r}_i) = d_i\ ,\quad i = 1,\ldots,n_e \tag{2.43b}$$

This is because of (2.41). In conclusion, the definition of the degree of freedom of these elements is the value of the function at the nodes, $w(\bar{r}_i)$, $1 \le i \le n_e$. Consider a FEM discretization of a weak integral form generated via any variational method. Assume that the strong solution exists and that the only error incurred in the computation of the weak solution is the interpolation error. This is equivalent to saying that errors of any other types are zero (e.g., geometric discretization, numerical integration, round-off, and truncation errors). Then an approximation of the function w with zero error at the nodes can be obtained.

A discretization by means of the Lagrange elements provides an approximation of the variable that will be continuous over the complete domain. Its first derivatives will be discontinuous at the interfaces between elements. Namely, the derivatives along the tangential direction to the interface will be continuous, while the derivative along the normal direction will be discontinuous [Dhatt and Touzot 1981; Ch. 2]. A necessary condition for the fulfillment of (2.43) is that the number of degrees of freedom and nodes, n_e, must coincide with the number of independent functions of the space P, that is, with the dimension of the space (see Appendix A.1.1).

The spaces of functions that generate rectangular and triangular Lagrange elements are the 2D space of polynomial functions in two variables, called $Q_p^{(2)}$ and $P_p^{(2)}$, respectively (see Appendix A.1.1). $P_p^{(2)}$ is the space of polynomials in two coordinates x,y containing monomials of total degree less than or equal to p, where $p \in \mathbb{N}$. Thus, it contains polynomials complete up to the degree p. The space $Q_p^{(2)}$ is formed by polynomials in two coordinates x,y containing monomials of degree less than or equal to p with respect to each coordinate. Thus, it contains polynomials of total degree $2p$ complete up to degree p. The space of polynomial functions $Q_p^{(2)}$ may be viewed as the result of multiplying a polynomial function of the one-dimensional space $P_p^{(1)}(x)$ with another polynomial function of the one-dimensional space $P_p^{(1)}(y)$. Figure 2.14 shows the Pascal triangle and the way in which it may be used to generate the base for the space $P_p^{(2)}$ and $Q_p^{(2)}$, respectively.

Three-dimensional elements may be generated from the corresponding linear spaces $P_p^{(3)}$ (tetrahedral elements) and $Q_p^{(3)}$ (cubic elements) in an analogous fashion. A Pascal tetrahedron is applied to generate the base for each space in that case.

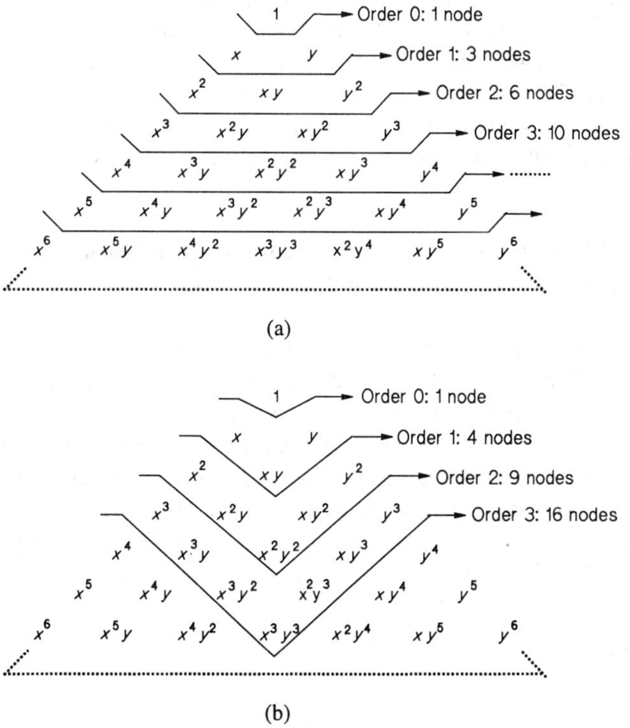

Figure 2.14 Pascal triangle. Generation of the base of the spaces (a) $P_p^{(2)}$ and (b) $Q_p^{(2)}$.

An interpolation through Lagrange elements of order p requires to be complete up to order p (i.e., it should contain a polynomial complete up to order p). This condition means that the basis functions must expand the space of polynomials or order $\leq p$. This condition is a requirement for having an interpolation error for the variable tending to zero when the size of the element tends to zero [Oden and Reddy 1976; Ch. 6], [Zienkiewicz 1977; Ch. 2].

Once the base for a given order has been determined, Lagrange element basis functions may be generated according to (2.41). They may also be generated through the Lagrange interpolating polynomial in one variable. Because of this, they are called Lagrange elements. The Lagrange interpolating polynomials, $L^p{}_k(x)$, are polynomials of degree p, having unity value at the kth point of coordinate x_k and zero value at the other p points of coordinates x_i, $1 \leq i \leq p{+}1$, $i \neq k$. The Lagrange polynomial in one dimension is given by

$$L_k^p(x) = \frac{(x-x_1)(x-x_2)\cdots(x-x_{k-1})(x-x_{k+1})\cdots(x-x_{p+1})}{(x_k-x_1)(x_k-x_2)\cdots(x_k-x_{k-1})(x_k-x_{k+1})\cdots(x_k-x_{p+1})} \tag{2.44}$$

An expression of the type of (2.44) fulfills conditions (2.41) and (2.42), so it may be used to generate the desired basis functions. For example, if (2.44) is used for each component of a Cartesian system, the basis functions for rectangular elements (2D) or cubic elements (3D) will be obtained for each element of the mesh [Raviart and Thomas 1983; Ch. 4], [Zienkiewicz 1977; Ch. 7]. In order to avoid the need for performing such computations, element by element, of a given mesh, one can generate a whole family of elements from just one, called the parent, master, or the reference element. This technique is described in the following subsection.

Equivalent Elements. Parent Element and Real Element. Straight and Curved Elements. Subparametric, Isoparametric, and Superparametric Elements.

Let $(\hat{\Omega}, \hat{P}, A)$ be a Lagrange element given by a compact region $\hat{\Omega}$ of $\mathbb{R}^n$, a space $\hat{P}$ of functions $\hat{p}(\{\hat{r}\})$, and a set A of n_e degrees of freedom to which a set of n_e nodes $\hat{\Sigma}$, $\{\hat{r}_i\}$, $1 \leq i \leq n_e$, $i \in \mathbb{N}$ (where $\{\hat{r}\}$ stands for a point of coordinates $\hat{x}$, $\hat{y}$ in the 2D case or $\hat{x}$, $\hat{y}$, $\hat{z}$ in the 3D case) may be associated. Given a compact region Ω^e of $\mathbb{R}^n$, a space P of scalar functions $p(\bar{r})$, and a set A of n_e degrees of freedom to which a set Σ of n_e nodes, $\bar{r}_i$, $1 \leq i \leq n_e$, $i \in \mathbb{N}$, is associated, it is said that the triplet (Ω^e, P, A) is a Lagrange element equivalent to $(\hat{\Omega}, \hat{P}, A)$ if a bidirectional transformation or parametric mapping F^e of $\mathbb{R}^n$ into $\mathbb{R}^n$ exists, so the following expressions are fulfilled

$$\Omega^e = F^e(\hat{\Omega}) \tag{2.45a}$$

$$\Sigma = F^e(\hat{\Sigma}) \tag{2.45b}$$

$$p(\{r\}) = \hat{p}(\{\hat{r}\}) \tag{2.45c}$$

The previous expressions essentially imply that there is an affine transformation of the real element into the parent element in its geometric sense and that the values of the degrees of freedom are invariant (see (2.45c)). More details on this topic are available in Appendix C.1. The transformation F^e that maps the element (Ω^e, P, A) into the element $(\hat{\Omega}, \hat{P}, A)$ depends on the shape and the location of the element Ω^e and thus on the coordinates of its vertices and eventually other geometric points defining the transformation. There will be a different transformation, F^e, for each different element Ω^e of a mesh equivalent to $\hat{\Omega}$, that is, for $e = 1, ..., N_e$, where N_e is the total number of elements of the mesh. Each transformation is characterized by its Jacobian matrix $[J^e]$. A necessary condition for the transformation to be bidirectional, that is, invertible, is that a point of the real element Ω^e will correspond to only one point of the parent element $\hat{\Omega}$ so that the determinant of the Jacobian matrix $|J^e|$ will not be zero. Furthermore, by ordering the vertices of the element in a trigonometric sense, $|J^e|$ should always be positive. Some properties of these transformations are summarized in Appendix C.1.

Assume that the basis functions of the parent element $\hat{\Omega}$ are known; then the computation of F^e and its inverse will allow one to obtain the basis function corresponding to any element Ω^e. In practice, it is not necessary to compute the inverse transformation because all computations may be performed for the parent element as shown in Appendix C.1. This technique is typical of FEM. It allows one to precompute any quantity related to the basis functions so that computations over each real element requires only the computation of its Jacobian matrix and its determinant. For these reasons, the CPU time required to fill the FEM local matrices (the stiffness matrix $[k^e]$, the mass matrix $[m^e]$, and the column matrix of excitations $\{f^e\}$) is rather low.

Usually, an element of very simple geometry and definition is selected as the element from which a complete family of elements will be generated. The generating element is called the parent, master, or reference element. Tables 2.1 to 2.3 summarize the characteristics of the Lagrange rectangular, triangular, and tetrahedral parent elements employed in this book. The geometry, the node coordinates, and the basis functions are also shown. The basis functions have been computed according to (2.41), as indicated in Appendix C.2. For the triangular and tetrahedral elements, natural coordinates have been used (see Appendix C.2.2). They are also called area (in 2D) or volume (in 3D) coordinates. Figure 2.15 illustrates the natural coordinates together with the triangular and tetrahedral parent element. Lagrange tetrahedral elements are not used in this book for interpolation of the unknown but for mapping only; that is, they have been used to perform the geometric transformation of other 3D tetrahedral elements from the corresponding parent element to the real element.

The elements (Ω^e, P, A) of a mesh that are equivalent to a parent element are called real elements. When the transformation F^e is linear, the element generated from the parent element will have straight contours or faces. When F^e is of higher order, it may give either a straight or a curved element, with quadratic (second-order) or cubic (third-order), contours or faces, for example. A parent element may generate a complete family of equivalent elements with straight or curved contours or faces. Figure 2.16 illustrates this point with a 2D example.

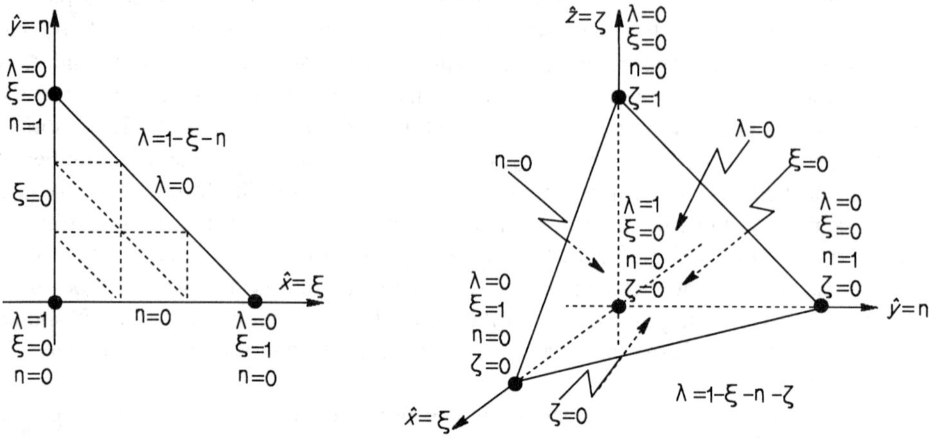

(a) (b)

Figure 2.15 Natural coordinates. (a) Triangular and (b) tetrahedral parent elements.

Table 2.1
Rectangular Lagrange parent elements

$Q(1)$	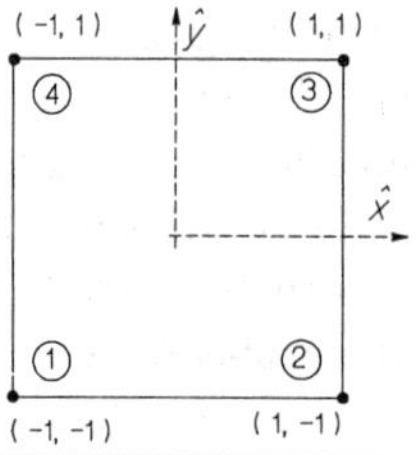	$\hat{N}_1 = a\,c \quad \hat{N}_3 = b\,d$ $\hat{N}_2 = b\,c \quad \hat{N}_4 = a\,d$
$Q(2)$	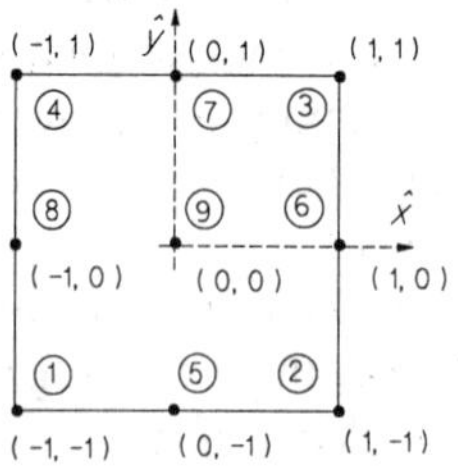	$\hat{N}_1 = a\,c\,\hat{x}\,\hat{y} \quad \hat{N}_5 = -4\,a\,b\,c\,\hat{y}$ $\hat{N}_2 = -b\,c\,\hat{x}\,\hat{y} \quad \hat{N}_6 = 4\,a\,c\,d\,\hat{x}$ $\hat{N}_3 = b\,d\,\hat{x}\,\hat{y} \quad \hat{N}_7 = 4\,a\,b\,d\,\hat{y}$ $\hat{N}_4 = -a\,d\,\hat{x}\,\hat{y} \quad \hat{N}_8 = -4\,b\,c\,d\,\hat{x}$ $\hat{N}_9 = 16\,a\,b\,c\,d$

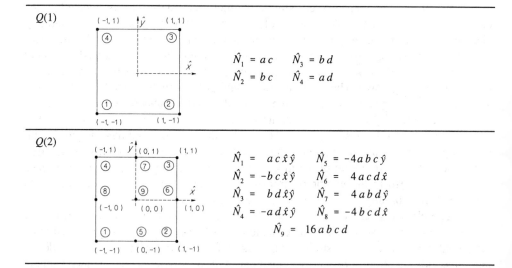

Table 2.1 (Continued)
Rectangular Lagrange parent elements

$Q(3)$

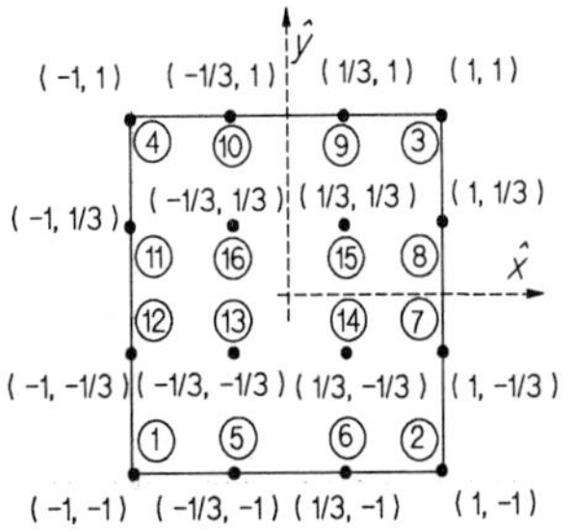

$$\hat{N}_1 = P_1(\hat{x})\,P_1(\hat{y}) \qquad \hat{N}_9 = P_3(\hat{x})\,P_4(\hat{y})$$
$$\hat{N}_2 = P_4(\hat{x})\,P_1(\hat{y}) \qquad \hat{N}_{10} = P_3(\hat{x})\,P_4(\hat{y})$$
$$\hat{N}_3 = P_4(\hat{x})\,P_4(\hat{y}) \qquad \hat{N}_{11} = P_2(\hat{x})\,P_3(\hat{y})$$
$$\hat{N}_4 = P_1(\hat{x})\,P_4(\hat{y}) \qquad \hat{N}_{12} = P_1(\hat{x})\,P_2(\hat{y})$$
$$\hat{N}_5 = P_2(\hat{x})\,P_1(\hat{y}) \qquad \hat{N}_{13} = P_2(\hat{x})\,P_2(\hat{y})$$
$$\hat{N}_6 = P_3(\hat{x})\,P_1(\hat{y}) \qquad \hat{N}_{14} = P_3(\hat{x})\,P_2(\hat{y})$$
$$\hat{N}_7 = P_4(\hat{x})\,P_2(\hat{y}) \qquad \hat{N}_{15} = P_3(\hat{x})\,P_3(\hat{y})$$
$$\hat{N}_8 = P_4(\hat{x})\,P_3(\hat{y}) \qquad \hat{N}_{16} = P_2(\hat{x})\,P_3(\hat{y})$$

$Q(2')$

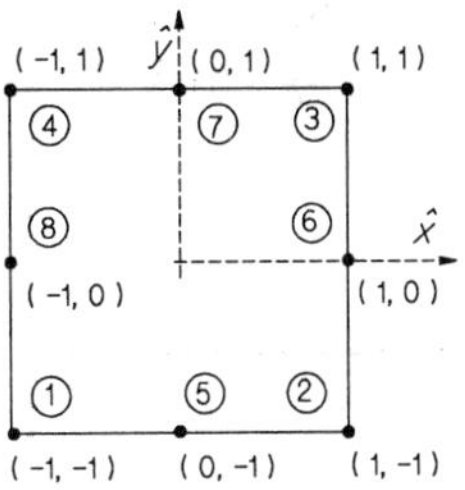

$$\hat{N}_1 = -ac(1+\hat{x}+\hat{y}) \qquad \hat{N}_5 = 4abc$$
$$\hat{N}_2 = -bc(1-\hat{x}+\hat{y}) \qquad \hat{N}_6 = 4bcd$$
$$\hat{N}_3 = -bd(1-\hat{x}-\hat{y}) \qquad \hat{N}_7 = 4abd$$
$$\hat{N}_4 = -ad(1+\hat{x}-\hat{y}) \qquad \hat{N}_8 = 4acd$$

$Q(3')$

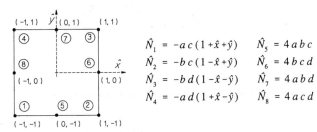

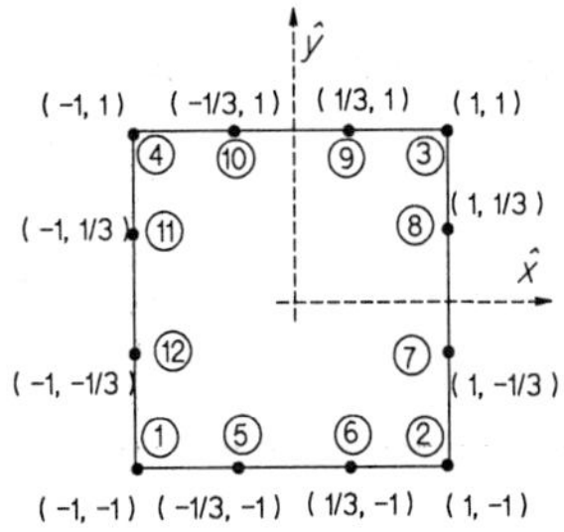

$$\hat{N}_1 = (9/8)ace \qquad \hat{N}_7 = (9/4)(1-3\hat{y})bcd$$
$$\hat{N}_2 = (9/8)bce \qquad \hat{N}_8 = (9/4)(1+3\hat{y})bcd$$
$$\hat{N}_3 = (9/8)bde \qquad \hat{N}_9 = (9/4)(1+3\hat{x})abd$$
$$\hat{N}_4 = (9/8)ade \qquad \hat{N}_{10} = (9/4)(1-3\hat{x})abd$$
$$\hat{N}_5 = (9/4)(1-3\hat{x})abc \qquad N_{11} = (9/4)(1+3\hat{y})abd$$
$$\hat{N}_6 = (9/4)(1+3\hat{x})abc \qquad \hat{N}_{12} = (9/4)(1-3\hat{y})abd$$

$$a = (1-\hat{x})/2, \quad b = (1+\hat{x})/2, \quad c = (1-\hat{y})/2, \quad d = (1+\hat{y})/2, \quad e = -10/9+\hat{x}^2+\hat{y}^2$$

$$P_1(w) = -(1/16)(1-w)(1-9w^2), \qquad P_2(w) = (6/16)(1-w^2)(1-3w)$$
$$P_3(w) = (9/16)(1-w^2)(1+3w), \qquad P_4(w) = (1/16)(1+w)(1-9w^2)$$

Table 2.2
Triangular Lagrange parent elements

$P(1)$

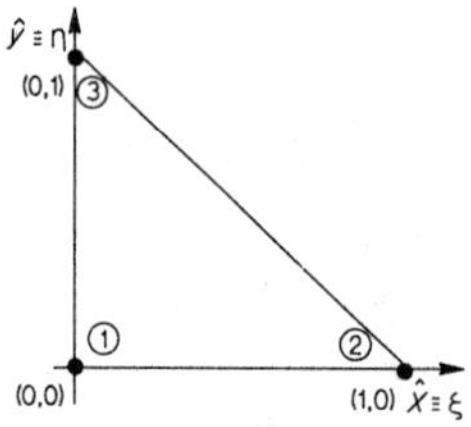

$$\hat{N}_1 = \lambda$$
$$\hat{N}_2 = \xi$$
$$\hat{N}_3 = \eta$$

$P(2)$

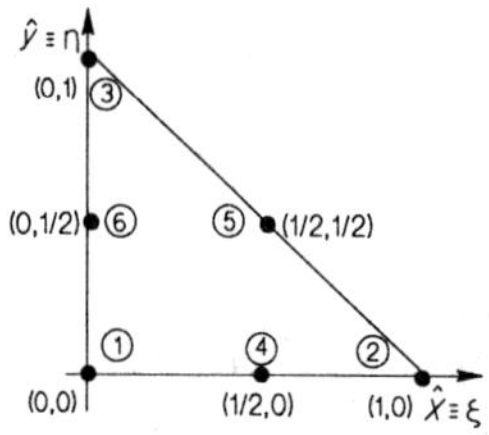

$$\hat{N}_1 = -(1-2\lambda)\lambda \qquad \hat{N}_4 = 4\xi\lambda$$
$$\hat{N}_2 = -(1-2\xi)\xi \qquad \hat{N}_5 = 4\xi\eta$$
$$\hat{N}_3 = -(1-2\eta)\eta \qquad \hat{N}_6 = 4\eta\lambda$$

$P(3)$

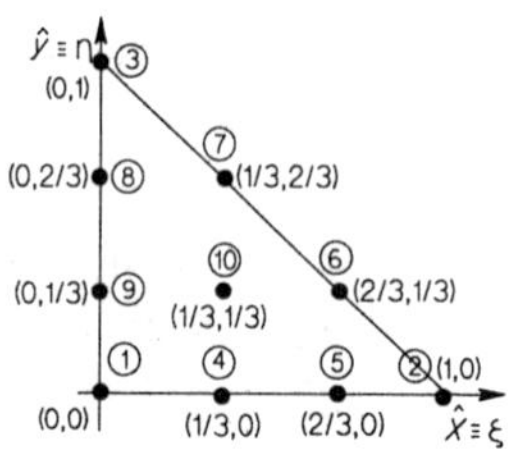

$$\hat{N}_1 = \frac{1}{2}\lambda(-1+3\lambda)(-2+3\lambda) \qquad \hat{N}_6 = \frac{9}{2}\xi\eta(-1+3\xi)$$
$$\hat{N}_2 = \frac{1}{2}\xi(-1+3\xi)(-2+3\xi) \qquad \hat{N}_7 = \frac{9}{2}\xi\eta(-1+3\eta)$$
$$\hat{N}_3 = \frac{1}{2}\eta(-1+3\eta)(-2+3\eta) \qquad \hat{N}_8 = \frac{9}{2}\lambda\eta(-1+3\eta)$$
$$\hat{N}_4 = \frac{9}{2}\lambda\xi(-1+3\lambda) \qquad \hat{N}_9 = \frac{9}{2}\lambda\eta(-1+3\lambda)$$
$$\hat{N}_5 = \frac{9}{2}\lambda\xi(-1+3\xi) \qquad \hat{N}_{10} = 27\xi\eta\lambda$$

$P(3')$

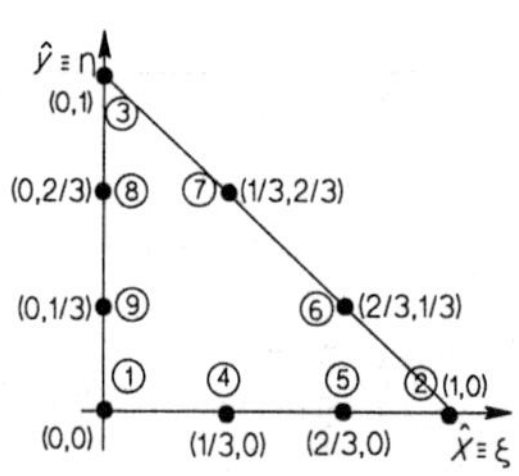

$$\hat{N}_i = \hat{N}_i^{(3)} + a_i \hat{N}_{10}^{(3)}, \quad 1 \le i \le 9, \quad i \in \mathbb{N}$$
$$\hat{N}_i = \text{basis functions } P(3)$$
$$a_i = -\frac{1}{6}, \qquad 1 \le i \le 3$$
$$a_i = \frac{1}{4}, \qquad 4 \le i \le 9$$

Table 2.3
Tetrahedral Lagrange parent elements

T(1)

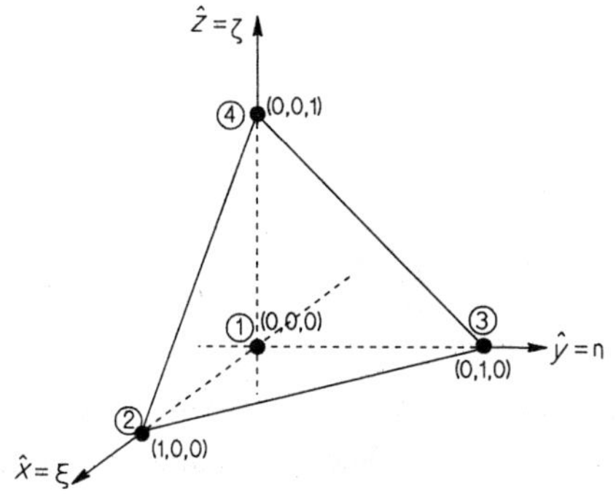

$$\hat{N}_1 = \lambda$$
$$\hat{N}_2 = \xi$$
$$\hat{N}_3 = \eta$$
$$\hat{N}_4 = \zeta$$

T(2)

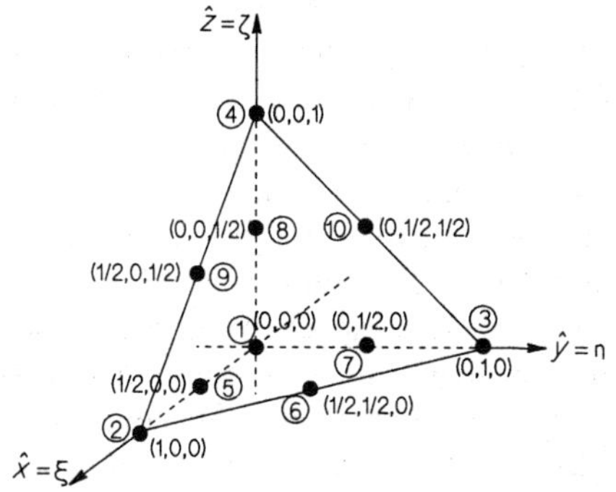

$$\hat{N}_1 = -(1-2\lambda)\lambda \qquad \hat{N}_6 = 4\xi\eta$$
$$\hat{N}_2 = -(1-2\xi)\xi \qquad \hat{N}_7 = 4\eta\lambda$$
$$\hat{N}_3 = -(1-2\eta)\eta \qquad \hat{N}_8 = 4\zeta\lambda$$
$$\hat{N}_4 = -(1-2\zeta)\zeta \qquad \hat{N}_9 = 4\xi\zeta$$
$$\hat{N}_5 = 4\xi\lambda \qquad \hat{N}_{10} = 4\eta\zeta$$

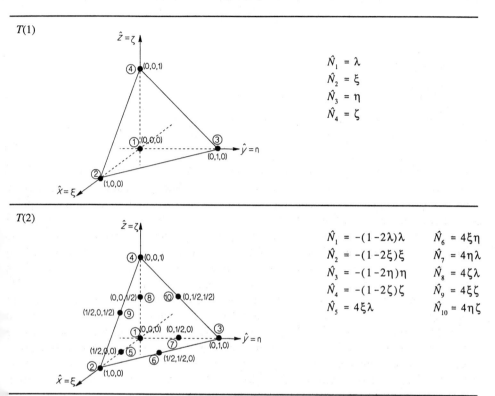

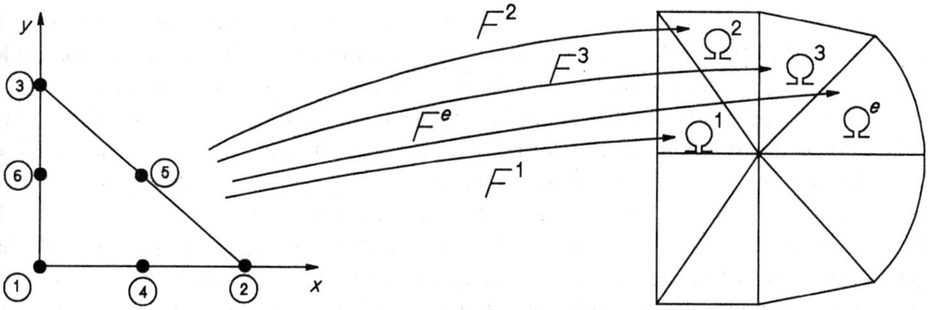

Figure 2.16 Generation of a family of curved and straight quadratic elements from a quadratic parent element.

Among the possible expressions for the geometric transformation F^e of degree s, it is easy to demonstrate that the following fulfills all the conditions mentioned previously:

$$\{r\} = \begin{bmatrix} \{r_1\} & \{r_2\} & \cdots & \{r_m\} \end{bmatrix} \begin{Bmatrix} \hat{N}_1^g(\{\hat{r}\}) \\ \hat{N}_2^g(\{\hat{r}\}) \\ \vdots \\ \hat{N}_m^g(\{\hat{r}\}) \end{Bmatrix} \tag{2.46}$$

In (2.46), the coefficients of the left-hand column matrix $\{r\}$ are the coordinates of any point within the real element, as defined previously. The jth column of the matrix of the right-hand term corresponds to the column matrix of the coordinates of the jth point of the real element. The number of points involved in the geometric transformation (also called geometric points) is assumed to be m. It is not required that it should be equal to n_e, the number of degrees of freedom and nodes. The coefficients of the column matrix of the right-hand term are the m geometric transformation basis functions of order s. Usually they are chosen as the Lagrange basis functions of order s.

A real element is said to be subparametric if its geometric transformation functions are of lower order than the interpolation functions. It is called isoparametric when the same order is used. Finally, superparametric elements are those utilizing geometric transformation functions of higher degree than the interpolation functions. Isoparametric elements are frequently used. The subparametric elements are a specific case of an isoparametric element. Figure 2.17 illustrates the generation of subparametric, isoparametric, and superparametric elements. The latter ones are not frequently used because they have convergence problems. In the example of Figure 2.17 the upper element is a subparametric second-order element; that is, second-order basis functions are used for the interpolation (there are six nodes), while only three points (the vertices) have been required for the linear transformation performed. The next element is an isoparametric second-order element; that is, second-order basis functions have been utilized for both the interpolation and the mapping. Finally, the lower element is a superparametric first-order element (thus, there are three nodes at the vertices), while second-order polynomials are utilized for the mapping (six geometric points).

The condition that F^e must be invertible ($|J^e|$ nonzero) may be translated into geometric terms by the requirement that there must not be any degenerate element. For elements with curved contours or faces, certain conditions should be fulfilled, related to the location of the geometric points in the edges. They must not deviate too much from the position they will have when the edge is straight. The position of the geometric points or nodes for the curved edges or faces is computed by projecting on the curved boundary the geometric point corresponding to the straight element as it is shown in Figure 2.18 for the cases of quadratic and cubic elements. The discretization error will be reduced when the degree of the mapping function is increased.

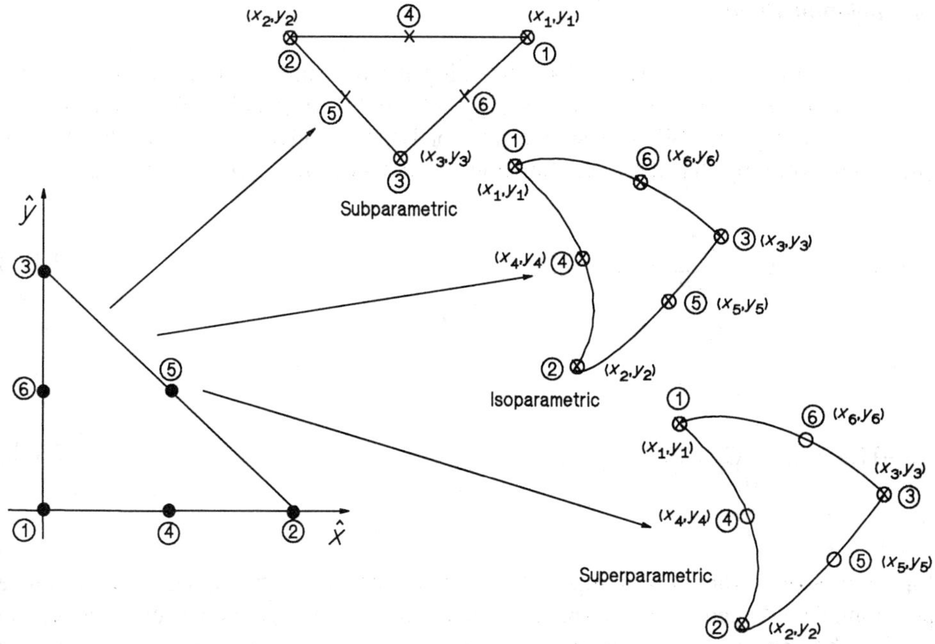

Figure 2.17 Generation of subparametric, isoparametric, and superparametric elements from a parent element with identical vertices and geometric points. ×: Points used for the interpolation (nodes). o: Points employed for the geometric transformation.

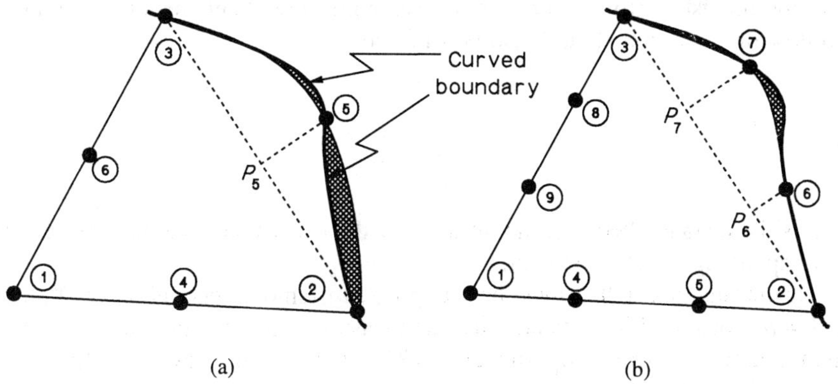

Figure 2.18 Position of the geometric points in a triangular element for the discretization of a curved contour. The shaded surface shows the geometric discretization error.
(a) Quadratic transformation. (b) Cubic transformation.

Interpolation Error

This subsection summarizes results concerning the measure of the interpolation error over Ω^e. For the different measures see Appendix A.1.3. In [Ciarlet 1973] and [Raviart and Thomas 1983; Ch. 4], it is shown that for Lagrange elements of type $P(p)$, $Q(p)$, $Q(p')$ and assuming that the exact function w belongs to $H^{(r+1)}(\Omega^e)$, $r \geq 1$, one gets:

$$|w - \Pi w|_{m,\Omega^e} \leq C_1 \frac{h_e^{k+1}}{\rho_e^m} |w|_{k+1,\Omega^e} \tag{2.47a}$$

$$\|w - \Pi w\|_{m,\Omega^e} \leq C_2 \frac{h_e^{k+1}}{\rho_e^m} |w|_{k+1,\Omega^e} \tag{2.47b}$$

for $0 \leq m \leq k+1$, where $k = \min(p,r)$. Here p is the degree of the higher order complete polynomial contained in the interpolation functions. C_1 and C_2 are constants that depend only on the degree p and the type of element; h_e is the diameter of the element Ω^e, and ρ_e is its roundness. It is assumed that Ω^e is small enough, that is, both h_e and ρ_e are assumed to be < 1.

For the case of the element $P(3')$, the interpolation will be complete only up to second order. Thus, for that case in (2.47) it should be $k = \min(2,r)$ [Ciarlet 1973].

Expression (2.47) may be extended to curved elements (although substituting in the right-hand side the seminorm by the norm [Ciarlet 1973]), if the position of each node is not too distant from that of the corresponding straight element (see Figure 2.18). The following condition should also be fulfilled:

$$\sum_{i=1}^{n_e} d_i^2 \leq C_3 h_e^p \tag{2.48}$$

Here d_i is the distance between the ith node of the curved element and the ith node of the corresponding straight element.

If the function w that must be interpolated is smooth enough over Ω^e, then $r > p$ and $k = p$. Thus, (2.47a) indicates that the measure of the interpolation error incurred for the function ($m = 0$) is proportional to $h_e^{(p+1)}$; while for the first derivatives ($m = 1$), it is proportional to $\sigma_e h_e^p$, where σ_e is the aspect ratio of the element. For triangular elements, σ_e is inversely proportional to $\sin\theta_e$, where θ_e is the lowest internal angle of the element. Hence, elements with low aspect ratio or with shapes close to the ideal equilateral triangle should be used. Conversely, distorted elements should be avoided.

It is interesting to note that for rectangular elements, although one is using polynomials having monomials of total degree higher than p, the error is proportional to $h_e^{(p+1)}$ or h_e^p. In other words, the error is fixed by the degree of the complete polynomial of higher order, p, contained in the interpolation. This is one of the reasons that the higher order rectangular elements are not frequently used. An increment on the order of the interpolation implies a considerable increase on the number of nodes (thus, on the size of the global system of equations to be solved) without a corresponding significant improvement on the interpolation error. It is also understandable that elements of type $Q(p')$ should be preferred over $Q(p)$.

If the exact function w is not smooth enough over Ω^e, then, $r < p$ and $k = r$, with r an integer number. The measure of the interpolation error is proportional to h_e^{r+1} for $m = 0$ and to h_e^r for $m = 1$. Thus, it is equivalent to the use of elements of degree r.

In cases where the derivatives of the exact function w are singular but still square integrable, expressions (2.47) give errors proportional to $h_e^{\alpha+1}$ (for $m = 0$) and $\sigma_e h_e^{\alpha}$ (for $m = 1$), where α $(0 < \alpha < 1)$ is a measure of the singularity ($r = \alpha$). If the dependence on p in those expressions is explicitly taken into account, it may be shown that for $m = 1$ the error is proportional to $\sigma_e h_e^{\alpha} p^{-\alpha+\varepsilon}$, where $\varepsilon > 0$ [Babuška and Dorr 1981], [Babuška and Suri 1987a,b], [Barnhill and Whiteman 1973], [Zienkiewicz and Craig 1986]. These results indicate that in order to obtain a prespecified interpolation error of similar magnitude over each element, regions where the solution is not smooth enough should be discretized with smaller elements (or with elements of higher order) than regions where the solution is regular enough.

Utilization of Lagrange Elements in Scalar and Vector Formulations

A formulation is said to be scalar if there is only one degree of freedom per node. Thus, expression (2.42) gives the approximation of w over the element Ω^e of n_e degrees of freedom and nodes of coordinates $\bar{r}_i$, $i = 1, \ldots, n_e$

$$\tilde{w}(\bar{r}) = \Pi\, w(\bar{r}) = \sum_{i=1}^{n_e} N_i(\bar{r})\, w(\bar{r}_i) = \{N\}^T \begin{Bmatrix} w(\bar{r}_i) \\ \cdot \\ \cdot \\ \cdot \\ w(\bar{r}_{n_e}) \end{Bmatrix} = \{N\}^T \{d\} \tag{2.49}$$

In a vector formulation there are two or more degrees of freedom per node. This is the case for problems with several unknowns requiring an approximation of the same type, namely they must belong to $H^1(\Omega)$. Then Lagrange elements with two or more degrees of freedom per node may be used. Let $\Pi w^{(j)}$ be the jth approximate function and $w_k^{(j)}$ the value at the kth node ($k = 1, \ldots, n_e'$) of the jth variable, $w^{(j)}$; m is the number of degrees of freedom per node, and n_e' is the total number of nodes. The total number of degrees of freedom is $n_e = m\, n_e'$ [Dhatt and Touzot 1981; Ch. 2]. Hence,

$$\{\tilde{w}\} = \{\Pi w\} = \begin{Bmatrix} \Pi w^{(1)}(\bar{r}) \\ \Pi w^{(2)}(\bar{r}) \\ \vdots \\ \Pi w^{(m)}(\bar{r}) \end{Bmatrix} = \begin{bmatrix} N_1, 0...0 & N_2, 0...0 & ... & N_{n_e}, 0...0 \\ 0, N_1...0 & 0, N_2...0 & ... & 0, N_{n_e}...0 \\ .... & .... & ... & .... \\ 0, 0...N_1 & 0, 0...N_2 & ... & 0, 0...N_{n_e} \end{bmatrix} \begin{Bmatrix} d_1^{(1)} \\ \vdots \\ d_1^{(m)} \\ \vdots \\ d_{n_e}^{(1)} \\ \vdots \\ d_{n_e}^{(m)} \end{Bmatrix} \quad (2.50)$$

Formulations of this type are used in Chapter 3 for the discretization of the full-wave analysis of waveguiding structures in terms of the longitudinal components of both the electric and the magnetic fields. They have also been utilized in the literature for the full-wave analysis in terms of two or more Cartesian components of either the electric and/or the magnetic fields.

Lagrange Infinite Elements

In Section 2.3.1, these elements have been mentioned for the solution of open-region problems. Given a problem with an open region, the domain is subdivided into a near-field region and a far-field region. Ordinary finite elements are used to discretize the near-field region. Infinite elements are used to represent the far-field region. At the interface between both regions, finite and infinite elements are contiguous.

Infinite elements should have the following three characteristics:

1. They must be able to simulate subdomains of infinite length in a given direction.
2. They must provide a suitable approximation of the function for points well into the far-field region; that is, they should provide the proper asymptotic behavior of the fields.
3. They must be compatible with ordinary elements.

The second characteristic indicates that the formulation for this element depends on the type of problem to be analyzed. In this book infinite elements have been used only for the quasi-static analysis of transmission lines with one or two infinite ground planes. Furthermore, the formulation selected for such problems use the scalar electrostatic potential as an unknown function. In these types of 2D problems, the asymptotic behavior of the unknown may be considered to be proportional to $1/r$ or $1/r^2$, where r is the distance from the origin. The bibliography on this topic is quite voluminous; see, for example, [Akin 1982; Ch. 6], [Bettess 1977, 1980], [Bettess and Bettess 1984], [Curnier 1983], [Damjanic and Owen 1984], [Dhatt and Touzot 1981; Ch. 2], [Kardestuncer and Norrie 1987; P. 3, Chs. 2 and 5], [Lynn and Hadid 1981], [Pantic and Mittra 1986], [Pissanetzky 1984], [Salazar-Palma and Hernández-Gil 1995],

[Zienkiewicz 1977; Ch. 23], [Zienkiewicz et al. 1983] among other references. No attempt has been made in this book to use infinite elements for solution of other open-region problems. Instead, Chapter 6 (for 2D problems) and Section 7.5 (for 3D) are devoted to a hybrid finite element-boundary integral (FE-BI) technique.

A good presentation of the infinite elements that have been proposed in the past for static problems is available in [Bettess and Bettess 1984]. According to the method followed to obtain the asymptotic behavior at infinity, they may be classified as follows:

1. *Direct approximation methods*. They are based on the use of parent elements of infinite dimension in one, two, or three directions. The basis functions for the interpolation are the Lagrange basis functions, modified in some way (normally, multiplying them by a suitable expression) in order to obtain the desired asymptotic behavior. These methods require the use of a special numerical integration formula to compute the coefficients of the FEM local matrices. The integration must be done over each infinite element, that is, over the real infinite subdomains.
2. *Inverse approximation methods*. In these methods, the parent elements are the finite standard ones. The interpolation basis functions are also those of the standard Lagrange elements. The real element, infinite in one or more directions, with the proper asymptotic behavior at infinity is obtained by means of an adequate geometric mapping applied to the parent element. Methods in this category do not require special techniques for numerical integration. At the same time, the numerical integration may be performed in the finite parent element, following the usual FEM procedure. For these reasons inverse infinite elements look more attractive than the previous ones. From the programming point of view, inverse infinite elements only require a different computation of the coefficients of the Jacobian matrix $[J^e]$ of the related geometric mapping (see Appendix C.1).

In this book, 2D inverse infinite elements, compatible with rectangular and triangular ordinary elements, have been used. Examples of its application are given in Chapter 3. The element employed is due to O. C. Zienkiewicz, C. Emson, and P. Bettess [1983]. It provides asymptotic behavior proportional to $1/r$, $1/r^2$ or, in general, $1/r^m$, where m is an integer. The generation of 2D infinite elements is easily accomplished from one-dimensional (1D) infinite elements.

Figure 2.19 shows a one-dimensional finite parent element defined by three geometric points of coordinates $\hat{x}_1 = -1$, $\hat{x}_2 = 1$, $\hat{x}_3 = 0$. The element should be mapped into an infinite 1D element given by three geometric points of coordinates x_1, $x_2 \rightarrow \infty$, x_3, respectively. In Appendix C.3, it is shown how that mapping is performed using the general expression (2.46). The column matrix of geometric basis functions, $\{\hat{N}^g\}$, should have three coefficients corresponding to the three geometric points of the mapping. The pertinent basis functions called $\{\hat{N}^I\}$ are given in the upper part of Table 2.4, that is, the basis functions of the type $1/r$. Furthermore, it is shown in Appendix C.3 that the use of such a mapping provides:

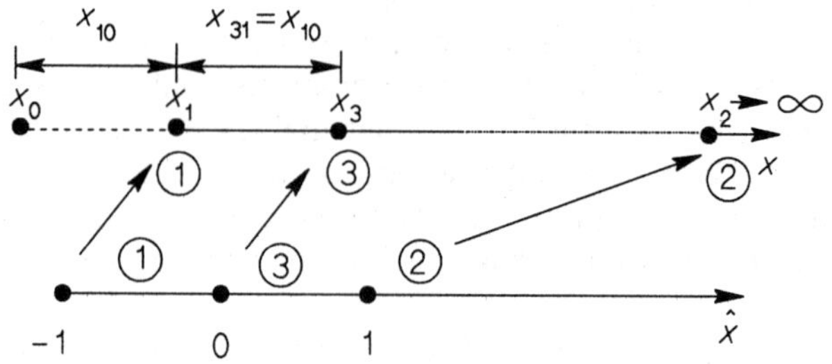

Figure 2.19 Transformation of a one-dimensional finite parent element into an element infinite in the x direction.

1. The transformation of that finite parent element into an infinite real element, defined by three geometric points of coordinates x_1, $x_2 = \infty$, x_3, respectively, as desired.
2. An expression for the function interpolating the variable, Πw, that has a dominant term proportional to $1/r$. Here r is the distance from any point of the real infinite element to a point, called *pole*, located outside it, in the near-field region. This result is obtained for any order of the Lagrange interpolation functions. Higher order interpolation functions will only improve the approximation, keeping the same asymptotic behavior.

The x coordinate of the pole is termed x_0. It is easy to see that $x_{10} = x_{31}$, where $x_{10} = x_1 - x_0$ and $x_{31} = x_3 - x_1$. The position of the pole is given by x_{10}, while x_{31} gives the so called *size* of the infinite element. By the size of the infinite element it is understood the distance between the origin of the infinite element (the vertex at the interface with the contiguous ordinary finite element) and its *central* geometric point (the point corresponding to $\hat{x} = 0$ on the parent element). The generalization of the previous results in order to obtain asymptotic behaviors like $(1/r)^m$ leads to the mapping basis functions given in the lower part of Table 2.4.

The geometric mapping of the basis functions for two-dimensional infinite elements can be obtained in a similar fashion. For 2D infinite elements in one direction, as in the example of Figure 2.20, the mapping basis functions may be defined by multiplying two sets of one-dimensional basis functions. The first one, for the coordinate of the infinite direction, is given by the infinite geometric basis functions shown in Table 2.4. The second one, for the coordinate of the finite direction, is given by the Lagrange first-order basis functions. Table 2.5 shows the 2D parent element $I(6)$ and the *geometric* basis functions for the generation of 2D elements, infinite in the x direction and with an asymptotic behavior as $1/r^m$. There are six geometric points, two of them placed at infinity in the real element.

That mapping may be used in conjunction with the interpolation basis functions

Table 2.4

One-dimensional parent element and geometric basis functions for the generation of infinite elements. The arrow indicates the point that will be mapped into infinity

One-dimensional parent element	Type	Geometric basis functions	Relation between the position of the pole and the size of the infinite element
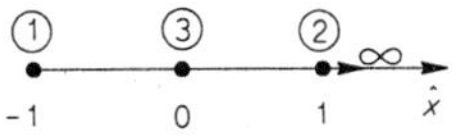	$1/r$	$\hat{N}_1^I = \dfrac{-2\hat{x}}{1-\hat{x}}$ $\hat{N}_2^I = 0$ $\hat{N}_3^I = \dfrac{1+\hat{x}}{1-\hat{x}}$	$x_{31} = x_{10}$
	$1/r^m$	$\hat{N}_1^I = \dfrac{k}{(1-\hat{x})^{1/m}} - k$ $\hat{N}_2^I = 0$ $\hat{N}_3^I = 1 - \dfrac{k}{(1-\hat{x})^{1/m}} + k$ where: $k = \dfrac{2^{1/m}}{1-2^{1/m}}$	$x_{31} = (2^{1/m}-1)x_{10}$

of any of the ordinary rectangular elements, $Q(p)$ or $Q(p')$ (see Table 2.1). Thus, the geometric basis functions of the element $I(6)$ may generate straight superparametric infinite elements of first order and straight subparametric infinite elements of second order, for example. Results of the analysis of open transmission lines using first- and second-order infinite elements are given in Chapter 3. These elements are compatible with ordinary 2D Lagrange elements of first- and second-order, respectively.

It is not necessary to generate 2D elements that are infinite in two directions, because an infinite domain of that type may be properly discretized using infinite elements in only one direction as shown in Figure 2.21.

As in the case of ordinary elements, infinite elements should not have shapes too distorted with respect to ideal ones. Hence, in order to avoid ill-conditioned matrices and the corresponding numerical problems, care should be taken not to generate infinite elements with internal angles deviating too much from 90 degrees. This issue is particularly important in the case of direct infinite elements [Bettess and Bettess 1984].

The pole position affects the accuracy of the FEM analysis. Results will not depend on the pole position (or on the size of the infinite element) if the location of the interface between the region where ordinary elements are used and that where infinite elements are used is well chosen. Actually, if that interface is deep in the far-field region, results will be quite accurate, but the number of elements in the near-field region will be very high. Hence, the FEM analysis will not be very efficient in terms of CPU time, memory, and storage requirements. If the interface is placed in the near-field region, the poles will be very close to the interface and the decay of the variable

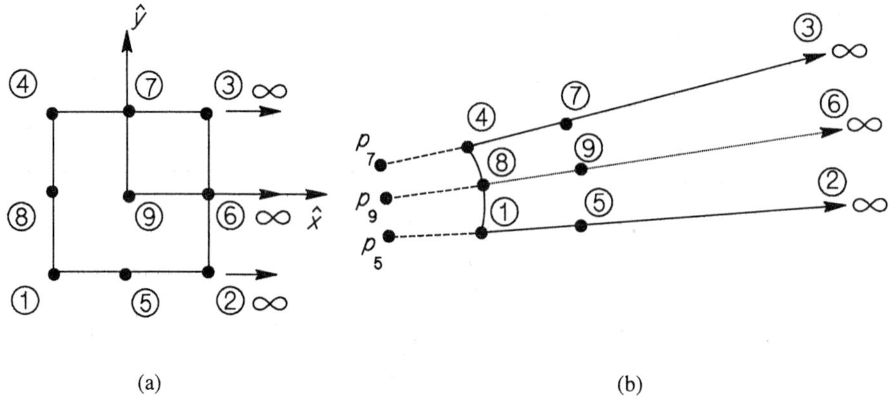

Figure 2.20 Example of a two-dimensional element infinite in a given direction. (a) Parent element. (b) Real element, p_i: pole corresponding to the ith central point, i = 5, 7, 9, along the infinite direction.

Table 2.5
Two-dimensional parent element and geometric basis functions for the generation of two-dimensional elements infinite in one direction. The arrows indicate the points mapped to infinity

Six geometric points $I(6)$	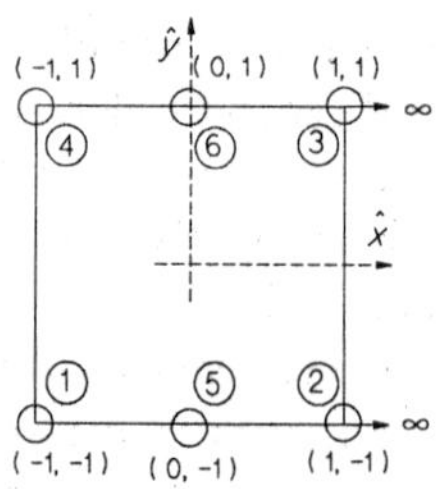	$\hat{N}_1^I = \hat{N}_1^{Im}(\frac{1}{2})(1-\hat{y})$ $\hat{N}_4^I = \hat{N}_1^{Im}(\frac{1}{2})(1+\hat{y})$ $\hat{N}_5^I = \hat{N}_3^{Im}(\frac{1}{2})(1-\hat{y})$ $\hat{N}_6^I = \hat{N}_3^{Im}(\frac{1}{2})(1+\hat{y})$	 $\hat{N}_2^I = 0$ $\hat{N}_3^I = 0$

$$\hat{N}_1^{Im} = k\left(\frac{1}{(1-\hat{x})^{1/m}} - 1\right), \quad \hat{N}_3^{Im} = 1 - k\left(\frac{1}{(1-\hat{x})^{1/m}} - 1\right), \quad k = \frac{2^{1/m}}{1-2^{1/m}}$$

to be approximated will not be adequately simulated. A compromise is needed. Thus, the analysis of open structures by infinite elements uses an iterative procedure to adapt the interface position to the optimum one.

Finally, it must be pointed out that when programming using infinite elements, care should be taken to preserve a specific local ordering of the nodes of each real infinite element in the domain. This is so because infinite elements (as well as singular or transition elements or any other special elements) are nonsymmetric elements. Ordinary elements are symmetric in the sense that any vertex of a real ordinary element Ω^e, may be considered to be the first one, for example, in the local numbering. Thus, any vertex of the real element may correspond to any vertex of the parent element. This is not the case for infinite elements where there is a specific and unique correspondence between the vertices placed at infinity and the vertices of the parent element corresponding to them. Thus, a given order must be followed in the computation of an infinite element local matrices. This local order will be changed if renumbering techniques are employed, with the scope of reducing the size of the band of the FEM matrices. Hence, techniques, like labeling, must be employed to restore the original local numbering for local matrices computation and postprocesses.

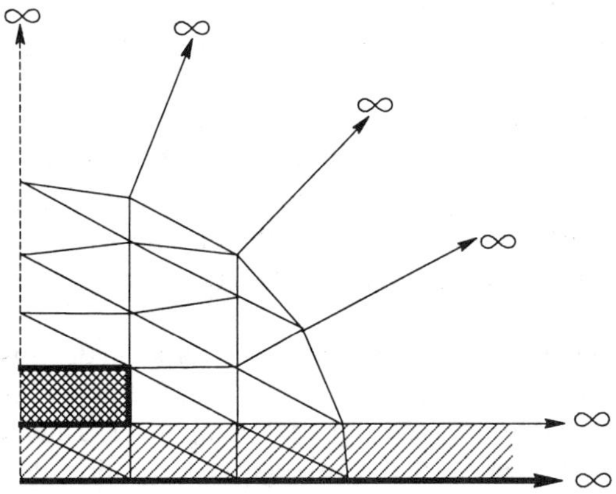

Figure 2.21 Mesh for the analysis of an open microstrip line. The symmetry of the problem allows one to consider just one half of the domain.

Div-Conforming Elements

The finite elements that are described in this subsection have been used in Section 3.2.3.3 to discretize the integral form obtained from a mixed formulation of the quasi-static analysis of transmission lines. In this formulation, the unknowns are the scalar electrostatic potential and the displacement vector over the 2D cross section of the transmission line. By means of a penalty function method, the scalar variable may be eliminated, so the only unknown left is the vector variable. As mentioned in Section 2.3.3.1, the vector solution should belong to the Hilbert space $H(\text{div},\Omega)$ (see Appendix A.1.4). Elements of the type described here will provide such solutions.

These elements approximate a vector variable by a linear combination of vector basis functions. For this reason they fall under the generic name of *vector* elements. The term vector is chosen as opposed to elements that give an approximation of a scalar variable by a linear combination of scalar basis functions. These latter elements may be generically called *scalar* elements. The Lagrange elements previously presented are scalar elements. Other scalar elements are, for example, those of the Hermite family [Ciarlet and Raviart 1972a], [Dhatt and Touzot 1981; Ch. 2], [Israel and Miniowitz 1987, 1990], [Oden and Reddy 1976; Ch. 6]. Nevertheless, this nomenclature may not be complete since scalar elements may also be used in vector approximation as described previously.

Historically, the first vector elements to be used were 2D first-order triangular elements. In this case, the values of the coefficients of the linear combination (the degrees of freedom) of vector basis functions were defined as the flux of the vector variable across each edge, that is, the contour integral extended to each edge of the component of the vector variable normal to the edge. For this reason these elements were called *edge* elements. This term was selected as opposed to *nodal* elements that were identified with scalar elements. In the latter case, the degrees of freedom coincide with the value of the scalar variable or its derivatives at nodal points. Nevertheless, as it will be seen, nodes may also be defined in connection with the definition of the degrees of freedom of edge elements. On the other hand, higher order 2D elements of the type under discussion require the definition of additional degrees of freedom related to the surface of the element. In the 3D case the degrees of freedom are related to the faces and the volume of the elements. Thus, the term edge does not properly encompass these elements.

Linear and higher order elements of the type under discussion impose the continuity of the normal component of the vector variable across element interfaces and do not impose the continuity of the tangential component. Thus, the approximation is differentiable only in the sense of the divergence operation. Hence, they are called divergence-conforming or *div-conforming* elements. The term conforming refers to the fact that the normal component of the vector variable across element interfaces is continuous, and this is a condition required to define the divergence of the vector variable. The divergence is then square integrable. The denomination div-conforming has been selected to refer to these elements. The component of the vector variable tangential to element interfaces is not continuous across element interfaces. Thus, the curl operation may not be defined. It may also be mentioned that the curl-conforming elements, to be described in the next subsection, fall also under the generic terms *vector* and *edge* elements. Thus, the denominations div-conforming and curl-conforming serve also to better distinguish between these two types of vector elements.

Div-conforming elements were first presented in [Raviart and Thomas 1977a]. A complete analysis may be found in [Thomas and Joly 1981; Ch. 3]. Out of all the conclusions that these authors draw, the following are important:

1. In order to construct a subspace of $H(\mathrm{div},\Omega)$ the condition

$$(\bar{a}_{n_{\Omega^1}} \cdot \bar{t}_{|\Omega^1}) + (\bar{a}_{n_{\Omega^2}} \cdot \bar{t}_{|\Omega^2}) = 0 \tag{2.51a}$$

must be satisfied by any element $\bar{t}$ in the subspace, for all contiguous elements Ω^1, Ω^2 at their common interface. Here, $\bar{a}_{n_{\Omega^i}}$ is the unit vector normal to the interface, toward the outside of element Ω^i, as it is expressed by Figure 2.22 for the 2D case. In 3D, the interface will be the common face of the two contiguous elements. In other words, because $\bar{a}_{n_{\Omega^1}} = -\bar{a}_{n_{\Omega^2}}$, (2.51a) indicates that the component of $\bar{t}$, normal to the element interfaces, must be continuous.

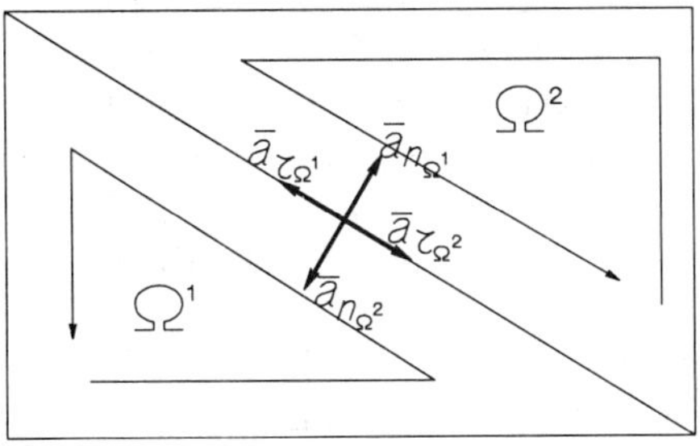

Figure 2.22 Definition of the unit vectors normal and tangential to an edge.

2. For a simplex (the only type of elements to be considered here: triangles for 2D and tetrahedrons for 3D), the proper space to perform the approximation for a vector variable belonging to $\mathbb{R}^n$ is given by $D_p^{(n)}$ (see Appendix A.1.1), that is, the space of vectors $\bar{d}$ in which components t_x, t_y (for 2D) and t_x, t_y, t_z (for 3D) are polynomials of order p containing complete polynomials of degree p-1. That space may also be described by

$$D_p^{(n)} = \left(P_{p-1}^{(n)}\right)^n \oplus P_{0,p-1}^{(n)}\, \bar{r} \tag{2.51b}$$

where $P_{p-1}^{(n)}$ is the linear space of polynomials of n variables of order less than or equal to p-1 (see Appendix A.1.1). $P_{0,\,p-1}^{(n)}$ is the space of homogeneous polynomials of degree p-1 (i.e., having only terms of degree p-1), and $\bar{r}$ is the position vector of a point of coordinates x,y (2D) and z (3D). The symbol $\oplus$ stands for the direct sum of two subspaces.

The number of degrees of freedom of an element of order p must be equal to the dimension of the space $D_p^{(n)}$. Here one is interested in 2D triangular elements of first and second order. Tetrahedral div-conforming elements have not been used in the book, but they will be described for completeness, for both first and second orders.

From (A.2d) the dimension of the space $D_p^{(n)}$ for the cases of interest is given by

- For $n = 2$ (triangles)
  - for $p = 1$, $\dim D_1^{(2)} = 3$,
  - for $p = 2$, $\dim D_2^{(2)} = 8$;
- For $n = 3$ (tetrahedrons)
  - for $p = 1$, $\dim D_1^{(3)} = 4$,
  - for $p = 2$, $\dim D_2^{(3)} = 15$.

3. Any element $\bar{t}$ of $D_p^{(n)}$ has the following properties:

- $\mathrm{div}\bar{t} \in P_{p\text{-}1}^{(n)}$, where $\mathrm{div}\bar{t}$ is defined in Appendix A (A.22), for $n = 2$ (i.e., 2D vectors in two variables, x and y); a similar definition holds for $n = 3$ (i.e., for 3D vectors in three variables, x, y, and z) substituting the 2D transverse operator by the 3D operator ;
- if $\mathrm{div}\bar{t} = 0$, then $\bar{t} \in (P_{p\text{-}1}^{(n)})^n$;
- for $\bar{\alpha}\cdot\bar{r} =$ constant (i.e., a straight line, for $n = 2$, or a plane for $n = 3$) then $\bar{\alpha}\cdot\bar{t}$ is a polynomial or order $\leq p\text{-}1$.

4. If Ω^e is a n-simplex of $\mathbb{R}^n$ (i.e., a triangle for $n = 2$, or a tetrahedron, for $n = 3$), a vector function $\bar{t}$ of $D_p^{(n)}$ is determined in an unique way by the quantities:

$$\int_{\Gamma_k^e} \bar{t}\cdot\bar{a}_n\, q\, \mathrm{d}\Gamma\ ,\quad \forall q \in P_{p-1}^{(n-1)}\ ,\quad \text{for } k = 1, \ldots, n+1 \tag{2.51c}$$

where Γ^e stands for the boundary of the element Ω^e, that is, its 3 edges ($k = 1$, 2, 3) for a triangle, or its 4 faces ($k = 1, 2, 3, 4$) for a tetrahedron, and

$$\int_{\Omega^e} t_{x_i}\, q\, \mathrm{d}\Gamma\ ,\quad \forall q \in P_{p-1}^{(1)}\ ,\quad \text{for } i = 1, \ldots, n \tag{2.51d}$$

where x_i stands for the components of $\bar{t}$ on the $x_1 = x$ and $x_2 = y$ directions, for $n = 2$, and $x_3 = z$ direction for $n = 3$.

The quantities (2.51c) are a way of specifying a polynomial of degree $\leq p\text{-}1$ over an edge ($n = 2$) or a face ($n = 3$) of Ω^e. An alternate way of determining such a polynomial is by giving its value at the appropriate number of points on the edges or in the faces of the element [Thomas and Joly 1981; Ch. 3]. That is, to sample $\bar{t}\cdot\bar{a}_n$ (that will be a polynomial of order p-1) at such points over the edges (for a triangular element) or the faces (for a tetrahedral element). Thus, these sampling points play the role of nodes, since the quantities required to specify a vector function of the space where the solution of the problem is sought for are related to them. It may be observed that contiguous elements will share these nodes and degrees of freedom (except for a sign, due to the outward definition of the unit vector normal to the boundary of an element: see Appendix C.4.3).

The quantities (2.51d) serve to complete the required number of degrees of freedom. Since they are related to the element (i.e., to its surface, for triangles, or to its volume, for tetrahedrons), they are not shared by contiguous elements. Thus, it is not necessary to relate them to any specific point. However, for practical implementation, nodes placed at the interior of the element are usually allocated for each degree of freedom of this type, or alternately, they may all be ascribed to a node placed at the centroid of the element.

A consequence of conclusions 1 and 4 is that for a div-conforming element of order p, the definition of the degrees of freedom should be a quantity proportional to the component of the vector variable, normal to its boundary. Actually, degrees of freedom related to boundaries are chosen as

$$C_k^e \int_{\Gamma_k^e} \bar{t} \cdot \bar{a}_n \, \mathrm{d}\Gamma \, , \quad k = 1, \ldots, n+1 \tag{2.51e}$$

sampled at the required points. Here, Γ_k^e stands for the kth boundary of the element Ω^e and C_k^e is a constant utilized for normalization purposes (the length of the edge for the 2D case, $n = 2$, or the area of the face for the 3D case, $n = 3$). Raviart and Thomas chose as sampling points the Gauss-Legendre integration points. The reason for selecting those points is that other choices, like the Newton-Cotes points of numerical integration, will include the vertices of the element where there will be a discontinuity of the normal vector. For the same reason, in the case of tetrahedrons, the Gauss-Legendre points that are chosen are those inside the faces, but not on the edges of the faces (see Table 2.11). This way of reasoning will provide the number and location of the nodes with which the degrees of freedom are associated. If more degrees of freedom are required, to fulfill the condition of conclusion 2, expression (2.51d) will provide their definition. For the elements one is interested in, the degrees of freedom are:

- For $n = 2$ (triangles)
  - for $p = 1$, that is, first-order polynomials, for the first type of degrees of freedom only one sampling point per edge is required. Thus, 3 nodes may be defined at the midpoints of each edge. This is the total number of degrees of freedom required for this case. Thus, they are defined as the length of each edge multiplied by the component of the vector variable normal to that edge at its midpoint; the sign allocated to the normal vector completes the definition of the degree of freedom;

- for $p = 2$, that is, second-order polynomials, for the first type of degrees of freedom two sampling points per edge are required. Thus, 6 nodes may be defined with two points per edge located at the two corresponding Gauss-Legendre points, with the same definition for the related degree of freedom. However, two more degrees of freedom are required. They are defined as

$$\int_{\Omega^e} \bar{t} \cdot \bar{a}_x \, \mathrm{d}\Omega \ , \quad \int_{\Omega^e} \bar{t} \cdot \bar{a}_y \, \mathrm{d}\Omega \tag{2.51f}$$

according to (2.51d). They may be related to the centroid of the element.

- For $n = 3$ (tetrahedrons)
  - for $p = 1$, for the first type of degrees of freedom one sampling point per face is required located at its centroid. Thus, 4 nodes may be defined, one per face. This is the total number of degrees of freedom required. The definition of the degree of freedom is done as for the 2D element except for substituting the length of the edge by the area of the face;
  - for $p = 2$, for the first type of degrees of freedom three sampling points are required. Thus, 12 nodes may be defined, three per face, located at the Gauss-Legendre points of integration placed inside the element (see Table 2.11). However, three more degrees of freedom are required. They are defined by (2.51f) plus a third one with an analogous definition except for using $\bar{a}_z$, the unit vector on the z-direction.

For higher order elements, the degrees of freedom and associated nodes may be easily defined following a similar procedure.

The general and formal description of div-conforming elements is given next [Nédélec 1980], [Raviart and Thomas 1977a], [Thomas and Joly 1981; Ch. 3]. A div-conforming simplex element used to discretize the space $H(\mathrm{div},\Omega)$ is defined by:

1. A compact region Ω^e of $\mathbb{R}^n$, connected and not empty, where for $n = 2$, Ω^e will be a triangle and for $n = 3$ a tetrahedron;
2. The space of vector functions $D_p^{(n)}$ described by (2.51b);
3. A set of degrees of freedom A (and eventually a set of sampling points Σ, or nodes $\bar{r}_i$, defined over Ω^e).

The degrees of freedom are of two types:

a. The first type is defined by

$$\int_{\Gamma_k^e} \bar{t} \cdot \bar{a}_n q \, \mathrm{d}\Gamma \ , \quad \forall \ q \in P_{p-1}^{(n-1)}(\Gamma_k^e) \ , \quad \text{for } k = 1,\ldots,n+1 \tag{2.51g}$$

where Γ_k^e stands for the boundaries of the element Ω^e (edges for 2D or faces for

3D) and $P_{p-1}^{(n-1)}(\Gamma_k^e)$ is the space of scalar polynomials of order $\leq p$-1 defined over the edges or the faces of Ω^e. The definition of the degree of freedom must include the sign allocated to the normal vector, i.e., to the edge or face.

b. The second type is defined by

$$\int_{\Omega^e} \bar{t} \cdot \bar{q}_i \, d\Omega \ , \quad \forall \ \bar{q}_i \in \left(P_{p-2}^{(n)}\right)^n (\Omega^e) \ , \quad i = 1,\ldots,n \tag{2.51h}$$

where $P_{p-2}^{(n)}(\Omega^e)$ is the space of functions of order $\leq p$-2 defined over the surface (for triangular elements) or the volume (for tetrahedral elements) of Ω^e. Vectors $\bar{q}_i$ may be any 2 or 3 vectors defining the surface or the volume of the element (for the 2D or 3D case, respectively). Usually, the Cartesian unit vectors.

Once the space of vector functions are selected and the degrees of freedom are defined, a vector function $\bar{t}(\bar{r})$ may be approximated by a function $\Pi\bar{t}(\bar{r})$ as

$$\{\tilde{t}(\bar{r})\} = \Pi\{t(\bar{r})\} = \sum_{i=1}^{n_e} \{N_i(\bar{r})\} d_i \tag{2.52a}$$

where n_e is the total number of degrees of freedom; d_i, $1 \leq i \leq n_e$, $i \in \mathbb{N}$, are the degrees of freedom; and $\{N_i(\bar{r})\}$ or $\bar{N}_i(\bar{r})$ are the vector basis functions of the approximation that must belong to the space $D_p^{(n)}$.

The computation of the vector basis functions associated with each degree of freedom may be easily accomplished by taking into account that the values of the degrees of freedom defined for each vector basis function must be equal to unity for the associated basis function and equal to zero for the rest. This is a property analogous to (2.41). To be more specific, for the cases in which one is interested here (i.e., div-conforming triangular elements of first and second orders), the relevant expressions are

- For $p = 1$

$$\left.\begin{array}{ll} L_k^e\left(\bar{a}_{nk} \cdot \bar{N}_r(x_k,y_k)\right) = 1, & k=r \\ L_k^e\left(\bar{a}_{nk} \cdot \bar{N}_r(x_k,y_k)\right) = 0, & k\neq r \end{array}\right\} \quad \begin{array}{l} k=1,2,3 \\ r=1,2,3 \end{array} \tag{2.52b}$$

- For $p = 2$

$$\left.\begin{array}{ll} L_k^e\left(\bar{a}_{nk} \cdot \bar{N}_r(x_k,y_k)\right) = 1, & k=r \\ L_k^e\left(\bar{a}_{nk} \cdot \bar{N}_r(x_k,y_k)\right) = 0, & k\neq r \end{array}\right\} \quad \begin{array}{l} k=1,\ldots,6 \\ r=1,\ldots,8 \end{array} \tag{2.52c}$$

$$\left.\begin{aligned} \int_{\Omega^e} N_{xr}(x,y)\,\mathrm{d}\Omega &= 1, & r&=7 \\ \int_{\Omega^e} N_{xr}(x,y)\,\mathrm{d}\Omega &= 0, & r&\neq 7 \end{aligned}\right\} \quad 1 \le r \le 8, \quad r \in \mathbb{N} \tag{2.52d}$$

$$\left.\begin{aligned} \int_{\Omega^e} N_{yr}(x,y)\,\mathrm{d}\Omega &= 1, & r&=8 \\ \int_{\Omega^e} N_{yr}(x,y)\,\mathrm{d}\Omega &= 0, & r&\neq 8 \end{aligned}\right\} \quad 1 \le r \le 8, \quad r \in \mathbb{N} \tag{2.52e}$$

where L_k^e stands for the length of the edge where the kth node of the element Ω^e is located; $\bar{a}_{nk}$ is the unit vector, normal to the edge at that point, directed in the outward direction; $\bar{N}_r(x_k,y_k)$ is the rth vector basis function sampled at the kth node of the element; and $N_{xr}(x,y)$, $N_{yr}(x,y)$ are the x, y components of $\bar{N}_r(x,y)$, respectively.

It is useful also in this case to make the computation over a parent element from which the expression corresponding to each real element may be obtained. The definition of equivalent elements is analogous to the case of Lagrange elements. The basis functions employed for the geometric mapping are, as in the previous case, the Lagrange basis functions. If straight real elements are desired, the first-order Lagrange basis functions should be used. If curved elements are required, higher order basis functions should be employed. In order to have an invariant space of vector functions $D_p^{(n)}$, the relation between a vector function $\{\hat{t}\}$ (in matrix notation) in the parent element $\hat{\Omega}$ and the vector function $\{t\}$ in the real element Ω^e must be given by

$$\{t(\{r\})\} = \frac{1}{|J^e|}[J^e]\{\hat{t}(\{\hat{r}\})\} \tag{2.53}$$

as pointed out by Raviart and Thomas [1977a] and Nédélec [1980]. Expression (2.53) should be added to expression (2.45). In Appendix C.4.1, details on the computation of the basis functions for the triangular parent elements of first and second orders are given. Table 2.6 provides the summary. It may be observed that the two internal nodes of the second order element are placed at the same point, the centroid of the triangle. Thus, strictly speaking there are two degrees of freedom assigned to the same node.

It is important to observe that for an element of order p the normal component of the vector variable at the edges (triangles) or the faces (tetrahedrons) will be a function of order p-1 with zero interpolation error at the nodes placed on edges or faces.

In summary, the relevant features of these elements are:

- They preserve the continuity of the normal component of the vector variable at element interfaces.

Table 2.6
Triangular div-conforming parent elements

First-order element

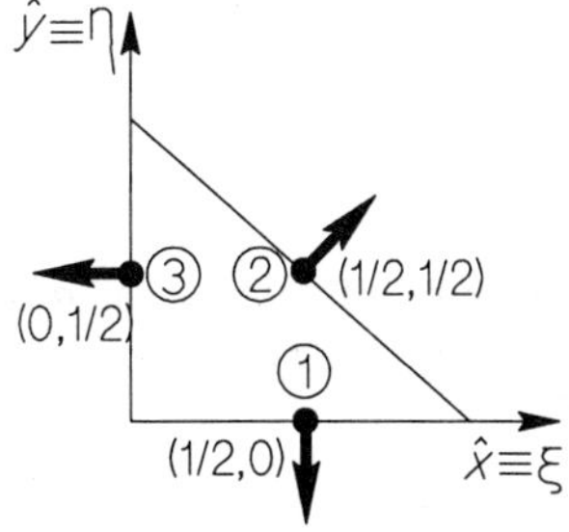

$$\{\hat{N}_1\} = \begin{Bmatrix} \xi \\ -1+\eta \end{Bmatrix}$$

$$\{\hat{N}_2\} = \begin{Bmatrix} \xi \\ \eta \end{Bmatrix}$$

$$\{\hat{N}_3\} = \begin{Bmatrix} -1+\xi \\ \eta \end{Bmatrix}$$

Second-order element

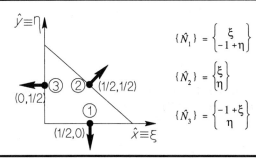

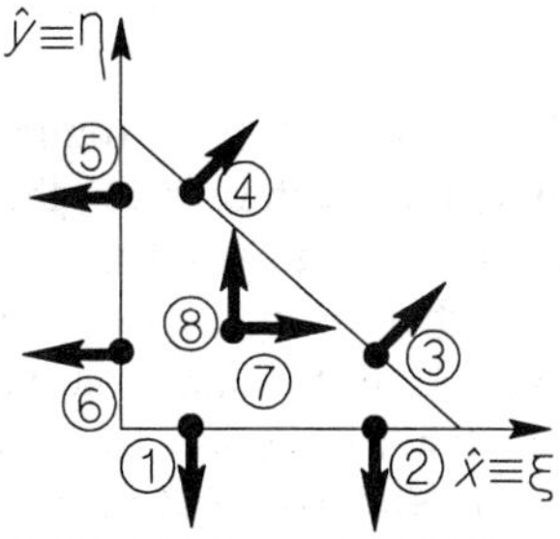

$$\{\hat{N}_i\} = \begin{Bmatrix} a_1^{(i)} + a_2^{(i)}\xi + a_3^{(i)}\eta + \xi(\alpha^{(i)}\xi + \beta^{(i)}\eta) \\ b_1^{(i)} + b_2^{(i)}\xi + b_3^{(i)}\eta + \eta(\alpha^{(i)}\xi + \beta^{(i)}\eta) \end{Bmatrix}$$

i	a_1	a_2	a_3	b_1	b_2	b_3	α	β
1	0	3-5a	0	-2+3a	3-6a	6-7a	-4+8a	-4+4a
2	0	-2+5a	0	1-3a	-3+6a	-1+7a	4-8a	-4a
3	0	-2+a	0	0	0	-1-a	4-4a	4a
4	0	-1-a	0	0	0	-2+a	4a	4-4a
5	1-3a	-1+7a	-3+6a	0	0	-2+5a	-4a	4-8a
6	-2+3a	6-7a	3-6a	0	0	3-5a	-4+4a	-4+8a
7	0	16	0	0	0	8	-16	-8
8	0	8	0	0	0	16	-8	-16

$a = 0.5(1-\sqrt{3}/3)$

Node	ξ	η
1	a	0
2	1-a	0
3	1-a	a
4	a	1-a
5	0	1-a
6	0	a
7, 8	1/3	1/3

- The tangential component of the vector variable at element interfaces is discontinuous;
- The nodes are not placed at the vertices of the elements. This is an important advantage in alleviating convergence problems when there are singularities of the vector variable at corners.

Interpolation Error

The interpolation error over the div-conforming element Ω^e of order p for a vector function $\bar{t}$ belonging to the space $(H^r(\Omega^e))^n$ with $r \geq 1$ is given by

$$\|\bar{t} - \Pi\bar{t}\|_{0,\Omega^e} \leq C_1 \frac{h_e^{k+1}}{\rho_e} |\bar{t}|_{k,\Omega^e} = C_1 \sigma_e h_e^k |\bar{t}|_{k,\Omega^e} \tag{2.54a}$$

where $k = \min(r,p)$, h_e is the diameter of the element, ρ_e is its roundness, and σ_e is the aspect ratio. When the exact solution $\bar{t}$ is smooth enough over Ω^e, then $k = p$, obtaining a measure of the error proportional to $\sigma_e h_e^p$. When that is not the case, then $k = r \geq 1$ and the error will be proportional to $\sigma_e h_e^r$, as when using only an element of order $r < p$ [Nédélec 1980], [Thomas and Joly 1981; Ch. 3]. If the exact solution has singularities but it is still square integrable (and, thus, the expression $|\bar{t}|_{0,\Omega^e}$ is bounded), a measure of the singularity, α, may be obtained [Babuška and Suri 1987a,b]. There are no works defining the interpolation error of these elements in such cases, however the work of [Křížek and Neittaanmäki 1984] for similar problems suggests that it will be proportional to $\sigma_e h_e^\alpha$, with $0<\alpha<1$.

For the divergence of $\bar{t}$, div$\bar{t}$, for div$\bar{t} \in H^r(\Omega^e)$, measures of the interpolation error are given by [Thomas and Joly 1981; Ch. 3] as

$$\|\text{div}\,\bar{t} - \text{div}\,\Pi\bar{t}\|_{0,\Omega^e} \leq C_2 h_e^k \,|\,\text{div}\,\bar{t}\,|_{k,\Omega^e} \tag{2.54b}$$

and by [Nédélec 1980] as

$$|\text{div}\,\bar{t} - \text{div}\,\Pi\bar{t}\,|_{0,\Omega^e} \leq C_3 \frac{h_e^{k+1}}{\rho_e} |\bar{t}|_{k+1,\Omega^e} \tag{2.54c}$$

Curl-Conforming Elements

In this section a family of finite elements that provides solutions belonging to the Hilbert space $H(\text{curl},\Omega)$ (see Appendix A.1.5) is described. It is the dual of the div-conforming element family.

Two-dimensional elements of this type combined with two-dimensional Lagrange elements provide a type of element that is suitable for the discretization of 3D vector variables with the following properties. The vector may be decomposed into two orthogonal components: one, in the x,y plane of a 3D Cartesian system of coordinates and the other in the z direction. The values of both the components depend only on the x,y coordinates but not on z. This is the case for the electric and magnetic field vectors in a uniform waveguiding structure. The dependence with distance along the direction of propagation (longitudinal direction), z, is through the factor $e^{\gamma z}$ (see Appendix D.3.2), where γ is the propagation constant. Then, the x,y-plane may be identified with the plane containing the cross section (transverse plane) of the waveguide structure. The transverse component of the electric or magnetic field vector may be approximated by means of 2D curl-conforming elements, while the amplitude of the longitudinal component is approximated by Lagrange elements. This approximation provides solutions belonging to the product space $H^1(\Omega)\times H(\text{curl},\Omega)$ (see Appendix A.1.6). This type of element is used in Chapters 3 to 5 for the full-wave analysis of waveguiding structures.

Three-dimensional curl-conforming elements are used in Chapter 7 for the analysis of 3D electromagnetic problems: cavity problems, discontinuities in waveguiding structures, and scattering and antenna problems. For these cases the vector solution belongs to the Hilbert space $H(\text{curl},\Omega)$.

The curl-conforming elements of linear and higher order that are going to be presented impose the continuity of the tangential component of the vector variable across element interfaces and do not impose the continuity of the normal component. Thus, the approximation is differentiable only in the sense of the curl operation. Hence, they are called *curl-conforming*. The term conforming refers to the fact that the tangential component of the vector variable across element interfaces is continuous, and this is a condition for defining the curl of the vector variable. The curl is then square integrable. The component of the vector variable, normal to element interfaces, is not continuous across element interfaces. Thus the divergence operation may not be defined.

J. C. Nédélec presented the curl-conforming elements in 1980 [Nédélec 1980]. Assume a vector variable, $\bar{t}$, that belongs to the space $H(\text{curl},\Omega)$. In order to construct a subspace of $H(\text{curl},\Omega)$, the following condition must be fulfilled for any 2D or 3D vector $\bar{t}$ of the subspace for any Ω^1 and Ω^2 elements (2D or 3D) sharing a common interface:

$$(\bar{a}_{\tau_{\Omega^1}} \cdot \bar{t}_{|\Omega^1}) + (\bar{a}_{\tau_{\Omega^2}} \cdot \bar{t}_{|\Omega^2}) = 0 \tag{2.55a}$$

where $\bar{a}_\tau$ stands for the unit vector parallel to the edges of a finite element. For the case of 3D elements the following condition should also be fulfilled:

$$(\bar{a}_{n_{\Omega^1}} \times \bar{t}_{|\Omega^1}) + (\bar{a}_{n_{\Omega^2}} \times \bar{t}_{|\Omega^2}) = 0 \tag{2.55b}$$

where $\bar{a}_{n_{\Omega^i}}$, $i = 1, 2$, is the unit vector normal to the common face of two contiguous 3D elements in the outward direction of each element, and $\times$ represents the vector product. Because $\bar{a}_{\tau_{\Omega^1}} = -\bar{a}_{\tau_{\Omega^2}}$ and $\bar{a}_{n_{\Omega^1}} = -\bar{a}_{n_{\Omega^2}}$, expressions (2.55a) and (2.55b) indicate that the component of $\bar{t}$ along a common edge (for 2D and 3D elements) or tangential to the face (3D) must be continuous. As a consequence, it is convenient to select values proportional to those quantities as the definition of the degrees of freedom.

For a simplex (which are the only elements to be considered here: triangles for 2D and tetrahedrons for 3D), the proper space to perform the approximation for a vector variable belonging to $\mathbb{R}^n$ is given in Appendix A.1.1 where its dimension is also shown. The space may be defined as

$$R_p^{(n)} = (P_{p-1}^{(n)})^n \oplus S_p^{(n)} \tag{2.55c}$$

where $\oplus$ represents the direct sum of the two spaces, $P_{p-1}^{(n)}$ is the space of polynomials of degree $\leq p$-1, and $S_p^{(n)}$ is the space of vectors whose components are homogeneous polynomials of degree p, having the property that for any vector $\bar{t}_s\,(\bar{r})$ belonging to it, the following condition holds: $\bar{t}_s\,(\bar{r}) \cdot \bar{r} = 0$ [Monk 1992a], [Nédélec 1980]. It may be seen that $R_p^{(n)}$ is not the space of complete polynomials of order p. It lies in between such space and the space of polynomials p-1. The reason for reducing it through the mentioned condition is to eliminate some degrees of freedom. This reduction is not arbitrary. In Section 7.2 it will be seen that it is performed in such a way that the elements of the space that have been eliminated belong to the null space of the curl operator. That condition is equivalent to imposing on the vector functions belonging to $R_p^{(n)}$ the following constraints, called the Nédélec constraints:

1. For 2D elements:
   a. for $p = 1$

$$\frac{\partial t_x}{\partial x} = 0\ , \qquad \frac{\partial t_y}{\partial y} = 0\ , \qquad \frac{\partial t_x}{\partial y} + \frac{\partial t_y}{\partial x} = 0 \tag{2.55d}$$

so that the space is given by

$$R_1^{(2)} = \left(P_0^{(2)}\right)^2 \oplus S_1^{(2)} \tag{2.55e}$$

where $S_1^{(2)}$ is the space generated by the vector

$$\begin{Bmatrix} -y \\ x \end{Bmatrix} \tag{2.55f}$$

The dimension of the space is 3 (see (A.2k));

b. for $p = 2$

$$\frac{\partial^2 t_x}{\partial x^2} = 0 \,, \quad \frac{\partial^2 t_y}{\partial y^2} = 0$$

$$\frac{\partial^2 t_x}{\partial y^2} + 2\frac{\partial^2 t_y}{\partial x \partial y} = 0 \,, \quad \frac{\partial^2 t_y}{\partial x^2} + 2\frac{\partial^2 t_x}{\partial x \partial y} = 0 \tag{2.55g}$$

so that the space is given by

$$R_2^{(2)} = \left(P_1^{(2)}\right)^2 \oplus S_2^{(2)} \tag{2.55h}$$

where $S_2^{(2)}$ is the space generated by the two vectors

$$\begin{Bmatrix} y^2 \\ -xy \end{Bmatrix} \,, \begin{Bmatrix} -xy \\ x^2 \end{Bmatrix} \tag{2.55i}$$

The dimension of the space is 8 (see (A.2k));

2. For 3D elements
 a. for $p = 1$

$$\frac{\partial t_x}{\partial x} = 0 \,, \quad \frac{\partial t_y}{\partial y} = 0 \,, \quad \frac{\partial t_z}{\partial z} = 0$$

$$\frac{\partial t_x}{\partial y} + \frac{\partial t_y}{\partial x} = 0 \,, \quad \frac{\partial t_x}{\partial z} + \frac{\partial t_z}{\partial x} = 0 \,, \quad \frac{\partial t_y}{\partial z} + \frac{\partial t_z}{\partial y} = 0 \tag{2.55j}$$

so that the space is given by

$$R_1^{(3)} = \left(P_0^{(3)}\right)^3 \oplus S_1^{(3)} \tag{2.55k}$$

where $S_1^{(3)}$ is the space generated by the vector $\bar{\beta} \times \bar{r}$ where $\bar{\beta}$ is a vector belonging to the space $(P_0)^3$; that is, the space $S_1^{(3)}$ is generated by

$$\begin{Bmatrix} 0 \\ -z \\ y \end{Bmatrix} \,, \begin{Bmatrix} z \\ 0 \\ -x \end{Bmatrix} \,, \begin{Bmatrix} -y \\ x \\ 0 \end{Bmatrix} \tag{2.55l}$$

The dimension of this space is 6 (see (A.2k));

b. for $p = 2$

$$\frac{\partial^2 t_x}{\partial x^2} = 0 \,, \quad \frac{\partial^2 t_y}{\partial y^2} = 0 \,, \quad \frac{\partial^2 t_z}{\partial z^2} = 0$$

$$\frac{\partial^2 t_x}{\partial y^2} + 2\frac{\partial^2 t_y}{\partial x \partial y} = 0 \,, \quad \frac{\partial^2 t_y}{\partial x^2} + 2\frac{\partial^2 t_x}{\partial x \partial y} = 0$$

$$\frac{\partial^2 t_x}{\partial z^2} + 2\frac{\partial^2 t_z}{\partial x \partial z} = 0 \,, \quad \frac{\partial^2 t_z}{\partial x^2} + 2\frac{\partial^2 t_x}{\partial x \partial z} = 0 \tag{2.55m}$$

$$\frac{\partial^2 t_y}{\partial z^2} + 2\frac{\partial t_z}{\partial y \partial z} = 0 \,, \quad \frac{\partial^2 t_z}{\partial y^2} + 2\frac{\partial^2 t_y}{\partial y \partial z} = 0$$

$$\frac{\partial^2 t_x}{\partial y \partial z} + \frac{\partial^2 t_y}{\partial x \partial z} + \frac{\partial^2 t_z}{\partial x \partial y} = 0$$

so that the space is given by

$$R_2^{(3)} = (P_1^{(3)})^3 \oplus S_2^{(3)} \tag{2.55n}$$

where $S_2^{(3)}$ is the space generated by the eight vectors

$$\begin{Bmatrix} y^2 \\ -xy \\ 0 \end{Bmatrix}, \begin{Bmatrix} 0 \\ -yz \\ y^2 \end{Bmatrix}, \begin{Bmatrix} -xy \\ x^2 \\ 0 \end{Bmatrix}, \begin{Bmatrix} -xz \\ 0 \\ x^2 \end{Bmatrix}$$
$$\begin{Bmatrix} z^2 \\ 0 \\ -xz \end{Bmatrix}, \begin{Bmatrix} 0 \\ z^2 \\ -yz \end{Bmatrix}, \begin{Bmatrix} yz \\ -xz \\ 0 \end{Bmatrix}, \begin{Bmatrix} 0 \\ xz \\ -xy \end{Bmatrix} \tag{2.55o}$$

The dimension of this space is 20 (see(A.2k)).

For a vector of order p belonging to this space its component along a straight line or tangential to a plane will be a polynomial of order $\leq p$-1.

The number of degrees of freedom of an element of order p must be equal to the dimension of the space $R_p^{(n)}$. As a consequence, besides the degrees of freedom proportional to the tangential component to edges or faces there will be the need to define other types of degrees of freedom. There are two types of degrees of freedom for 2D elements, while for 3D elements, degrees of freedom of three types are required. They are described in the following [Monk 1992a], [Nédélec 1980].

A curl-conforming element that may be employed to discretize the space $H(\text{curl},\Omega)$ is defined by:

1. A compact region, Ω^e, of $\mathbb{R}^n$, connected and nonempty;
2. The space of vector functions $R_p^{(n)}$;
3. A set of degrees of freedom, A, and eventually a set of sampling points, Σ, or nodes, defined over Ω^e.

The $p(p+2)$ (see Appendix A.1.1) degrees of freedom of 2D triangular curl-conforming elements are of two types:

a. The first one is defined by

$$\int_{\Gamma_k^e} \bar{t} \cdot \bar{a}_\tau q \, d\Gamma \ , \quad \forall \ q \in P_{p-1}^{(1)}(\Gamma_k^e) \ , \quad k = 1,...,3 \tag{2.55p}$$

where Γ_k^e stands for the kth edge of the triangle Ω^e and $P_{(p-1)}^{(1)}(\Gamma^e)$ is the space of polynomial functions of order $\leq p$-1 defined over the edges of Ω^e. As with div-conforming elements, expression (2.55p) is equivalent to sampling $\bar{t}\cdot\bar{a}_\tau$ at the adequate points of the edges to define a p-1 polynomial along them. Thus, the definition of the degree of freedom of this type is such value multiplied, for normalization purposes, by the length of the edge. The sign of the tangential vector, i.e., of the edge, must be added to the local definition. The sampling points may be considered as nodes associated to these degrees of freedom. There will be a total of $3p$ degrees of freedom of this type, p of them on each edge, chosen at the Gauss-Legendre points.

b. The second type is defined by

$$\int_{\Omega^e} (\bar{t} \cdot \times \bar{a}_n) \cdot \bar{q}_i \, d\Omega \ , \quad \forall \ \bar{q}_i \in (P_{p-2}^{(2)})^2(\Omega^e) \ , \quad i = 1,2 \tag{2.55q}$$

where $\bar{a}_n$ is the unit vector normal to the surface of the element and the space $(P_{(p-2)}^{(2)})^2(\Omega^e)$ contains polynomial functions of order $\leq p$-2 defined over the surface of Ω^e. Vectors $\bar{q}_i$ may be any two vectors defining the surface of the element. Usually, they are chosen on the Cartesian directions. Thus, the integrand of (2.55q) is the projection of the vector $\bar{t}\times\bar{a}_n$ (that is tangential to the surface of the element) along the direction of $\bar{q}_i$. This type of degrees of freedom will not exist for $p = 1$. For $p = 2$, two degrees of freedom of this type are required. For $p > 2$, $p(p$-1) degrees of freedom are required, that is, $p(p$-1)/2 for each Cartesian direction. These degrees of freedom are not shared by contiguous elements. Thus, there is no need to assign nodes to them. However, they may be defined at inner points of the element or at its centroid.

The $(p+3)(p+2)p/2$ degrees of freedom of tetrahedral elements are of three types:

a. The first type is defined by

$$\int_{e_i} \bar{t} \cdot \bar{a}_\tau \, q \, \mathrm{d}e \ , \quad \forall \ q \in P_{p-1}^{(1)}(e_i) \ , \quad \text{for } i = 1,...,6 \tag{2.55r}$$

where e_i stands for the ith edge of Ω^e and $P_{p-1}^{(1)}(e_i)$ is the space of functions of order $\leq p$-1 defined over the ith edge of Ω^e. This type of degree of freedom is defined as the first type of the 2D case, so similar considerations may be done. There are a total of $6p$ degrees of freedom of this type, p per edge. They may be associated to p sampling points at each edge (the Gauss-Legendre points).

b. The second type is defined by

$$\int_{\Gamma_k^e} \bar{t} \times \bar{a}_n \cdot \bar{q}_i \, \mathrm{d}\Gamma \ , \quad \forall \ \bar{q}_i \in \left(P_{p-2}^{(2)}\right)^2 (\Gamma_k^e) \ , \quad k = 1,...,4 \ , \quad i = 1,2 \tag{2.55s}$$

where $(P_{p-2}^{(2)})^2(\Gamma_k^e)$ is the space of functions of order $\leq p$-2 defined over the kth face of Ω^e. This type of degrees of freedom will not exist for $p = 1$. The vectors $\bar{q}_i$ are chosen on any two independent directions of each face Γ^e. The sign of the normal vector must be added to the local definition. It is important to note that these degrees of freedom will be shared by contiguous elements. Thus, in order to facilitate the assembly process it is convenient to associate nodes with them. There are a total of $4p(p$-1) degrees of freedom of this type, $p(p$-1) per face, i.e. $p(p$-1)/2 on each direction. The degrees of freedom on each direction may be associated to corresponding nodes. For $p = 2$ only two degrees of freedom of this type are required, one on each direction. The corresponding nodes may be placed at the centroid of the face. Care must be taken to properly distinguish one degree of freedom from the other when performing the assemblage of the FEM matrices (see Appendix C.4.3).

c. The third type is defined by

$$\int_{\Omega^e} \bar{t} \cdot \bar{q}_i \, \mathrm{d}\Omega \ , \quad \forall \ \bar{q}_i \in \left(P_{p-3}^{(3)}\right)^3 (\Omega^e) \ , \quad i = 1,2,3 \tag{2.55t}$$

where $(P_{p-3}^{(3)})^3(\Omega^e)$ is the space of functions of order $\leq p$-3 defined over the volume of Ω^e. This type of degrees of freedom will not exist for $p \leq 2$. Again, vectors $\bar{q}_i$ are chosen on three independent directions in the volume of Ω^e, usually the Cartesian directions. There are a total of $p(p$-1)(p-2)/2 degrees of freedom of this type, that is $p(p$-1)(p-2)/6 on each direction. They will not be shared by contiguous elements. Thus, it is not required to associate nodes with them. However, inner points or the centroid of the element may be assigned.

Any vector function $\overline{t}(\overline{r})$ defined over Ω^e may be interpolated by the function $\Pi\{t(\overline{r})\}$ given by (2.52a). The vector basis functions must belong to the space $R_p^{(n)}$. The computation of the basis functions is easily done from the definition of each degree of freedom and the expression of a vector belonging to the specified space.

It is useful to make the computation over a parent element from which the expression corresponding to each real element may be obtained. The definition of equivalent elements is analogous to the case of Lagrange elements. The basis functions employed for the geometric mapping are, as in the previous case, the Lagrange basis functions. If straight real elements are desired, the first-order Lagrange basis functions should be used. If curved elements are required, higher order basis functions should be employed. For an invariant space of vector functions, the relation between a vector function $\{\hat{t}\}$ in $\hat{\Omega}$ and the vector $\overline{t}$ (or $\{t\}$) in the real element, Ω^e, must be given by

$$\left\{t(\{r\})\right\} = \left(\left[J^e\right]^T\right)^{-1}\left\{\hat{t}(\{\hat{r}\})\right\} \tag{2.56}$$

The basis functions may be computed following a procedure analogous to the one described for div-conforming elements (see (2.52b) to (2.52e)). In Appendix C.4.2, details are given for simplex parent elements of first and second orders. Tables 2.7 and 2.8 describe them. The internal nodes of the second-order elements are placed at the centroid of the element (2D) or its faces (3D). The location of the nodes are the same for 2D triangular div-conforming and curl-conforming elements. It is observed that the 2D vector basis functions of the triangular curl-conforming element may be obtained from the 2D vector basis functions of the triangular div-conforming element by a 90° rotation. This is equivalent to performing the vector product of the unit vector in the longitudinal direction by the div-conforming vector basis functions. Instead, for 3D elements there is no such relation [Nédélec 1980]. Table 2.8 shows the basis function for second-order tetrahedral elements. There is some freedom in selecting the independent vectors $\overline{q}_i$ of (2.55s). Possible candidates are listed in Table 2.8. Depending on the concrete choice, the last eight vector basis functions are different. They are listed in Table 2.8. Appendix C.4.2.2 and C.4.3 provide more details on this topic.

It is important to observe that for an element of order p, the components of the approximate vector variables tangential to the edges (2D or 3D) or tangential to the faces (3D) are functions of order p-1 and they result in zero interpolation error at the nodes placed on edges (2D or 3D) or faces (3D). The relevant features of these elements are dual of those listed for div-conforming elements.

Interpolation Error

The interpolation error over the element Ω^e for a vector function t belonging to the

Table 2.7
Triangular curl-conforming parent elements

First-order element

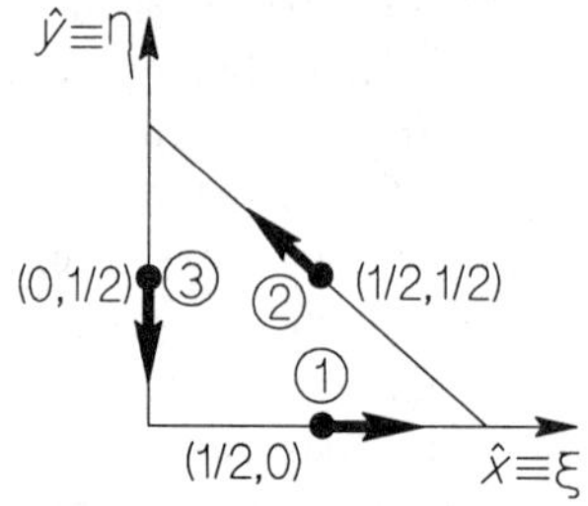

$$\{\hat{N}_1\} = \begin{Bmatrix} 1-\eta \\ \xi \end{Bmatrix}$$

$$\{\hat{N}_2\} = \begin{Bmatrix} -\eta \\ \xi \end{Bmatrix}$$

$$\{\hat{N}_3\} = \begin{Bmatrix} -\eta \\ -1+\xi \end{Bmatrix}$$

Second-order element

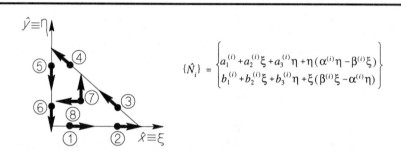

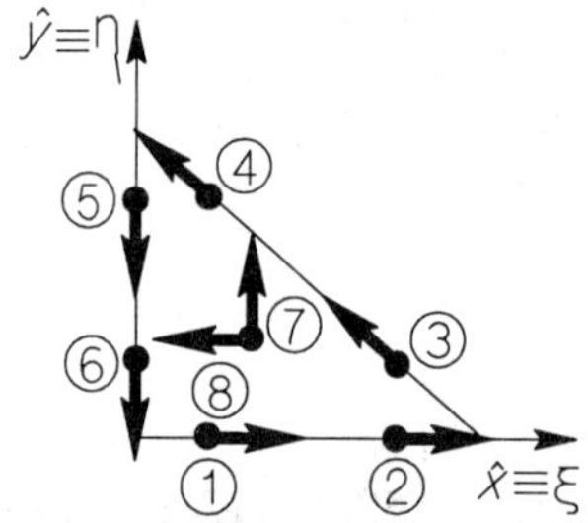

$$\{\hat{N}_i\} = \begin{Bmatrix} a_1^{(i)} + a_2^{(i)}\xi + a_3^{(i)}\eta + \eta(\alpha^{(i)}\eta - \beta^{(i)}\xi) \\ b_1^{(i)} + b_2^{(i)}\xi + b_3^{(i)}\eta + \xi(\beta^{(i)}\xi - \alpha^{(i)}\eta) \end{Bmatrix}$$

i	a_1	a_2	a_3	b_1	b_2	b_3	α	
1	2-3a	-3+6a	-6+7a	0	3-5a	0	4-4a	-4
2	-1+3a	3-6a	1-7a	0	-2+5a	0	4a	4
3	0	0	1+a	0	-2+a	0	-4a	4
4	0	0	2-a	0	-1-a	0	-4+4a	
5	0	0	2-5a	1-3a	-1+7a	-3+6a	-4+8a	
6	0	0	-3+5a	-2+3a	6-7a	3-6a	4-8a	-4
7	0	0	-8	0	16	0	8	
8	0	0	-16	0	8	0	16	

$a = 0.5(1-\sqrt{3}/3)$

Node	ξ	η
1	a	0
2	1-a	0
3	1-a	a
4	a	1-a
5	0	1-a
6	0	a
7, 8	1/3	1/3

Table 2.8
Tetrahedral curl-conforming parent elements

1st o.

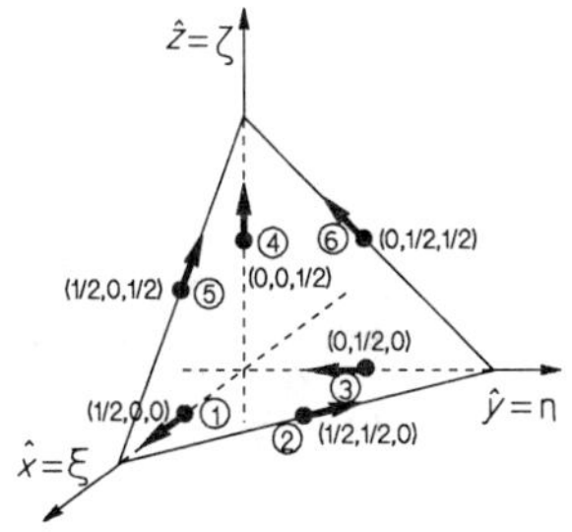

$$\{\hat{N}_1\} = \begin{Bmatrix} 1-\eta-\zeta \\ \xi \\ \xi \end{Bmatrix} \qquad \{\hat{N}_4\} = \begin{Bmatrix} \zeta \\ \zeta \\ 1-\xi-\eta \end{Bmatrix}$$

$$\{\hat{N}_2\} = \begin{Bmatrix} -\eta \\ \xi \\ 0 \end{Bmatrix} \qquad \{\hat{N}_5\} = \begin{Bmatrix} -\zeta \\ 0 \\ \xi \end{Bmatrix}$$

$$\{\hat{N}_3\} = \begin{Bmatrix} -\eta \\ -1+\xi+\zeta \\ -\eta \end{Bmatrix} \qquad \{\hat{N}_6\} = \begin{Bmatrix} 0 \\ -\zeta \\ \eta \end{Bmatrix}$$

2nd o.

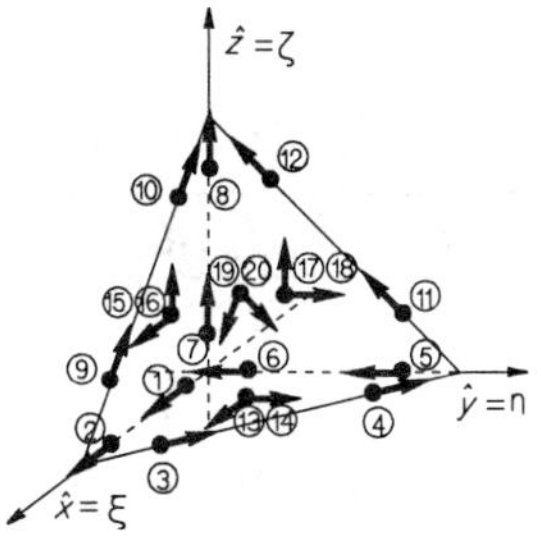

$$\{\hat{N}_i\} = \begin{Bmatrix} a_1^{(i)}+a_2^{(i)}\xi+a_3^{(i)}\eta+a_4^{(i)}\zeta+D^{(i)}\eta^2-F^{(i)}\xi\eta-G^{(i)}\xi\zeta+H^{(i)}\zeta^2+J^{(i)}\eta\zeta \\ b_1^{(i)}+b_2^{(i)}\xi+b_3^{(i)}\eta+b_4^{(i)}\zeta-D^{(i)}\xi\eta-E^{(i)}\eta\zeta+F^{(i)}\xi^2+I^{(i)}\zeta^2-J^{(i)}\xi\zeta+K^{(i)}\xi\zeta \\ c_1^{(i)}+c_2^{(i)}\xi+c_3^{(i)}\eta+c_4^{(i)}\zeta+E^{(i)}\eta^2+G^{(i)}\xi^2-H^{(i)}\xi\zeta-I^{(i)}\eta\zeta-K^{(i)}\xi\eta \end{Bmatrix}$$

Vectors for the degrees of freedom related to faces

	Face 1	Face 2	Face 3	Face 4
$\{\hat{q}_1\}$	$[1,0,0]^T$	$[0,1,0]^T$	$[1,0,0]^T$	$[0,-1,1]^T$
$\{\hat{q}_2\}$	$[0,1,0]^T$	$[0,0,1]^T$	$[0,0,1]^T$	$[-1,1,0]^T$
$\{\hat{q}_3\}$	$[-1,1,0]^T$	$[0,-1,1]^T$	$[-1,0,1]^T$	$[-1,0,1]^T$

$a = 0.5(1-\sqrt{3}/3)$
$b = 2-3a$
$c = 1-2a$
$d = 7a-6$
$e = 1-7a$
$f = 1+a$
$g = 2-a$
$h = 5a-2$
$j = 3-5a$
$k = 4-4a$
$l = -1+3a$

Node	ξ	η	ζ
1	a	0	0
2	1-a	0	0
3	1-a	a	0
4	a	1-a	0
5	0	1-a	0
6	0	a	0
7	0	0	a
8	0	0	1-a
9	1-a	0	a
10	a	0	1-a
11	0	1-a	a
12	0	a	1-a
13,14	1/3	1/3	0
15,16	1/3	0	1/3
17,18	0	1/3	1/3
19,20	1/3	1/3	1/3

Table 2.8 (Continued)

Tetrahedral curl-conforming parent elements

Coefficients of the first twelve basis functions of the second-order element

i	a_1	a_2	a_3	a_4	b_1	b_2	b_3	b_4	c_1	c_2	c_3	c_4	D	E	F	G	H	I	J	K
1	b	-3c	d	d	0	j	0	0	0	j	0	0	k	0	-4c	-4c	k	0	2k	k
2	l	3c	e	e	0	h	0	0	0	h	0	0	4a	0	4c	4c	4a	0	8a	4a
3	0	0	f	0	0	-g	0	0	0	0	0	0	-4a	0	k	0	0	0	0	0
4	0	0	g	0	0	-f	0	0	0	0	0	0	-k	0	4a	0	0	0	0	0
5	0	0	h	0	l	e	3c	e	0	0	h	0	4c	4c	4a	0	0	4a	-4a	4a
6	0	0	j	0	b	d	-3c	d	0	0	j	0	-4c	-4c	k	0	0	k	-k	k
7	0	0	0	j	0	0	0	j	b	d	d	-3c	0	k	0	k	-4c	-4c	-k	-2k
8	0	0	0	h	0	0	0	h	l	e	e	3c	0	4a	0	4a	4c	4c	-4a	-8a
9	0	0	0	f	0	0	0	0	0	-g	0	0	0	0	0	k	-4a	0	0	0
10	0	0	0	g	0	0	0	0	0	-f	0	0	0	0	0	4a	-k	0	0	0
11	0	0	0	0	0	0	0	f	0	0	-g	0	0	k	0	0	0	-4a	0	0
12	0	0	0	0	0	0	0	g	0	0	-f	0	0	4a	0	0	0	-k	0	0

Coefficients of the last eight basis functions of the second-order element when $\{\hat{q}_1\}$ and $\{\hat{q}_2\}$ are selected

i	a_1	a_2	a_3	a_4	b_1	b_2	b_3	b_4	c_1	c_2	c_3	c_4	D	E	F	G	H	I	J	K
13	0	0	-8	0	0	16	0	0	0	0	0	0	8	0	-16	0	0	0	8	-8
14	0	0	-16	0	0	8	0	0	0	0	0	0	16	0	-8	0	0	0	16	8
15	0	0	0	0	0	0	0	-8	0	0	16	0	0	-16	0	0	0	8	8	16
16	0	0	0	0	0	0	0	-16	0	0	8	0	0	-8	0	0	0	16	-8	8
17	0	0	0	8	0	0	0	0	0	-16	0	0	0	0	0	16	-8	0	-8	-16
18	0	0	0	16	0	0	0	0	0	-8	0	0	0	0	0	8	-16	0	-16	-8
19	0	0	0	0	0	0	0	0	0	0	0	0	0	0	0	0	0	0	-8	0
20	0	0	0	0	0	0	0	0	0	0	0	0	0	0	0	0	0	0	0	8

Coefficients of the last eight basis functions of the second-order element when $\{\hat{q}_1\}$ and $\{\hat{q}_3\}$ are selected

i	a_1	a_2	a_3	a_4	b_1	b_2	b_3	b_4	c_1	c_2	c_3	c_4	D	E	F	G	H	I	J	K
13	0	0	-24	0	0	24	0	0	0	0	0	0	24	0	-24	0	0	0	24	0
14	0	0	-16	0	0	8	0	0	0	0	0	0	16	0	-8	0	0	0	16	8
15	0	0	0	0	0	0	0	-24	0	0	24	0	0	-24	0	0	0	24	0	24
16	0	0	0	0	0	0	0	-16	0	0	8	0	0	-8	0	0	0	16	-8	8
17	0	0	0	24	0	0	0	0	0	-24	0	0	0	0	0	24	-24	0	-24	-24
18	0	0	0	16	0	0	0	0	0	-8	0	0	0	0	0	8	-16	0	-16	-8
19	0	0	0	0	0	0	0	0	0	0	0	0	0	0	0	0	0	0	-8	-8
20	0	0	0	0	0	0	0	0	0	0	0	0	0	0	0	0	0	0	0	8

Coefficients of the last eight basis functions of the second-order element when $\{\hat{q}_2\}$ and $\{\hat{q}_3\}$ are selected

i	a_1	a_2	a_3	a_4	b_1	b_2	b_3	b_4	c_1	c_2	c_3	c_4	D	E	F	G	H	I	J	K
13	0	0	-24	0	0	24	0	0	0	0	0	0	24	0	-24	0	0	0	24	0
14	0	0	8	0	0	-16	0	0	0	0	0	0	-8	0	16	0	0	0	-8	8
15	0	0	0	0	0	0	0	-24	0	0	24	0	0	-24	0	0	0	24	0	24
16	0	0	0	0	0	0	0	8	0	0	-16	0	0	16	0	0	0	-8	-8	-16
17	0	0	0	24	0	0	0	0	0	-24	0	0	0	0	0	24	-24	0	-24	-24
18	0	0	0	-8	0	0	0	0	0	16	0	0	0	0	0	-16	8	0	8	16
19	0	0	0	0	0	0	0	0	0	0	0	0	0	0	0	0	0	0	8	8
20	0	0	0	0	0	0	0	0	0	0	0	0	0	0	0	0	0	0	-8	0

space $(H^r(\Omega^e))^n$, with $r \geq 1$, by means of curl-conforming elements of degree p, is

$$\|\bar{t}-\Pi\bar{t}\|_{H(\mathrm{curl},\Omega^e)} \leq C_1 \frac{h_e^{k+1}}{\rho_e} |\bar{t}|_{k+1,\Omega^e} = C_1 \sigma_e h_e^k |\bar{t}|_{k+1,\Omega^e} \tag{2.57a}$$

where k, h_e, and ρ_e have been defined in (2.54). The considerations discussed regarding the interpolation error for div-conforming elements are also valid in this case [Nédélec 1980].

The measure of the interpolation error for the curl of $\bar{t}(\bar{r})$, $\bar{\nabla}\times\bar{t}(\bar{r})$, which in 2D is a vector in the longitudinal direction and in 3D is a vector with three components, has not been given explicitly by Nédélec. However the work of P. Monk [1991a,b, 1992a,b] suggests

$$\| \bar{\nabla}\times\bar{t} - \bar{\nabla}\times\Pi t \|_{0,\Omega^e} \leq C_2 h_e^k | \bar{\nabla}\times\bar{t} |_{k,\Omega^e} \tag{2.57b}$$

if $\bar{\nabla}\times\bar{t} \in (H^r(\Omega^e))^1$ in the 2D case, and $\bar{\nabla}\times\bar{t} \in (H^r(\Omega^e))^3$ in the 3D case. This is analogous to (2.54b) [Nédélec 1980].

Lagrange/Curl-Conforming Elements

Let us assume a vector variable $\bar{t}(x,y)$ with Cartesian components $t_x(x,y)$ and $t_y(x,y)$ so that the vector $t_x(x,y)\bar{a}_x + t_y(x,y)\bar{a}_y$ representing the transverse component must belong to the space $H(\mathrm{curl},\Omega)$. Assume that the vector variable also has a Cartesian component $t_z(x,y)$, or longitudinal component, that must belong to the space $H^1(\Omega)$. It is evident that the discretization of this vector variable may be done by means of a two-dimensional element that is a combination of two types of elements: the Lagrange element for the discretization of the longitudinal component and the curl-conforming element for the discretization of the transverse component. The definition of a 2D Lagrange/curl-conforming element of a given order makes use of the concepts corresponding to the two types of elements already mentioned. Thus, a formal definition will not be done. It is evident that in order to use a parent element it will be necessary to transform the scalar as well as the vector variables according to (2.45) and (2.56), respectively.

The parent elements for first and second orders are shown in Table 2.9. For the first-order element, 6 nodes are defined. The nodes located at the vertices are used for the discretization of the longitudinal component by means of the first-order triangular Lagrange basis functions (see Table 2.2). The nodes placed at the midpoints of the edges are used for the discretization of the transverse component by means of the first-order curl-conforming basis functions (see Table 2.7). For the second-order elements, 14 nodes are defined. The nodes are located at the vertices and at the two Gauss-Legendre integration points on the edges. The first ones are used for the discretization of the longitudinal component by means of the second-order triangular Lagrange basis functions (see Table 2.2). The rest are used for the discretization of the transverse component by means of the second-order curl-conforming basis functions (see Table 2.7). These elements will provide an approximation for the unknown vector that will exhibit continuity of the vector component tangential to the edges of the elements. The continuity of the longitudinal component (over the whole domain) as well as the continuity of the tangential component of the transverse vector across element interfaces will be ensured. The component of the transverse vector normal to the element interfaces will be discontinuous. Thus, the solution will belong to the product space $H^1(\Omega)\times H(\mathrm{curl},\Omega)$.

Conclusions concerning the interpolation error may be derived from the expressions corresponding to the Lagrange and curl-conforming elements.

Table 2.9
Triangular Lagrange/curl-conforming parent elements

First-order element	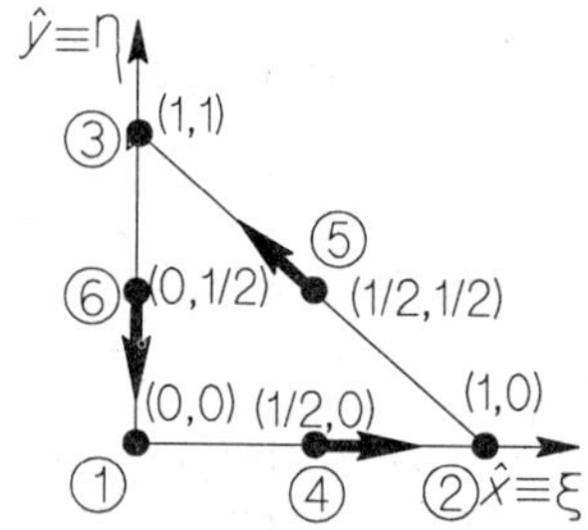	$\hat{N}_i = \hat{N}_i$, Lagrange $P(1)$ (see Table 2.2), $i=1,2,3$ $\{\hat{N}_{i+3}\} = \{\hat{N}_i\}$, curl-conforming first-order (see Table 2.7), $i=1,2,3$
Second-order element	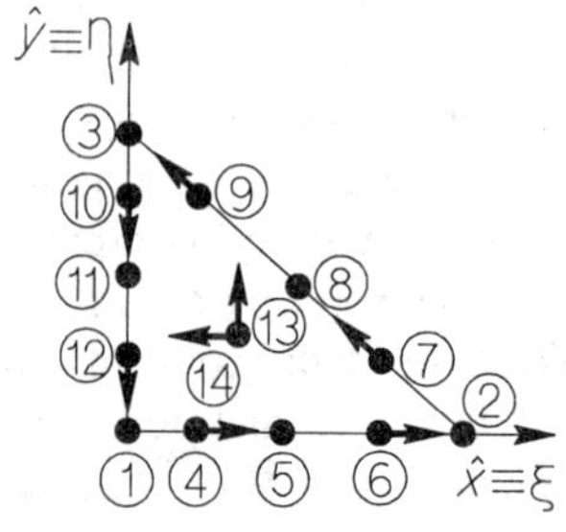	$\hat{N}_i = \hat{N}_i$, Lagrange triangle $P(2)$ (see Table 2.2), $i=1,2,3$ $\hat{N}_{3(i-1)+5} = \hat{N}_{i+3}$, idem $\{\hat{N}_4\} = \{\hat{N}_1\}$, curl-conforming second-order (see Table 2.7) $\{\hat{N}_6\} = \{\hat{N}_2\}$, idem $\{\hat{N}_7\} = \{\hat{N}_3\}$, idem $\{\hat{N}_9\} = \{\hat{N}_4\}$, idem $\{\hat{N}_{10}\} = \{\hat{N}_5\}$, idem $\{\hat{N}_{11+i}\} = \{\hat{N}_{5+i}\}$, idem, $i=1,2,3$

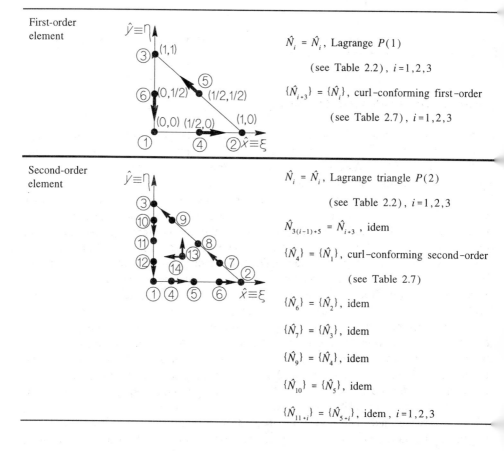

2.5.2 Discretization by the Finite Element Method

The approximation of the variables by means of FEM may be used in any of the variational methods of approximation described in Sections 2.4.2 and 2.4.3. In some applications the method of collocation (point matching) with variations have been used [Kardestuncer and Norrie 1987; P. 2, Ch. 6 and P. 3, Ch. 7] as well as the method of collocation by subdomains [Kardestuncer and Norrie 1987; P. 3, Ch. 2] and the least squares method [Kardestuncer and Norrie 1987; P. 3, Ch. 2], [Mitchell 1973], [Oden 1973], [Zlámal 1973]. In this book the Galerkin and the Ritz methods have been used. The general variational formulation in Section 2.3.3 is now discretized by FEM.

The standard variational formulation of the problems treated in this book admits a general expression of the following types:

$$W(\{u\},\{v\}) = \int_\Omega a_\Omega(\{u\},\{v\})\,\mathrm{d}\Omega - \int_\Gamma a_\Gamma(\{u\},\{v\})\,\mathrm{d}\Gamma - \int_\Omega l(\{v\})\,\mathrm{d}\Omega = 0 \tag{2.58a}$$

for deterministic problems or

$$W(\{u\},\{v\}) = \int_\Omega a_{1\Omega}(\{u\},\{v\})\,\mathrm{d}\Omega - \int_\Gamma a_\Gamma(\{u\},\{v\})\,\mathrm{d}\Gamma - \lambda \int_\Omega a_{2\Omega}(\{u\},\{v\})\,\mathrm{d}\Omega = 0 \tag{2.58b}$$

for eigenvalue problems.

The unknowns are the elements of $\{u\}$. The column vector $\{v\}$ stands either for the test functions in the Galerkin method or for the variation of the unknowns $\{\delta u\}$ in the Ritz method. In the boundary integrals the product of $\{v\}$ by the conormal derivative of $\{u\}$ (see (2.9)) is involved. Expressions (2.58) are valid for any domain Ω with any boundary Γ. So far no conditions have been imposed.

When boundary conditions of the Dirichlet and Neumann types should be used over Γ, the contour integrals are either zero (because $\{v\}$ will be zero for Dirichlet type boundaries and the conormal derivative will be zero for homogeneous Neumann type boundaries, as in the case of eigenvalue problems) or they are reduced to integrals extended only to the portion of the boundary in which nonhomogeneous Neumann conditions are present. Then, when the Neumann condition (2.11) is substituted, the corresponding integrand will turn out to be a function only of $\{v\}$.

Expression (2.58) becomes

$$W(\{u\},\{v\}) = \int_{\Omega} a_{\Omega}(\{u\},\{v\})\,\mathrm{d}\Omega - \int_{\Gamma_N} l_{\Gamma}(\{v\})\,\mathrm{d}\Gamma - \int_{\Omega} l(\{v\})\,\mathrm{d}\Omega = 0 \tag{2.59a}$$

for the case of deterministic problems, where

$$l_{\Gamma}(\{v\}) = a_{\Gamma}(\{u\},\{v\})\big|_{\Gamma = \Gamma_N} \tag{2.59b}$$

or

$$W(\{u\},\{v\}) = \int_{\Omega} a_{1\Omega}(\{u\},\{v\})\,\mathrm{d}\Omega - \lambda \int_{\Omega} a_{2\Omega}(\{u\},\{v\})\,\mathrm{d}\Omega = 0 \tag{2.59c}$$

in the case of eigenvalue problems.

The mixed variational formulation utilizing a penalty function used in this book for the analysis of deterministic quasi-static problems admits a general expression of the following type:

$$W(\bar{q},\bar{t}) = \int_{\Omega} a_{\Omega}(\bar{q},\bar{t})\,\mathrm{d}\Omega - \int_{\Gamma} a_{\Gamma}(\bar{q},\bar{t})\,\mathrm{d}\Gamma = 0 \tag{2.60}$$

where both the unknown $\bar{q}$ and the test function $\bar{t}$ are vector functions. The integrand of the boundary integral of (2.60) involves the product of the penalty function of the scalar variable $u = -(1/\alpha)(\bar{\nabla}_t \cdot \bar{q})$ (where α is a penalty factor) by the component of $\bar{t}$ normal to Γ, directed outward. The expression (2.60) is valid for any domain Ω with any boundary Γ in which, for the time being, no conditions have been imposed.

At the boundary Γ there are portions with Dirichlet Γ_D or Neumann Γ_N conditions (using the terminology of the direct formulation). The boundary integral extended to Neumann portions will be identically zero because these boundaries are now those in which the essential boundary conditions should be imposed. Thus the weighting functions are chosen to be zero along them (see Appendix E.3.1). Then the boundary integral terms extends only to the portions of the boundary that are of Dirichlet type Γ_D, on which $u = u_d$ may be substituted. Thus it depends only on $\bar{t}$, resulting in

$$W(\bar{q},\bar{t}) = \int_{\Omega} a_{\Omega}(\bar{q},\bar{t})\,\mathrm{d}\Omega - \int_{\Gamma_D} l(\bar{t})\,\mathrm{d}\Gamma = 0 \tag{2.61a}$$

where

$$l(\bar{t}) = a_{\Gamma}(\bar{q},\bar{t})\big|_{\substack{u = -(1/\alpha)(\nabla_t \cdot \bar{q}) \\ \Gamma = \Gamma_D}} \tag{2.61b}$$

Because in FEM the domain Ω, with boundary $\Gamma = \Gamma_N \cup \Gamma_D$, is subdivided into a set of elements, the variational formulation must be imposed, not on the total domain Ω of boundary Γ but in each element Ω^e of boundary Γ^e that consists of a closed boundary given by the edges (2D) or faces (3D), Γ_m^e (with $1 \leq m \leq 3$ for triangular elements or $1 \leq m \leq 4$ for rectangular or tetrahedral ones). In general, the boundary Γ^e does not coincide with Γ. Thus, in each subdomain Ω^e, the general expressions (2.58) or (2.60) should be applied. Then the contribution from each element should be added up.

For the sake of clarity only the case of the direct formulation for the deterministic problem is going to be considered. The strong formulation is given by (2.1), where u is the scalar variable. The application of (2.58a) to element Ω^e leads to

$$\int_{\Omega^e} a_\Omega(u,v)\,\mathrm{d}\Omega - \int_{\Omega^e} l(v)\,\mathrm{d}\Omega - \sum_{m=1}^{3\,\mathrm{or}\,4} \int_{\Gamma_m^e} a_\Gamma(u,v)\,\mathrm{d}\Gamma = 0 \tag{2.62}$$

The extension of (2.62) to the N_e elements of the domain will lead to

$$\sum_{e=1}^{N_e} \int_{\Omega^e} a_\Omega(u,v)\,\mathrm{d}\Omega - \sum_{e=1}^{N_e} \int_{\Omega^e} l(v)\,\mathrm{d}\Omega - \sum_{e=1}^{N_e} \left(\sum_{m=1}^{3\,\mathrm{or}\,4} \int_{\Gamma_m^e} a_\Gamma(u,v)\,\mathrm{d}\Gamma \right) = 0 \tag{2.63}$$

When both v and the absolute value of the conormal derivative of u are continuous across the interface of contiguous elements, then the contribution from two contiguous elements to the boundary integral may cancel. However, this is not the case with FEM. While v will be continuous, the absolute value of the conormal derivative of u is discontinuous across element interfaces. However, if small enough elements are assumed, because of the bilinear character of a_Γ, and the continuity of v, the absolute value of the contribution from each element is equal amongst them in a distributional sense but of opposite sign. For a better understanding of this point, see the detailed description of Section 7.3.1.1, where the formulation for the 3D cavity problem is developed. Going back to the general description, assuming that the contribution on element interfaces may be equated to zero, (2.63) results in

$$\sum_{e=1}^{N_e} \int_{\Omega^e} a_\Omega(u,v)\,\mathrm{d}\Omega - \sum_{e=1}^{N_e} \int_{\Omega^e} l(v)\,\mathrm{d}\Omega - \sum_{\substack{\Gamma_m^e \subset \Gamma_N \\ 1 \leq m \leq 3\,\mathrm{or}\,4 \\ 1 \leq e \leq N_e}} \int_{\Gamma_m^e} a_\Gamma(u,v)\,\mathrm{d}\Gamma = 0 \tag{2.64}$$

where, the boundary integral is extended now only to the element boundaries of Neumann type. Taking into account (2.59b), (2.64) may be written as

$$\sum_{e=1}^{N_e} \int_{\Omega^e} a_\Omega(u,v)\, \mathrm{d}\Omega - \sum_{e=1}^{N_e} \int_{\Omega^e} l(v)\, \mathrm{d}\Omega - \sum_{\substack{\Gamma_m{}^e \subset \Gamma_N \\ 1 \le m \le 3\,\mathrm{or}\,4 \\ 1 \le e \le N_e}} \int_{\Gamma_m{}^e} l_\Gamma(v)\, \mathrm{d}\Gamma = 0 \tag{2.65}$$

If u and v are continuous in Ω, this expression must coincide with (2.59a). The formulation given by (2.65) is equivalent to considering for each element Ω^e the integral

$$W^e(u,v) = \int_{\Omega^e} a_\Omega(u,v)\, \mathrm{d}\Omega - \int_{\Omega^e} l(v)\, \mathrm{d}\Omega - \sum_{\substack{\Gamma_m{}^e \subset \Gamma_N \\ 1 \le m \le 3or4}} \int_{\Gamma_m{}^e} l_\Gamma(v)\, \mathrm{d}\Gamma \tag{2.66a}$$

instead of (2.62). Then the global integral form would be given by

$$W(u,v) = \sum_{e=1}^{N_e} W^e(u,v) = 0 \tag{2.66b}$$

Expressions (2.66) constitute the variational formulation by subdomains that is used in FEM for the solution of deterministic problems of second order. Analogous expressions may be obtained for eigenvalue problems as well as for the mixed formulation for deterministic problems [Dhatt and Touzot 1981; Ch. 2], [Reddy 1984; Ch. 4], [Reddy 1993; Chs. 3 and 8].

Continuing with the direct or standard formulation of deterministic problems, the approximate solution by means of FEM implies seeking an approximate function $\tilde{u}$ such that the following conditions are fulfilled:

1. In each element, the solution will belong to the proper polynomial space.
2. It will fulfill the continuity conditions that are required for the equivalence between (2.63) and (2.65).
3. It will solve (2.66b) for any v (or $\delta\tilde{u}$) belonging to the set of basis functions utilized in the approximation of u that will fulfill the specified continuity conditions, and will have homogeneous Dirichlet type conditions on Γ_D.
4. It must fulfill the Dirichlet type conditions, required on Γ_D.

Condition (1) implies the selection of the element to be employed, its number of degrees of freedom, and its set of basis function $\{N\}$. This selection will determine the continuity conditions that will be fulfilled by the approximate function $\tilde{u}$ and the test function v along with its derivatives in each element and on the interfaces between them. In FEM methodology, there has to be continuity of the degrees of freedom of the problem, but it is not always possible to guarantee the exact fulfillment of the rest of the continuity conditions across element interfaces required by condition 2. Thus, (2.63) and (2.66) will not be exactly equivalent (for example, in the case of Lagrange elements the continuity of the conormal derivative of u in the direction normal to the edges (2D) or faces (3D) is not guaranteed). In those cases, (2.63) and (2.66) are equivalent only in a distributional sense. In fact, in expression (2.66b) a term for each interior edge or face of the geometric discretization of Ω will be neglected. This term is the difference (or jump) between the value of $a_\Gamma(\tilde{u},v)$ at each side of the interior interface between elements (see (2.63)) [Patterson 1973]. This introduces a new type of error in FEM. In these cases the method will converge if when the number of degrees of freedom is increased the continuity conditions are better satisfied. Then the element or elements employed are said to be conformal. Usually *conformality* is understood only in the sense of the continuity of the primary variable. The *patch test* technique allows one to verify if a mesh, with a given type of element (or with elements of different type, but compatible between them), is conformal [Oden and Reddy 1976; Ch. 6], [Strang and Fix 1973; Ch. 1], [Zienkiewicz 1977; Chs. 2 and 11], [Zienkiewicz and Taylor 1989; Chs. 2 and 11]. Next, one needs to compute for each element the discretized integral forms corresponding to each basis function $v = N_j$, $1 \le j \le n_e$, for the Galerkin method (see Section 2.4.2) or $v = \delta\tilde{u} = \{N\}^T\{\delta d\}$ for the Ritz method (see Section 2.4.3) following (2.66a). Once that computation is done,the local equations can be assembled according to (2.66b). Finally, taking into account condition 4, the Dirichlet conditions should be imposed and a system of equations will be obtained whose solution will provide the degrees of freedom over the whole domain.

The previous procedure can be extended to eigenvalue problems for the cases when Lagrange, curl-conforming, or Lagrange/curl-conforming elements are employed. An analogous study may be done for the case of deterministic problems employing curl-conforming elements or a mixed formulation discretized by div-conforming elements. When using div-conforming or curl-conforming elements,the conformal mesh involves the jump of the normal component of the vector dual variable or the jump of the tangential component of the vector dual variable across interfaces, respectively.

In the following subsection, the various mathematical and numerical aspects related to the processes presented are given with more detail.

2.5.2.1 *Calculation of the Local Integral Forms: Numerical Integration*

Once a type of element is selected, its number of degrees of freedom, n_e, and the corresponding basis functions are known. Then, the computation of the integral form for each element (2.66a) is done according to Section 2.4.2 for the Galerkin method or Section 2.4.3 for the Ritz method. As an example, consider a deterministic problem with just one scalar variable u given by (2.1), then for $v = N_j$ one obtains

$$W_j^e(\{N\}^T\{d\},N_j) = \int_{\Omega^e} a_\Omega(\{N\}^T\{d\},N_j)\,d\Omega - \int_{\Omega^e} l(N_j)\,d\Omega$$
$$- \sum_{\substack{\Gamma_m^e\subset\Gamma_N \\ 1\le m\le 3\text{or}4}} \int_{\Gamma_m^e} l_\Gamma(N_j)\,d\Gamma\,, \qquad 1\le j\le N_e\,, \qquad j\in\mathbb{N} \tag{2.67}$$

Taking into account the bilinear character of a_Ω, (2.67) may be written in matrix form as

$$\{W^e\} = [k^e]\{d^e\} - \{f^e\} \tag{2.68a}$$

where

$$k_{ij}^e = \int_{\Omega^e} a_\Omega(N_i, N_j)\, \mathrm{d}\Omega\ , \qquad 1 \le i \le n_e\ , \quad 1 \le j \le n_e\ , \quad i,j \in \mathbb{N} \tag{2.68b}$$

$$f_i^e = \int_{\Omega^e} l(N_i)\, \mathrm{d}\Omega \ + \sum_{\substack{\Gamma_m^e \subset \Gamma_N \\ 1 \le m \le 3 or 4}} \int_{\Gamma_m^e} l_\Gamma(N_i)\, \mathrm{d}\Gamma\ , \qquad 1 \le i \le n_e\ , \qquad i \in \mathbb{N} \tag{2.68c}$$

When the operator a_Ω is symmetric, it is evident that $k_{ij}^e = k_{ji}^e$, and the matrix $[k^e]$ will be symmetric.

For eigenvalue problems, expression (2.68a) will lead to

$$\{W^e\} = [k^e]\{d^e\} - \lambda[m^e]\{d^e\} \tag{2.69a}$$

where

$$k_{ij}^e = \int_{\Omega^e} a_{1\Omega}(N_i, N_j)\, \mathrm{d}\Omega\ , \qquad 1 \le i \le n_e\ , \qquad 1 \le j \le n_e\ , \qquad i,j \in \mathbb{N} \tag{2.69b}$$

$$m_{ij}^e = \int_{\Omega^e} a_{2\Omega}(N_i, N_j)\, \mathrm{d}\Omega\ , \qquad 1 \le i \le n_e\ , \qquad 1 \le j \le n_e\ , \qquad i,j \in \mathbb{N} \tag{2.69c}$$

where, in general, matrix $[m^e]$ is symmetric, and in many cases, matrix $[k^e]$ will also be symmetric.

Because the basis functions are known, the computation of the coefficients of matrices $[k^e]$, $[m^e]$, and $\{f^e\}$ can be done easily. It is not necessary to compute the integrals involved in (2.68) and (2.69) for each real element Ω^e since they can be performed for the parent element $\hat{\Omega}$. When the basis function appears in the integrand, they may be written in terms of the corresponding functions of the parent element according to (C.5), (C.7), and (C.9). The derivatives of the basis functions may be written in terms of the derivatives of the basis function of the parent element according to (C.10) using if necessary, (C.15) and (C.16f). And finally, the domain of integration may be transformed to that corresponding to the parent element according to (C.13), (C.14) and (C.16b). It is convenient to take into account (C.17) whenever it is possible.

The technique of performing all the integrations by transformation to the parent element constitutes one of the attractive features of FEM because the expressions for the computation of the coefficients of $[k^e]$, $[m^e]$, and $\{f^e\}$ only differ from one element to the other in the Jacobian matrix $[J^e]$ and/or its determinant $|J^e|$ corresponding to the mapping from the parent element $\hat{\Omega}$ to each real element Ω^e (see Appendix C.1). Because the coefficients of $[J^e]$ only depend on the coordinates of the parent element and/or the coordinates of the vertices and geometrical points of each real element Ω^e, matrices $[k^e]$, $[m^e]$ and $\{f^e\}$ of all the elements may be computed using little CPU time.

Depending on the type of elements and its order, the integrations can be done in an analytical way or they may require a numerical integration. The numerical integration may be exact if the integrands are polynomials. This is the case for subparametric triangular and tetrahedral elements because the coefficients of $[J]$ and $|J|$ will not be polynomials but real numbers. In general, when dealing with rectangular (or cubic) elements as well as for curved elements, the numerical integration will give approximate coefficients for $[k^e]$, $[m^e]$, and $\{f^e\}$.

Whenever the integrand turns out to be a polynomial, the following analytical expressions may be used for one-dimensional, rectangular, and cubic elements

$$\int_{-1}^{1}\int_{-1}^{1}\int_{-1}^{1} \hat{x}^i \hat{y}^j \hat{z}^k \, d\hat{x}\, d\hat{y}\, d\hat{z} = \begin{Bmatrix} 0 & \text{for } i \text{ or } j \text{ or } k \text{ odd} \\ \dfrac{2^n}{(i+1)(j+1)(k+1)} & \text{for } i \text{ and } j \text{ and } k \text{ even} \end{Bmatrix} \tag{2.70}$$

where n is the dimension of the element. For the one-dimensional parent element a single integral stands and $j = k = 0$. For the rectangular parent element a double integral holds and $k = 0$. For triangular and tetrahedral elements one has

$$\int_{0}^{1}\int_{0}^{1}\int_{0}^{1-\hat{x}} \hat{x}^i \hat{y}^j \hat{z}^k \, d\hat{x}\, d\hat{y}\, d\hat{z} = \frac{i!\, j!\, k!}{(i+j+k+n)!} \tag{2.71}$$

where for triangular elements a double integral holds and $k = 0$.

The numerical integration that is usually employed is that of Gauss (also called the Gauss-Legendre quadrature) because it provides the highest precision for a given number of samples of the integrand. In one dimension the integration formula is

$$\int_{-1}^{1} f(\hat{x})\, d\hat{x} = \sum_{k=1}^{n_p} w_k f(\hat{x}_k) \tag{2.72}$$

where n_p is the number of integration points, $\hat{x}_k$ is the coordinate of the point where the function is sampled, and w_k is the corresponding weight. This formula is based on the approximation of the integrand by a polynomial of degree $2n_p - 1$. Thus, polynomials of degrees less than or equal to that value will be integrated in an exact way. Table 2.10 shows the maximum order of the polynomials that will be integrated exactly when the number of integration points is n_p. The values of the abscissa where the function is to be sampled are $\hat{x}_k$, and the corresponding weights are w_k, for $1 \leq k \leq n_p$, $k \in \mathbb{N}$.

Table 2.10
Gauss numerical integration in one dimension

Maximum order of polynomials exactly integrated	n_p	$\hat{x}_k$	w_k
1	1	0	2
3	2	±0.577350269189626 = ±(1/√3)	1
5	3	0 ±0.774596669241483 = ±(√.6)	0.888888888888889 0.555555555555556
7	4	±0.339981043584856 $= \pm\sqrt{(3-2\sqrt{1.2})/7}$ ±0.861136311594053 $= \pm\sqrt{(3+2\sqrt{1.2})/7}$	0.652145154862546 = (1/2)+(1/(6√1.2)) 0.347854845137454 = (1/2)-(1/(6√1.2))

In two and three dimensions, the integration on the parent rectangular or cubic element is usually performed using a one-dimensional Gauss integration in each direction. The corresponding formula is

$$\int_{-1}^{1}\int_{-1}^{1}\int_{-1}^{1} f(\hat{x},\hat{y},\hat{z})\, \mathrm{d}\hat{x}\, \mathrm{d}\hat{y}\, \mathrm{d}\hat{z} = \sum_{i=1}^{n_p}\sum_{j=1}^{n_p}\sum_{k=1}^{n_p} w_i\, w_j\, w_k\, f(\hat{x}_i,\hat{y}_j,\hat{z}_k) \tag{2.73}$$

For the rectangular element only two summations will hold. The function needs to be evaluated at $n_p \times n_p$ or at $n_p \times n_p \times n_p$ points, for the 2D or 3D case, respectively. The integration would be exact for polynomials of degree $2n_p$-1 in each variable.

When the parent element is a triangle or a tetrahedron, a direct method may be employed, so that

$$\int_{\hat{\Omega}} f(\{\hat{r}\})\, \mathrm{d}\Omega = \sum_{l=1}^{n_p} w_l\, f(\{\hat{r}_l\}) \tag{2.74}$$

Tables 2.11 and 2.12 refer to the direct integration formula over a triangular and a tetrahedral parent element, respectively, that have been used in this book. The tables show the total maximum degree, m (also called order of integration), of the polynomials $\Sigma\, \hat{x}^i\, \hat{y}^j\, \hat{z}^k$, ($i+j+k \leq m$), that may be exactly integrated through the use of n_p sampling points ($\hat{x}_l$, $\hat{y}_l$, $\hat{z}_l$) and the corresponding weights w_l, $1 \leq l \leq n_p$.

The numerical integration allows the exact computation of the coefficients of the local matrices whenever the integrands of the elements are polynomials. In this case it is enough to select the minimum number of points according to Tables 2.10 and 2.11. For cases in which the integrand is not a polynomial, the computation is only approximate, giving rise to an error that should be minimized. This situation arises, for example, in the computation of the coefficients of the matrix $[k^e]$ whenever rectangular elements as well as curved triangular or tetrahedral elements are employed. In these cases the derivatives of the basis functions over the real element appear in the integrand. Thus, after performing the mapping to the parent element, according to (C.10), the determinant of the Jacobian matrix $|J^e|$ will appear dividing a polynomial expression. Here $|J^e|$ is also a polynomial. Thus, the integrand turns out to be a rational expression. In these cases, the integration by means of a numerical procedure requires a careful selection of the number of points of integration in order to obtain a minimum integration error. For a given number of points, the accuracy of the integration is reduced whenever the distortion of the element increases. In fact, if the inverse of $|J|$ is developed as an infinite series of monomials an infinite-order polynomial will be obtained. The polynomial should be truncated from a monomial of high order. This order should be higher for a greater distortion of the element because the integrand will be approximated by the product of the polynomial of the numerator and the truncated series. In conclusion, it is convenient to avoid very distorted elements, so a formula employing a minimum number of integration points may be used, reducing the CPU time but not the rate of convergence. In some cases the integration may be performed using a reduced number of points. Details on this topic may be found in [Carey and Oden 1984], [Dhatt and Touzot 1981; Ch. 5], [Kardestuncer and Norrie 1987; P. 2, Ch. 3], [Ottosen and Peterson 1992; Ch. 20], [Raviart and Thomas 1983; Ch. 5], [Silvester and Ferrari 1983; Ch. 8], [Zienkiewicz 1977; Chs. 8 and 11], [Zienkiewicz and Taylor 1989; Chs. 8 and 11].

The numerical integration in the parent elements greatly simplifies the programming of the modules of computation of the local matrices.

2.5.2.2 *Computation of the Global Integral Form: Assembly Process*

The operation that leads to a discretized global integral form, starting from the discretized local integral forms, is called the assembly process. It may be written as

$$\{W(u,v)\} = \sum_{e=1}^{N_e} \{W^e(u,v)\} \tag{2.75}$$

Table 2.11
Integration on a triangular domain

	Maximum order of the polynomial exactly integrated	n_p	$\hat{x}_l$	$\hat{y}_l$	w_l
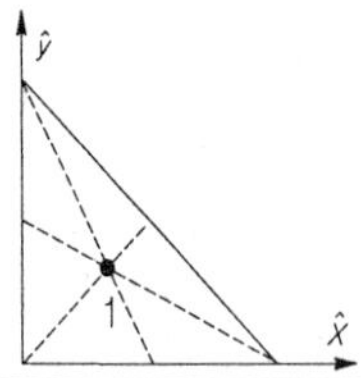	1	1	1/3	1/3	1/2
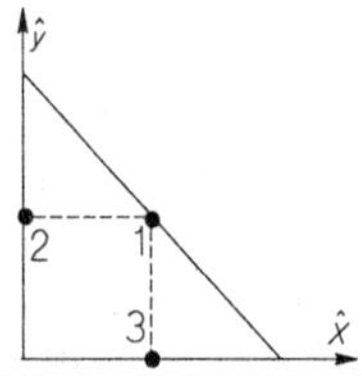	2	3	1/2 0 1/2	1/2 1/2 0	1/6
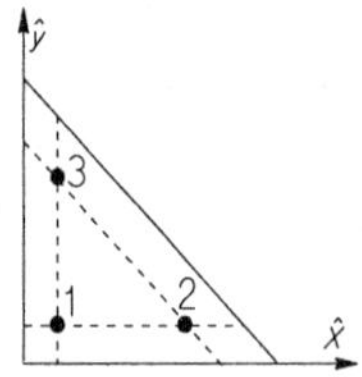	2	3	1/6 2/3 1/6	1/6 1/6 2/3	1/6
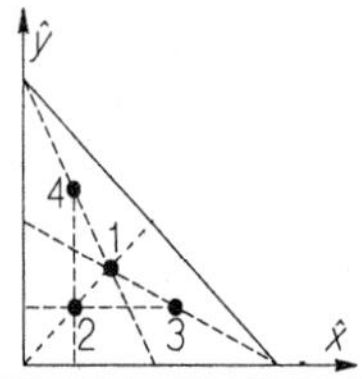	3	4	1/3 1/5 3/5 1/5	1/3 1/5 1/5 3/5	-27/96 25/96 25/96 25/96
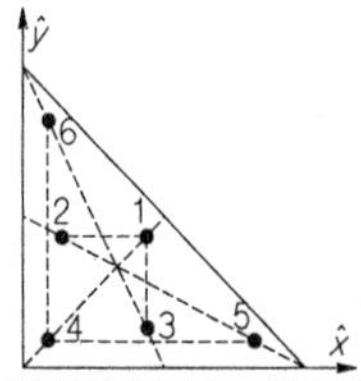	4	6	a 1-2a a b 1-2b b	a a 1-2a b b 1-2b	0.111690794839005 0.111690794839005 0.111690794839005 0.054975871827661 0.054975871827661 0.054975871827661

a = 0.445948490915965 b = 0.091576213509771

Table 2.11 (Continued)
Integration on a triangular domain

Maximum order of the polynomial exactly integrated	n_p	$\hat{x}_l$	$\hat{y}_l$	w_l
		c	c	0.025422453185103
6	12	$1-2c$	c	0.025422453185103
		c	$1-2c$	0.025422453185103
		d	d	0.058393137863189
		$1-2d$	d	0.058393137863189
		d	$1-2d$	0.058393137863189
		e	f	0.041425537809187
	$c = 0.063089014491502$	f	e	0.041425537809187
	$d = 0.249286745170910$	$1-(e+f)$	e	0.041425537809187
	$e = 0.310352451033785$	$1-(e+f)$	f	0.041425537809187
	$f = 0.053145049844816$	e	$1-(e+f)$	0.041425537809187
		f	$1-(e+f)$	0.041425537809187

Table 2.12
Integration on a tetrahedral domain

Order	n_p	$\hat{x}_l$	$\hat{y}_l$	$\hat{z}_l$	w_l
1	1	1/4	1/4	1/4	1/6
2	4	a	a	a	1/24
		a	a	b	1/24
	$a = (5-\sqrt{5})/20$	a	b	a	1/24
	$b = (5+3\sqrt{5})/20$	b	a	a	1/24
3	5	a	a	a	-2/15
		b	b	b	3/40
	$a = 1/4$	b	b	c	3/40
	$b = 1/6$	b	c	b	3/40
	$c = 1/2$	c	b	b	3/40
5	15	a	a	a	8/405
		b_i	b_i	c_i	w_i
		b_i	b_i	b_i	w_i
		b_i	c_i	b_i	w_i
	$a = 1/4$	c_i	b_i	b_i	w_i
	$b_i = (7\pm\sqrt{15})/34,\ i=1,2$	d	d	e	5/567
	$c_i = (13\mp3\sqrt{15})/34,\ i=1,2$	d	e	d	5/567
	$d = (5-\sqrt{15})/20$	e	d	d	5/567
	$e = (5+\sqrt{15})/20$	d	e	e	5/567
		e	d	e	5/567
	$w_i = (2665\mp14\sqrt{15})/226800,\ i=1,2$	e	e	d	5/567

where

$$\{W\} = [K]\{D\} - \{F\} \tag{2.76a}$$

for a deterministic problem or

$$\{W\} = [K]\{D\} - \lambda [M]\{D\} \tag{2.76b}$$

for an eigenvalue problem. Here $\{D\}$ is the vector of global degrees of freedom or unknowns of the discretized problem; $[K]$, $[M]$, and $\{F\}$ are the stiffness, mass and excitation global matrices obtained from the local matrices $[k^e]$, $[m^e]$, $\{f^e\}$; and λ represents the eigenvalue.

The assembly process should not be understood as just an addition. In each discretized local form given by (2.68) or (2.69), the vector $\{d^e\}$ refers to the local degrees of freedom. Similarly, each row of the local matrices corresponds to a local degree of freedom. In FEM the value of a degree of freedom is continuous across element interfaces. Thus, a degree of freedom may be shared by two or more elements, like in the case of nodes placed at element interfaces. Thus, it is necessary to consider a global vector $\{D\}$ that is constructed using all the degrees of freedom over the domain of the problem. In order to obtain (2.76) it is enough to do the following.

1. Express the discretized local integral form in terms of the global number assigned to each global degree of freedom $\{D\}$. This operation is known as an expansion of the local matrices and vectors. Each expanded local matrix will be a $N_d \times N_d$ matrix, where N_d is the total number of degrees of freedom over the domain of the problem. Each expanded local vector will be a vector of N_d coefficients. Let $[K^e]$, $[M^e]$, and $\{F^e\}$ be the expanded matrices and vector of element Ω^e. The coefficients K_{ij}^e and M_{ij}^e will be zero if the ith and/or the jth global degree of freedom do not belong to the element under consideration. The coefficient F_i will be zero if the ith global degree of freedom does not belong to element Ω^e. Assume that the ith global degree of freedom corresponds to the lth local degree of freedom of Ω^e and that the jth global degree of freedom corresponds to the mth local degree of freedom of Ω^e. Then $K_{ij}^e = k_{lm}$, $K_{ji}^e = k_{ml}$, $M_{ij}^e = m_{lm}$, $M_{ji}^e = m_{ml}$, $F_i^e = f_l$ and $F_j^e = f_m$ [Dhatt and Touzot 1981; Ch. 4].
2. Add all the assembled local matrices to obtain the global matrices

$$[K] = \sum_{e=1}^{N_e} [K^e]\,, \quad [M] = \sum_{e=1}^{N_e} [M^e]\,, \quad \{F\} = \sum_{e=1}^{N_e} \{F^e\} \tag{2.76c}$$

The assembly process involves different programming techniques depending on the type of problems under consideration [Dhatt and Touzot 1981; Ch. 4].

2.5.2.3 *Enforcement of the Essential Boundary Conditions: Global System of Equations*

In order to obtain the discretized final system of equations, it is necessary to impose on the global form (2.75) the boundary conditions that have not yet been taken into account either because they are not natural to the formulation employed (see Appendix A.2) or because they have not been directly included in the formulation (see Section 2.3.3.1) by means of correction functions (see Section 2.3.3.2) or by Lagrange multipliers (see Appendix A.3.2). In other words, what is left is to explicitly enforce the boundary conditions that are called essential (Dirichlet type) for the given formulation employed. This step consists of giving to each degree of freedom located on the Dirichlet boundary the value of the corresponding Dirichlet condition. The values may be different over different portions of the boundary. Essential conditions may also be provided as a linear combination. Thus, given coefficients of the global vector of degrees of freedom, $\{D\}$ will be set to known values.

After the essential boundary conditions are enforced, a discretized global system of equations is obtained

$$[K]\{D\} - \{F\} = \{0\} \tag{2.77a}$$

or

$$[K]\{D\} - \lambda\,[M]\{D\} = \{0\} \tag{2.77b}$$

where the different matrices are obtained from (2.76). The techniques used to enforce the essential boundary conditions depend on the type of problem: deterministic (2.77a) or eigenvalue (2.77b).

For *deterministic problems* with essential boundary conditions constant over certain portions of the boundary, the methods most frequently used are [Akin 1982; Ch. 8], [Dhatt and Touzot 1981; Ch. 4], [Reddy 1984; Ch. 3], [Reddy 1993; Chs. 3 and 8] as follows:

1. *The method of the dominant diagonal term.* The diagonal term K_{ii} corresponding to the degree of freedom that should be enforced is substituted by $K_{ii}+\alpha$ or αK_{ii}, α being a very high number in comparison with the coefficients of $[K]$; similarly the term F_i corresponding to the degree of freedom to be enforced, is substituted by $\alpha \underline{D}_i$ or $\alpha K_{ii}\underline{D}_i$, respectively, where $\underline{D}_i$ is the value to which the ith degree of freedom must be set. This method is easy to program but may have numerical problems if the matrix $[K]$ is not well conditioned or if any of the coefficients of $\{D\}$ is very high.
2. *The method of the unit (or constant) term over the diagonal.* It consists of substituting K_{ii} by unity (or by a number α) and F_i by $\underline{D}_i$ (or αD_i). The terms

K_{ij}, $j \neq i$, are set to a zero value, and the coefficients F_j, $j \neq i$, are substituted by the quantity $F_j - K_{ij}D_i$. This method is more complicated to program, but it does not present numerical problems. This is the method that has been implemented in solving the problems discussed in this book.

3. *The method of elimination of equations*. It consists of restructuring matrices [*K*] and {*F*} by eliminating the row and the column corresponding to the *i*th degree of freedom that should be enforced through an essential condition. The coefficients F_j, $j \neq i$, are modified as in the previous method. This method has the advantage of reducing the number of unknowns of the system, but the operations involved are costly.

In *eigenvalue problems*, the essential conditions are zero. The method that is usually employed consists of substituting K_{ii} by a high value, α, and M_{ii} by unity. The value of α depends on the solver to be employed. The coefficients K_{ij}, M_{ij}, $i \neq j$, either remain the same or are set equal to zero [Bernadou et al. 1985].

2.5.2.4 *Matrix Storage and Solution of the Global System of Equations*

Once the boundary conditions are enforced, the system of equations (2.77) is obtained. Its solution will provide the degrees of freedom of the problem {*D*} or the possible eigenvalues λ and the associated eigenvectors {*D*}. A number of solvers for deterministic or eigenvalue problems are available. Nevertheless, if the number of unknowns is large, in order to reduce storage, memory, and CPU time requirements, it is necessary to employ specific algorithms for the solvers that will take advantage of the sparsity and other characteristics of the FEM matrices [Duff 1977], [Duff et al. 1986], [Evans 1973], [George and Liu 1981].

In the solution of systems of equations corresponding to *deterministic problems* (2.77a) the methods that are usually employed are the following:

1. *Direct methods*. Gaussian elimination is the best known method. There are other methods derived from it. In all of them the original system is substituted by a sequence of systems, easier to solve, employing a factorization of the matrix [*K*] into triangular forms. Depending on the type of factorization that rely on the characteristics of [*K*], different methods are available (such as Doolittle, LDU, Crout, Choleski) [Akin 1982; Ch. 9], [Bernadou et al. 1985], [Carey and Oden 1984; Ch. 3], [Dhatt and Touzot 1981; Ch. 5], [Duff et al. 1986], [Evans 1973], [Kardestuncer and Norrie 1987; P. 4, Ch. 1], [Segerlind 1976; Chs. 7 and 18], [Silvester and Ferrari 1983; Ch. 10].

   The accuracy of the solution for a direct method is very sensitive to the condition number of the matrix [*K*] and, thus, is affected by the round-off error (accuracy of the arithmetic employed) and truncation error (number of digits employed in the representation of a number). A measure of the condition number of the matrix is given by the ratio between the highest and the lowest eigenvalues [Fried 1973] or, to be mathematically correct, its singular values. It will be high (ill-conditioned matrix) when the coefficients of the matrix, and in particular, those of the main diagonal, do not have absolute values that are comparable between them. This will happen when element diameters are quite different from each other or when the transition between regions of elements of different sizes is not smooth or when the physical characteristics of the different media that are included in the domain of the problem change considerably in an abrupt way. The condition number of $[K]$ is worse when a given mesh is refined. Thus, refined meshes may lead to a degradation of the accuracy [Fried 1973]. At the same time when the number of unknowns is high, the truncation and round-off errors after each operation may lead to poor results.

In direct methods the number of operations required to solve the system of equations is known *a priori*. It does not depend on the condition number of matrix $[K]$. The use of algorithms tailored for sparse matrices and the use of storage methods especially designed for such matrices allow one to reduce the number of operations and, thus, the CPU time.

The Gauss method requires more operations than the rest of the direct methods. This is the technique employed in the frontal method in which, at each instance, only a submatrix is used. Thus, the memory requirement is reduced. Actually, it depends on the size of the front. This may be minimized with a good numbering of the elements. Instead, the computation time required by the frontal method is high because it uses file storage and requires accessing file operations [Hinton and Owen 1977; Ch. 2], [Johnson 1992; Ch. 6].

All the other methods employ band or skyline storage. The width of the band (that depends on the maximum difference between the global number of the degrees of freedom of the same element) may be minimized with a good strategy of node numbering. If the matrices are symmetric, it is enough to store the upper or the lower triangular matrix. Skyline storage saves memory. Figure 2.23(a) shows an example of a band matrix. Nonzero coefficients are marked with $\times$. In the example of Figure 2.23 the matrix is symmetric. Figures 2.23(b),(c) represent the coefficients that should be stored if the lower triangular part is selected in a band or skyline storage, respectively. In both cases, in addition to a vector in which the coefficients of each row are sequentially stored, the width of the band and a vector indicating the location of the coefficient of the diagonal should be stored. If the upper triangular part is selected, the storage is done by columns. It may be concluded that in the case of sparse matrices with band structure and large dimensions the skyline storage will economize storage requirements in comparison to band storage [Akin 1982; Ch. 3], [Bathe and Wilson 1976; Ch. 6], [Hinton and Owen 1977; Ch. 2], [Segerlind 1976; Ch. 18], [Zienkiewicz 1977; Ch. 24]. When matrix $[K]$ is symmetric and positive definite the Choleski method minimizes the CPU time.

2. *Iterative methods*. In these methods, one starts with an initial guess for $\{D\}$ and refines the solution at each iteration step. These methods are called linear when the solution at a given iteration is a linear combination of the previous solutions.

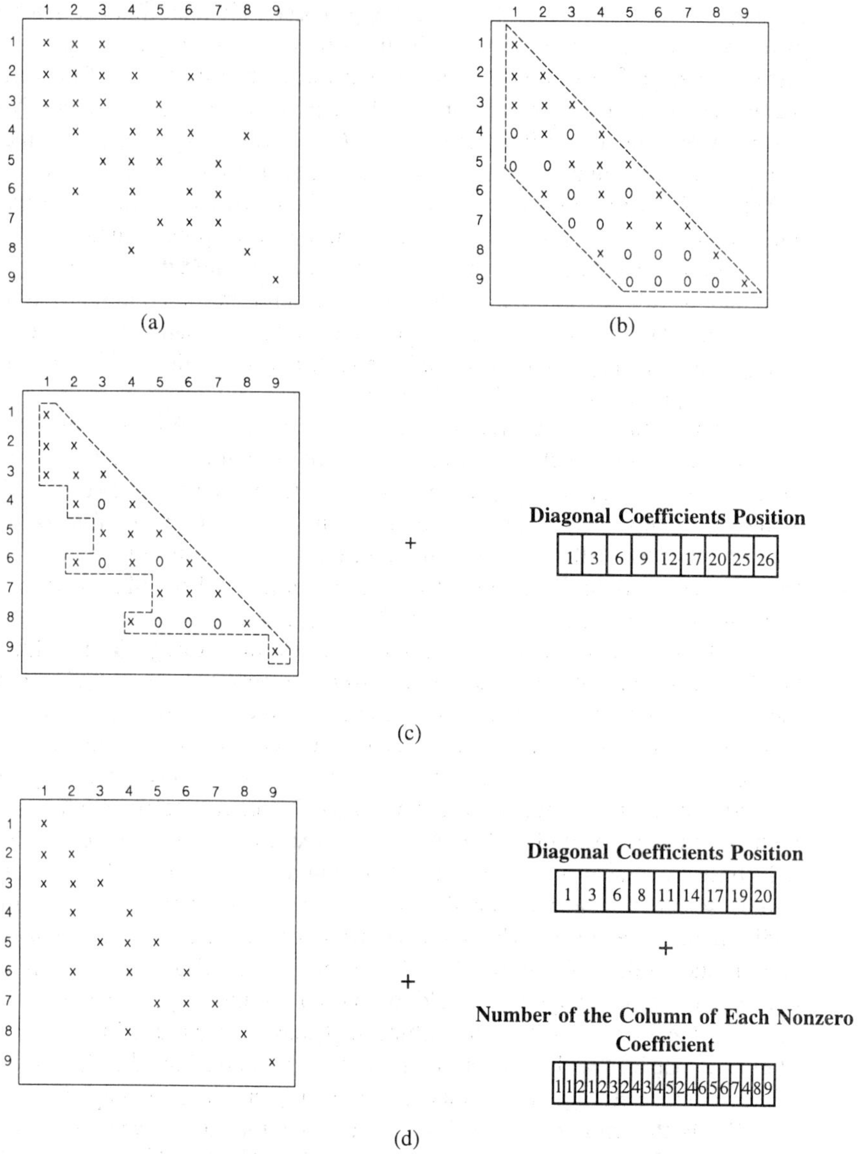

Figure 2.23 (a) A symmetric sparse matrix, (b) band storage, (c) skyline storage, and (d) nonzero coefficient storage.

Otherwise, they are called nonlinear. Among the early iterative methods are those of Gauss, Jacobi, Seidel, successive over relaxation (SOR), and all the methods derived from them. Among the nonlinear there is the method of steepest-descent and the conjugate gradient method [Bathe and Wilson 1976; Ch. 7], [Carey and Oden 1984; Ch. 3], [Kardestuncer and Norrie 1987; P. 4, Ch. 1], [Silvester and Ferrari 1983; Ch. 7], [Tortschonoff 1984].

Iterative procedures are easier to program than the direct ones. In the first case it is enough to store exclusively only the nonzero coefficients. Figure 2.23(c) shows an example of the coefficients that should be stored in the case of a symmetric matrix from which the lower triangular part has been selected. In addition to the vector containing the nonzero coefficients, the position of the coefficients of the diagonal, and that of the column corresponding to each nonzero coefficient should be stored. For sparse matrices of large dimensions this method is much more economic than those of band and skyline storage.

Any iterative method converges toward the correct solution in a number of iterations that depends on the condition number of the matrix. This means that the number of operations and the CPU time cannot be predicted. In order to accelerate the convergence, some type of preconditioning is usually employed. There are a number of preconditioners available, like those of relaxation and incomplete Choleski factorization or Crout factorization [Bernadou et al. 1985].

The accuracy of the solution is not sensitive to the condition number of the matrix. Truncation and round-off errors will slightly affect the final result in the case of systems with a very large number of unknowns.

The iterative method employed in this book is the conjugate gradient method either without a preconditioner or with different types of preconditioners, like relaxation and incomplete Choleski factorization. There exist different variations of the method that have not been employed to solve the problems analyzed in this book. For some iterative methods when the initial guess for $\{D\}$ (that initialize the iterative procedure) is a good approximation of the final solution, the rate of convergence is greatly enhanced. However, this does not seem to be the case for the conjugate gradient method.

Depending on the situation, it will be preferable to choose either a direct method or an iterative procedure. In the references given in this section, some comparisons between the various types of methods and different procedures may be found. An additional reference is [Seitelman 1973]. In Chapter 3 some comparisons are also given.

In the solution of *eigenvalue problems* (2.77b), there exists a variety of methods and algorithms. Most of them are especially adapted for the case of sparse matrices, like those of direct or inverse iteration. Included in this category will be the methods due to Jacobi, Lanczos, QR, Ritz, subspace, conjugate gradient, and the methods derived from them [Bathe and Wilson 1976; Chs. 10 to 12], [Bernadou et al. 1985], [Carey and Oden 1984; Ch. 4], [Dhatt and Touzot 1981; Ch. 5], [Kardestuncer and Norrie 1987; P. 4, Ch. 1], [Yang et al. 1989]. In this book the method of inverse iteration by subspaces is used in conjunction with the Rayleigh-Ritz method. The method assumes that the matrices $[K]$ and $[M]$ are symmetric and positive definite. For some of the problems to be treated in this book, the matrix $[K]$ does not always fulfill that condition. For those cases, a translation in λ (the eigenvalue) will be needed in order to transform the problem into one where both matrices $[K]$ and $[M]$ are positive definite (see Appendix F.2.1.2). In all cases, the skyline storage method has been used.

2.5.2.5 *Postprocessing of the Solution*

Once a solution $\{D\}$ is obtained, the value of the variational unknown at any point in the domain or on the boundary or any other quantity related to it may be computed.

For example, in the quasi-static analysis of transmission lines, the unknown is the electrostatic potential. It may be necessary to compute the value of the potential at certain points of the domain (in order to draw equipotential lines, for example) or the corresponding electric or magnetic field. Similarly, quantities like the stored energy, transmission line parameters, particularly characteristic impedance matrices, and losses in the different media and conductors, may be the final objective of a given analysis. Of special interest is the computation of estimates or indicators of the error incurred in the computation of the solution or any other quantity. Because these estimates are obtained from the solution, they are called *a posteriori* estimates. The term error indicator is used when it only serves as a comparative quantity from one element to the other (or between analysis of the same problem with different discretizations) of the error incurred. Error estimates and indicators allow one to design automatic refinement processes that will lead to solutions that are accurate up to a prespecified degree [Carey and Oden 1984; Ch. 1], [Reddy 1984; Ch. 3], [Reddy 1993; Ch. 5].

Assume that the quantity of interest $\varphi(\bar{r})$ is a function related to the unknown, u, $\varphi(u)$, by means of a linear differential expression. Taking into account that in each element the following expression holds

$$\tilde{u}^e(\bar{r}) = \{N^e(\bar{r})\}^T\{d^e\} \tag{2.78}$$

one obtains

$$\tilde{\varphi}^e(\bar{r}) = \tilde{\varphi}^e(\{N^e(\bar{r})\}^T\{d^e\}) = \{c^e(\bar{r})\}^T\{d^e\} \tag{2.79}$$

where $\{c^e(\bar{r})\}$ (or, in general, $[c^e(\bar{r})]$) is the local postprocess column matrix (or local postprocess matrix) whose coefficients depend on the basis functions, on the coordinates of the points where the quantity should be computed (thus, on the element Ω^e to which each given point belongs), and on the mathematical relation between φ and u. The coefficients of the postprocess matrices may be computed at the same time as the coefficients of the other local matrices $[k^e]$, $[m^e]$, and $\{f^e\}$ are computed. In this way it would not be necessary to repeat, during the postprocess step, computations that may have been done previously [Akin 1982; Chs. 9 and 14]. Then, at the postprocess step, if φ should be evaluated at one or more points $\overline{r}_i$ (or $\{r_i\}$), it is enough to identify the element Ω^e to which that point belongs, recover the corresponding matrix $[c^e]$ (that it is assumed to have been evaluated precisely at points $\overline{r}_i$), and apply expression (2.79). In general, for each real element the points of interest $\{r_i\}$ correspond to the same point $\{\hat{r}_i\}$ of the parent element. Thus, when computing the local matrices it is necessary only to evaluate the postprocess matrix $[c^e]$ at each point of interest of the parent element.

When performing the postprocesses, the possible discontinuities of the approximate variable $\tilde{\varphi}$ over the domain Ω of the problem, must be taken into account. For example, let us consider a deterministic problem formulated in terms of a scalar variable u, through a direct formulation discretized with Lagrange elements. Assume that the gradient of the scalar variable is also of interest. The Cartesian components of $\overline{\nabla}_t\, \tilde{u}$ will be discontinuous at points on element interfaces, while, inside each element, they will be continuous. In this case it is not convenient to compute $\overline{\nabla}_t\, \tilde{u}$ at points on the element boundaries [Zienkiewicz 1977; Ch. 11], [Zienkiewicz and Taylor 1989; Ch. 11], because different values will be obtained for a point shared by two or more elements. It is common practice to perform this computation at the points of numerical integration. Then, if the values at points on element boundaries are required, an interpolation from the value obtained at points of numerical integration may be performed. The interpolation should be of one order less than the interpolation employed for $\tilde{u}$ because the approximation of the first derivatives of u involve the derivatives of polynomials that are of one less order. This process is called the *recovery process*. Other procedures, like local least squares techniques, may be employed as recovery processes. The approximate solution for $\overline{\nabla}_t\, \tilde{u}$ will still be discontinuous from one element to the other. In order to obtain a globally smooth solution, the mean of the values computed from each element sharing a particular point may be assigned to that point, that is, if the given Cartesian component should be continuous across element interfaces, when the point is not located at interfaces between different media. If the point is placed at a media interface, two values will be assigned to the component that should be discontinuous across an interface. The assigned mean value will be computed for elements belonging to the same physical medium. This process is called a *smoothing process* [Lakshmipathi et al. 1989]. Also a weighted average can be carried out using the element areas as weights. In this book all those methods are used. Global least squares methods are also used as well as other more sophisticated smoothing techniques [Lowther et al. 1993], [Tessler et al. 1994], [Zienkiewicz and Zhu 1992a,b].

For some cases, the quantity of interest may be the sum of integral expressions or integral forms that may be linear or bilinear with respect to the local column matrix of the degrees of freedom, according to

$$F = \Sigma \{c^e\}^T \{d^e\} \tag{2.80a}$$

or

$$U = \Sigma \{d^e\}^T [c^e] \{d^e\} \tag{2.80b}$$

where the coefficients of $\{c^e\}$ or $[c^e]$ may be obtained during the computation of the local matrices. The sum of expressions (2.80) may be extended to all the elements (like in the case of the stored energy or the global error estimate), or just to some of them (like for the computation of the power lost in a lossy medium), or for certain boundaries of the domain (like in the case of the computation of losses in conductors).

In addition to the postprocesses, graphic facilities in two or three dimensions are usually employed in order to visualize the mesh, the boundary conditions, the matrix structure, the equipotential lines, or the field lines. These graphic displays are very useful in detecting possible errors as well as in the presentation of the results.

2.5.3 Convergence of the Finite Element Method

To understand convergence, it is necessary to identify the various sources of errors in a FEM procedure. They are:

1. The geometrical discretization error, arising from the subdivision of a domain into a set of elements, because when boundaries and interfaces are curved it is not possible in general to perform an exact subdivision;
2. The physical discretization error, arising from the approximation used to represent the physical characteristics of the different media and the specific value chosen for each element or at the integration points of each element;
3. The interpolation error, from the interpolation of the unknowns in each element;
4. The conforming error, arising from the discontinuity at element boundaries of the values of the secondary variables, or the approximate fulfillment of natural conditions, or the use of nonconforming elements, and from discontinuities of the primary variable at some nodes due to the use of noncompatible elements;
5. The numerical integration error, if nonexact numerical integration is used;
6. The numerical solution error, arising from the use of numerical procedures to solve the global system of equations:
   a. for direct methods, the truncation and round-off errors due to ill-conditioned matrices;
   b. for iterative methods, the convergence error inherent to any iterative technique using a limited number of iteration steps;
7. The truncation and round-off errors, arising from the use of finite precision arithmetic, and finite precision representation.

Out of all errors, 3 and 4 are more relevant for the study of FEM convergence because they are more specific to the method. In fact, both of them may be considered as just one, the interpolation error, because 4 results from the type of elements chosen, and the interpolation error may be measured by norms including the error incurred in the variable as well as the error in their first derivatives. For this reason, the interpolation error is of primary concern. However, any attempt to reduce the error of a FEM analysis should consider all possible sources of error. For example, it will not be convenient to excessively refine a mesh with the idea of reducing errors 3 and 4, if a direct solver is being used, because the truncation and round-off errors may be comparatively high [Carey and Oden 1984; Ch. 4].

Going back to the analysis of the interpolation error, the goal is to obtain global *a priori* expressions of the error valid for the whole domain Ω from the interpolation error over each element Ω^e. Any convergence analysis is related to the type of problem under consideration (in this book second-order boundary value deterministic or eigenvalue elliptic problems), the formulation used, and the finite elements selected.

Our objective is to extend the interpolation error given by expressions (2.47), (2.54), and (2.57), for the *e*th Lagrange, div-conforming, or curl-conforming element Ω^e, respectively, to the whole domain of the problem Ω. Then, it is necessary to impose certain conditions to the mesh and on the nature of the exact solution of the problem.

Consider a deterministic second-order boundary value problem, regular in the domain Ω and with an exact solution belonging to $H^{r+1}(\Omega)$, where $r \geq 1$. These assumptions imply that the domain is convex without re-entrant corners on its boundary [Raviart and Thomas 1983; Ch. 5]. Let the problem be discretized with a quasi-uniform and regular mesh, that is, with elements of similar size, fulfilling $h/h_e < \tau$, and bounded aspect ratio, $\sigma_e < \sigma$, where σ and τ are finite real numbers, h_e is the diameter of the *e*th element, and $h = \max(h_e)$, for $e = 1, ..., N_e$, (for the definition of the aspect ratio and the diameter of an element see Section 2.5.1.1). Let us assume that Lagrange elements of a single type (or compatible types) have been used and that the interpolation of the unknown contains polynomials complete up to degree p. By means of the patch test technique it can be shown that this discretization is conformal. This means that the primary variable will be continuous over the domain, and that the discontinuity of the secondary variable at element interfaces will tend to zero when the size of the elements is reduced [Patterson 1973]. Then, assuming $h < 1$, the approximate solution $\tilde{u}$ converges toward the exact solution over the whole domain Ω according to

$$\|u - \tilde{u}\|_{m,\Omega} \leq C_1 h^{k+1-m} |u|_{k+1,\Omega} , \quad k = \min(p,r) , \quad 0 \leq m \leq k+1 \tag{2.81a}$$

C_1 is a constant independent of h and u, but dependent on the type and order of the element, and on the nature of the solution. The rest of the variables have been already defined [Barnhill and Whiteman 1973], [Oden 1973], [Oden and Reddy 1976; Ch. 6], [Raviart and Thomas 1983; Ch. 5], [Strang and Fix 1973; Ch.3], [Zienkiewicz 1977; Chs. 2, 3, and 11], [Zienkiewicz and Taylor 1989; Chs. 2, 8, 9 and 11], [Zlámal 1973].

For second-order problems the norms to be considered are those corresponding to $m = 0$ and $m = 1$. Then, (2.81a) indicates that over domains in which the exact solution u is regular and smooth enough (that is, when $r > p$), if the size of the elements is refined the error in the approximation of u will tend to zero as h^{p+1}, while the error for the first derivatives of u will decrease as h^p. Thus, the rate of convergence of the h-version is ruled by p. One can note that for the triangular element $P(3')$ that provides interpolation complete only up to second order, the rates of convergence will be given by h^{2+1} and h^2, respectively, as if the element $P(2)$ had been used.

The exact solution may not be sufficiently smooth, so $r < p$. This may be due to different reasons: nonsmooth boundaries, abrupt changes in the value of the physical constants characterizing different media of the domain of the problem, singular excitations, and discontinuities in the boundary conditions [Babuška 1977]. For electromagnetic problems, coupling and proximity effects between conductors also produce nonsmooth solutions. In those cases, if the condition $r \geq 1$ is fulfilled, then $k = r < p$ and the expression (2.81a) indicates that the rate of convergence would be that corresponding to the use of elements of only degree r.

When the domain is nonconvex, there are singularities of the derivatives of the solution, and if those derivatives are square integrable, that is, the problem has finite energy, then $r = \alpha$, where α is a real positive number lower than unity. Then, expression (2.81a) with $r = \alpha < 1$ may be used providing a rate of convergence for the function as $h^{\alpha+1}$ and for the first derivatives as h^{α}, showing that the h-version will have a slow convergence ruled by the measure ($\alpha < 1$) of the smoothness of the solution. Apparently no improvement may be obtained by increasing the order of the elements. However, if the dependence on p of the constant in (2.81a) is made explicit, an expression for the error has been obtained in [Babuška and Suri 1987a,b] and [Babuška and Dorr 1981] as

$$\| u - \tilde{u} \|_{1,\Omega} \leq C_3 h^{\min(r,p)} p^{-r} \| u \|_{r,\Omega} , \quad r > 0 \tag{2.81b}$$

where the dependence of C_2 is as mentioned previously. This expression shows that the accuracy of the solution may be improved by increasing p, even if the convergence is ruled by $r = \alpha < 1$ [Barnhill and Whiteman 1973], [Zienkiewicz and Craig 1986].

For elliptic problems the energy norm turns out to be a measure of the error that is more natural and convenient than (2.81a)(see Appendix B.1). It is given by

$$\| u - \tilde{u} \|_E \leq C_3 h^k , \quad k = \min(p,r) , \quad r > 0 \tag{2.81c}$$

where C_3 is a constant independent of h, but dependent on u and the type and order p, of the elements employed. Note that (2.81b) is valid for values of r lower than unity (i.e., for strongly nonsmooth problems).

From (2.81) it is also possible to deduce the order of convergence of any other quantity computed by means of a postprocess. For example, the rate of convergence of the stored energy in an electrostatic or quasi-static problem (related to the square of the first derivatives) will be given by $h^{2\min(p,r)}$ [Zienkiewicz and Taylor 1989; Ch. 2].

The error on the ith eigenvalue $\tilde{\lambda}_i$ of regular eigenvalue problems of order $2s$ is

$$|\lambda_i - \tilde{\lambda}_i| \leq C_4 h^{2(k+1-s)} \lambda_i^{(k+1)/s} \tag{2.82}$$

Thus, for second-order problems one gets

$$|\lambda_i - \tilde{\lambda}_i| \le C_5 h^{2k} \lambda_i^{(k+1)} \tag{2.83}$$

Hence, eigenvalues have a rate of convergence that is twice that for the first derivatives. From (2.83) it may also be deduced that the convergence for eigenvalues larger in magnitude is slower than for the smaller ones [Carey and Oden 1984; Ch. 4], [Murty and Rao 1973], [Raviart and Thomas 1983; Ch. 6], [Strang and Fix 1973; Ch.2].

All the expressions given previously are applicable to elements with curved boundaries if condition (2.48) is fulfilled over the whole domain Ω. Thus, element shapes should not be too distorted [Ciarlet 1973]. They are also applicable for meshes with elements of different geometry or type as long as they are compatible amongst them [Raviart and Thomas 1983; Ch. 3].

In the case of div-conforming and curl-conforming elements, expressions (2.54) and (2.57) may be extended to the whole domain of the problem Ω as long as the problem is regular, the unknown $\bar{t}$ belongs to $(H^r(\Omega))^n$, with $r \ge 1$ and the mesh is regular and quasi-uniform. Under these conditions, if elements of order p are used a conformal discretized problem is obtained. Then, the following expression holds

$$\| \{t\} - \{\tilde{t}\} \|_0 \le C_6 h^k \tag{2.84}$$

where the parameter k has the meaning previously given to it.

For Lagrange/curl-conforming elements, expression (2.81) will correspond to the scalar unknown and (2.84) to the vector unknown. Conclusions analogous to those mentioned previously hold for the measure of the error over the whole domain.

When the exact solution is not sufficiently regular (i.e, smooth), an important conclusion may be derived from expressions (2.47), (2.54), and (2.57). Namely, in a quasi-uniform mesh the interpolation error is higher in those elements located in regions where the solution is not smooth enough. Thus, a better approach should be to generate a mesh with smaller elements in those regions (e.g., in the vicinity of corners and discontinuities) than in the rest, trying to obtain a nonuniform mesh with local errors of the same order of magnitude. These meshes are called equilibrated meshes because the error is said to be equilibrated. Then, in order to draw valid conclusions regarding convergence, any measure of the error should be expressed in terms of the total number of degrees of freedom N_d instead of using the parameter related to the size of the elements h that will not be meaningful for nonuniform meshes. Then, for the case of uniform or quasi-uniform meshes, the following expression is obtained:

$$\|e\|_E \le C_7 N_d^{-\frac{1}{n}\min(p,r)} \quad , \quad r > 0 \tag{2.85}$$

where e stands for the error $e = \tilde{u}$-u; n is the dimensionality of the problem (i.e., 1 for 1D elements, 2 for 2D elements, or 3 for 3D elements). Expression (2.85) again indicates that in problems with a strong singularity ($r < 1$) the rate of convergence for uniform meshes is determined by α.

If equilibrated meshes are used, the following expression holds:

$$\|e\|_E \leq C_8 N_d^{-\frac{p}{n}} \tag{2.86}$$

It may be observed that the previous expression is independent of the regularity or smoothness of the solution. It has been demonstrated by many authors [Babuška and Dorr 1981], [Babuška and Szabó 1982], [Babuška et al. 1979, 1981], [Babuška and Suri 1987a,b], [Dorr 1984, 1986], [Gui and Babuška 1986a,b,c], [Babuška and Guo 1988], [Guo 1988], [Lee and Lo 1992], [Rank 1986, 1993], [Szabó 1986], [Suri 1990], [Zienkiewicz et al. 1989].

Expressions (2.85) and (2.86) are the basis for the so-called *h*-version of FEM, in which the number of degrees of freedom is increased by subdividing the elements of the original mesh either uniformly or selectively. In particular, self-adaptive techniques using selective *h* refinements are based on (2.86), that indicates that if equilibrated meshes are obtained through an appropriate refinement, then the maximum rate of convergence of the *h*-version will be achieved as if quasi-uniform meshes were used for a smooth enough problem, regardless the actual regularity of the solution.

If instead of increasing the number of degrees of freedom by subdividing the elements of a particular mesh, the order of the elements is increased (*p*-version), the following expression holds:

$$\|e\|_E \leq C_9 N_d^{-\beta} \tag{2.87}$$

where C_9 is a constant that depends on the type of elements used, on the parameter β, and on the regularity of the exact solution u. The parameter β is a positive real number that depends on the quality of the mesh and the regularity of the solution. For quasi-uniform meshes and smooth solutions, the rate of convergence of the *p*-version is at least that of the *h*-version. For a problem with re-entrant corners, if they are located at the vertices of the elements and quasi-uniform meshes are used, then $\beta = r$, giving a rate of convergence at least n times that of the quasi-uniform *h*-version. For nonuniform meshes the rate of convergence of the *p*-version may be even higher, depending on the quality of the mesh and on the type of *p* refinement used, uniform or nonuniform.

The combination of *h* refinements and *p* refinements will lead to

$$\|e\|_E \leq C_{10} e^{-\delta N_d^{\theta}}, \quad \theta \geq 1/3 \tag{2.88}$$

where δ and θ are constants. Thus, exponential rates of convergence are achieved.

If the number of integration points is selected carefully, the expressions for the rate of convergence that have been given previously do not deteriorate because of numerical integration. This is illustrated in [Raviart and Thomas 1983; Ch. 5] using the direct formulation of a second-order deterministic problem discretized by a straight triangular element of the Lagrange type. If $n_p = p$ points of integration is selected so that polynomials of order $q = 2n_p$-1 are integrated exactly, the same rate of convergence is achieved as when using analytical integration. The use of numerical integration in the case of the rectangular elements or curved elements would not deteriorate the rate of convergence if the elements are not distorted excessively. Criteria relative to the way of eliminating the effects of other sources of error have been mentioned previously. Additional references are [Ciarlet and Raviart 1972c], [Strang and Fix 1973; Ch.4], [Cangellaris 1996], [Monk et al. 1994], [Warren and Scott 1994].

In summary, if a reasonable discretization of the domain is done in addition to a careful choice of the types of elements, the number of integration points, and the method to solve the matrix equations using double-precision arithmetic, then the previous expressions will constitute *a priori* estimates of the error. They indicate that a monotonic rate of convergence will be obtained if the size of the elements is reduced (*h* version of FEM) or if the order of the interpolation is increased (*p* version of FEM) or if both procedures are employed simultaneously (*h-p* version of FEM) [Carey and Oden 1984; Ch. 2]. A limitation for all FEM methods is an excessive increase of the number of degrees of freedom resulting in a large number of operations. This will increase numerical errors as well as memory and CPU time requirements. The *h* refinement may be limited also by the increase of the condition number of the matrices. Actually the condition number is proportional to h^{-2s}, where s = 1 for the problems considered here [Carey and Oden 1984; Ch. 4], [Fried 1973], [Zienkiewicz 1977; Ch. 2], [Zienkiewicz and Taylor 1989; Ch. 2]. Thus, truncation and round-off errors will increase with decreasing values of h if a direct solver is used. If iterative solvers are utilized a longer CPU time will be required. The refinement in p is limited by the widening of the band and an increase on memory requirements. The problem of how significantly the elements should be refined is a complicated one since the constants appearing in expressions (2.81)-(2.88) are not computable. Moreover, if the refinements are not uniform they do not have the same value for all the meshes.

For problems with a regular solution discretized by regular uniform meshes, Richardson's method can extrapolate the exact value of the unknown from solutions obtained over an initial mesh and a reduced number of uniformly refined meshes [Heise 1973], [Zienkiewicz 1977; Ch. 2], [Zienkiewicz and Taylor 1989; Ch. 2].

For most problems the regularity conditions are not fulfilled. The domain may not be convex. The physical constants of the media involved may have abrupt changes. The boundary conditions may be discontinuous, and, in general, the solution may not be smooth enough because of coupling or proximity effects. The value of r is difficult to obtain. Thus, *a priori* estimates are not useful in determining the refinement strategy. And, on many occasions, they will not provide a realistic estimate of the error. Then, *a posteriori* estimates or indicators of the local (over each element) and global (over the whole domain) errors incurred are useful. These estimates are computed from the solution obtained for a given mesh. When the global error estimate is greater than the desired accuracy, the number of degrees of freedom will be increased. The local error estimates are used to decide which elements should be subdivided (h refinement) or support a higher order interpolation (p refinement) or both. The process will continue until a prespecified degree of accuracy is achieved. These methods automatically generate meshes well adapted to the regularity conditions of the solution of the problem. They guarantee a good accuracy in the description of the local behavior of the solution. These techniques are termed self-adaptive mesh generators. Here they have been implemented by means of selective h refinements, eventually followed by uniform p refinement. In Chapter 4 a self-adaptive procedure is analyzed. Appendix G provides the computation of error estimates. Chapter 5 presents examples utilizing this procedure.

In summary, FEM provides accurate solutions with a high rate of convergence if the mesh is well adapted to the nature of the exact solution of the problem. In addition, if advantage is taken of the characteristics of the discretized problem, moderate storage, memory, and CPU time requirements may be obtained.

2.6 FLOW DIAGRAM OF A FINITE ELEMENT ANALYSIS

Figure 2.24 shows the general flow diagram of the program used to solve the examples of this book. The task of developing a FEM computer program is quite complex and long because specific techniques should be employed in order to obtain high efficiency in terms of rate of convergence, accuracy, CPU time, storage, and memory requirements. For this reason it is convenient to use existing algorithms for all those tasks that are not specific to given problems as for example mesh generation, numbering techniques, assembly procedures, and solvers. Of course, all the algorithms should exploit specific FEM characteristics (such as sparse matrices, band, and skyline or nonzero coefficients storage). In this way it would be necessary only to program the specific modules required, like those computing the local matrices (including the postprocess matrices), the final postprocesses, and the interfaces between them.

The MODULEF library of Finite Elements [Bernadou et al. 1985] is very well adapted to this situation. The modular conception of this package allows one to use it in a very flexible way. In Figure 2.24 a continuous line stands for the MODULEF routines that have been used with a minimum modification. A dashed line indicates new modules that have been programmed. Lines of both types indicate MODULEF routines that have been substantially modified or where interfaces have been added. A dynamic treatment of the memory has been used to reduce the memory requirements.

2.7 PUBLIC DOMAIN AND COMMERCIAL SOFTWARE PACKAGES FOR THE ANALYSIS OF ELECTROMAGNETIC PROBLEMS UTILIZING THE FINITE ELEMENT METHOD

A large number of public domain commercial FEM-based software packages are available. Because in the electromagnetic field this method has had a relatively recent

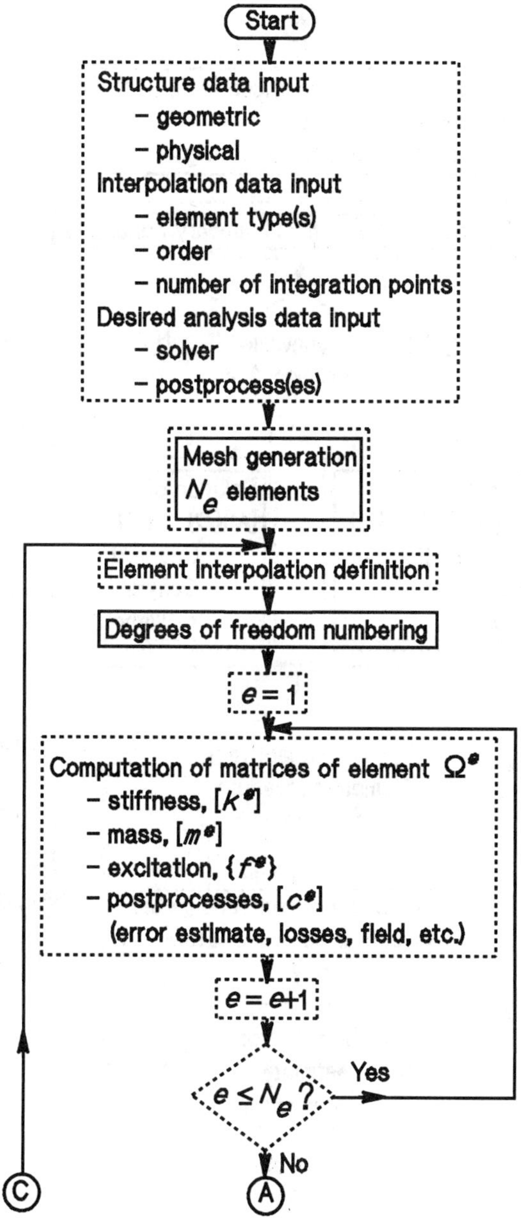

Figure 2.24 General flow diagram of a FEM program.

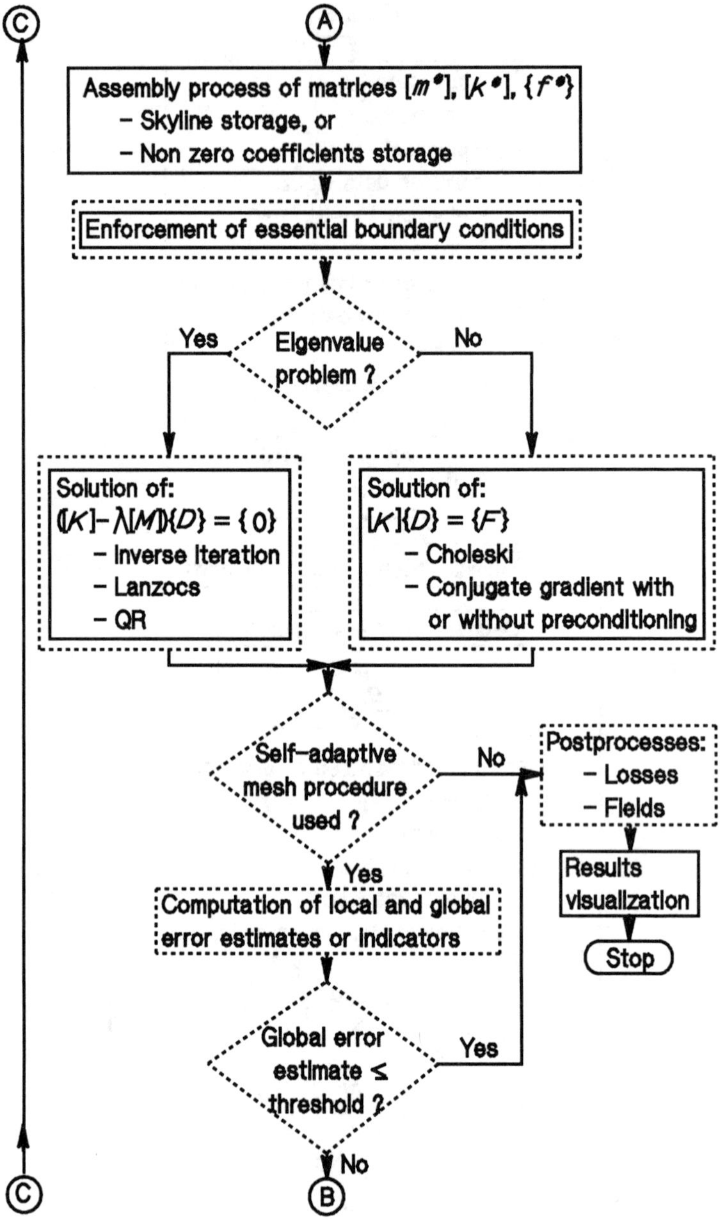

Figure 2.24 (Continued.)

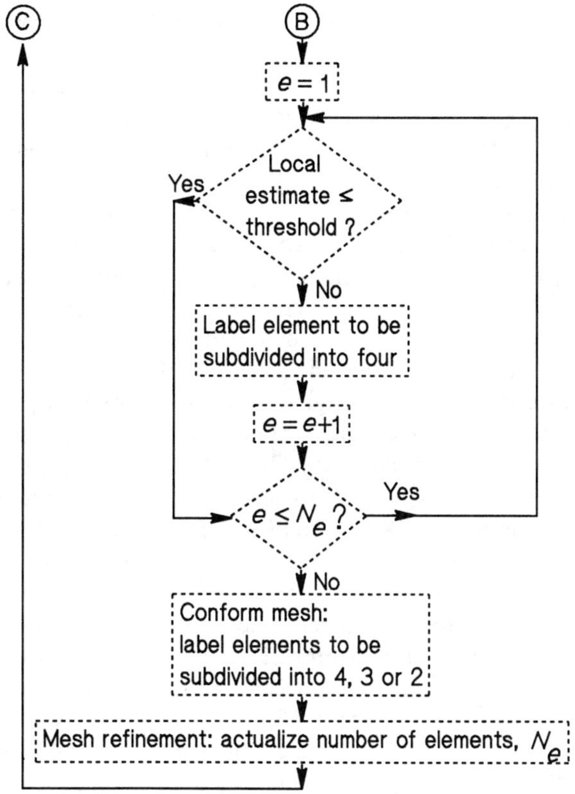

Figure 2.24 (Continued.)

development, most of these packages are not well adapted to the specific needs of the electromagnetic community. In [Kardestuncer and Norrie 1987; P. 4, Ch. 5] an exhaustive list of the FEM packages available in the market as of 1987 is given. Some of them are of general purpose, and they are either too complex or too large. Some of them pay special attention to nonlinear problems and are not as efficient for other types of problems. Some of them are specific for mechanical, structural, and fluid problems. ANSYS, from Swanson Analysis Systems Inc., has been employed with success for the full-wave analysis of some microwave waveguiding structures [Navarro et al. 1991], but it does not cover all the needs in the electromagnetic and microwave fields.

During the last two decades, new packages have appeared in the market that are devoted to the study of electromagnetic problems [Coulomb et al. 1985], [Heighway and Biddlecombe 1982], [Sabonnadière et al. 1982]. A summary of various packages may be found in [Tortschonoff 1984]. The latest versions of MAXWELL, from Ansoft Corp., and MAGNET, from Infolytica Corp., can perform both electrostatic and full-wave analyses of microwave structures [Csendes 1990].

In the field of microwave engineering, the earlier packages were devised for scalar problems and waveguide analysis [Konrad and Silvester 1971], [Silvester 1969b]. A package for planar structures was soon available [Silvester and Konrad 1977]. Since then, numerous packages with one or other characteristics have been developed by researchers and software companies,and have been updated, or improved every year. Some of them are limited to specific problems while other have a broader scope. Some of them are public domain packages and may even be obtained through the Internet, while others are commercially available. To give a list and an evaluation of all of them is out of the scope of this section. Here, some will be mentioned and some references will be given. The interested reader is encouraged to keep track on their development through the Internet network.[2] A recent update of commercial packages for electromagnetic problems is available in [Silvester 1996b]. There, a number of FEM packages relevant for the microwave engineering community is described together with other tools that are based on different numerical procedures (like MAFIA, from CST—Computer Simulation Technology; PROFI, from Profi Engineering Systems GmbH; and Sonnet, from Sonnet Software, Inc). The FEM-based packages for microwave applications are: FLUX2D, and FLUX3D, from CEDRAT S.A., that perform static, transient, and dynamic analysis of 2D and 3D structures, respectively; EMP, E3, and Superfish 3.1, from Field Precision: the first two packages perform 2D and 3D static analysis, while the third one computes the resonant modes of cylindrical or linear structures; HFSS, from HP, that performs 2D and 3D simulations of electromagnetic problems, including open-region ones; ELECTRO (2D/RS), and COULOMB (3D electrostatic), from Integrated Engineering Software, that are electrostatic field solvers; FEMAX, FEMAX3, FEMAXT, maintained by Gerrit Mur, that perform 3D time harmonic and arbitrary time dependence analysis of electromagnetic fields; a package for static analysis has been added recently [Lager and Mur 1996]; and PDE2D, from Granville Sewell, that analyzes 2D and 3D time dependent and eigenvalue problems. The mentioned reference give details about the FEM implementation of each package. Other software modules not included in that reference are: STINGRAY, from ArguMens, for quasi-static analysis of transmission lines [Waldow 1989]; PC-OPERA, from Vector Fields, in the market since 1992, that is a 2D electrostatic and full-wave solver; MSC/EMAS, from MacNeal-Schwendler Corp., in the market from 1989, an electromagnetic simulator for zero-, one-, two-, and three-dimensions, including open-region problems; and LAGRANGEINE 2D, from Technische Universität Berlin, that performs a quasi-static analysis of transmission lines, among others [Anger 1990], [Csendes 1990], [Sabonnadière and Konrad 1992]. The authors of this book have developed a software code for 2D and 3D quasi-static and full-wave analysis of microwave structures including open-region problems (scattering and radiation). Its features are described throughout the book.

---

2 Information related to FEM resources in general, and FEM packages in particular, may be found in http://www.engr.usask.ca%7Emacphed/finite/fe_resources/fe_resources.html.

Chapter 3

Application of the Finite Element Method to the Analysis of Waveguiding Problems

3.1 INTRODUCTION

In the study of electromagnetic devices it is of paramount importance to analyze systems for the transmission of electromagnetic energy between two points on the device. At high frequency, namely, at microwave and millimeter waves, the sizes of passive and active devices are of the same order of magnitude as the wavelengths corresponding to the frequency band used. This fact implies that the propagation time of the electromagnetic response due to an excitation between two points of the device, is comparable to the periods of the harmonic signals in which the excitation may be decomposed by means of a Fourier series. Thus, the description of the electromagnetic behavior of high-frequency devices may not be done in terms of circuital concepts as in the case of low-frequency devices.

The term microwave generally refers to the frequency band in which the free-space wavelengths are between centimeter (cm) and millimeter (mm). Some authors suggest that the microwave/millimeter wave spectrum corresponds to the band between 1 GHz (30 cm) and 300 GHz (1 mm). However, the usual agreement is to give the name microwave/millimeter wave to the frequency band between 300 MHz (100 cm) and 300 GHz (1 mm). The actual development of high-frequency technology predicts an extension of such bands up to 3 THz (100 μm). The microwave band occupies from 300 MHz (100 cm) to 30 GHz (1 cm). The millimeter wave band extends from 30 GHz (10 mm) to 300 GHz (1 mm). Finally, the rest of the band is called the submillimeter wave band [Bhal and Bhartia 1988; Ch. 1], [Collin 1966; Ch. 1], [Elliot 1993; Ch. 1], [Wiltse 1984].

All systems utilized for electromagnetic energy transmission between two points of a high-frequency device have a common characteristic. Namely, their geometry may be defined in a plane perpendicular to the direction of propagation of the electromagnetic energy. Actually, the cross section of the energy transmission system is invariant along the direction of propagation. This type of symmetry is called a translational symmetry along a given axis. In this case, the axis coincides with the direction of propagation. This characteristic defines a waveguiding structure or a transmission line in a broad sense. In the following, the direction of propagation will be identified with the z-direction of a Cartesian coordinate system. That direction will also be called the longitudinal direction. Thus, the cross section of a waveguiding structure is in the x-y plane.

In this chapter the finite element method (FEM) is applied to the analysis of waveguiding structures. The analysis is performed in the frequency domain, and hence the solutions are time-harmonic electric and magnetic fields. Under such conditions the propagating electromagnetic field can be described in terms of the following electric and magnetic field phasors:

$$\underline{\bar{E}}(x,y,z) = \bar{E}(x,y)e^{\mp\gamma z} \tag{3.1a}$$

$$\underline{\bar{H}}(x,y,z) = \bar{H}(x,y)e^{\mp\gamma z} \tag{3.1b}$$

where $\underline{\bar{E}}$ stands for the electric field intensity (volts/meter) and $\underline{\bar{H}}$ represents the magnetic field intensity (amperes/meter) at a point given by the cartesian coordinates (x,y,z). $\gamma = \alpha + j\beta$ represents the complex propagation constant along the longitudinal direction, where α represents the attenuation constant (nepers/meter) and β depicts the phase constant (rad/m). Because of the translational symmetry, the objective of the analysis is to solve the two-dimensional (2D) electromagnetic problem, namely, to obtain, besides the value of the propagation constant γ, the description of the electric and magnetic fields at each point (x,y) of the cross section. In (3.1) the complex phasor notation has been used, so the time variation of the form $e^{-j\omega t}$ is assumed. However, for ease of notation it has been suppressed since phasors are used. ω is the angular frequency (rad/sec) and t is the time variable (sec). The electric and magnetic field intensities are complex vectors having nonzero components in all three Cartesian directions for the general case. It is possible to have propagation in both the positive and the negative z-directions by means of the terms $e^{\mp\gamma z}$, respectively. This implies that the structure is infinite in the longitudinal direction, or equivalently, the sources of the electromagnetic fields are located at infinity.

The type of waveguiding structure dealt with in this chapter is an arbitrarily shaped multiconductor structure embedded in a multidielectric medium. The permittivity and permeability of the dielectric media may be described, in general, by a tensor. The transmitted electromagnetic energy takes place through travelling waves that are solutions of Maxwell's equations in a source-free region (i.e., fields have the form shown in (3.1)). The fields are time-harmonic, and the nature of the waves is dictated by the geometrical cross section of the waveguiding structure which is expressed partly through the boundary conditions. The guided wave problem is, in general, an eigenvalue problem where, for a given frequency, each eigenvalue provides the propagation constant of a corresponding spatial field distribution described by the associated eigenvector. Each solution is called a mode.

For structures shielded by a perfect conductor the set of modes will have discrete eigenvalues that are infinite in number for each value of the frequency variable f (Hz)

A number of modes having cut-off frequencies lower than f will be propagating inside the structure while all the rest (infinite in number) will not be propagating or, equivalently they will be evanescent. The cut-off frequency of a mode is the value of the frequency that limits the propagation of the mode. For a working frequency under the cut-off frequency of a given mode, this implies that such a mode will not propagate along the waveguiding structure. Assuming, for the moment, that ohmic losses are not present (hence, that all conductors are perfect, and the dielectric media are ideal, with real permittivity and permeability), the propagation constant, γ, for a propagating mode will be purely imaginary, $\gamma_i = j\beta_i$, while for an evanescent mode it is real, $\gamma_i = \alpha_i$. Conjugate pairs of complex modes may also be present depending on the nature of the waveguiding structure and the working frequency [Sarkar and Salazar-Palma 1994]. A pair of complex conjugate modes transmit zero power along the waveguiding structure. If the waveguiding structure is lossy (imperfect conductors and/or non ideal dielectric media) the propagation constant will be complex in general, with $\beta > \alpha$, for propagating modes, $\alpha > \beta$, for evanescent modes and complex conjugate pairs for complex modes. If the cross section of the waveguiding structure is finite and the structure is not completely shielded then there will be a continuous spectrum of eigenvalues, since there will be radiation in addition to an infinite set of discrete eigenvalues corresponding to the infinite propagating modes. An important requirement of all waveguiding structures is that the radiation losses be kept as small as possible. The electromagnetic field propagating along a waveguiding structure at a given frequency is a linear combination of all the propagating modes (possibly with a complex propagation constant).

At this point it may be interesting to note that microwave technology is essentially a mono-mode technology. To be more precise, devices are generally designed so that along the waveguiding structures only the fundamental or dominant mode (that having the lowest cut-off frequency) will be propagating, or in the case of multiconductor lines (with N_c+1 conductors) only the N_c fundamental modes. Microwave technology is also a broadband technology, which means that the gap between the working frequency, at which the fundamental mode propagates, and the first higher order mode should be large.

There are different types of modes inside a waveguiding structure. They may be either transverse electromagnetic (TEM), transverse magnetic (TM), transverse electric (TE) or hybrid. In the case of the TEM modes, the longitudinal components of both the electric and the magnetic fields are identically zero. For the TM modes, the longitudinal component of the magnetic field is zero. For the TE modes, the longitudinal component of the electric field is zero. Finally, for the hybrid modes, the longitudinal components of both the electric and the magnetic fields are finite and never identically zero. In the microwave area one often deals with quasi-TEM modes that are essentially hybrid modes in which the longitudinal components of the fields are much smaller in magnitude than the transverse components. Then the longitudinal components may be neglected. Such an assumption often greatly simplifies the analysis in many situations without sacrificing accuracy. In fact, as in the case of pure TEM modes, the fundamental modes of the waveguiding problem may be obtained by solving a deterministic problem. However, the computation of higher order modes (as well as the exact computation of the fundamental modes) requires the solution of an eigenvalue problem. The latter analysis is often called a full-wave analysis.

Pure TEM modes can propagate only in structures consisting of at least two or more perfect conductors embedded in a homogeneous isotropic medium, eventually lossy. A structure with N_c+1 conductors will have N_c fundamental TEM modes with no cut-off frequency. Higher order modes will be TM or TE modes. Structures of this type are called TEM transmission lines.

If the multiconductor structure is embedded in an inhomogeneous or anisotropic media or if the conductors are not perfect (i.e., the conductivity of at least one conductor is finite), then the fundamental modes will be hybrid modes as well as the higher order modes. As mentioned before, the number of the fundamental modes will be equal to the total number of conductors minus one and their cut-off frequency will be zero. At low frequency the fundamental modes will be quasi-TEM modes. These structures are often called quasi-TEM transmission lines.

A unique property of TEM and quasi-TEM transmission lines is the possibility of associating with their fundamental modes voltage and current waves, which allows one to describe TEM and quasi-TEM transmission lines employing low frequency circuital concepts in terms of distributed parameters, i.e., *per-unit-length* (p.u.l.) capacitance, inductance, conductance, and resistance matrices. This allows the definition of multiport parameters, in terms of impedance $[z]$, admittance $[y]$, transmission, A,B,C,D, or scattering $[s]$ parameters.

Waveguiding structures with just one conductor and a homogeneous isotropic medium will propagate TE and TM modes with nonzero cut-off frequency. They are usually called waveguides or, more precisely, metallic waveguides.

In all other cases (structures with just one conductor and several dielectric media or structures consisting of just two or more dielectric media), modes are hybrid in general. Nevertheless, special symmetry conditions will allow the propagation of TE or TM modes. In general, the cut-off frequency of the fundamental mode is nonzero. Some waveguiding structures of the first class (structures with one conductor and several dielectric media) are improperly called non-TEM transmission lines. When a conductor encloses different dielectric media in a transverse plane, then they are called partially or completely filled (inhomogeneous) metallic waveguides. Finally, some structures of the first class and all of the waveguiding structures of the second class (two or more dielectric media) are called dielectric waveguides.

A characteristic of microwave waveguiding structures is that the physical dimensions of the cross section are of the same order (or smaller) as the wavelength related to the operating frequency and the phase velocity of the fundamental propagating mode. Namely, the cross-sectional dimensions of the TEM and the quasi-TEM transmission lines are much smaller than the wavelength corresponding to the phase velocity of the fundamental mode. Metallic waveguides, on the other hand, have cross-sectional dimensions comparable to the wavelength of the fundamental mode. This characteristic is common to microwave and millimeter wave dielectric waveguiding structures. Instead, dielectric waveguides used in the optical band (optic fibers) have transverse dimensions much greater than the wavelength of the fundamental mode.

This chapter is divided into two main sections. The first one deals with the FEM analysis of TEM and quasi-TEM transmission lines employing the quasi-static approach. This implies that FEM will be applied to solve 2D deterministic problems. The result of this analysis will be the distributed parameters mentioned previously as well as the propagation parameters (the complex propagation constant and the characteristic impedance matrix). From these parameters the electric and magnetic field distributions of the fundamental modes of the multiconductor transmission line are computed.

The next section deals with the finite element full-wave analysis of general waveguiding structures. This implies that FEM will be applied to solve 2D eigenvalue problems. The result will be, for the general case, the frequency and the electric and magnetic field distribution of a mode with a given propagation constant, or the propagation constant and the electromagnetic field distribution of any desired mode for a given frequency. For the case of quasi-TEM transmission lines it is possible to define also, in an approximate way, frequency-dependent primary and secondary parameters. The chapter finishes with some conclusions.

3.2 QUASI-STATIC ANALYSIS OF TRANSMISSION LINES

In this section, first the type of structure to be analyzed is defined. Appendix D.2 provides further details on this subject. The theory related to the electromagnetic and circuital characterizations of multiconductor TEM and quasi-TEM transmission lines is summarized next. Its is shown that the characterization of such structures requires the solution of a given number of quasi-static (electrostatic) problems with different dielectric constant characterizing the medium and different boundary conditions.

The following section explains the various quasi-static formulations of the electromagnetic problem and the FEM applied for their discretization. The direct (also called standard) formulations in terms of the scalar electric potential (primal formulation) or the vector magnetic potential (dual formulation) that use a FEM based on Lagrange ordinary and infinite elements is described. Appendix E.2 provides the details. The rate of convergence of the method is analyzed. A variety of structures with different characteristics (such as geometries that do not lead to field singularities, with curved boundaries, re-entrant corner structures that lead to field singularities of different order, and open configurations) is considered. Results are compared with the exact solution (in cases where it is available) and/or with those obtained by other methods. A conclusion from this research is the need for self-adaptive procedures.

The mixed formulation utilizing FEM and div-conforming elements is also described. Details may be found in Appendix E.3. This formulation provides a bound for the computed values of the same type that the bound obtained through the direct dual formulation. The bound obtained with the direct primal formulation is of an opposite nature. It will be shown that while the implementation of the dual formulation has certain disadvantages, the mixed formulation is always easy to implement. Hence the mixed formulation may be used together with the primal formulation providing upper and lower bounds of the parameters of the waveguiding structure.

3.2.1 Description of the Structures To Be Analyzed

Figure 3.1(a) is a general representation of the transverse section of the structures to be analyzed. These configurations have a total of N_{ct} ($N_{ct} = N_c+1$) conductors embedded in one or more different media whose physical characteristics (electric permittivity and magnetic permeability) can be homogeneous or inhomogeneous, isotropic or anisotropic. The structure is invariant along the direction of propagation as suggested by Figure 3.1(b). This type of symmetry allows one to perform an electromagnetic analysis in two dimensions. From the N_{ct} conductors, at least two of them must be nonconnecting, thus enabling the propagation of at least one TEM mode (for the case of an isotropic homogeneous medium and ideal conductors) or quasi-TEM mode. One of these conductors is usually chosen as the reference conductor and the remainder N_c non-connecting conductors are the active conductors. In Figures 3.1(a,b), the general structure is enclosed by the reference conductor. However, open structures with boundaries placed at infinity are also considered. Figure 3.1(c) provides an example.

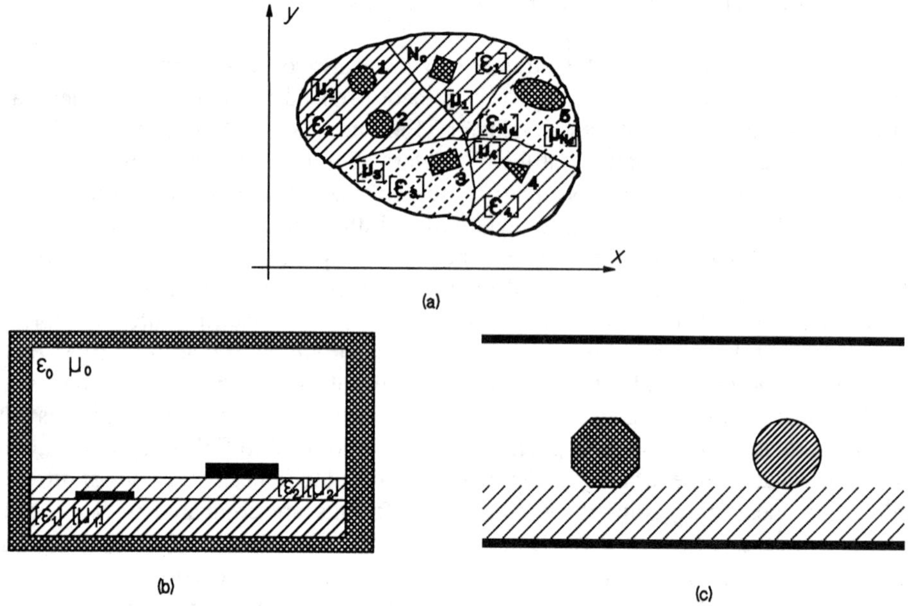

Figure 3.1 Structures which allow the propagation of TEM or quasi-TEM modes. (a) General cross section. (b) Closed structure. (c) Example of open structure.

The N_c+1 conductors are electromagnetically characterized by their conductivity σ that in principle, can vary with frequency f. In general, these conductors can be ideal (infinite conductivity under any angular frequency ω), or they may have a finite conductivity. For the electromagnetic analysis the conductors have always been assumed to be ideal. The effect of imperfections can be taken into account by a perturbational technique provided σ is reasonably high. In other words, the ohmic losses are assumed to be reasonably low.

The material media in which the conductors are embedded are characterized electromagnetically by their electric permittivity and their magnetic permeability. These are described by three-dimensional tensors, $\underline{\underline{\varepsilon}}$ and $\underline{\underline{\mu}}$, that may be represented by 3×3 matrices, $[\varepsilon]$ and $[\mu]$, respectively, as illustrated in Appendix D.2. Thus, the media (or medium) can be isotropic or anisotropic. However, the type of anisotropy or material orientation with respect to the axis of propagation is such so that it may be characterized by means of matrices of relative permittivity and permeability, $[\varepsilon_r]$ and $[\mu_r]$ ($[\varepsilon] = \varepsilon_0[\varepsilon_r]$ and $[\mu] = \mu_0[\mu_r]$), given by

$$[\varepsilon_r] = \begin{bmatrix} \varepsilon_{rxx} & \varepsilon_{rxy} & 0 \\ \varepsilon_{rxy} & \varepsilon_{ryy} & 0 \\ 0 & 0 & \varepsilon_{rzz} \end{bmatrix}, \quad [\mu_r] = \begin{bmatrix} \mu_{rxx} & \mu_{rxy} & 0 \\ \mu_{rxy} & \mu_{ryy} & 0 \\ 0 & 0 & \mu_{rzz} \end{bmatrix} \tag{3.2}$$

Here, ε_0 and μ_0 are the permittivity and permeability of vacuum, respectively.

The media can be homogeneous or inhomogeneous, however in the latter case, the coefficients in (3.2) are only permitted to vary along the transverse section of the structure. This preserves the translational symmetry. In principle, these media may have electric or magnetic losses, so the coefficients in (3.2) are generally complex.

Summarizing, the media will be characterized by matrices of relative permittivity and permeability of the following type:

$$[\varepsilon_r(x,y)] = \begin{bmatrix} \varepsilon_{rxx}(x,y) & \varepsilon_{rxy}(x,y) & 0 \\ \varepsilon_{rxy}(x,y) & \varepsilon_{ryy}(x,y) & 0 \\ 0 & 0 & \varepsilon_{rzz}(x,y) \end{bmatrix} = \begin{bmatrix} [\varepsilon_{rt}] & & 0 \\ & & 0 \\ 0 & 0 & \varepsilon_{rzz} \end{bmatrix} \tag{3.3a}$$

$$[\mu_r(x,y)] = \begin{bmatrix} \mu_{rxx}(x,y) & \mu_{rxy}(x,y) & 0 \\ \mu_{rxy}(x,y) & \mu_{ryy}(x,y) & 0 \\ 0 & 0 & \mu_{rzz}(x,y) \end{bmatrix} = \begin{bmatrix} [\mu_{rt}] & & 0 \\ & & 0 \\ 0 & 0 & \mu_{rzz} \end{bmatrix} \tag{3.3b}$$

where

$$[\varepsilon_{rt}] = \begin{bmatrix} \varepsilon_{rxx}(x,y) & \varepsilon_{rxy}(x,y) \\ \varepsilon_{rxy}(x,y) & \varepsilon_{ryy}(x,y) \end{bmatrix}, \quad [\mu_{rt}] = \begin{bmatrix} \mu_{rxx}(x,y) & \mu_{rxy}(x,y) \\ \mu_{rxy}(x,y) & \mu_{ryy}(x,y) \end{bmatrix} \tag{3.4}$$

are 2 × 2 symmetric matrices of relative transverse permittivity and permeability, which represent the bidimensional tensors of relative permittivity and permeability, $\underline{\underline{\varepsilon}}_{rt}$ and $\underline{\underline{\mu}}_{rt}$.

Multiconductor waveguiding systems may have arbitrary shapes for the transverse sections of the conductors as well as for the different media surrounding them. As it will be seen, the FEM flexibility allows handling of a variety of geometries and configurations without modifying the formulation of the problem. In fact, conductors of rectangular cross sections (with arbitrary aspect ratio) as well as circular, and trapezoidal shapes as well as other shapes may be present.

The basic structures mainly used in microwave engineering as energy transmission systems have cross sections that correspond to Figure 3.2. For hybrid technology usually one dielectric layer is present. Figure 3.3 is an example of the cross section for monolithic technology, where several dielectric layers are used.

From the configurations of Figure 3.2, only the ideal stripline (with perfect conductors and a homogeneous and isotropic dielectric) transmits a TEM mode (zero longitudinal field components) without any cut-off frequency. The other configurations, due to their intrinsic inhomogeneity (to which possibly anisotropy and conductor losses can be added), transmit a quasi-TEM mode (with small, though nonzero, longitudinal field components) that again does not have a cut-off frequency. The analysis of TEM

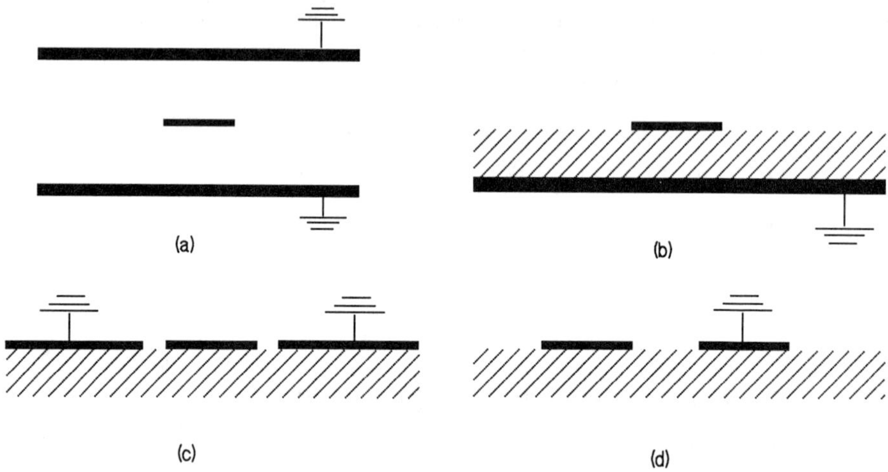

Figure 3.2 Basic transmission lines: (a) stripline, (b) microstrip line, (c) coplanar waveguide, and (d) coplanar lines.

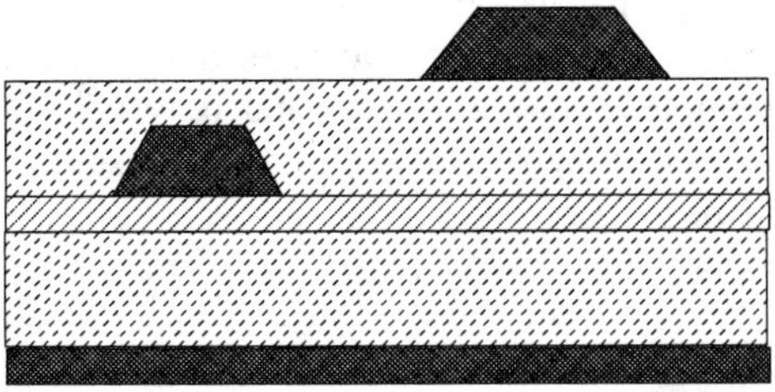

Figure 3.3 Cross section of microstrip transmission line in monolithic technology.

modes does not require a full-wave formulation. The use of a static formulation is much simpler, allowing an accurate characterization of these modes. For a quasi-TEM mode, the use of a quasi-static analysis allows one to obtain (in a simpler way than using full-wave formulations) solutions for the field valid up to a certain high frequency. Beyond such a frequency it would be necessary to perform a full-wave analysis. However, in general, the upper limit of the working band is usually below that maximum frequency.

It is also worth mentioning that due to the high-speed nature of digital circuits and the short rise-time of pulses involved with those devices, it is necessary to model interconnects as multiconductor transmission lines because a lumped element model is not applicable for the analysis of the transmission of energy containing very high frequencies.

3.2.2 Circuital and Electromagnetic Characterization of TEM and Quasi-TEM Multiconductor Transmission Lines

The classic definition of a transmission line refers to a couple of ideal conductors surrounded by a homogeneous isotropic, possibly lossy medium, with translational symmetry with respect to an axis. Through this system propagates a mode whose electric field components are contained in the plane transverse to the direction of propagation, i.e., a TEM mode. This mode has no cut-off frequency. The relevant theory is summarized first and the analysis is extended to treat inhomogeneous, anisotropic structures with imperfect conductors. Finally, general multiconductor quasi-TEM transmission lines are considered.

3.2.2.1 *Two-Conductor Transmission Line*

Homogeneous and Isotropic Transmission Line with Perfect Conductors

Figure 3.4(a) shows a transmission line of two nonconnecting ideal conductors in which the electromagnetic field propagates in a homogeneous and isotropic medium. The electric permittivity $\varepsilon = \varepsilon_0 \varepsilon_r$ and the magnetic permeability $\mu = \mu_0 \mu_r$ are constant in the medium but may be complex, $\varepsilon_r = \varepsilon_r' - j\varepsilon_r''$ and $\mu_r = \mu_r'' - j\mu_r''$. This implies that dielectric and magnetic losses may be present. Such a transmission line could be open, as shown in Figure 3.4(b).

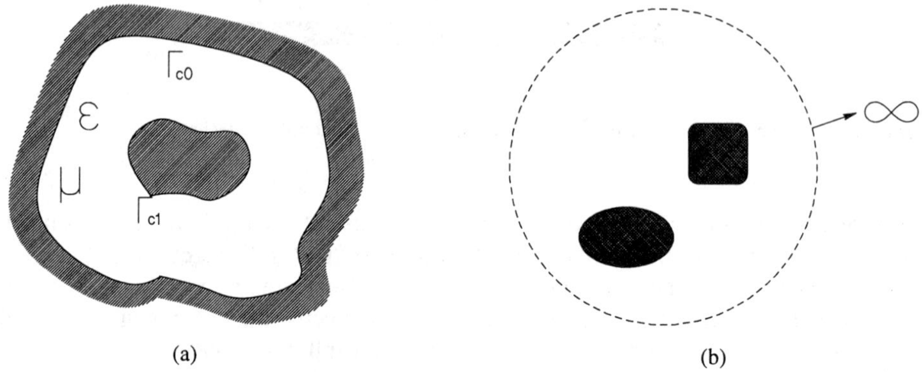

Figure 3.4 Two-conductor transmission line: (a) general closed line and (b) an open line.

According to Appendix D.3, field solutions with harmonic time variation are sought. Hence the phasor or complex notation will be used to avoid the factor $e^{j\omega t}$. In Appendix D.3, Maxwell's equations in a source free medium are specialized to a time variation of this type. The fundamental mode propagating through this structure is TEM, and therefore it is possible to write the electric field $\underline{\bar{E}}(x,y,z)$ and the magnetic field $\underline{\bar{H}}(x,y,z)$ as

$$\underline{\bar{E}}\,(x,y,z) = \bar{E}_t(x,y)\; e^{\mp\gamma z} \tag{3.5a}$$

$$\underline{\bar{H}}\,(x,y,z) = \bar{H}_t\,(x,y)\; e^{\mp\gamma z} \tag{3.5b}$$

Here γ is the propagation constant in the positive ($-\gamma z$) or negative ($+\gamma z$) z-direction along the axis with respect to which the structure has translational symmetry. Vectors $\bar{E}_t(x,y)$ and $\bar{H}_t(x,y)$ are contained in the plane transverse to the direction of propagation.

The field $\bar{E}_t(x,y)$ satisfies

$$\bar{E}_t = -\bar{\nabla}_t \phi(x,y) \quad , \quad \text{in } \Omega \tag{3.6a}$$

$$\bar{D}_t = \varepsilon_0\, \varepsilon_r\, \bar{E}_t \quad , \quad \text{in } \Omega \tag{3.6b}$$

$$\bar{\nabla}_t \cdot \bar{D}_t = 0 \quad , \quad \text{in } \Omega \tag{3.6c}$$

where $\phi(x,y)$ is a scalar potential function defined in the transverse section Ω of the structure. $\bar{\nabla}_t$ is defined in (D.11f), and $\cdot$ stands for the scalar product. Boundary conditions are $\phi = V_1$ on the contour Γ_{c1} of the conductor C_1, $\phi = 0$ on the contour Γ_{c0} of the reference conductor C_0, and $\bar{a}_n \cdot \bar{D}_t = 0$ along the symmetry walls, if any. Here $\bar{a}_n$ is the unity vector normal to the walls. V_1 is the potential difference between the trace, Γ_{c0}, in the x-y plane, of the reference conductor C_0 and the trace, Γ_{c1}, of conductor C_1. The propagation constant γ and the phase velocity v_f in the medium are:

$$\gamma = \alpha + j\beta = j\omega(\mu_0\varepsilon_0)^{1/2}\,(\mu_r\varepsilon_r)^{1/2} \tag{3.7a}$$

$$v_f = \frac{\omega}{\beta} = \frac{c_0}{\mathrm{Re}[(\mu_r\varepsilon_r)^{1/2}]} \tag{3.7b}$$

respectively, where c_0 is the velocity of light in vacuum, $c_0 = 1/(\mu_0\varepsilon_0)^{1/2}$, and Re[$\cdot$] denotes the real part of the quantity between brackets.

The magnetic field may be obtained from the electric field through

$$\bar{H}_t = \mp\, \mathrm{j}\frac{\gamma}{\omega\mu_0}\mu_r^{-1}\bar{a}_z \times \bar{E}_t \tag{3.8}$$

$\bar{a}_z$ is the unit vector in the longitudinal direction and $\times$ is the vector product. Alternatively one could solve a magnetostatic problem in terms of $\bar{H}_t$. Then $\bar{E}_t$ will be computed from $\bar{H}_t$.

Substituting (3.6a) into (3.6b) and (3.6c) leads to the direct (also called standard) formulation of the problem, given by the Laplace's equation

$$-\bar{\nabla}_t^2\,\phi(x,y) = 0 \tag{3.9}$$

with the boundary conditions $\phi = V_1$ on Γ_{c1} and $\phi = 0$ on Γ_{c0} and $\partial\phi/\partial n = 0$ along the symmetry walls. Dual or mixed formulations may also be used.

Voltage and current waves may be associated with the fields (3.5) as given by

$$V(z) = V_1 e^{\mp\gamma z} \tag{3.10a}$$

$$I(z) = I_1 e^{\mp\gamma z}\ , \quad I_1 = \oint_{\Gamma_{c1}} \bar{H}_t \cdot \mathrm{d}\bar{\Gamma} = \oint_{\Gamma_{c1}} (\bar{a}_n \times \bar{H}_t)\cdot\bar{a}_z\,\mathrm{d}\Gamma \tag{3.10b}$$

It is seen that I_1 is the current flowing along conductor C_1 at the reference cross section. Here $\bar{a}_n$ is the unit vector normal to Γ_{c1}, toward the inside of conductor C_1.

A p.u.l. admittance of the line may be defined as

$$Y = G + j\omega C \tag{3.11}$$

where C is the capacity p.u.l. of the line, given by

$$C = \varepsilon_0 \frac{\int_\Omega \bar{E}_t^* \cdot \varepsilon_r' \bar{E}_t \, d\Omega}{V_1 V_1^*} = \varepsilon_0 \frac{-\oint_{\Gamma_{c1}} \bar{a}_n \cdot \varepsilon_r' \bar{E}_t d\Gamma}{V_1} \tag{3.12}$$

and G is the conductance p.u.l., given by

$$G = \omega \varepsilon_0 \frac{\int_\Omega \bar{E}_t^* \cdot \varepsilon_r'' \bar{E}_t \, d\Omega}{V_1 V_1^*} = \omega \varepsilon_0 \frac{-\oint_{\Gamma_{c1}} \bar{a}_n \cdot \varepsilon_r'' \bar{E}_t d\Gamma}{V_1} \tag{3.13a}$$

Because the structure is homogeneous, G is given by

$$G = \omega \frac{\varepsilon_r''}{\varepsilon_r'} C \tag{3.13b}$$

In (3.12) and (3.13a), the first definitions are deduced from the electric energy stored and dissipated in the structure, respectively, while the second definitions are obtained from the p.u.l. induced electric charge.

The p.u.l. impedance of the line may be defined by

$$Z = R_m + j\omega L \tag{3.14}$$

where L is the inductance p.u.l. according to the following definitions:

$$L = \mu_0 \frac{\int_\Omega \bar{H}_t^* \cdot \mu_r' \bar{H}_t \, d\Omega}{I_1 I_1^*} = \mu_0 \frac{-\int_{\Gamma_{c0}}^{\Gamma_{c1}} \mu_r' \bar{H}_t \cdot (\bar{a}_z \times \overline{dl})}{I_1} \tag{3.15a}$$

Hence, one obtains

$$L = \mu_0 \varepsilon_0 \mu_r' \varepsilon_r' C^{-1} \tag{3.15b}$$

R_m is the p.u.l. resistance due to magnetic losses, given by

$$R_m = \omega\mu_0 \frac{\int_\Omega \bar{H}_t^* \cdot \mu_r'' \, \bar{H}_t \, \mathrm{d}\Omega}{I_1 I_1^*} = \omega\mu_0 \frac{-\int_{\Gamma_{c0}}^{\Gamma_{c1}} \mu_r'' \, \bar{H}_t \cdot (\bar{a}_z \times \overline{\mathrm{d}l})}{I_1} \tag{3.16a}$$

and therefore

$$R_m = \omega \frac{\mu_r''}{\mu_r'} L \tag{3.16b}$$

The preceding circuit parameters satisfy the telegrapher's equation:

$$\frac{\partial V(z)}{\partial z} = -Z\,I(z) \tag{3.17a}$$

$$\frac{\partial I(z)}{\partial z} = -Y\,V(z) \tag{3.17b}$$

Substituting (3.10) into (3.17) yields

$$\gamma V_1 = Z I_1 \,, \quad \gamma I_1 = Y V_1 \tag{3.18}$$

from which an alternative expression for the propagation constant γ can be obtained

$$\gamma = (ZY)^{1/2} \tag{3.19}$$

The line characteristic impedance Z_c is defined as the ratio between the forward voltage and current waves, yielding

$$Z_c = \left(\frac{Z}{Y}\right)^{1/2} \tag{3.20}$$

The parameters L, C, G, and R_m are called the line primary parameters. They can be understood as distributed circuital parameters. L and C do not depend explicitly on the angular frequency ω, while R_m and G show an explicit linear dependence.

The parameters γ, α, β, v_f, and Z_c are the secondary parameters of the line; v_f and Z_c do not depend explicitly on the angular frequency; while γ, α, and β have an explicit linear dependence on the angular frequency.

The line characterization can be obtained just by the solution of the electrostatic problem (3.9) and the calculation of the capacity C. For the lossless case this provides the various parameters:

$$\gamma = j\beta = j\omega(\mu_0\varepsilon_0)^{1/2}(\mu_r'\varepsilon_r')^{1/2} = j\omega(LC)^{1/2} \tag{3.21}$$

$$v_f = \frac{1}{(\mu_0\varepsilon_0)^{1/2}(\mu_r'\varepsilon_r')^{1/2}} = \frac{c_0}{(\mu_r'\varepsilon_r')^{1/2}} \tag{3.22}$$

$$L = \frac{1}{v_f^2 C} \tag{3.23}$$

$$Z_c = \frac{1}{v_f C} \tag{3.24}$$

The theory on this subject may be found in [Arabi et al. 1991a,b], [Collin 1960; Ch. 4, 1966; Ch. 3], [Elliot 1993; Chs. 2 and 3], [Gardiol 1987; Chs. 2 and 4], [Goosen and Hammond 1989], [Gupta et al. 1981; Chs. 2 and 11], [Harrington 1961; Ch. 2], [Kraus 1992; Ch. 12], [Ramo et al. 1956; Chs. 5 and 7], [Schelkunoff 1945; Ch. 7].

TEM Mode Perturbations

Next to be considered are the cases where the fundamental mode propagating along a structure having a translational symmetry with two conductors is not a TEM mode, i.e., the electromagnetic field has nonzero longitudinal field components. Thus, it becomes a hybrid mode. This mode, however, has no cut-off frequency.

In most practical cases, the magnitude of the longitudinal components of these modes is very small compared to the magnitude of the transverse components. Then these modes are termed quasi-TEM modes. In such cases the analysis of these structures by a static approach, where the longitudinal component of the fields are neglected, provides valid results up to a certain frequency.

The cases to be considered are when the medium is anisotropic and/or inhomogeneous, and when the conductors are lossy.

Anisotropic and Inhomogeneous Medium

When the medium through which the electromagnetic wave propagates (see Figure 3.4) is anisotropic or inhomogeneous, any mode propagating along the transmission line at nonzero frequency must have longitudinal components with non-trivial values.

Here anisotropy is of the type given by (3.2) and there is only transverse inhomogeneity as indicated by (3.3) and (3.4). Then, up to a certain value of the frequency, the longitudinal components are much lower in magnitude than the transverse ones. The approach that assumes zero longitudinal components can then be used as a zero-order approach. As a result, the transverse components of the electric and magnetic fields can be obtained by solving the static problems related to each other.

Anisotropic medium. Consider a lossless homogeneous medium, with isotropic permeability and having a tensor of electric permittivity characterized by means of a diagonal matrix, according to (D.8) of Appendix D. Imposing the condition that the propagating mode is TEM ($E_z = H_z = 0$) and with $\gamma = j\beta$, (D.28c) and (D.28d) yield

$$\beta \bar{a}_z \times \bar{E}_t = \omega \mu_0 \mu_r \bar{H}_t \tag{3.25a}$$

$$\beta \bar{a}_z \times \bar{H}_t = -\omega \varepsilon_0 \underline{\underline{\varepsilon}}_{rt} \bar{E}_t \tag{3.25b}$$

Identifying the components in the x- and y-directions, (3.25a) would yield

$$-\beta E_y = \omega \mu_0 \mu_r H_x \tag{3.26a}$$

$$\beta E_x = \omega \mu_0 \mu_r H_y \tag{3.26b}$$

while (3.25b) leads to

$$-\beta H_y = -\omega \varepsilon_0 \varepsilon_{rxx} E_x \tag{3.26c}$$

$$\beta H_x = -\omega \varepsilon_0 \varepsilon_{ryy} E_y \tag{3.26d}$$

Thus, (3.26b) and (3.26c) provide a phase constant given by

$$\beta_x = \omega \sqrt{\varepsilon_0 \mu_0} \sqrt{\varepsilon_{rxx} \mu_r} \tag{3.27a}$$

while (3.26a) and (3.26d) give a different value

$$\beta_y = \omega \sqrt{\varepsilon_0 \mu_0} \sqrt{\varepsilon_{ryy} \mu_r} \tag{3.27b}$$

Therefore, it must be concluded that the propagation of a TEM mode in the longitudinal direction is not possible. In order to obtain a unique definition of the phase constant, so that (D.28c) and (D.28d) can be fulfilled simultaneously, it is necessary to assume the existence of nonzero longitudinal components of the electric and magnetic fields.

Inhomogeneous medium. Consider a lossless, isotropic and inhomogeneous medium. If the propagating mode is TEM and $\gamma = j\beta$, then (D.28c) and (D.28d) do not lead to a unique value for the phase constant but to

$$\beta = \omega \sqrt{\varepsilon_0 \mu_0} \sqrt{\varepsilon_r(x,y) \, \mu_r(x,y)} \tag{3.28}$$

which is a function of the (x,y) coordinates. This result indicates that under these conditions it is not possible to assume zero longitudinal field components.

If the inhomogeneity is due to a discontinuity of the medium, then in order to meet the conditions (D.32) of Appendix D along two media interfaces, the existence of nonzero longitudinal components of the field is necessary. A simple example is the configuration shown in Figure 3.2(b) that represents a microstrip line with two dielectric media. Suppose that the substrate (medium 1) has a relative permittivity ε_r and relative permeability unity and that medium 2 is air with relative permittivity and permeability equal to unity. At the interface one should have

$$E_{x1} = E_{x2} = E_{x12} \,, \quad H_{y1} = H_{y2} = H_{y12} \tag{3.29}$$

Hence from (D.28d) one obtains

$$\frac{\partial H_z}{\partial y} \Big|_1 - \varepsilon_r \frac{\partial H_z}{\partial y} \Big|_2 = \beta(\varepsilon_r - 1) H_{y12} \tag{3.30}$$

Since $H_{y12} \neq 0$, (3.30) implies that, in general, $H_z \neq 0$. Proceeding in the same manner with (D.32b) and (D.32c), the conclusion that $E_z \neq 0$ is obtained. The existence of those longitudinal components is related to the components E_x and H_x of the transverse fields. However, the longitudinal components E_z and H_z are much lower in magnitude than the transverse components. Hence, the propagating mode may be characterized as a quasi-TEM mode [Gupta et al. 1979].

Quasi-TEM Approximation for Anisotropic and Inhomogeneous Two-Conductor Lines. Similar to (3.6), the transverse electric field, $\bar{E}_t$, is given by

$$\bar{E}_t = -\bar{\nabla}_t \phi(x,y) \quad , \quad \text{in } \Omega \tag{3.31a}$$

$$\bar{D}_t = \varepsilon_0 \underline{\underline{\varepsilon}}_{rt}(x,y) \bar{E}_t \quad , \quad \text{in } \Omega \tag{3.31b}$$

$$\bar{\nabla}_t \cdot \bar{D}_t = 0 \quad , \quad \text{in } \Omega \tag{3.31c}$$

with $\phi = 0$ on Γ_{c0}, $\phi = V_1$ on Γ_{c1}, and $\bar{a}_n \cdot \bar{D}_t = 0$ along symmetry walls.

Equations (3.31) lead to the generalized harmonic equation or direct formulation

$$\bar{\nabla}_t \cdot \underline{\underline{\varepsilon}}_{rt} \bar{\nabla}_t \phi = 0 \quad , \quad \text{in } \Omega \tag{3.32}$$

with the outlined Dirichlet conditions on perfect electric walls (ideal conductors), Γ_{c0} and Γ_{c1}, and $\bar{a}_n \cdot \underline{\underline{\varepsilon}}_{rt} \bar{\nabla}_t \phi = 0$ along perfect magnetic walls (symmetry walls).

The transverse magnetic field may be obtained from a vector potential $A(x,y)\bar{a}_z$

$$\bar{B}_t(x,y) = \bar{\nabla}_t \times A(x,y) \bar{a}_z \quad , \quad \text{in } \Omega \tag{3.33a}$$

$$\bar{B}_t = \mu_0 \underline{\underline{\mu}}_{rt}(x,y) \bar{H}_t \quad , \quad \text{in } \Omega \tag{3.33b}$$

$$\bar{\nabla}_t \times \bar{H}_t = 0 \quad , \quad \text{in } \Omega \tag{3.33c}$$

with $A = 0$ on Γ_{c0} and $A = \psi_1$ on Γ_{c1}, where ψ_1 is the lateral magnetic flux between Γ_{c0} and Γ_{c1}, and $\bar{a}_n \times H_t = 0$ on symmetry walls. Here, ψ_1 is given by

$$\psi_1 = \int_{\Gamma_{c0}}^{\Gamma_{c1}} \mu_0 \underline{\underline{\mu}}_{rt} \bar{H}_t \cdot (\overline{dl} \times \bar{a}_z) \tag{3.33d}$$

Expressions (3.33a) to (3.33c) lead to the generalized harmonic equation

$$\bar{\nabla}_t \cdot |\mu_{rt}|^{-1} \underline{\underline{\mu}}_{rt}^T \bar{\nabla}_t A = 0 \quad , \quad \text{in } \Omega \tag{3.34}$$

with the previously mentioned conditions on perfect electric walls and $\bar{a}_n \cdot |\mu_{rt}|^{-1} \underline{\underline{\mu}}_{rt}^T \bar{\nabla}_t A$

= 0 on symmetry walls. Here $\underline{\underline{\mu}}_{rt}^{T}$ stands for a tensor that may be described by a 2 × 2 matrix that is the transpose of the 2 × 2 matrix describing the tensor $\underline{\underline{\mu}}_{rt}$ while $|\mu_{rt}|$ stands for the determinant of those matrices.

There is a formal analogy between the problem (3.32) with its boundary conditions and problem (3.34) with its own. In fact, between these two problems there exists an isomorphism, which allows to formulate the magnetostatic problem as an equivalent electrostatic problem in which the permittivity tensor is given by $\underline{\underline{\varepsilon}}_{rteq} = |\mu_{rt}|^{-1}\underline{\underline{\mu}}_{rt}^{T}$. This method is a generalization of the usual method for treating homogeneous isotropic magnetic media. When dealing with homogeneous nonmagnetic medium, $\varepsilon_{rteq} = 1$ [Djordjević et al. 1989, 1996], [Harrington and Wei 1984], [Venkataraman et al. 1985], [Wei et al. 1984]. For the case of isotropic permeability μ_r, one has $\varepsilon_{rteq} = \mu_r^{-1}$. The formulation of the isomorphism for the general case is due to Horno et al. [1990]. Table 3.1 shows the quantities related in this isomorphism.

Therefore, for the quasi-TEM electromagnetic field characterization, the electrostatic problem (3.31) must be solved either by the direct formulation (3.32) or by any other method that may be deduced from (3.31). If the magnetostatic problem is to be solved it can be done either directly through (3.32) or by any other formulation, or by the mentioned isomorphism. However, in this case (two conductor) there is no need to solve the magnetostatic or the equivalent problem to obtain $\overline{H}_t$ since it can be calculated from $\overline{E}_t$ by means of the expression (3.8) and replacing μ_r by $\underline{\underline{\mu}}_{rt}$.

Table 3.1

Quantities related in the isomorphism between the electrostatic and magnetostatic problems

Electrostatic problem	Magnetostatic problem
$\phi(x,y)$	$A(x,y)$
$\underline{\underline{\varepsilon}}_{rt}$	$\lvert\mu_{rt}\rvert^{-1}\underline{\underline{\mu}}_{rt}^{T}$
Dirichlet b.c. at perfect conductors: V_i, $i = 0, 1$	Dirichlet b.c. at perfect conductors: ψ_i, $i = 0, 1$
Neumann b.c. at symmetry walls: $\overline{a}_n \cdot \underline{\underline{\varepsilon}}_{rt}\overline{\nabla}_t\phi = 0$	Neumann b.c. at symmetry walls: $\overline{a}_n \cdot \lvert\mu_{rt}\rvert^{-1}\underline{\underline{\mu}}_{rt}^{T}\overline{\nabla}_t A = 0$
Normalized surface charge density: $\rho_{sn} = \rho_s/\varepsilon_0$, on Γ_{ci} (see (D.21a))	Normalized surface current density: $J_{sn} = \mu_0 J_s$, on Γ_{ci} (see (D.21b))
Normalized charge: $Q_{ni} = \oint_{\Gamma_{ci}} \rho_{sn} d\Gamma$, $i = 0, 1$, $Q_0 = -Q_1$	Normalized current: $I_{ni} = \oint_{\Gamma_{ci}} J_{sn} d\Gamma$, $i = 0, 1$, $I_0 = -I_1$

Primary parameters may be defined through expressions (3.11), (3.12), (3.13a), (3.14), (3.15), and (3.16a), substituting ε_r' and ε_r'' by $\underline{\underline{\varepsilon}}_{rt}'$ and $\underline{\underline{\varepsilon}}_{rt}''$, respectively; and μ_r' and μ_r'' by $\underline{\underline{\mu}}_{rt}'$ and $\underline{\underline{\mu}}_{rt}''$, respectively. Expressions (3.19) and (3.20) give the propagation constant γ and the characteristic impedance; while the phase velocity is given by, $v_f = \omega/\beta$, where $\beta = \text{Im}[\gamma]$ and Im[·] stands for the imaginary part of the complex quantity between brackets. The expression (3.7b) leads to the use of the parameter termed effective permittivity $\varepsilon_{eff} = (\beta/k_0)^2$, where $k_0 = 2\pi/\lambda_0 = \omega/c_0$ is the wave number in vacuum (or the propagation constant in vacuum for an angular frequency ω), and λ_0 is the corresponding wavelength.

Imperfect conductors

Equations (D.33) to (D.35) of Appendix D provide the boundary conditions that must be satisfied at the surface between an ideal conductor (with infinite conductivity, σ) and a dielectric medium. In that case the electromagnetic field inside the conductor is zero and the current flows on its surface, with a surface density related to the magnitude of the tangential component of the magnetic field on the conductor surface.

When the conductors are imperfect (σ finite) those conditions are no longer valid, since (D.32) has to be satisfied. Therefore, the electromagnetic field penetrates inside the conductor. This indicates that there is a distribution of the current density in the longitudinal direction, i_z. From Ohm's law, this implies that there must be a non-zero longitudinal component of the electric field.

However, for good conductors, the TEM perturbation should be small and the electromagnetic field in the dielectric media may be assumed to be quasi-TEM with approximately zero longitudinal components.

The main consequences of imperfect conductors are ohmic losses (due to the existence of i_z inside the conductors) and magnetic effects (due to the existence of the magnetic field).

Ohmic losses may be characterized by the electric resistance p.u.l., R_c:

$$R_c = \frac{\sum_{k=0}^{1} \int_{\Gamma_{ck}} \text{Re}[Z_s]\bar{J}_s \cdot \bar{J}_s^* \, d\Gamma}{I_1 I_1^*} \tag{3.35}$$

and the magnetic effects by the internal inductance p.u.l., L_c:

$$L_c = \frac{1}{\omega} \frac{\sum_{k=0}^{1} \int_{\Gamma_{ck}} \text{Im}[Z_s]\bar{J}_s \cdot J_s^* \, d\Gamma}{I_1 I_1^*} \tag{3.36}$$

Here $\bar{J}_s = -\bar{a}_n \times \bar{H}$, where $\bar{H}$ is the magnetic field on the surface of the conductor. It has been assumed that the well-known semi-infinite plane conductor approximation may be used (see, e.g., [Ramo et al. 1965; Ch. 4]). This implies that the current density distribution inside the conductors has an exponential variation. Then, Z_s, the p.u.l. and width surface impedance, defined as the ratio between the electric field at the conductor surface and the current density per unit width is given by

$$Z_s = \frac{\tau}{\sigma} \tag{3.37a}$$

where τ is the exponent related to the variation of the current density. For normal conductors one gets

$$Z_s = \frac{1+j}{\delta_s \sigma} \tag{3.37b}$$

where δ_s is the penetration depth given by expression (D.1) of Appendix D. For superconductors Z_s may be considered to be reactive, given by

$$Z_s = j\omega L_s \,, \quad L_s = \mu_c \delta_{ssc} \tag{3.37c}$$

where μ_c stands for the permeability of the conductor (usually considered to be equal to that of vacuum) and δ_{ssc} is the frequency-invariant penetration depth for superconductors (see Appendix D.2) [How et al. 1992].

Since one is dealing with good conductors, a perturbational technique can be used to compute J_s. It consists of assuming ideal conductors. Then one applies a quasi-static approach to compute the electromagnetic fields. Then from the ideal magnetic field H_t at the conductor surface, J_s is obtained.

Once R_c and L_c, are computed, the p.u.l. impedance of the dissipative transmission line, Z, is defined by

$$Z = R_m + R_c + j\omega(L + L_c) \tag{3.38}$$

3.2.2.2 *Multiconductor Transmission Line in an Inhomogeneous Anisotropic Medium with Dielectric and Magnetic Losses and Imperfect Conductors*

The previous formulation may be extended to the case of a line with N_c+1 imperfect conductors surrounded by an inhomogeneous, anisotropic medium with dielectric and magnetic losses [Lindell 1981], [Marqués and Horno 1985], [Santos and Figanier 1975].

The electromagnetic characterization requires the solution of the electrostatic problem given by (3.31) satisfying the conditions $\phi = V_i$ on Γ_{ci}, $i = 0, ..., N_c$ (where V_i

is the potential difference between the ith and the reference conductor) as well as the magnetostatic problem (3.33) satisfying the conditions $A = \psi_i$ on Γ_{ci}, $i = 0, ..., N_c$ (where ψ_i is the magnetostatic flux p.u.l. between the ith and the reference conductor). Both problems are mutually related. Such a relation is determined by the value of the propagation constant γ, according to

$$\gamma \{V\} = j\,\omega \{\psi\} \tag{3.39}$$

where $\{V\}$ and $\{\psi\}$ are column matrices whose ith coefficients are the voltage and magnetostatic flux associated with the ith conductor for the mode being analyzed. Hence $\{V\}$ and $\{\psi\}$ cannot be chosen arbitrarily.

The propagation constant γ of each mode is obtained by solving the eigenvalue problem

$$\gamma^2 \{V\} = [Z][Y]\{V\} \tag{3.40a}$$

or

$$\gamma^2 \{I\} = [Y][Z]\{I\} \tag{3.40b}$$

where $\{I\}$ is the column matrix of currents corresponding to $\{V\}$. The p.u.l. impedance matrix $[Z]$ and the p.u.l. admittance matrix $[Y]$ are given by

$$[Z] = [R_m] + [R_c] + j\omega([L] + [L_c]) \tag{3.41a}$$

$$[Y] = [G] + j\omega[C] \tag{3.41b}$$

Matrices $[C]$, $[L]$, $[G]$, $[R_m]$, $[R_c]$, and $[L_c]$ are symmetric. They are the generalization to the case of the multiconductor line of the primary or distributed parameters C, L, G, R_m, R_c, and L_c of the two-conductor line.

The calculation of the matrices of primary parameters requires the solution of various static problems with the boundary conditions corresponding to the definition of each coefficient.

Calling C_{nij} the C_{ij} coefficient normalized with respect to ε_0, i.e., $C_{nij} = C_{ij}/\varepsilon_0$, and W_{en} the normalized electrostatic energy stored, i.e., $W_{en} = W_e/\varepsilon_0$, one has

$$W_{en} = \frac{1}{2}\{V\}^T[C_n]\{V\} = \frac{1}{2}\int_\Omega \bar{\nabla}_t\phi \cdot \underline{\underline{\varepsilon}}'_{rt}\, \bar{\nabla}_t\phi \; d\Omega \tag{3.42a}$$

where

$$C_{nii} = 2W_{en}\Big|_{\substack{V_0=0\ ,\ V_i=1,\\ V_k=0\ ,\ k\neq i\ ,\ k=1,\cdots,N_c}} , \quad \text{for } i = 1,\cdots,N_c \tag{3.42b}$$

$$C_{nij} = -\frac{1}{2}(C_{nii}+C_{njj}) + W_{en}\Big|_{\substack{V_0=0\ ,\ V_i=1\ ,\ V_j=1\\ V_k=0,\ k\neq i\ ,\ k\neq j\ ,\ k=1,\cdots,N_c}} , \quad \text{for } i,j = 1,\cdots,N_c\ ,\ i\neq j \tag{3.42c}$$

Using a perturbational method, coefficients of $[G_n] = (1/\omega\varepsilon_0)[G]$ is computed from

$$P_{den} = \frac{1}{2}\{V\}^T[G_n]\{V\} = \frac{1}{2}\int_\Omega \bar{\nabla}_t\phi\cdot\underline{\underline{\varepsilon}}''_{rt}\ \bar{\nabla}_t\phi\ d\Omega \tag{3.43}$$

where ϕ is the ideal electrostatic potential obtained by solving problem (3.31) with the boundary conditions corresponding to the definitions of G_{nii} and G_{nij}. These coefficients have expressions formally identical to (3.42b) and (3.42c), respectively, substituting W_{en} by P_{den}. Hence for the computation of matrices $[C]$ and $[G]$ one has to solve a set of $N_c(N_c+1)/2$ electrostatic problems of type (3.31). The first N_c problems require a unity volt excitation on the ith conductor and zero on the rest in order to compute the stored energy and the dissipated power and, thus, C_{ii} and G_{ii}. For the computation of C_{ij} and G_{ij}, $i \neq j$, the rest of the electrostatic problems should have unity excitation on the ith and jth conductors and zero on the rest. Then the stored energy and the dissipated power must be computed in order to obtain the desired coefficients through (3.42c).

Matrix $[L_n] = (1/\mu_0)[L]$ may be obtained using the isomorphism between the electrostatic and the magnetostatic problem by computing matrix $[C_{neq}]$ of the equivalent electrostatic problem. Then

$$[L_n] = [C_{neq}]^{-1} \tag{3.44}$$

Thus, another set of $N_c(N_c+1)/2$ electrostatic equivalent (i.e., with the equivalent permittivity tensor) problems should be solved with identical excitations.

For the computation of matrices $[R_m]$, $[R_c]$, and $[L_c]$ the isomorphism between static problems may be used as well as the perturbational approach. Using the normalization $[R_{mn}] = (1/\omega\varepsilon_0)[R_m]$ one gets

$$R_{mnii} = 2P_{deneq}\Big|_{\substack{V_0=0\\ V_k=L_{nki}\ ,\ k=1,\cdots,N_c}} , \quad \text{for } i = 1,\cdots,N_c \tag{3.45a}$$

$$R_{mnij} = -\frac{1}{2}(R_{mnii}+R_{mnjj}) + P_{deneq}\Big|_{\substack{V_0=0\ ,\\ V_k=L_{nki}+L_{nkj},\ k\neq i\ ,\ k\neq j\ ,\ k=1,\cdots,N_c}} , \quad \text{for } i,j=1,\cdots,N_c\ ,\ i\neq j \tag{3.45b}$$

where $P_{d\varepsilon neq}$ is given by the right hand side of (3.43) when the equivalent tensor is used.

Let $P_{dcneq} = (1/\varepsilon_0\mu_0)P_{dceq}$ be the normalized dissipated power on conductors that may be computed from the equivalent electrostatic problem as

$$P_{dcneq} = \frac{1}{2}\oint_{\Gamma_c} \mathrm{Re}[Z_{sn}]\rho_{sn}^2 \, d\Gamma \tag{3.46}$$

For conductors (i.e., not superconductors) the normalization $[R_{cn}] = a(2/\mu_0 f^{1/2})[R_c]$ is used, where a is a geometric normalization factor. Then the expressions for the computation of R_{cnii} and R_{cnij} are formally identical to (3.45a) and (3.45b) with $P_{d\varepsilon neq}$ substituted by P_{dcneq}.

Let $P'_{dcneq} = (1/\varepsilon_0\mu_0)P'_{dceq}$ be the reactive power on conductors given by

$$P'_{dcneq} = \frac{1}{2}\oint_{\Gamma_c} \mathrm{Im}[Z_{sn}]\rho_{sn}^2 \, d\Gamma \tag{3.47}$$

As $[X_c] = \omega[L_c]$, and using the normalization $[X_{cn}] = (a/\omega\mu_0 b)[X_c]$, where $b = 1/(4\pi f^{1/2})$ for normal conductors and $b = 1$ for superconductors, the expressions for the computation of X_{cnii} and X_{cnij} are formally identical to (3.45a) and (3.45b) with $P_{d\varepsilon neq}$ substituted by P'_{dcneq}.

Thus, to compute $[R_m]$, $[R_c]$, and $[L_c]$ one needs to solve another set of $N_c(N_c+1)/2$ electrostatic equivalent problems (i.e., with the equivalent permittivity tensor). Now the first N_c problems should have excitations given by $V_0 = 0$ and $V_k = L_{nki}$, for $k,i = 1, ..., N_c$, in order to compute the iith diagonal coefficients. The rest of the problems should have excitations given by $V_0 = 0$ and $V_k = L_{nki}+L_{nkj}$, for $k,i,j = 1, ..., N_c$, $k \neq i$, $k \neq j$, $i \neq j$, in order to obtain the ijth coefficients.

The number of problems to be solved for the computation of the primary parameters may be reduced when special conditions hold, like for a lossless structure with homogeneous and isotropic permittivity and permeability (in which the N_c eigenvalues are identical and one has $[L][C] = 1/v_f^2$) or for structures with one (or more) symmetry axis. It may be noticed that the definitions for energy or dissipated power utilizing potentials have been used in (3.42a) and (3.43). This is the proper definition for the direct formulation in terms of the scalar potential. However the number of problems in each set may also be reduced if the definition for charge is utilized. This is also the proper setting for the mixed formulation in terms of the displacement vector. The reader is referred to [Djordjević et al. 1989, 1996] where details are given about the relevant expressions for this case.

Having obtained the matrices of primary parameters, the eigenvalue problem (3.40a), can be solved computing the eigenvector $\{V\}$ corresponding to each eigenvalue, γ^2. The matrix $[\gamma]$ is defined as a diagonal matrix in which the nonzero coefficient of the jth column is the square root of the jth eigenvalue, i.e., the jth propagation constant (i.e., the propagation constant of the *j*th mode). The voltage matrix $[M_v]$ is defined as the matrix in which the *j*th column is the voltage eigenvector.

The characteristic impedance matrix $[Z_c]$ is defined as

$$[Z_c] = [M_v][\gamma][M_v]^{-1}[Y]^{-1} \tag{3.48a}$$

In a similar way, as in the case of a two-conductor line, the following relation holds:

$$[M_v] = [Z_c][M_i'] \tag{3.48b}$$

where $[M_i']$ is the current matrix associated with the voltage matrix $[M_v]$.

One may prefer to solve the eigenvalue problem (3.40b), obtaining $[\gamma]$ and $[M_i]$. Then $[Z_c]$ is defined as

$$[Z_c] = [Z][M_i][\gamma]^{-1}[M_i]^{-1} \tag{3.49a}$$

and the following relation holds:

$$[M_v'] = [Z_c][M_i] \tag{3.49b}$$

where $[M_v']$ is the voltage matrix associated with the current matrix $[M_i]$.

If the field configuration of the *j*th mode is required, it could be obtained by solving the electrostatic problem (3.32) with the boundary conditions given by the *j*th eigenvector $\{V\}$ of the eigenvalue problem (3.40a). The magnetic field $\bar{H}_t$ may be obtained from (3.8), by replacing μ_r with $\underline{\mu}_{rt}$ and using the propagation constant γ of the *j*th mode.

Alternatively the magnetostatic problem (3.34) can be solved with the boundary conditions given by the eigenvector $\{\psi\}$ as computed from (3.39) for the *j*th mode. Then the electric field $\bar{E}_t$ can be obtained from the magnetic field.

From $[\gamma]$, $[Z_c]$, and the length l of the transmission line, any network parameter, $[z]$, $[y]$, *ABCD* or $[s]$, or the matrices of modal impedance, $[Z_m]$, or diagonalized modal impedance, $[Z_{md}]$, may be computed. In [Gentili and Salazar-Palma 1995] a study of the various definitions of these matrices may be found.

3.2.3 Application of the Finite Element Method to the Quasi-Static Analysis of Transmission Lines

The electromagnetic and circuital characterizations of a multiconductor transmission line require the analysis of one or more sets of electrostatic problems with specific values for the dielectric permittivity of the medium and for the boundary conditions. As pointed out previously for the solution of the electromagnetic problem the medium as well as the conductors are considered to be lossless.

This section deals with the application of various FEMs to the quasi-static analysis of transmission lines. The numerical techniques are validated by applying them to different test structures. First, the direct formulation that uses a FEM based on Lagrange elements is presented. The unknown variable is the electric scalar potential. The rate of convergence is analyzed by applying the method to several transmission line structures using different meshes consisting of straight and curved triangular and/or rectangular elements, finite and infinite elements (for the case of open structures), and different orders of the polynomial approximation. Questions related to the computational methods such as the number of integration points and solvers are also addressed.

With the purpose of finding error bounds, the dual direct formulation of FEM which uses the magnetic vector potential as unknown, has also been considered. This formulation is easily applicable to a particular class of transmission line structures. However, it is difficult to apply to general geometries. It also uses a Lagrange-based FEM.

Next, the mixed formulation, where the unknowns are the scalar electric potential and the displacement vector, is presented. It uses a FEM based on div-conforming elements. This formulation is an alternative to the dual formulation, since these two formulations provide bounds of the same nature. However, the mixed formulation is easily applicable to any geometry. Hence, the use of both the primal direct and the mixed formulations allows one to obtain upper and lower bounds of the parameters characterizing a transmission line.

3.2.3.1 *Application of the Finite Element Method to the Direct Formulation*

The direct formulation of the quasi-static analysis of a transmission line with N_c+1 conductors embedded in an inhomogeneous and anisotropic medium characterized by $\underline{\underline{\varepsilon}}_{rt}(x,y)$ is given by (3.32). The unknown variable in this case is the electrostatic potential $\phi(x,y)$. For a given cross section of the structure Ω (i.e., the 2D domain of the problem), its boundary Γ is defined by $\Gamma = \Gamma_{pe} \cup \Gamma_{pm}$, where $\Gamma_{pe} = \Gamma_c = \cup \Gamma_{ci}$ for $i = 0, 1, \ldots, N_c$, where Γ_{ci} is the trace in the *x-y* plane of the *i*th conductor, and Γ_{pm} stands for perfect magnetic walls or symmetry walls. The mathematical formulation of the problem is as follows:

Given $\underline{\underline{\varepsilon}}_{rt}$ *in* Ω, *find* ϕ *so that* $\phi = V_i$ *on* Γ_{ci}, $i = 0, \ldots, N_c$, *and*

$$-\bar{\nabla}_t \cdot \varepsilon_0 \underline{\underline{\varepsilon}}_{rt} \bar{\nabla}_t \phi = 0 \ , \quad in\ \Omega \tag{3.50a}$$

$$\bar{a}_n \cdot \varepsilon_0 \underline{\underline{\varepsilon}}_{rt} \bar{\nabla}_t \phi = 0 \ , \quad on\ \Gamma_{pm} . \tag{3.50b}$$

Appendix E.2.1 details the classical (also called strong) and the weak formulation. The latter may be obtained minimizing the functional

$$W(\phi) = \frac{1}{2}\int_{\Omega} \bar{\nabla}_t\phi \cdot \varepsilon_0 \underline{\underline{\varepsilon}}_{rt}\, \bar{\nabla}_t\phi \; d\,\Omega \tag{3.50c}$$

which represents the electrostatic stored energy. In Appendix E.2.1 references of FEM works based on this functional may be found.

Appendix E.2.2 describes the discretization of such a formulation using a Lagrange element-based FEM after applying the Rayleigh-Ritz method to the functional (3.50c). The choice of Lagrange elements is justified by the fact that $\phi(x,y)$ must be a continuous function over Ω. Its first derivatives must be square integrable. In fact, Lagrange elements provide a continuous solution with continuous tangential derivatives and discontinuous normal derivatives at element interfaces. Thus, the electric field that is derived from the electrostatic potential will be square integrable with continuous tangential components and discontinuous normal components across element interfaces. Thus, the solution will provide the continuity of the tangential component of the electric field across media interfaces, while allowing the normal component to be discontinuous at media interfaces. The continuity of the normal component of the displacement vector across symmetry walls will not be exactly accomplished because it is a natural condition for the formulation, according to (3.50b). Similarly, the continuity of the normal component of the displacement vector across elements and media interfaces is not exactly accomplished, but the method will conform more and more with refined meshes.

Discretization Through FEM

The discretization of the problem is carried out according to the procedure detailed in Sections 2.5 and 2.6. The mesh generation is performed by means of an improved version of the MODULEF semiautomatic 2D mesh generator [Bernadou et al. 1985]. A simplified automatic mesh generator has been used, just requiring the introduction of the structure geometry and configuration data, through a graphical interface. In both cases, structured and unstructured meshes may be generated. The second type is based on Delaunay-Voronoi procedures [Delaunay 1934], [Watson 1981].

All the rectangular and triangular elements in Tables 2.1 and 2.2 of Chapter 2, respectively, are used in their straight and curved versions, except for the first-order approximation where only straight elements have been utilized to avoid the convergence problems for superparametric elements. The infinite elements generated by the geometric mappings given in Table 2.5 of Chapter 2 are used for open structures. Thus, a total of nineteen Lagrange elements have been used.

After generating the mesh and defining the interpolation polynomials, the computation of the stiffness, excitation, and postprocesses matrices of each element is carried out. The possible postprocesses are: calculation of the electrostatic stored energy, dissipated power in the dielectric media, dissipated power on the conductors, components of the electric field, and, eventually, the components of the magnetic field as well as the local error estimate. The latter will be dealt with in Chapter 4 and Appendix G. The rest of the postprocesses are described in the following. All computations related to the elements are carried out through the parent element (see Appendix C.1). Thus, most of the calculations may be precomputed. For each element, only factors related to the coefficients of the Jacobian matrix need to be computed and multiplied with the precomputed ones. Table 3.2 shows the number of integration points used to program the three sets of elementary computations for each of the nineteen elements utilized. Analytical integration has been used in the case of all triangular straight elements (first- order isoparametric element and higher order subparametric elements).

Table 3.2

Number of points of integration for the calculation of the coefficients of the stiffness matrix

Ordinary finite elements	Triangular elements	P(2) isoparametric	3 points
		$P(3')$ isoparametric	6 points
		$P(3)$ isoparametric	6 and 7 points
	Rectangular elements	$Q(1)$ isoparametric	2×2 points
		$Q(2')$ sub and isoparametric	2×2 and 3×3 points
		$Q(2)$ sub and isoparametric	2×2 and 3×3 points
		$Q(3')$ sub and isoparametric	3×3 and 4×4 points
		$Q(3)$ sub and isoparametric	3×3 and 4×4 points
Infinite elements	$Q(1)$ with I(6)		2×2 and 3×3 points
	$Q(2')$ with I(6)		2×2 and 3×3 points
	$Q(2)$ with I(6)		2×2 and 3×3 points

For the assembly of the stiffness matrix and the excitation vector as well as for the imposition of the essential boundary conditions, MODULEF routines, conveniently modified, have been used. Renumbering techniques for nodes and elements have been applied in order to reduce the band width of the global matrix. Since the stiffness matrix is symmetric, solvers especially designed for sparse symmetric matrices from the MODULEF library have been applied. These are Choleski and conjugate gradient methods (without preconditioning and with preconditioning, through an incomplete Choleski factorization or through relaxation). The first method uses skyline storage. The second method is based on nonzero coefficient storage. The solution of the global system of equations provides the degrees of freedom for ϕ at the nodes. The value of the electrostatic potential at any point of the domain is obtained through interpolation in the element to which the given point belongs.

Postprocessing the Solution

Once the degrees of freedom of the unknown have been solved, the postprocessing of the solution may be carried out. The following computations have been programmed for straight and curved triangular elements of first and second order.

1. The electrostatic stored energy normalized with respect to ε_0 for the computation of the coefficients of matrices $[C]$ and $[L]$ (in the equivalent electrostatic problem) is obtained from

$$W_{en} = \frac{1}{2}\{D\}^T[K_{nbc}]\{D\} \tag{3.51}$$

where $[K_{nbc}]$ is the assembled stiffness matrix before the imposition of the essential boundary conditions and $\{D\}$ is the global vector of degrees of freedom. Hence, this computation does not require local postprocess matrices but only the global matrix $[K_{nbc}]$.

2. The calculation of the dissipated power in the dielectric media (hence, over Ω) normalized with respect to $\omega\varepsilon_0$, for the computation of the coefficients of matrices $[G]$ and $[R_m]$ (for the equivalent electrostatic problem), is obtained from

$$P_{den} = \sum_{e=1}^{N_e} P_{dene} \tag{3.52}$$

where N_e is the total number of elements and P_{dene} is the dissipated power in the eth element, given by

$$P_{dene} = \frac{1}{2}\{d^e\}^T[k''^e]\{d^e\} \tag{3.53a}$$

where $\{d^e\}$ are the degrees of freedom of the eth element and $[k''^e]$ is

$$[k''^e] = \left[\int_{\Omega^e}[\partial N]^T[\varepsilon''_{rt}][\partial N]\,d\Omega\right] \tag{3.53b}$$

Here $[\varepsilon_{rt}'']$ is the matrix of the imaginary part of $[\varepsilon_{rt}]$, and $[\partial N]$ is the matrix of the derivatives of the basis functions defined in Appendix E, (E.15d). The postprocess matrix that must be calculated for each element is formally identical to the local stiffness matrix $[k^e]$ given by Appendix E, (E.15b), where matrix $[\varepsilon_{rt}]$ must be substituted by $[\varepsilon_{rt}'']$. Computations are done in the parent element.

3. The power dissipated in imperfect conductors normalized with respect to $a/(\omega\mu_0 b)$ (a being a geometric normalization parameter, eventually used for the structure description, and $b = 1$, for superconductors or $b = 1/(4\pi f^{1/2})$ for normal conductors) for the computation of matrix $[R_c]$ is calculated from

$$P_{dcn} = \sum_{i=1}^{N_{edc}} P_{dcni} \tag{3.54}$$

where N_{edc} is the total number of edges in the mesh that are conductors and P_{dcni} is the dissipated power on the ith edge/conductor belonging to the eth element

$$P_{dcni} = \frac{1}{2}\{d^e\}^T[c_{ci}^e]\{d^e\} \tag{3.55a}$$

where $[c_{ci}^e]$ is given by

$$[c_{ci}^e] = R_{si}\left[\int_{\Gamma_{edi}^e}[\partial N]^T[\varepsilon_{rteq}]\{a_n\}\{a_n\}^T[\varepsilon_{rteq}][\partial N]\,\mathrm{d}\Gamma\right] \tag{3.55b}$$

and R_{si} is the surface resistance for the ith edge/conductor (set equal to zero, if the edge is a perfect conductor). $\{a_n\}$ is the matrix notation for the unit vector normal to the edge and $[\varepsilon_{rteq}]$ is the equivalent relative dielectric constant of the element to which the edge belongs. The coefficients of the matrix $[c_{ci}^e]$ are calculated using numerical integration at one or two Gauss-Legendre points depending on whether it is a first- or second-order element.

4. A formally identical procedure is carried out for the calculation of the coefficients of the matrix $[L_c]$ (the internal inductance matrix) in order to obtain ωW_{mcn}, where W_{mcn} is the magnetic energy stored inside the conductors using the same normalizations. R_{si} is substituted by $\mathrm{Im}[Z_{si}]$, the imaginary part of the surface impedance for the ith edge. For the case of normal conductors $\omega W_{mcn} = P_{dcn}$, and hence it is not necessary to perform any additional postprocess.

5. The following expressions are used to calculate the electric and magnetic fields at the eth element, according to (3.6a) and (3.8), respectively:

$$\{\bar{E}_t^e\} = -[\partial N^e]\{d^e\} \tag{3.56}$$

$$\{\bar{H}_t^e\} = \frac{1}{v_f\mu_0}[\mu_{rt}^e]^{-1}\begin{bmatrix}0 & -1\\ 1 & 0\end{bmatrix}\{\bar{E}_t^e\} \tag{3.57}$$

where v_f is the phase velocity of the mode whose field is being computed. As it can be noticed these expressions do not involve any integration but derivatives. They may be computed at any point of the domain. However, they will provide discontinuous results across element interfaces.

In many cases the value of the field components is required at the nodes of the mesh. Then, in order to provide a continuous field (except at dielectric interfaces, where the requirements are that the tangential component of the electric field as well as the normal component of the displacement vector should be continuous), the following smoothing algorithms may be applied:

a. For each element, compute the electric field given by (3.56) at the nodes. Assign to each node the mean value obtained by averaging the local value computed for each element sharing it. This average may be simple or weighted by the area of each element. This procedure is implemented enforcing the continuity of the tangential components of the electric field and the normal components of the smoothed electric displacement vector at dielectric interfaces.
b. For each element, compute the electric field given by (3.56) at the optimum points (the Gauss-Legendre integration points). Then, from those values, interpolate the field values at the nodes of the element. Then assign to each node the simple or weighted average of the values obtained from each element sharing it as for the previous algorithm.
c. For each element, apply a smoothing by means of a local least square method that consists of calculating $\{E_t^e\}$, not by (3.56), but through

$$\{\bar{E}_t'^{\,e}\} = -[N_{ext}^e]\{d_E^e\} \tag{3.58a}$$

where $[N_{ext}^e]$ is the following extended matrix of basis functions:

$$[N_{ext}^e] = \begin{bmatrix} N_1^e & 0 & N_2^e & 0 & N_3^e & 0 & \cdots \\ 0 & N_1^e & 0 & N_2^e & 0 & N_3^e & \cdots \end{bmatrix} \tag{3.58b}$$

and $\{d_E^e\}$ are the degrees of freedom of this local problem:

$$\{d_E^e\} = \left[\, d_{Ex_1}^e \;\; d_{Ey_1}^e \;\; d_{Ex_2}^e \;\; d_{Ey_2}^e \;\; d_{Ex_3}^e \;\; d_{Ey_3}^e \;\; \cdots \right]^T \tag{3.58c}$$

The unknown $\{d_E^e\}$ is calculated using a least squares approach as

$$\left[\int_{\Omega^e} [N_{ext}]^T [N_{ext}]\, d\Omega\right]\{d_E^e\} = \left[-\int_{\Omega^e} [N_{ext}]^T [\partial N]\, d\Omega\right]\{d^e\} \tag{3.59}$$

Then $\{d_E^e\}$ is obtained from the following equations:

$$\{d_E^e\} = -[m_E^e]^{-1}[p^e]\{d^e\} \tag{3.60a}$$

where matrices $[m_E^e]$, $[p^e]$ may be computed over the parent element (see Appendix C.1)

$$[m_E^e] = \int_{\Omega^e}[N_{ext}]^T[N_{ext}]\,\mathrm{d}\Omega = \int_{\hat{\Omega}}[\hat{N}_{ext}]^T[\hat{N}_{ext}]|J^e|\,\mathrm{d}\hat{\Omega} \tag{3.60b}$$

$$-[p^e] = \int_{\Omega^e}[N_{ext}]^T[\partial N]\,\mathrm{d}\Omega = \int_{\hat{\Omega}}[\hat{N}_{ext}]^T([J^e]^T)^{-1}[\partial\hat{N}]\,|J^e|\,\mathrm{d}\hat{\Omega} \tag{3.60c}$$

Here the symbol ^ refers to the parent element. In summary, the postprocess matrix $-[m_E^e]^{-1}[p^e]$ may be precomputed and stored when performing the local computations. Then, once the degrees of freedom of the solution are computed, the local degrees of freedom for each element, $\{d^e\}$, may be obtained. Thus, the column matrix $\{d_E^e\}$ may be computed using expression (3.60a). The local least-square smoothed values of the field $\{E_t'^e\}$ are computed from (3.58a). Finally, a simple or weighted average value is assigned to each node.

Rate of Convergence

In this formulation, since one minimizes the energy functional (3.50), it seems natural to measure the error incurred by means of the energy norm. Then, according to Section 2.5.3, because Lagrange elements are used, the rate of convergence is given by h^p provided that the potential function ϕ meets the requirements of smoothness and that regular and quasi-uniform meshes are used. Here, h is the maximum diameter of the elements and p is the order of the polynomial approximation. However, in most of the transmission line configurations, the potential ϕ is not sufficiently smooth. In Section 2.5.3 it was pointed out that in such cases it is better to use the total number of degrees of freedom, N_d, as the parameter for the calculation of the rate of convergence. The following expression holds for 2D quasi-uniform meshes:

$$\|e\|_E \leq C_1 N_d^{-\frac{1}{2}\min(p,\alpha)} \tag{3.61}$$

where $\|e\|_E$ is the energy norm of the error, C_1 is a constant that does not depend on N_d nor on p. α is the measure of the regularity or smoothness of the exact solution, which may be lower than unity. Hence, if the function is smooth enough, $p < \alpha$, then the rate of convergence will be given by the order of the polynomials used. Conversely, if it is determined by α, the rate of convergence can turn out to be quite low.

For such cases, equilibrated meshes should be used. These are meshes in which the error at each element is approximately the same. Thus, in regions with higher local error, elements should be smaller than in regions where the solution is smooth enough. The measure of the error is given by

$$\|e\|_E \leq C_2 N_d^{-p/2} \tag{3.62}$$

that is, the same rate of convergence as that which would be obtained with quasi-uniform meshes and a smooth solution, i.e., the maximum possible rate of convergence for the h version of FEM.

If the order of polynomials is increased (p version), then one obtains

$$\|e\|_E \leq C_3 N_d^{-\alpha} \tag{3.63}$$

where α measures the regularity of the exact solution. Thus, for nonsmooth solutions, it will be at least twice that of the h version for quasi-uniform meshes.

If both types of enhancement of the mesh are combined, it is possible to reach exponential rates of convergence as detailed in Section 2.5.3.

The most important sources of lack of smoothness in transmission lines are the corners in conductors and dielectrics, discontinuities in permittivity or permeability, discontinuities in boundary conditions, coupling, and proximity effects. In the literature, attention has been focused only on the effects of corners associated with conductors and dielectrics. However, it is important to consider all the other sources of lack of smoothness that are present in most of the structures being addressed.

The literature about the order of the regularity of the solution of a problem and, more precisely, of the singularity, when conductor and dielectric corners are present is very broad. Different geometries have been approximately analyzed, like metallic edges of different angles surrounded by various anisotropic dielectrics, possibly lossy, and dielectric edges [Beker 1991], [Bressan and Gamba 1994], [Burais et al. 1985], [Chan et al. 1988], [Geisel et al. 1992], [Hurd 1976], [Lang 1973], [Marchetti and Rozzi 1991a,b], [Meixner 1972], [Mittra and Lee 1971], [Miyata et al. 1989], [Seshadri et al. 1980], [Van Bladel 1991]. In any case, results from these studies are only approximations to the problem, since edges are considered to be semi-infinite and isolated and proximity effects between conductors are not included. On the other hand, it is not always possible to obtain the analytical solution of the transcendental equations involved that would provide the order of regularity of the solution. Moreover, in some cases the solution is not unique.

For some simple configurations, the order of the regularity can be estimated. For instance, in quasi-static problems having a metallic corner of internal angle $\varphi = \pi/2$, the potential ϕ has an asymptotic behavior at points close to the corner of the type $r^{2/3}$ (i.e., $\alpha = 2/3$). Here, r is the distance from the corner to the point where the potential is measured.This indicates that its derivative (and therefore, the electric field) will have a behavior like $r^{-1/3}$. Hence, it becomes singular as r tends to zero in the proximity of the corner with an order given by -1/3. If the internal angle is zero, the behavior is of type $r^{1/2}$. This is the worst case, since its derivative will have a behavior of type $r^{-1/2}$. If the corner is surrounded by one or more dielectrics the order of the singularity will be different. The order of the singularities of dielectric corners may be estimated similarly. Singular elements are based on these estimates. The idea is to model the asymptotic behavior near corners either by modifying appropriately the basis functions or by using suitable geometric transformations. The use of singular elements does not deal with the effect of the singularity in regions distant from it. Additionally, they fail to consider other sources of nonsmoothness, namely, coupling and proximity effects, for example. In Section 2.5.1.2 references on the topic are given. The self-adaptive procedure proposed in Chapter 5 considers all the sources of nonsmoothness. Its rate of convergence is given by (3.62).

Convergence Study

In this section the FEM is applied to the analysis of various structures in order to perform a study of the numerical convergence of the procedure. The run-time shown in the tables to follow refers to various computers. Namely two different PC 486, at 2×50 MHz, one with 8 Megabytes of RAM, the other with 16 Megabytes of RAM, and a Pentium at 150 MHz with 16 Megabytes of RAM. In all cases swapping is used. Thus, no attempt has been made to compare time results from one example to the other. Nevertheless the run-time data are reported for the relative information they provide.

a) *Circular coaxial line*

First a structure having a smooth solution for ϕ has been considered. It is a circular coaxial cable having an inner conductor of radius a, shielded by an external conductor of radius $2a$. The medium in between is vacuum and the conductors are assumed to be perfect. Because there are only two conductors and the medium is homogeneous and isotropic, it is only necessary to obtain the capacity p.u.l., C, in order to completely characterize the structure electrically. The phase velocity, v_f, is that of vacuum, $v_f = c_0$. The inductance p.u.l. is given by $L = 1/(v_f^2 C)$, and the characteristic impedance may be written as $Z_c = 1/(v_f C)$. The exact result for the capacitance is $C = 2\pi\varepsilon_0/(\ln 2)$ F/m = $9.064720284\varepsilon_0$ F/m [Kraus 1992; Ch. 4]. For the analysis, triangular and rectangular elements of first-, second- and third- order have been used. One-fourth of the structure has been analyzed. In all cases, the FEM procedure starts from a coarse mesh with few elements. Then it is uniformly refined by subdividing each element into four, until an error $\leq 0.1\%$ is achieved.

Figure 3.5 shows the initial triangular mesh as well as the mesh corresponding to the third step of uniform refinement. Table 3.3 gives the results. Figure 3.6 shows the rate of convergence for h refinements for different degrees of the polynomial approximation. The vertical axis represents, in logarithmic scale, the energy norm of the relative error. It is computed from

$$(e_r)_E = \frac{\| e \|_E}{\| \phi \|_E} = \left| \frac{\tilde{C} - C}{C} \right|^{1/2} \tag{3.64}$$

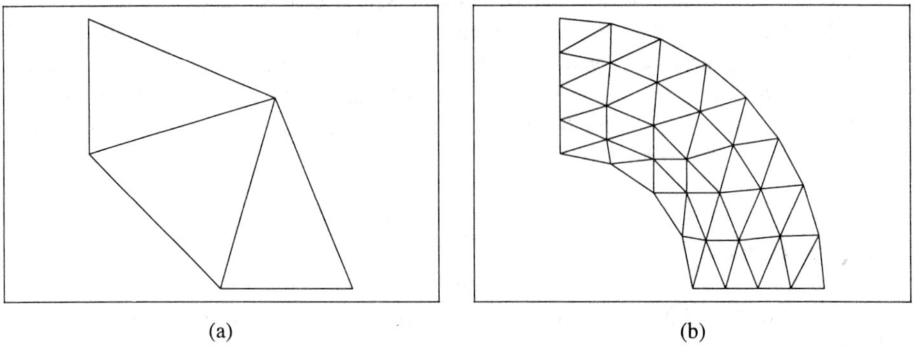

Figure 3.5 Mesh of one-fourth of the circular coaxial line utilizing triangular elements: (a) initial mesh and (b) mesh after the third iteration using uniform refinements.

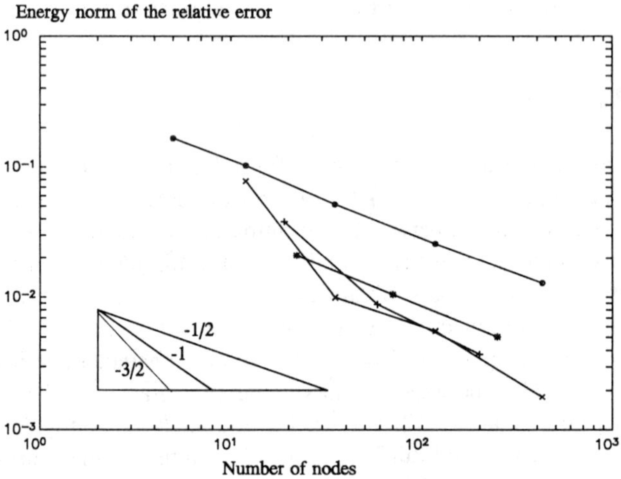

Figure 3.6 Rate of convergence for h refinements for the analysis of the circular coaxial line using triangular elements. o: first-order (straight elements); ×: second-order (straight and curved elements); +: serendipity third-order (straight and curved elements); *: complete third-order (straight and curved elements).

Table 3.3
Results for the analysis of the circular coaxial line using triangular elements and uniform refinements

Order	Mesh step⇒ No. elements	No. nodes	C/ε_0	Relative error (%)	Run-time (sec)
First (straight elements)	1⇒3 e.	5	8.815086	-2.754	0.99
	2⇒12 e.	12	8.969689	-1.048	0.99
	3⇒48 e.	35	9.040583	-0.266	1.21
	4⇒192 e.	117	9.058740	-0.066	2.09
	5⇒768 e.	425	9.063235	-0.016	16.26
Second (straight and curved elements)	1⇒3 e.	12	9.009589	-0.608	1.26
	2⇒12 e.	35	9.065617	0.01	1.37
	3⇒48 e.	117	9.065000	0.003	1.92
	4⇒192 e.	425	9.064749	0.0003	18.01
Third serendipity (s. and c. e.)	1⇒3 e.	19	9.051597	-0.144	1.54
	2⇒12 e.	58	9.065426	0.007	1.92
	3⇒48 e.	199	9.064846	0.001	3.79
Third complete (s. and c. e.) (6 int. p.)	1⇒3 e.	22	9.048844	-0.175	2.58
	2⇒12 e.	70	9.063414	-0.014	4.34
	3⇒48 e.	247	9.064603	-0.001	5.21
Third complete (s. and c. e.) (7 int. p.)	1⇒3 e.	22	9.060779	-0.043	2.65
	2⇒12 e.	70	9.063728	-0.011	4.65
	3⇒48 e.	247	9.064490	-0.002	5.50

where C is the exact value of the capacitance and $\tilde{C}$ is the calculated one. The horizontal axis, also in logarithmic scale, represents the total number of degrees of freedom N_d. The inset gives the theoretical slopes according to (3.61) and (3.62) for first-, second-, and third-order of the polynomial approximation. Thus, the slope of the straight lines of the inset are, respectively, -1/2, -1, -3/2. The asymptotic rate of convergence is reached except for the case of the complete third-order element, where a saturation effect may be observed. It should be mentioned that although the elements of the meshes of Figure 3.5 are depicted as straight elements, curved (isoparametric) second-order and third-order elements are used along the curved boundaries. It may be observed that the convergence of the serendipity third-order element is of second-order type, as mentioned in Section 2.5.1.2. For the complete third-order element, it is shown that more accurate results are obtained with seven integration points for a low number of nodes while six points of integration give a more accurate result for a higher number of unknowns. An interesting result is that in problems with curved boundaries, the FEM solution does not always provide an upper bound of the exact value. It depends on the type of elements and on the total number of degrees of freedom. Figure 3.7(a) shows the results for the rate of convergence for p refinements. The curves correspond to the initial mesh (3 triangles), where first-, second-, and third- (serendipity and complete) order elements have been used successively. In agreement with (3.63), the rate of convergence is higher than unity ($\alpha>1$, for this problem) and is better than for *h* refinements. Figure 3.7(b) shows results for the rate of convergence of the *h-p* version. It has been obtained starting with the initial mesh of Figure 3.6 and first-order elements. Then a uniform mesh refinement is performed. Second-order elements are used over this refined mesh. A uniform mesh refinement takes place and finally third-order serendipity and complete elements are utilized. The expected exponential rate of convergence is obtained.

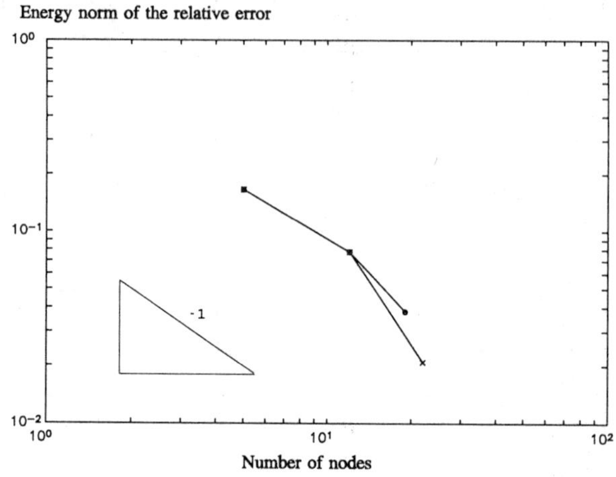

(a)

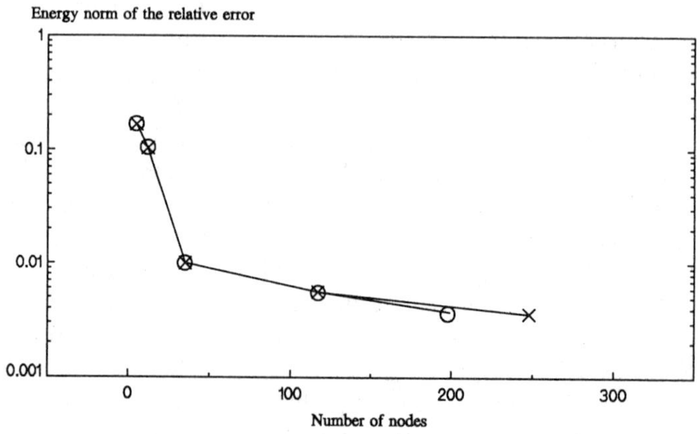

(b)

Figure 3.7 Rate of convergence for the analysis of the circular coaxial line: (a) *p*-version, and (b) *h-p* version. Final data: o: third-order serendipity; ×: third-order complete.

Figure 3.8 shows the initial mesh for one-fourth of the circular coaxial structure using rectangular elements and the mesh obtained with uniform refinements, after third iteration. Results using elements of first-, second- (serendipity and complete), and third- (serendipity and complete) order are shown in Table 3.4. For each case, a comparison for two different number of points of integration (4 and 9 for first- and second-order, 9 and 16 for third-order) is made for the last mesh.

Figure 3.9 gives the asymptotic rate of convergence for h refinements. It can be observed that the theoretical predictions are verified, except for third-order elements,

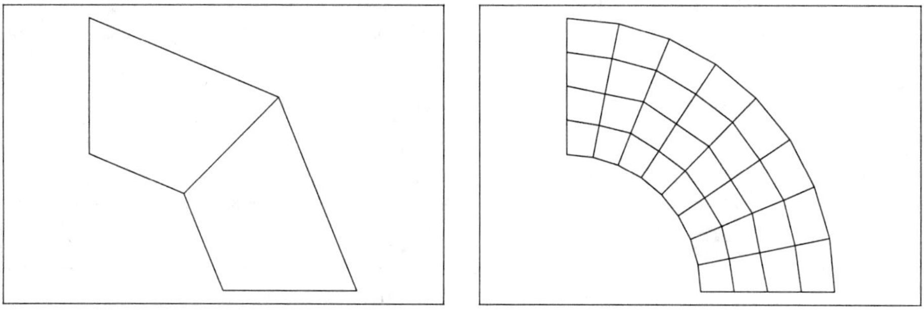

(a) (b)

Figure 3.8 Mesh of one-fourth of the circular coaxial line utilizing rectangular elements: (a) initial mesh and (b) mesh after the third iteration using uniform refinements.

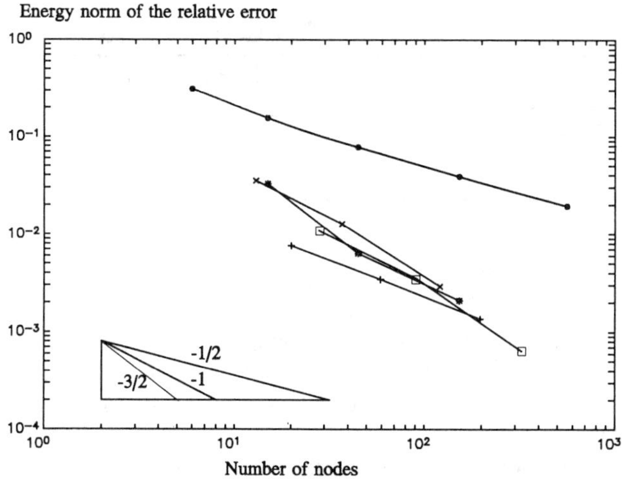

Figure 3.9 Rate of convergence for h refinements for the analysis of the circular coaxial line utilizing rectangular elements. o: first-order; ×: serendipity second-order; *: complete second-order; +: serendipity third-order; □: complete third-order.

Table 3.4
Results for the analysis of the circular coaxial line using rectangular elements and uniform refinements

Order	Mesh step⇒ No. elements	No. nodes	C/ε_0	Relative error (%)	Run-time (sec)
First o.	1⇒2 e.	6	9.941125	9.67	1.05
	2⇒8 e.	15	9.282577	2.403	1.15
	3⇒32 e.	45	9.118995	0.59	1.27
	4⇒128 e.	153	9.078263	0.15	2.25
4 int. p.	5⇒512 e.	561	9.068106	0.037	32.95
9 int. p.	5⇒512 e.	561	9.068104	0.037	33.3
Second o.	1⇒2 e.	13	9.075829	0.122	1.48
Serendipity	2⇒8 e.	37	9.066177	0.016	1.53
4 int. p.	3⇒32 e.	121	9.064797	0.0008	2.53
9 int. p.	3⇒32 e.	121	9.064797	0.0008	2.69
Second o.	1⇒2 e.	15	9.074317	0.106	1.54
Complete	2⇒8 e.	45	9.065083	0.004	1.65
4 int. p.	3⇒32 e.	153	9.064688	0.0003	2.03
9 int. p.	3⇒32 e.	153	9.06476	0.0004	2.15
Third o.	1⇒2 e.	20	9.065241	0.005	1.54
Serendipity	2⇒8 e.	59	9.064613	-0.001	1.81
9 int. p.	3⇒32 e.	197	9.064736	0.0002	4.67
16 int. p.	3⇒32 e.	197	9.064737	0.0002	4.85
Third o.	1⇒2 e.	28	9.065776	0.011	1.59
Complete	2⇒8 e.	91	9.064832	0.0012	2.08
9 int. p.	3⇒32 e.	325	9.064724	0.00004	3.3
16 int. p.	3⇒32 e.	325	9.064727	0.00007	3.79

in which case a saturation effect is also shown.

Except for first-order, rectangular elements give more accurate results than triangular ones. They seem to be better adapted to the symmetry of the solution (see Section 2.5.1.1).

From the previous results it could be concluded that the use of rectangular elements of an order higher than two does not signify a noticeable improvement since the relative error obtained with second order is sufficiently low. Using third-order elements results in a considerable increase of the degrees of freedom, without a considerable improvement in accuracy. This confirms what was previously mentioned in Section 2.5.1.2.

It can also be concluded that it is enough to choose the lower number of integration points for each order, namely, 4 for first- and second-order and 9 for third-order elements.

b) ***Square coaxial line***

Figure 3.10 shows the cross section of the structure analyzed next. It is a square coaxial line whose inner conductor shape is a square of side a and outer conductor is another square of side $2a$. The dielectric in between is free space. Like the previous case it is enough to calculate the parameter C. The exact value, obtained through a conformal transformation, is C/ε_0 = 10.23408862 [Terakado 1976]. The potential has a behavior of type $r^{2/3}$ near the corners. Hence, with uniform mesh refinements, the rate of convergence will be proportional to $N_d^{-1/3}$, independent of the order of the polynomial approximation.

A thorough study of this structure with different meshes and elements is carried out. In Chapter 4 the structure is analyzed using the self-adaptive mesh algorithm verifying that the maximum rate of convergence for h refinements, given by the order of the element used, is achieved.

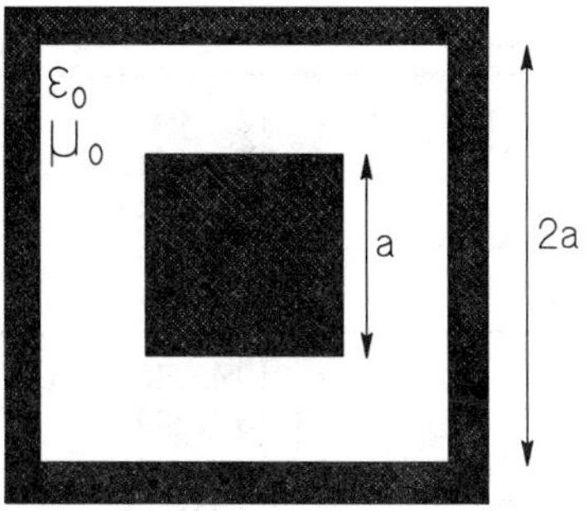

Figure 3.10 Square coaxial line.

Figure 3.11 shows the initial and the mesh after the third iteration, using triangular elements and uniform refinements. One-eighth of the structure has been analyzed. Table 3.5 gives the results for different orders of approximation. Figure 3.12 summarizes the rate of convergence for h refinements.

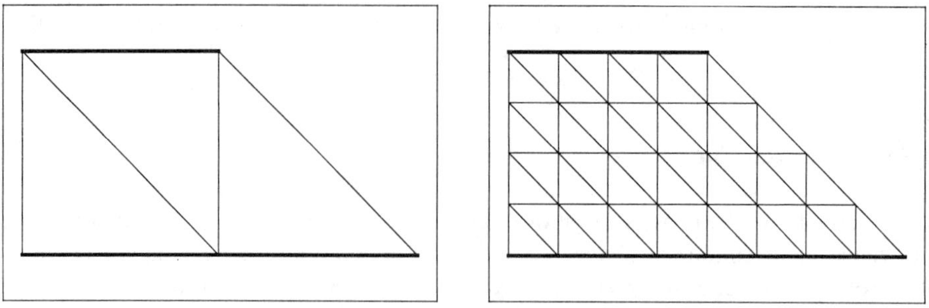

(a) (b)

Figure 3.11 Mesh of one-eighth of the square coaxial line utilizing triangular elements: (a) initial mesh and (b) mesh after the third iteration using uniform refinements.

Table 3.5
Results for the analysis of the square coaxial line using triangular elements and uniform refinements

Order	Mesh steps⇒ No. elements	No. nodes	C/ε_0	Relative error (%)	Run-time (sec)
First o.	1⇒3 e.	5	12	17.25	0.99
	2⇒12 e.	12	10.84444	5.96	0.99
	3⇒48 e.	35	10.46224	2.23	1.21
	4⇒192 e.	117	10.32194	0.86	2.14
	5⇒768 e.	425	10.26837	0.33	16.25
Second o.	1⇒3 e.	12	10.45926	2.2	1.05
	2⇒12 e.	35	10.32649	0.903	1.14
	3⇒48 e.	117	10.27083	0.35	1.70
	4⇒192 e.	425	10.24867	0.14	18.57
Third o. Serendipity	1⇒3 e.	19	10.39838	1.61	1.15
	2⇒12 e.	58	10.29764	0.62	1.31
	3⇒48 e.	199	10.25935	0.24	3.52
Third o. Complete	1⇒3 e.	22	10.32855	0.92	1.15
	2⇒12 e.	70	10.27111	0.36	1.42
	3⇒48 e.	247	10.24875	0.14	5.00

From the figure it can be observed that it agrees with the predicted behavior. Elements of any order give rates of convergence that asymptotically have the same slope, that is -1/3, as imposed by the regularity of the solution. The serendipity third-order element provides results identical to that of second-order elements.

Figure 3.13 shows the rate of convergence for p refinements using the first mesh and elements of first-, second-, and third-order (serendipity and complete), successively. It can be noticed that the rate of convergence is higher than twice that obtained with h refinements in accordance with the theoretical predictions.

The behavior of structured meshes of the criss-cross type has also been investigated. This type of mesh leads in some cases to nonconvergent results. Figure 3.14 shows the initial mesh and the mesh after the third step when using uniform refinements. Table 3.6 summarizes the results for elements of different orders. Figure 3.15 shows the rate of convergence for h refinements. They agree with the predicted results. However, comparing Figures 3.12 and 3.15, it can be observed that slightly worse results are obtained with this type of mesh than with the previous one.

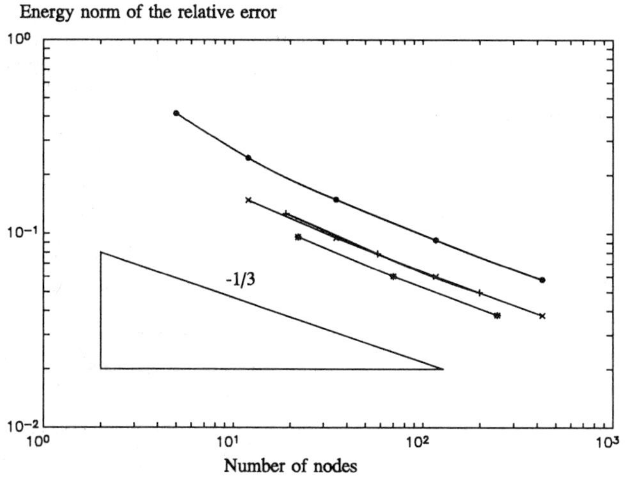

Figure 3.12 Rate of convergence for h refinements for the analysis of the square coaxial line utilizing triangular elements. o: first-order; ×: second-order; +: serendipity third-order; *: complete third-order.

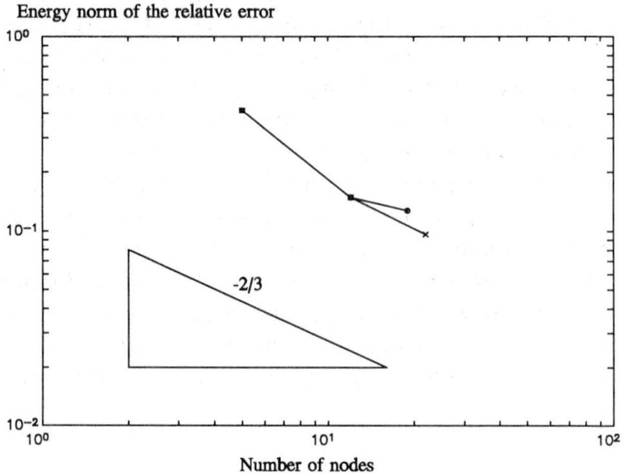

Figure 3.13 Rate of convergence for p refinements for the analysis of the square coaxial line using the initial mesh of Figure 3.11 and first-, second-, and third-order (o: serendipity; ×: complete) elements.

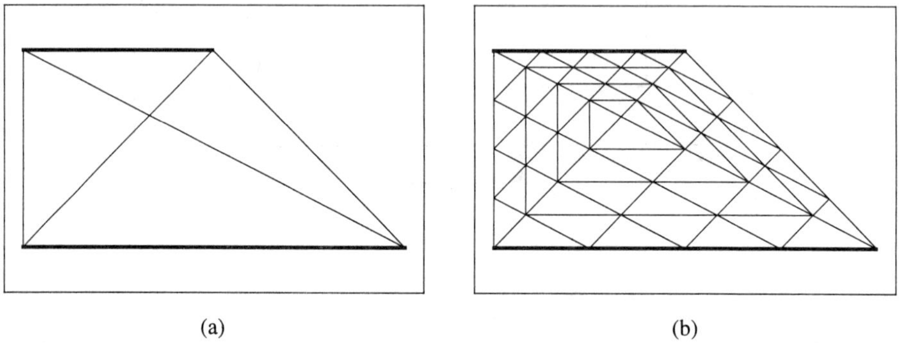

Figure 3.14 Triangular criss-cross mesh of one-eighth of the square coaxial line: (a) initial coarse mesh and (b) mesh after the third iteration using uniform refinements.

Table 3.6
Results for the analysis of the square coaxial line using uniformly refined triangular criss-cross meshes

Order	Mesh steps⇒ No. elements	No. nodes	C/ε_0	Relative error (%)	Run-time (sec)
First o.	1⇒4 e.	5	12.0	17.25	0.99
	2⇒16 e.	13	10.89012	6.41	1.04
	3⇒64 e.	41	10.4804	2.41	1.32
	4⇒256 e.	145	10.32828	0.92	2.64
	5⇒1024 e.	545	10.27061	0.35	30.76
Second o.	1⇒4 e.	13	10.51062	2.7	0.99
	2⇒16 e.	41	10.33444	0.981	1.1
	3⇒64 e.	145	10.27314	0.38	2.09
	4⇒256 e.	545	10.24151	0.15	36.37
Third o. Serendipity	1⇒4 e.	21	10.37504	1.37	1.15
	2⇒16 e.	69	10.27905	0.44	1.43
	3⇒64 e.	249	10.25132	0.17	5.11
Third o. Complete	1⇒4 e.	25	10.33185	0.95	1.15
	2⇒16 e.	85	10.27265	0.377	1.49
	3⇒64 e.	313	10.24934	0.150	8.19

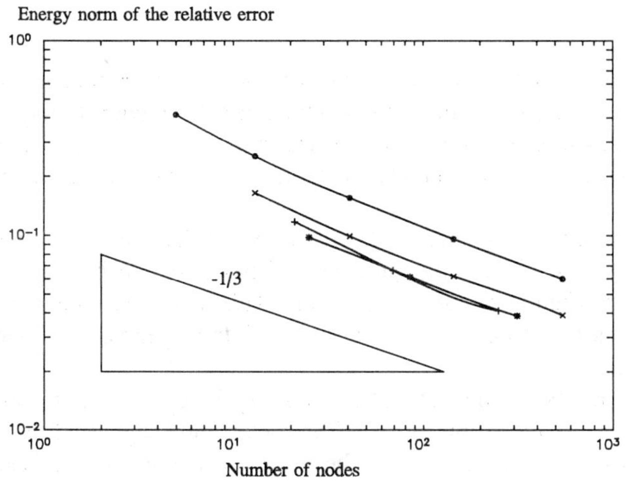

Figure 3.15 Rate of convergence for h refinements for the analysis of the square coaxial line with criss-cross triangular meshes and elements of different orders. o: first-order; ×: second-order; +: serendipity third-order; *: complete third-order.

The structure has also been analyzed using rectangular elements. Figure 3.16 shows the initial mesh and the mesh after the third iteration utilizing uniform refinements. Table 3.7 summarizes the results.

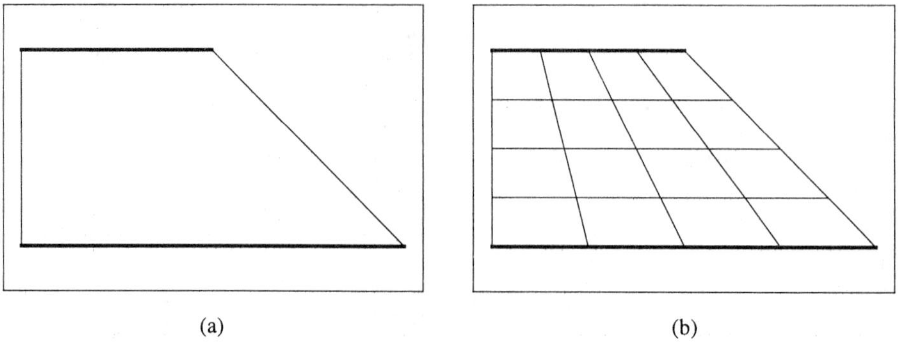

Figure 3.16 Mesh of one-eighth of the square coaxial line utilizing rectangular elements: (a) initial mesh and (b) mesh after the third iteration using uniform refinements.

Figure 3.17 shows the rate of convergence for h refinements and elements of different orders. They follow the predicted theoretical behavior. It may be observed that for the same number of unknowns second-order elements provide more accurate results than those obtained with serendipity third-order elements. Again there is a saturation effect.

If compared with Figure 3.12, it is observed that triangular elements are more efficient in this case than rectangular ones. They seem to be better adapted to the symmetry or configuration of the solution. This result has been observed for all structures having re-entrant corners. Thus, for arbitrary geometries, triangular elements are preferred.

Finally, a mesh with rectangular and triangular elements has also been used. The idea was to check if the combination of both elements, suggested by the geometry of the structure was more efficient than the use of just triangular elements. Figure 3.18 shows the initial mesh and the mesh after three uniform refinements. For this analysis, elements compatible with each other must be used. Table 3.8 shows the results for elements of different orders. Figure 3.19 shows the rate of convergence for h refinements. It may be seen that again the rate of convergence agrees with the theoretical predictions.

Table 3.7
Results for the analysis of the square coaxial line using rectangular elements and uniform refinements

Order	Mesh steps⇒ No. elements	No. nodes	C/ε_0	Relative error (%)	Run-time (sec)
First o.	1⇒1 e.	4	12	17.25	0.93
	2⇒4 e.	9	11.14063	8.86	0.94
	3⇒16 e.	25	10.62623	3.83	1.04
	4⇒64 e.	81	10.39355	1.56	1.43
	5⇒256 e.	289	10.29846	0.63	6.43
Second o. Serendipity	1⇒1 e.	8	10.78017	5.337	1.04
	2⇒4 e.	21	10.40934	1.71	1.05
	3⇒16 e.	65	10.30453	0.69	1.43
	4⇒64 e.	225	10.25915	0.246	3.85
Second o. Complete	1⇒1 e.	9	10.44462	2.06	1.09
	2⇒4 e.	25	10.32891	0.927	1.15
	3⇒16 e.	81	10.27338	0.38	1.54
	4⇒64 e.	289	10.25007	0.16	4.25
Third o. Serendipity	1⇒1 e.	12	10.75148	5.05	1.1
	2⇒4 e.	33	10.39397	1.56	1.26
	3⇒16 e.	105	10.29718	0.62	1.87
	4⇒64 e.	369	10.25899	0.244	4.55
Third o. Complete	1⇒1 e.	16	10.33425	0.98	1.25
	2⇒4 e.	49	10.27560	0.406	1.65
	3⇒16 e.	169	10.25104	0.16	3.52
	4⇒64 e.	625	10.23421	0.6	6.55

Comparing the results obtained with those of Figure 3.12, where triangular elements are used, no improvement is observed in terms of accuracy. Thus it could be concluded that triangular elements are preferable. Second-order elements are efficient and provide quite accurate results.

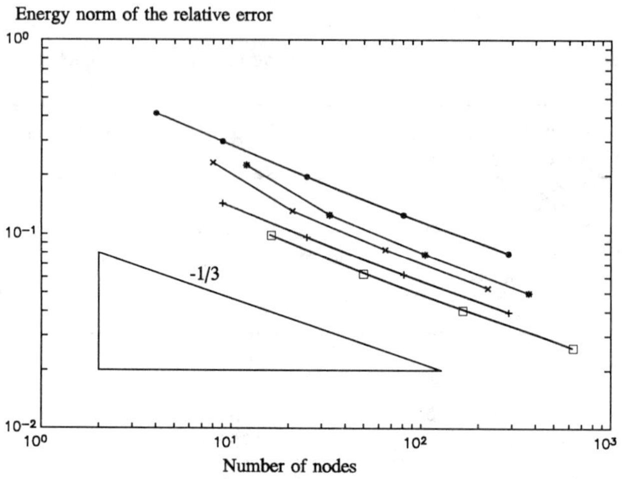

Figure 3.17 Rate of convergence for h refinements for the analysis of the square coaxial line with rectangular elements of different orders. o: first-order; ×: serendipity second-order; +: complete second-order; *: serendipity third-order; □: complete third-order.

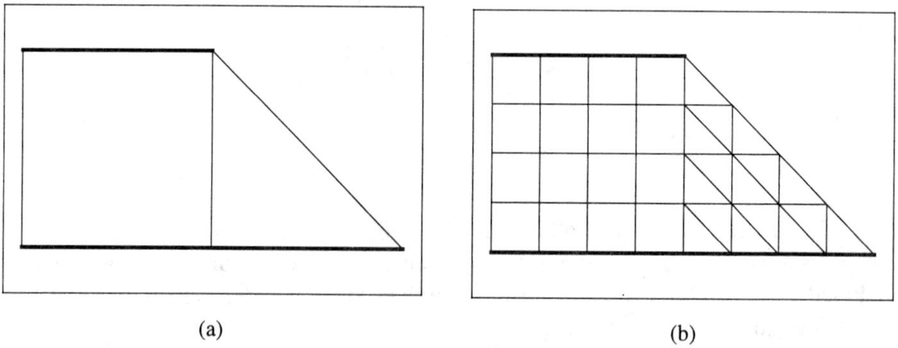

Figure 3.18 Mesh of one-eighth of the square coaxial line, utilizing triangular and rectangular elements: (a) initial mesh and (b) mesh after the third iteration using uniform refinements.

Table 3.8

Results for the analysis of the square coaxial line using rectangular and triangular elements and uniform refinements

Order	Mesh steps⇒ No. elements	No. nodes	C/ε_0	Relative error (%)	Run-time (sec)
-Triangular e.:	1⇒2 e.	5	12	17.25	0.99
first-order	2⇒8 e.	12	10.84444	5.96	0.99
-Rectangular e.:	3⇒32 e.	35	10.46224	2.23	1.26
first-order	4⇒128 e.	117	10.32194	0.86	2.10
	5⇒512 e.	425	10.26837	0.33	15.9
-Triangular e.:	1⇒2 e.	11	10.45783	2.18	1.26
second-order	2⇒8 e.	31	10.32198	0.86	1.32
-Rectangular e.:	3⇒32 e.	101	10.269	0.25	1.81
Serendipity	4⇒512 e.	361	10.24794	0.13	12.1
second-order					
-Triangular e.:	1⇒2 e.	12	10.42854	1.9	1.16
second-order	2⇒8 e.	35	10.31397	0.78	1.31
-Rectangular e.:	3⇒32 e.	117	10.26592	0.32	1.98
Complete	4⇒512 e.	425	10.24672	0.12	18.24
second-order					
-Triangular e.:	1⇒2 e.	17	10.41617	1.78	1.21
Serendipity	2⇒8 e.	50	10.29957	0.64	1.43
third-order	3⇒32 e.	167	10.25990	0.25	3.07
-Rectangular e.:	4⇒512 e.	605	10.24432	0.1	50.2
Serendipity					
third-order					
-Triangular e.:	1⇒2 e.	18	10.38022	1.43	1.2
Complete	2⇒8 e.	54	10.28839	0.53	1.43
third-order	3⇒32 e.	183	10.25551	0.21	3.24
-Rectangular e.:	4⇒512 e.	669	10.24258	0.08	65.53
Serendipity					
third-order					
-Triangular e.:	1⇒2 e.	21	10.36924	1.32	1.26
Serendipity	2⇒8 e.	66	10.28346	0.48	1.82
third-order	3⇒32 e.	231	10.25359	0.19	4.89
-Rectangular e.:					
Complete					
third-order					
-Triangular e.:	1⇒2 e.	22	10.31630	0.804	1.21
Complete	2⇒8 e.	70	10.26644	0.317	1.7
third-order	3⇒32 e.	247	10.2469	0.126	5.28
-Rectangular e.:					
Complete					
third-order					

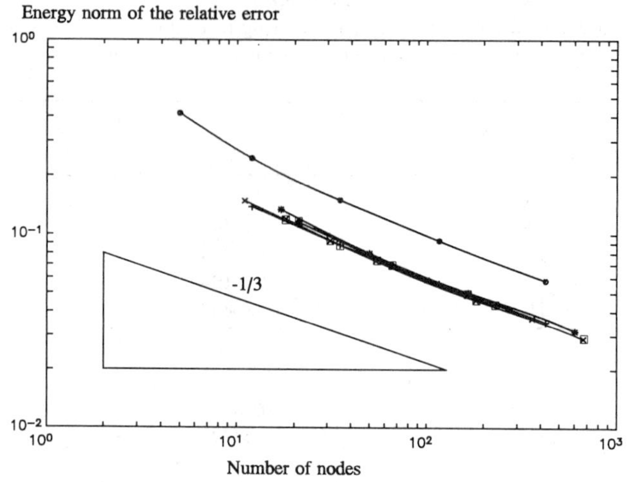

Figure 3.19 Rate of convergence for h refinements for the analysis of the square coaxial line with rectangular and triangular elements of different compatible orders. o: first-order; ×: second-order triangles, serendipity second-order rectangles; +: second-order triangles, complete second-order rectangles; *: serendipity third-order; ⊠: complete third-order triangles, serendipity third-order rectangles; ⊞: serendipity third-order triangles, complete third-order rectangles; ⊠: complete third-order.

c) ***Shielded symmetric stripline***

The next structure that has been analyzed is the symmetric stripline that is shown in Figure 3.20. It consists of a conductor of width a shielded by a rectangular waveguide

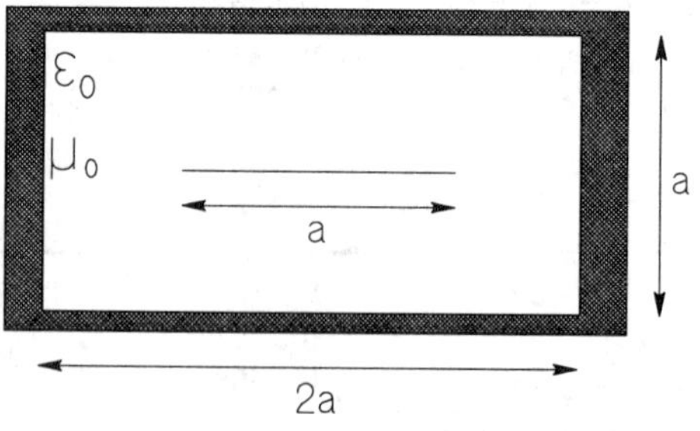

Figure 3.20 Symmetric stripline.

of dimensions $2a \times a$. The thickness of the strip is zero. The behavior of the potential in the proximity of the edge of the strip is of the type $r^{1/2}$. Because of this, the electric field shows a strong singularity of the type $r^{-1/2}$. As in the two preceding cases just the calculation of parameter C is needed. The exact solution is given in [Daly 1973a] and [Tippet and Chang 1976] and is $C/\varepsilon_0 = 5.87688$.

First a mesh with triangular elements is used. Figure 3.21 shows the initial mesh of one-fourth of the structure and the third mesh utilizing uniform refinements. Table 3.9 summarizes the results of the analysis employing elements of different orders.

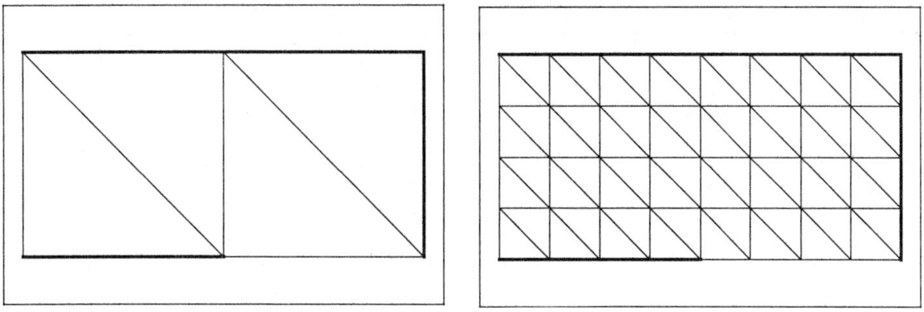

(a) (b)

Figure 3.21 Mesh of one-fourth of the symmetric stripline utilizing triangular elements: (a) initial mesh and (b) mesh after the third iteration using uniform refinements.

Figure 3.22 shows the rate of convergence for h refinements for elements of different order. It can be observed that the slope is the same for all cases. If compared with the results corresponding to the square coaxial structure it can be observed that the slope is smaller (the convergence is slower). In fact, in this case the parameter α that measures the regularity of the solution is $\alpha = 1/2$, (i.e., in the corner region, the potential behaves as $r^{1/2}$ and its gradient as $r^{-1/2}$) so the convergence is now proportional to $N_d^{-1/4}$ instead of $N_d^{-1/3}$ as it was in the previous case. Again it can be seen how the rate of convergence cannot be improved by using higher order elements if uniform refinements are utilized. In fact, higher order elements only provide an increased accuracy. Figure 3.22 clearly shows how the serendipity third-order triangular element behaves as the second-order one.

Figure 3.23 gives the rate of convergence for p refinements using the initial mesh and first-, second-, and third-order (serendipity and complete) elements. It can be observed that the rate of convergence is more than twice that of h refinements.

A triangular-based structured mesh has also been used in which the triangles have been obtained by subdividing the initial rectangles by the diagonals in the opposite direction to that of Figure 3.21. Results are identical to the previous ones except for the serendipity third-order element.

Table 3.9
Results for the analysis of the symmetric stripline using triangular elements and uniform refinements

Order	Mesh step⇒ No. elements	No. nodes	C/ε_0	Relative error (%)	Run-time (sec)
First o.	1⇒4 e.	6	8.0	36.12	0.99
	2⇒16 e.	15	6.761905	15.06	1.15
	3⇒64 e.	45	6.288339	7.00	1.32
	4⇒256 e.	153	6.075963	3.39	2.69
	5⇒512 e.	561	5.97485	1.67	32.9
Second o.	1⇒4 e.	15	6.349206	8.03	0.93
	2⇒16 e.	45	6.112856	4.01	1.16
	3⇒64 e.	153	5.993612	1.98	2.58
	4⇒256 e.	561	5.934886	0.98	43.2
Third o. Serendipity	1⇒4 e.	24	6.251910	6.38	1.32
	2⇒16 e.	75	6.063583	3.17	1.54
	3⇒64 e.	261	5.969647	1.58	6.31
Third o. Complete	1⇒4 e.	28	6.118267	4.11	1.16
	2⇒16 e.	91	5.995278	2.01	1.81
	3⇒64 e.	325	5.935666	1.00	13.62

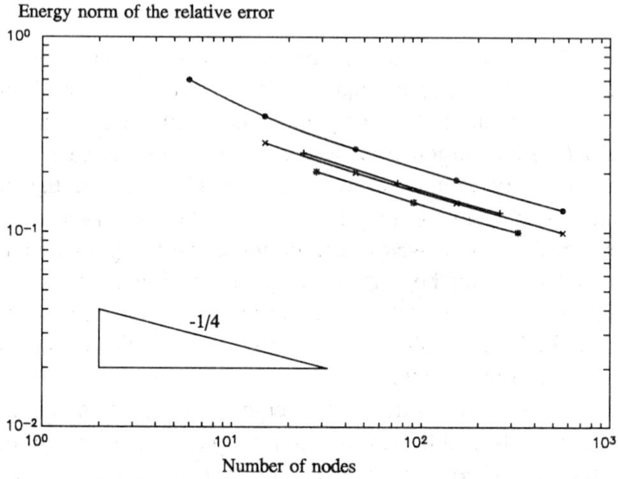

Figure 3.22 Rate of convergence for h refinements for the analysis of the symmetric stripline using triangular elements. o: first-order; ×: second-order; +: serendipity third-order; *: complete third-order.

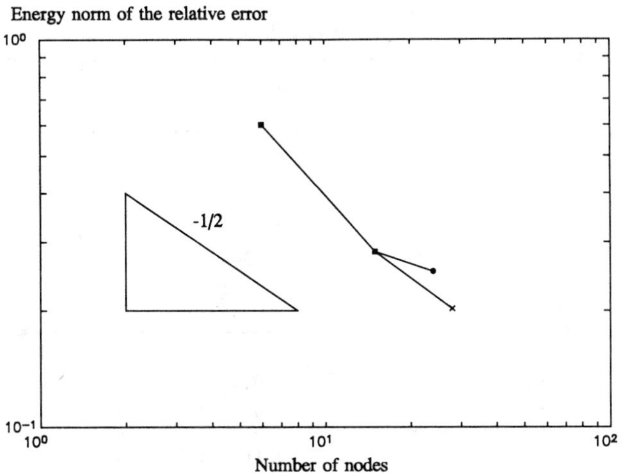

Figure 3.23 Rate of convergence for p refinements for the analysis of the symmetric stripline with the initial mesh of Figure 3.21 with first-, second-, and third-order (o: serendipity and ×: complete) elements.

The criss-cross mesh of this structure is shown in Figure 3.24. It provides results similar to the previous ones, as can be seen from Table 3.10 and Figure 3.25.

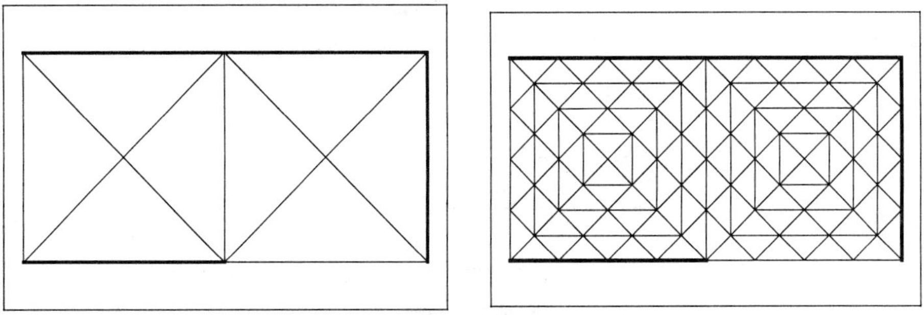

(a) (b)

Figure 3.24 Triangular criss-cross mesh of one-fourth of the symmetric stripline: (a) initial mesh and (b) mesh after the third iteration using uniform refinements.

Table 3.10
Results for the analysis of the symmetric stripline with criss-cross triangular meshes and uniform refinements

Order	Mesh step⇒ No. elements	No. nodes	C/ε_0	Relative error (%)	Run-time (sec)
First o.	1⇒8 e.	8	7.0	19.11	1.16
	2⇒32 e.	23	6.434807	9.49	1.21
	3⇒128 e.	77	6.153168	4.70	1.76
	4⇒512 e.	281	6.014198	2.33	8.68
Second o.	1⇒8 e.	23	6.219436	5.83	1.25
	2⇒32 e.	77	6.043124	2.83	1.43
	3⇒128 e.	281	5.959147	1.4	7.25
Third o. Serendipity	1⇒8 e.	38	6.118474	4.11	2.36
	2⇒32 e.	131	5.997869	2.06	2.64
	3⇒128 e.	485	5.937068	1.02	28.61
Third o. Complete	1⇒8 e.	46	6.044855	2.86	2.8
	2⇒32 e.	163	5.960221	1.42	2.96
	3⇒128 e.	613	5.918366	0.706	52.9

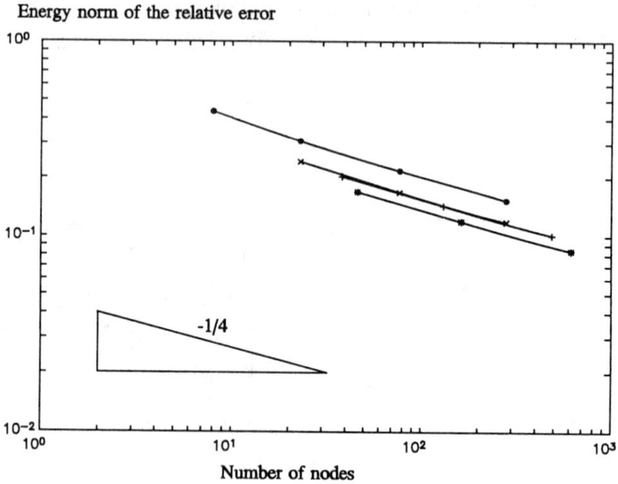

Figure 3.25 Rate of convergence for the analysis of the symmetric stripline with triangular criss-cross meshes, using h refinements. o: first-order; ×: second-order; +: serendipity third-order; *: complete third-order.

In this case, the use of higher-order rectangular elements has given a higher efficiency than triangular elements of the same order. Figure 3.26 shows the mesh used. Table 3.11 provides the results for different orders of elements. Figure 3.27 plots the rate of convergence for h refinements.

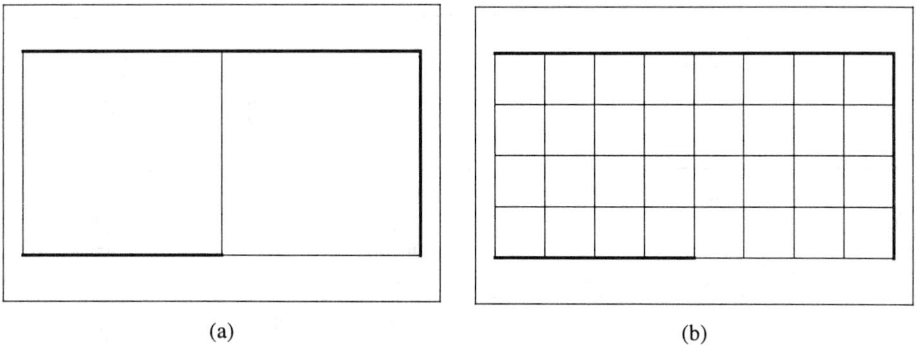

Figure 3.26 Mesh of one-fourth of the symmetric stripline utilizing rectangular elements: (a) initial mesh and (b) mesh after the third iteration using uniform refinements.

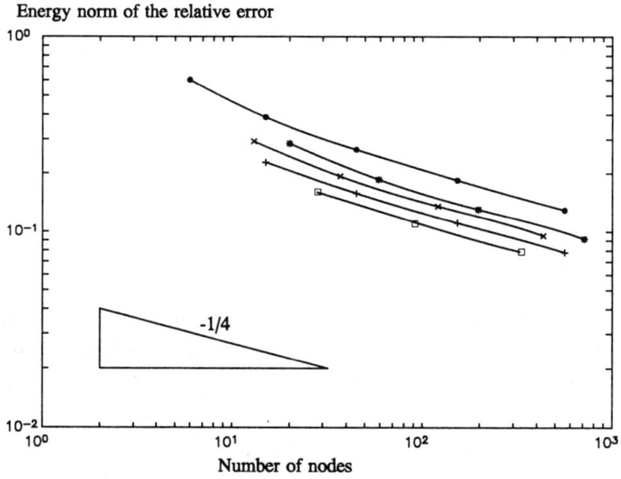

Figure 3.27 Rate of convergence for h refinements for the analysis of the symmetric stripline with rectangular elements. o: first-order; ×: serendipity second-order; +: complete second-order; *: serendipity third-order; □: complete third-order.

Table 3.11

Results for the analysis of the symmetric stripline with rectangular elements and uniform refinements

Order	Mesh step⇒ No. elements	No. nodes	C/ε_0	Relative error (%)	Run-time (sec)
First o.	1⇒2 e.	6	8.0	36.12	0.87
	2⇒8 e.	15	6.761905	16.06	0.99
	3⇒32 e.	45	6.288339	7.00	1.16
	4⇒128 e.	153	6.075963	3.387	2.31
	5⇒512 e.	561	5.97485	1.67	33.4
Second o. Serendipity	1⇒2 e.	13	6.377054	8.51	1.16
	2⇒8 e.	37	6.095055	3.71	1.26
	3⇒32 e.	121	5.984007	1.83	2.03
	4⇒128 e.	433	5.926401	0.84	18.7
Second o. Complete	1⇒2 e.	15	6.180976	5.17	1.16
	2⇒8 e.	45	6.022628	2.48	1.31
	3⇒32 e.	153	5.949085	1.23	2.37
	4⇒128 e.	561	5.879494	0.044	37.3
Third o. Serendipity	1⇒2 e.	20	6.353044	8.1	1.21
	2⇒8 e.	59	6.079523	3.45	1.48
	3⇒32 e.	197	5.976293	1.69	4.17
	4⇒128 e.	713	5.926104	0.84	108.0
Third o. Complete	1⇒2 e.	28	6.026363	2.54	1.31
	2⇒8 e.	91	5.950392	1.25	1.87
	3⇒32 e.	325	5.882117	0.09	9.6

Finally a mesh using rectangular and triangular elements has been utilized. Results are similar to those obtained with the triangular meshes. Figure 3.28 shows the initial mesh and the mesh after the third iteration. Table 3.12 provides the results for the capacitance. Figure 3.29 depicts the rate of convergence for h refinements.

Next, the improvement in the rate of convergence is studied when refinements are made at re-entrant corners. The initial mesh of Figure 3.21 is refined first at the corner. In successive steps it is further refined uniformly. Table 3.13 shows the results.

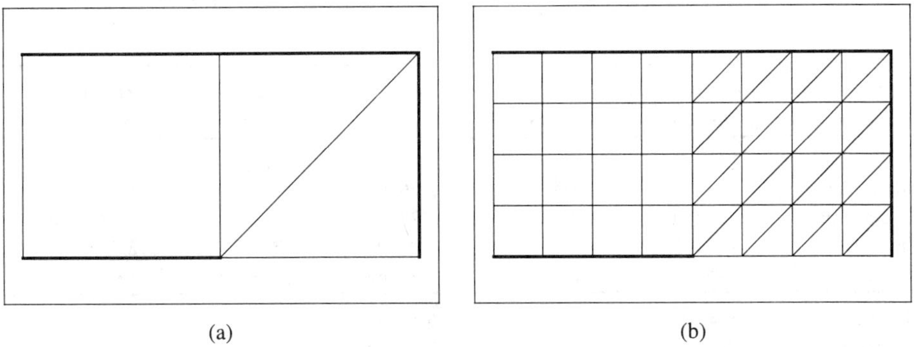

Figure 3.28 Mesh of one-fourth of the symmetric stripline utilizing rectangular and triangular elements: (a) initial mesh and (b) mesh after the third iteration using uniform refinements.

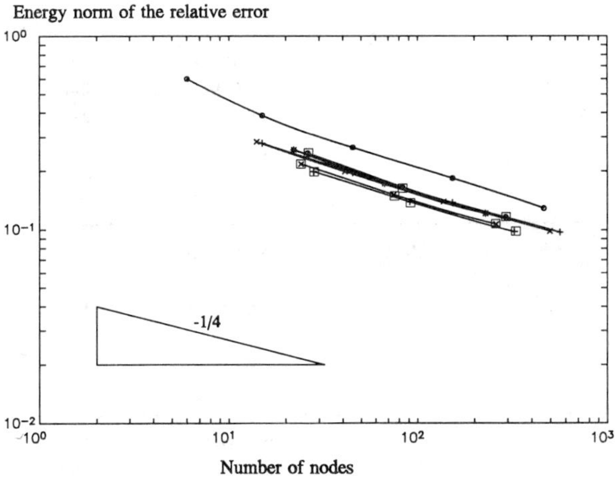

Figure 3.29 Rate of convergence for *h* refinements for the analysis of the symmetric stripline using rectangular and triangular elements of different compatible orders. o: first-order; ×: second-order triangles, serendipity second-order rectangles; +: second-order triangles, complete second-order rectangles; *: serendipity third-order; ⊠: complete third-order triangles, serendipity third-order rectangles; ⧇: serendipity third-order triangles, complete third-order rectangles; ⊞: complete third-order.

Table 3.12

Results of the analysis of the symmetric stripline using rectangular and triangular elements and uniform refinements

Order	Mesh step⇒ No. elements	No. nodes	C/ε_0	Relative error (%)	Run-time (sec)
-Triangular e.:	1⇒3 e.	6	8.0	36.12	1.04
first-order	2⇒12 e.	15	6.761905	15.06	1.21
-Rectangular e.:	3⇒48 e.	45	6.288339	7.00	1.37
first-order	4⇒192 e.	153	6.075963	3.387	2.75
	5⇒768 e.	561	5.974850	1.667	44.43
-Triangular e.:	1⇒3 e.	14	6.348387	8.023	1.43
second-order	2⇒12 e.	41	6.109664	3.96	2.53
-Rectangular e.:	3⇒48 e.	137	5.991895	1.957	2.64
Serendipity	4⇒192 e.	497	5.933990	0.971	30.21
second-order					
-Triangular e.:	1⇒3 e.	15	6.331733	7.74	1.58
second-order	2⇒12 e.	45	6.103625	3.86	2.59
-Rectangular e.:	3⇒48 e.	153	5.988755	1.904	2.69
Complete	4⇒192 e.	561	5.932364	0.944	42.3
second-order					
-Triangular e.:	1⇒3 e.	22	6.266516	6.63	1.59
Serendipity	2⇒12 e.	67	6.052969	2.996	2.59
third-order	3⇒48 e.	229	5.963755	1.48	5.11
-Rectangular e.:					
Serendipity					
third-order					
-Triangular e.:	1⇒3 e.	24	6.155148	4.735	1.69
Complete	2⇒12 e.	75	6.011980	2.298	2.65
third-order	3⇒48 e.	261	5.944306	1.15	6.54
-Rectangular e.:					
Serendipity					
third-order					
-Triangular e.:	1⇒3 e.	26	6.236403	6.12	1.96
Serendipity	2⇒12 e.	83	6.038687	2.75	2.59
third-order	3⇒48 e.	293	5.956357	1.35	8.23
-Rectangular e.:					
Complete					
third-order					
-Triangular e.:	1⇒3 e.	28	6.109406	3.95	1.75
Complete	2⇒12 e.	91	5.990647	1.936	2.81
third-order	3⇒48 e.	325	5.933258	0.96	10.44
-Rectangular e.:					
Complete					
third-order					

Figure 3.30 compares the rate of convergence with the previous computations with uniform refinements. The inset shows the slope of the theoretical rate of convergence corresponding to uniform refinements (ruled by the order of the singularity) and that

Table 3.13

Results for the analysis of the symmetric stripline using triangular elements and local refinements at the corner of the conductor and successive uniform refinements

Order	No. nodes	C/ε_0	Relative error (%)	Run-time (sec)
First o.	10	6.710679	14.18	1.15
	29	6.231548	6.03	1.2
	97	6.037343	2.73	1.92
	353	5.952673	1.29	10.38
Second o.	29	6.092999	3.67	1.26
	97	5.971945	1.61	14.9
	353	5.922442	0.775	17.1

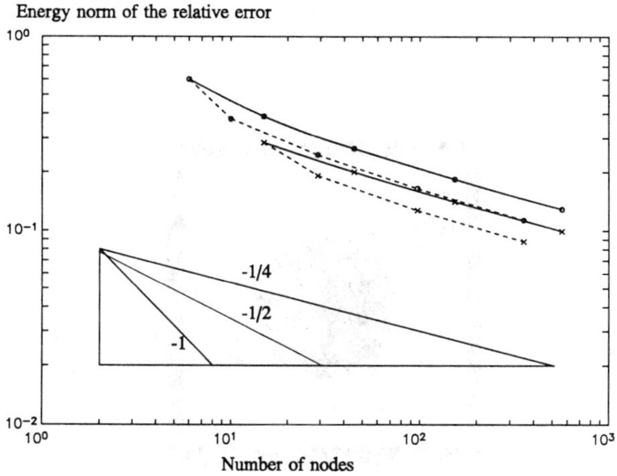

Figure 3.30 Rate of convergence for h refinements for the analysis of the symmetric stripline with triangular mesh uniformly (—o—: first-order; —x—: second-order) and locally (--o--: first-order; --x--: second-order) refined.

corresponding to the use of first and second order elements in a problem with a smooth enough solution. An improvement of the rate of convergence can be noticed when local refinements are carried out.

d) *Square-circular coaxial line partially filled with dielectric*

Next, the structure of Figure 3.31 is considered. The central conductor is cylindrical of radius $a/2$, and the outside conductor has a square cross section of side $2a$. The material between the conductors has permittivity $\varepsilon_r > 1$ and permeability $\mu_r = 1$. The contour of that medium is a square of side $1.5a$. The rest of the space between both conductors is occupied by vacuum. In this case, parameters C and L would have to be calculated, since the domain of the problem is an inhomogeneous one. The parameter L would be $L = c_0 / C_{eq}$, where C_{eq} is the capacity of the structure with the inhomogeneous media replaced by free space. For the calculation of C, the analysis of the geometry of Figure 3.31 is required. The dielectric corners cause the solution of the problem to have a non-smooth behavior. An estimate of the measure of the regularity of the solution may be computed from the expression $\varepsilon_r \tan(\alpha\theta) = -\tan(\alpha(\pi-\theta))$, where θ is half of the inner angle of the dielectric (here, $\theta = \pi/4$ rad). Two cases have been considered for $\varepsilon_r = 15$ (thus, $\alpha = 0.6945$) and 25 (hence, $\alpha = 0.71$). Thus, the rate of convergence for h uniform refinements should be given by $N_d^{-\alpha/2}$. The analysis is done using straight and curved second-order triangular elements starting with the mesh of Figure 3.32.

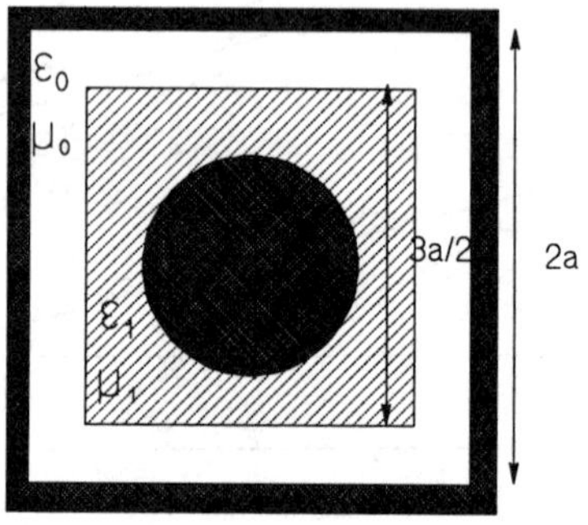

Figure 3.31 Coaxial line partially filled with dielectric.

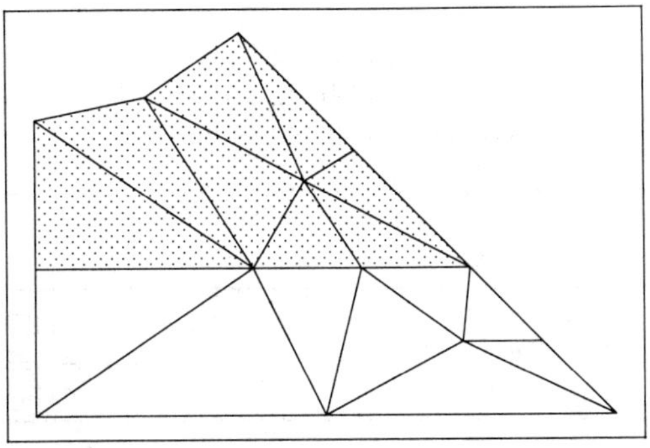

Figure 3.32 Initial mesh of one-eighth of the structure of Figure 3.31.

First, an analysis with uniform refinements is performed. Table 3.14 shows the results. The sècond analysis consists of further refinement at the dielectric corner with a total of 115 nodes. Figure 3.33 shows the mesh after the fourth iteration of the local refinement process. Then, in order to ensure a good meshing of the rest of the structure, each element was further subdivided into four other elements, resulting in a total of 489 nodes. The solution obtained from this last mesh has been taken as a good approximation of the exact value. Table 3.15 shows the results.

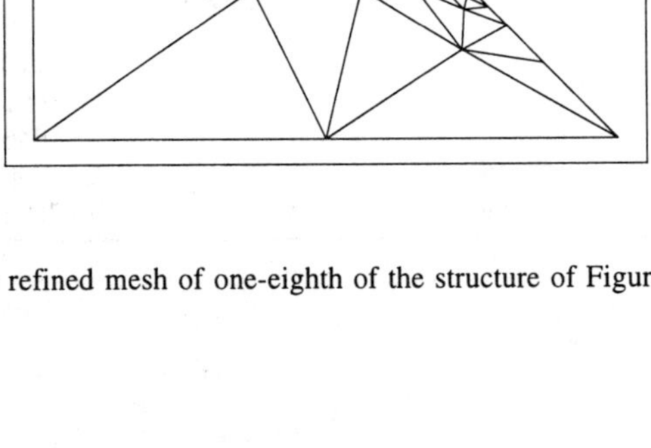

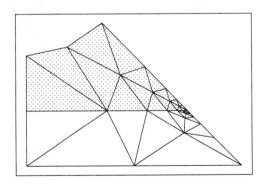

Figure 3.33 Locally refined mesh of one-eighth of the structure of Figure 3.31.

Table 3.14
Results for the analysis of the coaxial line partially filled with dielectric using the initial mesh of Figure 3.32 and uniform refinements

ε_r	No. nodes	C/ε_0
	25	22.61494
15	81	22.47092
	289	22.42748
	25	24.04513
25	81	23.85898
	289	23.80112

Table 3.15
Results for the analysis of the coaxial line partially filled with dielectric using the initial mesh of Figure 3.32 and successive refinements at the corner

ε_r	No. nodes	C/ε_0
	25	22.61494
	43	22.46068
	61	22.43602
15	79	22.42715
	97	22.42390
	115	22.42269
	489	22.40456
	25	24.04513
	43	23.84379
25	61	23.81106
	79	23.79913
	97	23.79469
	115	23.79300
	489	23.76922

Figure 3.34 plots the results of both analyses. The theoretical slope of the rate of convergence for the case of quasi-uniform meshes (ruled by the order of the singularity) and the slope for second-order elements for the case of a smooth enough problem are also provided. The improvement in the rate of convergence using local refinements is evident.

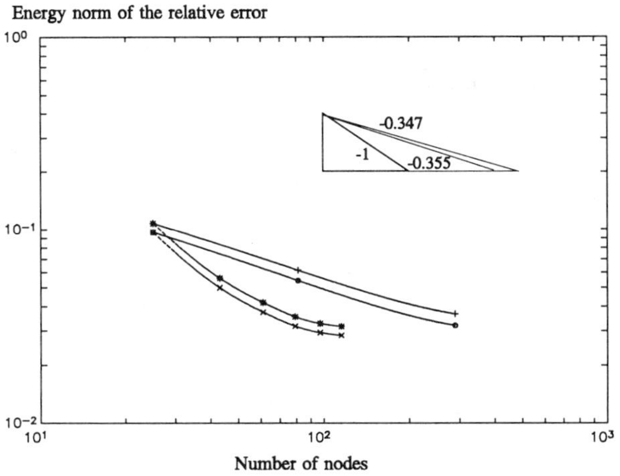

Figure 3.34 Comparison between the rate of convergence for h refinements for the analysis of the coaxial line with dielectric corner using uniform and local refinements along with second-order triangular elements. +: uniform refinements, $\varepsilon_r = 25$; *: local refinements, $\varepsilon_r = 25$; o: uniform refinements, $\varepsilon_r = 15$; ×: local refinements, $\varepsilon_r = 15$.

e) *Circular conductor over an infinite ground plane*

The next structure to be analyzed is open. The goal is to study the application of the infinite elements [Salazar-Palma and Hernández-Gil 1995]. The structure of Figure 3.35 has been chosen since it does not produce field singularities. This allows one to focus the attention on the behavior of the infinite elements. The radius of the line is a and its center is located at a height of $2a$ over an infinite ground plane. The surrounding medium is vacuum. The analytical value for the capacitance p.u.l. is given by $C = 2\pi\varepsilon_0/\log(2 + \sqrt{3}) = 4.770984\varepsilon_0$ [Kraus 1992; Ch. 4].

Figure 3.36 shows two different meshes for the structure. The mesh for Figure 3.36(b) is more refined than that of Figure 3.36(a). In both cases, finite triangular elements and rectangular elements, either finite or infinite, have been used.

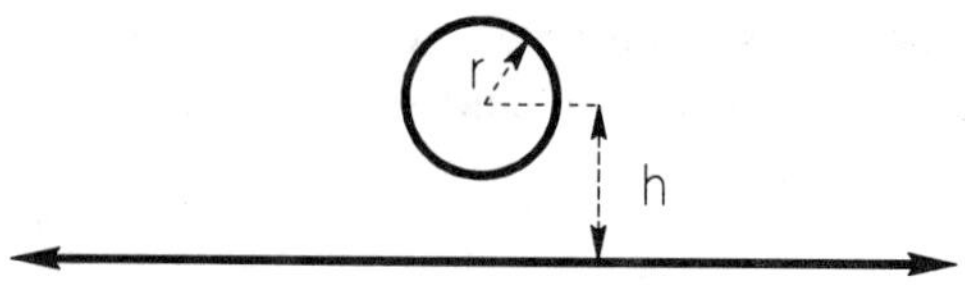

Figure 3.35 Line of circular cross section over an infinite ground plane.

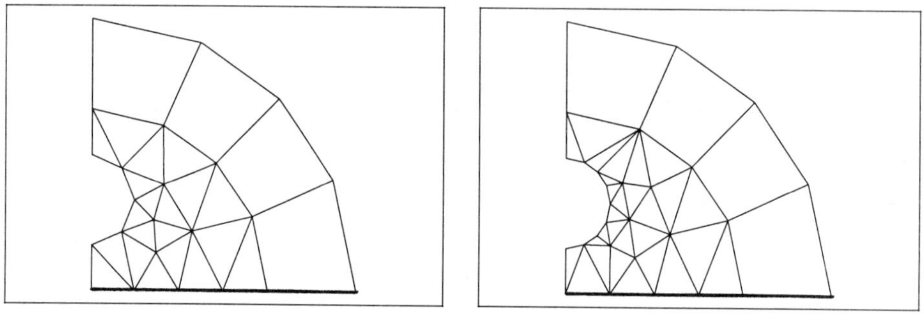

(a) (b)

Figure 3.36 Meshes for half of the structure of Figure 3.35: (a) initial mesh and (b) refined mesh.

Table 3.16 compares the results obtained only with finite elements (triangles and rectangles) or with finite elements (triangles) and infinite elements (rectangles). Different orders of approximation have been used. The efficiency of the analysis employing infinite elements (mainly when using second-order approximation) is quite noticeable even though they are located quite close to the near-field region.

Computational Aspects

Computational techniques. Other subjects have also been investigated such as the use of meshes where most of the elements are equilateral and node reordering techniques. The proper use of these tools increases the efficiency and accuracy of the results and reduces the computational time. Table 3.17 shows the results obtained using the example of Table 3.4 with and without node reordering.

Table 3.16
Comparison of the analysis of a circular conductor over an infinite ground plane using only finite elements or using both finite and infinite elements

	Order	No. nodes	No. unknowns	C/ε_0	Relative error (%)	Comput. time (sec)
Finite elements only. Mesh (a)	1	23	8	5.809291	21.76	1.7
	Serend. 2nd	69	41	5.346444	12.06	1.85
	Compl. 2nd	73	45	5.345681	12.04	2
Finite and infinite elements. Mesh (a)	1	23	8	5.331488	11.74	1.98
	Serend. 2nd	69	41	4.679494	-1.92	2.62
	Compl. 2nd	73	45	4.678005	-1.95	2.71
Finite and infinite elements. Mesh (b)	Serend. 2nd	89	53	4.762961	-0.17	3.29
	Compl. 2nd	93	57	4.761533	-0.2	3.25

Table 3.17
Comparison of storage space needed and computational time for the analysis of the circular coaxial line with complete third-order rectangular elements with and without node reordering techniques

No. nodes = 325	Half-band length	Sky-line profile	Half sky-line profile	Computational time (sec)
No Ordering	320	48723	149.917	11.1
With Ordering	49	9733	29.948	3.67

Solvers. Comparisons between the methods for the solution of the global system of equations have also been made. Figure 3.37 shows a shielded microstrip line. Half of the structure has been analyzed using various meshes and Lagrange elements of different order. Table 3.18 shows that the conjugate gradient method is more efficient than the direct Choleski method when the number of equations is large. A conjugate gradient method with an incomplete Choleski factorization appears to be the most efficient one for all the closed structures analyzed. If infinite elements are used then a relaxation preconditioner is preferable.

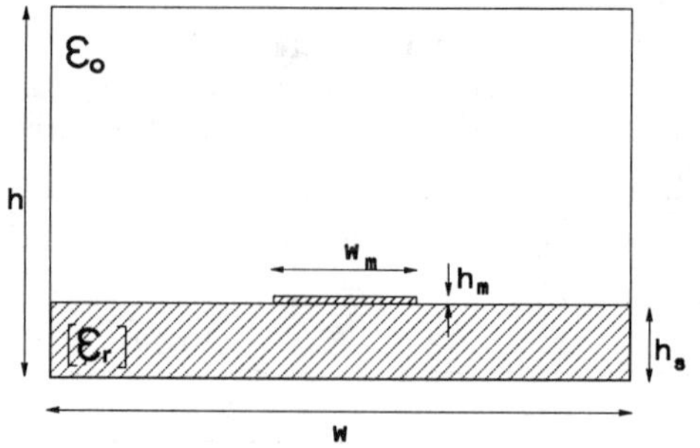

Figure 3.37 Shielded microstrip line: $w = 3.56$ mm; $h_s = 0.254$ mm; $h_m = 0.018$ mm; $w_m = 0.8$ mm; $\varepsilon_r = 2.17$.

Table 3.18
Comparison between the methods of solution of a microstrip line

No. unknowns	Charac. Impedance Z_c (Ω)	Computational time (sec): Choleski	Computational time (sec): Conjugate Gradient, Choleski precond.	Conjugate Gradient, Relaxation precond.	Conjugate Gradient, No precond.	Number of iterations: Conjugate Gradient, Choleski precond.	Conjugate Gradient, Relaxation precond.	Conjugate Gradient, No precond.
270(1)	47.39	9	9	10	11	10	19	74
1521(2)	48.00	39	29	38	51	14	40	174

(1) for $h = 2.272$ mm
(2) for $h = 3.272$ mm

Postprocesses. In order to validate losses and electromagnetic field postprocesses, a coaxial circular line made of copper conductors ($\sigma = 5.8\times10^7$ mhos/m) with an isotropic dielectric medium of $\varepsilon_r = \varepsilon_r'(1 - j10^{-3})$ and $\mu_r = \mu_r'(1 - j10^{-3})$ is considered. The radius of the inner conductor is 1 mm and that of the outer conductor is 2 mm. Second order triangular elements have been used. The structure is considered to be lossless, homogeneous, and anisotropic (i.e., calculating parameters L, R_m, G, R_c, and L_c as in the general case, but with $\varepsilon_r' = 1$ and $\mu_r' = 1$). Then the postprocesses were performed. Table 3.19 shows the results utilizing 91 unknowns and the exact solution [Ramo et al. 1965; Ch. 8]. Computational time for the complete solution, including postprocesses, was 19.62 sec.

Figure 3.38 shows the electric field over the cross section. Table 3.20 provides the value of the electric field along the axis $y = 0$ for various values of the x coordinate. A comparison is made with the exact results.

Table 3.19
Results for the lossy coaxial cable

$f = 10$ GHz	C (pF/m)	L (nH/m)	G (mho/m)	R_m (Ω/m)	R_c (Ω/m)	L_c (nH/m)
Computed	80.26	138.63	$0.504 \cdot 10^{-2}$	8.710	6.3	0.1002
Exact	80.14	138.6	$0.503 \cdot 10^{-2}$	8.708	6.23	0.0991
Error (%)	0.15	0.02	0.2	0.02	0.01	1.11

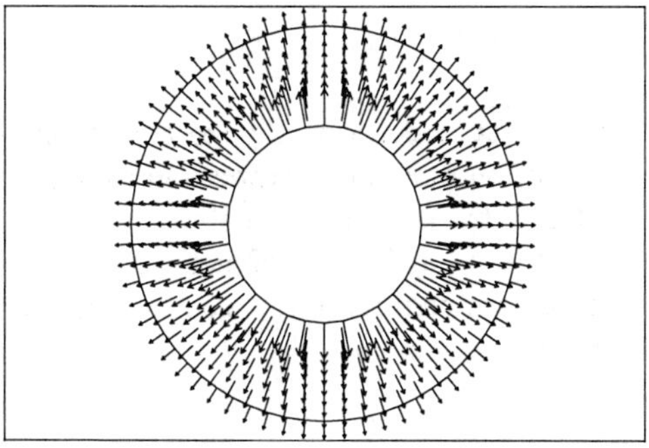

Figure 3.38 Coaxial cable: electric field.

Table 3.20
Electric field along the horizontal axis inside a coaxial cable

x (mm)	Computed (volt/m)		Exact value (volt/m)
	Mean	Weighted mean	
1.0	1.48712	1.47490	1.44269
1.125	1.28889	1.28889	1.23315
1.25	1.16233	1.16233	1.09135
1.375	1.05105	1.05105	0.98853
1.5	0.966846	0.968495	0.91024
1.625	0.889995	0.889990	0.84842
1.75	0.827905	0.828312	0.79823
1.875	0.770058	0.770058	0.756569
2.0	0.716639	0.716639	0.721347

3.2.3.2 *Application of the Finite Element Method to the Dual Standard Formulation*

The dual formulation for electrostatic problems in order to provide error bounds was first used by C. T. Carson and G. K. Cambrell [1966]. The method assumes that the electric displacement vector $\bar{D}_t$ is derived from a vector potential

$$\bar{D}_t = -\bar{\nabla} \times \bar{\varphi} \tag{3.65}$$

where $\bar{\varphi} = \varphi \bar{a}_z$. According to (3.31), and taking into account that the curl of vector $\bar{E}_t$ must vanish over the cross section, it is easy to demonstrate that the potential function φ must fulfill the equation

$$\bar{\nabla}_t \cdot \underline{\underline{\varepsilon}}_{rt}^T \left|\varepsilon_{rt}\right|^{-1} \bar{\nabla}_t \varphi = 0 \tag{3.66}$$

with $\bar{a}_n \cdot \left|\varepsilon_{rt}\right|^{-1} \underline{\underline{\varepsilon}}_{rt}^T \bar{\nabla}\varphi = 0$ on perfect electric walls and $\varphi = \varphi_i$ on symmetry walls.

It can be observed that, on one hand, expression (3.66) is formally identical to (3.32) with $\left|\varepsilon_{rt}\right|^{-1} \underline{\underline{\varepsilon}}_{rt}^T$ replaced by $\underline{\underline{\varepsilon}}_{rt}$, while on the other hand, the role of perfect electric walls and symmetry walls has been interchanged.

Between the primal scalar potential ϕ (the electrostatic potential) and its dual φ, there exists a relationship that is given by

$$\underline{\underline{\varepsilon}}_{rt} \bar{\nabla}_t \phi + \bar{a}_z \times \bar{\nabla}_t \varphi = 0 \tag{3.67}$$

This relationship implies that if the dual problem is solved, the expression

$$W_{dn} = \int_\Omega \bar{\nabla}_t \varphi \cdot \underline{\underline{\varepsilon}}_{rt}^T \left|\varepsilon_{rt}\right|^{-1} \bar{\nabla}_t \varphi \, d\Omega \tag{3.68}$$

meets the condition

$$W_n W_{dn} = \varepsilon_0^2 \tag{3.69a}$$

where W_n is

$$W_n = \int_\Omega \bar{\nabla}_t \phi \, \underline{\underline{\varepsilon}}_{rt} \bar{\nabla}_t \phi \, d\Omega \tag{3.69b}$$

If W_{dn} is calculated by means of an approximate method, it is clear that its computed approximate value will be a bound for the exact value of a different nature than that given by W_n. Thus, upper and lower bounds of the exact value may be computed through the analysis of the problem employing both the primal and dual formulations.

The application of FEM to the dual formulation does not require any other modification with respect to the primal direct formulation except that of using the variable $|\varepsilon_{rt}|^{-1}\underline{\underline{\varepsilon}}_{rt}^{T}$ instead of $\underline{\underline{\varepsilon}}_{rt}$ and interchanging the role of the symmetry walls and the electric walls. Hence, the previous methodology can be used for this formulation. However, an implementation problem arises because at least two symmetry walls are needed (which is not the case for most of the structures) together with the condition that the definition of the potential assigned to these boundaries must be unique. These are the same types of difficulties that arise in a static formulation using the magnetic field as derived from a scalar potential. In fact, if there are no two symmetry walls in which potentials $\varphi = 0$ and $\varphi = 1$ can be enforced (in the case of a two-conductor line), the use of double-faced boundaries becomes necessary since φ is not a single-valued function. In [Daly 1973a, 1984, 1985] and [Shepherd and Daly 1985] the dual formulation has been developed based on this approach. For the case of multiconductor lines, the number of double faced boundaries must be equal to the number of active conductors. It is evident that such a requirement complicates the generation of a mesh. Thus, in the context of this book, this formulation has been applied only to some simple structures, together with the primal direct formulation in order to compute lower and upper bounds of the exact values of the primary and secondary line parameters.

As an example of the application of this procedure, consider the stripline of Figure 3.20. One-fourth of the structure has been analyzed. Thus, two walls of symmetry exist. Therefore, the previous method can be directly applied. Table 3.21 shows results for several meshes generated by the self-adaptive mesh technique

Table 3.21

Error bounds for the computation of the capacity of a symmetric stripline using the primal and dual standard formulation with second-order triangular elements

Mesh steps	Upper bound primal formulation $C/(4\varepsilon_0)$	Lower bound dual formulation $C/(4\varepsilon_0)$	Mean value
4	1.48409	1.4545572	1.4693236
5	1.4768863	1.4616349	1.4692601
6	1.4733027	1.4651872	1.4692449
[Daly 1984]	1.51899	1.42244	1.47071
[Daly 1973a]	1.47107	1.46737	1.46922
Exact: [Daly 1973a], [Tippet and Chang 1976]	1.46922	1.46922	1.46922

explained in Chapter 4. The mean value of both the bounds approximates the exact value with high accuracy [Tippet and Chang 1976]. Comparison with the results of [Daly 1973a, 1984] is also provided.

3.2.3.3 *Application of the Finite Element Method to the Mixed Formulation*

The mixed formulation for the quasi-static analysis of transmission lines is obtained from (3.31) by expressing the unknowns in terms of the scalar electric potential $\phi(x,y)$ and the electric displacement vector $\bar{D}_t(x,y)$. The classical and weak formulation are detailed in Appendix E.3.1. Using the same terminology of Section 3.2.3.1, the problem is mathematically written as:

Given $\underline{\underline{\varepsilon}}_{rt}$ *in* Ω, *find* $\bar{D}_t$ *and* ϕ *with* $\bar{a}_n \cdot \bar{D}_t = 0$ *on* Γ_{pm}, *and* $\phi = \phi_{ci}$ *on* Γ_{ci}, *so*

$$\bar{D}_t + \varepsilon_0 \underline{\underline{\varepsilon}}_{rt} \bar{\nabla}_t \phi = 0 \,, \quad in\ \Omega \tag{3.70a}$$

and

$$\bar{\nabla}_t \cdot \bar{D}_t = 0 \,, \quad in\ \Omega \tag{3.70b}$$

Appendix E.3.2 deals with the FEM discretization of this problem. It is set up from the weighted residual method applied to the vector and the scalar equations (3.70a) and (3.70b), respectively. Then the scalar unknown, ϕ, is eliminated via a penalty function method. The integral expression obtained is discretized by the Galerkin method using div-conforming elements. They provide an approximate solution for $\bar{D}_t(x,y)$ which belongs to $H(\text{div},\Omega)$ (see Section 2.5.1.2). Thus, the displacement vector will be square integrable and it will have continuous normal components and discontinuous tangential components across element interfaces. These characteristics justify the choice of div-conforming elements because the computed solution vector will fulfil the conditions prescribed by Maxwell's equations. Triangular div-conforming elements of first and second order are used (see Table 2.6). Second-order div-conforming elements have been applied to analysis of transmission lines [Salazar-Palma et al. 1991].

The application of FEM is carried out in the usual way. This method has the following significant features:

(1) The nodes are placed at the midpoint of the edges (first order) or at the two points of numerical integration (Gauss-Legendre points) of each edge (second order), but never at element vertices. Thus, if the structure under analysis has re-entrant corners which coincide with element vertices, the solution will be more accurate for the discrete problem than for the Lagrange elements. This is because no nodes are placed at points where theoretically the value of the degree

of freedom is going to be infinite.

(2) For second-order elements, in addition to the six nodes related to edges, there are also two internal degrees of freedom. However, these two degrees of freedom may be eliminated at a local level. Hence, for an element, the increment in the number of unknowns from first to second order is the same as when using Lagrange elements.

(3) The sign assigned to each edge must be taken into account during the assembly process of the local matrices into the global one (see Appendix C.4.3).

(4) Once the global column matrix of the degrees of freedom is obtained, in order to carry out any postprocess, the sign of the local degrees of freedom as well as the value of the internal local degrees of freedom must be recovered.

The use of div-conforming elements provides a lower connectivity between nodes or degrees of freedom than when using Lagrange elements. For the former case, each node or degree of freedom is shared at most by two elements because nodes are placed at edges. For Lagrange elements, the same happens for nodes placed at edges, but nodes located at vertices are shared by as many elements sharing the same vertex. As a consequence on one hand, the total number of unknowns of a discretization performed by div-conforming elements is higher than when using Lagrange elements. But, on the other hand, the sparsity of global matrices arising from div-conforming elements is higher than when using Lagrange elements. Then, in terms of computation time, the overall solution of the global system of equations for a div-conforming discretization may be of similar efficiency (if not more) as for a Lagrange FEM.

It is interesting to note that the value of the degrees of freedom for nodes located at conductor edges represent the charge density, more precisely, the degree of freedom is its sampled value at the points of numerical integration. Therefore, the computation of the capacitance matrix coefficients is reduced to the weighted addition of the degrees of freedom over each conductor, simplifying the postprocess required.

Another advantage is that the calculation of each set of transmission line parameters requires the solution of a set of only N_c electrostatic problems instead of $N_c(N_c+1)/2$ as for the standard formulation.

Since the formulation is obtained from a complementary functional of (3.50c), the values of the parameters obtained through this formulation represent bounds of the exact values of an opposite nature than the bounds obtained from the direct formulation. In this book, the mixed formulation has been used for this purpose. However, as it has been already mentioned, the mixed formulation has several advantages with respect to the standard or direct formulation that justifies its use by itself. Appendix C.5.1 details the advantages and disadvantages of both formulations.

The div-conforming/mixed formulation has been validated using the same test structures as in the case of the Lagrange/direct formulation. A convergence study has been performed. The rate of convergence obtained agrees in all cases with the theoretical predictions for div-conforming elements (see Chapter 2). Further details are not given herewith. The following applications illustrate the use of both the direct and the mixed formulation in order to obtain the upper and lower bounds of the transmission line parameters.

Table 3.22 provides results for the analysis of a square coaxial line. Refinement of the meshes is obtained through the self-adaptive process detailed in Chapter 4. It may be seen how both methods converge to the exact solution. The direct method provides an upper bound and the mixed formulation, a lower bound of the solution. The mean of the values obtained at each step of the mesh using both the procedures gives a more accurate value. In this example, the accuracy of the solution obtained using averaging is one order of magnitude better than each of the bounds separately. When this procedure is used, one can reduce the number of steps required to obtain a solution with a prespecified degree of accuracy. Of course the structure must be analyzed twice.

Table 3.22

Results for the normalized capacity C/ε_0 of a square coaxial line

Mesh Steps	First-order elements			Second-order elements		
	Direct	Mixed	Mean	Direct	Mixed	Mean
1	12.000	9.1428	10.5714	10.4592	10.000	10.2296
2	10.8461	9.6986	10.2723	10.3273	10.1059	10.2166
3	10.4718	9.9818	10.2668	10.2725	10.1823	10.2274
4	10.3697	10.0928	10.2315	10.2500	10.2130	10.2315
Exact: [Daly 1984], [Carson and Cambrell 1966]	10.2341					

As shown in the previous section, the square coaxial line can also be analyzed using the dual direct formulation. The next example illustrates a case in which the dual formulation is not adequate. Figure 3.39 shows an asymmetric structure. The analyzed transmission line structure is composed of two zero-thickness lines coupled in an asymmetric fashion. They are embedded in a dielectric and placed between two ground planes separated by 3 units. If this structure is to be analyzed through the dual direct formulation, two double-faced artificial boundaries will be required. Each one of them will be located between each active conductor and the ground plane. A jump of 1 V for the potential function should be defined across each double-faced boundary. That will complicate the mesh generation of the structure and, eventually, the self-adaptive mesh procedure. Instead, the structure is easily analyzed through the mixed formulation.

Table 3.23 shows the results obtained using second-order triangular div-conforming elements and its comparison with those obtained in [Djordjević et al. 1996] using the method of moments (MOM) (the software package, LINPAR) and by [Medina and Horno 1987] using a spectral method. A good agreement can be observed. The table gives the values of the coefficients of matrix $[C]$ and matrix $[C_{eq}]$ obtained by replacing the dielectric medium by vacuum.

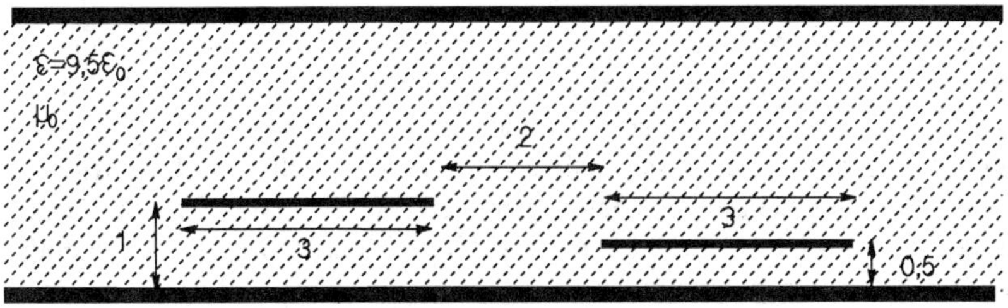

Figure 3.39 Asymmetric coupled lines between two infinite ground planes.

Table 3.23
Coefficients of matrices $[C]$ and $[C_{eq}]$ of the structure of Figure 3.39

	Direct FEM	Mixed FEM	Mean value	[Djordjević et al. 1996]	[Medina and Horno 1987]
C_{11}(*1.E-9)	0.5568	0.5089	0.5329	0.5356	0.5320
C_{12}(*1.E-10)	-0.1205	-0.0826	-0.1015	-0.0925	-0.1000
C_{22}(*1.E-9)	0.8153	0.7468	0.7810	0.7834	0.7790
C_{eq11}(*1.E-10)	0.5861	0.5357	0.5609	0.5466	0.5600
C_{eq12}(*1.E-11)	-0.1269	-0.0869	-0.1069	-0.0943	-0.1062
C_{eq22}(*1.E-10)	0.8582	0.7861	0.8221	0.7994	0.8200

3.2.4 Conclusions

The application of the FEM to the quasi-static analysis of transmission lines has been described. The primary and secondary parameters characterizing TEM and quasi-TEM transmission lines as well as the electromagnetic fields of TEM and quasi-TEM fundamental modes propagating along the line may be computed through FEM. The direct formulation that employs Lagrange elements has been described. A study of the rate of convergence has been performed.

The dual direct formulation allows one to obtain bounds of opposite nature to those obtained through its primal counterpart. However, the straightforward application of this formulation is limited to structures with particular symmetry conditions.

The mixed formulation that employs div-conforming elements has been proposed as an alternate procedure for bounding the error of a FEM solution. By itself, it is an attractive alternative to Lagrange element-based FEM.

The structures analyzed through this validation procedure of the FEMs presented have been of simple geometry. Nevertheless, they have been useful in illustrating the weak points of FEM. Namely, if the mesh of the structure is not well adapted to the exact electromagnetic field behavior, a poor rate of convergence is obtained. This is an important issue if one takes into account that the characteristics of the geometry and the configuration of most transmission lines involve conductor or dielectric edges and corners, abrupt changes in the constants characterizing the materials, and discontinuities in boundary conditions, coupling, and proximity effects leading to nonsmooth enough solutions for the electromagnetic field.

It has been shown how the rate of convergence may be improved by introducing refinements in the mesh for regions where the fields are not smooth. However, this selective refinement has been carried out in an intuitive way and without an error estimation or indication of which elements are responsible for the higher errors within a given mesh. It is known that if equilibrated meshes can be provided (meshes where all elements produce similar errors), then the maximum rate of convergence will be achieved. Hence, it seems most convenient to look for a method that will compute the local error estimates or indicators so that, taking into account this information, the mesh could be automatically refined in selected regions. Hence, equilibrated meshes will result, and the highest rate of convergence provided by the order of the element used will be obtained. This is the objective of Chapter 4.

3.3 FULL-WAVE ANALYSIS OF WAVEGUIDING STRUCTURES UTILIZING FINITE ELEMENT METHOD

This section deals with the application of the FEM to the computation of the TE, TM, and hybrid modes that propagate through general waveguiding structures. These modes either are higher order modes of TEM transmission lines or the fundamental and higher order modes of quasi-TEM transmission lines, metallic waveguides, dielectric waveguides, and finlines, for example.

First, the waveguiding structures to be analyzed are described. A brief summary of the FEM/full-wave formulations follows. Next, the formulations developed in the chapter are presented. Their details may be found in Appendix F. The first formulation makes use of the longitudinal components of the electric and magnetic fields and utilizes Lagrange elements. Although this may be used for general waveguiding structures, however, because of inherent numerical problems, it is mainly utilized for isotropic homogeneous structures. Some results of its application are given. The variables of the second formulation are the transverse and longitudinal components of the electric or the magnetic field. It uses a curl-conforming/Lagrange element. A study of the rate of convergence is carried out. The section finishes with some conclusions regarding the need for developing self-adaptive mesh procedures for full-wave analysis.

3.3.1 Description of the Geometry and Configuration of the Structures To Be Analyzed

The TEM or quasi-TEM fundamental modes that propagate along a waveguiding structure with two or more conductors embedded in one or more dielectrics have been discussed in Section 3.2 using a quasi-static approximation. In this section a full-wave analysis is used that allows the study of higher order (TE or TM) modes of TEM structures and the frequency dependence of the fundamental modes (or quasi-TEM modes) and the higher order hybrid modes propagating in multiconductor inhomogeneous anisotropic structures. In other words, the first type of waveguiding structures to be considered is the same as that described in Section 3.2.1. What is different in this section is the type of analysis, in which no approximation is made, so the frequency dependence of the electromagnetic field of any desired mode may be investigated.

Additionally the TE, TM, or hybrid modes that propagate along waveguiding structures, either with just one conductor and one or more dielectrics (such as metallic waveguides, non-TEM transmission lines, and dielectric waveguides) or just consisting of several dielectrics (dielectric waveguides) with no conductors involved, may be analyzed through the same formulations described.

In all cases, the propagating medium can be homogeneous and isotropic or inhomogeneous and anisotropic. For the latter case, the type of anisotropy to be considered has been described in Appendix D.2. The propagating medium as well as the conductors may not be ideal, i.e., lossy. However, for the computation of the electromagnetic field configuration they will be assumed to be perfect. A perturbational method will be used for the calculation of losses.

3.3.2 Survey of Various Formulations of the Waveguiding Problem Utilizing the Finite Element Method

In Appendix D.3.2, Maxwell's equations in a source-free medium specialized for waveguiding structures have been described. There is a wide variety of possibilities for the formulation of the waveguiding problem, since some of the field components can be expressed as functions of others.

In some structures, the number of unknowns can be reduced to one, as in the case of the analysis of homogeneous isotropic metallic waveguides propagating TE or TM modes. For this case, the longitudinal component of the magnetic field H_z or the electric field E_z is usually chosen [Ahmed 1968], [Ahmed and Daly 1969a], [Alvarez-Melcón et al. 1994], [Daly 1974], [García-Castillo et al. 1994, 1995], [Gil and Zapata 1994, 1995, 1997], [Israel and Miniowitz 1987], [Jin 1993; Ch. 7], [Lagasse and Van Bladel 1972], [Laura et al. 1980], [Sarkar et al. 1994], [Silvester 1969a,b], [Silvester and Ferrari 1983; Ch. 3]. These types of formulations are called scalars. Exact or approximate scalar formulations can also be obtained for the study of modes that propagate along structures in which, due to some particular type of symmetry, some field components may be set equal to zero or have a negligible value [Koshiba and

Suzuki 1982], [Koshiba et al. 1982b, 1984b, 1985c], [Mabaya et al. 1981], [Vandenbulcke and Lagasse 1976].

Besides these cases, the analysis of waveguiding structures with general configuration and geometry requires formulations with two or more field components as unknowns. These types of formulations are usually called vector formulations. This terminology must be understood in the sense that the problem is formulated with more than one variable as unknown (see Section 2.5.1.2). A summary of the different possibilities may be found in [Koshiba 1992; Ch. 3].

Some authors have used the longitudinal component of the electric field E_z and that of the magnetic field H_z [Ahmed and Daly 1969b], [Aubourg et al. 1983], [Chang et al. 1990, 1991], [Csendes and Silvester 1970], [Daly 1971], [Eswarappa et al. 1989], [Hernández-Figueroa and Pagiatakis 1993], [Ikeuchi et al. 1981], [Jin 1993; Ch. 7], [Mabaya et al. 1981], [McAulay 1977], [Okamoto and Okoshi 1978], [Tzuang and Itoh 1986], [Vandenbulcke and Lagasse 1976], [Welt and Webb 1985], [Wu and Chen 1985a].

Others employ the three components of the electric field $\bar{E}$ and of the magnetic field $\bar{H}$ [Svedin 1989]; the transverse components of the electric field $\bar{E}_t$ and of the magnetic field $\bar{H}_t$ [Angkaew et al. 1987], [García-Castillo 1992], [García-Castillo and Salazar-Palma 1992a], [Madrangeas et al. 1994]; the field components $\bar{E}_t$ and H_z [Ohtaka and Kobayashi 1990]; the electric field $\bar{E}$ or the magnetic field $\bar{H}$ [Alam et al. 1993, 1994b, 1997], [Bárdi et al. 1993], [Bermúdez de Castro and Gómez-Pedreira 1992], [Blanc-Castillo 1994], [Blanc-Castillo et al. 1995], [García-Castillo 1992], [García-Castillo and Salazar-Palma 1992a,b], [Hano 1984], [Hayata et al. 1986], [Helal et al. 1994], [Israel and Miniowitz 1990], [Jin 1993; Ch. 8], [Koshiba and Inoue 1992], [Koshiba et al. 1984a, 1985a, 1986, 1994], [Lee et al. 1991a], [Lee 1994], [Nuño et al. 1997], [Rahman and Davies 1984a,b], [Salazar-Palma et al. 1994], [Valor and Zapata 1995, 1996]; and any one of the transverse components $\bar{E}_t$ or $\bar{H}_t$ [Fernández and Lu 1991], [Fernández et al. 1993], [Israel and Miniowitz 1990], [Lu and Fernández 1993a, 1994].

Also there is the possibility of using the scalar potential ϕ and the vector potential $\bar{A}$ as unknowns [Bárdi and Biro 1991], [Bárdi et al. 1994], [Boyse et al. 1992, 1993], [Paulsen et al. 1992].

The choice of the formulation depends on the type of analysis desired. Thus, for instance, if the losses are not required, it will be convenient to look for formulations that lead to real symmetric matrices. In that case, the problem may be positive definite. This allows the use of highly effective solvers. For similar reasons, sparse matrices are quite convenient. Thus, whenever possible, formulations should lead to sparse matrices. However, the main issue is the type of elements to be used. A suitable combination of formulation and type of elements gives highly efficient codes.

Besides Hermite elements [Israel and Miniowitz 1987, 1990], and other special basis functions (such as *wavelet*-like [García-Castillo et al. 1994, 1995], [Sarkar et al. 1994]), the most widely used elements are Lagrange elements and those belonging to the family that may be called vector elements. The div-conforming and curl-conforming elements described in Section 2.5.1.2 belong to the vector family. Their properties as well as those of Lagrange elements and Lagrange/curl-conforming elements are discussed in Section 2.5.1.2 and Appendix C.

Classical FEMs use Lagrange elements in which the unknowns are scalar quantities (i.e., in the context of this book, each of the Cartesian components acting as unknowns) and approximated by a linear combination of Lagrange basis functions. The coefficients of the linear combination are the degrees of freedom of the discrete problem. The degrees of freedom are defined as the values of the unknowns at the nodes. Element vertices are a subset of nodes. This procedure leads to continuous solutions over the problem domain belonging to the space $H^1(\Omega)$. Therefore, for waveguiding problems, Lagrange elements are suitable for formulations based on longitudinal field components. The transverse components will be obtained through a postprocess. Thus, they will be calculated with less accuracy than the longitudinal ones. However, one is usually more interested in the transverse components than on the longitudinal components. In fact, the magnitude of the former ones is greater than that of the latter ones. In this sense, this formulation may not be too convenient. Additionally, if a vector formulation (more than one unknown) is used for given values of the propagation constant, spurious modes will appear. The generation of spurious modes is due to the undefined character of the formulation and to the nonfulfillment of field continuity conditions at media interfaces. Spurious modes may be discriminated from the real physical ones since the former have a chaotic nature and do not always satisfy the boundary conditions. A tedious but possible procedure is to plot them and eliminate those modes showing chaotic field components. For scalar formulations (one unknown only), spurious modes will not occur.

Lagrange elements have also been used in formulations based on E_x, E_y and/or H_x, H_y components (and, eventually, E_z and/or H_z) [Fernández and Lu 1991], [Hayata et al. 1986, 1988, 1989], [Koshiba et al. 1985a,b],[Paulsen and Lynch 1991], [Rahman and Davies 1984a,b], [Svedin 1989]. Depending on the specific formulation, it becomes necessary to force additional conditions at media interfaces in order to avoid incoherences of the behavior of the electric field across them. This technique does not only complicate the formulation but reduces or destroys the sparsity of the matrices of the discrete problem and its band structure. The convergence in cases of structures with re-entrant corners may not be good enough. Besides, spurious modes pollute the spectrum of the physical modes. This is the worst problem that arises from these formulations. The generation of spurious modes has been attributed to the nonfulfillment of Maxwell's divergence equation that, for a source-free region, implies divergence-free fields. The problem of spurious modes is addressed in detail in Section 7.2, in the broader context of both 2D and 3D problems. The origin of these spurious modes is an incorrect modeling of the discrete solutions corresponding to the null space of the operator. Researchers have used different methods or modifications of the original formulations to avoid or, at least, to differentiate real modes from spurious ones. Among them (see Section 7.2), it may be mentioned penalty function methods [Koshiba et al. 1984b], [Rahman and Davies 1984a,b] and the elimination of the longitudinal components using the divergence equation either at the stage of the continuous problem formulation or after the global matrices of the discrete problem have been obtained

[Fernández and Lu 1991], [Hayata et al. 1986], [Israel and Miniowitz 1990], [Koshiba 1992; Ch. 3]. Most of the techniques utilized to solve in one manner or the other the spurious modes problem, modify given characteristics of the discrete problem. This fact may represent a serious inconvenience, such as resulting in non-self adjoint problems, a higher number of unknowns, nonsparse matrices, asymmetric or undefined matrices.

Vector elements approximate vector unknowns using a linear combination of vector basis functions. The degrees of freedom are related to the tangential component of the vector unknown at nodes placed at the edges of the curl-conforming elements (or related to the normal component for div-conforming elements). For higher order elements, degrees of freedom with a different definition are required. The vertices of the element are not a subset of the nodes. These elements provide solutions belonging to the space $H(\text{curl},\Omega)$ (or $H(\text{div},\Omega)$) with continuity of the tangential component across media interfaces (or of the normal component). They are therefore consistent with the electromagnetic field behavior for $\bar{E}_t$ and $\bar{H}_t$ (or for $\bar{D}_t$ and $\bar{B}_t$). The fact that the vertices of elements are not nodes provides a better convergence in cases of structures with re-entrant corners. Also the imposition of boundary conditions is easily done. The Lagrange/curl-conforming elements combine curl-conforming elements (that discretize transverse vector components) and Lagrange elements (that approximate longitudinal components). They give solutions belonging to the space $H(\text{curl},\Omega)\times H^1(\Omega)$.

From 1977, researchers in the field of applied mathematics have introduced various types of vector elements (as opposed to the elements discretizing scalar unknowns). The lower order ones may be seen as particular cases of a wider family of finite element shape functions, first described by H. Whitney [Whitney 1957]. The first vector elements are the elements of P. A. Raviart and J. M. Thomas [1977a]. Then follow the basis functions of the first type of J. C. Nédélec [Nédélec 1980], those of F. Brezzi, J. Douglas, and L. D. Marini [Brezzi et al. 1985], and those of the second type of J. C. Nédélec [Nédélec 1986].

In microwave engineering, M. Hano in 1984, was the first to publish results with a vector element, different from the previous ones, that suppressed spurious modes [Hano 1984]. Since then, different vector elements have been proposed, like [Mur and Hoop 1985], [Angkaew et al. 1987], [Barton and Csendes 1987], [Csendes 1991], [Lee et al. 1991a,b], [Miniowitz and Webb 1991], [Bermúdez de Castro and Gómez-Pedreira 1992], [Koshiba and Inoue 1992], [Wang and Ida 1993], [Blanc-Castillo 1994], [Helal et al. 1994], [Koshiba et al. 1994], [Peterson 1994], [Blanc-Castillo et al. 1995], [Peterson and Wilton 1996], [Graglia et al. 1997]. Some of them are based on the elements proposed by Raviart and Thomas, Nédélec, or Brezzi and coworkers, while others use different definition for the space where the solution is sought, the degrees or freedom, or the location of nodes. For obvious reasons, the lower order ones (using first-order polynomials) were used at the beginning. The first works utilizing higher order vector elements for both $H(\text{div})$ and $H(\text{curl})$ spaces appear in 1991, like, for example [Salazar-Palma et al. 1991], [Lee et al. 1991a,b]. Curved elements (for linear and higher order approximations) have been used in [Blanc-Castillo 1994] and [Blanc-Castillo et al. 1995], for example. All authors using vector elements report the suppression of the spurious modes or, to be more precise, the correct modeling of them (see Section 7.2).

The vector elements utilized in this book and in particular those used for the full-wave analysis of waveguiding structures are the 2D Nédélec elements of the first type of first and second order [Nédélec 1980]. A curved version for both orders has been developed. Following the work of [Raviart and Thomas 1977a] and [Thomas and Joly 1981; Ch. 3], the degrees of freedom related to edges are defined at the Gauss-Legendre points of integration. This definition has certain advantages from a practical implementation point of view and for the postprocessing of the solution. Other authors do not follow this approach [Lee et al. 1991a,b]. The definition of the degrees of freedom related to the surface of the element is also that given in [Nédélec 1980], and again is different from that used by other authors [Lee et al. 1991a,b], [Peterson 1994]. As a consequence, the basis functions of higher order elements used in this book differ from those used by the authors just mentioned and others (see Section 2.5.1.2 and Appendix C.4). To be more precise, the elements used in this chapter for the discretization of a formulation based on the longitudinal and transverse components of the electric or the magnetic field are the Lagrange/curl-conforming elements described in Section 2.5.1.2. The first order element is the same one used in [Bermúdez de Castro and Gómez-Pedreira 1992] except for a normalization constant in the definition of the degree of freedom. These authors demonstrate the correct modeling of the spurious modes by curl-conforming elements. It is also the same element used in [Koshiba and Inoue 1992], however in their implementation, these authors do not retain the sparsity of the FEM matrices since they suppress the degrees of freedom related to the longitudinal component at the global level. The higher order elements for both 2D and 3D (see Chapter 7) problems implemented in this book differ from other higher order elements used by other authors in the field of microwave engineering.

3.3.3 Full-Wave Analysis of Waveguiding Structures Using the Finite Element Method

Two full-wave formulations are going to be described. The first one uses the longitudinal component of the electric field E_z and/or the magnetic field H_z as unknowns. The related finite element procedure is based on Lagrange elements. The second one utilizes a formulation based on the electric field $\bar{E}$ or the magnetic field $\bar{H}$ and more specifically, on their transverse and longitudinal components. It uses curl-conforming/Lagrange elements. Details about the strong and weak formulations and their FEM discretization are given in Appendix F.2 and F.3, respectively. For the second formulation, in the multiconductor case, a postprocessing module has been developed that allows one to compute the approximate frequency-dependent primary parameter matrices (the p.u.l. inductance, capacitance, conductance, and resistance matrices) and secondary parameter matrices (the characteristic impedance and attenuation constant matrices). The following subsections describe these formulations and their validation through the analysis of test structures.

3.3.3.1 *Formulations Using Longitudinal Field Components: Lagrange Elements*

Consider a waveguide structure with an arbitrary number of conductors and dielectric media (one or more conductors and one or more dielectric media, or two or more dielectric media). The waveguide problem can be formulated using the longitudinal components of the electric and magnetic fields as described in Appendix F.2. The formulation can be specialized to the case of an homogeneous and isotropic medium, obtaining two different formulations: one corresponding to the general case (Appendix F.2.1) and another for the homogeneous and isotropic case (Appendix F.2.2).

Inhomogeneous and Anisotropic Structures

In Appendix D.3, the classical formulation leading to the weak formulation discussed in Appendix F.2.1.1 has been presented. It results in an eigenvalue problem where k_0^2 (the square of the wave number in vacuum) is the eigenvalue, while the eigenvector has two coefficients, E_{zn} and H_{zn} (the longitudinal components, E_z and H_z, normalized with respect to a given factor). The quantity $\beta_n = \beta/k_0$ is the input. However, the system matrices are not definite for all the values of β_n. For certain values, the matrices involved become singular. Thus, the input data must be carefully chosen.

First- and second-order straight and curved Lagrange triangular elements have been used for its discretization, following the normal procedure that is described in detail in Appendix F.2.1.2. As indicated in this section, the generation of spurious modes in this formulation is unavoidable. However, their discrimination is relatively simple since the eigenvectors introduce a chaotic behavior. To validate this formulation several inhomogeneous structures have been analyzed.

a) *Rectangular waveguide half-filled with dielectric*

Figure 3.40 shows the structure of dimensions $b = a/2$, $d = b/2$, with $\varepsilon_r = 4$ and $\mu_r = 1$. Half of it has been discretized. A magnetic wall is first placed along the central plane and, then, an electric wall. In this way the first modes with even or odd symmetry are obtained. A uniform mesh of 72 first-order elements (49 nodes) and second-order elements (169 nodes) has been used. Figures 3.41 and 3.42 plot β_n^2, which is related to the effective relative permittivity ε_{ref}. A comparison with analytic results is given in [Jin 1993; Ch. 7], [Harrington 1961; Ch. 4]. The effect of the singularities, intrinsic to this formulation, may be observed. There are convergence problems in the proximity of $\beta_n = 1$. Dashed lines represent the spurious modes. This is a drawback of this formulation. As seen from Figures 3.41 and 3.42 they appear as modes without dispersion, polluting all the spectrum.

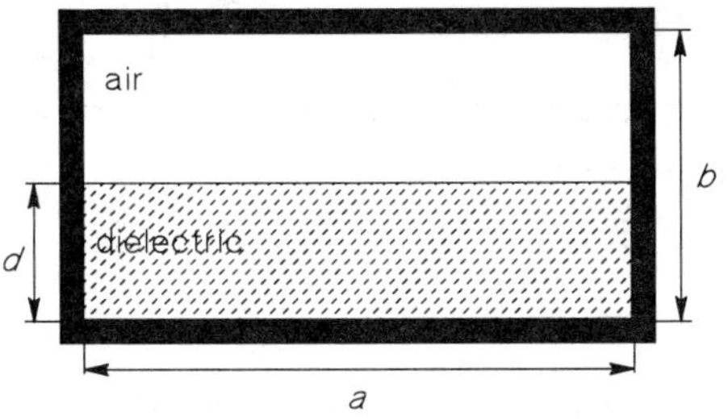

Figure 3.40 Rectangular waveguide half-filled with dielectric.

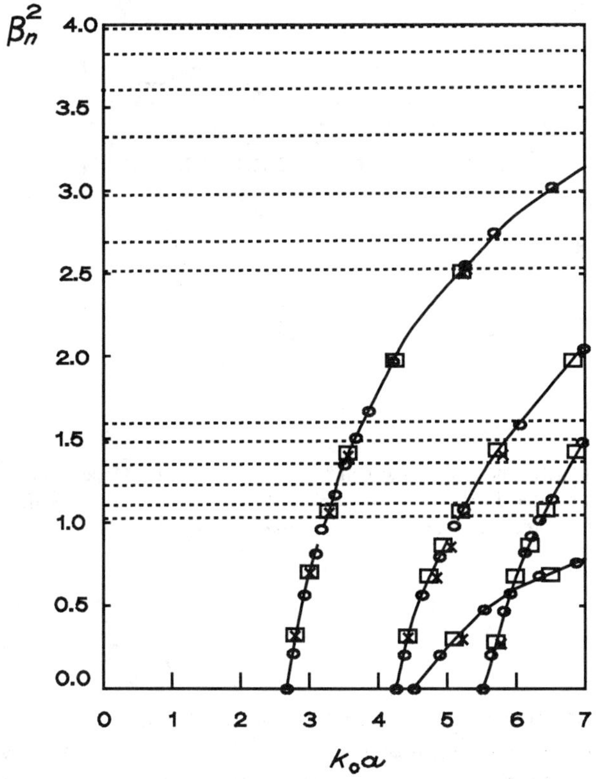

Figure 3.41 Dispersion for the waveguide half-filled with dielectric. Modes with even symmetry. —: [Harrington 1961; Ch. 4]; o: [Jin 1993; Ch. 7]; - -: spurious modes [Jin 1993; Ch. 7]; ×: first-order elements; □: second-order elements.

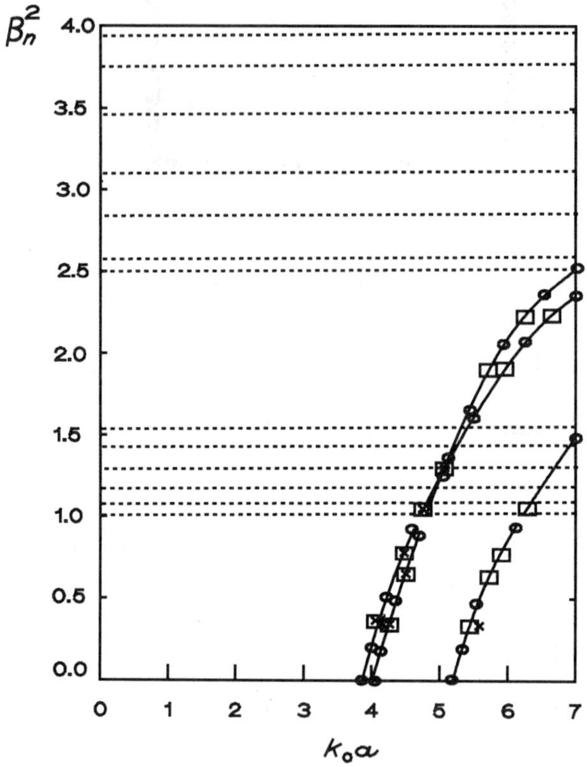

Figure 3.42 Dispersion for the waveguide half-filled with dielectric. Modes with odd symmetry. —: [Harrington 1961; Ch. 4]; o: [Jin 1993; Ch. 7]; - -: spurious modes [Jin 1993; Ch. 7]; ×: first-order elements; □: second-order elements.

b) *Microstrip line*

Figure 3.43 shows a microstrip line. Its dimensions are $b = a/2$, $d = b/2$, $w = b$, with $\varepsilon_r = 4$ and $\mu_r = 1$. Half of the line has been analyzed utilizing a magnetic wall along the symmetry plane. A local refinement of the mesh at the corners of the conductors has been used to enhance the rate of convergence of the second-order elements. Figure 3.44 shows the dispersion of β_n^2 and compares it with the results of [Jin 1993; Ch. 7]. Slow convergence of the higher order modes, erroneous results in regions close to the singularities, and the presence of spurious modes are problems associated with this approach. Because of these problems, this formulation has been discarded.

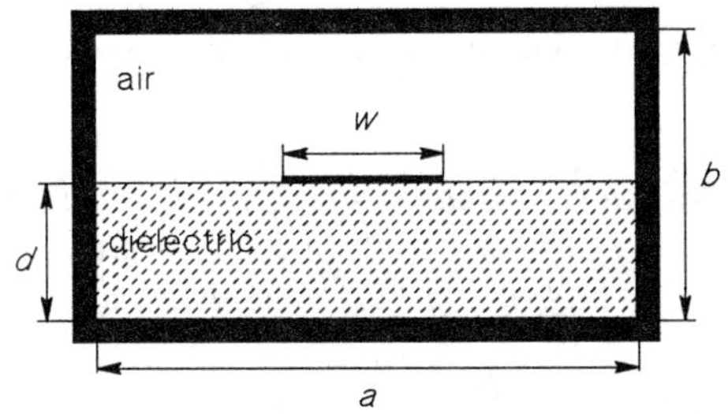

Figure 3.43 Microstrip line.

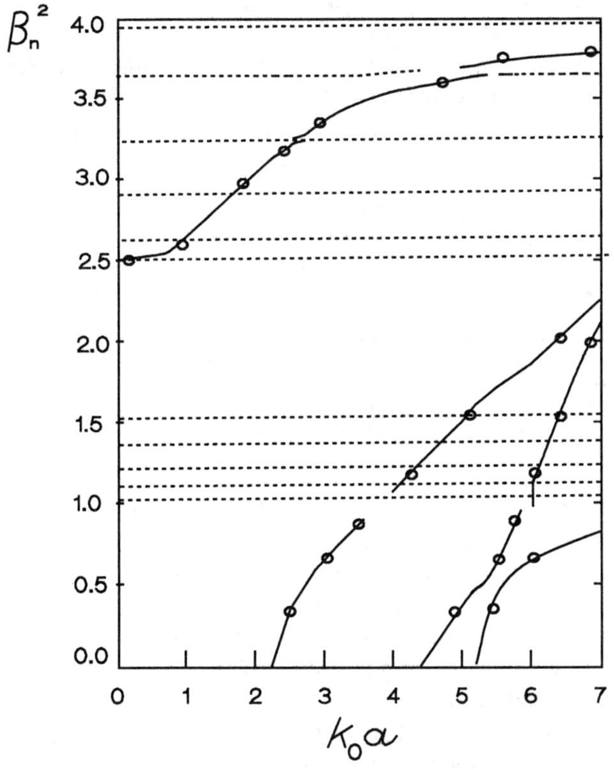

Figure 3.44 Dispersion for a microstrip line. Modes with even symmetry. —: [Jin 1993; Ch. 7]; -: Spurious modes [Jin 1993; Ch. 7]; o: this method.

Homogeneous and Isotropic Structures

The weak formulation for this class of problems is discussed in Appendix F.2.2.1 and can be summarized as:

Let $H_0^1(\Omega)$ be the space of scalar functions φ such that $\varphi \in H^1(\Omega)$ and $\varphi = 0$ on Γ_D, where Γ_D is composed of those parts of the boundary Γ of the domain Ω with Dirichlet conditions. Find k_c and $\phi \in H_0^1(\Omega)$ such that the functional

$$F = \frac{1}{2}\left(\int_\Omega \bar{\nabla}_t \phi \cdot \bar{\nabla}_t \phi \, d\Omega - k_c^2 \int_\Omega \phi^2 \, d\Omega\right) \tag{3.71}$$

is made stationary, where k_c is the cut-off wave number $k_c^2 = k_0^2\,(\varepsilon_r\,\mu_r - \beta_n^2)$.

Table 3.24 provides the interpretation for ϕ and Γ_D depending on whether one is analyzing TE or TM modes. The weak formulation leads to a positive semidefinite eigenvalue problem where k_c^2 is the eigenvalue and ϕ is the eigenvector. Once the problem is solved, using k_c and ϕ any of the field components corresponding to a particular propagation constant may be obtained, as shown in Appendix F.2.2.1.

Table 3.24
Interpretation for ϕ and Γ_D for TM and TE mode analysis

	TM	TE
ϕ	E_z	H_z
Γ_D	Electric walls	Magnetic walls

For the discretization of the continuous problem straight and curved, first- and second-order Lagrange triangular elements are used. The method is described in detail in Appendix F.2.2.2. The efficiency, convergence, and performance of this scalar formulation is well known. Table 3.25 shows the relative error obtained in the calculation of the square of the normalized cut-off wave number of the first five modes of a rectangular $2a \times a$ waveguide using first-order elements and meshes of 144 and 576 elements (91 and 325 nodes, respectively). Figure 3.45 shows the rate of convergence and compares it with the theoretical prediction. Since the solution is smooth, the rate of convergence should be ruled by the order of the approximation. Thus, for the eigenvalue, it is given by N_d^{-p}. It may be observed that the cut-off wave numbers of the degenerate modes (TE_{20} and TE_{01}, in this case) are numerically distinguished. Also, they are obtained with lower accuracy than the rest due to the numerical difficulty of computing two close eigenvalues. Superconvergence effects are observed for the TE_{11} and TM_{11} modes.

Table 3.25
Relative error, in %, for the computation of the s_c^2 of the first five modes of a rectangular waveguide

	TE_{10}	TE_{20}	TE_{01}	TE_{11}	TM_{11}
144 elements: 91 nodes	0.72	2.86	2.78	0.11	0.18
576 elements: 325 nodes	0.2	0.8	0.75	0.03	0.05

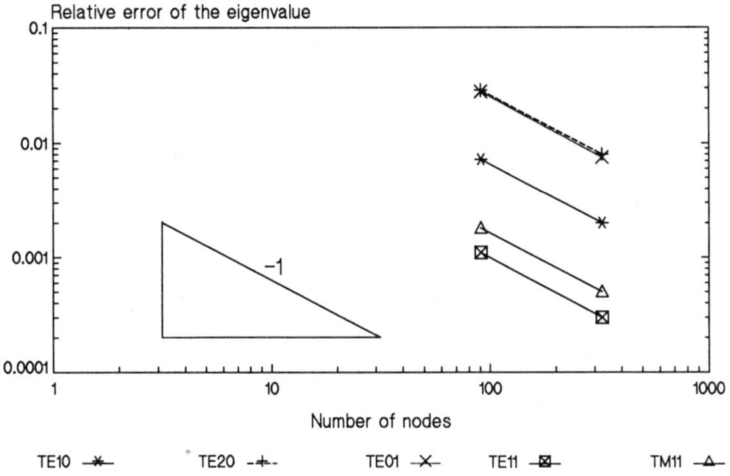

Figure 3.45 Rate of convergence for the square of the cut-off wave numbers of the rectangular waveguide.

3.3.3.2 *Formulations Using Transverse and Longitudinal Field Components: Lagrange/Curl-Conforming Elements*

The double curl equations arising from Maxwell's equations for source-free media (see Appendix D) results in a selfadjoint positive semidefinite system that is derived from the functional

$$F = \int_\Omega \left(\bar{\nabla} \times \underline{\bar{V}}^*\right) \cdot \left(\underline{\underline{f}}^{-1}\, \bar{\nabla} \times \underline{\bar{V}}\right) d\Omega - \omega^2 \int_\Omega \underline{\bar{V}}^* \cdot \underline{\underline{p}}\, \underline{\bar{V}}\, d\Omega \tag{3.72}$$

where $\underline{\bar{V}}$, $\underline{\underline{f}}$, and $\underline{\underline{p}}$ are described in Table 3.26, depending on whether the problem is analyzed in terms of the electric or the magnetic field. One should recall that the functional dependence of $\underline{\bar{V}}$, is given by $\underline{\bar{V}}(x,y,z) = \bar{V}(x,y)e^{-j\beta z}$.

Specializing functional (3.72) for waveguiding problems and utilizing the normalizations detailed in Appendix F.3.1.1 one arrives at the integral form

Table 3.26
Interpretation for the variables of the general functional (3.72) depending on whether it is an electric field formulation or a magnetic field one

$\bar{V} = \bar{V}(x,y,z)$	$\bar{E} = \bar{E}(x,y,z)$	$\bar{H} = \bar{H}(x,y,z)$
$\underline{\underline{f}}$	$\underline{\underline{\mu}}$	$\underline{\underline{\varepsilon}}$
$\underline{\underline{p}}$	$\underline{\underline{\varepsilon}}$	$\underline{\underline{\mu}}$

$$F_n = \frac{c_0^2}{2}\left[A_n - k_0^2 B_n\right] \tag{3.73a}$$

where

$$\begin{aligned} A_n = {} & \int_\Omega \bar{\nabla}_t \times V_{zn}\bar{a}_z \cdot \underline{\underline{f}}_{rt}^{-1}\, \nabla_t \times V_{zn}\bar{a}_z \, d\Omega + \beta^2 \int_\Omega \bar{a}_z \times \bar{V}_{tn} \cdot \underline{\underline{f}}_{rt}^{-1}\, \bar{a}_z \times \bar{V}_{tn} \, d\Omega \\ & - 2\beta \int_\Omega \bar{\nabla}_t \times V_{zn}\bar{a}_z \cdot \underline{\underline{f}}_{rt}^{-1}\, \bar{a}_z \times \bar{V}_{tn} \, d\Omega + \int_\Omega \bar{\nabla}_t \times \bar{V}_{tn} \cdot f_{rzz}^{-1}\, \bar{\nabla}_t \times \bar{V}_{tn} \, d\Omega \end{aligned} \tag{3.73b}$$

$$B_n = \int_\Omega \left(\bar{V}_{tn} \cdot \underline{\underline{p}}_{rt}\bar{V}_{tn} + V_{zn}\, p_{rzz}\, V_{zn}\right) d\Omega \tag{3.73c}$$

and c_0 is the velocity of light in a vacuum.

The functional obtained is real, symmetric, and positive semidefinite. Table 3.27 defines the different variables depending on whether one is using an electric field or a magnetic field formulation. The variable $\bar{U}$ refers to the dual variable (the magnetic field or the electric field, respectively, and vice versa).

The weak formulation can be stated as follows:

Let $X^0(\Omega)$ be the vector function space consisting of elements $\bar{q}$, in the 2D domain Ω, such that $\bar{q} \in H(\mathrm{curl},\Omega)$ and $\bar{a}_n \times \bar{q} = 0$ on Γ_D. Let $H_0^1(\Omega)$ be the space of the scalar functions consisting of elements ψ such that $\psi \in H^1(\Omega)$ and $\psi = 0$ on Γ_D, where $\Gamma = \Gamma_D \cup \Gamma_N$, is the boundary of Ω. Given β, $\underline{\underline{f}}_r$ and $\underline{\underline{p}}_r$ in Ω, where both the tensors may be decomposed into their transverse ($\underline{\underline{f}}_{rt}$ and $\underline{\underline{p}}_{rt}$, respectively) and longitudinal (f_{rz} and p_{rz}, respectively) components. Assume that the inverse of $\underline{\underline{f}}_{rt}$ exists. Then find the triads k_0^2, $\bar{V}_{tn}$, and V_{zn} with $\bar{V}_{tn} \in X^0$ and $V_{zn} \in H_0^1(\Omega)$ which makes the functional (3.73) stationary.

The discretization of the weak formulation is carried out using a FEM that is based on triangular Lagrange/curl-conforming elements. Details are given in Appendix

F.3.1.2. The parent elements are described in Section 2.5.1.2, and Table 2.9. Straight and curved elements of first- and second- orders have been used. This leads to an eigenvalue problem, in which the input is the phase constant β (more specifically, the

Table 3.27

Interpretation for the variables involved in the formulation in terms of the electric field or the magnetic field

$\bar{V}$	$\bar{E}$	$\bar{H}$
$\bar{V}_t$	$\bar{E}_t$	$\bar{H}_t$
V_z	E_z	H_z
$\bar{V}_n = \sqrt{p_0}\ \bar{V}$	$\bar{E}_n = \sqrt{\varepsilon_0}\ \bar{E}$	$\bar{H}_n = \sqrt{\mu_0}\ \bar{H}$
$\bar{V}_{tn} = \sqrt{p_0}\ \bar{V}_t$	$\bar{E}_{tn} = \sqrt{\varepsilon_0}\ \bar{E}_t$	$\bar{H}_{tn} = \sqrt{\mu_0}\ \bar{H}_t$
$V_{zn} = \sqrt{p_0}\ V_z$	$E_{zn} = \sqrt{\varepsilon_0}\ E_z$	$H_{zn} = \sqrt{\mu_0}\ H_z$
$\bar{U}$	$\bar{H}$	$\bar{E}$
$\bar{U}_t$	$\bar{H}_t$	$\bar{E}_t$
U_z	H_z	$\bar{E}_t$
$\bar{U}_n = \sqrt{f_0}\ \bar{U}$	$\bar{H}_n = \sqrt{\mu_0}\ \bar{H}$	$\bar{E}_n = \sqrt{\varepsilon_0}\ \bar{E}$
$\bar{U}_{tn} = \sqrt{f_0}\ \bar{U}_t$	$\bar{H}_{tn} = \sqrt{\mu_0}\ \bar{H}_t$	$\bar{E}_{tn} = \sqrt{\varepsilon_0}\ \bar{E}_t$
$U_{zn} = \sqrt{f_0}\ U_z$	$H_{zn} = \sqrt{\mu_0}\ H_z$	$E_{zn} = \sqrt{\varepsilon_0}\ E_z$
f_0	μ_0	ε_0
$\underline{\underline{f}}_{rt}$	$\underline{\underline{\mu}}_{rt}$	$\underline{\underline{\varepsilon}}_{rt}$
f_{rzz}	μ_{rzz}	ε_{rzz}
p_0	ε_0	μ_0
$\underline{\underline{p}}_{rt}$	$\underline{\underline{\varepsilon}}_{rt}$	$\underline{\underline{\mu}}_{rt}$
p_{rzz}	ε_{rzz}	μ_{rzz}
sign	+1	-1
Dirichlet boundaries	Electric walls	Magnetic walls
Neumann boundaries	Magnetic walls	Electric walls

normalized value, $\beta_a = a\beta$, where a is a geometric factor as explained in Appendix F.3.1.2) and the eigenvalue is related to the square of the wave number k_0^2 (more specifically, the normalized one, $s^2 = a^2k_0^2$). The resulting matrices $[K]$ and $[M]$ are sparse, real, symmetric, and positive definite (more precisely, $[K]$ is positive semidefinite). This allows the use of highly efficient solution algorithms, especially adapted to these types of matrices. This formulation allows one to calculate the cut-off frequency of the modes just by introducing $\beta_a = 0$.

A variant of the previous formulation can be made that leads to an eigenvalue problem in which s is the input and β_a^2 the eigenvalue to be computed [Lee et al. 1991a]. However, for the latter case, the matrices are no longer positive definite. Details about this variant can be found in Appendix F.3.2.

Postprocesses

Once the discrete eigenvalue problem is solved for a given mode, its frequency dependent phase constant $\beta(\omega)$ and its electromagnetic field at any point of the waveguiding structure may be computed. Any related quantity, such as dielectric or conductor losses at a given frequency, may also be obtained.

Appendix C.3.2 gives some details with respect to the curl-conforming elements. Namely, the sign corresponding to the local degrees of freedom must be recovered in order to perform any local postprocess. Once the sign is recovered, the approximate transverse and longitudinal components of the unknown variable may be obtained through the evaluation of the expressions (F.75) and (F.76). The dual variable may be obtained through (F.58). Smoothing procedures similar to those described for the quasi-static analysis are used to obtain the electric and magnetic fields over the domain.

The power loss is calculated so that the value of the frequency-dependent attenuation constant, $\alpha(\omega)$, may be obtained for each mode, leading to the modal complex propagation constant, $\gamma(\omega) = \alpha(\omega) + j\beta(\omega)$. These postprocesses assume low losses so that perturbational methods may be applied.

A two conductor line may be characterized by a frequency-dependent characteristic impedance together with the propagation constant. In many cases, a complete matrix of frequency-dependent coefficients, called characteristic impedance matrix, $[Z_c(\omega)]$, is applied to characterize a multiconductor waveguiding structure together with the frequency-dependent diagonal propagation constant matrix, $[\gamma(\omega)] = [\alpha(\omega)] + j[\beta(\omega)]$. The aim is to use those two frequency-dependent concepts as in the case of a quasi-static multiconductor structure having an impedance matrix $[Z_c]$ and a diagonal propagation constant matrix $[\gamma] = [\alpha] + j[\beta]$. The diagonal coefficients are the frequency-dependent complex propagation constants of each of the N_c dominant modes of the waveguiding structure and may be easily computed for each frequency point. However, for the case of the characteristic impedance its definition is not unique because the static concepts of voltage and current no longer have a meaning. Several definitions are given in the literature. Here the definition using the transmitted power and the current is utilized since they are easily computed using FEM [Amari 1993]

[Mesa et al. 1992], [Slade and Webb 1991, 1992], [Wiemer and Jansen 1988]. Thus, a postprocessing module for the frequency-dependent characterization of multiconductor waveguiding structures must be used to obtain such parameters [Salazar-Palma et al. 1994].

Study of Convergence

The previous method has been applied to different test structures for its validation.

a) ***Rectangular waveguide***

Figure 3.46 shows the initial mesh for a $2a \times a$ rectangular waveguide and given steps of the uniform refinement process. First- and second-order elements have been used. Table 3.28 shows results corresponding to first-order elements for the eigenvalue, s^2, with $\beta_a = 0$ (i.e., the s_c^2 value corresponding to k_c^2) of the first five modes. The exact values are provided. The analysis has been done with both formulations in terms of the electric field $\overline{E}$ and in terms of the magnetic field $\overline{H}$. Figure 3.47 shows the rate of convergence that coincides with the theoretical prediction. The exact value of the solution is bounded from below and above by the results of both the formulations. It has been observed that the TE modes are obtained with more accuracy using the $\overline{E}$ formulation, while the TM modes are obtained with more accuracy with the $\overline{H}$ formulation. To check the expression $s^2 = s_c^2 + \beta_a^2$, s^2 has been calculated for different values of β_a. Figures 3.48 and 3.49 show the field plots for the first two modes. Tables 3.29 to 3.31 provide the error in the computation of the field for the TE_{10} mode. Table 3.32 shows results corresponding to second-order elements. Figure 3.50 shows the rate of convergence. Again the theoretical prediction is obtained. A superconvergent rate is obtained for some modes.

Table 3.28

$s_c^2 = (ak_c)^2$ obtained for the rectangular waveguide using first-order elements

Mode	45 nodes	325 nodes	1225 nodes	Exact value
TE_{10}	2.4247 $(\overline{E})$ 2.5842 $(\overline{H})$	2.4627 $(\overline{E})$ 2.4813 $(\overline{H})$	2.4662 $(\overline{E})$ 2.4709 $(\overline{H})$	2.4674
TE_{20}	9.1372 $(\overline{E})$ 11.7154 $(\overline{H})$	9.7934 $(\overline{E})$ 10.0910 $(\overline{H})$	9.8507 $(\overline{E})$ 9.9250 $(\overline{H})$	9.8696
TE_{01}	9.1960 $(\overline{E})$ 11.8514 $(\overline{H})$	9.7947 $(\overline{E})$ 10.0935 $(\overline{H})$	9.8508 $(\overline{E})$ 9.9257 $(\overline{H})$	9.8696
TE_{11}	12.2904 $(\overline{E})$ 16.9178 $(\overline{H})$	12.3334 $(\overline{E})$ 12.7962 $(\overline{H})$	12.3359 $(\overline{E})$ 12.4527 $(\overline{H})$	12.3370
TM_{11}	12.2508 $(\overline{H})$ 16.7405 $(\overline{E})$	12.3319 $(\overline{H})$ 12.8061 $(\overline{E})$	12.3358 $(\overline{H})$ 12.4535 $(\overline{E})$	12.3370

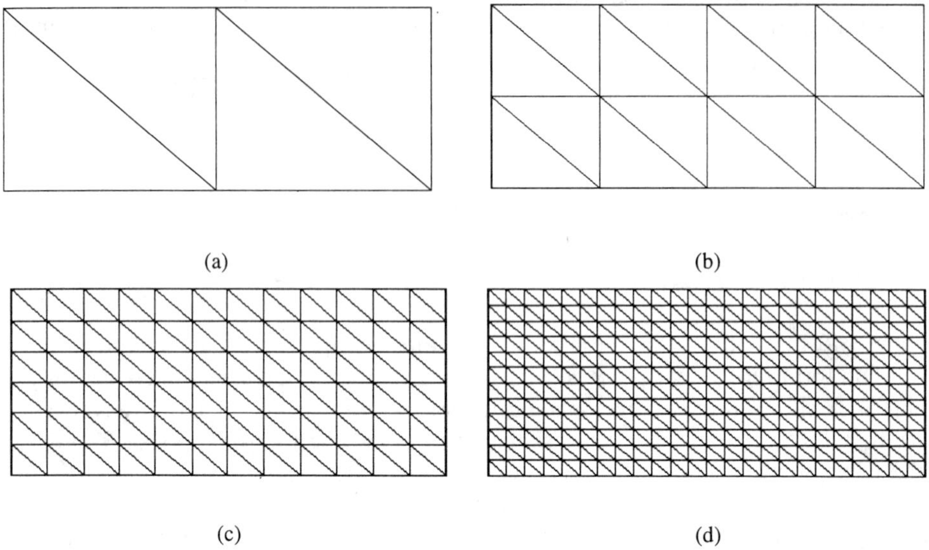

Figure 3.46 Meshes for the analysis of the rectangular waveguide: (a) 4 elements; first-order: 15 nodes, second-order: 45 nodes, (b) 16 elements; first-order: 45 nodes, second-order: 137 nodes, (c) 144 elements; first-order: 325 nodes, second-order: 1081 nodes, and (d) 576 elements, first-order: 1225 nodes.

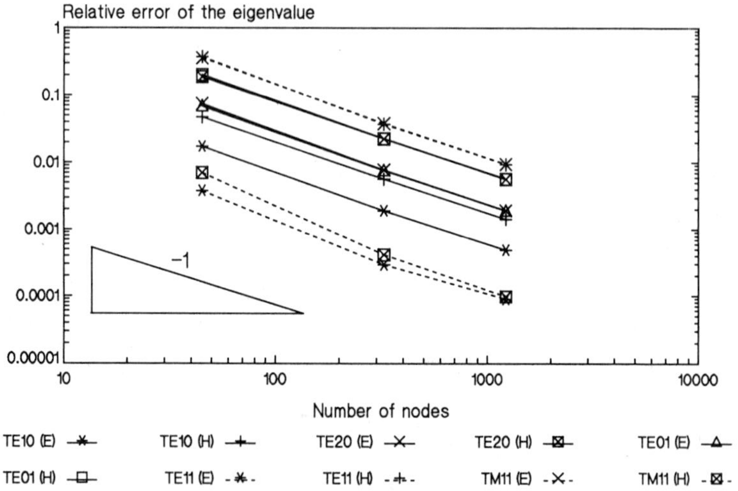

Figure 3.47 Rate of convergence for the square of the cut-off wave number of the first five modes of a rectangular waveguide using first-order elements.

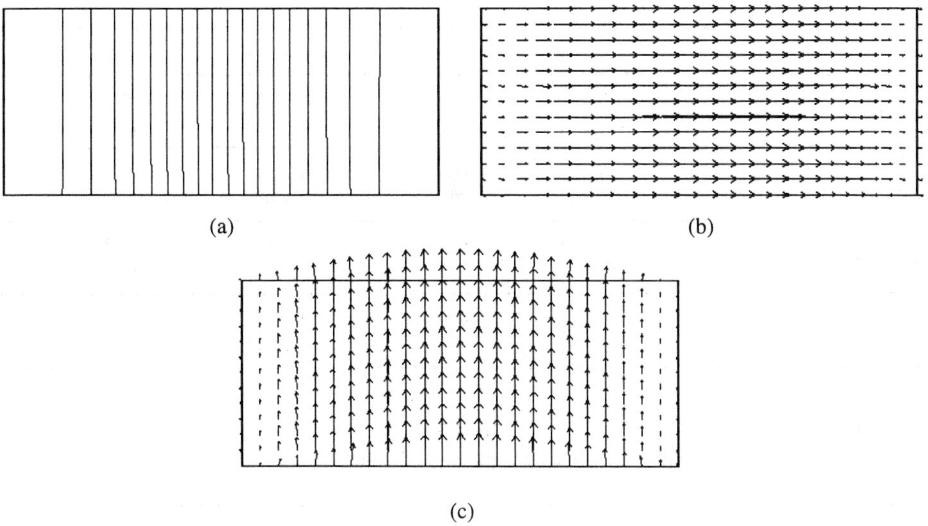

Figure 3.48 Field plots of the TE_{10} mode: (a) H_z; b) $\overline{H}_t$; (c) $\overline{E}_t$.

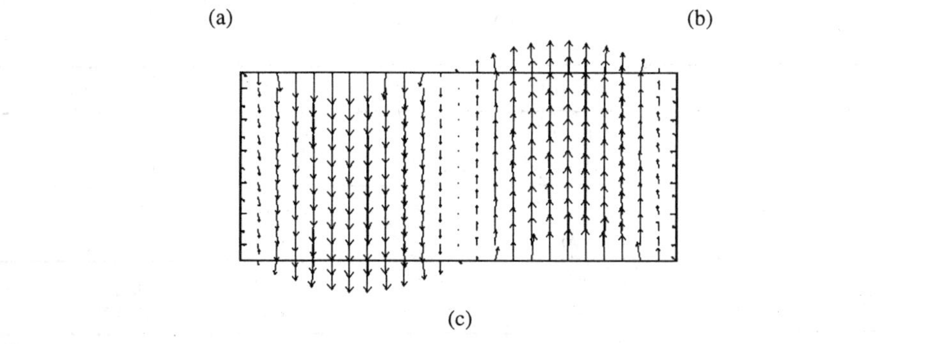

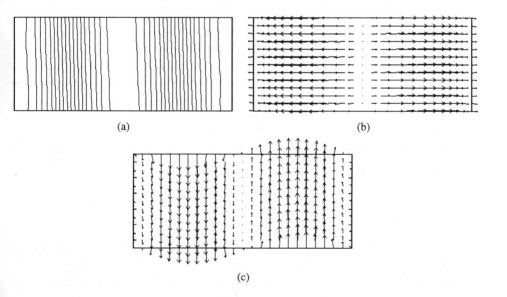

'igure 3.49 Field plots of the TE_{20} mode: (a) H_z; (b) $\overline{H}_t$; (c) $\overline{E}_t$.

Table 3.29

TE_{10} mode. Normalized computed (and exact) values for H_z (x,y)

	$x = 0$	$x = 0.5$	$x = 1$	$x = 1.5$	$x = 2$
y = 0	1 (1)	0.715 (0.707)	0.003 (0)	-0.712 (-0.707)	-1.019 (-1)
y = 0.33	1.008 (1)	0.714 (0.707)	0.001 (0)	-0.714 (-0.707)	-1.011 (-1)
y = 0.66	1.011 (1)	0.714 (0.707)	0.001 (0)	-0.714 (-0.707)	-1.008 (-1)
y = 1	1.019 (1)	0.713 (0.707)	0.004 (0)	-0.715 (-0.707)	-1 (-1)

Table 3.30

TE_{10} mode. Normalized computed (and exact) values for H_x (x,y)

	$x = 0.083$	$x = 0.583$	$x = 1.083$	$x = 1.583$	$x = 1.916$
y = 0	-0.068 (-0.083)	-0.503 (-0.505)	-0.632 (-0.631)	-0.393 (-0.387)	-0.098 (-0.083)
y = 0.33	-0.082 (-0.083)	-0.504 (-0.505)	-0.632 (-0.631)	-0.388 (-0.387)	-0.084 (-0.083)
y = 0.66	-0.084 (-0.083)	-0.507 (-0.505)	-0.632 (-0.631)	-0.386 (-0.387)	-0.082 (-0.083)
y = 1	-0.098 (-0.083)	-0.509 (-0.505)	-0.631 (-0.631)	-0.383 (-0.387)	-0.069 (-0.083)

Table 3.31

TE_{10} mode. Normalized computed (and exact) values for E_y (x,y)

	$x = 0.083$	$x = 0.583$	$x = 1.083$	$x = 1.583$	$x = 1.916$
y = 0	0.065 (0.130)	0.749 (0.793)	0.995 (0.991)	0.658 (0.609)	0.195 (0.130)
y = 0.33	0.13 (0.130)	0.789 (0.793)	0.987 (0.991)	0.606 (0.609)	0.13 (0.130)
y = 0.66	0.13 (0.130)	0.789 (0.793)	0.987 (0.991)	0.606 (0.609)	0.13 (0.130)
y = 1	0.195 (0.130)	0.83 (0.793)	0.978 (0.991)	0.553 (0.609)	0.065 (0.130)

Table 3.32

$s_c^2 = (ak_c)^2$ obtained for the rectangular waveguide using second-order elements

Mode	41 nodes	137 nodes	497 nodes	Exact value
TE_{10}	2.467748 $(\bar{E})$ 2.481668 $(\bar{H})$	2.467454 $(\bar{E})$ 2.468522 $(\bar{H})$	2.4674050 $(\bar{E})$ 2.4674770 $(\bar{H})$	2.4674
TE_{20}	9.120798 $(\bar{E})$ 11.420010 $(\bar{H})$	9.871542 $(\bar{E})$ 9.935223 $(\bar{H})$	9.8698210 $(\bar{E})$ 9.8743660 $(\bar{H})$	9.8696
TE_{01}	9.127655 $(\bar{E})$ 11.620380 $(\bar{H})$	9.871573 $(\bar{E})$ 9.938880 $(\bar{H})$	9.8698220 $(\bar{E})$ 9.8744980 $(\bar{H})$	9.8696
TE_{11}	12.454910 $(\bar{E})$ 14.069390 $(\bar{H})$	12.377510 $(\bar{E})$ 12.515170 $(\bar{H})$	12.3399300 $(\bar{E})$ 12.3505600 $(\bar{H})$	12.3370
TM_{11}	13.232330 $(\bar{H})$ 14.327700 $(\bar{E})$	12.38119 $(\bar{H})$ 12.53454 $(\bar{E})$	12.3400500 $(\bar{H})$ 12.3512100 $(\bar{E})$	12.3370

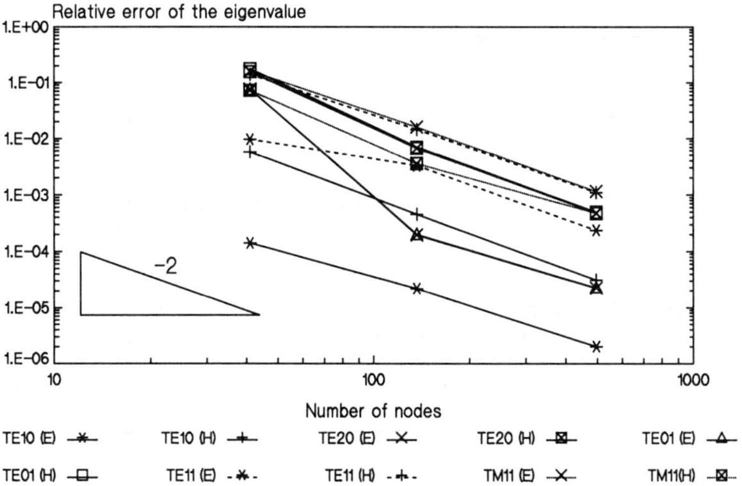

Figure 3.50 Rate of convergence for the analysis of the rectangular waveguide using second-order elements.

b) *Rectangular waveguide half-filled with dielectric*

The next structure to be analyzed is a rectangular waveguide of dimensions $2a \times a$, partially filled with dielectric of permittivity $\varepsilon_r = 2.25$. Figure 3.51 shows the structure and its discretization using a mesh of 144 first-order elements (325 nodes). Tables 3.33 and 3.34 provide the computed eigenvalues for the first four modes for different β_a. Figure 3.52 shows the normalized dispersion curve for the first four modes and their comparison with the exact values [Marcuvitz 1951, 1986] and with those obtained by Angkaew et al. [1987] through a FEM that uses a different type of vector element. It can be observed that for the higher order modes a finer mesh would be required. The accuracy for the first and second modes is excellent. The third and fourth modes are better approximated with a particular formulation ($\bar{H}$ formulation for the third mode; and $\bar{E}$ formulation for the fourth mode). Spurious modes did not pollute the spectrum since they all have a numerically zero eigenvalue. Figures 3.53 and 3.54 show the magnetic field for the first two modes.

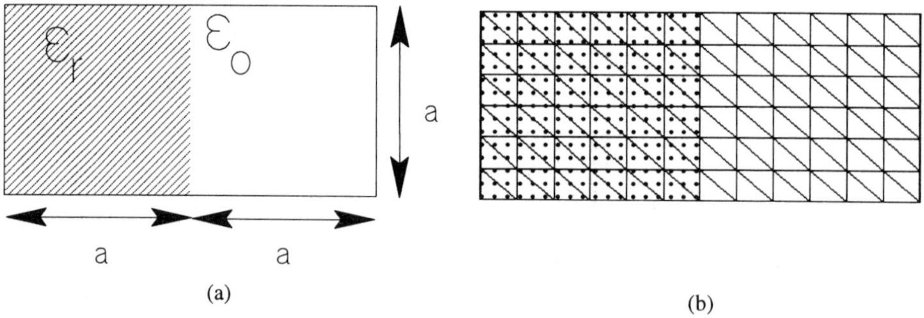

Figure 3.51 Rectangular waveguide half-filled with dielectric: (a) structure and (b) mesh.

Table 3.33
Rectangular waveguide half-filled with dielectric. Computed eigenvalues for the first four modes and different β_a ($\bar{H}$ formulation). First-order element

Mode	$\beta_a = 0$	$\beta_a = 1$	$\beta_a = 2$
1	1.4750	2.0435	3.6799
2	5.3025	5.7609	7.1268
3	6.6675	7.1689	8.6492
4	6.8984	7.6290	9.8113

Table 3.34
Rectangular waveguide half-filled with dielectric. Computed eigenvalues for the first four modes and different β_a ($\bar{E}$ formulation). First-order element

Mode	$\beta_a = 0$	$\beta_a = 1$	$\beta_a = 2$
1	1.4644	2.0332	3.6706
2	5.1215	5.5831	6.9563
3	6.9719	7.3539	8.9187
4	6.6360	7.4628	9.4957

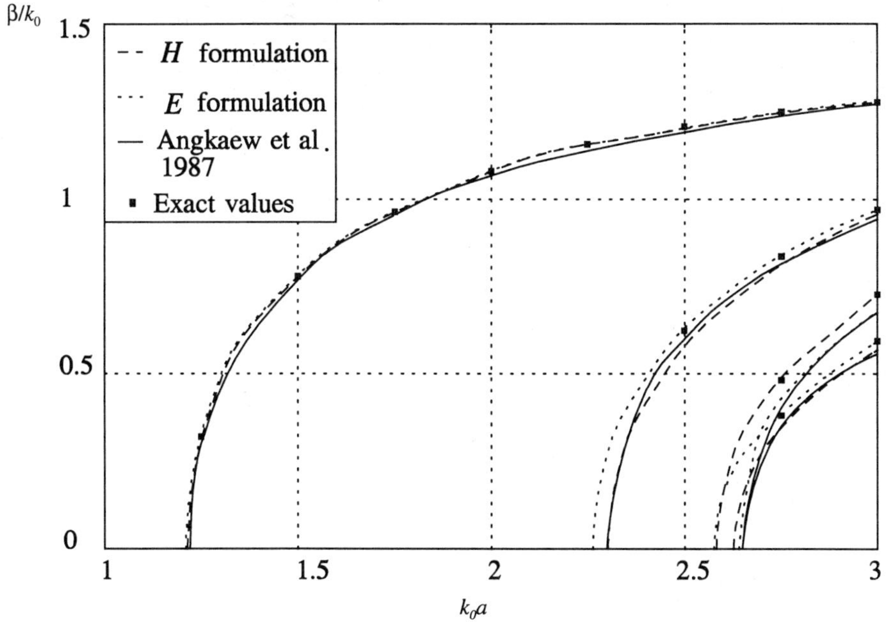

Figure 3.52 Dispersion curve for the rectangular waveguide half-filled with dielectric.

Second-order elements have also been used. Figure 3.55 compares the rate of convergence obtained for the first mode, with $\beta_a = 1$, using first- and second-order elements and uniform refinements. Again the theoretical results are obtained, showing the efficiency of the second-order element.

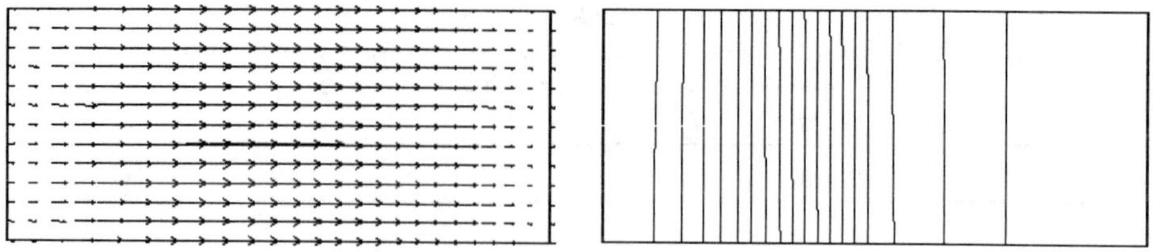

(a) (b)

Figure 3.53 Magnetic field of the first mode: (a) $\bar{H}_t$ and (b) H_z.

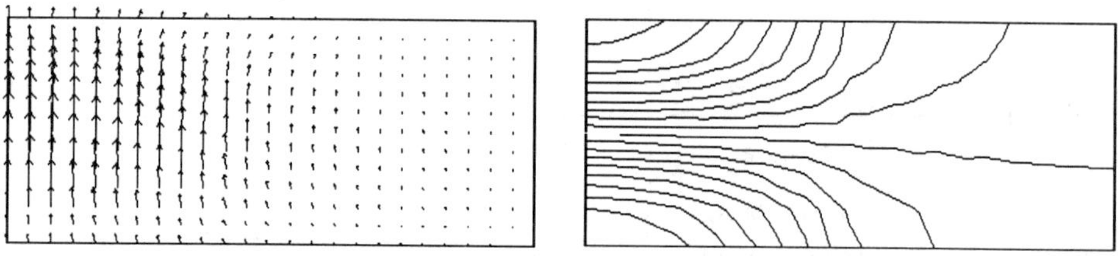

(a) (b)

Figure 3.54 Magnetic field of the second mode: (a) $\bar{H}_t$ and (b) H_z.

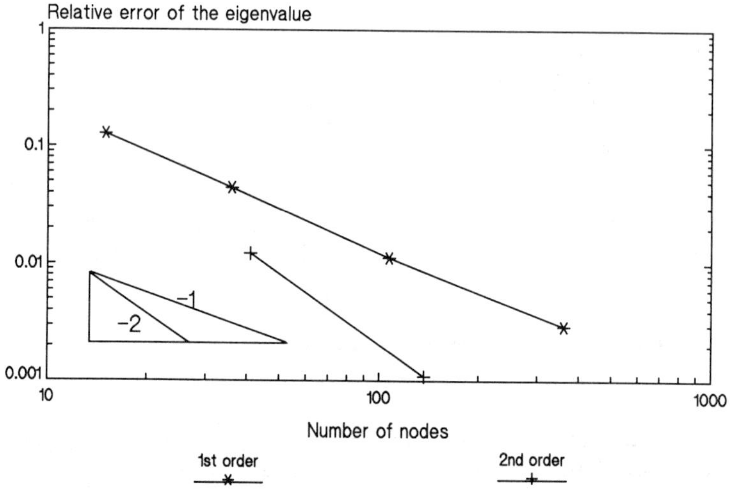

Figure 3.55 Rate of convergence for the analysis of the first mode ($\beta_a = 1$) using first- and second-order elements.

c) *Ridge waveguide*

In order to study the behavior of structures with re-entrant corners, the ridge waveguide of Figure 3.56 has been analyzed using two meshes. One has 160 elements (369 nodes) and the other has 448 elements (985 nodes). Tables 3.35 and 3.36 provide s_c^2 for the first eight modes for both meshes. They are compared with results from [Bermúdez de Castro and Gómez-Pedreira 1992] using first-order Nédélec curl-conforming elements with a slightly different implementation. By comparing Tables 3.35 and 3.36, it is observed that there are convergence problems for some modes depending on the formulation used. The TE or TM character of each mode has been derived from the field plots [García-Castillo 1992].

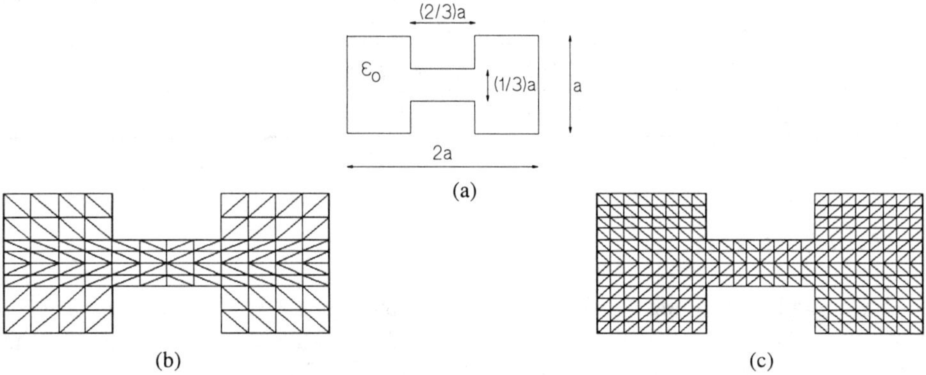

Figure 3.56 Ridge waveguide: (a) structure, (b) mesh 1: 369 nodes, and (c) mesh 2: 985 nodes.

Table 3.35
s_c^2 for the first eight modes of the ridge waveguide, with a 369 nodes mesh

Mode	$\bar{H}$ formulation	$\bar{E}$ formulation	[Bermúdez de Castro and Gómez-Pedreira 1992]
1 (TE)	2.048557	1.94132	2.01912
2 (TE)	11.84285	11.04284	11.52805
3 (TE)	11.85219	11.04625	11.53311
4 (TE)	12.63136	10.83795	12.03664
5 (TE)	24.31541	23.00215	23.48747
6 (TE)	31.63223	27.28184	29.52791
7 (TM)	29.12342	34.08824	
8 (TM)	29.21388	34.14274	

Table 3.36

s_c^2 for the first eight modes of the ridge waveguide, with a 985 nodes mesh

Mode	$\bar{H}$ formulation	$\bar{E}$ formulation	[Bermúdez de Castro and Gómez-Pedreira 1992]
1 (TE)	2.019123	1.96859	2.01912
2 (TE)	11.52805	11.22316	11.52805
3 (TE)	11.53312	11.22611	11.53311
4 (TE)	12.03664	11.247671	12.03664
5 (TE)	23.48746	23.11106	23.48747
6 (TE)	29.52791	27.90759	29.52791
7 (TM)	29.84113	31.70562	29.89188
8 (TM)	29.89185	31.74023	

d) *Microstrip line*

A shielded microstrip line (see Figure 3.57) of dimensions w = 12.18 mm and h = 3.14 mm with ε_r = 11.7 is analyzed. A mesh of half the structure is used to study the fundamental mode. The number of elements is 162 with 369 nodes. The magnetic field formulation has been used and compared with [Lee et al. 1991a], which uses a FEM with second-order tangential elements, and with [Shih et al. 1988], which utilizes a variational method with a conformal transformation. Figure 3.58 compares the various results.

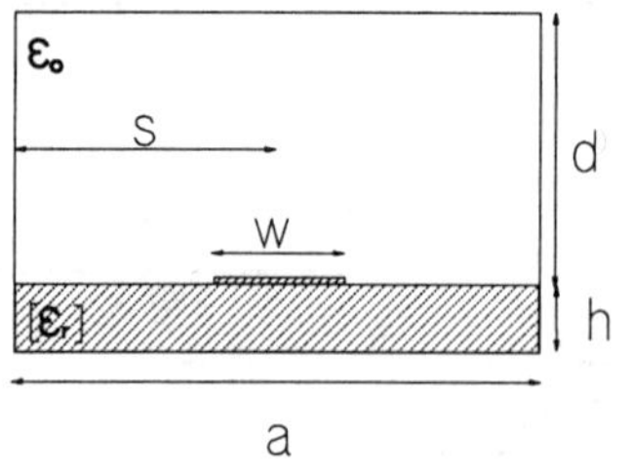

(a)

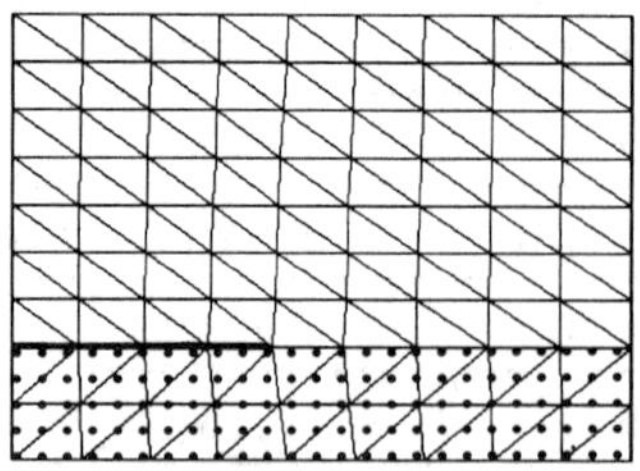

(b)

Figure 3.57 Microstrip line: (a) structure and (b) mesh of one-half of the structure.

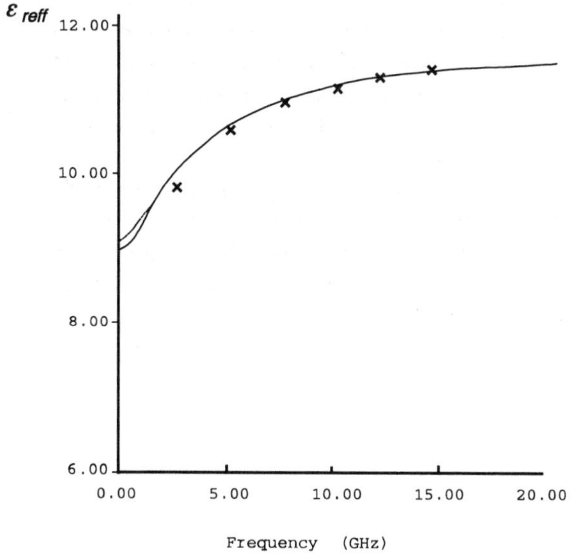

Figure 3.58 Dispersion curve for the fundamental mode of a microstrip line. —: [Lee et al. 1991a], ····: [Shih et al. 1988], ×: this method.

e) *Circular waveguide and other structures with curved contours*

The performance of curved elements has also been analyzed. First, the behavior of the curved first-order element has been studied as compared with the straight version of same order. Then, the curved second-order element has been used. Figure 3.59 shows the meshes employed for the analysis of a circular waveguide. The number of elements of the first mesh was not enough to obtain convergence for the first-order element. The last mesh was not used for the second-order analysis. Table 3.37 compares the results for s_c^2 of the TE_{11} mode using only straight elements, and both straight and curved first- and second-order elements, utilizing $\bar{E}$ and $\bar{H}$ formulations. Figure 3.60 shows the rate of convergence. The superconvergent result of the second-order curved element may be observed. Figure 3.61 shows the field plots for the TE_{11} mode. Table 3.38 shows the results for the TM_{01} mode. Figure 3.62 plots the rate of convergence. Figure 3.63 depicts the field plots. For the TE_{21} mode, convergence was not achieved with the second mesh and first-order elements (41 nodes). This confirms the need to use more refined meshes when computing the higher order modes. Table 3.39 shows the results using $\bar{E}$ formulation. Figure 3.64 shows the rate of convergence.

Other structures such as a circular coaxial cable, a cigar-shaped guide, and semicircular ridge guides have been analyzed, obtaining in all cases an optimum behavior for the curved element as opposed to the straight element for the discretization of curved boundaries.

The method has also been applied to several circular guides with re-entrant corners. For these structures, some of the modes have nonsmooth fields in regions close to the corners. Hence, the mesh should be refined in those regions where the fields display a nearly singular behavior. The use of curved elements has a secondary effect, since the main source of error is due to the nonsmoothness of the field. Figure 3.65 shows a circular waveguide with two fins. The meshes shown in the figure have 353 and 1509 nodes, respectively. Table 3.40 compares the computed values for s_c^2 of the first mode using curved elements and the $\underline{\bar{E}}$ formulation using two different meshes. Figure 3.66 shows the dispersion curve for the first mode computed with the more refined mesh and compares it with [Thorburn et al. 1990], which uses the method of lines. Good agreement may be observed.

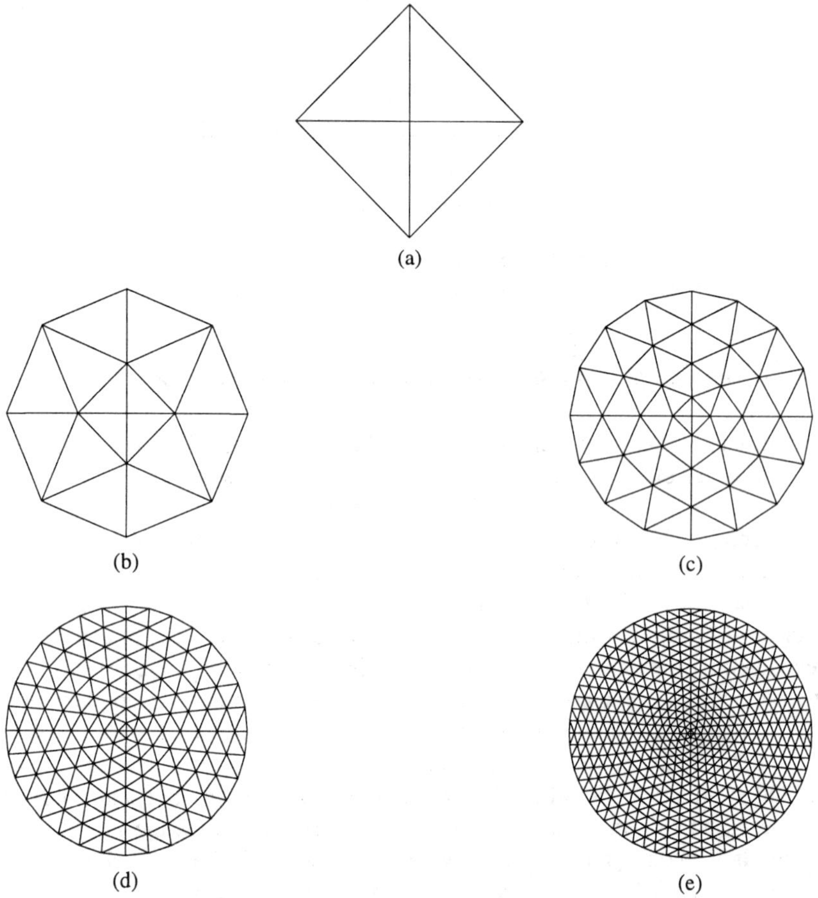

Figure 3.59 Homogeneous circular waveguide: meshes with (a) 4 elements (first-order: 13 nodes; second-order: 37 nodes), (b) 16 elements (first-order: 41 nodes; second-order: 129 nodes), (c) 64 elements (first-order: 145 nodes; second-order: 481 nodes), (d) 256 elements (first-order: 545 nodes; second order: 1857 nodes), and (e) 1024 elements (first-order: 2113 nodes).

Table 3.37
Relative error on the computation of s_c^2 of the TE_{11} mode for the circular waveguide

Number of elements	First-order						Second-order			
	Number of nodes	CPU time (sec)	$\bar{E}$ formulation		$\bar{H}$ formulation		Number of nodes	CPU time (sec)	$\bar{E}$ formulation	
			Straight	Curved	Straight	Curved			Straight	Curved
4	-	-	-	-	-	-	37	5.44	44.59	2.23
16	41	5.5	14.0	4.70	19.6	7.60	129	10.6	0.43	0.003
64	145	6.05	3.30	0.97	4.70	2.00	481	38.77	0.13	-
256	545	24.73	0.84	0.23	1.10	0.48	-	-	-	-
1024	2113	145.4	0.21	0.07	0.29	0.13	-	-	-	-

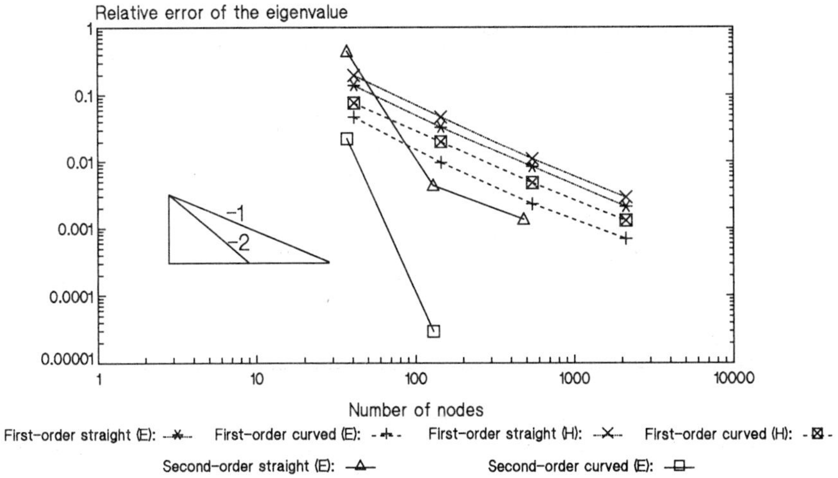

Figure 3.60 Rate of convergence for the square of the cut-off wave number of the TE_{11} mode of the circular waveguide.

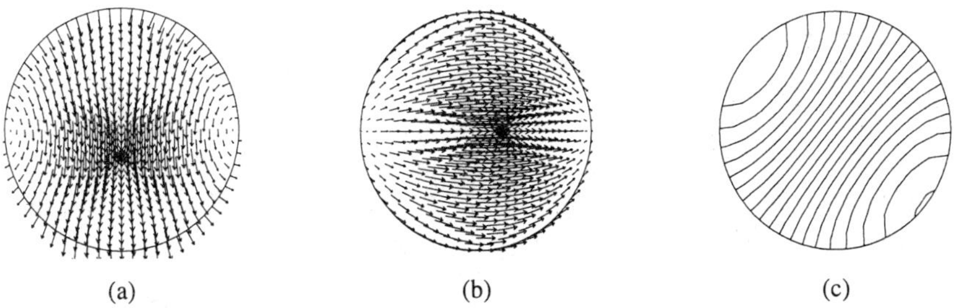

Figure 3.61 Field plot of the TE_{11} mode for the circular waveguide: (a) $\bar{E}_t$, (b) $\bar{H}_t$, and (c) H_z.

Table 3.38
s_c^2 for the TM_{01} mode of the circular waveguide

Number of elements	First-order					Second-order	
	Number of nodes	$\bar{E}$ formulation		$\bar{H}$ formulation		Number of nodes	$\bar{H}$ formulation
		Straight	Curved	Straight	Curved		Curved
4	-	-	-	-	-	37	5.4240
16	41	6.8957	6.0446	7.0245	6.3169	129	5.7800
64	145	6.0603	5.8765	6.0525	5.9011	481	5.7841
256	545	5.8542	5.8126	5.8445	5.8073	-	-
Exact value				5.7840			

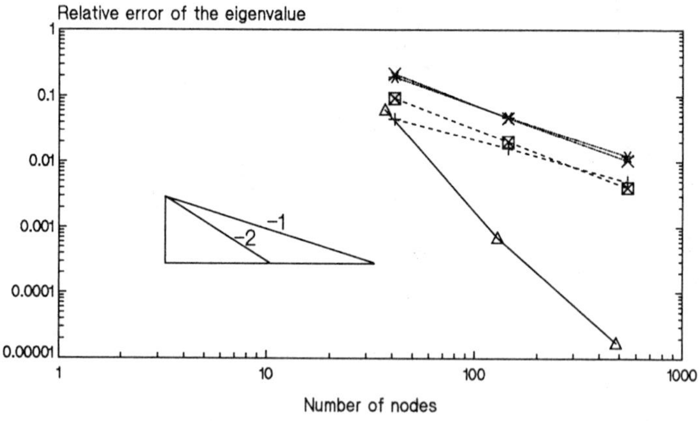

Figure 3.62 Rate of convergence for the square of the cut-off wave number of the TM_{01} mode of the circular waveguide.

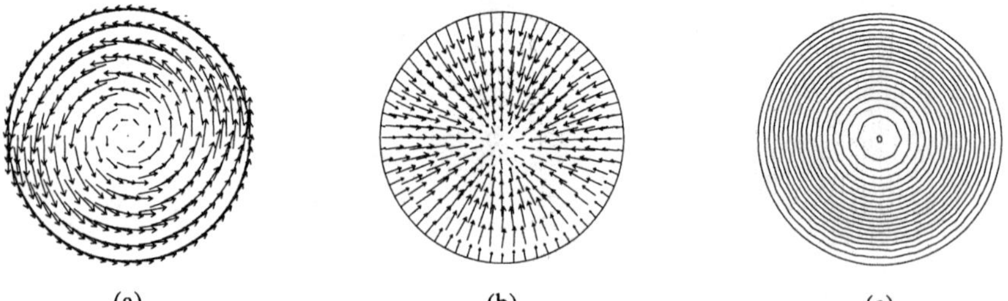

(a) (b) (c)

Figure 3.63 Field plot of the TM_{01} mode for the circular waveguide: (a) $\bar{H}_t$, (b) $\bar{E}_t$, and (c) E_z.

Table 3.39

s_c^2 for the TE_{21} mode of the circular waveguide using $\bar{E}$ formulation

Number of elements	First-order			Second-order	
	Number of nodes	Straight	Curved	Number of nodes	Curved
16	-	-	-	129	9.2461
64	145	9.7976	9.5297	481	9.3236
256	545	9.4403	9.3816	-	-
Exact value			9.3269		

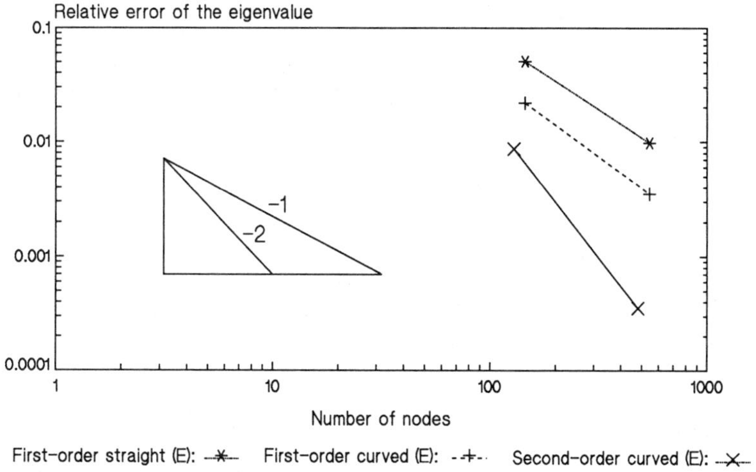

Figure 3.64 Rate of convergence for the square of the cut-off wave number of the TE_{21} mode of the circular waveguide.

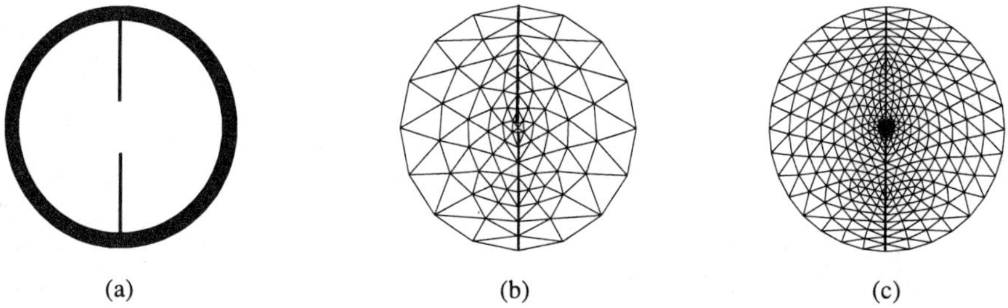

Figure 3.65 Circular waveguide with fins: (a) structure, (b) 353 nodes mesh, and (c) 1509 nodes refined mesh.

Table 3.40
s_c^2 for the first mode of the circular finline

Number of nodes	s_c^2
353	0.0279
1509	0.03249

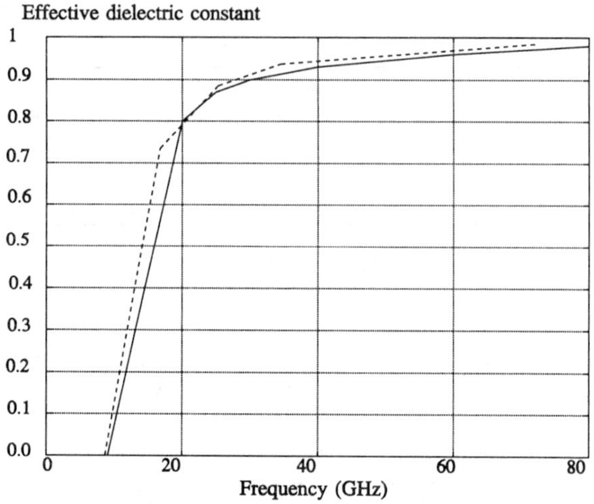

Figure 3.66 Dispersion curve for the first mode of the finline in a circular waveguide: —: [Thorburn et al. 1990]; - -: this method.

3.3.4 Conclusions

Two formulations for the full-wave analysis of waveguiding structures have been presented. The difficulties arising from the first formulation when applied to inhomogeneous structures have been highlighted. However, for isotropic homogeneous structures, such a formulation deals with unknowns that are scalars. It uses the longitudinal component of either the electric field (for TM modes) or the magnetic field (for TE modes). When discretized with Lagrange elements the computation is very efficient. For structures leading to nonsmooth solutions, the use of locally refined meshes is advisable.

The unknowns of the second formulation are the longitudinal and the transverse components of either the electric or the magnetic field. For its discretization straight and curved, first- and second-order Lagrange/curl-conforming elements are utilized. The method leads to convergent results. However, it has been observed that higher order modes require more refined meshes. On the other hand, modes with a nonsmooth electromagnetic field configuration require meshes well adapted to them in order to ensure a fast rate of convergence. Thus, self-adaptive mesh procedures are again highly desired. The next chapter deals with this subject.

Chapter 4

Self-Adaptive Mesh Algorithm

4.1 INTRODUCTION

A unique feature of the Finite Element Method (FEM) is the way in which the approximating functions are defined. The potential of FEM lies in the fact that the contribution of each finite element is calculated and added in a global assembly independently from its neighbors. Therefore, a high density of elements or degrees of freedom may be utilized in regions where field variation is abrupt, while a smaller number may be assigned to regions with relatively small field variation. In fact, the accuracy of the method greatly depends on a proper discretization of the domain. In general, the method is used in an iterative way applying it to a sequence of meshes with an increasing number of degrees of freedom until a convergent solution is obtained.

Once a finite element code is available, the most cumbersome task for the user is to create an appropriate mesh. Hence, if this technique were to be utilized to its fullest potential, a methodology should be available particularly related to the refinement of the mesh to improve the accuracy of the solution. This methodology should be transparent to a user [Shepard 1985, 1986]. The capability to decide which parts of the domain present the greatest error in the approximation procedure and the capability to ameliorate the approximation in these parts is called *adaption* [Babuška 1986], [Ewing 1990]. Nowadays adaption is considered *inherent* in FEM and an *indispensable* part of it [Golias and Tsiboukis 1996].

This chapter addresses the implementation of an automatic self-adaptive mesh technique that provides numerical results to a prespecified degree of accuracy. This procedure may be briefly described as follows. An initial coarse mesh is generated from the definition of the geometry of the problem. The finite element method is applied to it and the error of the computed solution is estimated both locally (in each element) and globally. Those estimates are used to decide if the mesh should be refined and, if that is the case, which regions should be enriched. A refined mesh is then created, by subdividing the selected elements (*h*-refinement), or increasing their order (*p*-refinement), or combining both techniques, for example. The procedure will continue until a stopping criterion is met. This technique provides a powerful tool to automatically generate meshes with local errors of the same order of magnitude. Meshes of this type are called equilibrated or optimal. They ensure a high rate of convergence of FEM, minimizing the overall computational time required to obtain the final result.

The objective of a self-adaptive scheme should be to achieve at least the same rate of convergence that would be obtained for problems with smooth enough exact solutions discretized by means of uniformly refined meshes. In Section 2.5.3, there were given expressions for *a priori* estimates of the error incurred in an FEM analysis by means of uniform and nonuniform (equilibrated) discretizations. They make use of problem dependent unknown constants that preclude their practical utilization. The use of singular elements is not enough to solve for the lack of smoothness of the exact solution of a problem that may be due to nonsmooth boundaries or media interfaces, abrupt changes in the value of the physical constants of the medium, singular excitations, and discontinuities of the boundary conditions, for example [Babuška 1977]. Singular elements take into account just the first source of lack of smoothness mentioned, which indeed is particularly important in electromagnetics. However, they underestimate electromagnetic coupling and proximity between conductors that are important sources of lack of smoothness in electromagnetic field problems. Hence, any procedure devised to ensure highly accurate results should consider not just one source of lack of smoothness, but the local and global behavior of the field on itself. Hence, self-adaptive procedures based on *a posteriori* error estimates are the best candidates.

A self-adaptive mesh scheme relies on the computation of *a posteriori* error estimates or indicators and an appropriate strategy for the refinement of the mesh. In this chapter, a brief overview of the various procedures utilizing different error criteria is given first. Then the *a posteriori* error estimate selected for this book is presented. It is based on the residual of the differential equation defining the problem. The energy norm is used to give a measure of the error (see Appendix B). The procedure chosen to enrich the mesh is a selective *h*-refinement, eventually followed by a uniform *p*-refinement. The self-adaptive algorithm is presented next. Its has been applied to the quasi-static analysis of transmission lines. Its application to several examples demonstrates the efficiency of the method. The extension of the self-adaptive scheme to the full-wave analysis of waveguiding structures is shown next. The analysis of several simple structures illustrates the results obtained by using this technique. Details about the computation of the various error estimates are given in Appendix G. Results from the analysis of more complicated waveguiding structures are given in Chapter 5.

In summary, this chapter describes a specific self-adaptive procedure for the analysis of waveguiding structures, that is, two-dimensional (2D) problems. However, its general features are common to any self-adaptive algorithm. Extension to one- and three-dimensional (1D and 3D) problems is easily done. These procedures turn out to be essential tools in obtaining user-friendly software codes for electromagnetics.

4.2 SELF-ADAPTIVE TECHNIQUES, ERROR ESTIMATES, AND REFINEMENT PROCEDURES

The origins of self-adaptive procedures are related on one hand to the analysis of problems with singularities. Local *h*-refinements were successfully utilized as alternate solutions to singular elements (see, for example, [Daly 1973]). The mathematical justification was addressed and systematic procedures were proposed [Thatcher 1976], [Schatz and Wahlbin 1978, 1979], [Parks 1979]. On the other hand, the study of self-adaptive procedures requires an understanding of the *a priori* interpolation errors and rates of convergence that researchers have obtained towards the second part of the 1970s and during the 1980s. In fact, the idea of adaptivity may be found in the conclusions of many papers devoted to the study of *a priori* errors for the *h*-, *p*-, and *hp*-versions [Babuška et al. 1981], [Babuška and Dorr 1981], [Babuška and Szabó 1982], [Dorr 1984, 1986], [Gui and Babuška 1986a,b,c], [Babuška and Suri 1987a,b], [Guo 1988], [Babuška and Guo 1988], [Suri 1990].

The basis for self-adaptivity was posed by I. Babuška during the 1975 MAFELAP Conference [Babuška 1976]. However, in the area of electromagnetics there were some pioneering works. The main concepts of self-adaptive procedures (computation of *a posteriori* error estimates, and *h*-refinement based on the information provided by these estimates) may be recognized in [Kinsner and Torre 1974], for example. From his 1976 seminal paper, Babuška together with his coworkers developed the mathematical theory for the definition of *a posteriori* error criteria and the norms for their measurement [Babuška and Rheinboldt 1978a,b, 1979]. In [Babuška et al. 1979] it was explicitly demonstrated that "a proper mesh refinement gives the same rate of convergence of the error as in the case of smooth solutions and a quasi-uniform mesh." The theory was further studied and the concept of optimal meshes was analyzed [Rheinboldt 1980, 1981], [Babuška and Rheinboldt 1981], [Carey and Humphrey 1981]. Those initial papers were concerned mainly with *h*-refinements but papers extending the approach to *p*-refinements immediately followed [Dunavant and Szabó 1983].

An important two-part paper dealing with second-order boundary value problems was published in 1983. In the first part [Kelly et al. 1983], the authors analyze an *a posteriori* error estimate based on the residual of the differential equation both in the domain of the problem (i.e., in the domain of each element, and across element interfaces), and on the boundaries (i.e., on the boundaries with natural conditions). The distinction between error estimators (quantities that approximate the actual error) and error indicators (relative quantities that serve to indicate how large the error is in a given element or region in comparison with other elements or regions) is formally done. The global error estimate is obtained as the sum of the contributions of each element (local error estimates, that serve also as local error indicators). The robustness of this error estimate (i.e., its reliability to indicate when the error is small) is also demonstrated. The proposed error estimate is essentially the same one described in [Babuška and Rheinboldt 1978a,b] for 1D problems. In the second part of the paper [Gago et al. 1983], the authors address the question of *h*-adaptivity and optimal meshes (i.e., meshes in which the error is equally distributed among the elements) to achieve a given accuracy. Other papers followed [Babuška and Vogelius 1984], [Babuška and Miller 1984] with variants for the computation of the error estimate.

For 2D and 3D problems the *a posteriori* error estimate proposed in [Kelly et al. 1983] requires rather involved programming. Also, different error criteria may be more adequate for different problems. In any case, since 1983 this area of research has been very active and a number of different error estimates have been proposed. In the following papers an overview of possible methods is done [Ainsworth and Craig 1992], [Babuška et al. 1994], [Beckers and Zhong 1992]. In the area of electromagnetics, partial reviews may be found in [Adamiak 1992], [Csendes and Shenton 1985], [Hoole et al. 1989], [Shepard 1985]. The different approaches may be classified as follows.

- Estimates based on the residual of the strong formulation. Variants are:
  - Complete residual, as explained previously [Belytschko and Tabbara 1993], [Rivara 1986], [Zienkiewicz and Craig 1986]. Works in electromagnetic engineering are [Fernandes et al. 1988, 1990a, 1991], [Meyer and Davidson 1994], [Salazar-Palma and Hernández-Gil 1989], [Salazar-Palma et al. 1992, 1994], [Salazar-Palma and Recio-Peláez 1996], [Salazar-Palma and García-Castillo 1996].
  - Self-equilibrated residual, in which the computation of the error estimate although based on the complete residual is partitioned into self-equilibrating systems [Ainsworth and Oden 1993], [Bank and Weiser 1985], [Kelly 1984, 1986], [Kelly et al. 1987], [Ladeveze 1991], [Ohtsubo and Kitamura 1992], [Oden et al. 1990], [Yang et al. 1993]. The following approaches are possible:
    - Local-Dirichlet analysis, in which the error for each point of the mesh is obtained from a Dirichlet problem defined over a patch of a number of elements surrounding that point. This approach has been used for electromagnetic problems in [Csendes and Shenton 1985], [Fernandes et al. 1992a,b], for example.
    - Local-Neumann analysis, in which the error for each element is computed as the solution of a local Neumann problem [Bank 1986], [Verfürth 1991]. In computational electromagnetics, examples of this approach are [Chellamuthu and Ida 1994, 1995], [Drago et al. 1992], [Fernandes et al. 1988, 1990a,b, 1991].
  - Incomplete residual, in which only the residual in the domain of each element is taken into account [Baehmann et al. 1992]. In the area of magnetics it has been implemented in [Fernandes et al. 1988].
  - Interface residual, in which only the residual across element interfaces is considered [Rank 1986, 1993]. This type of error estimate has been used for magnetic problems in [Golias and Tsiboukis 1992].
- Estimates based on the violation of the continuity conditions of the variables of the problem (i.e., the field vectors $\bar{E}$, $\bar{H}$, $\bar{D}$, or $\bar{B}$, the current density $\bar{J}$, the vector potential $\bar{A}$, or the scalar potential ϕ) or the constitutive relations [Ladeveze et al. 1986]. In electromagnetic engineering, examples are [Csendes and Shenton 1985], [Golias and Tsiboukis 1991, 1996], [Hahn et al. 1988], [Kim et al. 1991], [Remacle et al. 1995, 1996]. This method usually coincides with the interface residual technique mentioned previously.
- Estimates based on complementary, dual, mixed or hybrid methods. These methods provide either upper and lower bounds that may be used to compute

error estimates, or approximations of both the primary and the secondary variables of the problem, from which error estimates may be derived [Ferragut 1993], [Kelly et al. 1987]. Several authors have used this approach in electromagnetics [Csendes and Shenton 1985], [Golias et al. 1994], [Golias and Tsiboukis 1996], [Li et al. 1994], [Mitchell and Penman 1992], [Penman and Grieve 1985], [Pinchuk and Silvester 1985], [Rikabi et al. 1988].

- Estimates based on recovery, averaging or smoothing techniques. In these methods the error is computed as the difference between the actual FEM solution and a more accurate solution obtained through different recovery or smoothing procedures [Ainsworth et al. 1989], [Babuška and Rodríguez 1993], [Belytschko and Tabbara 1993], [Bugeda and Oñate 1992], [Burkley and Bruch 1991], [Donzelli et al. 1992], [Durán et al. 1991], [Fuenmayor and Oliver 1993], [Hinton et al. 1991], [Lee and Lo 1992], [Lewis et al. 1991], [Liu and Elmaraghy 1992], [Wiberg and Abdulwahab 1992], [Zeng et al. 1992], [Zienkiewicz and Zhu 1987, 1989, 1991, 1992a,b,c], [Zienkiewicz et al. 1988, 1989]. An example of the use of these types of estimates for electromagnetic problems is [Lowther et al. 1993].
- Estimates based on extrapolation techniques [Georges and Shepard 1991], [Szabó 1986], [Szabó and Babuška 1991; Ch. 4].
- Estimates based on interpolation theory. These methods require the computation (estimate) of higher-order derivatives over each element [Belytschko and Tabbara 1993], [Eriksson and Johnson 1988], [Oden et al. 1986]. In the field of electromagnetic engineering, examples of their use are [Biddlecombe et al. 1986], [Fernandes et al. 1988, 1991], [Girdinio et al. 1983].
- Estimates based on the Green integration formula. In this method the FEM solution at a node is compared with a value obtained from the Green integration along a contour enclosing the node [Adamiak 1992].
- Estimates based on the perturbation of some quantity between two finite element solutions where one is considered more accurate than the other [Field and Pressburger 1993], [Georges and Shepard 1991], [Zienkiewicz and Craig 1986]. Several authors have used this approach for electromagnetic problems [Golias and Tsiboukis 1996], [Hoole 1987, 1988, 1990], [Luomi and Rouhiainen 1988], [McFee and Webb 1992].
- Estimates based on the sensitivity of a variable to node position. An example of this approach in magnetics is [Henneberger et al. 1990].

In most cases, different norms have been proposed to measure the error computed through the various techniques summarized previously. Thus, a wide variety of choices is available nowadays for the implementation of *a posteriori* error estimates.

The mathematical analysis of most of the residual-based error estimates is available. Additionally, their robustness has been proven through numerical experiments. For second-order elliptic boundary value problems, the energy norm is usually employed. Under this norm it may be said that complete and incomplete residual-based error estimates (either in their original formulation or through self-equilibrated approaches) are adequate for problems in which global quantities are the variables of interest. Complete residual and interface residual-based estimates (as well as constitutive relation-based estimates) provide good performance for local behavior.

Error estimates based on complementary or dual principles usually require the FEM solution of two different problems, one in terms of the primary variable and the other in terms of the secondary variable (see Section 2.3.3.2). Thus, they may be too expensive. An alternate approach is the use of mixed or hybrid formulations that provide approximate solutions for both the primary and the secondary variables while solving only one problem, although much more involved.

The error estimates based on recovery techniques are quite popular nowadays. They were initially proposed on a heuristic basis as an easy alternative to residual-based error estimates [Zienkiewicz and Zhu 1987]. However, they rely on superconvergent recovery procedures and much research has been required to obtain methods providing satisfactory results. In practice their implementation is as involved as the residual-based procedures. The mathematical analysis of recovery or smoothing-based error estimates has been developed during the last few years. It has been demonstrated that they are equivalent to residual-based error estimates. Recovery-based error estimates may be adequate for problems in which the variable of interest is the one on which the recovery or smoothing is carried out (e.g, the electric or magnetic field).

The extrapolation-based error estimates are usually employed with *p*-refinements. Many of the error estimation procedures summarized previously are quite heuristic. However, for the problems to which they have been applied, their performance is good.

Historically, the first self-adaptive procedures used *h*-refinements. In fact, most of the works referenced so far are oriented towards them. However, in most cases they may be used to derive self-adaptive procedures employing any refinement method. Possible techniques are[1] [Shepard 1985]:

- Refinement in *h* [Biswas et al. 1991], [Mercader and Rivara 1992], [Nambiar et al. 1993], [Rivara 1984, 1987, 1989, 1992].
- Refinement in *p* [Daigang and Kexun 1994], [Dunavant and Szabó 1983], [Georges and Shepard 1991], [Mcfee and Webb 1992], [Meyer and Davidson 1994], [Szabó 1986], [Szabó and Babuška 1991; Ch. 4], [Zienkiewicz and Craig 1986].
- Refinement in *r* (i.e., relocation of nodes) [Shepard 1985].
- Possible combinations of the previous ones: *r-h* [Casas and Figueiredo 1992], [Hoole et al. 1988, 1989, 1990], *h-p* [Babuška 1986], [Chen and Rice 1992], [Field and Pressburger 1993], [Georges and Shepard 1991], [Giannacopoulos and Mcfee 1994], [Tang et al. 1994], [Zienkiewicz and Craig 1986], [Zienkiewicz et al. 1989], [Zienkiewicz and Zhu 1991].
- Generation of a new refined mesh [Bugeda and Oñate 1992], [Lee and Lo 1992], [Zienkiewicz and Zhu 1991, 1992].

---

1 In the following list, works already referenced will be again cited only for non *h*-refinement strategies.

The first procedure consists of subdividing the elements selected by means of the error analysis into two or more new elements. It is demonstrated that if optimal (equilibrated) meshes are obtained, the rate of convergence of the finite element procedure depends on the order of the polynomial approximation used and not on the regularity of the exact solution.

The second method consists of increasing the order of the basis functions of the elements selected. The procedure requires either the use of transition elements between elements of different order to obtain a conforming mesh, or the use of hierarchical basis functions. Hierarchical basis functions are those where an element of order p possesses a set of degrees of freedom that is compatible (see Section 2.5.1.1) with a neighboring element of order p-1 [Rossow and Katz 1978], [Sears 1988], [Wang et al. 1996], [Webb and Forghani 1993, 1994], [Webb and Abouchacra 1995], [Zienkiewicz et al. 1983]. The rate of convergence of a p-refinement procedure is higher (at least twice) than for h-refinements. However, it should be mentioned that there are certain limits to increasing the order of the basis functions (see Section 2.5.3).

The third technique consists of the relocation of the nodes of the original mesh by distributing them in a more appropriate fashion. The development of this procedure is quite time consuming. It is a very useful technique for time-dependent problems.

The combination of various methods is very efficient at the added cost of more complex implementation (e.g., h-p methods produce exponential rates of convergence).

The last technique consists of generating a new refined mesh in which the size of the elements (and, eventually, their order) would have been computed from the error analysis. This technique is not easy to incorporate in an existing mesh generator. Special procedures have been suggested in [Frey 1987], [Lo 1991].

In the field of electromagnetics, most of the research papers employ h-refinements, although some authors have used other techniques [Daigang and Kexun 1994], [Mcfee and Webb 1992], [Meyer and Davidson 1994], [Hoole et al. 1988, 1989, 1990], [Giannacopoulos and Mcfee 1994], [Tang et al. 1994].

The problems dealt with in this book are elliptic second-order boundary value problems. The parameters of interest are related to the energy norm of the problem (coefficients of capacitance, inductance, conductance, resistance and magnetic reluctance matrices for the quasi-static case; and the propagation constant in a full-wave analysis), although electric and magnetic fields are needed for certain postprocesses. However, the accuracy of the fields depends on how well the energy variational principle has been minimized. For these reasons, an *a posteriori* complete residual-based error estimate measured by the energy norm has been chosen. The mesh enrichment procedure chosen is a selective h-refinement eventually followed by a uniform p-refinement.

4.3 APPLICATION OF A SELF-ADAPTIVE MESH ALGORITHM TO THE QUASI-STATIC ANALYSIS OF TRANSMISSION LINES

The self-adaptive mesh algorithm developed for the quasi-static analysis of transmission lines is described next. In this case, the direct formulation in terms of the scalar electrostatic potential is used with first- and second-order, straight and curved Lagrange triangular elements. Next, the refinement criterion is discussed and the element subdivision algorithms are described. Finally, the method is validated by applying it to several test structures and studying the behavior of the error estimates.

4.3.1 Local and Global Error Estimates

In [Kelly et al. 1983] a model two-dimensional elliptic second-order boundary value problem is considered. Its strong formulation may be given as follows.

Given $a(x,y)$ and b in $\Omega \subset \mathbb{R}^2$, find $\phi(x,y)$ such that

$$-\bar{\nabla}_t \cdot a \bar{\nabla}_t \phi + b\phi + f = 0 \,, \quad \text{in } \Omega \tag{4.1a}$$

with

$$\phi = 0 \,, \quad \text{on } \Gamma_D \tag{4.1b}$$

$$\bar{a}_n \cdot a \bar{\nabla}_t \phi = g_N \,, \quad \text{on } \Gamma_N \tag{4.1c}$$

where $\Gamma = \Gamma_N \cup \Gamma_D$ is the boundary of Ω and $\bar{a}_n$ is its unit normal vector.

It is assumed that the solution of (4.1) belongs to the Hilbert space $H^1(\Omega)$. The space of functions fulfilling condition (4.1b) is called $H_0^{\,1}(\Omega)$. Multiplying (4.1a) by an arbitrary weighting function $v \in H_0^{\,1}(\Omega)$, and integrating by parts, one obtains

$$B(\phi, v) = \int_\Omega \bar{\nabla}_t \phi \cdot a \bar{\nabla}_t v \, \mathrm{d}\Omega + \int_\Omega b \, \phi \, v \, \mathrm{d}\Omega = - \int_\Omega f v \, \mathrm{d}\Omega + \int_{\Gamma_N} g_N \, v \, \mathrm{d}\Gamma \tag{4.2}$$

Then, (4.2) may be taken as the definition of (4.1) and one can deal now with functions which have discontinuous derivatives (C^0 continuity). The application of FEM will lead to the use of the approximate function

$$\tilde{\phi} = \sum_{i=1}^{n_e} N_i d_i \tag{4.3}$$

where N_i is the ith basis function, for $i = 1, \ldots, n_e$, and n_e is the number of degrees of freedom. The degrees of freedom d_i are obtained by substituting v with N_j, $j = 1, \ldots, n_e$, in the usual way. This will lead to the FEM equations for the determination of d_i

$$B(\tilde{\phi},v) = \int_{\Omega} \bar{\nabla}_t \tilde{\phi} \cdot a\, \bar{\nabla}_t N_j \,\mathrm{d}\Omega + \int_{\Omega} b\, \tilde{\phi}\, N_j \,\mathrm{d}\Omega = -\int_{\Omega} f N_j \,\mathrm{d}\Omega + \int_{\Gamma_N} g_N N_j \,\mathrm{d}\Gamma\,, \quad j=1,\ldots,n_e \tag{4.4}$$

It can be noticed that $\tilde{\phi}$ is an exact solution of the original problem with perturbed right hand terms f and g_N (given by (4.1c)). This is easily seen by integrating (4.4) by parts and substituting N_j by v which gives

$$\begin{aligned}&\sum_{i=1}^{N_e} \left(\int_{\Omega^e} (-\bar{\nabla}_t \cdot a\, \bar{\nabla}_t \tilde{\phi})\, v \,\mathrm{d}\Omega \right) + \sum_{\Gamma_k \not\subset \Gamma} \int_{\Gamma_k} [\![\bar{a}_n \cdot \bar{\nabla}_t \tilde{\phi}]\!]_{\Gamma_k} v \,\mathrm{d}\Gamma + \sum_{\Gamma_k \subset \Gamma_N} \int_{\Gamma_k} (\bar{a}_n \cdot a\, \bar{\nabla}_t \tilde{\phi})_{\Gamma_k} v \,\mathrm{d}\Gamma \\ &+ \int_{\Omega} b\, \tilde{\phi}\, v \,\mathrm{d}\Omega = \int_{\Omega} f v \,\mathrm{d}\Omega + \int_{\Gamma_N} g_N v \,\mathrm{d}\Gamma\end{aligned} \tag{4.5}$$

where $[\![\cdot]\!]_{\Gamma_k}$ denotes the jump of the quantity between double brackets at the kth interface between elements, and N_e is the number of elements. From (4.5) it is seen that one may write

$$-\bar{\nabla}_t \cdot a\, \bar{\nabla}_t \tilde{\phi} + b\, \tilde{\phi} + f + \rho = r\,, \quad \text{in } \Omega \tag{4.6a}$$

$$\bar{a}_n \cdot a\, \bar{\nabla}_t \tilde{\phi} = g_N - \xi\,, \quad \text{on } \Gamma_N \tag{4.6b}$$

that formally coincides with (4.1). In (4.6) one has for the residual

$$r = \bigcup r_s^e\,, \quad e = 1,\ldots,N_e \tag{4.6c}$$

where r_s^e is the residual distributed on the surface of the eth element Ω^e

$$r_s^e = -\bar{\nabla}_t \cdot a\, \bar{\nabla}_t \tilde{\phi} + b\, \tilde{\phi} + f\,, \quad in\ \Omega^e\,, \quad e = 1,\ldots,N_e \tag{4.6d}$$

ρ is a singular function given by

$$\rho = \rho_k \delta_k = [\![\bar{a}_n \cdot a\, \bar{\nabla}_t \tilde{\phi}]\!]_{\Gamma_k}\,, \quad \text{on } \Gamma_k\,, \quad k = 1,\ldots,n_{edi} \tag{4.6e}$$

where n_{edi} is the number of internal edges of Ω, and ξ is

$$\xi = g_N - \bar{a}_n \cdot a\, \bar{\nabla}_t \tilde{\phi}\,, \quad \text{on } \Gamma_N \tag{4.6f}$$

Denoting the error in ϕ by $e = \phi-\tilde{\phi}$, and subtracting (4.6a) from (4.1a) it is seen that e satisfies the original problem with $r+\rho$ and ξ instead of f and g_N. Thus, one has identified a distributed residual $r = \cup r_s^e$ over the surfaces of the elements, a lumped or singular residual ρ over the internal edges, and a lumped residual ξ over the edges located at the Neumann boundaries.

From (4.2), (4.4), and (4.5) it can be seen that the following

$$\int_\Omega \bar{\nabla}_t e \cdot \bar{\nabla}_t v = \sum_{i=1}^{N_e} \left(\int_{\Omega^e} r_s^e \, v \, \mathrm{d}\Omega \right) + \sum_{\Gamma_k \not\subset \Gamma} \int_{\Gamma_k} \rho_k v \, \mathrm{d}\Gamma = \sum_{\Gamma_k \subset \Gamma_N} \int_{\Gamma_k} \xi \, v \, \mathrm{d}\Gamma \tag{4.7}$$

is a measure of the error in self equilibration.

To measure the error incurred by the approximation (4.3) the natural choice is the energy norm given by

$$\| e \|_E^2 = \int_\Omega \left(\bar{\nabla}_t e \cdot a \bar{\nabla}_t e + b e^2 \right) \mathrm{d}\Omega \tag{4.8a}$$

that may be written in terms of the residuals as

$$\| e \|_E^2 = \sum_{i=1}^{N_e} \left(\int_{\Omega^e} r_s^e \, e \, \mathrm{d}\Omega \right) + \sum_{\Gamma_k \not\subset \Gamma} \int_{\Gamma_k} \rho_k e \, \mathrm{d}\Gamma + \sum_{\Gamma_k \subset \Gamma_N} \int_{\Gamma_k} \xi \, e \, \mathrm{d}\Gamma \tag{4.8b}$$

An estimate of (4.8) may be obtained in the form

$$\| e \|_E^2 \approx \epsilon_G^2 = \sum_{e=1}^{N_e} \epsilon_e^2 \tag{4.9a}$$

where ϵ_G is called the global error estimate and ϵ_e denotes the local error estimate associated with the Ω^e element as given by

$$\epsilon_e^2 = \alpha_e \int_{\Omega^e} r_s^2 \, d\Omega + \beta_e \sum_{\substack{\Gamma_k \subset \Gamma_{\Omega^e} \\ \Gamma_k \not\subset \Gamma_D}} \int_{\Gamma_k} r_{ed}^2 \, \mathrm{d}\Gamma \tag{4.9b}$$

Here, for ease of notation, r_s denotes the distributed residual r_s^e over the eth element and r_{ed} denotes the lumped residual over the interior edges ρ, or over the edges located at the Neumann boundaries ξ. Γ_{Ω^e} is the boundary of the eth element

The difficulty of obtaining an error estimate is to compute the values of the constants α_e and β_e in (4.9b). In [Kelly et al. 1983] the authors arrive to the following expression for Lagrange rectangular elements

$$\epsilon_e^2 = \frac{h_e^2}{24 a_e p} \int_{\Omega^e} r_s^2 \, d\Omega + \frac{h_e}{24 a_e p} \sum_{\substack{\Gamma_k \subset \Gamma_{\Omega^e} \\ \Gamma_k \not\subset \Gamma_D}} \int_{\Gamma_k} r_{ed}^2 \, d\Gamma \tag{4.9c}$$

where h_e is the diameter of the eth element, p is the order of the approximation used, and a_e is the value of a for the eth element.

The quasi-static analysis of transmission lines may be performed by solving a set of boundary value problems as given by (4.1) where the unknown ϕ is the scalar electrostatic potential, the quantity a stands for the transverse part of the permittivity tensor $\underline{\underline{\varepsilon}}_{rt}$, $b = 0$, $f = 0$, and $g_N = 0$ (see Sections 3.2.2.2). The FEM solution of the problem uses Lagrange elements (see Section 3.2.3.1). Expression (4.9c) may be extended to the anisotropic case by substituting a_e with the minimum eigenvalue of the 2×2 matrix representing the transverse permittivity tensor of the eth element. However, triangular elements are preferred over rectangular ones because they offer a greater flexibility to discretize complex geometries. Expression (4.9c) may be extended to triangular elements [Rivara 1986] in the following form

$$\epsilon_e^2 = f_s \frac{h_e^2}{\varepsilon_{rtmin}\, p} \int_{\Omega^e} r_s^2 \, d\Omega + f_{ed} \frac{h_e}{\varepsilon_{rtmin}\, p} \sum_{\substack{\Gamma_k \subset \Gamma_{\Omega^e} \\ \Gamma_k \not\subset \Gamma_D}} \int_{\Gamma_k} r_{ed}^2 \, d\Gamma \tag{4.9d}$$

where f_s and f_{ed} are weighting factors associated with the surface and the singular residual errors, respectively. These factors have been estimated numerically.

In summary, once a finite element analysis is carried out over a given mesh, the surface and singular residuals may be computed for each element. Then (4.9d) is applied to obtain the local error estimate. In Appendix G.2, details are given about these computations for the case of first- and second-order, straight and curved triangular elements, as well as for first- and second-order infinite elements. The global error estimate is obtained from (4.9a).

If a self-adaptive procedure works properly, the estimate of the error should approach the energy norm of the exact error as the refinement proceeds. Thus, the effectivity index defined as

$$i_e = \frac{\epsilon_G}{\|e\|_E} \tag{4.10}$$

should tend asymptotically to unity. Examples having analytical solutions are employed to analyze the robustness of an error estimate and of a self-adaptive procedure by calculating the behavior of the effectivity index.

4.3.2 Refinement Strategy

The indication of whether a mesh refinement should be carried out or not can be obtained just by comparing $\|e\|^2_E$ with $\|\phi\|^2_E$. The refinement process will take place whenever

$$\frac{\|e\|_E^2}{\|\phi\|_E^2} \geq \vartheta \tag{4.11a}$$

where ϑ is the prespecified degree of accuracy or global refinement parameter selected by the user. In practice, the exact values are not known, but they may be substituted by their approximations. Hence, the mesh will be enriched whenever the following expression holds:

$$\frac{\epsilon_G^2}{\|\tilde{\phi}\|_E^2} \geq \vartheta \tag{4.11b}$$

Typical values for ϑ are from 10^{-3} (0.1%) to 10^{-4} (0.01%).

Once the decision of refining a mesh has been taken, the next issue is to determine which elements should be enriched. For this purpose, a characteristic local error estimate ϵ_c is defined. For example, this can be the maximum local error estimate

$$\epsilon_M = \max(\epsilon_e) \ , \quad e = 1,\ldots,N_e \tag{4.12a}$$

or the optimal local error estimate defined as

$$\epsilon_o = \sqrt{\frac{\sum_{e=1}^{N_e} \epsilon_e^2}{N_e}} \tag{4.12b}$$

that is, the value of the local error estimate in each element of an equilibrated or optimal mesh.

The eth element having a local error estimate ϵ_e will be refined if

$$\frac{\epsilon_e^2}{\epsilon_c^2} \geq \gamma \tag{4.12c}$$

holds, where γ is the local refinement parameter between 0 and 1. This parameter is selected by the user. Elements fulfilling (4.12c) will be called primary elements.

In [Gago et al. 1983] it is shown that the selection of one or the other characteristic local error estimate leads asymptotically to the same optimal mesh. After experimenting with both characteristic local error estimates, it has been observed that its definition according to (4.12a), that is, $\in_c = \in_M$, accelerates the rate of convergence. The parameter γ only influences the speed at which the maximum rate of convergence is reached. A typical value is 0.5.

4.3.3 Element Subdivision Algorithms

As mentioned previously, an *h*-refinement procedure has been chosen. Thus each primary element must be subdivided into two or more elements. The simplest approach is to subdivide every primary triangle into three smaller triangles by inserting a point at its centroid and connecting it to the vertices of the original triangle (see Figure 4.1(a)). This method produces conforming meshes because all new nodes are inserted inside existing elements (at the new vertex or edges generated inside them), not at their boundaries. Thus, the resulting mesh will consist of compatible elements. However, by itself it creates elements with high aspect ratios (with obtuse angles) leading in a few iterations to degenerate elements that produce less accurate results (see Section 2). Hence, it must be used in conjunction with methods like diagonal swapping (see Figure 4.1(b)) between the new elements and their neighbors or vertex relocation (see Figure 4.1(c)) in order to obtain high quality triangles. The Delaunay method is quite appropriate [Csendes and Shenton 1985], [Delaunay 1934], [Golias and Tsiboukis 1991], [Henneberger et al. 1990], [Penman and Grieve 1985], [Watson 1981]. This complicates the procedure. Also, if the initial mesh does not have a large enough number of elements, the first few iterations of the self-adaptive procedure will not be free of distorted elements. This method may not produce a smooth enough transition between regions with elements of large and small sizes.

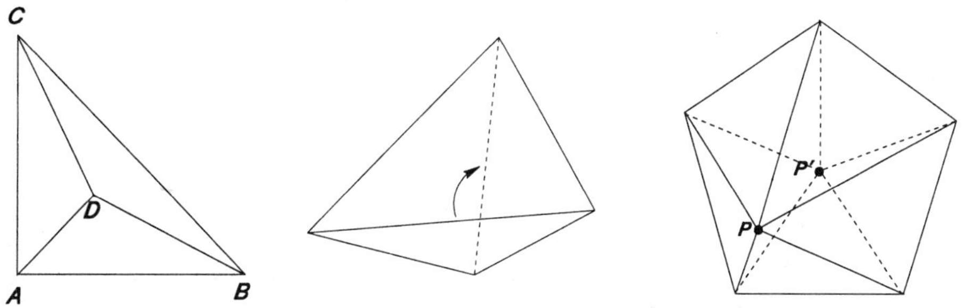

(a) (b) (c)

Figure 4.1 (a) Subdivision into three triangles. (b) Diagonal swapping. (c) Vertex relocation.

The next choice will be to bisect a primary triangle into two smaller elements by placing a point at the midpoint of an edge (possibly the longest one) and connecting it to the opposite vertex. This procedure by itself generates nonconforming meshes if the edge on which the point has been placed is not on a boundary of the domain of the problem but at an interface with a nonprimary element (i.e., an element that has not been selected for subdivision). Thus, the procedure requires the extension of the refinement process to those elements. Nonprimary elements that should be subdivided in order to obtain a conformal mesh are called secondary elements. Figure 4.2 illustrates this issue. This method of simple bisection is not adequate for several reasons. Even if for primary elements the longest edge is used for bisection, distorted elements may be generated in the process of subdivision of a given secondary element. This is because the new point may have been placed along any one of its edges. The same techniques explained previously may be used to alleviate this problem [Luomi and Rouhiainen 1988]. Another drawback is that bisecting a primary element into two elements may not be enough for making a good distinction between primary and secondary elements. Moreover, both primary and secondary elements may be forced to be subdivided into more than two elements in the process of conforming the mesh. Thus, it may happen that a primary element in which the error is high will be subdivided into two elements while a secondary element may be subdivided into three or four elements. Thus, the method will not provide a robust self-adaptive procedure. Other possible methods are [Fernandes et al. 1992] and [Tärnhuvud et al. 1990].

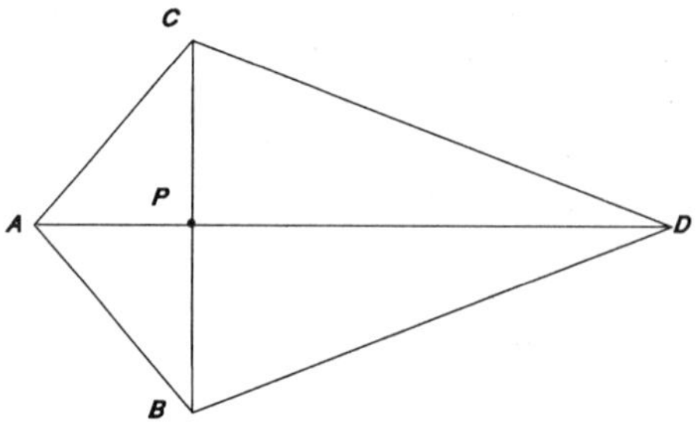

Figure 4.2 Simple bisection of the primary element *ABC* into two smaller elements *ABP* and *APC*. The secondary element *BDC* must also be subdivided.

Summarizing, one would look for subdivision algorithms in which primary elements would be divided into more than two elements. Assuming that the initial mesh does not contain high aspect ratio elements, the method should ensure that degeneracies will not occur in the process of refining both primary and secondary elements. The transition between regions with elements of large and small sizes should be smooth. The refinement should not be extended too far away from the primary elements. Finally, the subdivision algorithm in conjunction with the error estimate chosen and the refinement strategy selected should lead asymptotically to equilibrated meshes and to a good behavior of the effectivity index.

A triangle may be subdivided into four triangles by placing a point at the midpoint of each edge and connecting all three new points between them (see Figure 4.3(a)). It may be observed that all four new triangles are similar to the original one, so that the aspect ratio has been preserved (this was the procedure utilized in Section 3.2.3.1 for uniform refinements). Hence, no degeneracies will occur for the subdivision of primary elements [Kinsner and Torre 1974], [Löhner et al. 1986]. However, because the procedure requires the extension of the refinement process to nonprimary elements, a good method for their refinement should also be provided. The first subdivision algorithm that has been analyzed is based on the subdivision into two of each of the edges connected to any of the vertices of a primary element, by placing a point at their mid points. Then, if the three edges of an element should be subdivided (i.e., in the case of primary elements, or eventually secondary elements), their midpoints are connected according to Figure 4.3(a). If only two edges of an element should be subdivided (i.e., for secondary elements), their midpoints are connected between them. This will give a triangle similar to the original one, and a quadrilateral element that is further subdivided into two triangles by a diagonal subdivision. The diagonal giving triangles with better aspect ratio is selected for this partition (i.e., the diagonal subdividing the largest angle of the quadrilateral). Figure 4.3(b) shows the subdivision of a secondary element whose vertex C is also a vertex of a primary element. It is observed that this algorithm will subdivide all secondary elements having a common edge with a primary one into four elements, while secondary elements with just a common vertex will be subdivided into three elements. No element will be subdivided into two elements. Thus, the refinement may be excessive and may extend too far away from the primary elements.

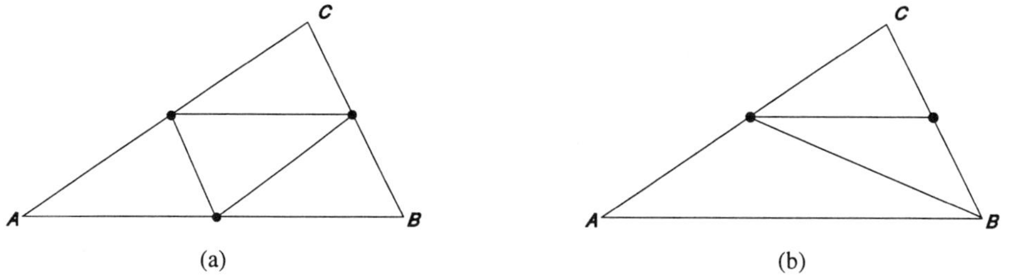

Figure 4.3 First subdivision algorithm: (a) primary elements (and eventually secondary elements) and (b) secondary elements.

A triangle may also be subdivided into four elements by placing a point on its longest edge and connecting it to the opposite vertex and to the midpoints of the other two edges (see Figure 4.4(a)). Hence, at least two of the new elements will be similar to the original one [Hahn et al. 1988], [Iribarren and Rivara 1992], [Rivara 1984, 1986,

1987, 1989, 1992], [Rivara and Inostroza 1997]. Secondary elements are divided into two, three, or eventually four elements as follows. If a new vertex created in a primary element happens to belong to the longest side of a contiguous secondary element, then the latter is subdivided into two elements connecting the midpoint of its longest side to the opposite vertex. If the new vertex created in a primary element happens to belong to any other edge of a contiguous secondary element, then the latter is subdivided into three elements by connecting that point to the midpoint of its longest side, and this latter point to the opposite vertex (see Figure 4.4(c)). Figure 4.5 illustrates the procedure. *ABC* is a primary element that is subdivided into four elements creating the vertices *f*, *g*, and *h*. *ACE* is a secondary element where *f* appears on its longest edge. Then it is subdivided into two elements. *BDC* is a secondary element where *g* appears on an edge that is not the longest one. Then it is subdivided into two elements. A secondary element may require to be subdivided into four elements if at the midpoint of both nonlongest edges there happens to appear a new vertex.

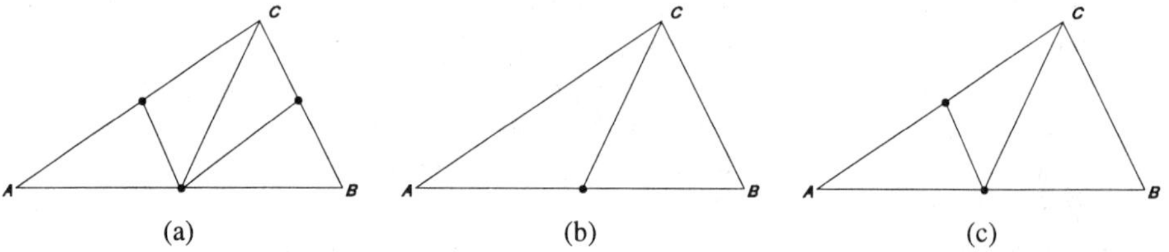

Figure 4.4 Second subdivision algorithm: (a) primary element and (b),(c) secondary elements.

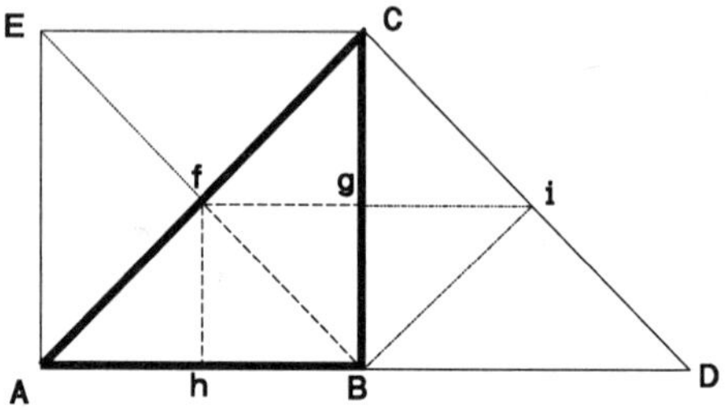

Figure 4.5 Illustration of the second subdivision algorithm.

This second algorithm has been analyzed in [Iribarren and Rivara 1992], [Mercader and Rivara 1992], [Rivara 1984, 1986, 1987, 1989, 1992]. It guarantees that no angle of the final mesh will be smaller than one half the smallest angle of the initial mesh. The refinement procedure does not result in an excessive number of elements.

In addition, the transition between regions of triangles of different sizes is smooth.

A third algorithm combines the two former algorithms, using the method in Figure 4.3(a) for primary elements and those in Figures 4.4(b) and 4.4(c) for secondary elements. This third algorithm is used either by itself or as an alternate to the second one for structures with curved contours in order to avoid the degeneration of curved triangles [Mercader and Rivara 1992].

4.3.4 Self-Adaptive Algorithm

Figure 4.6 shows the flow diagram of the self-adaptive procedure as described in the previous sections. The stopping criterion may be controlled through the ϑ parameter (the prespecified degree of accuracy), through a prespecified degree of change in the computation of a given parameter, or through a maximum number of steps.

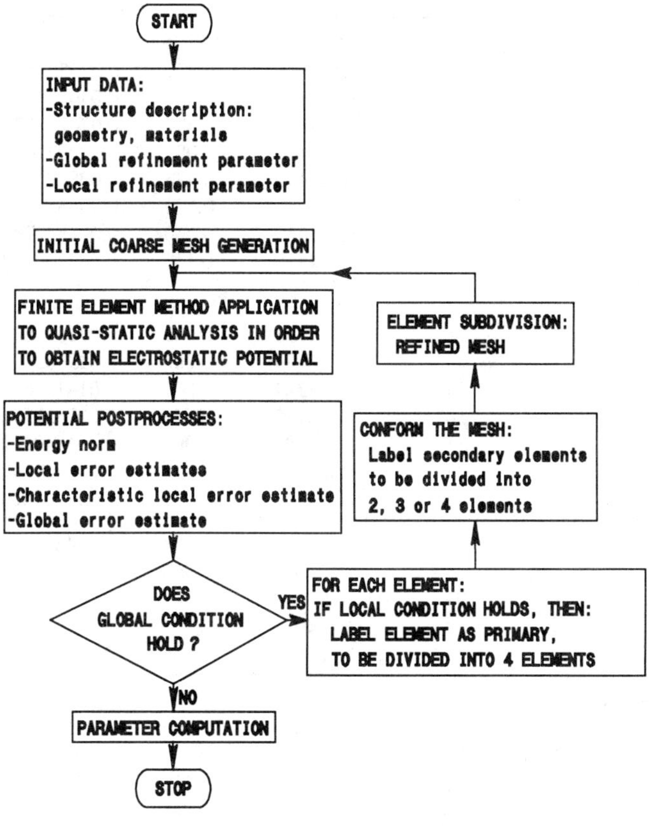

Figure 4.6 Flow diagram of the self-adaptive algorithm.

4.3.5 Validation of the Self-Adaptive Algorithm

The method has been validated through the analysis of the structures described in Section 3.2.3.1: the coaxial square line, the symmetric stripline, the coaxial line with dielectric corner, and the circular coaxial line.

Tables 4.1 to 4.3 show the results for the square coaxial line using first- and second-order elements using two different values of the γ parameter and for the three

Table 4.1
Results for the analysis of the coaxial square line using first- and second-order triangular elements and the first refinement algorithm

		Mesh step	Number of nodes	C/ε_0	Error (%)	Square of the global error estimate	CPU time (sec)
First order	$\gamma = 0.5$	1	5	12.0	17.256	0.1666667	2.36
		2	11	10.84615	5.981	0.07554663	2.91
		3	23	10.47819	2.386	0.03050300	4.83
		4	33	10.38002	1.427	0.02025857	6.42
		5	43	10.34465	1.081	0.01693567	8.17
		6	93	10.28442	0.493	0.008185401	12.41
		7	122	10.27288	0.380	0.006557072	16.48
		8	184	10.26065	0.260	0.004842088	22.46
		9	330	10.24855	0.142	0.002709698	31.36
	$\gamma = 0.1$	1	5	12.0	17.256	0.1666667	2.36
		2	11	10.84615	5.981	0.07554663	5.33
		3	25	10.46754	2.282	0.02998220	6.22
		4	54	10.34324	1.067	0.01419866	7.91
		5	103	10.28782	0.526	0.007529700	11.04
		6	155	10.26489	0.302	0.004840382	15.49
		7	235	10.25363	0.192	0.003217692	23.07
		8	426	10.24413	0.099	0.001708088	36.58
Second order	$\gamma = 0.5$	1	12	10.45926	2.201	0.034673930	1.49
		2	35	10.32649	0.903	0.014419420	3.62
		3	79	10.27142	0.365	0.005884058	6.31
		4	123	10.24951	0.151	0.002520779	9.23
		5	167	10.24082	0.066	0.001187407	12.69
	$\gamma = 0.1$	1	12	10.45926	2.201	0.03467393	1.54
		2	35	10.32649	0.904	0.01441942	3.68
		3	79	10.27142	0.365	0.005884050	6.04
		4	123	10.24951	0.151	0.002520779	9.40
		5	167	10.24082	0.066	0.001187407	12.74

refinement algorithms. One eighth of the structure was analyzed. Thus, the square of the global error estimate (equation (4.9c) has been used) refers to one eighth of the structure. However, the normalized capacitance refers to the complete structure. The error refers to the percentage error incurred in the computation of the normalized capacitance. The CPU time refers to the total time taken for the computations.

Table 4.2

Results for the analysis of the coaxial square line using first- and second-order triangular elements and the second refinement algorithm

		Mesh step	Number of nodes	C/ε_0	Error (%)	Square of the global error estimate	CPU time (sec)
First order	$\gamma = 0.5$	1	5	12.0	17.256	0.1666667	2.35
		2	9	10.84615	5.981	0.0590552	2.92
		3	15	10.52453	2.839	0.0331687	4.67
		4	24	10.39494	1.573	0.0203698	6.76
		5	36	10.33810	1.017	0.0138769	8.96
		6	57	10.29523	0.598	0.0088068	11.53
		7	83	10.27403	0.391	0.0056813	14.66
		8	107	10.26522	0.305	0.0043444	18.62
		9	150	10.25597	0.214	0.0031000	26.42
	$\gamma = 0.1$	1	5	12.0	17.256	0.1666667	2.36
		2	9	10.84615	5.981	0.05905522	3.03
		3	23	10.46675	2.274	0.02296155	4.89
		4	42	10.34896	1.123	0.01251221	7.19
		5	79	10.28852	0.533	0.006281040	9.99
		6	101	10.26951	0.347	0.004461660	14.17
		7	168	10.25511	0.206	0.002757531	19.66
		8	302	10.24447	0.105	0.00144504	28.12
		9	408	10.24152	0.073	0.001023945	41.68
Second order	$\gamma = 0.5$	1	12	10.45926	2.201	0.03467393	1.37
		2	29	10.32725	0.912	0.01189280	3.62
		3	59	10.27255	0.377	0.005200896	6.43
		4	89	10.25005	0.157	0.002297376	8.46
		5	119	10.24112	0.069	0.001146787	12.31
		6	149	10.23758	0.035	0.000694092	16.20
	$\gamma = 0.1$	1	12	10.45926	2.201	0.03467893	1.53
		2	29	10.32735	0.912	0.01189280	3.57
		3	73	10.27106	0.362	0.00459863	6.54
		4	111	10.24906	0.147	0.001912223	9.28
		5	149	10.24034	0.062	0.0008517209	12.80
		6	187	10.23650	0.035	0.000431835	19.17

Table 4.3
Results for the analysis of the coaxial square line using first- and second-order triangular elements and the third refinement algorithm

		Mesh step	Number of nodes	C/ε_0	Error (%)	Square of the global error estimate	CPU time (sec)
First order	$\gamma = 0.5$	1	5	12.0	17.25	0.1666667	1.38
		2	9	10.84615	5.98	0.07487965	3.13
		3	17	10.48475	2.45	0.03043302	4.72
		4	23	10.39494	1.57	0.02250222	6.86
		5	35	10.33810	1.02	0.01599649	8.95
		6	57	10.29413	0.58	0.009603545	11.60
		7	79	10.27369	0.39	0.006333431	14.22
		8	114	10.26146	0.27	0.004582760	18.78
		9	209	10.24857	0.14	0.002392792	24.44
	$\gamma = 0.1$	1	5	12.0	17.25	0.1666667	2.59
		2	9	10.84615	5.98	0.07487965	3.24
		3	23	10.46754	2.28	0.02879987	5.00
		4	45	10.34656	1.10	0.01446424	6.97
		5	84	10.28663	0.51	0.007061084	10.66
		6	105	10.26964	0.35	0.005360948	14.12
		7	209	10.25148	0.17	0.002639213	20.76
		8	297	10.24501	0.11	0.001757654	29.88
Second order	$\gamma = 0.5$	1	12	10.45926	2.20	0.03467393	1.48
		2	29	10.32735	0.91	0.01477757	3.57
		3	51	10.27286	0.38	0.006392315	6.04
		4	73	10.25119	0.17	0.003114485	9.06
		5	95	10.24264	0.08	0.001818313	11.53
	$\gamma = 0.1$	1	12	10.45926	2.20	0.03467393	1.71
		2	29	10.32735	0.91	0.01477757	3.68
		3	51	10.27286	0.38	0.006392315	6.59
		4	87	10.24975	0.15	0.002619213	9.12
		5	109	10.24116	0.07	0.001322503	11.97

Figures 4.7 to 4.9 show the evolution of the mesh with the first, second, and third algorithms, respectively. The mesh for the complete structure is shown. It may be seen how the mesh is more and more refined at the corners of the inner conductor. It is observed that the first algorithm generates a refined mesh over a larger area than the other two. The third algorithm extends the refinement region less than the second one

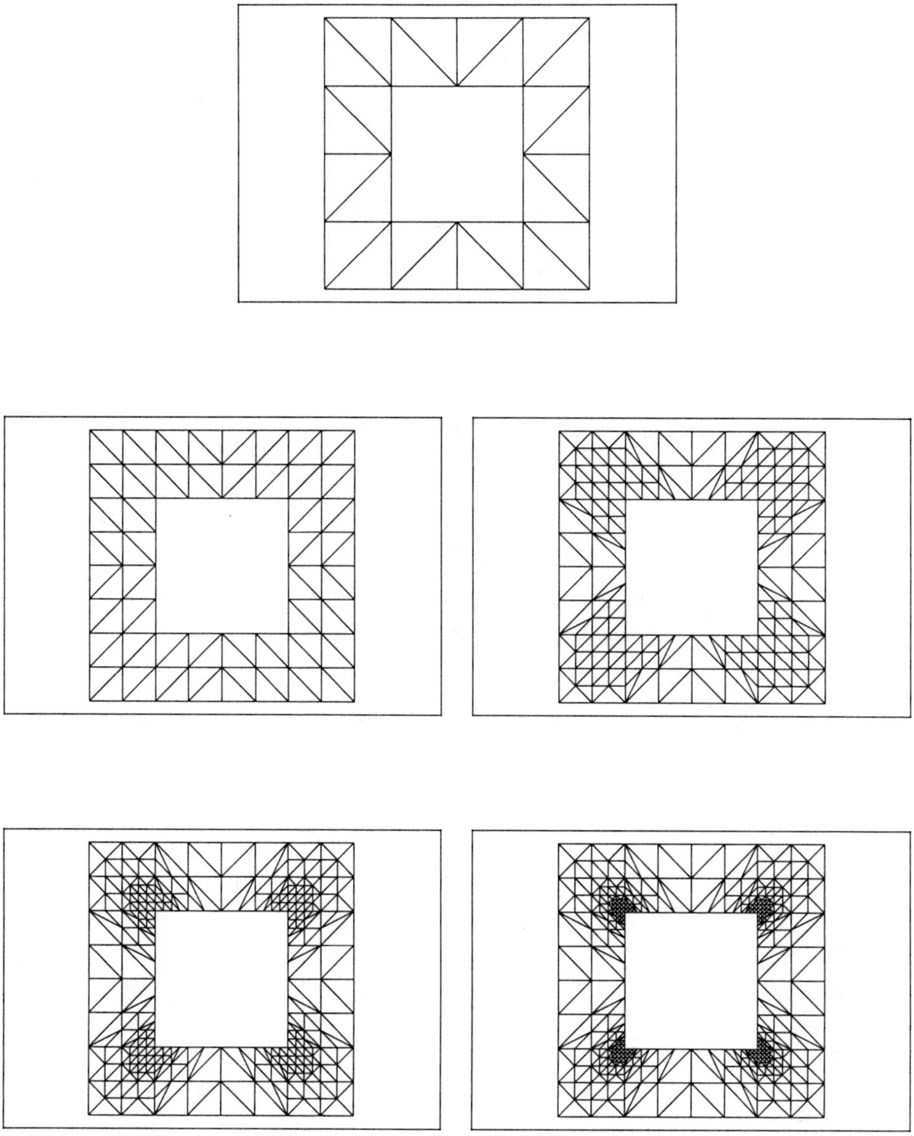

igure 4.7 Evolution of the mesh for the analysis of the rectangular coaxial structure with the first refinement algorithm.

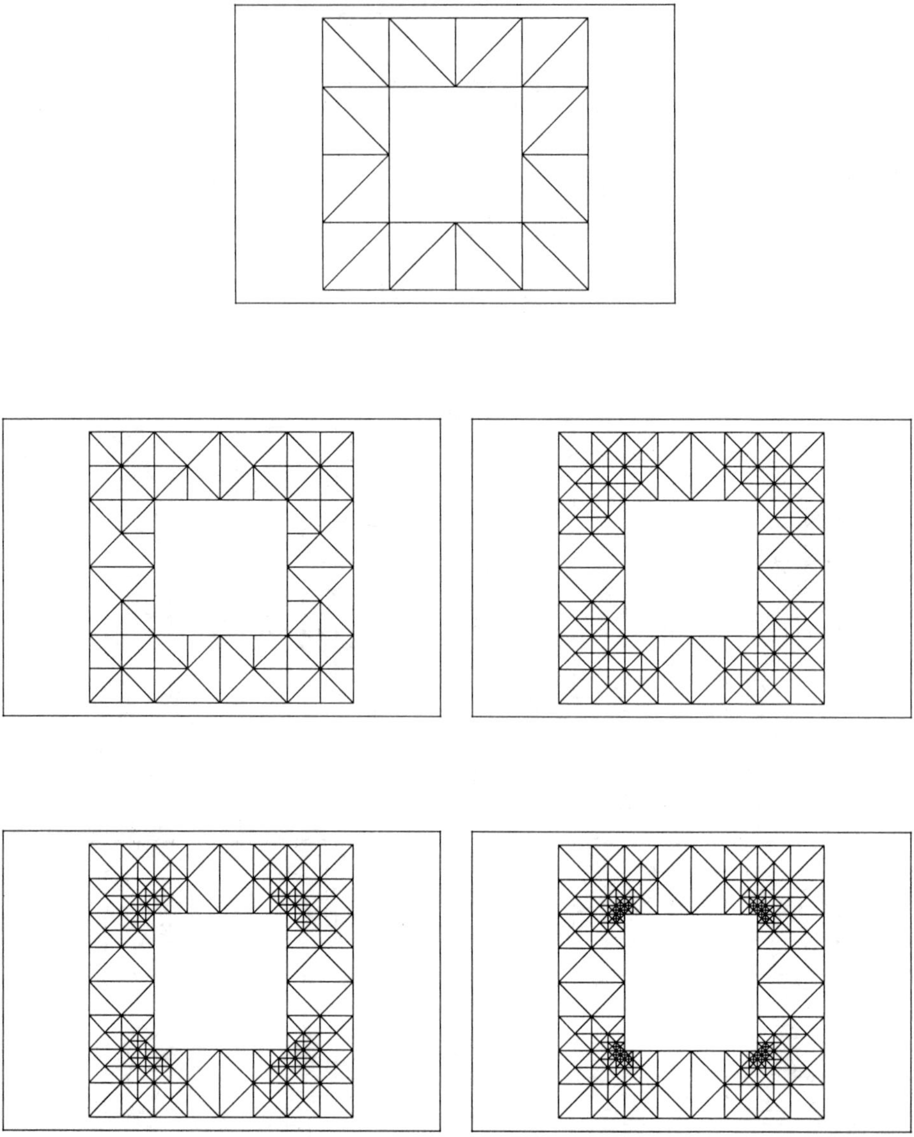

Figure 4.8 Evolution of the mesh for the analysis of the rectangular coaxial structure with the second refinement algorithm.

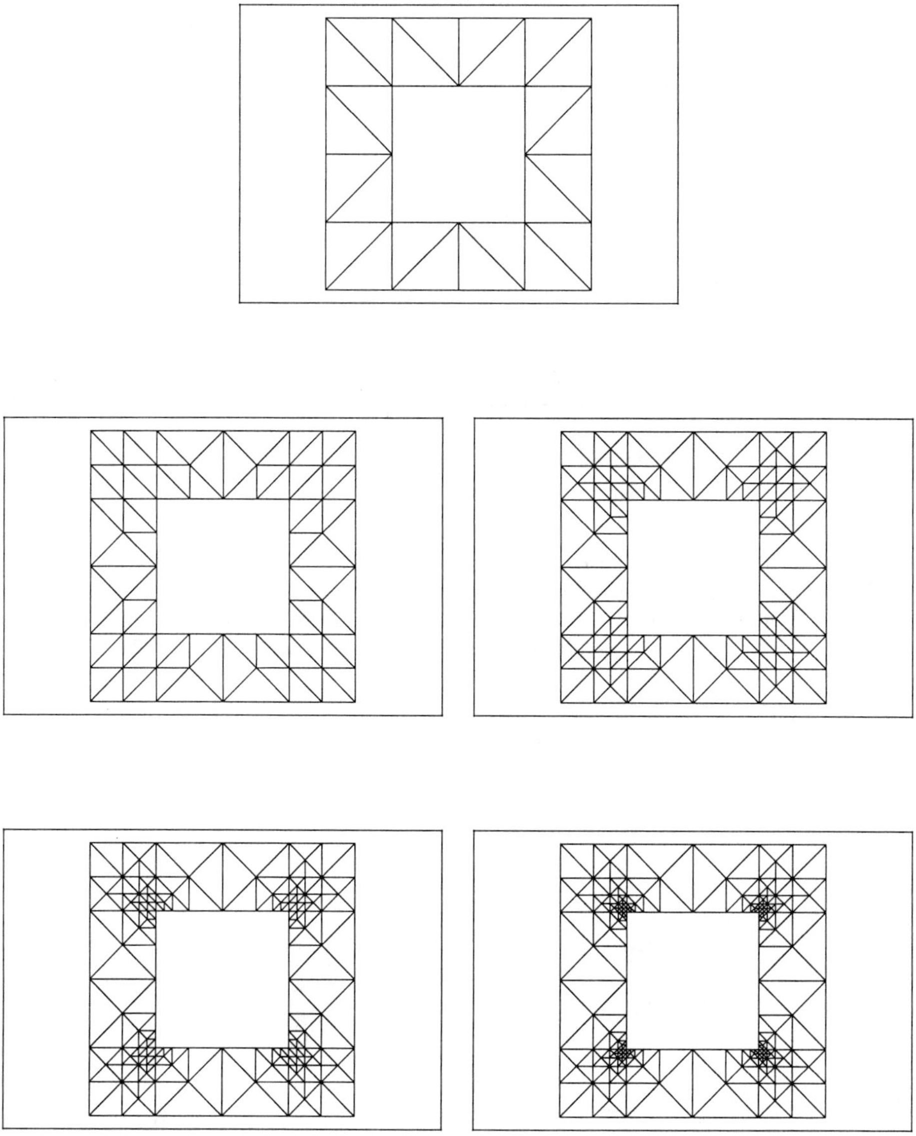

Figure 4.9 Evolution of the mesh for the analysis of the rectangular coaxial structure with the third refinement algorithm.

Figures 4.10 to 4.12 show the rate of convergence obtained in the three cases and compare it with that corresponding to uniform refinements. In all cases the maximum rate of convergence is reached, as if a quasi-uniform mesh has been utilized

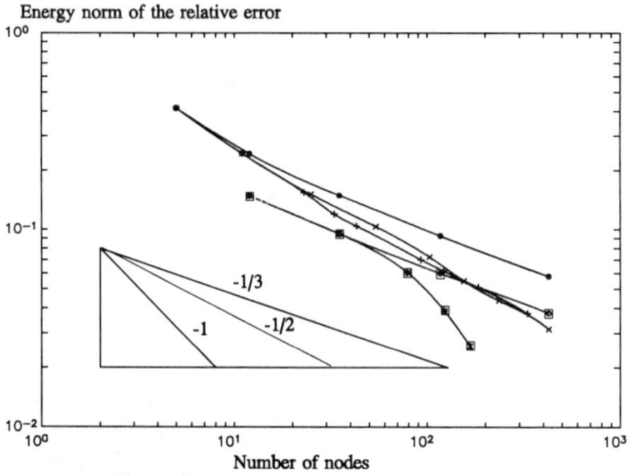

Figure 4.10 Convergence of the self-adaptive analysis with the first refinement algorithm compared with the uniform refinement for the rectangular coaxial line. o: first order, uniform refinement; ×: first-order, self-adaptive, with $\gamma = 0.5$; +: first-order, self-adaptive, with $\gamma = 0.1$; ⧇: second-order, uniform refinement; ⊞: second-order, self-adaptive, with $\gamma = 0.5$.

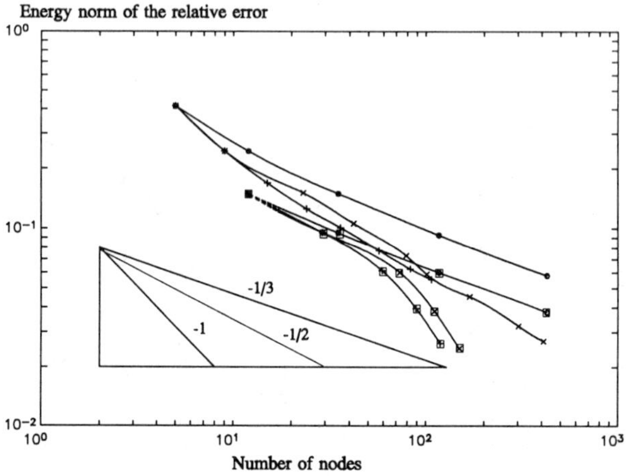

Figure 4.11 Convergence of the self-adaptive analysis with the second refinement algorithm compared with the uniform refinement for the rectangular coaxial line. o: first-order, uniform refinement; ×: first-order, self-adaptive, with $\gamma = 0.5$; +: first-order, self adaptive, with $\gamma = 0.1$; ⧇: second-order, uniform refinement; ⊠: second-order, self-adaptive, with $\gamma = 0.5$; ⊞: second-order, self-adaptive, with $\gamma = 0.1$.

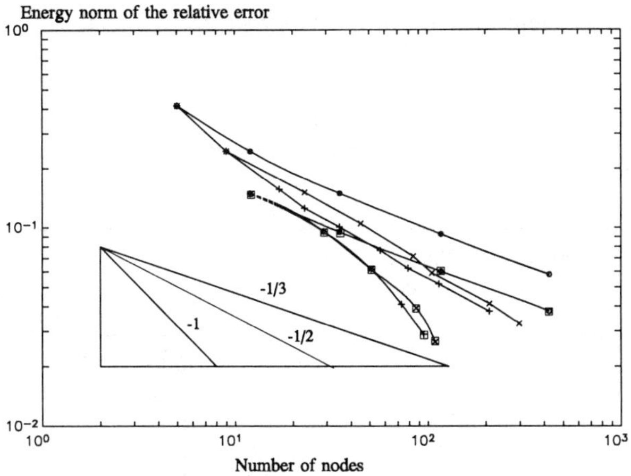

Figure 4.12 Convergence of the self-adaptive analysis with the third refinement algorithm compared with the uniform refinement for the rectangular coaxial line. o: first-order, uniform refinement; ×: first-order, self-adaptive, with $\gamma = 0.5$; +: first-order, self-adaptive, with $\gamma = 0.1$; ⊡: second-order, uniform refinement; ⊠: second-order, self-adaptive, with $\gamma = 0.5$; ⊞: second-order, self-adaptive, with $\gamma = 0.1$.

to solve a problem with a smooth solution. The first algorithm converges more slowly than the other two. The third algorithm converges faster than the rest. However, it is easy to check that the effectivity index of the third algorithm shows a worse behavior than that of the second algorithm. Figure 4.13 shows the evolution of the effectivity index using the second algorithm.

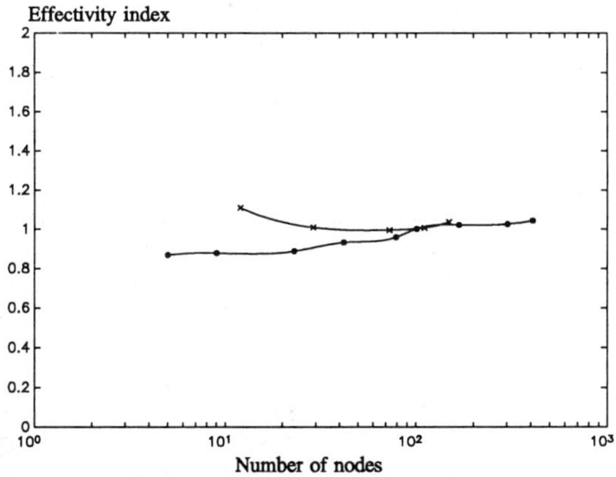

Figure 4.13 Effectivity index for the analysis of the rectangular coaxial line with the second refinement algorithm; o: first-order; x: second-order.

Tables 4.4 to 4.6 and Figures 4.14 to 4.20 show the results of the analysis of the symmetric stripline. Here one fourth of the structure was analyzed.

Table 4.4
Results for the analysis of the symmetric stripline using first- and second-order triangular elements and the first refinement algorithm

		Mesh step	Number of nodes	C/ε_0	Error (%)	Square of the global error estimate	CPU time (sec)
First order	$\gamma = 0.5$	1	6	8.0	36.13	0.6868867	1.09
		2	14	6.762887	15.08	0.2545903	2.97
		3	25	6.370205	8.39	0.1375158	4.89
		4	36	6.196409	5.43	0.09398648	6.98
		5	47	6.112855	4.01	0.0743300	9.12
		6	70	6.006142	2.20	0.04214842	12.25
		7	81	5.986564	1.87	0.03800873	15.21
		8	108	5.960993	1.43	0.02936727	19.44
		9	164	5.931347	0.93	0.01951620	24.16
	$\gamma = 0.1$	1	6	8.0	36.13	0.6868867	2.59
		2	15	6.761905	15.06	0.2547239	4.83
		3	36	6.314342	7.44	0.1142820	5.27
		4	55	6.130204	4.31	0.06613627	7.75
		5	86	6.019257	2.42	0.03993758	11.31
		6	105	5.977406	1.71	0.03020099	14.94
		7	157	5.939915	1.07	0.01931471	19.56
		8	232	5.918516	0.71	0.01350570	26.25
		9	316	5.905931	0.50	0.01012652	35.76
Second order	$\gamma = 0.5$	1	15	6.349206	8.04	0.1889882	1.43
		2	42	6.112896	4.05	0.09446052	4.07
		3	83	5.994537	2.00	0.04606543	6.09
		4	124	5.936360	1.01	0.02367290	9.61
		5	165	5.907546	0.52	0.01245588	13.13
		6	206	5.893211	0.28	0.006921180	16.75
		7	247	5.886065	0.16	0.004158405	20.93
		8	288	5.882498	0.10	0.002803548	25.81
	$\gamma = 0.1$	1	15	6.349206	8.04	0.1889882	1.53
		2	45	6.112856	4.05	0.09438592	3.90
		3	102	5.994277	2.00	0.04639373	6.99
		4	159	5.935959	1.00	0.02332097	10.05
		5	216	5.907059	0.51	0.01253006	13.72
		6	273	5.892675	0.27	0.006442322	18.57
		7	330	5.885500	0.15	0.003662529	24.00
		8	387	5.881919	0.09	0.002286498	30.05

Table 4.5
Results for the analysis of the symmetric stripline using first- and second-order triangular elements and the second refinement algorithm

		Mesh step	Number of nodes	C/ε_0	Error (%)	Square of the global error estimate	CPU time (sec)
First order	$\gamma = 0.5$	1	6	8.0	36.13	0.6868867	1.15
		2	14	6.762887	15.08	0.2545903	3.02
		3	25	6.370205	8.39	0.1375158	4.95
		4	36	6.196409	5.44	0.09398648	7.03
		5	47	6.112855	4.01	0.0743300	9.62
		6	70	6.006142	2.20	0.04214842	12.86
		7	81	5.986564	1.87	0.03800873	15.65
		8	108	5.960993	1.43	0.02936727	19.39
		9	164	5.931347	0.97	0.01951620	24.06
	$\gamma = 0.1$	1	6	8.0	36.13	0.6868867	2.41
		2	15	6.761905	15.06	0.2547239	4.61
		3	36	6.314342	7.44	0.1142820	5.38
		4	55	6.130204	4.31	0.06613627	7.91
		5	86	6.019257	2.42	0.03993758	11.26
		6	105	5.977406	1.71	0.03020099	14.45
		7	157	5.939915	1.07	0.01931471	19.94
		8	232	5.918516	0.71	0.01350570	26.97
		9	316	5.905931	0.50	0.01012652	36.80
Second order	$\gamma = 0.5$	1	15	6.349206	8.04	0.1889882	1.54
		2	30	6.115105	4.05	0.04724657	3.79
		3	70	5.996230	2.03	0.02375097	6.26
		4	104	5.937657	1.03	0.01271276	9.88
		5	138	5.908633	0.54	0.007322916	13.12
		6	172	5.894197	0.30	0.004659201	16.59
		7	206	5.887003	0.17	0.003337003	21.97
		8	254	5.882340	0.09	0.001924773	25.81
	$\gamma = 0.1$	1	15	6.349206	8.04	0.1889882	1.54
		2	33	6.115105	4.05	0.04709609	3.79
		3	88	5.995031	2.01	0.02248409	6.70
		4	138	5.936019	1.00	0.01159994	10.65
		5	188	5.906738	0.51	0.006004564	13.89
		6	238	5.892156	0.26	0.003220404	18.68
		7	228	5.884881	0.14	0.001833439	24.45
		8	338	5.881250	0.07	0.001142869	29.88

Table 4.6
Results for the analysis of the symmetric stripline using first- and second-order triangular elements and the third refinement algorithm

		Mesh step	Number of nodes	C/ε_0	Error (%)	Square of the global error estimate	CPU time (sec)
First order	$\gamma = 0.5$	1	6	8.0	36.12	0.6868867	1.15
		2	10	6.762887	15.07	0.2554753	3.00
		3	15	6.394909	8.81	0.1464729	3.80
		4	20	6.231598	6.03	0.1057946	4.91
		5	29	6.086007	3.56	0.06322833	6.15
		6	34	6.048638	2.42	0.05546367	7.35
		7	49	5.989155	1.91	0.03717622	10.50
		8	57	5.974473	1.66	0.03330106	12.75
		9	96	5.933078	0.95	0.01879003	20.25
	γ=0.1	1	6	8.0	36.12	0.6868867	1.15
		2	13	6.762887	15.07	0.2548837	4.72
		3	25	6.322173	7.57	0.1172320	5.15
		4	36	6.142899	4.52	0.07115166	6.05
		5	52	6.032374	2.64	0.04420787	8.05
		6	63	5.991561	1.95	0.03508612	9.95
		7	104	5.944304	1.15	0.02042756	14.80
		8	135	5.925067	0.82	0.01505822	20.01
		9	178	5.910949	0.58	0.01095252	24.89
Second order	$\gamma = 0.5$	1	15	6.349206	8.04	0.1889882	1.54
		2	30	6.115550	4.06	0.09616809	3.98
		3	48	5.997613	2.05	0.04844980	6.55
		4	66	5.939385	1.06	0.02543498	9.27
		5	84	5.910515	0.57	0.01419817	12.85
		6	102	5.896144	0.33	0.008651085	15.15
		7	120	5.888978	0.21	0.005896368	17.75
		8	138	5.885399	0.14	0.004528107	20.07
	$\gamma = 0.1$	1	15	6.349206	8.04	0.1889882	1.54
		2	33	6.115105	4.05	0.09590805	4.15
		3	59	5.996603	2.04	0.04778200	8.30
		4	85	5.938156	1.04	0.02458903	12.95
		5	111	5.909165	0.55	0.01325590	16.18
		6	137	5.894729	0.30	0.007658519	20.52
		7	163	5.887526	0.18	0.004876991	28.85
		8	203	5.882988	0.10	0.002907334	39.61

From Figures 4.14 to 4.16, it may be observed that again the first algorithm has extended too much the refinement process, while the third algorithm seems to confine the refinement to the region close to the corner of the stripline. It turns out that the behavior of the third algorithm is quite dependent on the initial mesh. The same structure has been analyzed starting from an initial mesh in which triangles were obtained by subdividing rectangles by means of the diagonal opposite to that used in Figures 4.14 to 4.16, obtaining a similar performance for the second algorithm and a quite different performance for the third one.

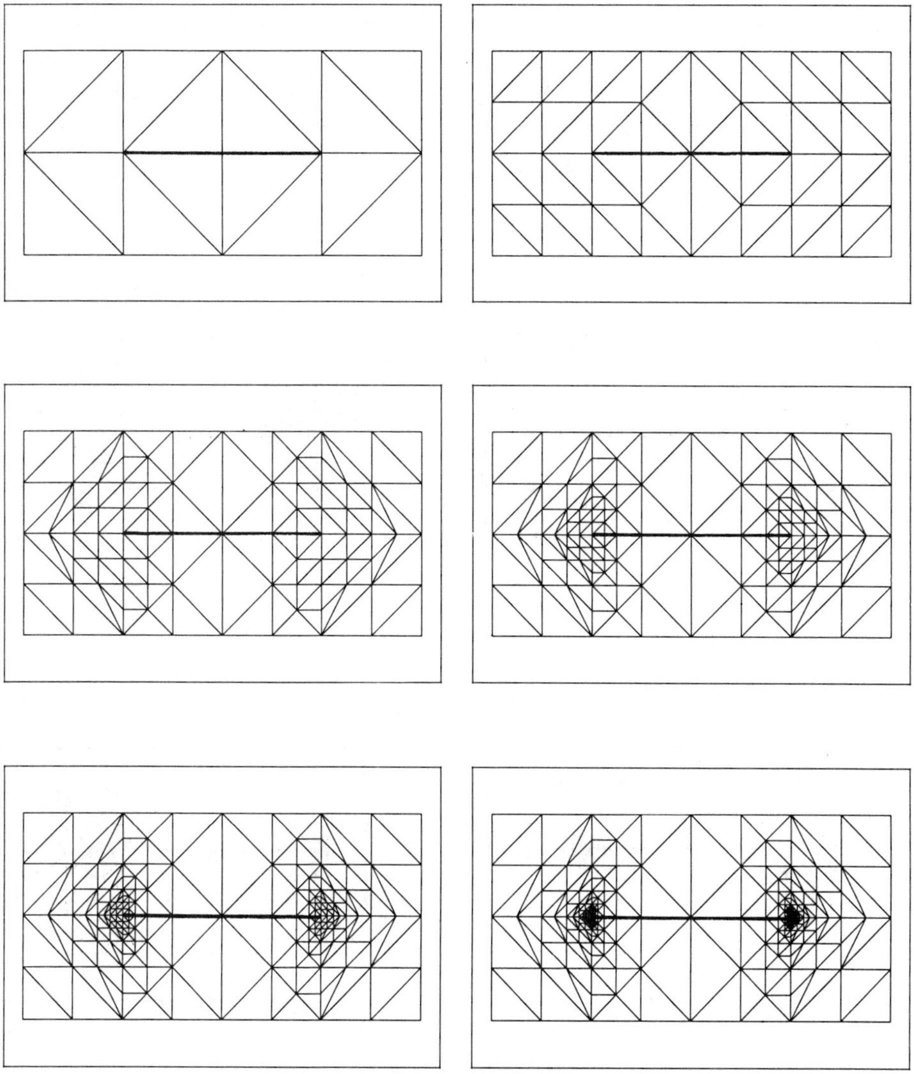

'igure 4.14 Evolution of the mesh for the analysis of the symmetric stripline with the first refinement algorithm.

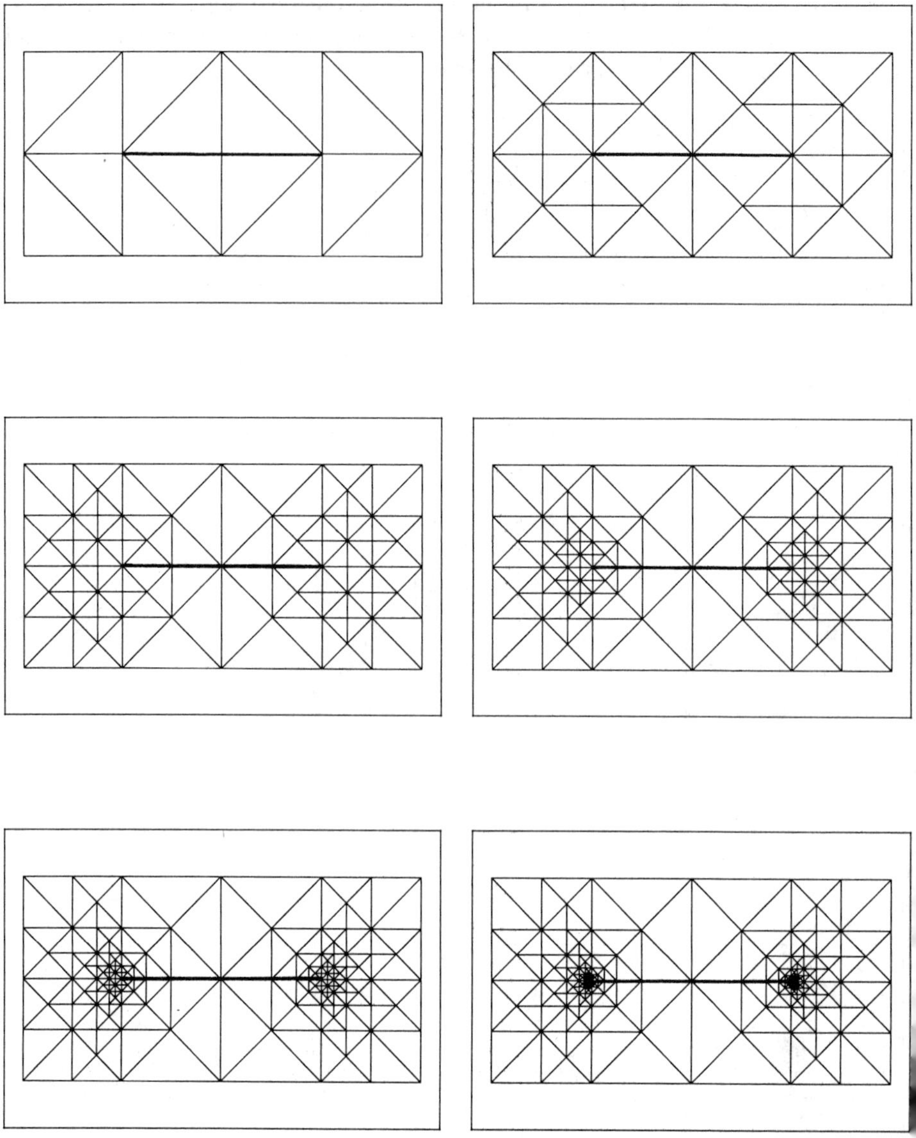

Figure 4.15 Evolution of the mesh for the analysis of the symmetric stripline with the secor refinement algorithm.

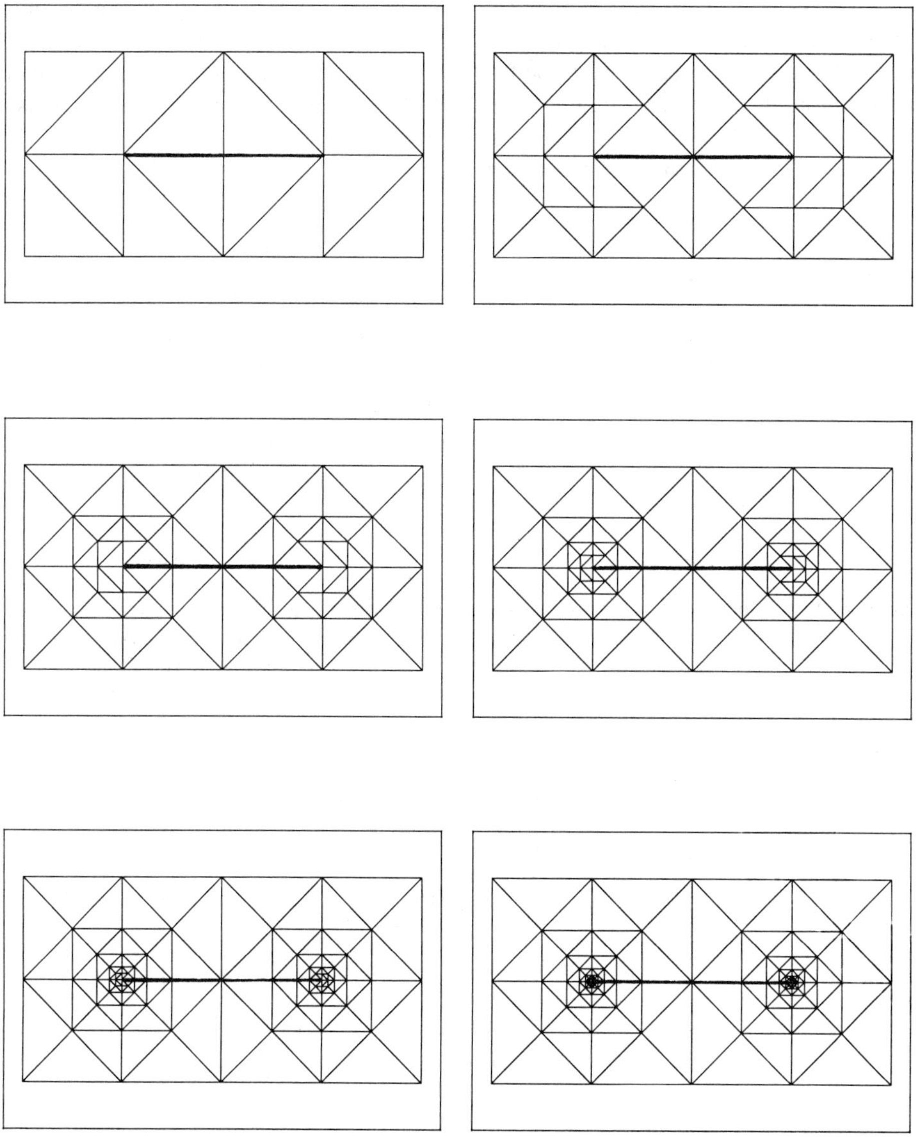

Figure 4.16 Evolution of the mesh for the analysis of the symmetric stripline with the third refinement algorithm.

Figures 4.17 to 4.19 show how the self-adaptive procedure gives the maximum rate of convergence for the h-version of FEM. The third algorithm converges faster than the other two. However, the behavior of this algorithm is highly dependent on the initial mesh. Also from Tables 4.5 and 4.6, it may be seen that the behavior of the effectivity index of the second algorithm is better than that of the third algorithm. The second algorithm is thus preferred. Figure 4.20 shows the evolution of its effectivity index.

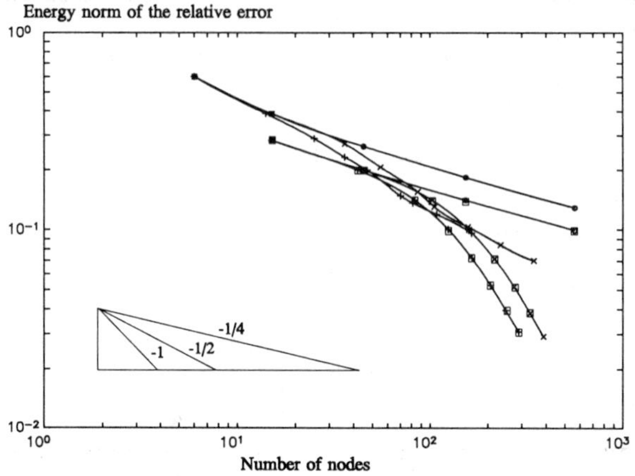

Figure 4.17 Convergence of the self-adaptive analysis with the first refinement algorithm compared with the uniform refinement for the symmetric stripline. o: first-order, uniform refinement; ×: first-order, self-adaptive, with $\gamma = 0.5$; +: first-order, self-adaptive, with $\gamma = 0.1$; ⧇: second-order, uniform refinement; ⊠: second-order, self-adaptive, with $\gamma = 0.5$; ⊞: second-order, self-adaptive, with $\gamma = 0.1$.

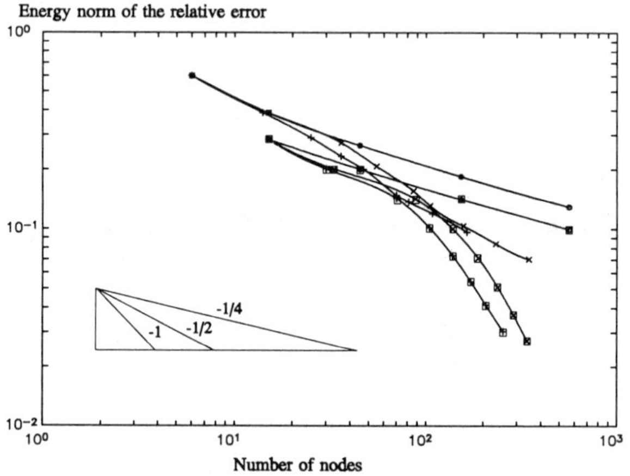

Figure 4.18 Convergence of the self-adaptive analysis with the second refinement algorithm compared with the uniform refinement for the symmetric stripline. o: first-order, uniform refinement; ×: first-order, self-adaptive, with $\gamma = 0.5$; +: first-order, self-adaptive, with $\gamma = 0.1$; ⧇: second-order, uniform refinement; ⊠: second-order, self-adaptive, with $\gamma =$ 0.5; ⊞: second-order, self-adaptive, with $\gamma = 0.1$.

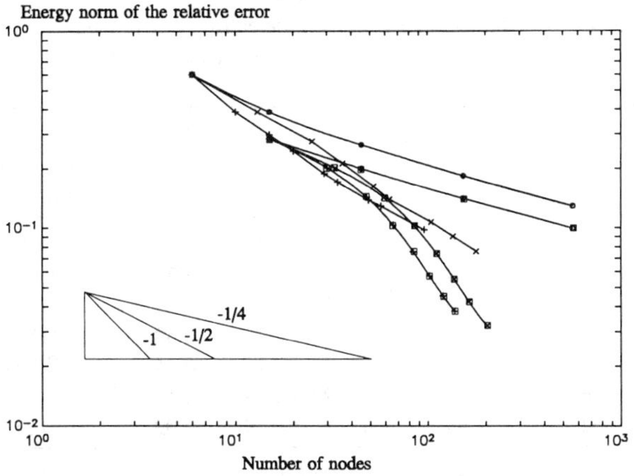

Figure 4.19 Convergence of the self-adaptive analysis with the third refinement algorithm compared with the uniform refinement for the symmetric stripline. o: first-order, uniform refinement; ×: first-order, self-adaptive, with $\gamma = 0.5$; +: first-order, self-adaptive, with $\gamma = 0.1$; ⊡: second order, uniform refinement; ⊠: second order, self adaptive, with $\gamma = 0.5$; ⊞: second-order, self-adaptive, with $\gamma = 0.1$.

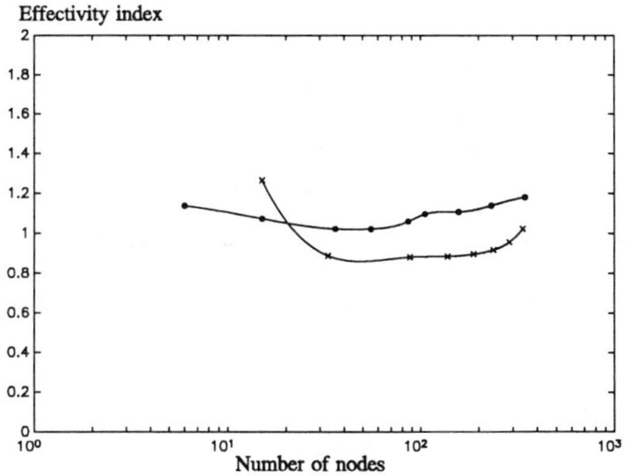

Figure 4.20 Effectivity index for the symmetric stripline analysis with the second refinement algorithm; o: first-order; ×: second-order.

A coaxial structure partially filled with a high permittivity dielectric is considered next. The corners of the dielectric produces a nonsmooth behavior of the solution. In this case, the third algorithm has been used because of the existence of curved contours. Table 4.7 provides the results of the analysis. Figure 4.21 shows the rate of convergence of the relative error of the energy norm and compares it with the case in which uniform refinements are used. Once again, it is observed that the maximum rate of convergence for the h-version is reached.

Table 4.7
Results for the analysis of the coaxial line with dielectric corners using first- and second-order triangles and the third refinement algorithm

		Mesh step	Number of nodes	C/ε_0	Error (%)	Square of the global error estimate	CPU time (sec)
First order	$\gamma = 0.5$	1	9	25.06975	5.47	0.2142707	1.32
		2	14	24.31888	2.31	0.1066141	3.24
		3	20	24.10662	1.42	0.7372717	5.16
		4	34	23.90675	0.58	0.03151618	7.80
		5	40	23.87519	0.44	0.02705186	9.83
		6	68	23.81398	0.19	0.01472982	13.35
		7	83	23.79179	0.09	0.01105726	16.20
	$\gamma = 0.1$	1	9	25.06975	5.47	0.2142707	2.80
		2	14	24.31888	2.31	0.1066141	3.57
		3	30	23.97884	0.88	0.04209408	5.55
		4	44	23.87720	0.45	0.02531727	8.13
		5	85	23.79840	0.122	0.01177041	11.37
		6	122	23.79757	0.119	0.007275507	15.49
Second order	$\gamma = 0.5$	1	25	24.04513	1.16	0.08897444	1.76
		2	43	23.86972	0.42	0.02688807	4.73
		3	65	23.80548	0.15	0.01028939	7.90
		4	87	23.78345	0.06	0.004646557	10.38
	$\gamma = 0.1$	1	25	24.04513	1.16	0.08897444	3.62
		2	43	23.86972	0.42	0.02688807	6.97
		3	65	23.80548	0.15	0.01028939	8.07
		4	87	23.78345	0.06	0.004646557	10.98

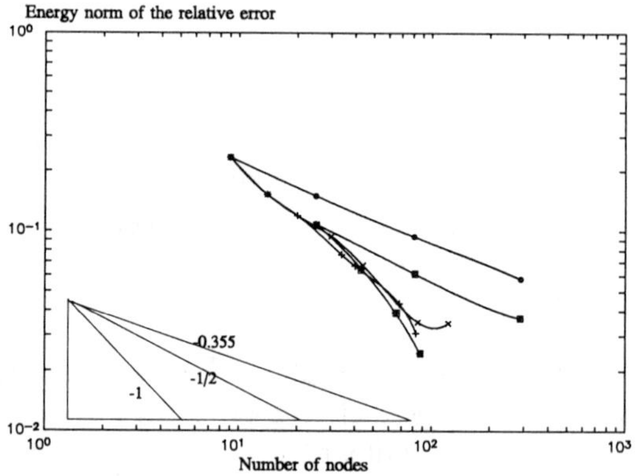

Figure 4.21 Convergence of the self-adaptive analysis compared with the uniform refinement for the coaxial line with dielectric corner. o: first-order, uniform refinement; +: first-order, self-adaptive, with $\gamma = 0.5$; ×: first-order, self-adaptive, with $\gamma = 0.1$; ⊡: second- order, uniform refinement; ⊠: second-order, self-adaptive, with $\gamma = 0.5$.

Figure 4.22 shows the evolution of the mesh. It is seen that in addition to the refinement at the dielectric corners, there is also more refinement along the air-dielectric interface in the air region where the strength of the electric field is higher. Thus, the adaptive procedure has taken into account the behavior of the electromagnetic field in the overall structure.

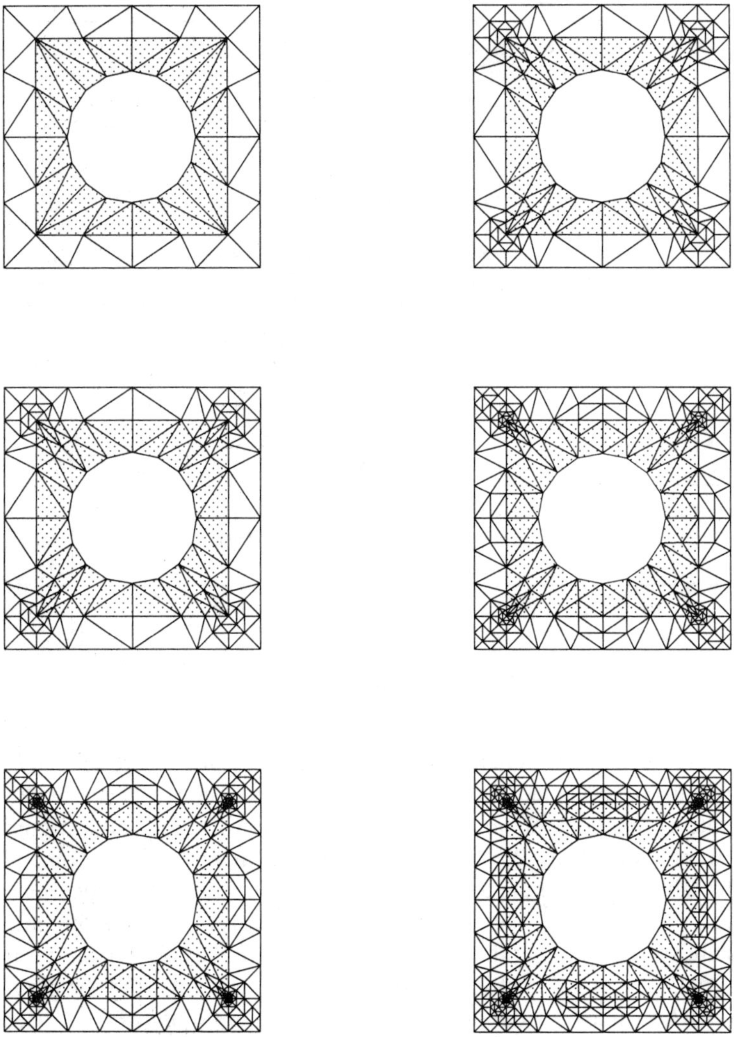

Figure 4.22 Evolution of the mesh for the analysis of the coaxial structure partially filled with dielectric.

Finally, Table 4.8 provides the results of the analysis of a circular coaxial line. If compared with Table 3.3, it may be seen that the use of the self-adaptive procedure also accelerates the convergence for the case when the solution is smooth.

Table 4.8
Results for the analysis of a coaxial circular line using first- and second-order triangular elements and the third subdivision algorithm

		Mesh step	Number of nodes	C/ε_0	Error (%)	Square of the global error estimate	CPU time (sec)
First order	$\gamma = 0.5$	1	5	8.815086	-2.750	0.1697789	1.27
		2	12	8.969689	-1.050	0.09097046	3.08
		3	29	9.140366	0.830	0.03572862	5.44
		4	78	9.106036	0.450	0.01473650	8.41
		5	99	9.076530	0.130	0.01127884	11.70
		6	200	9.083244	0.200	0.006013584	18.12
	$\gamma = 0.1$	1	5	8.815086	-2.750	0.1697789	2.69
		2	12	8.969689	-1.050	0.09097046	3.63
		3	35	9.040583	-0.260	0.03001181	5.44
		4	109	9.084201	0.210	0.008826048	9.45
		5	400	9.069779	0.060	0.002446393	5.48
		6	1330	9.066748	0.020	0.00077722	15.51
Second order	$\gamma = 0.5$	1	12	9.009589	-0.610	0.02143814	1.76
		2	35	9.065618	0.010	0.003007665	4.40
		3	90	9.066121	0.009	0.000428175	7.47
	$\gamma = 0.1$	1	12	9.009589	-0.610	0.02143814	2.91
		2	35	9.065618	0.010	0.003007665	4.50
		3	117	9.065000	0.003	0.000249943	7.85

The previous analyses demonstrate the efficiency of the self-adaptive procedure. Accurate results are obtained with the maximum rate of convergence of the *h*-version and moderate computational time. It must be pointed out that modest computational resources were utilized: a 50 MHz PC with a 486 processor. The computation time given in the tables refers to the total one (i.e., the time required for the total number of iterations at each step of the self-adaptive procedure). For the initial mesh, the unknowns are solved utilizing the Choleski method. For the subsequent mesh-steps, the conjugate gradient method is utilized. To initialize it, the solution of the previous mesh-step is used. The values for the new nodes are obtained through first- or second-order interpolation if first- or second-order elements are used, respectively. Thus, the time required by the conjugate gradient solver is minimized. In Chapter 5, additional examples are provided that further demonstrate the efficiency of the self-adaptive procedure for FEM postprocesses.

4.4 EXTENSION OF THE SELF-ADAPTIVE ALGORITHM TO THE FULL-WAVE ANALYSIS OF WAVEGUIDING STRUCTURES

This section deals with the extension of the self-adaptive procedure to the full-wave analysis of waveguiding structures. From [Babuška and Rheinboldt 1978], it may be deduced that for positive definite (or semi-definite) second-order elliptic eigenvalue problems once an eigenvalue $\tilde{\lambda}_i$ and its corresponding eigenvector $\{D^i\}$ has been computed so that an approximate solution of the unknown is available, then (4.9a) and (4.9b) also holds. Thus (4.9d) has been extended to give an estimate (or to be more precise an indicator) of the error in the eth element given by

$$\varepsilon_e^2 = f_s \frac{h_e^2}{a_{e_{\min}} p} \int_{\Omega^e} \{r_s^*\}^T \{r_s\}\, d\Omega + f_{ed} \frac{h_e}{a_{e_{\min}} p} \sum_{k=1}^{n_{ed}^e} \int_{\Gamma_k \not\subset \Gamma_D} \{r_{ed}^*\} \{r_{ed}\}\, d\Gamma \tag{4.13}$$

where $a_{e_{min}}$ depends on which formulation is being used. n_{ed}^e is the number of edges for the eth element. $\{r_s\}$ and $\{r_{ed}\}$ represent the distributed residual and the singular residual for the Ω^e elements, respectively. Appendix G.3 provides a detailed derivation for each of the terms of (4.13).

The self-adaptive mesh algorithm is essentially the same as that presented in Figure 4.6 except that $\|\phi\|^2_E$ in (4.11a), and $\|\tilde{\phi}\|^2_E$ in (4.11b) should be substituted by the eigenvalue λ_i, and the computed eigenvalue $\tilde{\lambda}_i$, respectively. Here the comparison is made for an indicator of the error. Another distinctive characteristic of this case with respect to the quasi-static analysis is that the self-adaptive process must be performed for each of the eigenvalues. Figure 4.23 shows the flow diagram for these classes of problems.

The full-wave self-adaptive mesh algorithm has been developed for the analysis of homogeneous isotropic waveguides employing the formulation in terms of the longitudinal component of the electric or the magnetic field, for the analysis of TM or TE modes, respectively. The operator in this case is positive definite. Straight Lagrange elements of first- and second-order are used. The computation of (4.13) is given in Appendix G.3.1. Here $a_{e_{min}} = 1$.

The method is also applicable for the analysis of general waveguiding structures using the formulation in terms of the transverse and the longitudinal components of the electric or the magnetic field, using the first-order Lagrange/curl-conforming element. Again the operator is positive semi-definite. Details of the computation of (4.13) are given in Appendix G.3.2. Here $a_{e_{min}}$ is the lowest eigenvalue of matrix $[f_r]$, according to Table 3.27. The surface and the singular residuals are vector quantities.

The full-wave self-adaptive approach has been validated utilizing the following examples: a rectangular homogeneous waveguide with smooth analytical solutions for any propagating mode; an L-shaped homogeneous waveguide in which some of the propagating modes exhibit a singular behavior at the corner, and a rectangular waveguide half-filled with dielectric.

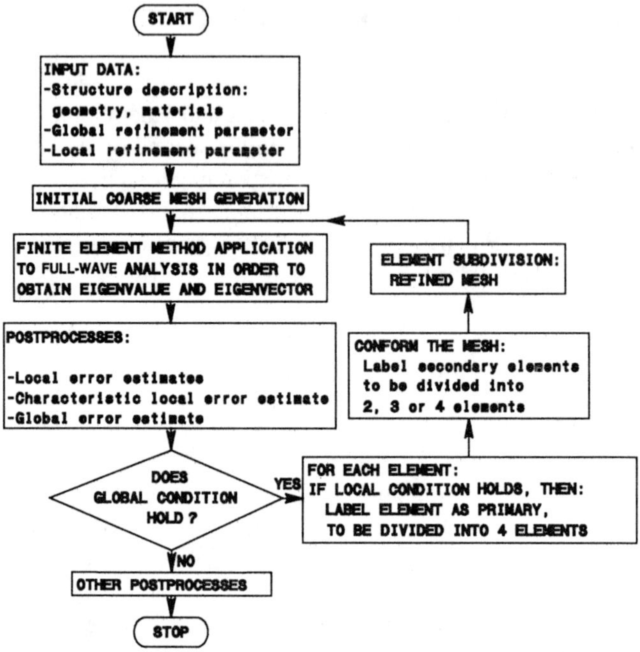

Figure 4.23 Flow diagram of the extension of the self-adaptive algorithm to an eigenvalue problem.

Figure 4.24 shows the evolution of the mesh for the computation of the first TE mode of a rectangular waveguide of dimensions $2a \times a$. The $\bar{E}$ field formulation has been used. Table 4.9 shows the corresponding results. Figure 4.25 compares the rate of convergence for the case of uniform refinements. Because the solution is smooth, the rate of convergence for uniform refinements follows the theoretical predictions. However, it may be seen that the self-adaptive procedure accelerates the rate of convergence.

Table 4.9
Results of the full-wave analysis of a rectangular waveguide

Mesh refinement	Number of nodes	s_c^2	Relative error (%)
1	45	2.424808	1.72
2	133	2.458323	0.37
3	489	2.465035	0.096
Exact		2.4674	

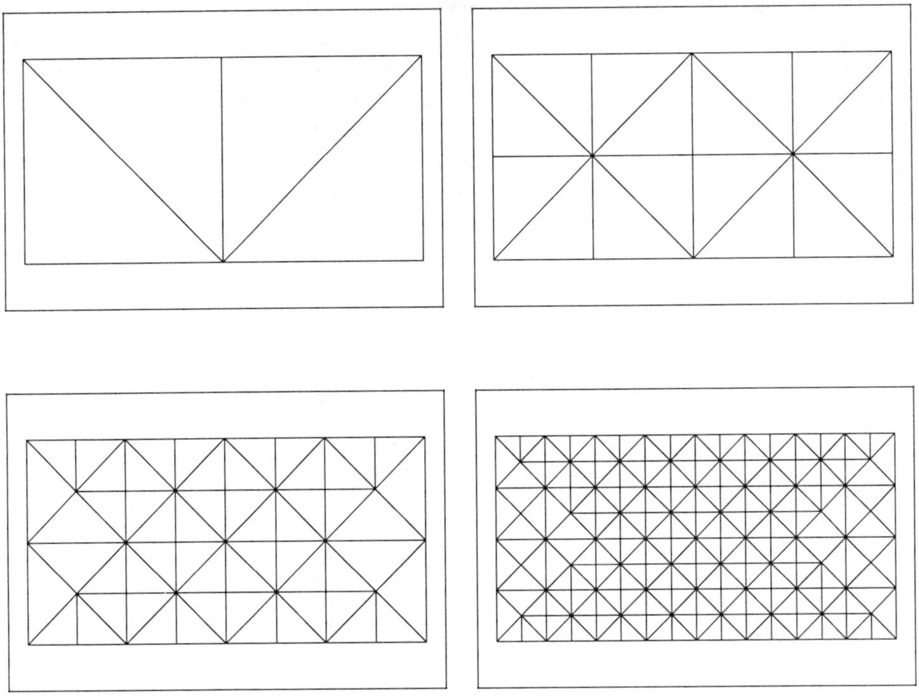

Figure 4.24 Evolution of the mesh for the analysis of the first TE mode of a rectangular waveguide.

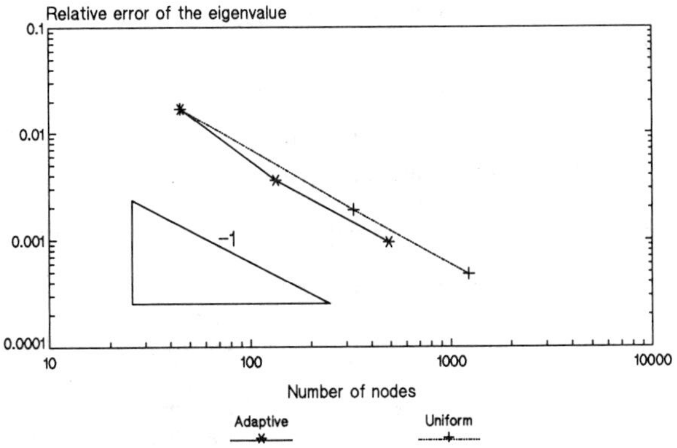

Figure 4.25 Rate of convergence of the eigenvalue (s_c^2) of the first TE mode of a $2a \times a$ rectangular waveguide.

Figure 4.26 shows the evolution of the mesh for the analysis of the first TE mode of a symmetric L-shaped waveguide and the transverse electric field that exhibits a singular behavior at the corner. The length of the longer sides is a, while that of the shorter sides is $a/2$. $\bar{E}$-field formulation has been used. Table 4.10 provides the corresponding results for $\gamma = 0.5$. Figure 4.27 shows the rate of convergence of the self-adaptive procedure. The relative error has been computed by assuming the exact value as the mean value of both the $\bar{E}$-and $\bar{H}$-field formulations over a mesh refined after 13 steps. It may be seen that the maximum rate of convergence is obtained.

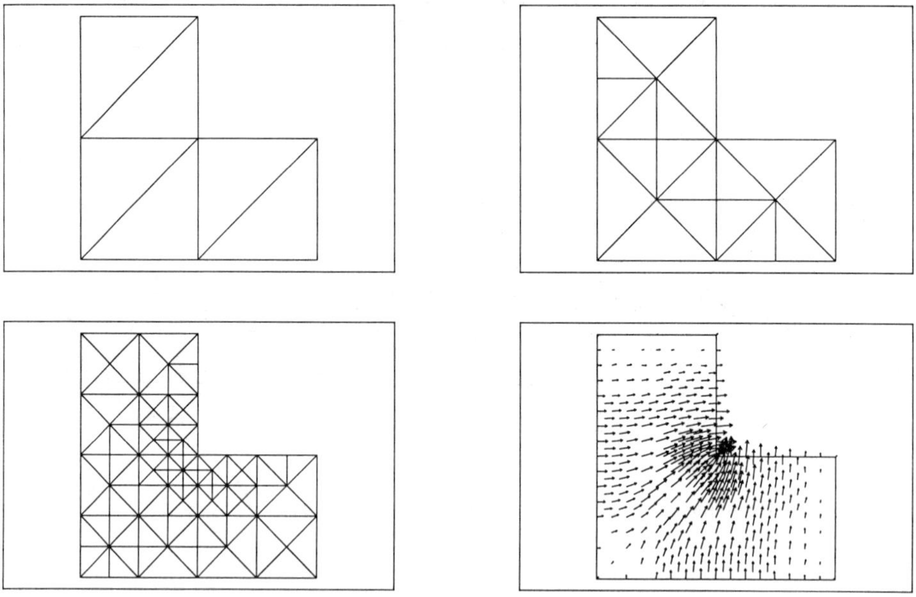

Figure 4.26 Evolution of the mesh for the first TE mode of a L-shaped waveguide along with the transverse component of the electric field.

Table 4.10

Results for the first TE mode of the L-shaped waveguide (geometric normalization factor a = 1.27)

Mesh step	Number of nodes	s_c^2
1	21	2.681323
2	41	3.303096
3	145	3.516722
4	227	3.574351
5	351	3.585899
[Swaminathan et al. 1990]		3.578528
[Sarkar et al. 1989]		3.534400

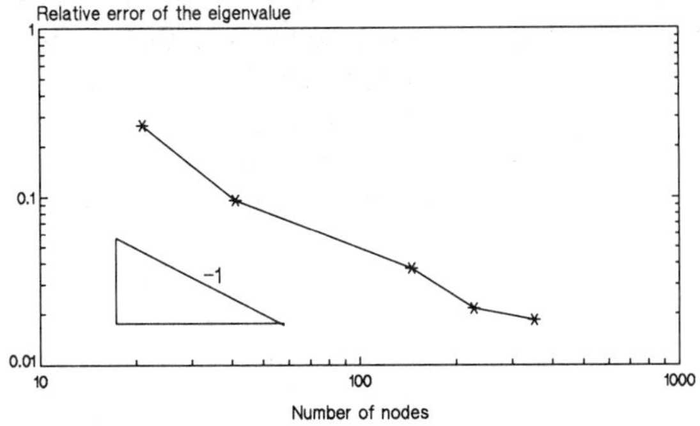

Figure 4.27 Rate of convergence of the eigenvalue (s_c^2) of the first TE mode of the L-shaped waveguide.

Figure 4.28 shows the evolution of the mesh for the second TE mode and the transverse electric field distribution for this mode. It can be observed how the mesh has adapted itself to the field configuration that does not exhibit a singular behavior at the corner in agreement with [Andersen and Solodukhov 1978].

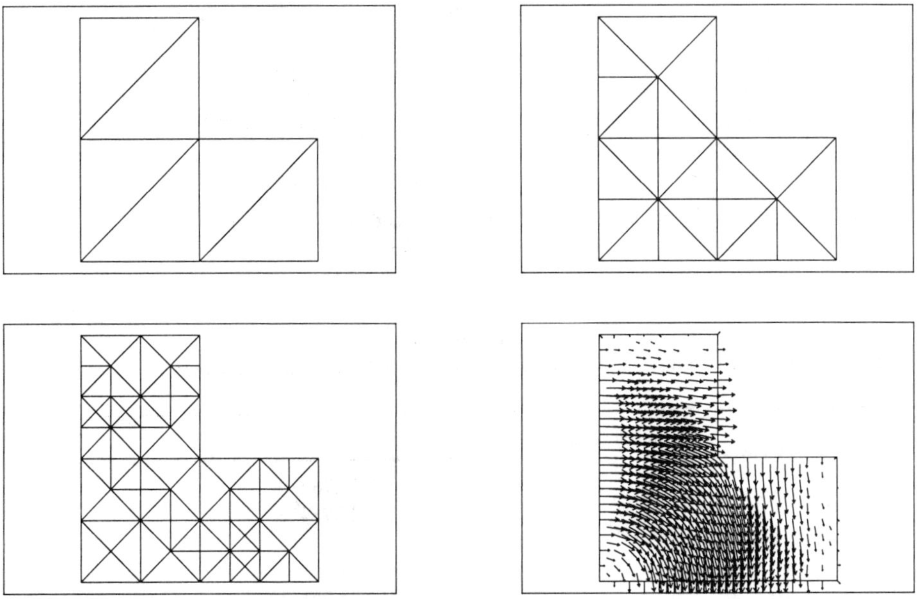

Figure 4.28 Evolution of the mesh for the second TE mode of a L-shaped waveguide along with the transverse component of the electric field.

Figure 4.29 shows the evolution of the mesh for the analysis of the first TM mode and the plot of the transverse magnetic field that again exhibits a singular behavior. The mesh adapts itself accordingly. The $\bar{H}$-field formulation and $\gamma = 0.3$ were used. Table 4.11 presents the results and compares them with those of [Swaminathan et al. 1990] and [Sarkar et al. 1989], which are based on an integral equation formulation and the frequency domain finite difference methods, respectively.

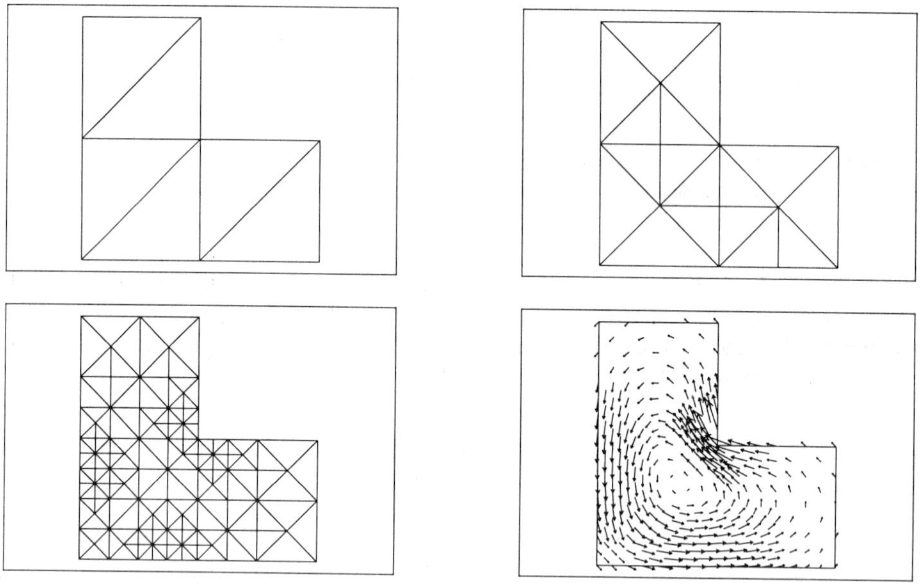

Figure 4.29 Evolution of the mesh for the first TM mode of a L-shaped waveguide along with the transverse component of the magnetic field.

Table 4.11
Results for first TM mode of the L-shaped waveguide (geometric normalization factor a = 1.27)

Mesh	Number of nodes	s_c^2
1	21	29.75980
2	65	25.73466
3	209	24.50415
4	605	24.44148
5	835	23.46113
[Sarkar et al. 1989]		23.04000
[Swaminathan et al. 1990]		23.694503

A rectangular waveguide half-filled with dielectric is analyzed next {see Figure 3.51(a)}. The waveguide dimensions are $2a \times a$, and $\varepsilon_r = 2.25$. The propagating modes will all have a smooth behavior. Figure 4.30 shows the evolution of the mesh for the first mode. The H field formulation and $\beta_a = 1$ have been used. Table 4.12 presents the results. Figure 4.31 compares the rate of convergence with that obtained for uniform refinements. It is seen that the convergence is accelerated by the adaptive procedure. Figure 4.32 refers to the second mode. Table 4.13 depicts the results.

Therefore the self-adaptive procedure leads to meshes that appropriately models the behavior of the electromagnetic field.

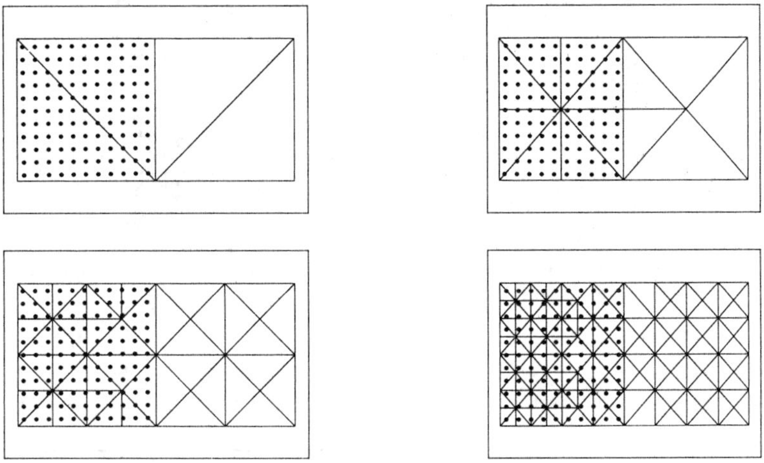

Figure 4.30 Evolution of the mesh for the rectangular waveguide half-filled with dielectric (first hybrid mode).

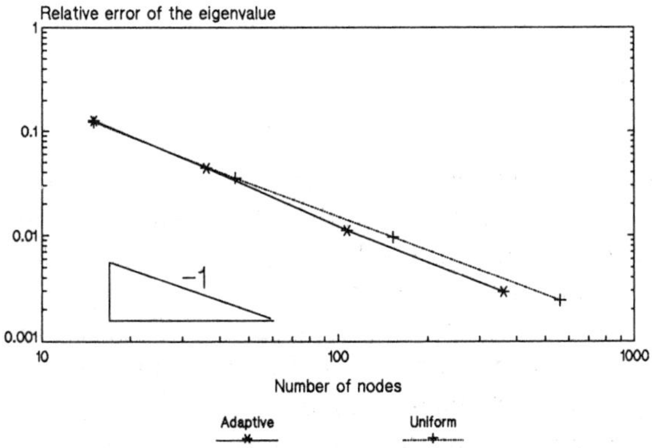

Figure 4.31 Rate of convergence of the eigenvalue (s^2) of the first hybrid mode of the rectangular waveguide half-filled with dielectric.

Table 4.12
Results for the first mode of a rectangular waveguide half-filled with dielectric $\beta_a = 1$

Mesh refinement	Number of nodes	s^2	CPU time (sec)
1	15	2.284481	1.21
2	36	2.124683	3.07
3	107	2.057605	6.16
4	361	2.040770	17.03

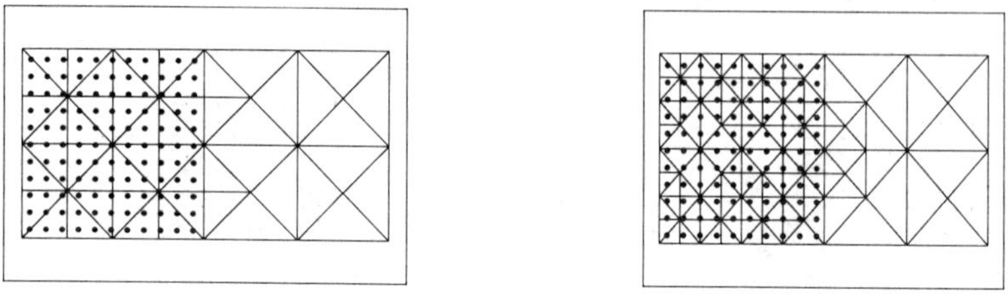

Figure 4.32 Third and fourth meshes of the analysis of the second hybrid mode of the rectangular waveguide half-filled with dielectric.

Table 4.13
Results for the second mode of a rectangular waveguide half-filled with dielectric $\beta_a = 1$

Mesh refinement	Number of nodes	s^2	CPU time (sec)
1	15	6.230510	1.10
2	36	6.172208	3.02
3	119	5.942790	7.36
4	309	5.719765	20.98

4.5 CONCLUSIONS

A self-adaptive mesh algorithm has been described that allows one to obtain results with a prespecified degree of accuracy. An algorithm has been developed for the quasi-static analysis using FEM with Lagrange elements. A reliable error estimate has been described along with three algorithms for the refinement of the mesh. The second is advantageous over the other two. The third avoids degenerate elements when discretizing structures with curved contours. The method has also been used for a full-wave analysis of transmission lines. Applications to several examples reveal that a maximum rate of convergence can be achieved for the *h*-version of FEM. The refined mesh adapts itself automatically to the appropriate behavior of the electromagnetic field and provides accurate results for arbitrary shaped complex waveguiding structures.

Chapter 5

Additional Examples

5.1 INTRODUCTION

This chapter deals with transmission lines and waveguides that are analyzed using the Finite Element Method (FEM) with a self-adaptive mesh. The first group of examples deals with transmission lines that are analyzed using the quasi-static approach. They are analyzed using the direct primary formulation with second-order Lagrange elements and the automatic mesh refinement algorithm.

The next group of examples deals with the full-wave analysis of transmission lines and waveguides. For this group, the solution is formulated in terms of the transverse and longitudinal components of the electric or the magnetic field using Lagrange/curl-conforming elements and the self-adaptive procedure. Use of the magnetic field formulation is preferred because the frequency dependent characteristic impedance is defined in terms of the power and current. This facilitates the calculation of the secondary parameters and losses, for example. In some examples, second order Lagrange/curl-conforming elements have been utilized.

The results are compared with those obtained by other authors and methods, commercial software packages and with experimental data.

5.2 QUASI-STATIC ANALYSIS OF TRANSMISSION LINES

5.2.1 Finite-Thickness Coupled Microstrip Lines

Figure 5.1 represents the cross section of two identical microstrips with finite-thickness conductors. The strips are located on top of a dielectric substrate above an infinite ground plane. Figure 5.1 shows the dimensions of the structure and the other details required for the analysis. The dimensions are in relative units, since the per-unit-length (p.u.l.) capacitance and inductance do not depend on the scale of the line cross section. This transmission line can represent a symmetric directional coupler. Thus, the phase velocities of the even and odd modes along with the corresponding modal characteristic impedances are calculated.

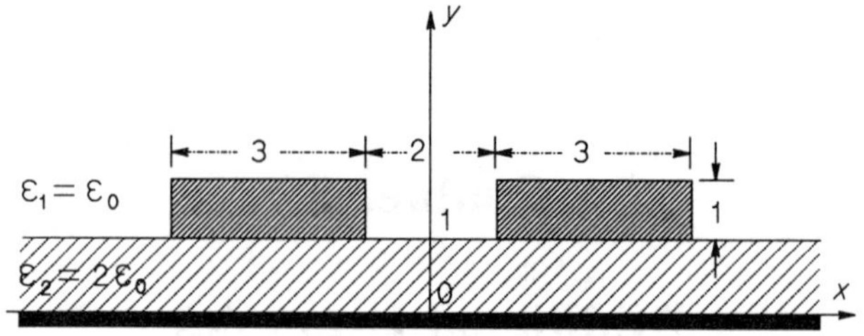

Figure 5.1 Finite-thickness coupled microstrip lines.

Figure 5.2 shows the mesh used for the analysis using the right half of the structure of Figure 5.1. In the analysis, the plane of symmetry is represented by a magnetic wall (for the analysis of the even mode) or by an electric wall (for the odd mode). It also depicts some of the intermediary steps of the self-adaptive mesh for the analysis of the odd mode. It is observed that there is an increased refinement around the lower corners of the strip, in particular at the lower-left corner that is close to both the ground plane and the electric wall. A coarser refinement can be observed around the upper edges, in particular around the upper-right corner. All of this agrees with the behavior of the electric field of the odd mode. The results for the p.u.l. capacitance, capacitance in vacuum (i.e., assuming the dielectric is absent), effective relative permittivity, phase velocity, and modal impedances for the odd and even modes are shown in Table 5.1.

Table 5.1
Results of the analysis of the structure of Figure 5.1

Mode	Capacitance (pF/m)	Capacitance in vacuum (pF/m)	ε_{reff}	Phase velocity ($\times 10^9$ m/s)	Z_m (Ω)
Even	87.05209	50.96271	1.708153	0.229381	50.11
Odd	10.25142	66.60098	1.539229	0.2416401	40.39

Table 5.2 compares the values of the capacitance matrix $[C]$ obtained using the present method along with the values obtained by Djordjević et al. [1996] using the method of moments (MOM) via the Galerkin method with an integral equation formulation (the LINPAR software package), Weeks [1970] using a least-squares method, and Khebir et al. [1990a] utilizing FEM with absorbing conditions. The results of the present analysis have been obtained after four automatic refinements of the initial

coarse mesh. The difference of 2.7% in the diagonal elements can be further reduced with a refined mesh as shown later on.

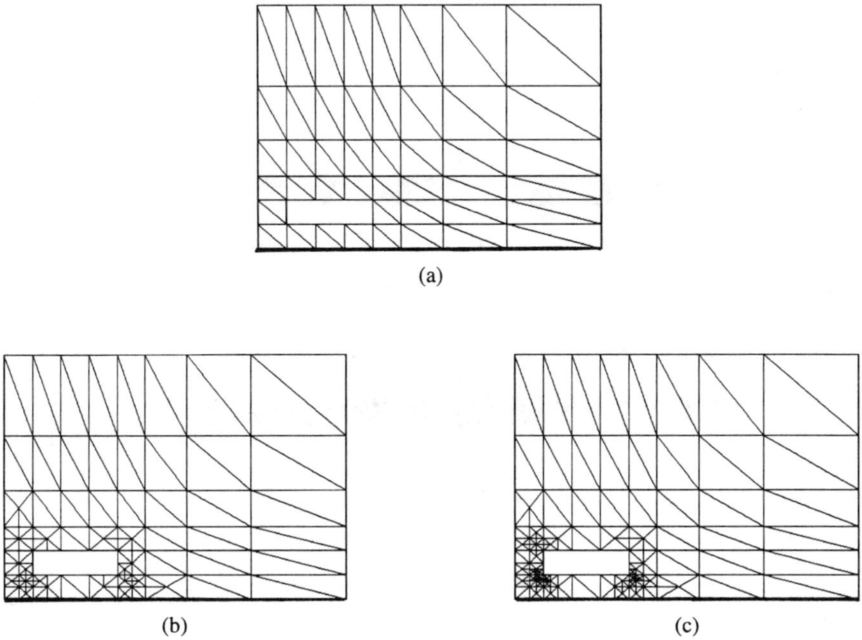

Figure 5.2 Mesh of the right half of the structure of Figure 5.1: (a) initial mesh and (b),(c) subsequent mesh refinement steps for the analysis of the odd mode.

Table 5.2
Comparison of capacitance coefficients obtained by the present method for the structure of Figure 5.1 with the results of other methods

(pF/m)	Present method	[Djordjević et al. 1996]: LINPAR	[Weeks 1970]	[Khebir et al. 1990a]
$C_{11} = C_{22}$	94.78	93.59	92.24	92.49
$C_{12} = C_{21}$	-7.731	-8.62	-8.504	-8.061

5.2.2 Finite-Thickness Coupled Striplines

Figure 5.3 shows the cross section of a stripline structure consisting of two identical strips of a finite thickness asymmetrically located between two infinitely wide ground planes. In this example, a meshing of the whole structure has been done, although symmetry could be exploited like in the previous example to effectively analyze only one half of the structure (with an appropriate boundary condition on the symmetry wall).

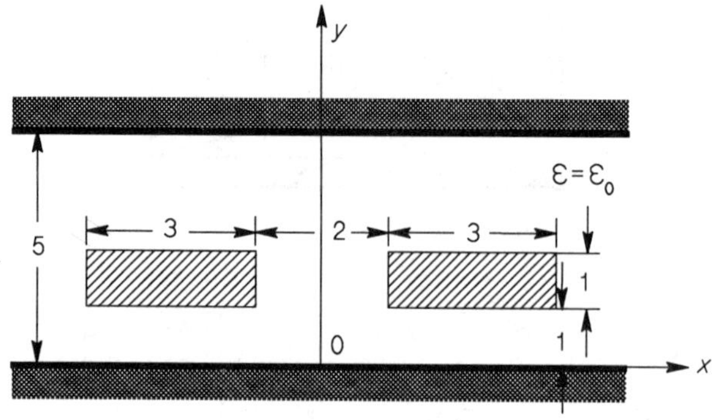

Figure 5.3 Finite-thickness coupled striplines.

Figure 5.4 shows the initial mesh for the right half of the structure and two mesh refinement steps corresponding to the odd mode. Figure 5.5 shows a zoom of a meshing step for the whole structure for the calculation of parameter C_{12} when the conductors are at opposite potentials (this corresponds to the odd mode). A symmetric mesh refinement with respect to the vertical plane of symmetry can be observed.

The results obtained for the odd and even modes are presented in Table 5.3. Table 5.4 shows the coefficients of matrix [*C*], comparing them with values obtained in [Djordjević et al. 1996] and [Weeks 1970]. It can be observed that the agreement is now better than 0.11% due to the increased accuracy of the analysis. This has been achieved by increasing the number of adaptive steps for the mesh.

Table 5.3

Results of the analysis of the structure of Figure 5.3

Mode	Capacity (pF/m)	Capacity in vacuum (pF/m)	ε_{reff}	Phase velocity ($\times 10^9$ m/s)	Z_m (Ω)
Even	57.51331	57.51331	1.0	0.2997925	58.04
Odd	69.31269	69.31269	1.0	0.2997925	48.16

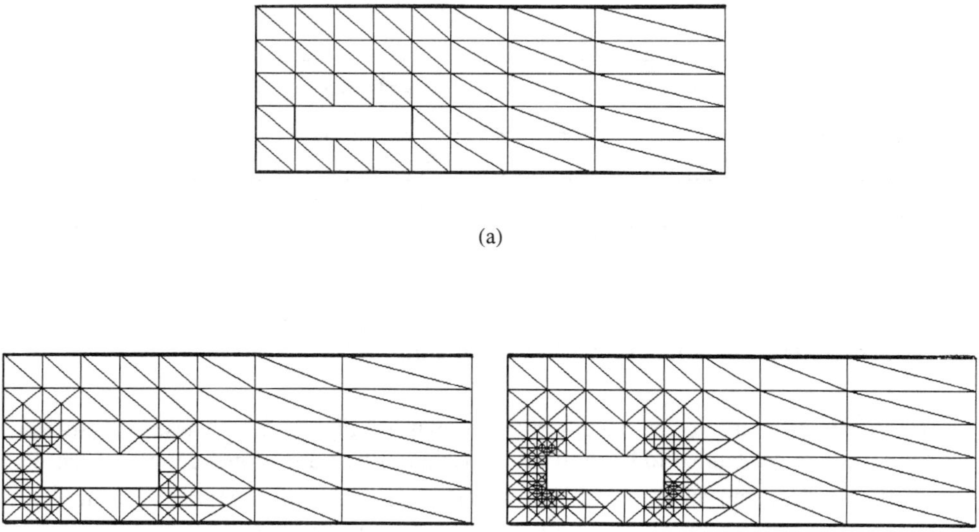

(a)

(b)

(c)

Figure 5.4 Mesh of the right half of the structure of Figure 5.3: (a) initial mesh and (b),(c) subsequent meshing steps for the analysis of the odd mode.

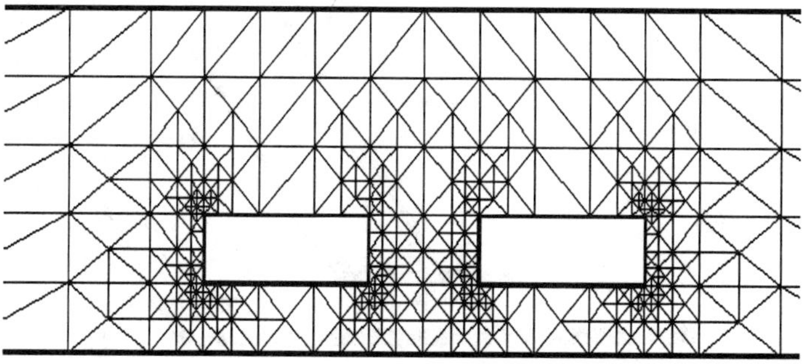

'igure 5.5 Zoom-in of a mesh for the analysis of the odd mode for the structure of Figure 5.3.

Table 5.4
Comparison of the coefficients of capacitance obtained by the present method for the structure of Figure 5.3 with the results of other methods

(pF/m)	Present method	[Djordjević et al. 1996]: LINPAR	[Weeks 1970]
$C_{11} = C_{22}$	63.41	63.31	63.07
$C_{12} = C_{21}$	-5.899	-5.902	-5.724

5.2.3 Zero-Thickness Coupled Microstrip Lines

The next structure to be analyzed is shown in Figure 5.6. These edge-coupled microstrips are intended to be used in the design of two directional couplers: one with -6 dB coupling and another with -10 dB coupling. The cross-sectional dimensions of the coupled microstrips on an alumina substrate, of relative permittivity $\varepsilon_r = 9.5$, have been calculated by Judd et al. [1970] with $w/h = 0.595$ and $s/h = 0.061$ for the 6 dB coupler and $w/h = 0.793$ and $s/h = 0.285$ for the 10 dB coupler. The port nominal impedances are supposed to be 50 Ω. The results for the transfer scattering parameter obtained using the self-adaptive mesh are compared in Figures 5.7 and 5.8 with the theoretically calculated results of [Judd et al. 1970] (using a semianalytical separation of variables method) and with experimental data. The difference between the experimental data and results obtained by the self-adaptive mesh is quite small.

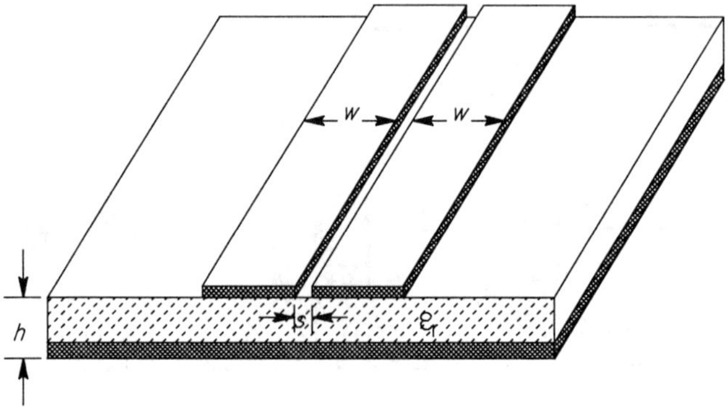

Figure 5.6 Coupled zero-thickness microstrip lines.

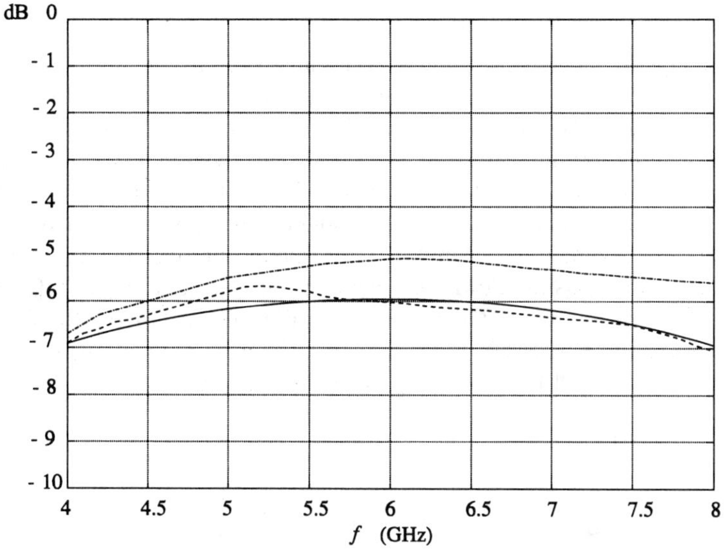

Figure 5.7 Transfer scattering parameter of the -6 dB coupler. Symbols: -.- , calculated in Judd et al. [1970]; --, experimental results; —, calculated using FEM.

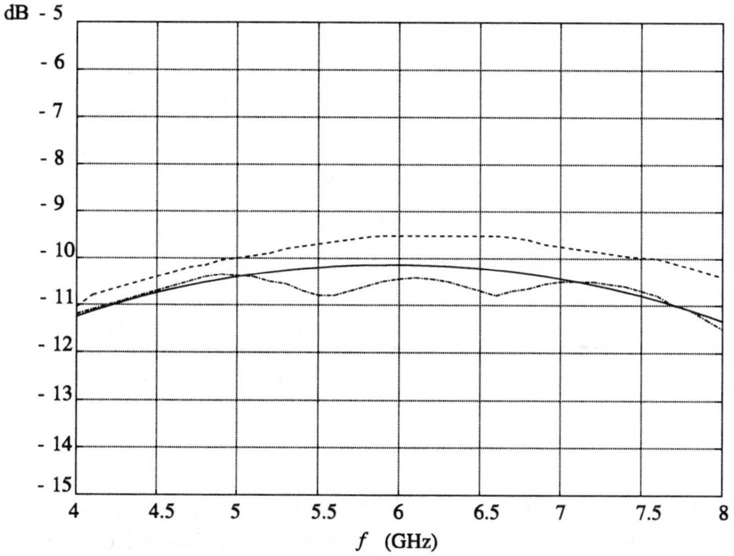

Figure 5.8 Transfer scattering parameter of the -10 dB coupler. Symbols: --, calculated in [Judd et al. 1970]; -.-, experimental results; —, calculated using FEM.

5.2.4 Three Coupled Microstrip Lines

The three coupled microstrips, shown in Figure 5.9, can be used to build a dual directional coupler. It can be used in measurements or for coupling two independent signals into a third conductor. For a nominal coupling of 10 dB, the dimensions of the coupler, built on an alumina substrate ($\varepsilon_r = 9.8$), are given in Table 5.5. The nominal impedances for the ports are 50 Ω. The length of the coupling structure, as in previous examples, has been calculated so that at the center frequency of the band, it is equal to one-fourth of the average wavelength of the modes propagating through the three multiconductor lines. Using FEM, the elements of matrices $[C]$ and $[C_{eq}]$ (for air-dielectric), obtained are: $C_{11} = C_{33} = 159.4434$ pF/m; $C_{22} = 174.2623$ pF/m; $C_{12} = C_{23} = -43.208$ pF/m; $C_{13} = -1.8960$ pF/m; $C_{eq11} = C_{eq33} = 27.3112$ pF/m; $C_{eq22} = 30.8658$ pF/m; $C_{eq12} = C_{eq23} = -9.1379$ pF/m; $C_{eq13} = -0.8482$ pF/m.

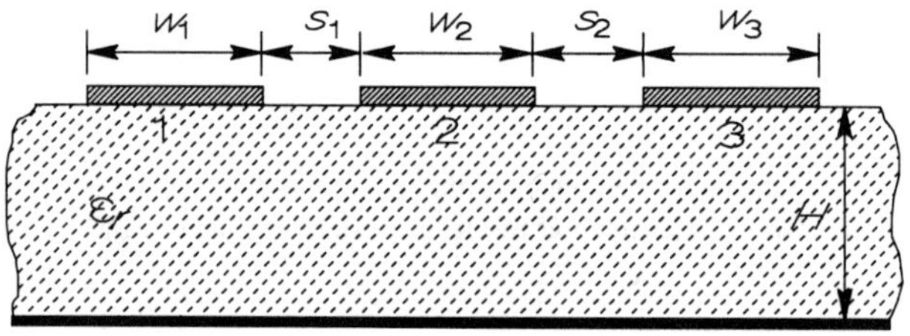

Figure 5.9 Three coupled microstrip lines.

Table 5.5

Dimensions of the structure of Figure 5.9 for a -10 dB coupler

Nominal coupling (dB)	$w_1 = w_2 = w_3 = w$ (μm)	H (μm)	$s_1 = s_2 = s$ (μm)	l (mm)
- 10	431.8	293.624	190.5	7.69

Figure 5.10 shows the coupling between the central conductor and the lateral conductors. The matrix parameters have been evaluated using FEM along with the computed results of Pavlidis and Hartnagel [1976] obtained using the finite difference method and experimental data. Again, a good agreement between the experimental results and the computed results using the self-adaptive mesh is seen.

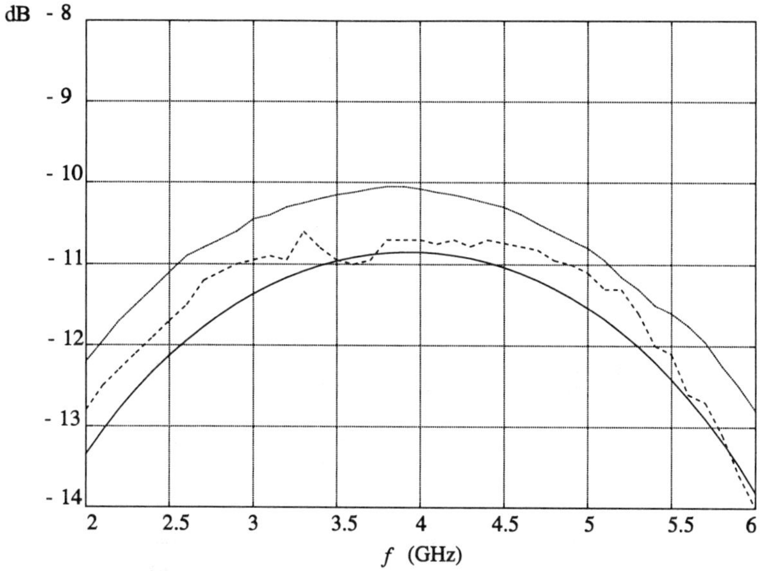

Figure 5.10 Transfer scattering parameter of the dual -10 dB coupler. Symbols: ..., theoretical results [Pavlidis and Hartnagel 1976]; --, experimental results; —, FEM results.

5.2.5 Microstrip Line with Undercutting

Figure 5.11 shows the cross section of an undercut shielded microstrip line. The undercutting is defined by the angles Φ_1 and Φ_2. For most transmission lines used in monolithic technology, the metallization processes generate overcutting or undercutting effects with typical angles of 30° to 45°. It is interesting to consider the corresponding effect on the impedance of the structure by comparing the results with the situation in which undercutting does not exist (i.e., when $\Phi_1 = \Phi_2 = 90°$). The dimensions of the structure are shown in Table 5.6. Here, the signal conductor is located very close to the right-side wall. This situation can be of interest for the study of the odd mode in a coupler or for the analysis of the proximity effects of the lateral grounded walls existing in an encapsulated monolithic circuit. The structure is also interesting because the solution has singularities of different orders. Table 5.7 shows the results for the characteristic impedance of the line using the present method and by Rizzoli [1979] using an integral formulation that includes special basis functions to model the singularities. A good agreement can be observed. It must be pointed out that there is no need for special functions for the case of FEM that utilizes only the self-adaptive mesh. From the results it is seen how important it is to properly take the metallization thickness and the undercutting into account.

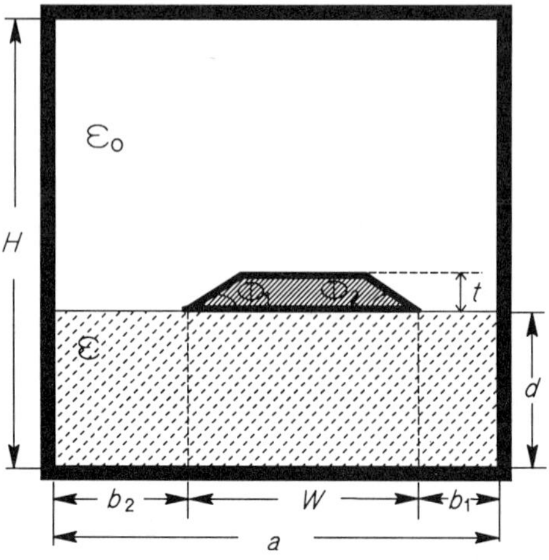

Figure 5.11 Cross section of a microstrip line with undercutting.

Table 5.6

Dimensions of the structure of Figure 5.11

a (mm)	H (mm)	d(μm)	ε_r	W(μm)	b_1 (mm)	b_2 (μm)
7.6	10	635	10	75	7.5	25

5.2.6 Symmetric Coplanar Waveguide with Broadside-Coupled Lines

Figure 5.12 shows the cross section of a symmetric pair of broadside-coupled coplanar waveguides. This structure can also be used to build directional couplers. The data for the structure are s_c = 50 μm, and the relative permittivity of the central dielectric (GaAs) is 12.9. The structure can be analyzed using its two axes of symmetry. Figure 5.13 shows the initial mesh of one-fourth of the structure and the mesh at a subsequent refinement step. In Figure 5.14, a zoom-in over the slot between the active conductor and the ground plane can be observed. It is noticed that the refinement is produced at the edges of the signal conductor as well as at the edges of the grounded conductor.

A parametric study is carried out for the modal impedances, effective permittivities, and phase velocities of the even and odd modes. These quantities are evaluated as a function of the metallization width (w) for different separations between the signal conductors and the coplanar ground planes (s_g) considering three different

Table 5.7

Characteristic impedance of the line of Figure 5.11

t (μm)	$\Phi_1 = \Phi_2$	Z_c (Ω)	
		Present method	[Rizzoli 1979]
0		45.37	45.4
10	90°	41.43	41.6
	45°	43.05	43.2
	30°	43.63	43.8
15	90°	40	40.1
	45°	42.43	42.6
	30°	43.24	43.3
20	90°	38.76	38.9
	45°	41.93	42
	30°	42.98	43

dielectric materials for the medium at both sides of the gallium arsenide layer. The values obtained for the coupling coefficient (C_c) and the ratio of the even and odd mode velocities (v_e / v_o) are given in Tables 5.8 to 5.10, corresponding to the metallization widths of 100 μm, 200 μm, and 300 μm, respectively. The results obtained by FEM are labeled (1) in these tables. Also shown are the values as evaluated in [Bedair and Wolff 1992] using a semianalytic method based on conformal transformations. These results are labeled (2). There is a good agreement between the two results.

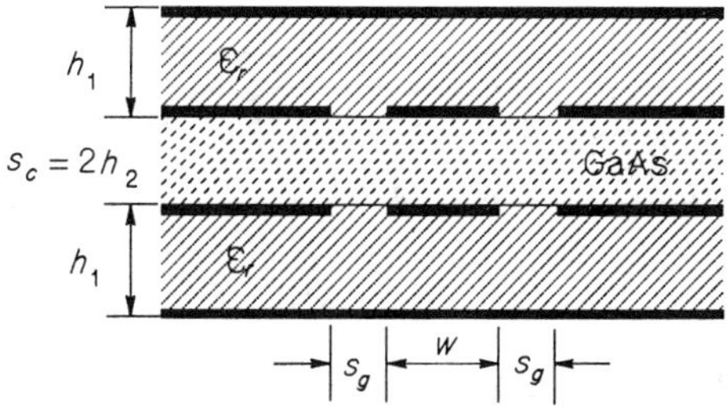

Figure 5.12 Cross section of a pair of symmetric broadside-coupled coplanar waveguides.

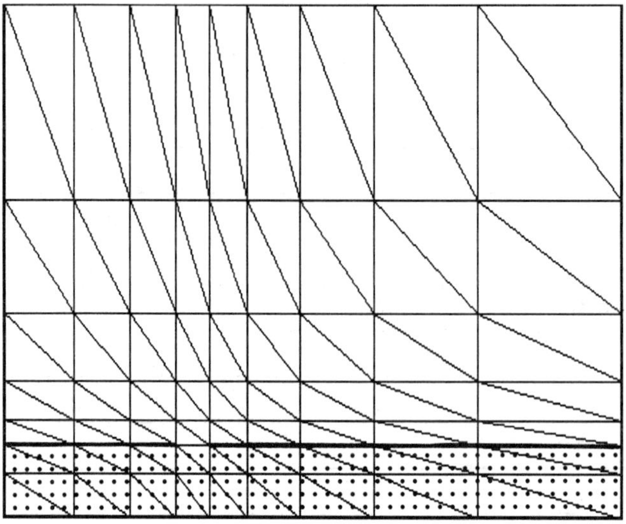

(a)

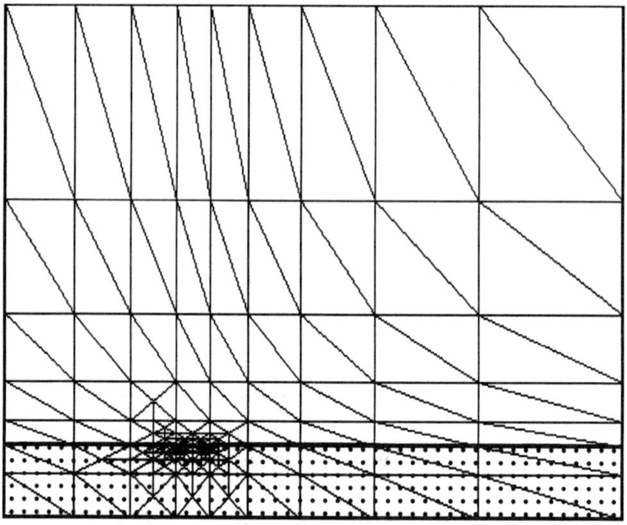

(b)

Figure 5.13 Mesh of the top-right quarter of the structure of Figure 5.12: (a) initial mesh and (b) subsequent refinement step for the analysis of the odd mode.

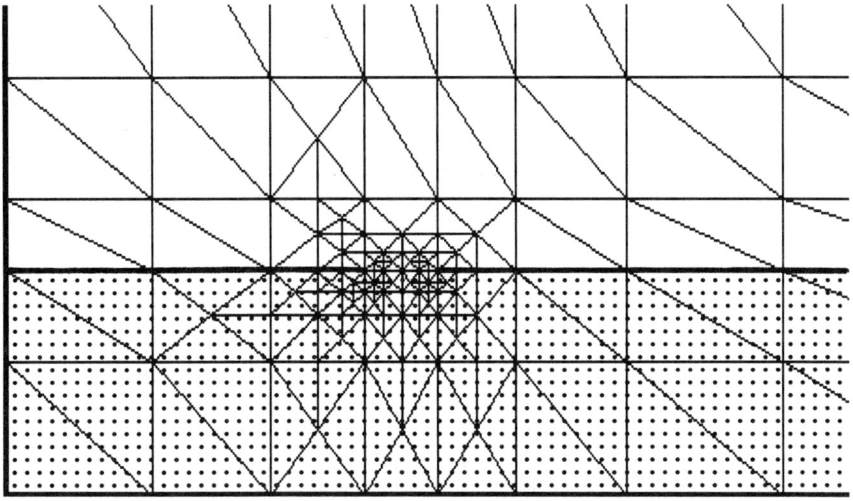

Figure 5.14 Zoom-in of the slot area of the mesh of Figure 5.13 (b).

Table 5.8
Coupling coefficient and the ratio of modal velocities for the structure of Figure 5.12 for w = 100 μm.

		s_g = 10 μm		s_g = 30 μm		s_g = 90 μm	
		(1)	(2)	(1)	(2)	(1)	(2)
$\varepsilon_r = 1$	C_c	0.396	0.40	0.531	0.54	0.682	0.70
	v_e / v_o	1.259	1.25	1.387	1.38	1.648	1.62
$\varepsilon_r = 3.78$	C_c	0.361	0.37	0.487	0.49	0.624	0.64
	v_e / v_o	1.159	1.15	1.231	1.23	1.345	1.34
$\varepsilon_r = 10$	C_c	0.312	0.32	0.424	0.43	0.548	0.58
	v_e / v_o	1.035	1.03	1.049	1.04	1.068	1.06

Table 5.9
Coupling coefficient and the ratio of modal velocities for the structure of Figure 5.12 for w = 200 μm.

		s_g = 10 μm		s_g = 30 μm		s_g = 90 μm	
		(1)	(2)	(1)	(2)	(1)	(2)
$\varepsilon_r = 1$	C_c	0.560	0.57	0.665	0.68	0.770	0.80
	v_e / v_o	1.400	1.38	1.562	1.52	1.874	1.78
$\varepsilon_r = 3.78$	C_c	0.517	0.53	0.616	0.63	0.710	0.75
	v_e / v_o	1.241	1.23	1.319	1.30	1.437	1.42
$\varepsilon_r = 10$	C_c	0.454	0.47	0.545	0.57	0.632	0.69
	v_e / v_o	1.052	1.05	1.065	1.06	1.082	1.08

Table 5.10
Coupling coefficient and the ratio of modal velocities for the structure of Figure 5.12 for w = 300 μm.

		s_g = 10 μm		s_g = 30 μm		s_g = 90 μm	
		(1)	(2)	(1)	(2)	(1)	(2)
$\varepsilon_r = 1$	C_c	0.643	0.67	0.729	0.76	0.809	0.85
	v_e / v_o	1.505	1.44	1.689	1.58	2.030	1.85
$\varepsilon_r = 3.78$	C_c	0.597	0.63	0.677	0.72	0.748	0.82
	v_e / v_o	1.294	1.27	1.375	1.34	1.492	1.46
$\varepsilon_r = 10$	C_c	0.529	0.57	0.604	0.66	0.671	0.75
	v_e / v_o	1.062	1.06	1.075	1.07	1.089	1.09

5.2.7 V-Grooved Microstrip Line

The V-grooved microstrip line (or valley line) has been proposed for use in monolithic technology. It takes advantages of the metallization process and tends to have reduced loss and dispersion [Hasegawa et al. 1991, 1992]. Figure 5.15 shows the general geometry of the line. There is a slot in the metallization used as a ground plane, which is placed on top of a gallium arsenide substrate. The metallization of the signal conductor is deposited over a polyamide layer.

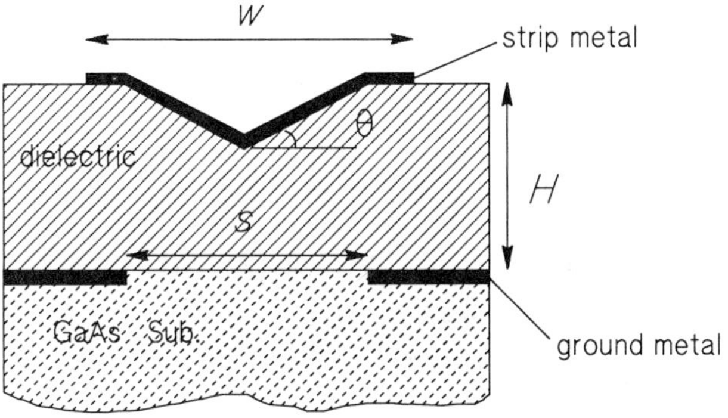

Figure 5.15 A V-grooved microstrip line.

In the present example, the thickness of the polyamide layer is H = 10 μm. The valley aperture angle is θ = 35° and the valley depth is 5 μm. Figure 5.16 shows the initial mesh used and a zoom-in of the valley line area where various refinements can be observed. Table 5.11 shows a comparison between the values obtained using FEM and those obtained by Hasegawa et al. [1992], also using FEM.

5.2.8 Suspended Stripline with Supporting Grooves

Shown here are results of a parametric analysis of the effect of substrate supporting bridles (grooves) in lines based on suspended microstrip technology. This type of situation is quite common not only in configurations with this geometry but also in finline and the general type of encapsulated structures. Figure 5.17 shows the geometry of the line. Its dimensions are a/b = 1, $h_1/b = h_3/b$ = 0.4, h_2/b = 0.2. Figure 5.18 shows the initial mesh of the right half of the structure and the meshing at the fourth step corresponding to ε_r = 2.22 for two different widths of the suspended strip, w/b = 0.2 and w/b = 0.6. Comparing the second case with the first one, it is seen that there is a finer refinement due to the proximity between the edges of the signal conductor and the bridle. The corresponding equipotential lines are also shown in Figure 5.18.

Tables 5.12 and 5.13 show the dependence of the characteristic impedance and the wavelength reduction factor λ/λ_0 for ε_r = 2.22 as a function of the bridle depth, d/b, and the width of the signal conductor, w/b. These results are also plotted in Figure 5.19 and Figure 5.20 together with the results of Yamashita et al. [1985] obtained using a

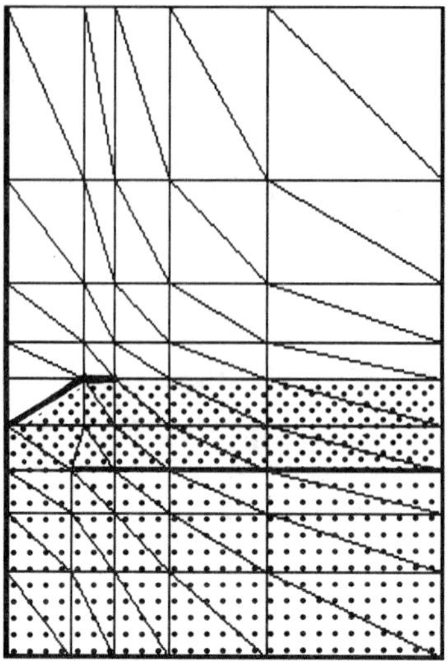

(a)

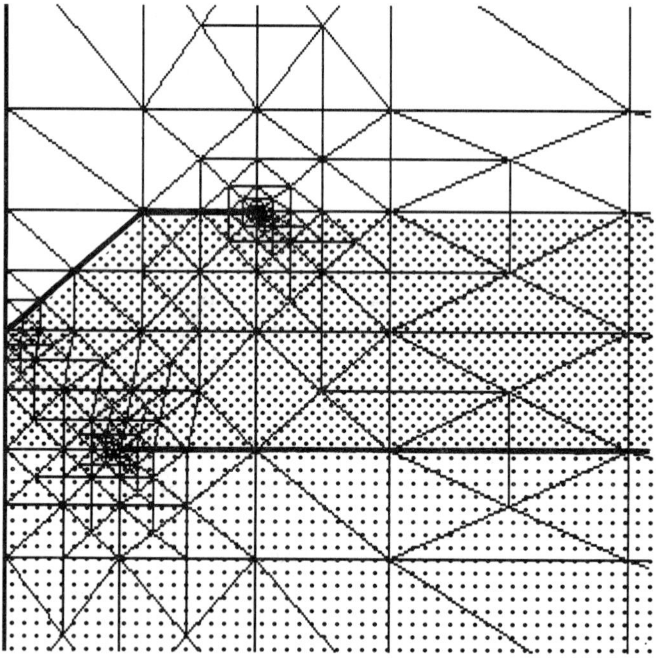

(b)

Figure 5.16 Mesh of the right half of the structure of Figure 5.15: (a) initial mesh and (b) zoom-in of the valley area at a subsequent meshing step.

Table 5.11

Characteristic impedance, normalized phase velocity, and effective relative permittivity for the structure of Figure 5.15 versus the cross-sectional dimensions

S (μm)	W (μm)	Z_c (Ω)		v_f / c_0	ε_{reff}
		Present method	[Hasegawa et al. 1992]	Present method	Present method
0	14.28	50.02	51.0	0.6149	2.6444
	24.00	39.90	41.0	0.6183	2.6157
	30.00	35.06	36.0	0.6175	2.6224
12	20.00	47.96	48.5	0.6003	2.7746
	26.00	41.61	42.0	0.6002	2.7760
	34.00	35.12	35.2	0.6000	2.7781
24	24.00	49.24	50.8	0.5806	2.9667
	34.00	39.14	40.8	0.5821	2.9508
	40.00	34.67	36.0	0.5835	2.9370

variational method. Results for the effect of the bridles and the substrate relative permittivity on the characteristic impedance and the wavelength reduction factor are summarized in Figures 5.21 and 5.22, respectively, together with the results of Yamashita et al. [1985]. Again the agreement is quite good.

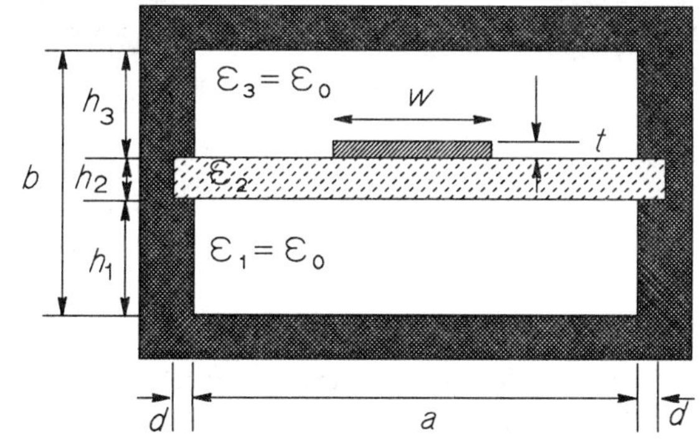

Figure 5.17 Suspended substrate stripline with supporting grooves.

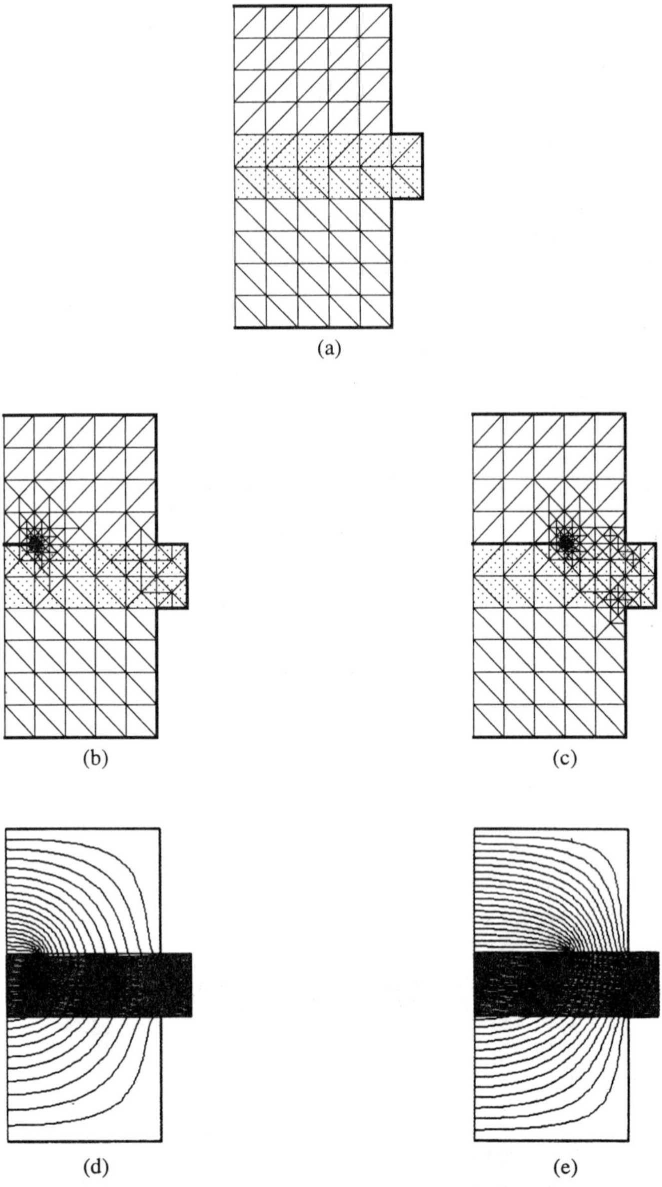

Figure 5.18 Mesh of the right half of the structure of Figure 5.17 with $d/b = 0.1$: (a) initial mesh; (b) mesh of the fourth step for $w/b = 0.2$; (c) mesh of the fourth step for $w/b = 0.6$; (d) equipotential lines for case (b); (e) equipotential lines for case (c).

Table 5.12
Characteristic impedance of the suspended substrate stripline with supporting grooves

w/b \ *d/b*	0.0	0.1	0.2	0.3	0.4
0.2	117.50	118.50	118.6	118.60	118.60
0.4	83.83	85.06	85.12	85.12	85.12
0.6	62.71	64.27	64.35	64.36	64.37
0.8	45.04	47.33	47.44	47.45	47.45

Table 5.13
Wavelength reduction factor for the structure of Figure 5.17 as a function of cross-sectional dimensions

w/b \ *d/b*	0.0	0.1	0.2	0.3	0.4
0.2	0.8387	0.8420	0.8422	0.8422	0.8422
0.4	0.8422	0.8533	0.8535	0.8535	0.8535
0.6	0.8483	0.8561	0.8566	0.8566	0.8566
0.8	0.8391	0.8524	0.8530	0.8531	0.8531

5.2.9 Microstrip Line Near a Dielectric Edge

Figure 5.23 shows a microstrip line where the strip is in the proximity of a dielectric edge. The abrupt termination of the substrate alters the electromagnetic field as compared with the case of a very wide substrate. A structure of dimensions $h/w = 1.382$ over a substrate of $\varepsilon_r = 12.9$ has been analyzed for different values of s/w. Figure 5.24 shows a zoom-in of the strip in the mesh refinement steps corresponding to the case when the dielectric edge does not exist (or far away from the microstrip, $s/w > 2.0$) and when the dielectric edge is near the strip ($s/w = 0.5$). Figure 5.25 shows the equipotential lines for these two cases, as well as for $s/w = 0.2$. Table 5.14 shows the calculated characteristic impedance using the present technique along with the results obtained by [Yamashita et al. 1989] using a modal method. The latter provides an upper bound, while FEM provides a lower bound.

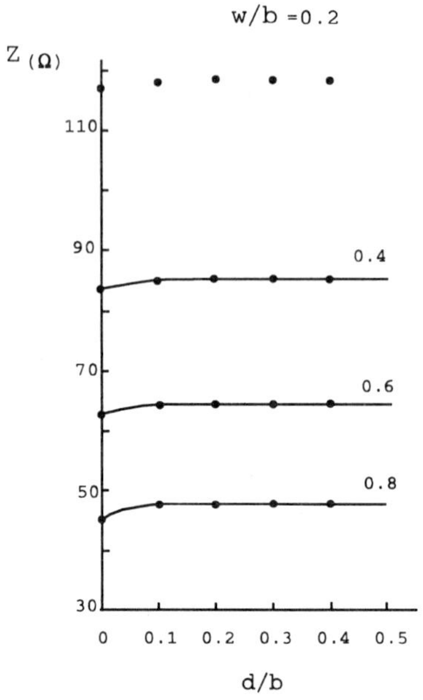

Figure 5.19 Variation in Z; $\varepsilon_r = 2.22$. Symbols: —, [Yamashita et al. 1985]; •, FEM.

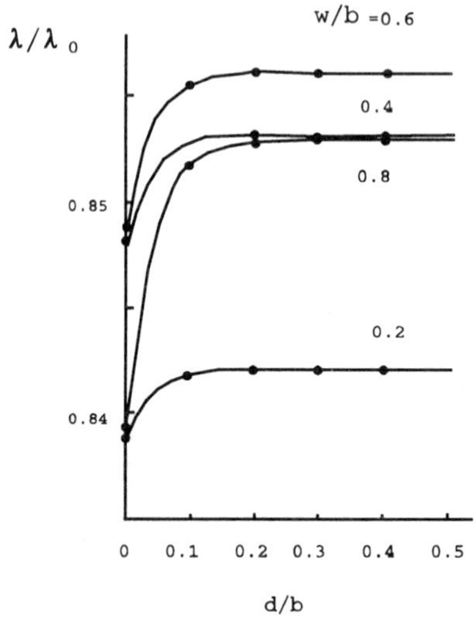

Figure 5.20 Variation of λ/λ_0; $\varepsilon_r = 2.22$. Symbols: —, [Yamashita et al. 1985]; •, FEM.

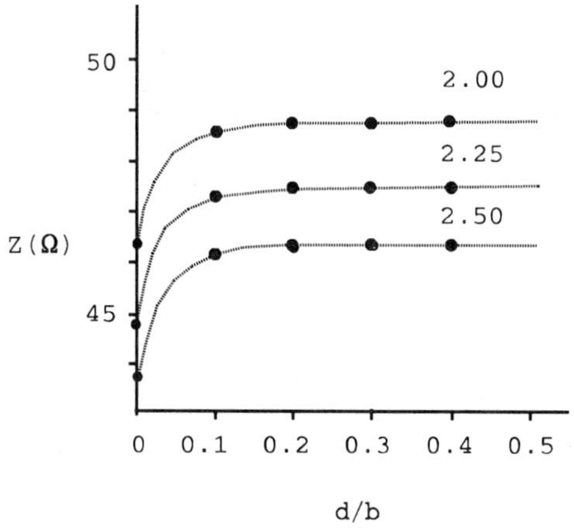

Figure 5.21 Variation in Z; $w/b = 0.8$; for different values of the dielectric constant of the suspended substrate. Symbols: --, [Yamashita et al. 1985]; •, FEM.

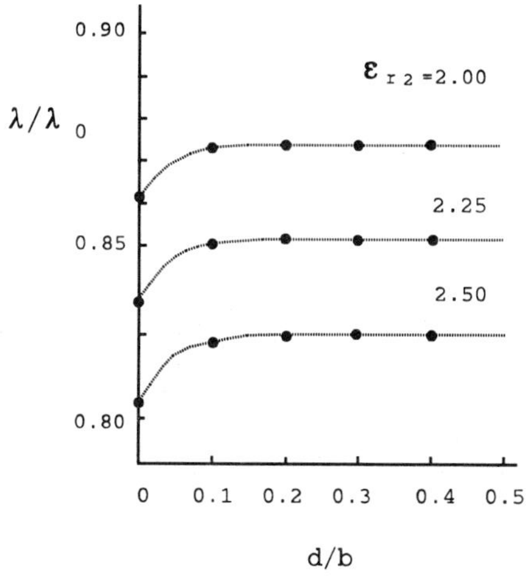

Figure 5.22 Variation of λ/λ_0; w/b = 0.8; for different values of the dielectric constant of the suspended substrate. Symbols: ---, [Yamashita et al. 1985]; •, FEM.

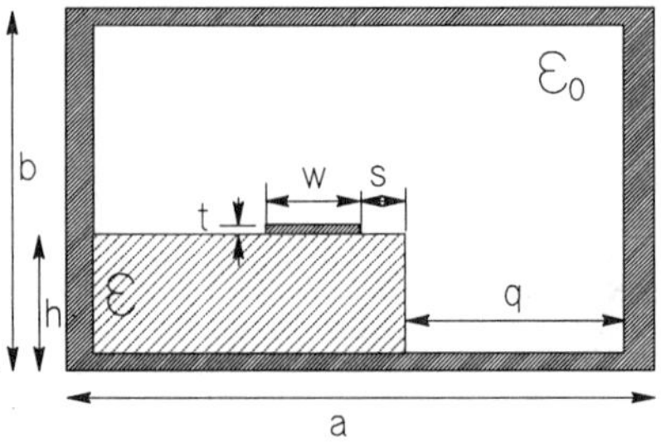

Figure 5.23 Microstrip line near a dielectric edge.

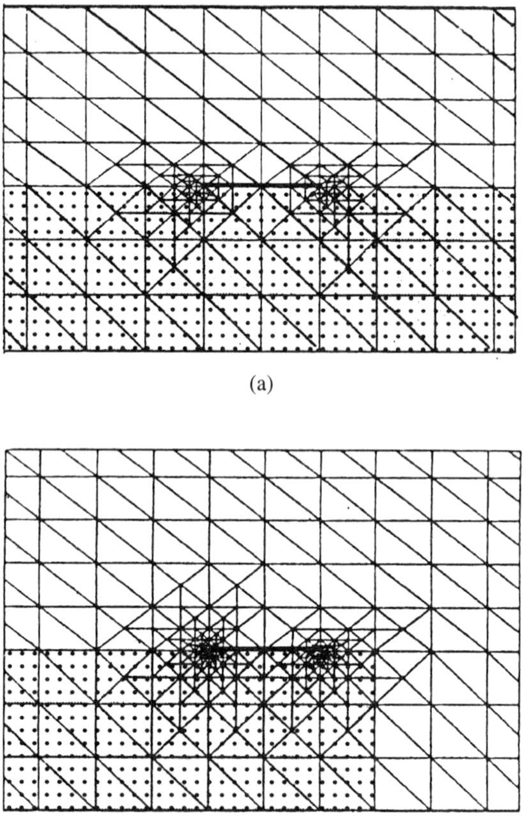

(a)

(b)

Figure 5.24 Zoom-in of the mesh in the strip area for: (a) $s/w = 2.0$ and (b) $s/w = 0.5$.

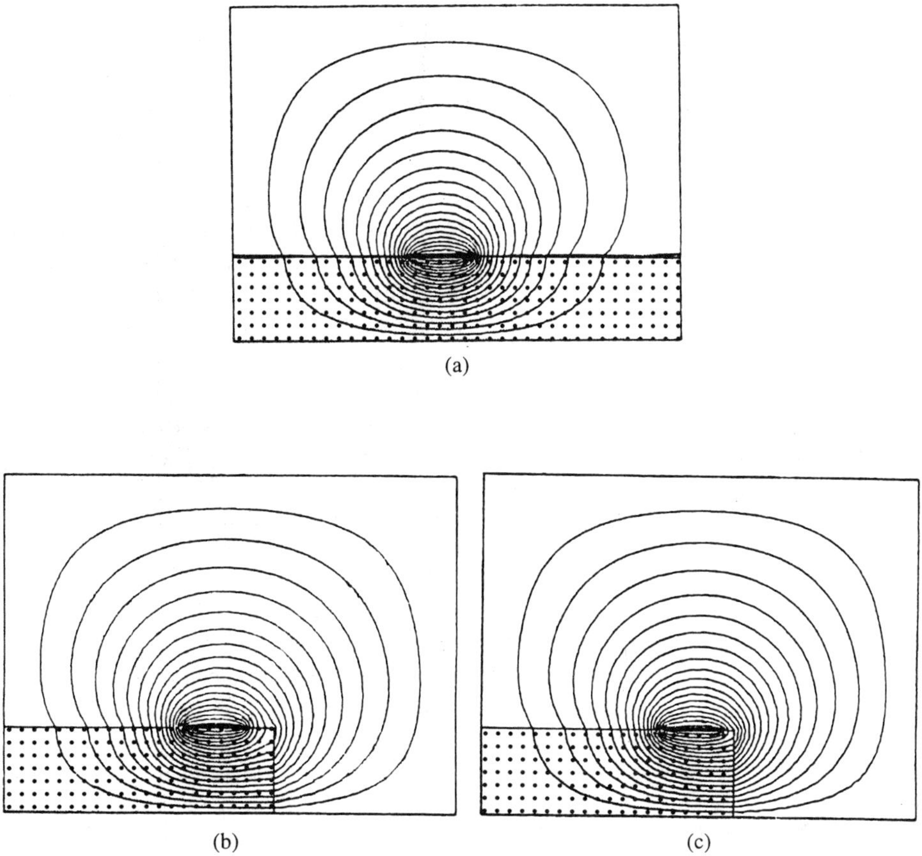

Figure 5.25 Zoom-in of the strip area, showing equipotential lines: (a) $s/w = 2.0$; (b) $s/w = 0.5$; and (c) $s/w = 0.2$.

Table 5.14

Characteristic impedance of the structure of Figure 5.23 as a function of the distance between the strip and the dielectric edge

s/w	0.2	0.5	> 2.0
FEM	52.93	50.93	48.56
[Yamashita et al. 1989]	54.50	52.40	50.00

5.2.10 Electro-Optical Coupler

The principle of operation of some of the optical devices, such as modulators, converters and couplers, is based on the interaction of light with an electrostatic field through the electro-optical effect. The electrostatic field is generated using electrodes situated on the device surface. They are located close to the optical guide. At each point of the device material, the refraction index is a function of the electro-optical tensor and the electric field at that point. The relationship between those quantities is given by the Poisson equation, in which the unknown is the inverse of the square of the refraction index and the right hand term is the product of the electro-optical tensor times the electrostatic field. At the same time, the intensity of light depends on the refraction index and is calculated through Maxwell's equations. In other words, applying a voltage between two electrodes produces a local variation of the refractive index of the optical guide. Hence, this modifies the optical field distribution and its propagation parameters. For a good design, it is necessary to perform a very accurate characterization of the electrostatic field in the zone between the electrodes. The main difficulty lies on the strong field intensity due to the sharp edges of the electrodes. Hence, three steps are required for the computation of the optical field: (1) the computation of the electrostatic field, (2) the calculation of the refractive index, and (3) the full-wave analysis of the optical waveguide.

First consider a hypothetical structure that consists of two semi-infinite electrodes, as shown in Figure 5.26. The electric field of this structure has an analytical solution. Figure 5.27 shows a zoom-in of the zone between the electrodes for the initial mesh and for a subsequent step in a self-adaptive process applied to the analysis of the structure. Figure 5.28 shows a comparison between the computed x component of the electric field and the analytical result as a function of the coordinate x for several values of the coordinate y. This field component is responsible for the change of the refractive index in the structure of Figure 5.26. Figure 5.29 shows the corresponding three-dimensional plots for the whole xy-plane.

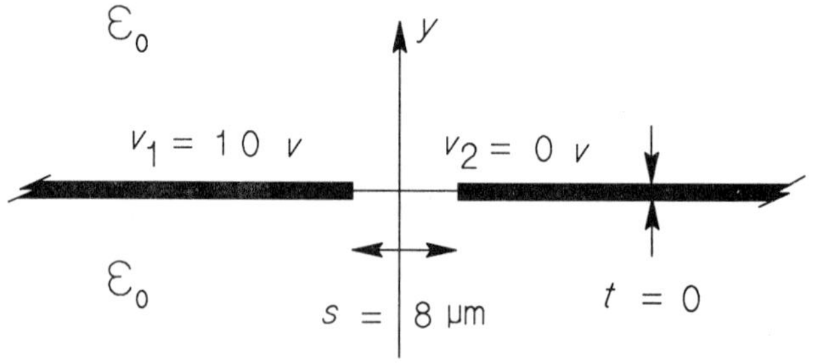

Figure 5.26 Electro-optical coupler with semi-infinite electrodes.

This illustrates that the adaptive procedure also produces accurate field components which are derived from a postprocess of the results.

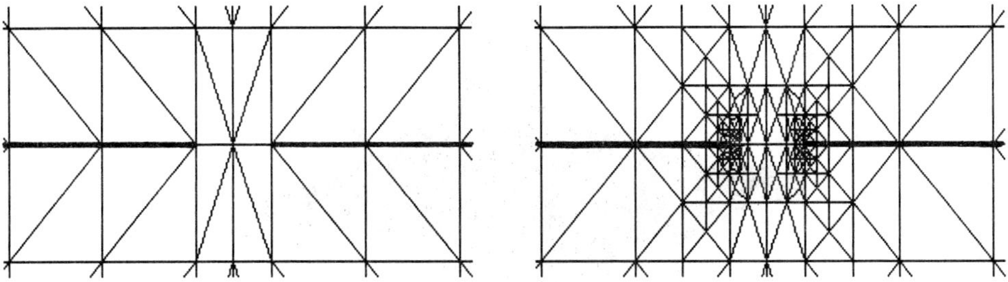

(a) (b)

Figure 5.27 Zoom-in of the zone between electrodes of the system of Figure 5.26: (a) initial mesh and (b) mesh refinement step.

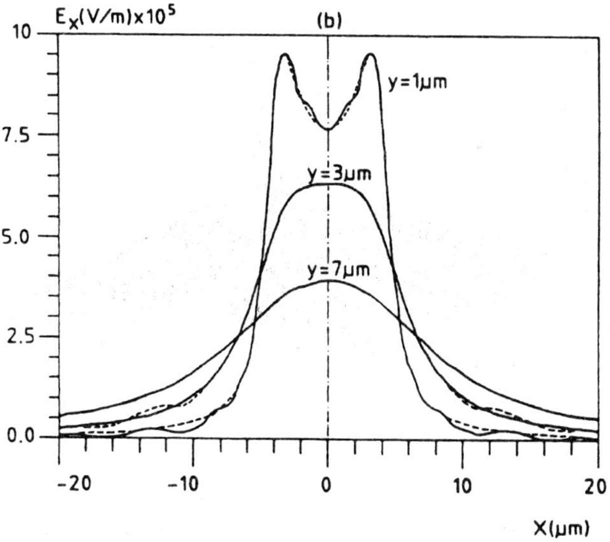

Figure 5.28 Comparison between the analytical and computed results for the *x*-component of the electric field in the system of Figure 5.27, with the *y* coordinate as a parameter. Symbols: - -, analytical; —, calculated for *y*=1 μm, and 7 μm; - -, calculated; —, analytical for *y*=3 μm.

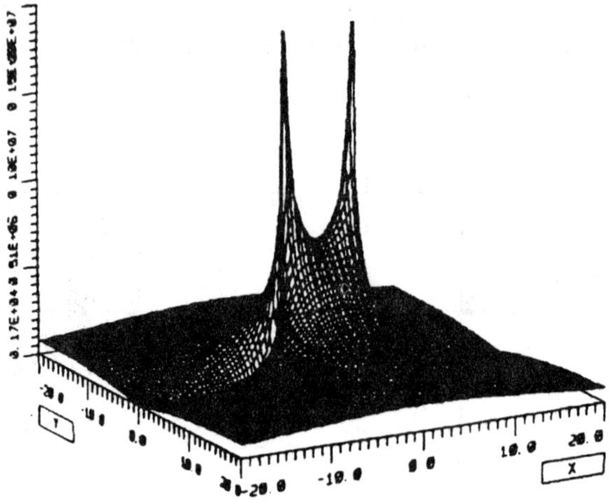

(a)

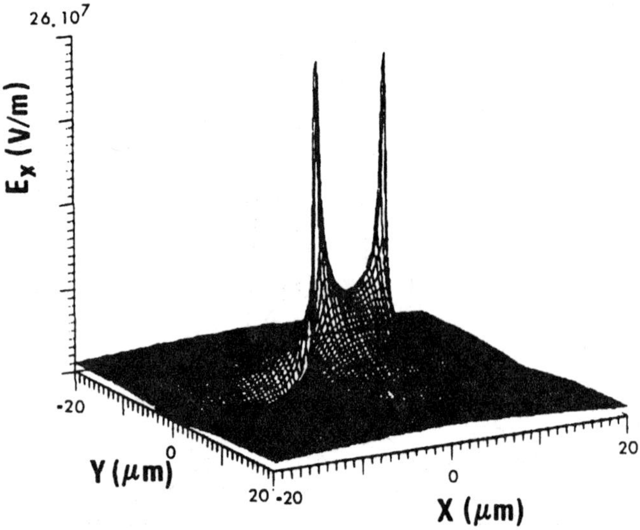

(b)

Figure 5.29 Three-dimensional plots of the x-component of the electric field for the structure of Figure 5.26: (a) analytical and (b) computed.

The self-adaptive algorithm is next applied to a $LiNbO_3$ electro-optical coupler, whose cross section is shown in Figure 5.30(a). Figures 5.30(b),(c) show the x- and y-component of the electric field generated by the two finite-width electrodes located on top of the coupler. These field components are calculated using the present self-adaptive algorithm.

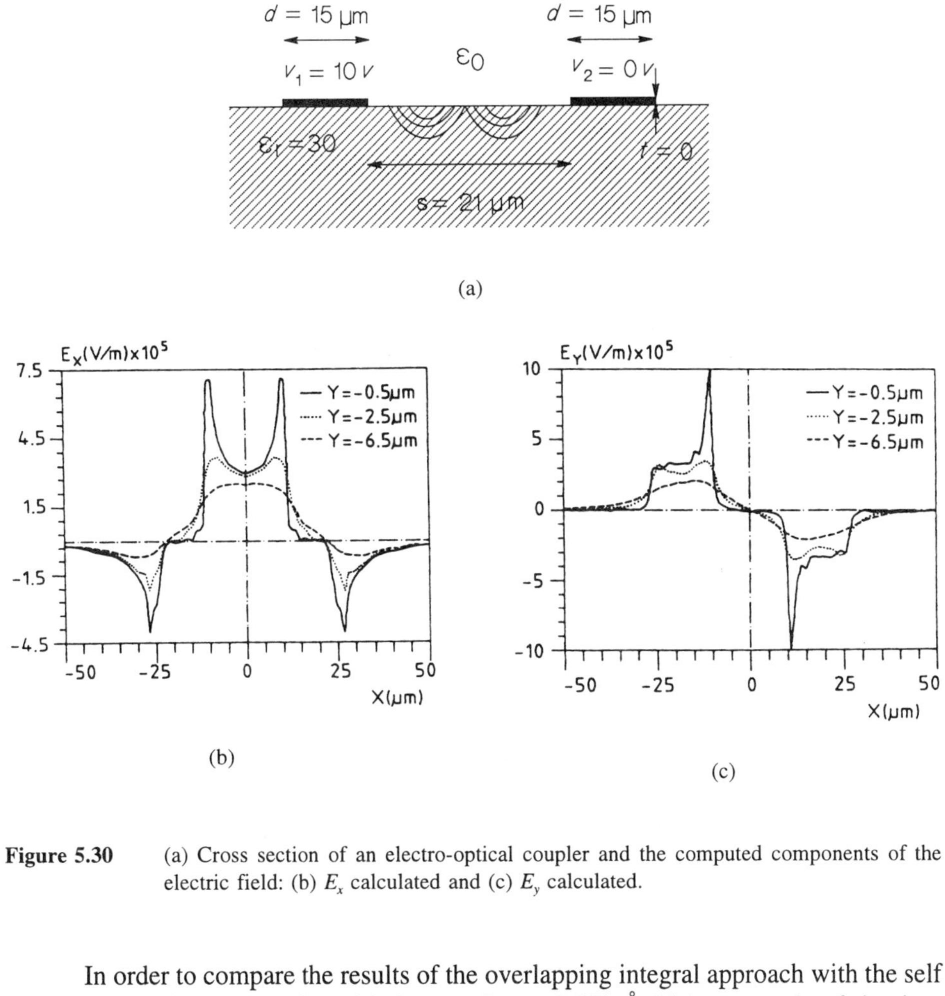

Figure 5.30 (a) Cross section of an electro-optical coupler and the computed components of the electric field: (b) E_x calculated and (c) E_y calculated.

In order to compare the results of the overlapping integral approach with the self adaptive procedure, a coupler with 8 μm strips and 600 Å thickness made of titanium is analyzed. The distance between the electrodes is 5 μm. The distance between the centers of the dielectric guides is 13 μm. Diffusion conditions were at 1050° C for 8.5 hours. The working wavelength is 1.55 μm. Figure 5.31 shows the results for the crossover efficiency as a function of the potential difference applied between the electrodes for two different lengths of the coupler. A comparison with results obtained through the approximate method and FEM shows differences of the order of 10 to 15%. This illustrates that use of FEM to optical devices can provide a more accurate solution.

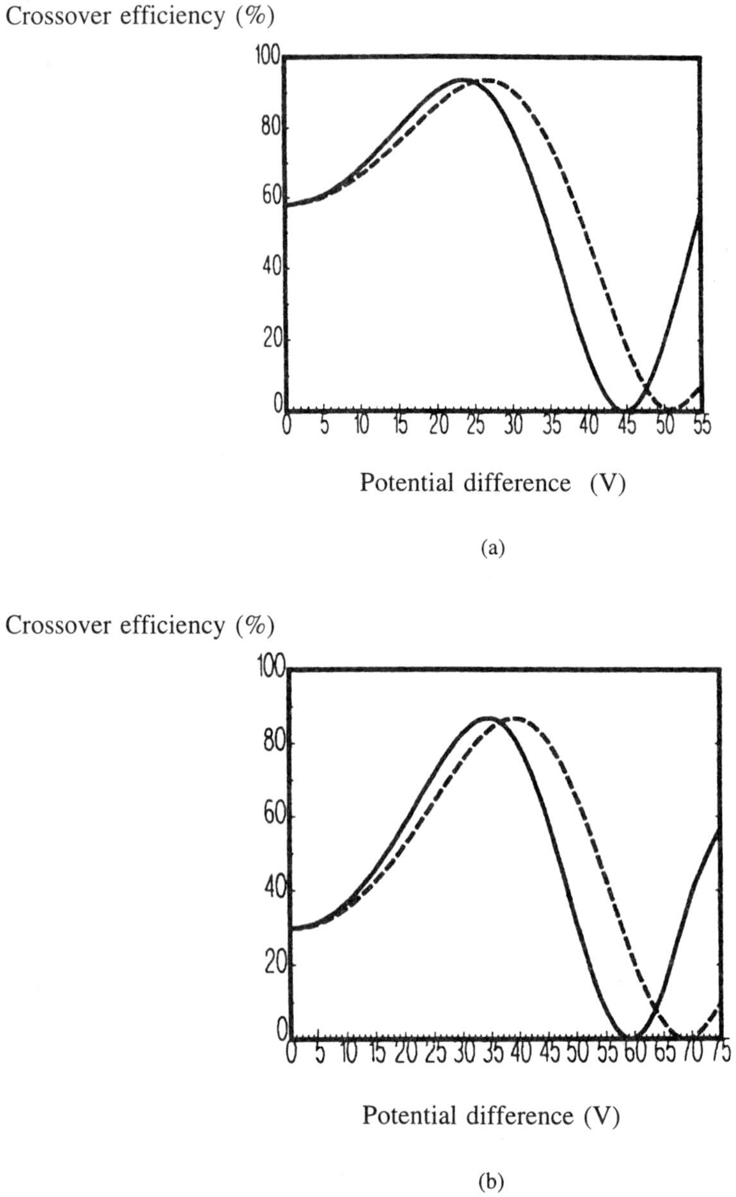

Figure 5.31 Crossover efficiency in a directional coupler as a function of the potential difference between electrodes, for two different lengths of the coupler: (a) 15 mm and (b) 10 mm. Symbols: —, overlapping integral approach; - -, present method.

5.2.11 Dissipative Structures

Next, conductor and dielectric losses are computed through a postprocessing of the solution. Here, comparison of the results are made with those obtained using commercial software packages.

a) *Finite-thickness coupled striplines*

Consider the geometry shown in Figure 5.3 but of different dimensions. The width of the strips is 2 mm. The thickness is 1 mm. The ground planes are 40 mm wide. The separation between them is 5 mm. The strips are coplanar and are located symmetrically with respect to the coordinate y but asymmetrically with respect to x. The coordinates of the lower-left corner of the first strip are (19 mm, 2 mm) and those of the second strip are (22 mm, 2 mm). The coordinate origin is at the lower-left edge of the bottom ground plane. The medium is vacuum with a loss tangent of 10^{-3}. Conductors are copper of conductivity $\sigma = 5.8\times10^{7}$ mho/m.

The comparison between the primary parameters calculated with the present method and those obtained using LINPAR [Djordjević et al. 1996] is shown in Table 5.15. A very good agreement is observed with accuracy better than 1% for the coefficients of matrices [C], [L], and [G], and better than 5% for the elements of matrix [R_c].

Table 5.15
Primary parameters of the finite-thickness coupled striplines

		Present method	LINPAR
[C] (pF/m)	C_{11}	48.50	48.476
	C_{22}	48.57	48.470
	$C_{12} = C_{21}$	-14.49	-14.47
[L] (μH/m)	L_{11}	251.87	252.0
	L_{22}	251.53	252.0
	$L_{12} = L_{21}$	75.17	75.2
[G] (mhos/m)	G_{11}	3.05	3.046
	G_{22}	3.05	3.046
	$G_{12} = G_{21}$	-0.91	-0.9093
[R_c] (ohm/m)	R_{c11}	7.28	7.592
	R_{c22}	7.36	7.592
	$R_{c12} = R_{c21}$	0.52	0.536

b) ***Two symmetric coupled microstrip lines***

Consider a pair of coupled identical microstrip lines. The strip width is 70 mm and the thickness is 3 μm. Their conductivity is 41 MSm, and they are placed on top of a lossless dielectric substrate of relative permittivity 12.9 and height 100 mm. The width of the gap between the strips is 30 mm. The structure is shielded. The distance between a strip and the adjacent side wall is 515 mm, and the distance between the strips and the metallic cover is 2 mm. Table 5.16 shows the results, at 10 GHz, for the attenuation coefficient for the odd and even modes. The difference in the results obtained using the present method and the Hewlett-Packard program HFFS [Anger 1990] is less than 5%.

Table 5.16
Attenuation coefficient of the coupled microstrip pair

α (Db/m)	Even mode	Odd mode
Present method	26.1	48.5
HFSS	27.0	51.0

5.3 FULL-WAVE ANALYSIS OF GUIDING STRUCTURES

5.3.1 Shielded Microstrip Line: Case A

A microstrip line whose general cross section is shown in Figure 5.32 is considered. Three cases (θ=45°, 90° and 135°) of the signal conductor are considered. The meshes for the last two cases are shown in Figures 5.33 and 5.34.

For this example, the goal is to compute the dispersion characteristics of the fundamental, quasi-TEM mode. Symmetry has been exploited to reduce the size of the FEM mesh by half. In the adaptive procedure, the mesh matches the field behavior. The electric field is more concentrated around the lower edge of the strip due to the presence of the dielectric and the proximity of the ground plane. The nature of the field singularity at the edges depends on the conductor shape, as in the trapezoidal case. First order Lagrange/curl conforming elements have been used. It is observed that the self adaptive procedure generates a finer mesh for Figure 5.34 than for Figure 5.33. This is expected as the singularity of the fields are more pronounced for θ = 135° than for θ = 90°.

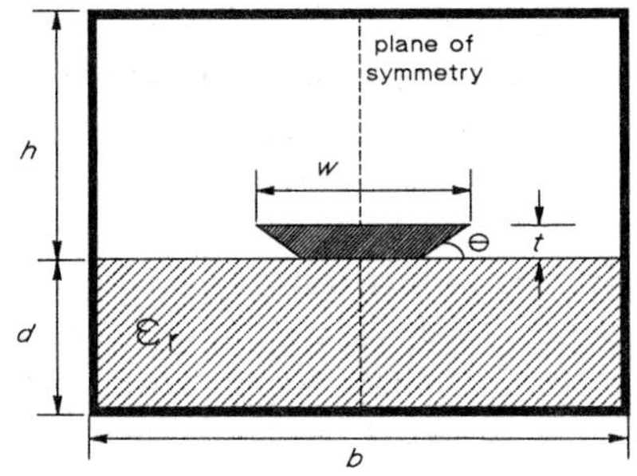

Figure 5.32 General cross section of a microstrip line: b = 10 mm, d = 0.635 mm, h = 6.35 mm, W = 3 mm, t = 0.3 mm, ε_r = 9.8.

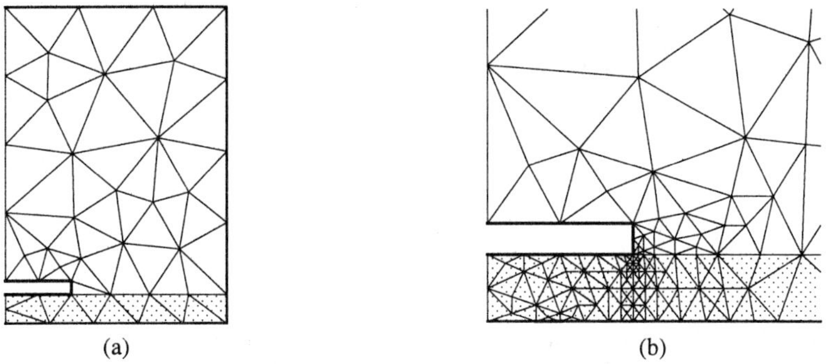

Figure 5.33 Microstrip line, θ = 90°. (a) Initial mesh of one half of the structure. (b) Zoom-in over the last mesh (fifth one, 501 nodes).

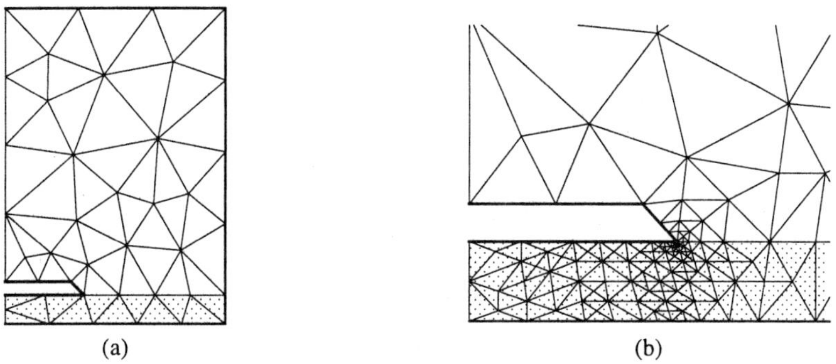

Figure 5.34 Microstrip line, θ = 135°. (a) Initial mesh of one half of the structure. (b) Zoom-in over the last mesh (sixth one, 537 nodes).

Results of the analysis for the three cross sections of the signal conductor are shown in Figures 5.35 and 5.36. These results are the normalized propagation constant and characteristic impedance. The normalization is with respect to the propagation coefficient (wave number) in vacuum and the intrinsic impedance of vacuum, respectively. Also shown for comparison are the results obtained by Alam et al. [1994b] and Olyslager et al. [1993]. The former results are obtained using a FEM with vector elements of higher order than used here. The latter results are obtained using the moment method in the spectral domain. The agreement between the results generated by the self adaptive mesh and the other two methods is excellent.

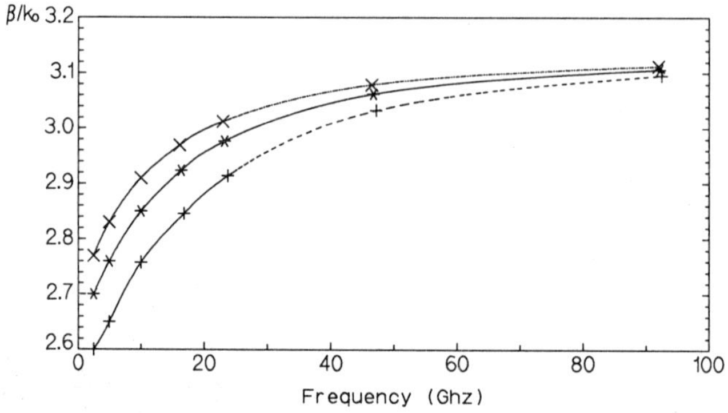

Figure 5.35 Dispersion of the normalized propagation constant. Symbols: -+-, $\theta = 45°$; —*-, $\theta = 90°$; -x-, $\theta = 135°$. Lines: [Alam et al. 1994b] and [Olyslager et al. 1993]. Markers: present method.

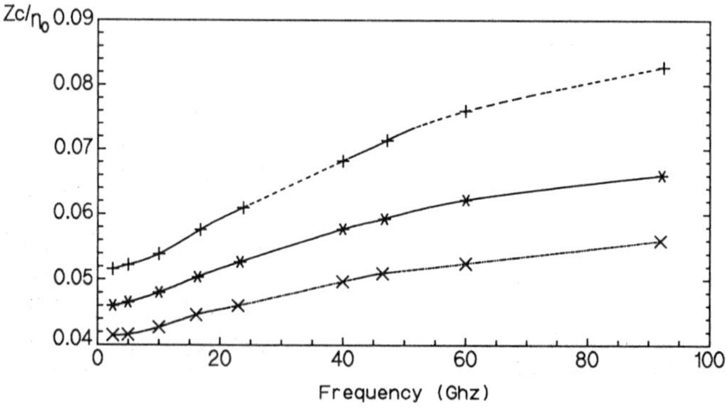

Figure 5.36 Dispersion of the normalized characteristic impedance. Symbols: -+-, $\theta = 45°$; —*-, $\theta = 90°$; -x-, $\theta = 135°$. Lines: [Alam et al. 1994b] and [Olyslager et al. 1993]. Markers: present method.

5.3.2 Shielded Microstrip Line: Case B

The cross-sectional dimensions are b = 10 mm, d = 1 mm, h = 4 mm, W = 1 mm and t = 0 mm, with θ = 90° as shown in Figure 5.32. The dielectric has a relative permittivity of ε_r = 10. Only one half of the structure is analyzed, as in the previous case. The results given are computed with a final mesh of 943 nodes using first order elements, as opposed to 4500 unknowns employed by Slade and Webb [1992] utilizing both Lagrange and singular elements. Figure 5.37(a) shows the normalized propagation coefficient for the quasi-TEM fundamental mode and for the first higher mode. Figure 5.37(b) shows the characteristic impedance of the quasi-TEM mode. It is clear that the self adaptive mesh with a much lower number of unknowns displays a better agreement with the results obtained using the moment method and the spectral domain analysis (SDA) [Slade and Webb 1992], as well as those obtained using commercial CAD packages Libra and Super Compact.

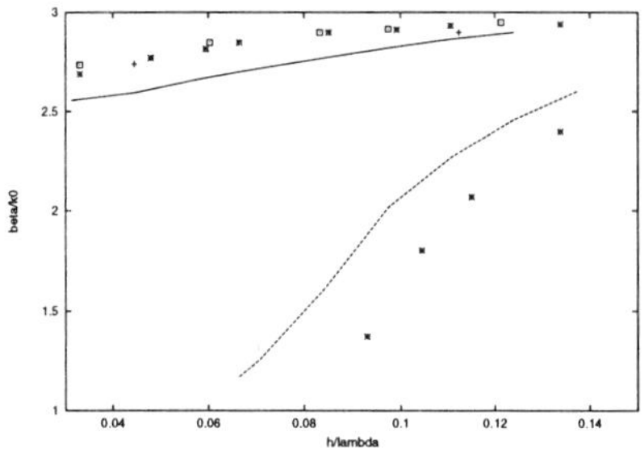

(a)

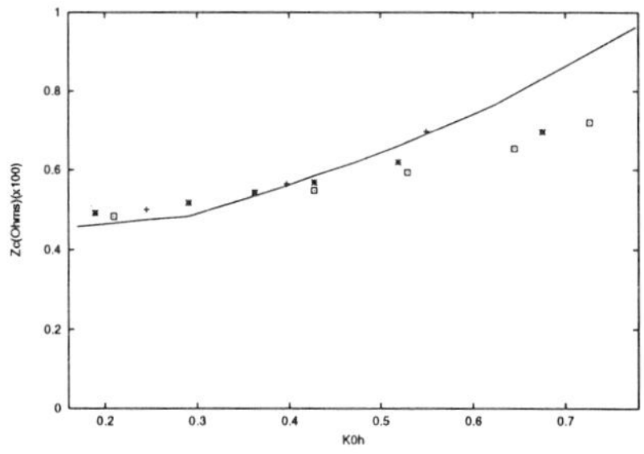

(b)

Figure 5.37 Results for the single microstrip line. (a) Normalized propagation constant: —, even mode; --, odd mode. (b) Characteristic impedance. Symbols: ×, present method; □, Libra-Super Compact; lines, [Slade and Webb 1992]; +, SDA (same reference).

5.3.3 Shielded Microstrip Line: Effect of Walls

The objective is to analyze the effect of the side walls and the top cover as shown in Figure 5.38, on the propagation characteristics of the fundamental mode. Figures 5.39 and 5.40 show the dispersion of the effective dielectric constant as a function of the location of the side walls and top cover, respectively. The results are compared with a finite difference time domain solution [Wu and Chang 1991].

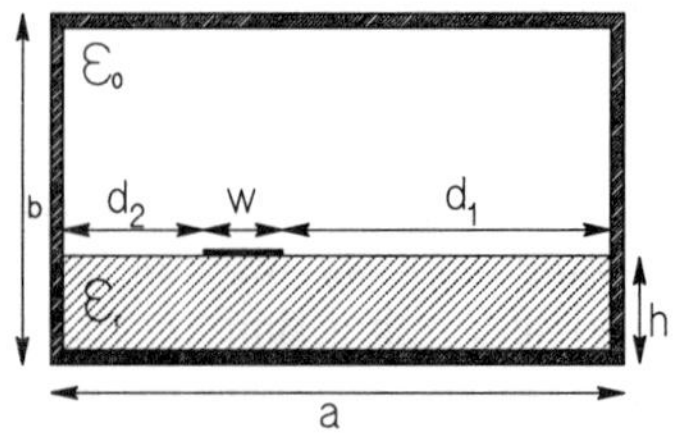

Figure 5.38 Asymmetric microstrip line with $w = h = 0.635$ mm and $\varepsilon_r = 9.7$.

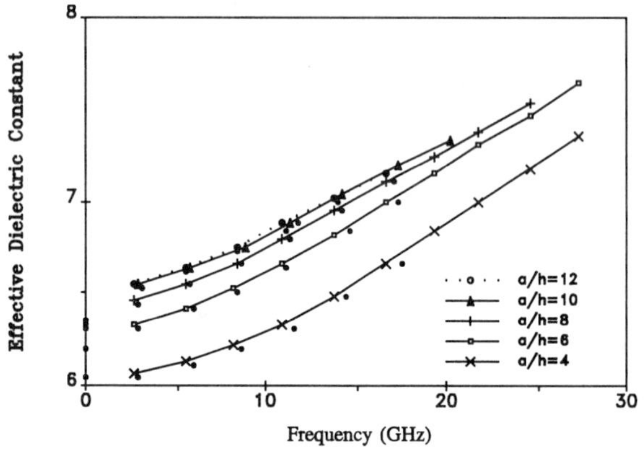

Figure 5.39 Effect of the side walls; $d_1/h = d_2/h = (a/h\text{-}w/h)/2$, $b/h = 7$. Symbols: —, [Wu and Chang 1991]; •, present method with nine mesh refinements.

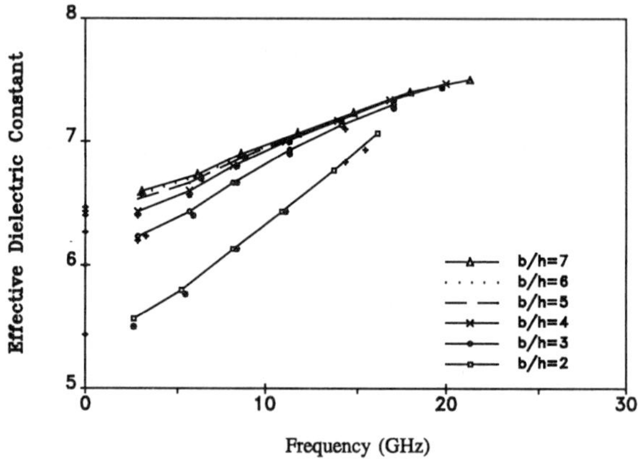

Figure 5.40 Effect of the top conductor; $a/h = 10$, $d_1/h = d_2/h = 4.5$. Symbols: —, [Wu and Chang 1991]; •, present method with nine mesh refinements.

5.3.4 Shielded Microstrip Line: Losses

The structure considered here corresponds to the one in Figure 5.32, now with b = 1.2 mm, h = 0.1 mm, d = 2 mm, W = 70 μm, t = 3 μm, θ = 90° and ε_r = 12.9. Figure 5.41 shows the results for the total attenuation coefficient as a function of frequency. The total attenuation is composed of the attenuation coefficient due to the substrate losses (assuming tan δ = 0.0003) and the attenuation coefficient due to conductor losses (because of its finite conductivity, assuming $\sigma = 4.1\times10^7$ mho/m). The results of the present method are compared with the values provided by Kitazawa [1993], utilizing SDA, as well as with the results of the quasi-static analysis and experimental data. The analysis has been performed over one half of the structure. The self-adaptive procedure has been used. The results given in Figure 5.41 have been obtained with a final mesh of 1325 nodes and using first order elements.

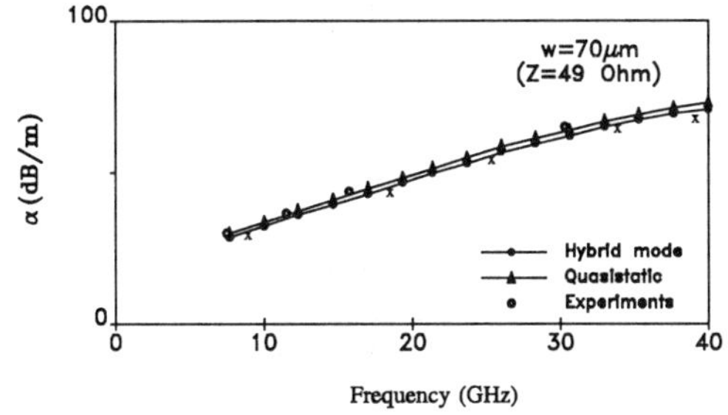

Figure 5.41 Total attenuation constant. Symbols: ×, present method; •, [Kitazawa 1993]; o, measurements.

5.3.5 Bilateral Circular Finline

The bilateral finline, shown in Figure 5.42, has been chosen to illustrate the use of curved Lagrange/curl-conforming elements for the analysis of an inhomogeneous structure with curved contours. The conductors are supported by a dielectric layer of relative permittivity ε_r = 2.2 shielded by a circular guide of radius 4.165 mm. The distance between the fins is 0.3 mm. The thickness of the dielectric layer is 0.254 mm. Figure 5.43 shows the mesh for 977 nodes utilizing first-order straight and curved elements.

The results for the structure of Figure 5.42 are summarized in Table 5.17 and Figure 5.44, where s is k_0a. k_0 is the free-space wave number, and a is the geometric normalization factor (i.e., the radius of the circle). The result of the present method are compared with the results obtained by Eswarappa et al. [1989] using a FEM-based formulation based on the longitudinal components of both the electric and the magnetic fields and the Lagrange elements.

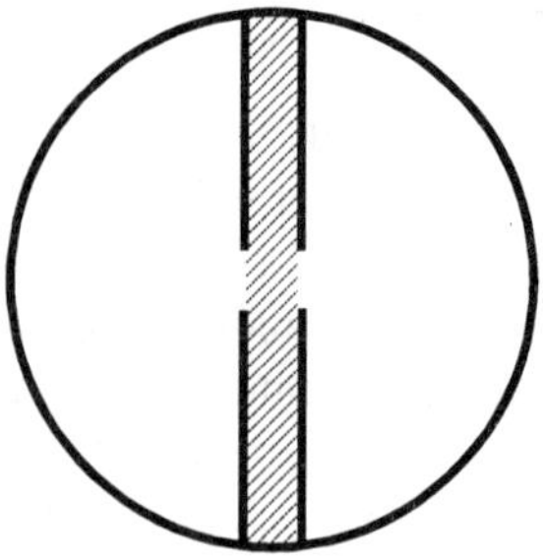

Figure 5.42 Bilateral circular finline.

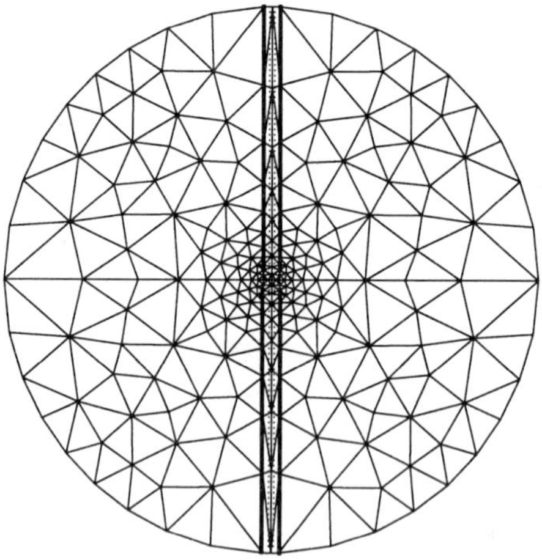

Figure 5.43 Mesh generated for the analysis of the structure of Figure 5.42.

Table 5.17
Results for the structure of Figure 5.42

β_a	s^2	ε_{reff}	f (GHz)
0.0	0.01946	0.0	6.66
0.2	0.05030	0.795	10.70
0.5	0.21233	0.177	22.00
1.3	1.37540	1.228	55.99

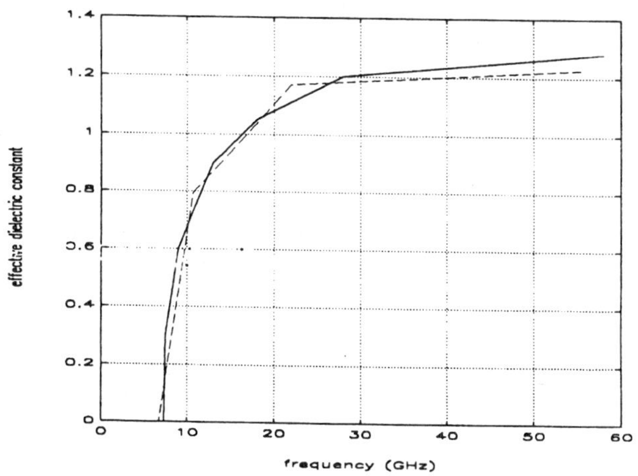

Figure 5.44 Dispersion curve for the structure of Figure 5.42. Symbols: —, Eswarappa et al. [1989]; ---, present method.

5.3.6 Double Semicircular Ridge Guide

A bilateral circular finline is analyzed using first-order curved edge elements. Figure 5.45 shows two meshes of the structure, assuming the width of the guide to be 10 cm and height 7.5 cm. The equivalent dimensions for a rectangular guide would be 10.6 cm in width and 5.9 cm in height according to [Meinke et al. 1963], where the guide has been analyzed using a conformal transformation. In Table 5.18, results are presented for the first four guided modes as computed using straight elements only and straight and curved elements. In this table, measurement data [Meinke et al. 1963] are also listed. Figure 5.46 shows the electric and the magnetic fields of the TE_{10} mode.

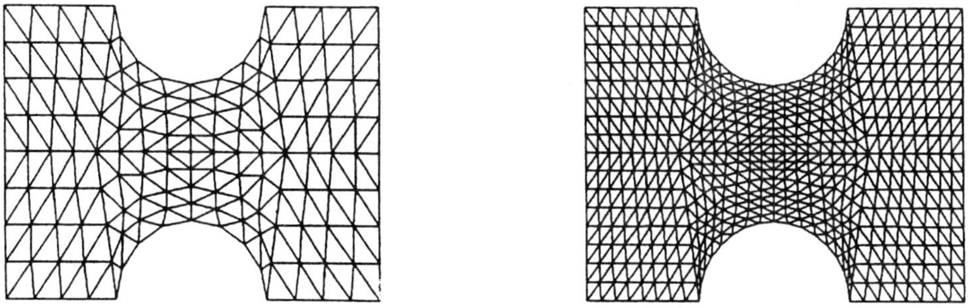

(a) (b)

Figure 5.45 Double semi-circular ridge waveguide. Mesh with (a) 561 nodes and (b) 2145 nodes.

Table 5.18
Results of the analysis of the double semicircular ridge waveguide

		TE_{10}	TE_{01}	TE_{11}	TE_{20}
Mesh 1	Straight	0.05755	0.2104	0.2209	0.4006
	Curved	0.05698	0.2108	0.2206	0.3984
	Measurements	0.06393	0.2143	0.2293	0.4240
Mesh 2	Straight	0.05774	0.2110	0.2212	0.4043
	Curved	0.05759	0.2211	0.2211	0.4039
	Measurements	0.06393	0.2143	0.2293	0.4240

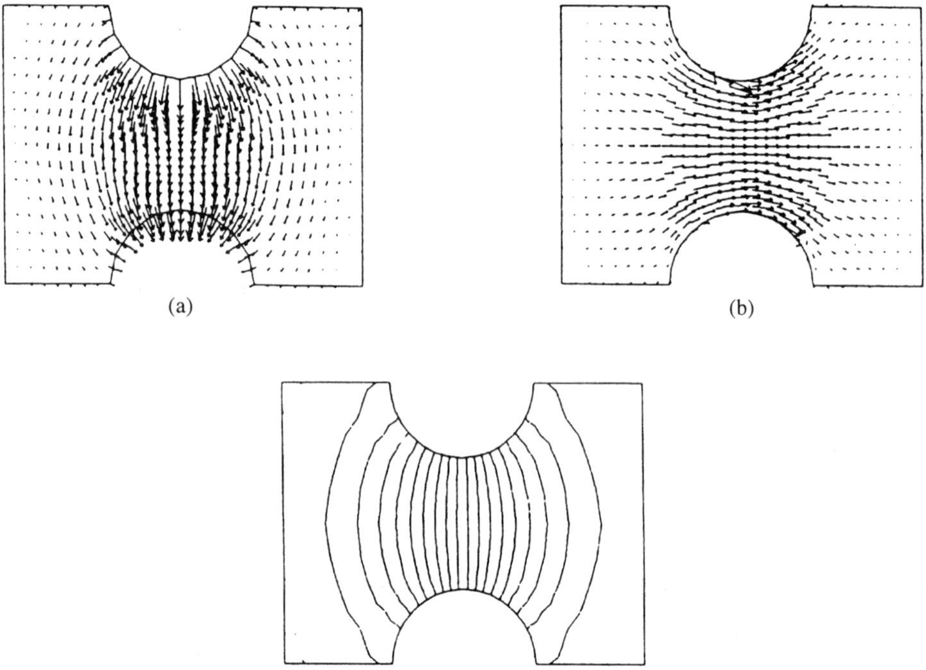

(a) (b) (c)

Figure 5.46 Electromagnetic field of the dominant TE mode. (a) transverse electric field, (b) transverse magnetic field, and (c) longitudinal magnetic field.

5.3.7 Coplanar Line with Anisotropic Substrate

The geometry and dimensions of the structure of interest are shown in Figure 5.47. Both the permittivity and permeability are considered to be anisotropic according to

$$[\varepsilon_r] = \begin{bmatrix} \varepsilon_x & 0 & 0 \\ 0 & \varepsilon_y & 0 \\ 0 & 0 & \varepsilon_z \end{bmatrix} \tag{5.1}$$

$$[\mu_r] = \begin{bmatrix} \mu_x & 0 & 0 \\ 0 & \mu_y & 0 \\ 0 & 0 & \mu_z \end{bmatrix} \tag{5.2}$$

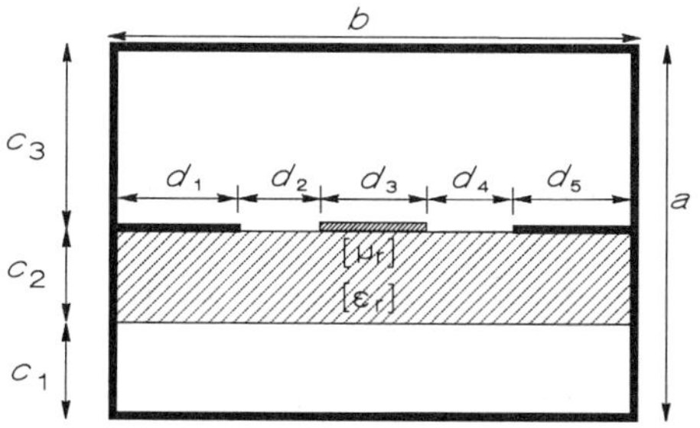

Figure 5.47 Coplanar line, $c_1/a = 0.4$; $c_2/a = 0.1$; $c_3/a = 0.5$; $d_1 = d_2 = d_3 = d_4 = d_5 = b/5$.

In Figure 5.48, results are given for the normalized propagation constant as a function of frequency for dielectric, magnetic, and both types of anisotropy. The results for the fundamental mode are compared with those of Mazé-Merceur et al. [1993], who used a spectral domain technique. The adaptive procedure starts initially with a mesh with 990 nodes using first order elements as shown in Figure 5.49. Figure 5.50 illustrates the results for both the $\bar{E}$ formulation and the $\bar{H}$ formulation showing a zoom-in over the strip region of a refined mesh and the nature of the electric and the magnetic fields of the fundamental mode for both types of anisotropy. By comparing Figure 5.50(b) and 5.50(e) one can observe the effect of anisotropy in the y-direction, $\varepsilon_y = 5$ in Figure 5.50(b) while $\mu_y = 1$ in Figure 5.50(e).

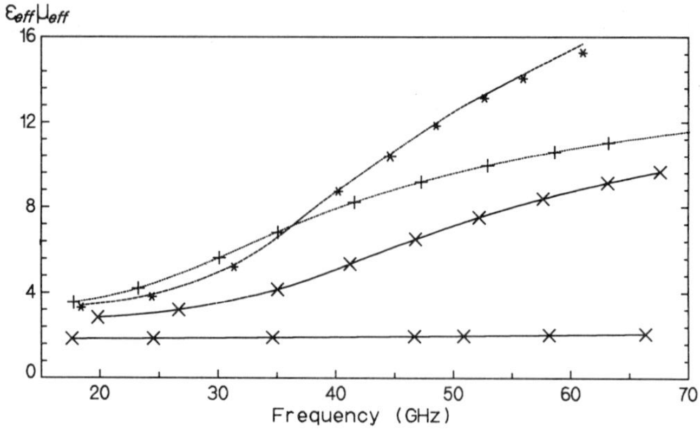

Figure 5.48 Square of the normalized propagation coefficient for the fundamental quasi-TEM mode. Symbols: *: $\boldsymbol{\varepsilon}_r$ = (3;5;3), μ_r = (5;1;1); +: $\boldsymbol{\varepsilon}_r$ = 3, μ_r = 5; ×: $\boldsymbol{\varepsilon}_r$ = 3, μ_r = (5;1;1); × (lower line): $\boldsymbol{\varepsilon}_r$ = 3, μ_r = 1. Lines: [Mazé-Merceur et al. 1993]. Markers: present method.

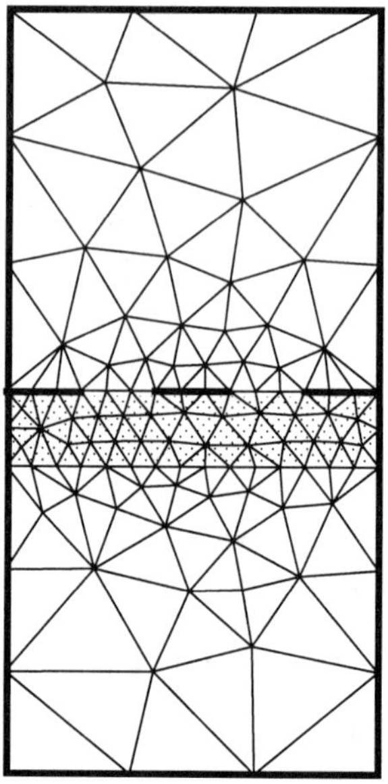

Figure 5.49 Initial mesh for the coplanar waveguide analysis.

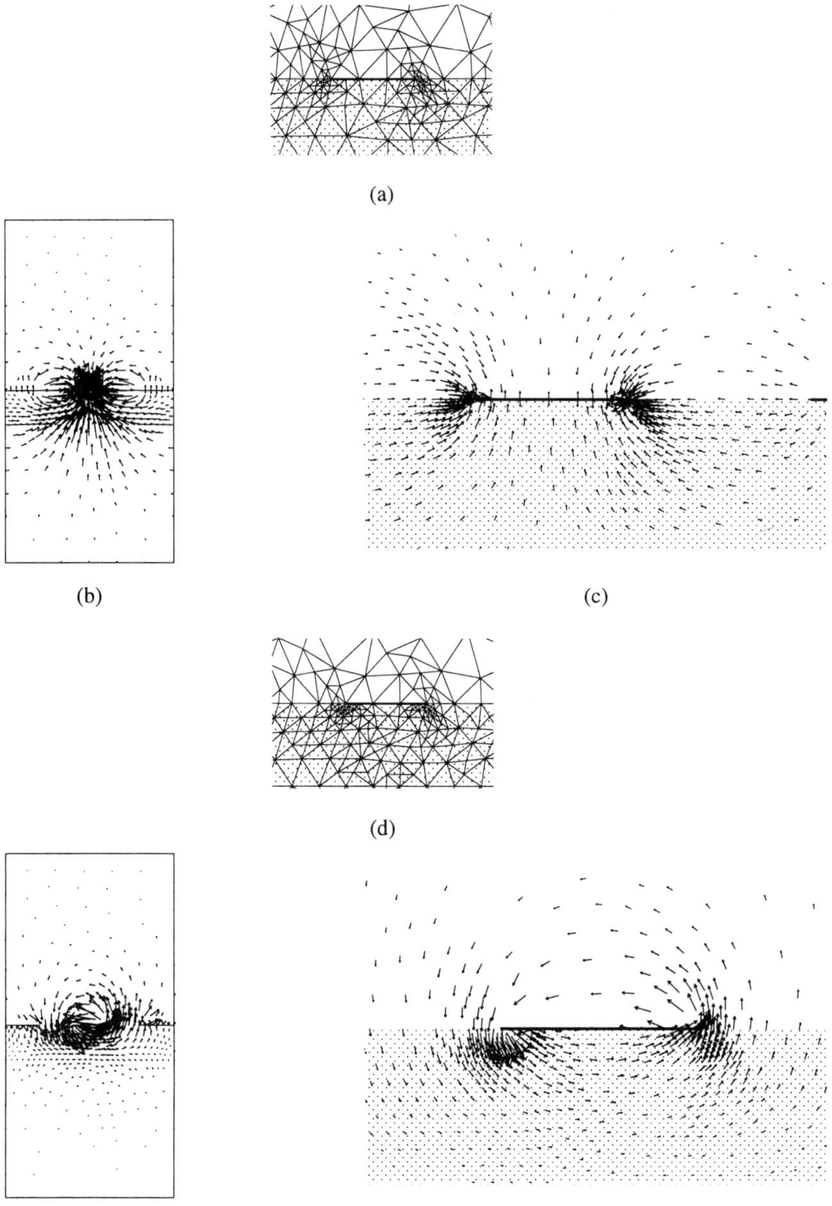

Figure 5.50 Analysis of the coplanar waveguide for $\varepsilon_r = (3;5;3)$ and $\mu_r = (5;1;1)$. $\bar{E}$ formulation: (a) zoom-in over the refined mesh, (b) transverse electric field, and (c) zoom-in over the strip area. $\bar{H}$ formulation: (d) zoom-in over the refined mesh, (e) transverse magnetic field, and (f) zoom-in over the strip area.

5.3.8 Microstrip Line with Anisotropic Substrate

Referring to Figure 5.32, the line dimensions are $w/d = 5$, $b/d = 30$, and $h/d = 30$, $t = 0$, $\theta = 90°$. The substrate has a relative dielectric constant given by (5.1) with $\varepsilon_x = 2.89$, $\varepsilon_y = 2.45$, and $\varepsilon_z = 2.95$ while the permeability is that of a vacuum. Figure 5.51 shows the dispersion curve for the fundamental mode. Two sets of results are shown using both the electric and the magnetic-field formulations. The results are compared with those from Cano et al. [1989], who used a method in the spectral domain. The results of the present method are obtained using first order elements and a mesh with 504 elements and 1079 nodes.

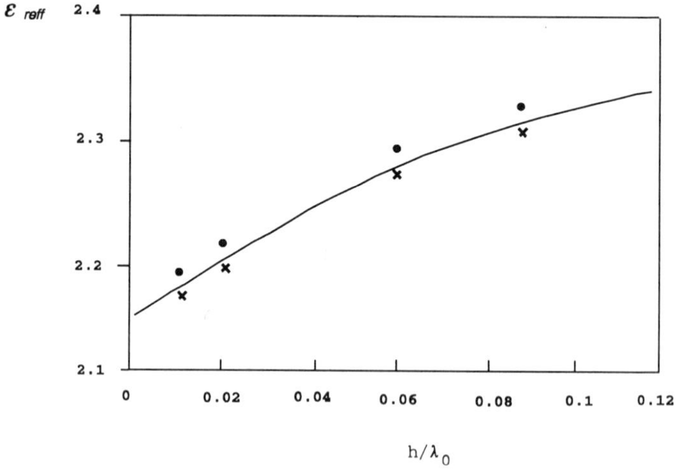

Figure 5.51 Dispersion of the relative effective dielectric constant of the fundamental mode of the microstrip line with anisotropic substrate. Symbols: •, $\overline{E}$ formulation; ×, $\overline{H}$ formulation; —, [Cano et al. 1989].

5.3.9 Finline with Anisotropic Substrate

The geometry of the finline is shown in Figure 5.52, with $a/2 = 3.556$ mm, $b = 3.556$ mm, $a_x = 0.127$ mm, and $d = 1$ mm. The substrate is teflon, with dielectric anisotropy, given by (5.1) with $\varepsilon_x = 2.89$, $\varepsilon_y = 2.45$, and $\varepsilon_z = 2.95$. Again, the permeability is that of vacuum. Figure 5.53 shows the dispersion of the effective dielectric constant of the fundamental mode computed using the electric-field and the magnetic-field formulations. The results of the present method are compared with those provided by Bulutay and Prasad [1993], who used the TLM method as well as a technique in the spectral domain (SDA).

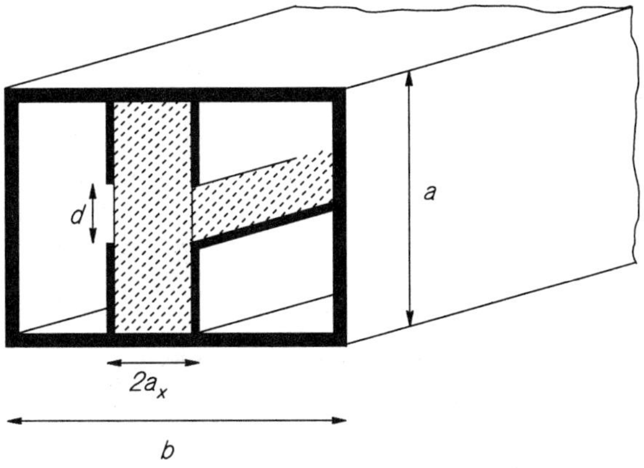

Figure 5.52 Geometry of the finline.

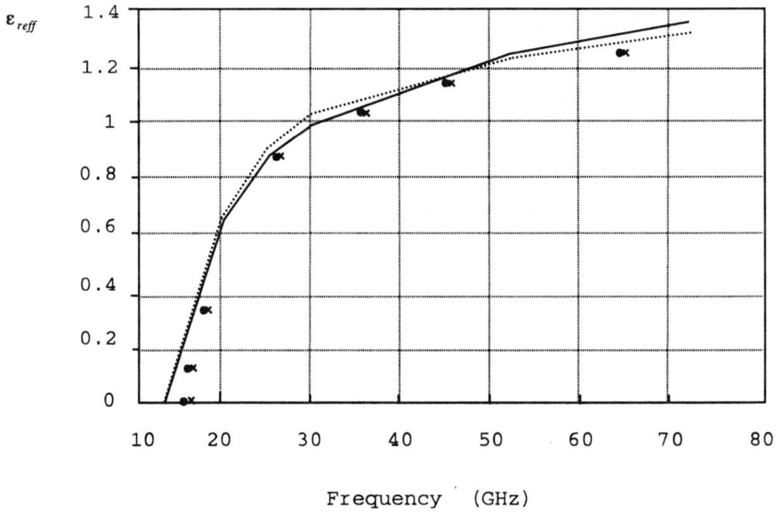

Figure 5.53 Dispersion of the effective dielectric constant of the fundamental mode of the anisotropic finline. Symbols: —: TLM [Bulutay and Prasad 1993]; •: FEM, $\bar{E}$ formulation; --: SDA [Bulutay and Prasad 1993]; ×: FEM, $\bar{H}$ formulation.

5.3.10 Suspended Coplanar Waveguide

The geometry of the suspended coplanar waveguide is shown in Figure 5.54. The data for the structure are $h_{L1} = h_{U1} = 4.5$ mm, $h_{L2} = 1$ mm, $A = 10$ mm, and $w = s = 2$ mm; and $\varepsilon_{rL1} = \varepsilon_{rU1} = 1$ and $\varepsilon_{rL2} = 9.35$. Figure 5.55 shows the dispersion curve obtained using both the electric-field and the magnetic-field formulations. First order elements have been used with a mesh of 716 nodes. The fundamental mode and the first higher order modes have been analyzed. The results are compared with those evaluated by Lyons et al. [1993] using a method in the spectral domain. These authors compare their results with those obtained by Mirshekar-Syahkal [1990], who also used SDA.

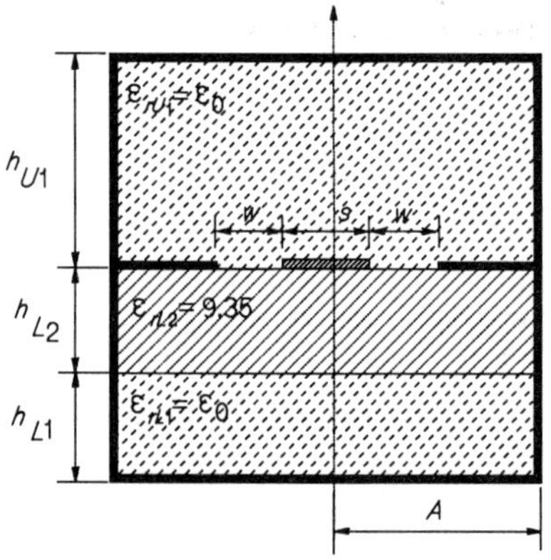

Figure 5.54 Suspended coplanar waveguide.

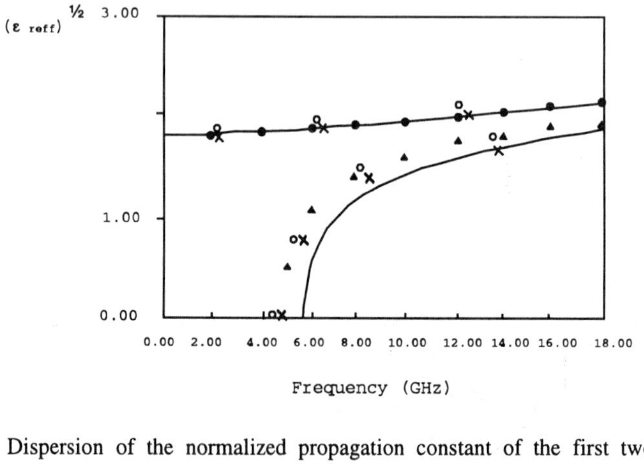

Figure 5.55 Dispersion of the normalized propagation constant of the first two modes of the structure shown in Figure 5.54. Symbols: —, [Lyons et al. 1993]; •, even mode [Mirshekar-Syahkal 1990]; ▴, odd mode [Mirshekar-Syahkal 1990]; o, $\bar{E}$ formulation, this method; ×, $\bar{H}$ formulation, this method.

5.3.11 Coupled Microstrip Lines

This is a structure with two signal conductors, and the geometry is shown in Figure 5.56. The results for the normalized phase coefficient for the odd and even modes, and the coefficients of the characteristic impedance matrix, obtained with first order elements and a mesh of 840 nodes, are shown in Figures 5.57 and 5.58. The results are compared with those provided by Slade and Webb [1992]. They used FEM based on Lagrange elements and singular elements with a total of 4500 unknowns. The results of the present method are also compared with those obtained using SDA [Slade and Webb 1992] and using the software package Super-Compact by Compact Software, Inc. The self-adaptive procedure uses a lower number of unknowns and yet produce quite accurate results as shown in Figures 5.57 and 5.58.

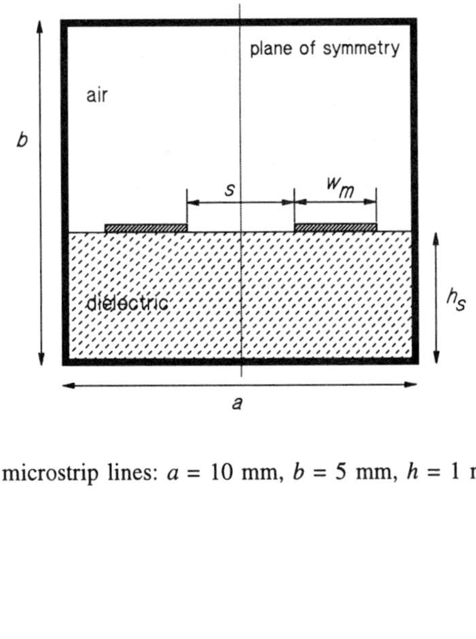

Figure 5.56 Coupled microstrip lines: a = 10 mm, b = 5 mm, h = 1 mm, w = 2 mm, s = 1 mm, ε_r = 4.

5.4 CONCLUSIONS

This chapter presents additional results for the self-adaptive mesh algorithms for the quasi-static and full-wave analysis of different guiding structures and passive devices. The variety of examples presented demonstrates the power and the flexibility of the self-adaptive mesh FEM. The results obtained by this analysis have been compared with the results published in the literature by other authors, commercial software packages, as well as experimental results. In all cases, a high degree of accuracy of the results of the self- adaptive FEM has been obtained.

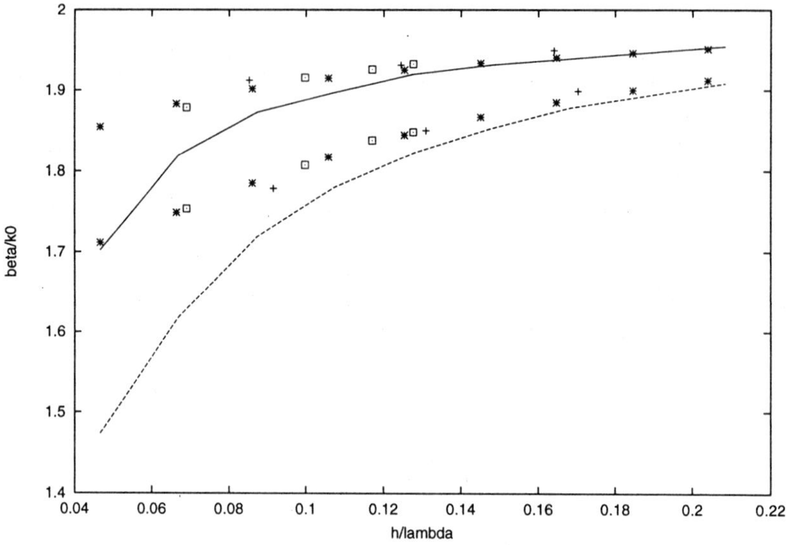

Figure 5.57 Dispersion of the normalized propagation constant of the even and odd modes. Symbols: lines (—, even mode; --, odd mode), [Slade and Webb 1992]; *, present method; □, Super-Compact; +, SDA [Slade and Webb 1992].

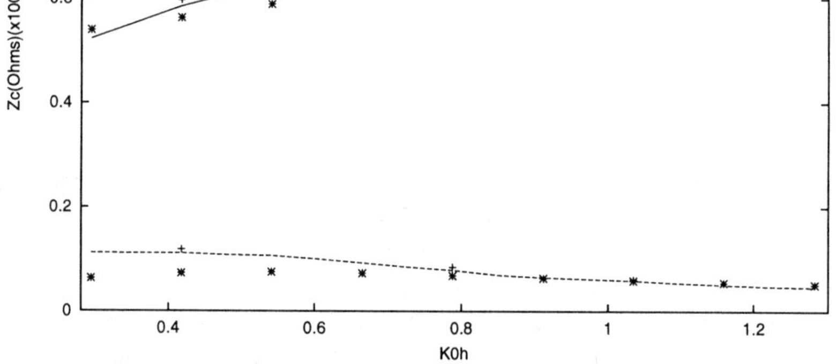

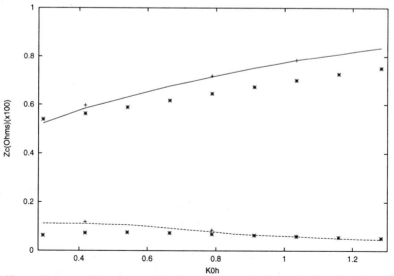

Figure 5.58 Elements of the characteristic impedance matrix. Symbols: lines (—, $Z_{11} = Z_{22}$; --, $Z_{12} = Z_{21}$), [Slade and Webb 1992]; *, present method; +, SDA [Slade and Webb 1992].

Chapter 6

Application of the Finite Element Method for the Solution of Open-Region Problems

6.1 INTRODUCTION

In this chapter, the Finite Element Method (FEM) is applied for the solution of open-region problems. Although the problem is stated generally, particular attention is given to two-dimensional electrostatic problems as well as scattering from conducting structures with incident fields having a transverse electric (TE) or transverse magnetic field (TM) polarization. Three-dimensional (3D) problems are treated in Section 7.5.

6.2 STATEMENT OF THE PROBLEM

6.2.1 Introduction

FEM is basically used to solve Maxwell's equations in differential form to evaluate the electromagnetic fields or potentials. As a consequence, and in contrast to other approaches that deal with the sources of the fields as the unknowns, FEM inherently involves a discretization of the whole region where the electromagnetic fields exist. In many practical problems, the fields are bounded in a finite region (e.g., by a closed perfectly conducting surface). In other problems, however, the fields exist in an unbounded (open) region. Typical examples are two- and three-dimensional antenna and scattering problems. The classical discretization by FEM of an infinite region is numerically impossible. Hence, modifications of FEM have to be made to enable treating open-region problems. There exist several such possibilities that are summarized in the next section, but the stress in this chapter is on the combination of FEM and integral-equation formulations (see Section 1.1.2.2) through an iterative scheme.

6.2.2 The Finite Element Method and Open-Region Problems

Different procedures may be used to adapt FEM for treating open-region problems. The simplest approach in problems where interest is focused in the region where field variations are significant is to enclose such a region by an artificial shielding placed sufficiently far away, so that a crude simulation of an infinite domain would be obtained. Then, the boundary conditions corresponding to infinity would be applied at the artificial shielding boundary. Hence, the problem results in a closed-region problem. This is, however, a dangerous and restrictive technique. On one hand the computational domain may be still too large resulting in large memory, storage, and CPU time requirements. Conversely, the shield may be placed not sufficiently far away significantly affecting the fields in the solution region.

All other methods are based on the subdivision of the infinite region problem into two regions. One is the region where field variations are significant (e.g., source regions and regions of inhomogeneities) up to a fictitious surface S. Usually, S encompasses all internal impressed sources and material discontinuities constituting the electromagnetically complex region of interest, as shown in Figure 6.1. The region that is outside S is homogeneous. All methods perform a standard discretization of the finite region enclosed by S. The methods vary with respect to treating the region outside S, and they can be classified into two groups.

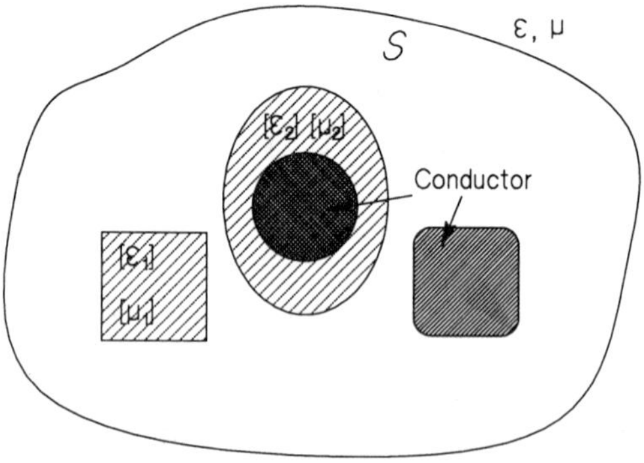

Figure 6.1 Problem of interest with surface S.

Methods of the first group are based on a nonclassical discretization of the unbounded region external to S. Approaches like special elements of infinite extension called *infinite elements* (see the infinite element presented in Section 2.5.1.2), *recursive condensation* [Silvester et al. 1977], or *infinitesimal scaling* [Hurwitz 1984], or the use of suitable *geometric transformations* such as conformal mappings (see Section 1.1.1.1), correspond to this group. These approaches are mainly applied to static and quasi-static problems. In all cases, either the special elements or the other techniques, must be able to properly describe the asymptotic field behavior as the field point moves away. This group of methods is popular in static and quasi-static field analysis.

Methods of the second group discretize only the interior of S, i.e., only a portion of the original infinite region. Hence they perform a classical discretization of a truncated domain. The field representation within S must then be related to a suitable representation of the fields in the unbounded portion of the region. The basic point is to impose appropriate boundary conditions on the surface S, which account for the infinite extent of the original geometric region. In other words, the solution of Maxwell's equations for the fields inside the fictitious surface S (i.e., the solution to the equivalent internal problem) along with these boundary conditions must yield a good approximation of the solution of the original unbounded problem. Some authors refer to these methods as *picture-frame methods* because, like in a picture of a landscape, the solution is obtained only in a portion (a frame) of the total. However, this does not mean that the solution corresponding to the external region, i.e., the exterior field, can not be obtained, as it will be clear later.

Several procedures have been reported in the literature for imposing suitable boundary conditions. The techniques can be classified following different criterions. One criterion that is relevant from the numerical implementation point of view is the one that attends to the nature of the boundary conditions imposed at the artificial boundary S that truncates the original unbounded domain. This results in dividing the methods into two categories: those based on local boundary conditions and those employing integral (nonlocal) boundary conditions.

Nonlocal boundary conditions provide a numerically exact mesh truncation scheme, valid for near and far fields. This is accomplished by making use of the equivalence principle and the integral representation of the exterior field in terms of the equivalent sources (charges and/or currents) and the Green's function of the problem. The nonlocal character of the boundary condition comes from the integral nature of the boundary condition, which relates the value of the field (or the potential) at some generic point on the external boundary S with the values of the field (or the potential) at all other points of the boundary. Therefore, the use of nonlocal boundary conditions together with FEM leads to a discretized problem where, in general, the sparsity and the banded structure of the original FEM matrices are destroyed yielding to full or at least partially full matrices. However, due to the exact character of the boundary condition valid for the near field, the external boundary may be placed very close to the original sources of the problem leading to meshes with a lower number of nodes. Thus, the number of unknowns can be kept within reasonable limits increasing the efficiency of the approach.

In this context, two families of methods may be distinguished depending on the kind of basis functions employed for the discretization of the equivalent sources (charges and/or currents) of the integral representation of the exterior field.

The first family makes use of identical basis functions to those employed in the FEM discretization of the internal region enclosed by S. They are commonly known as *hybrid finite element/boundary integral methods* (FE-BI method) or simply *hybrid finite element methods*. The first works on the application of the FEM to unbounded domain problems belong to this category ([Silvester and Hsieh 1971] for the Laplace equation and [McDonald and Wexler 1972] for the Helmholtz equation). Basically, as it has been already mentioned, the hybrid methods employ the equivalence principle to yield a boundary integral representation of the exterior field in terms of equivalent sources on the external artificial boundary S. These equivalent sources (charges and/or currents) are expressed as a function of the physical unknown of the problem, i.e., the potential or the field, which is discretized using finite elements. Thus, the equivalent sources and hence, the boundary integral on S, are actually FEM discretized. Therefore, due to the flexibility of the FEM method, the geometry of the boundary S may be, in general, arbitrary. In addition, due to the exact character of the radiation condition of the boundary integral, the external boundary S may be placed very close to the original sources of the problem, leading to systems with a low number of unknowns. However, because of the integral character (nonlocal) of the boundary condition, the sparsity and the banded structure of the matrices are destroyed. Specifically, partly full matrices are obtained. Considering the N_S degrees of freedom associated to the boundary S numbered consecutively, the FEM matrices have a dense matrix block of dimensions $N_S \times N_S$. This makes the process of solving the FEM system for large problems computationally expensive. Since the work of [Silvester and Hsieh 1971] and [McDonald and Wexler 1972] the method has been widely applied to a variety of problems with different formulations and type of elements, until nowadays, e.g., [Bossavit and Verité 1982], [Boyse and Seidl 1991, 1994], [Collins et al. 1990a,b, 1992], [Cristina and Di Napoli 1983], [Gong et al. 1994], [Janković et al. 1994], [Jin and Liepa 1988a,b], [Jin and Volakis 1991a,b,c], [Kagawa et al. 1982], [Lucas and Fontana 1995], [Onuki 1990], [Paulsen et al. 1988], [Ramahi and Mittra 1991a], [Zienkiewicz et al. 1977]. The FEM/BEM procedures fall under this category. In this context there are a number of works that are usually categorized as MOM/FEM techniques. Although they lead to formulations equivalent to those of the FE-BI methods, their philosophy is somehow different. In fact, the open-region problem is decoupled into an external and an internal problem, discretized by MOM and FEM respectively. The coupling of both formulations is performed through the appropriate continuity conditions [Angelini et al. 1993], [Cwik 1992], [Cwik et al. 1996a,b], [Gedney and Mittra 1991], [Gedney et al. 1992], [Gong and Glisson 1990], [Hoppe et al. 1994], [Pekel and Mittra 1995], [Yuan 1990], [Yuan et al. 1990], [Zuffada et al. 1997].

The second family of methods that employ nonlocal boundary conditions use an eigenfunction expansion for the discretization of the equivalent sources. The *unimoment method* [Mei 1974] (see also the review of [Mei 1987]) is the most popular method among the approaches employing an eigenfunction expansion. It was originally applied to 2D scattering and antenna problems and since then it has been widely employed for the analysis of antenna and scattering from complex objects (e.g., [Chang and Mei 1976], [Morgan 1980, 1981], [Stoval and Mei 1975]). In the unimoment method, the equivalent sources on S are approximated by a linear combination of N eigenfunctions Φ_m ($m = 1, ..., N$) of the Helmholtz operator. The field in the interior region enclosed by the boundary S is obtained by the FEM solution of N sparse system of equations corresponding to N interior problems with the value of the eigenfunctions Φ_m as a Dirichlet boundary condition on S. The interior field is then computed as a linear combination of the previously mentioned N FEM field solutions, where the coefficients of the linear combination are the same as those of the approximate equivalent sources

Now, the only thing left is to determine the values of these coefficients. They are obtained by enforcing the continuity of the field and the equivalent sources through S, which involves the solution of a dense system of equations of $N \times N$.

The advantage of the unimoment method is that the dimension of the dense system of equations (computationally expensive to solve and store) is much less than in the case of the so called FE-BI methods. This is due to the fact that a larger number of functions are needed to discretize the equivalent sources on S with conventional FEM basis functions than with an eigenfunction expansion. However, the unimoment method requires, in order to perform the eigenfunction expansion, a constant coordinate boundary S, on which the Helmholtz operator could be separable (typically circular in 2D and spherical in 3D). Then, the approach may not be efficient for arbitrarily shaped geometries (that is the case for elongated geometries) due to the need of meshing all the space within a circle, or a sphere, which encloses all the sources (like scatterers and radiating sources) of the problem. On the other hand, the FE-BI methods lead, in this sense, to a more efficient approach because of the flexibility provided by the possible choice of an arbitrarily shaped surface to truncate the FEM mesh. Several approaches have appeared in the literature in trying to overcome some of the limitations of the unimoment method, in particular, the need of a constant coordinate boundary as the external boundary. Examples of the latter are the so called *finite-elements with extended boundary conditions method* [Morgan et al. 1984], and the *bymoment method* [Cangellaris and Lee 1990, 1991].

The second category of picture-frame methods is formed by those procedures based on local boundary conditions. The idea underlining these methods is the following. To preserve the sparsity and the banded nature of the FEM matrix, it is necessary to provide a local relation between the fields on the boundary surface S, i.e., to relate the value of the field at some generic point on S with only its neighboring points. For example, if S is a sphere well in the radiation (far-field) zone of an antenna, the electric and magnetic fields on S are locally orthogonal and their intensities are related by the wave impedance. This amounts to Sommerfeld's radiation condition. In the FEM solution, for each finite element adjacent to S an equation is formulated enforcing this relation. Other types of radiation conditions have also been designed that allow for the closer placement of the surface S [Jin 1993; Ch. 1]. However, local boundary conditions are not exact, even in the numerical sense. The reason is that, theoretically, there can not be strictly local relation between the fields, except deeply in the far-field zone. However, local boundary conditions may still be obtained but at the expense of performing some approximations to the exact radiation boundary condition. In this sense, local boundary conditions only consider the fields corresponding to outgoing waves. The local boundary conditions must then cancel any reflection on S, i.e., S represents a boundary that virtually absorbs the outgoing electromagnetic wave. That is the reason why they are also known as *absorbing boundary conditions* (ABCs). The limitations of the ABCs come from the hypotheses made for their computation. In order to improve their accuracy, the external boundary S must be placed at some distance from the original sources of the problem (scatterer, inhomogeneities, etc) in order to obtain an accurate solution. This leads to FEM system of equations with a large number of unknowns. Also, there may be undesirable effects when the external boundary *S* has re-entrant corners, because the hypotheses relative to the presence only of outgoing waves is no longer valid.

The research on ABCs started in the applied mathematics field with the work of Engquist and Majda [1977], in which scalar ABCs were developed for 2D problems. Closely related to the work of Engquist and Majda is the second-order scalar ABC derived by Mur [Mur 1981a], valid also for 3D. Bayliss and Turkel [1980] derived an alternate approach to the one of Engquist and Majda, which was later applied to the scalar Helmholtz equation [Bayliss et al. 1982]. Other noteworthy references based on the previously mentioned pioneering work are those of [Bayliss et al. 1985], [Blaschak and Kriegsmann 1988], [Trefethen and Haalpern 1986], among others. It was not until the end of the 1980s that the ABCs left the mathematical forums to start to be widely known in the engineering arena and, in particular, in computational electromagnetics. In [Peterson and Castillo 1989], a 2D scattering problem was solved employing the FEM together with a Bayliss-Turkel ABC to truncate the mesh. Since then, extensive research has been made in the development and application of ABCs for the FEM solution of electromagnetic problems in both, scalar ABCs (e.g., [D'Angelo and Mayergoyz 1990], [Jin et al. 1992]], [Lee et al. 1990], [Ma 1991], [Povinelli and D'Angelo 1981], [Ramahi and Mittra 1989], [Ramahi et al. 1991], [Sumbar et al. 1991]) and also in vector ABCs (see [Chatterjee et al. 1993], [Chatterjee and Volakis 1995], [Kanellopoulos and Webb 1991], [Peterson 1988], [Webb and Kanellopoulos 1989]). Because the approaches utilizing the local relations are approximate mesh truncation schemes, many papers defining absorbing boundary conditions, published after the fundamental works of [Bayliss and Turkel 1980], [Engquist and Majda 1977], tend toward *a posteriori* checking of the validity of the absorbing boundary conditions themselves by means of numerical experiments.

Note that for any kind of mesh termination schemes (local or integral), a proper choice of the terminating surface *S* is essential for obtaining an accurate and efficient solution. On one hand, the terminating surface *S* should be chosen so as to have as small an area as possible. The size of the discretized region should be minimized to reduce the number of unknowns and speed-up computations. On the other hand, if *S* is placed too close, the local boundary conditions may be invalid while the numerical implementation of the integral boundary conditions may be harder.

As an attempt to circumvent the problem of deriving the ABCs, new approaches have been developed in the last years. *Numerical absorbing boundary conditions* (NABCs) and the *measured equation of invariance* (MEI) may be cited as the most significative methods. Basically, both approaches have in common the *a priori* postulate of the existence of linear relationships that relate the degrees of freedom associated to the boundary *S* with their neighbors and *a posteriori* numerical computation of the coefficients of the mentioned linear combination. They differ in the way these coefficients are obtained and also in some implementation issues.

In the NABCs approach [Boag et al. 1994a], [Gordon et al. 1993], [Mittra et al. 1994], the coefficients of the linear combination are derived by enforcing the satisfaction of the linear relationships for several known field wave solutions (e.g planar waves or line wave sources). With certain assumptions, NABCs may be shown to be equivalent to the existing analytical ABCs when h (the distance between the nodes of the mesh) tends to zero [Stupfel and Mittra 1995]. NABCs have been applied to scalar and vector 2D and 3D problems (e.g., see [Boag and Mittra 1995], [Pantic-Tanner et al. 1994], [Stupfel and Mittra 1996]).

On the other hand, the MEI method [Mei et al. 1992, 1994], [Prouty et al. 1993], also enforces the satisfaction of the linear relationships between the degrees of freedom associated with S and their neighbors for several field wave solutions, but in this case, the field wave solutions (denoted as *measuring functions*) are numerically generated by using *metrons* and a suitable Green's function acting on those metrons. The metrons are a set of surface current densities on S, i.e., in the general case, a set of equivalent sources on S. The strength of the method resides in the postulate of invariance with the field of excitation, i.e., the coefficients computed are nearly independent of the choice of the metrons. Thus, in contrast to the NABCs, geometric specific measuring functions are obtained. This is done through the employment of the Green's function. For instance, the analysis of an open microstrip line may be performed by discretizing only the region covering the strip and the area surrounding it. It may be not necessary to mesh the whole dielectric region between the ground plane and the strip, and the ground plane itself, if the Green's function employed incorporates already the ground plane information (e.g., [Prouty et al. 1993]). The MEI method has been the object of further research by its own authors and others (see [Jetvić and Lee 1994, 1995a,b]) and it has been applied to a variety of problems including electrostatic, planar waveguiding structures and discontinuities, scattering and radiation as illustrated in [Gothard et al. 1995], [Luk et al. 1995], for example.

It is also worth noting another relatively recent method employed in the development of ABCs. It is the so called *complementary operator method* [Ramahi 1995] (a review of the method is presented in [Ramahi 1997]).

An alternate approach to all of the previous ones is the use of a layer of lossy material to enclose the problem. Therefore, a metal wall may be placed at the outer boundary with no significant influence. The idea, which is not new, did not gain widespread use because of the high level of reflections produced at the interface of the lossy material. The choice of a low loss material in order to minimize the reflections had the serious drawback of the need of a lossy material region large enough to attenuate the wave and hence, it leads to meshes with a prohibitively large number of nodes. That was until the appearance of the so called *perfectly matched layer* (PML) concept, proposed by Berenger in 1994 [Berenger 1994] (see also [Berenger 1996]). Berenger's PML was presented in the context of a finite difference technique but it is also used nowadays in FEM approaches. The PML is a lossy material layer designed to absorb plane waves of arbitrary frequency and for all incidence angles. Berenger developed such a PML by introducing modifications to Maxwell's equations by allowing the specification of material properties that results in a reflection-less lossy material. Thus, a thin metal backed PML may be placed close to the original sources of the problem (e.g., the scatterer) acting as a physically absorbing boundary condition. Additional studies on Berenger's PML have been performed by several authors, [Chew and Weedon 1994], [Mittra and Pekel 1995], [Rappaport 1995], among others.

However, Berenger's PML is obtained through the modification of the original Maxwell equations and hence, this leads to a PML material that is non-Maxwellian. An alternative representation of a perfect absorber was developed by Sacks et al. [Sacks et al. 1995], yielding to an anisotropic, but Maxwellian, PML. A comparison and review of both absorbers, the Berenger's PML and the anisotropic PML may be found in [Wu et al. 1997]. Some recent topics in this area of active research are the introduction of conformal PMLs [Kuzuoglu and Mittra 1996a], causal PMLs [Kuzuoglu and Mittra 1996b], absorption of evanescent waves [Moerloose and Stuchly 1995], and the search for physically realizable PML absorbers [Ziolkowski 1997].

The method presented in this chapter and Section 7.5 makes use of the boundary integral representation of the exterior field to truncate the mesh. Thus it falls under the category of a FE-BI approach. Its practical implementation has been done in two different ways. The problems in this chapter have been solved by a first approach, in which a system of equations that may be partitioned in sparse matrices and a dense matrix is solved iteratively. In Section 7.5 the method is applied in such a way that the sparsity and banded structure of the FEM matrices is not disturbed. This is done at the expense of performing a number of iterations. The BI equations provide at each step the values that should be imposed at the nodes located at the boundary S. A similar procedure was used in conjunction with the finite difference method in [Sandy and Sage 1971], inspired by the work of [Cermark and Silvester 1968]. A FEM iterative scheme has been developed independently by Aiello and coworkers, who have applied it to static problems, electrical field computation, and skin-effect analysis [Aiello et al. 1992, 1993, 1994a,b, 1995, 1996a, 1997a,b]. In this book the FE-BI iterative method is presented as a tool for the analysis of any kind of open-region field problems, including static, quasi-static, and dynamic problems, like scattering and radiation.

6.2.3 Nonlocal Boundary Conditions

For near fields, there is no local relation between the fields. Enforcing local boundary conditions may thus lead to nonphysical solutions, in particular, for static (e.g., electrostatic) and quasi-static problems. Hence, nonlocal boundary conditions must be constructed to truncate the FEM mesh.

The integral (nonlocal) boundary conditions on S can properly take into account the influence of the region exterior to S, as described in the remainder of this chapter. Such procedures are referred to as exact mesh truncation schemes because they are theoretically exact and applicable to any kind of fields, including static ones. However, these techniques have a deficiency because they partly destroy the sparsity of the finite element matrices.

Consider an open-region problem, sketched in Figure 6.2. The finite elements are generated within the region encapsulated by the closed surface S. Another closed surface, S_o, is defined also. This surface S_o is assumed to lie completely interior to S. It is assumed that the region outside S_o is isotropic and homogeneous, of parameters ε

and μ (in the steady-state regime, losses can be incorporated by taking these parameters to be complex). In other words, all the material inhomogeneities, including conductors, dielectrics, and magnetic materials, are encapsulated by S_o (and thus also encapsulated by S).

The basis for the classical FEM approach is to approximate the field distribution (or a potential or another related quantity) within a closed region. This approximation is then enforced to satisfy the differential form of Maxwell's equations (more precisely, equations derived from them) within the region along with appropriate boundary conditions on the surface bounding the region. In the classical approach, treated in the previous chapters, these boundary conditions are always local in nature because they specify the values of the unknowns or their normal derivatives on the boundary.

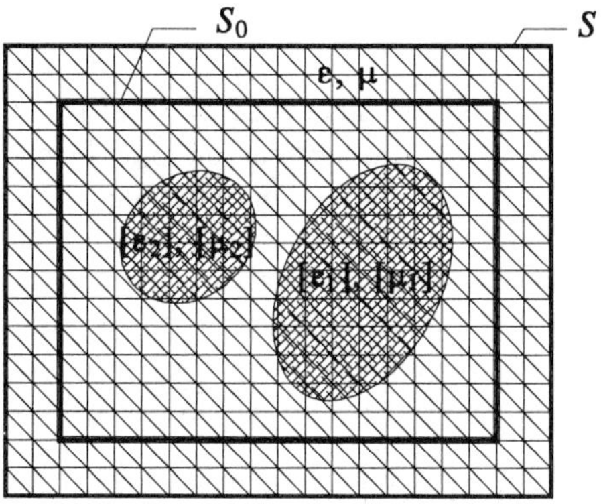

Figure 6.2 Problem of interest with finite elements and encapsulating surfaces.

In the present technique used for open-region problems, the classical FEM approach for the fields within the surface S in Figure 6.2 is followed. However, a special kind of boundary conditions for that surface is used. As explained later in this section, the boundary conditions are based on placing a set of equivalent field sources on the surface S_o and expressing the fields on S in terms of the sources on S_o.

Suppose, for a moment, that the FEM solution for the fields within S has been found. The fields on S are then defined by the degrees of freedom (unknowns) pertinent to the finite elements that are adjacent to S, i.e., by the elements that form a layer of elements touching S. For example, if one is solving for the electrostatic potential (V), the potential on S is defined by the degrees of freedom of nodes that are on S. For the sake of theory presented in this section, let us order the degrees of freedom in the FEM solution so that those defining the fields (potentials) on S come first. Let their total number be N_S, and let the set of these unknowns be denoted by $\{u_S\}$.

By enforcing the equivalence theorems [Harrington 1961], all field sources that are interior to S_o can be replaced with respect to the exterior region by an equivalent (surface) source located on S_o. By the interior field sources, currents and charges should be understood, including those modeling material inhomogeneities (e.g., polarization currents and charges) and excitation (e.g., impressed currents). The fields exterior to S_o can then be expressed in terms of the equivalent surface sources using surface integrals. In contrast to this, more involved volume integrals are generally needed to express the fields outside S_o in terms of the original sources located within S_o.

For dynamic problems, the equivalent sources can be surface electric and/or magnetic currents (Huygens's sources). The densities of these currents are

$$\bar{J}_s = \bar{a}_n \times \bar{H} \tag{6.1}$$

$$\bar{J}_{ms} = -\bar{a}_n \times \bar{E} \tag{6.2}$$

where $\bar{E}$ is the electric field on the outer face of S_o, $\bar{H}$ is the corresponding magnetic field, while $\bar{a}_n$ is the outward unit normal on S_o. Generally, these currents have associated surface charges related by the continuity equations presented later on.

For electrostatic problems, the equivalent sources can be surface charges and surface dipoles (a double-charge layer). The surface charges are related to the normal component of the electric field on S_o and the surface dipoles to the electrostatic potential. Instead of the surface dipoles, the surface magnetic currents from (6.2) can be used. However, the surface electric currents from (6.1) cannot be used because, for static fields, there is no continuity relation between electric currents and charges.

If this replacement is done in the equivalent system shown in Figure 6.3, the fields within the region bounded by S_o become zero (due to the equivalence theorems).

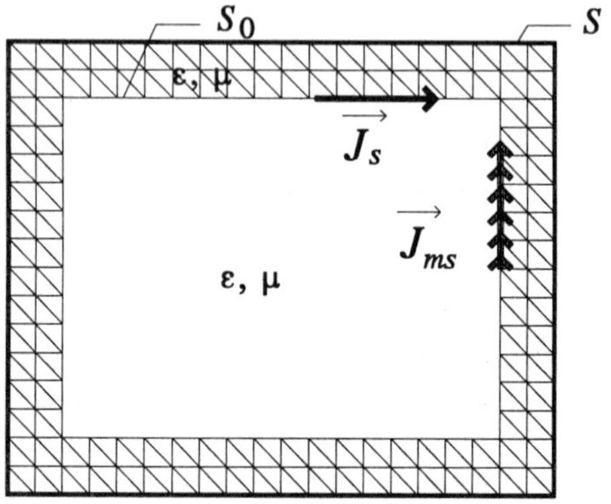

Figure 6.3 Equivalent system to that in Figure 6.2, where the field sources inside S_o are substituted by surface electric and magnetic currents.

Outside S_o, these equivalent sources produce fields (such as potentials) identical to those of the replaced sources (that are within S_o in Figure 6.2). The total fields (potentials) on S in the original system of Figure 6.2 are due to possible sources external to S_o (e.g., in the analysis of scattering) and all the internal sources. The fields of the latter sources are replaced in the system of Figure 6.3 by the fields of the surface sources. Hence, the total fields (potentials) on S are identical in the systems of Figures 6.2 and 6.3.

The electric and magnetic fields produced by the equivalent sources can be expressed in terms of the Lorentz potentials: the electric scalar-potential (V), the magnetic vector-potential ($\bar{A}$), the magnetic scalar-potential (V_m), and the electric vector-potential ($\bar{A}_m$), as

$$\bar{E} = -\bar{\nabla} V - j\omega\bar{A} - \frac{1}{\varepsilon}\bar{\nabla}\times\bar{A}_m \tag{6.3}$$

$$\bar{H} = \frac{1}{\mu}\bar{\nabla}\times\bar{A} - \bar{\nabla} V_m - j\omega\bar{A}_m \tag{6.4}$$

The fields within S_o in Figure 6.3 being zero, the medium within this surface can be replaced by the same medium as outside S_o. Hence, the equivalent sources on S_o can be assumed to produce their fields in an isotropic and homogeneous medium of parameters ε and μ. The potential can now be evaluated as

$$V(\bar{r}) = \frac{1}{\varepsilon}\int_{S_o} \rho_s(\bar{r}')\, G(\bar{r};\bar{r}')\, dS' \tag{6.5}$$

$$\bar{A}(\bar{r}) = \mu \int_{S_o} \bar{J}_s(\bar{r}')\, G(\bar{r};\bar{r}')\, dS' \tag{6.6}$$

$$V_m(\bar{r}) = \frac{1}{\mu}\int_{S_o} \rho_{ms}(\bar{r}')\, G(\bar{r};\bar{r}')\, dS' \tag{6.7}$$

$$\bar{A}_m(\bar{r}) = \varepsilon \int_{S_o} \bar{J}_{ms}(\bar{r}')\, G(\bar{r};\bar{r}')\, dS' \tag{6.8}$$

where $\bar{r}$ is the position-vector of the field point, $\bar{r}\,'$ the position-vector of the source point, and $G(\bar{r}\,;\bar{r}')$ is Green's function for a homogeneous space. The densities of the surface electric charges (ρ_s) and magnetic charges (ρ_{ms}) are related to the surface currents by the continuity equations

$$\rho_s = -\frac{j}{\omega}\,\bar{\nabla}_s \cdot \bar{J}_s \tag{6.9}$$

$$\rho_{ms} = -\frac{j}{\omega}\,\bar{\nabla}_s \cdot \bar{J}_{ms} \tag{6.10}$$

where $\bar{\nabla}_s$ denotes differentiation with respect to surface coordinates on S_o.

For three-dimensional (3D) dynamic problems, Green's function is

$$G(\bar{r};\bar{r}') = \frac{1}{4\pi}\,\frac{\exp(-jk|\bar{r}-\bar{r}'|)}{|\bar{r}-\bar{r}'|} \tag{6.11}$$

where

$$k = \omega\sqrt{\varepsilon\mu} \tag{6.12}$$

is the phase coefficient (wave number). This function naturally takes into account Sommerfeld's radiation condition.

For 3D electrostatic problems, Green's function is reduced to

$$G(\bar{r};\bar{r}') = \frac{1}{4\pi|\bar{r}-\bar{r}'|} \tag{6.13}$$

implying that the reference point for the potential V is at infinity.

For two-dimensional (2D) problems, where the system and all the fields are assumed to be uniform along an axis (e.g., the z-axis), Green's function can be integrated along that axis. As a consequence, in (6.5) to (6.8), the integration over the surface S_o is simplified to an integration along the contour associated with the cross section, i.e., the perimeter of that surface.

For 2D dynamic problems, Green's function reads

$$G(\bar{r};\bar{r}') = -\frac{j}{4}\,H_0^{(2)}(k|\bar{r}-\bar{r}'|) \tag{6.14}$$

where $H_0^{(2)}$ is Hankel's function of the second kind and order zero.

For 2D electrostatic problems, the reference point for V cannot be taken at infinity unless the total charge of the system is zero. So, generally, the reference point must be taken at a finite distance from the surface S (or within it). Green's function for such problems reads

$$G(\bar{r};\bar{r}') = \frac{1}{2\pi} \ln \left(\frac{K}{|\bar{r}-\bar{r}'|} \right) \tag{6.15}$$

where K is an arbitrary (positive) constant, which determines the position of the reference point. If the total charge is zero, one can formally set $K = 1$ (in units used to express $\bar{r}$ and $\bar{r}'$), yielding

$$G(\bar{r};\bar{r}') = -\frac{1}{2\pi} \ln(|\bar{r}-\bar{r}'|) \tag{6.16}$$

In most practical electrostatic problems, the free charges are localized only on conductor surfaces (but not within the volume of the dielectric). In such cases, the total charge of the system can be zero only if the system contains at least two conductors. For example, if there are exactly two conductors, they must be charged by opposite per-unit-length (p.u.l) charges. Equation (6.16) implies for such cases a reference point at infinity may be assumed [Djordjević et al. 1996].

The equivalent sources can be related to the finite elements in the layer that is adjacent to S_o, by expressing the fields on S_o in terms of the degrees of freedom corresponding to these elements and then using (6.1) and (6.2). In principle, this layer can be on either side of S_o (interior or exterior). Let us call the total number of these degrees of freedom N_{So} and let's order them immediately after the first N_S degrees. The set of these degrees of freedom is denoted by $\{u_{So}\}$. If the distance between S and S_o is small (e.g., smaller than two layers in some formulations), some degrees of freedom from $\{u_S\}$ may coincide with some degrees of freedom from $\{u_{So}\}$. The fields (potentials) produced by the equivalent sources on S_o can now be regarded as a function of the coefficients (degrees of freedom) of the column matrix $\{u_{So}\}$. To achieve a higher accuracy, it is possible to involve more than one layer of finite elements adjacent to S_o and S in formulating the boundary conditions, but this does not change the essence of the technique.

Suppose, now, that one is actually solving the problem using FEM. Then, for all the finite elements inside S, one formulates equations in the usual way. An exception is made for equations for the nodes (or edges) that are on S. For these nodes (edges), one imposes equations based on the boundary conditions. According to the uniqueness theorem for solutions of Maxwell's equations [Harrington 1961], one has to impose the condition that the tangential components of $\bar{E}$ or $\bar{H}$ on S are identical in the systems of Figures 6.2 and 6.3. (In electrostatic problems, one may just equate potentials on S.) On one hand, one expresses the fields in terms of the degrees of freedom of $\{u_{So}\}$, through the corresponding integral equations. On the other hand, one expresses the fields in terms of the degrees of freedom of $\{u_S\}$. Finally the corresponding field components are equated. Thus, a system of equations is obtained expressing the degrees of freedom of

$\{u_S\}$ in terms of the elements of $\{u_{So}\}$. The number of such equations (N_b) is related to N_S but may not be equal to N_S depending on the actual approximations used in FEM and the actual boundary conditions used in formulating the problem.

These equations are linear because the medium in the equivalent system is linear even if nonlinearities exist within S_o. Once the discretization of the domain is performed, the FE-BI formulation together with the appropriate boundary conditions is applied. After all the integrations are carried out (often, numerically) and substitutions are made, one obtains a system of linear equations with constant (and known) coefficients that is compatible with the remaining FEM equations. However, unlike the FEM system, which is sparse, the equations that describe the boundary conditions on S are dense (although not completely full). In problems involving scalar unknowns (e.g., electrostatics and TE and TM two-dimensional scattering), which are associated with the nodes of the elements and if S and S_o are separated by at least two layers, each equation of this form has typically a total of N_{So}+1 nonzero coefficients. Here, N_{So} coefficients come from the evaluation of the equivalent sources and the remaining one coefficient from the node where the Dirichlet boundary condition for the scalar function on S is imposed. So, the overall sparsity of the system is somewhat spoiled, but sparse-matrix solvers may still be applicable. The banded structure of the FEM systems of equations and matrices may also be spoiled. However, there is always a possibility of using sparse iterative solvers. The storage of the elements depends on the nonzero coefficients and not on the banded structure of the matrix. The equations based on the boundary condition on S are similar to those arising in the solution of electromagnetic field problems using integral equations involving surface sources. Hence, the preceding procedure can be considered as a hybrid FE-BI method.

So far, it has been assumed that the surfaces S and S_o are distinct, but they may also coincide. In that case, some of the degrees of freedom of $\{u_S\}$ and $\{u_{So}\}$ (or even all of them, depending on the formulation) coincide. However, this choice may lead to more pronounced numerical problems (compared with cases when S and S_o are distinct) for several reasons. First, the equivalent sources are evaluated only approximately, and the corresponding errors may yield more erratic fields near the sources (in particular, at interfaces between adjacent finite elements) than at remote points. Second, the boundary conditions on S involve an integration of the equivalent sources multiplied by the appropriate Green's functions. These functions have singularities when the source and field points coincide (i.e., when $\bar{r} = \bar{r}\,'$). Hence, in the general case, the resulting integrals are harder to evaluate numerically when S and S_o coincide than when these surfaces are separated by several layers of finite elements. The roles of S and S_o can also be interchanged; i.e., the equivalent sources can be calculated on S and the boundary conditions imposed on S_o. In this case, one has the so-called extended boundary conditions.

The surfaces S and S_o should be selected to minimize the number of unknowns, in particular those involved in the integral boundary condition. Hence, S_o should tightly encompass all the discontinuities of the system and have as small an area as possible. S should closely follow S_o at a distance of several layers. In most examples presented in this chapter, exactly two layers have been used.

6.2.4 Comments on Solution of Linear Equations

If the medium within S_o is linear, the hybrid technique described in Section 6.2.3 results in a system of linear equations. This system has to be solved numerically. Assume that the total number of degrees of freedom in the FEM solution is N_d. Also, consider that $N_b = N_S$ and place the N_b equations based on the boundary conditions first. The remaining N_d - N_b equations are classical FEM equations. (This results in a determined system of equations. However, for an overdetermined system, one may take more than N_d - N_b FEM equations.) If S and S_o are separated so that the elements of $\{u_S\}$ and $\{u_{So}\}$ are distinct, the equations based on the boundary conditions have the common form

$$u_i = b_i(\{U_2\}) \ , \quad i = 1, \ldots, N_S \tag{6.17}$$

where $u_i \in \{u_S\}$, $i = 1, \ldots, N_S$; $\{U_2\}$ is the column matrix (vector) of elements of $\{u_{So}\}$; and b_i denotes a linear function of the coefficients of $\{U_2\}$. Define the column matrix $\{U_1\}$ of the elements of $\{u_S\}$. Then the set of equations of the form (6.17) can be put in the following matrix form if there are no source terms

$$\{U_1\} + [B]\{U_2\} = \{0\} \tag{6.18}$$

For the general case, the entire system of equations including source terms can be obtained in the following matrix form:

$$\begin{bmatrix} [1] & [B] & [0] \\ [F_{21}] & [F_{22}] & [F_{23}] \end{bmatrix} \begin{Bmatrix} \{U_1\} \\ \{U_2\} \\ \{U_3\} \end{Bmatrix} = \begin{Bmatrix} \{E_1\} \\ \{E_2\} \end{Bmatrix} \tag{6.19}$$

where [1] is a unit submatrix, [0] is a zero submatrix; $[F_{21}]$, $[F_{22}]$, and $[F_{23}]$ are highly sparse submatrices that come from FEM; $\{U_3\}$ is a column matrix of those degrees of freedom not affected by the boundary conditions; while $\{E_1\}$ and $\{E_2\}$ are column matrices that describe the excitation to the system.

The system (6.19) can be solved in a variety of ways. Generally, direct or iterative techniques can be applied or even their combinations. Direct procedures, like Gaussian elimination and lower-upper (LU) decomposition can be highly efficient if care is taken about the sparsity of the system matrix.

Among a variety of iterative techniques, the classical relaxation formulation may be used. The linear equations constituting (6.19) can be modified (by proper divisions and renumbering) so that all elements on the main diagonal of the system matrix equal 1. The resulting system of linear equation can be represented in matrix form as

$$[S]\ \{U\} = \{E\} \tag{6.20}$$

which can further be rearranged as

$$\{U\} = \{E\} - ([S] - [1])\ \{U\} \tag{6.21}$$

where [1] is a unit matrix. In the relaxation process, the elements of $\{U\}$ on the right-hand side are given some initial values and the new values are evaluated on the left-hand side. Thereby, the column matrix $\{U\}$ can be updated for each element or after computing the whole matrix. This fully corresponds to the classical, simplest, iterative techniques (Gauss-Seidel and Jacobi's, respectively).

To speed up the convergence, an over-relaxation process can be used in the following way. The new value of an element of $\{U\}$ is computed as described previously. The old value is subtracted from the new value and the resulting difference is multiplied by the over-relaxation factor and added to the old value. This becomes the final new value. If the over-relaxation factor is 1, one has a pure relaxation. If this factor is greater than 1 (usually between 1 to 2), one has an over-relaxation. If this factor is smaller than 1, one effectively has an under-relaxation. For the system (6.21) obtained by combining FEM and BI equations, it was found that under-relaxation is desirable for the BI equations (with a factor of typically between 0.1 and 1) and over-relaxation for the FEM equations (of a factor of typically between 1 and 2).

A different approach is to split the system (6.19) into the following two matrix equations:

$$\{U_1\} + [B]\ \{U_2\} = \{E_1\} \tag{6.22}$$

$$[F_{21}]\ \{U_1\} + [F_{22}]\ \{U_2\} + [F_{23}]\ \{U_3\} = \{E_2\} \tag{6.23}$$

which can be rearranged as

$$[F_{22}]\ \{U_2\} + [F_{23}]\ \{U_3\} = \{E_2\} - [F_{21}]\ \{U_1\} \tag{6.24}$$

$$\{U_1\} = \{E_1\} - [B]\ \{U_2\} \tag{6.25}$$

Note that (6.24) can be regarded as a system of linear equations in the elements of $\{U_2\}$

and $\{U_3\}$. The matrix of this system is sparse, coming from FEM equations. Then, one starts with an arbitrary $\{U_1\}$ in (6.24) and then solves this equation by a sparse-matrix solver to get $\{U_2\}$ and $\{U_3\}$. Then the new values of the elements of $\{U_1\}$ are evaluated from (6.25). These new values are used in (6.24) and a new computation takes place. This technique is the basis of the approach described in Section 7.5 for the solution of open-region 3D problems.

6.2.5 Applications

In the remainder of this chapter, examples are presented of the application of the previously described hybrid technique to 2D electrostatic problems (in Section 6.3) and TM and TE scattering (in Sections 6.4 and 6.5, respectively). In all cases, the results are compared with MOM solutions, demonstrating a reasonable agreement. In some cases of objects with sharp edges, the agreement is somewhat worse, calling out for the adaptive FEM approach described in previous chapters. In all the examples, relatively simple structures are analyzed, consisting of only one perfectly conducting cylinder. The cylinder surface is taken to coincide with S_o. This reduces the demand to only one kind of equivalent sources on S_o (surface charges for electrostatic problems; and surface electric currents, possibly with associated charges, for the TM and TE scattering). The surface magnetic currents are zero in all these cases, since the tangential component of the electric field on S_o is zero. Also, the interior of the space bounded by S_o in Figure 6.2 need not be discretized using FEM, since there are no fields in that region. However, in more complicated problems, S_o need not coincide with the surface of a perfect conductor and two kinds of equivalent surface sources would be required.

6.3 TWO-DIMENSIONAL ELECTROSTATIC PROBLEMS

6.3.1 Introduction

In this section, the FEM is used in conjunction with the integral boundary conditions to solve Laplace's equation for two-dimensional electrostatic problems involving open regions. The analyzed system consists of one infinitely long cylinder of an arbitrary cross section situated in a vacuum. A Cartesian coordinate system is associated with the cylinder so that the cylinder generatrix is parallel to the z-axis, so that the x and y represent transverse coordinates. Referring to Figures 6.2 and 6.3, the conductor surface is denoted by S_o.

In the open region surrounding the cylinder, the electrostatic potential, V, satisfies Laplace's equation:

$$\nabla^2 V \equiv \frac{\partial^2 V(x,y)}{\partial x^2} + \frac{\partial^2 V(x,y)}{\partial y^2} = 0 \tag{6.26}$$

subject to the boundary conditions on the conductor surface. On this surface, the potential is constant and equal to a prescribed value, V_o (Dirichlet condition).

Since the conductor is infinitely long, the reference point for the potential cannot be taken at infinity. In all the examples presented in this section, the reference point is taken to be far away from the conductor at a (finite) distance from its centroid that is several orders of magnitude larger than the cross-sectional dimensions of the conductor.

6.3.2 Formulation

Nonoverlapping triangular finite elements are used in the solution of (6.26), because they can easily and accurately fit arbitrary boundaries. The electric field within the conductor is zero, so the potential is constant inside and on its surface. There is no need to place finite elements inside the conductor, but only outside. However, when one deals with an open-region problem, the discretization has to be terminated at some finite distance from the conductor. The terminating surface is denoted by S. In most examples presented in this section, only two layers of finite elements around the conductor are taken in order to keep the size of the computational domain small.

According to FEM, to find the potential distribution, $V(x,y)$, for the two-dimensional solution region, the region is divided into triangular finite elements. Seek an approximation for the potential, $V^e(x,y)$, within the eth element; and then interrelate that potential distribution to the surrounding elements so that the potential is continuous across interelement boundaries [Roy et al. 1995]. Hence, the solution for the whole discretized region is given by

$$V(x,y) \approx \sum_{e=1}^{N_e} V^e(x,y) \tag{6.27}$$

where N_e is the number of finite elements into which the solution region is divided. The most simple form of approximation for V^e within an element is a linear polynomial approximation, namely,

$$V^e(x,y) = a_e + b_e x + c_e y \tag{6.28}$$

for the eth triangular element. The constants a_e, b_e, and c_e are unknowns to be determined. In other words, linear Lagrange elements have been used (see Section 2.5.2.1). In (6.27), the potential V^e is generally nonzero within the eth element, but it is assumed to be zero outside the element.

The eth element has its vertices (nodes) located at (x_{e1},y_{e1}), (x_{e2},y_{e2}), and (x_{e3},y_{e3}). The potentials (which are unknowns) at the nodes are V_1^e, V_2^e , and V_3^e , respectively. They can be expressed as

$$\begin{Bmatrix} V_1^e \\ V_2^e \\ V_3^e \end{Bmatrix} = \begin{bmatrix} 1 & x_{e1} & y_{e1} \\ 1 & x_{e2} & y_{e2} \\ 1 & x_{e3} & y_{e3} \end{bmatrix} \begin{Bmatrix} a_e \\ b_e \\ c_e \end{Bmatrix} \tag{6.29}$$

Expressing a_e, b_e, and c_e in terms of V_1^e, V_2^e, and V_3^e, the potential at any point within the eth element is given by

$$V^e(x,y) = \sum_{i=1}^{3} N_i^e(x,y)\ V_i^e \tag{6.30}$$

where

$$N_1^e = \frac{1}{2A_e}\left(x_{e2}y_{e3} - x_{e3}y_{e2} + (y_{e2} - y_{e3})x + (x_{e3} - x_{e2})y\right) \tag{6.31}$$

$$N_2^e = \frac{1}{2A_e}\left(x_{e3}y_{e1} - x_{e1}y_{e3} + (y_{e3} - y_{e1})x + (x_{e1} - x_{e3})y\right) \tag{6.32}$$

$$N_3^e = \frac{1}{2A_e}\left(x_{e1}y_{e2} - x_{e2}y_{e1} + (y_{e1} - y_{e2})x + (x_{e2} - x_{e1})y\right) \tag{6.33}$$

and A_e is the area of the eth element, which is given by

$$2A_e = (x_{e1}y_{e2} - x_{e2}y_{e1}) + (x_{e3}y_{e1} - x_{e1}y_{e3}) + (x_{e2}y_{e3} - x_{e3}y_{e2}) \tag{6.34}$$

which is the determinant of the Jacobian matrix of the mapping from the real element to the parent element. The area A_e is positive if the nodes of the eth element are numbered counterclockwise (as observed from the top of the z-axis). The element shape functions N_i^e are linear interpolation functions and have the following properties (see Section 2.5.1.2):

$$N_i^e(x_j,y_j) = \begin{cases} 1, & i=j \\ 0, & i \neq j \end{cases} \tag{6.35}$$

and

$$\sum_{i=1}^{3} N_i^e(x,y) = 1 \tag{6.36}$$

The classical and the weak formulation of the problem are formally identical to that discussed in Section 3.2.3.1. Because Neumann boundary conditions are homogeneous, a variational principle given by (3.50c) may be used. Thus, a functional W^e, which is the p.u.l. energy associated with the eth element, is given by

$$W^e = \frac{1}{2} \int_{S^e} \varepsilon_0 \, |\bar{\nabla} V^e|^2 \, \mathrm{d}S \tag{6.37}$$

where S^e is the surface of the eth element. From (6.30),

$$\bar{\nabla} V^e(x,y) = \sum_{i=1}^{3} V_i^e \, \bar{\nabla}_t N_i^e \tag{6.38}$$

Substituting (6.38) into (6.37) gives

$$W^e = \frac{1}{2} \sum_{i=1}^{3}\sum_{j=1}^{3} \varepsilon_o V_i^e k_{ij}^e V_j^e \tag{6.39}$$

where

$$k_{ij}^e = \int_{S^e} \bar{\nabla} N_i^e \cdot \bar{\nabla} N_j^e \, \mathrm{d}S \tag{6.40}$$

Equation (6.39) can be written in matrix form as

$$W^e = \frac{1}{2} \varepsilon_o \{V^e\}^T [k^e] \{V^e\} \tag{6.41}$$

where the superscript ^{T} denotes the transpose of a matrix,

$$\{V^e\} = \begin{Bmatrix} V_1^e \\ V_2^e \\ V_3^e \end{Bmatrix} \tag{6.42}$$

and

$$[k^e] = \begin{bmatrix} k_{11}^e & k_{12}^e & k_{13}^e \\ k_{21}^e & k_{22}^e & k_{23}^e \\ k_{31}^e & k_{32}^e & k_{33}^e \end{bmatrix} \tag{6.43}$$

The coefficients of the matrix $[k^e]$ may be regarded as the coupling coefficients between nodes i and j. Using analytical integration these coefficients are given by

$$k_{11}^e = \frac{1}{4A_e}\left((y_{e2} - y_{e3})^2 + (x_{e3} - x_{e2})^2\right) \tag{6.44}$$

$$k_{12}^e = k_{21}^e = \frac{1}{4A_e}\left((y_{e2} - y_{e3})(y_{e3} - y_{e1}) + (x_{e3} - x_{e2})(x_{e1} - x_{e3})\right) \tag{6.45}$$

$$k_{13}^e = k_{31}^e = \frac{1}{4A_e}\left((y_{e2} - y_{e3})(y_{e1} - y_{e2}) + (x_{e3} - x_{e2})(x_{e2} - x_{e1})\right) \tag{6.46}$$

$$k_{22}^e = \frac{1}{4A_e}\left((y_{e3} - y_{e1})^2 + (x_{e1} - x_{e3})^2\right) \tag{6.47}$$

$$k_{23}^e = k_{32}^e = \frac{1}{4A_e}\left((y_{e3} - y_{e1})(y_{e1} - y_{e2}) + (x_{e1} - x_{e3})(x_{e2} - x_{e1})\right) \tag{6.48}$$

$$k_{33}^e = \frac{1}{4A_e}\left((y_{e1} - y_{e2})^2 + (x_{e2} - x_{e1})^2\right) \tag{6.49}$$

Assembling all such elements in the solution region, the total energy of the assemblage can be expressed as

$$W = \sum_{e=1}^{N_e} W^e = \frac{1}{2}\,\varepsilon_o\{V\}^T[K]\{V\} \tag{6.50}$$

where

$$\{V\} = \begin{Bmatrix} V_1 \\ V_2 \\ \vdots \\ V_{N_d} \end{Bmatrix} \tag{6.51}$$

N_d is the total number of nodes, whereas N_e denotes the number of finite elements in the solution region. The global coefficient matrix, $[K]$, is an assemblage of all the individual coefficient matrices $[k^e]$.

There are three types of nodes in the finite element mesh. The first set of nodes consists of those on the outer boundary of the finite element mesh, i.e., the nodes residing on the surface S. As in Sections 6.2.3 and 6.2.4, these nodes are ordered first. They are labeled "*ob*." For these nodes, equations based on the integral boundary conditions are formed, as explained later. The second set of nodes are those on the inner boundary of the finite element mesh, i.e., the nodes residing on the conductor surface (S_o). For these nodes, equations based on the Dirichlet condition ($V = V_o$) are formed, so that the potentials at these nodes are fixed. These nodes are ordered second and labeled "*ib*." Collectively, these two sets of nodes are referred to as boundary (or fixed) nodes and labeled "*b*." The third set of nodes are all the remaining nodes of the finite element mesh. They are referred to as free (or internal) nodes, labeled "*f*," and they are ordered last. One can now rewrite (6.50) by expressing the matrices by the correspon-ding submatrices as

$$W = \frac{1}{2}\,\varepsilon_o \begin{Bmatrix} \{V_b\} \\ \{V_f\} \end{Bmatrix}^T \begin{bmatrix} [K_{bb}] & [K_{bf}] \\ [K_{fb}] & [K_{ff}] \end{bmatrix} \begin{Bmatrix} \{V_b\} \\ \{V_f\} \end{Bmatrix} \tag{6.52}$$

where subscripts "*b*" and "*f*" refer to the boundary and free nodes, respectively. As explained in previous chapters, Laplace's equation is satisfied when the total energy in the solution region is minimal. Hence, one requires that the partial derivatives of W with respect to the potential of each free nodal potential be zero, i.e.,

$$\frac{\partial W}{\partial V_f} = 0 \tag{6.53}$$

In general, this yields

$$\left[[K_{fb}]\ [K_{ff}]\right] \begin{Bmatrix} \{V_b\} \\ \{V_f\} \end{Bmatrix} = \{0\} \tag{6.54}$$

This equation can be written as

$$[K_{ff}]\{V_f\} = -[K_{fb}]\{V_b\} \tag{6.55}$$

Formally, equations expressing the Dirichlet boundary condition on S_o should be appended to the system of equations in (6.54). These added equations contain the excitation to the system. In (6.19), the only nonzero terms are those in $\{E_2\}$ corresponding to the Dirichlet condition.

The basic attempt in the mesh termination boundary conditions is to simulate a homogeneous space surrounding the region covered by the mesh. Suppose one knows the equivalent field sources on S_o. In this case, these sources are the free charges located on the conductor surface. (In the general case, the equivalent sources in electrostatic problems must contain an additional term: surface dipoles or magnetic currents.) These sources produce fields in a free space. Based on Green's function (6.15), their potential at the terminating surface S can be evaluated using the integral

$$V = \frac{1}{\varepsilon_0} \sum_i \int_{\Delta l_i} \frac{\rho_{si}}{2\pi} \ln\left(\frac{K}{|\bar{r}-\bar{r}'|}\right) dl' \tag{6.56}$$

where ρ_{si} is the surface-charge density over the ith segment of the conductor surface (mathematically represented by its cross-sectional dimension, Δl_i). An example is the segment between points labeled "a" and "b" in Figure 6.4. Referring to Figure 6.3, (6.56) relates the potential on S to the equivalent sources (surface charges) on S_o (coinciding with the conductor surface). The union of all elements Δl_i represents the cross-sectional contour of the conductor.

Note that (6.56) can be used to evaluate the potential at any point, not only at the terminating surface.

When solving an electrostatic system, it is usual to *a priori* fix the position of the reference point. In that case, the constant K should be considered as an additional unknown, which can be evaluated from the condition that the potential of the reference point is zero. Mathematically, this condition is obtained by setting $\bar{r}$ in (6.56) so as to correspond to the location of the reference point. This equation is appended to the system (6.54) in addition to equations formulated from the boundary conditions on the mesh terminating surface (explained later in this section). Note that the potential is zero not only at the specified reference point, but also on an equipotential surface that contains the reference point.

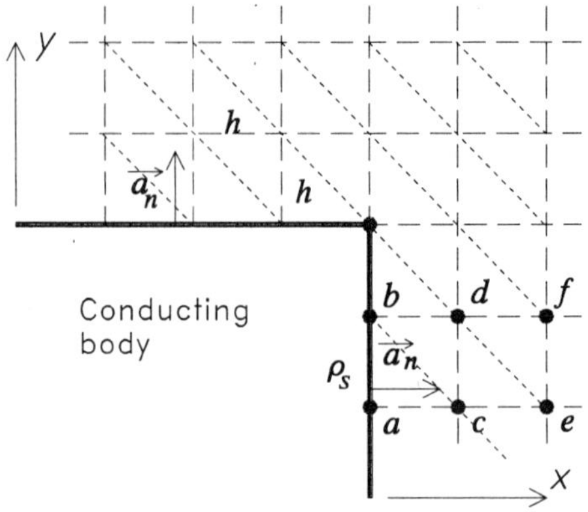

Figure 6.4 Calculation of the fields and charges for a subsection.

The cross section of an equipotential surface in the *xy*-plane is a contour, which is referred to as an equipotential line. One such line is the contour of the conductor cross section. Close to the conductor surface, equipotential lines resemble that contour. Far away from the cylinder, equipotential lines resemble circles, centered at the centroid of the cylinder cross section. Let us assume that the centroid is at the coordinate origin or very close to it while the reference point is far away. To specify the position of the reference point, it is sufficient to give its distance from the cylinder centroid (i.e., $|\bar{r}|$), while the direction in which this point is located is irrelevant (since the equipotential lines are circles). In this case, from (6.56), one gets $K = |\bar{r}|$ because $|\bar{r}| \gg |\bar{r}'|$. For example, if the cross-sectional dimensions of the cylinder are smaller than 1 m (one meter), setting $K = 1000$ m, fixes the reference point at a distance of 1 km from the cylinder centroid.

For a circular cylinder, the interpretation of K is even simpler. In this case, the equipotential lines are circles at any distance from the cylinder surface. Hence, an arbitrary K greater than the cylinder radius defines the reference point at a distance K from the cylinder axis.

The surface-charge density on the conductor surface is expressed in terms of the potential in the following way. The normal component of the electric field on the surface of the conducting body and the local surface-charge density are related by the boundary conditions as

$$\bar{a}_n \cdot \bar{E} = \frac{\rho_s}{\varepsilon_0} \tag{6.57}$$

where $\bar{E}$ is the electric field vector and $\bar{a}_n$ the unit normal on the conductor surface. The normal component of the electric field is related to the normal derivative of the potential as

$$\bar{a}_n \cdot \bar{E} \equiv E_n = -\frac{dV}{dn} \tag{6.58}$$

The field component at a point can be evaluated from this derivative, which can be computed from the degrees of freedom (the potential values) of the finite element. Consider, for simplicity, the case of Figure 6.4. (A more general case is mentioned in Section 6.3.3.3.) For example, the normal component of the electric field at the node labeled "a" in Figure 6.4 can be evaluated using the expression

$$E_{na} = E_x \approx -\sum_{i=1}^{3} \frac{\partial N^e}{\partial x} V_i^e \tag{6.59}$$

where the subscripts i = 1, 2, 3 correspond to the nodes labeled "a," "c," "b," respectively. Please note that because the basis functions utilized are linear, their derivatives are constant over the corresponding element. Now, the node labeled "a" is shared by other elements. Then expression (6.59) may be computed for all elements sharing node "a." In general, the result will be different from one element to the other. Thus, to have a smooth field, the value assigned to node "a" is computed as

$$E_{na} \approx \frac{1}{N_a} \sum_{j=1}^{N_a} E_{xj} \tag{6.60}$$

where N_a stands for the number of elements sharing node "a." Similar expressions can be written for other points on the conductor surface and, in particular, for the normal component of the electric field at the point labeled "b" (E_{nb}) in Figure 6.4. Considering the segment between points "a" and "b," the simplest approximation is to assume the surface charge to be uniform. The normal component along this segment can be evaluated as the average of E_{na} and E_{nb}. Finally, let us note that $h = \Delta l_i$ for all the segments on the conductor. Now, (6.56), (6.57), and (6.59) or (6.60) completely define the potential at an arbitrary point on the outer boundary in terms of the nodal potentials. Hence, the equations for the boundary integral extended to the terminating surface S can be formulated with ease by expressing the charge density in (6.56) in terms of the potentials at nodes adjacent to S_o. The integration in (6.56) can be carried out analytically, as explained in [Djordjević et al. 1996]. If two layers of finite elements are placed on top of the conducting surface, as shown in Figures 6.3 and 6.4, the nodes on the terminating surface are not involved in the boundary conditions on S_o.

The resulting system of linear equation that follows from FEM and boundary-condition equations is solved iteratively. In a relaxation process, the system is arranged according to (6.21). Thereby, the potentials of nodes on the inside boundary are fixed to a given value (10 V (Volt) in all the examples of Section 6.3.3) and are excluded from iterations. The iterations start by assigning a potential of 0 V to all the remaining nodes. In each step, a new set of potential values on the terminating surface replaces the old values in the column matrix of potentials of the nodes on the terminating surface ($\{V_{ob}\}$) and the free nodes ($\{V_f\}$). If the potentials in the kth iteration are denoted by the superscript k, then the potentials on the terminating surface in the iteration (k+1) are given by

$$\{V_{ob}^{k+1}\} \leftarrow \{V_{ob}^{k}\} + \alpha\left(\{V_{ob}^{k+1}\} - \{V_{ob}^{k}\}\right) \tag{6.61}$$

An under-relaxation factor, α, is chosen between 0.1 and 1.0 to get a convergence for the nodes on the terminating surface. The potentials of the free nodes are updated at each iteration by

$$\{V_f^{k+1}\} \leftarrow \{V_f^{k}\} + \beta\left(\{V_f^{k+1}\} - \{V_f^{k}\}\right) \tag{6.62}$$

For the free nodes, an over-relaxation factor, β, is chosen between 1.0 and 1.9 for a good convergence.

The progress of iterations is monitored by the error in potentials. This error is defined as the largest absolute difference of nodal potentials in two successive iterations. As the iteration count increases in a convergent procedure, the error should reduce in magnitude. The process stops when the error becomes less than a user-specified value (10^{-6} for all examples in Section 6.3.3). This procedure is simple to use and the convergence is achieved for all the structures to which this method has been applied.

Direct solution techniques using sparse-matrix solvers have been found to be more efficient than the relaxation procedure for all the examples presented in Section 6.3.3. As explained in Section 6.2.4, the solution can alternatively be carried out by combining an iterative and a direct technique. The process starts as for the relaxation procedure, by assigning fixed potentials to the nodes on the inner boundary ($\{V_{ib}\}$) and 0 V to the nodes on the outer boundary ($\{V_{ob}\}$). Equation (6.55) is then solved (by a sparse-matrix algorithm, e.g., based on Gaussian elimination) for the potentials of the free nodes ($\{V_f\}$). With the potentials known everywhere in the computational domain, the charge densities on the segments of the conductor surface may be evaluated. Hence, using (6.56) one computes the potentials at nodes located on the outer boundary, thus updating the elements of $\{V_{ob}\}$. This terminates the iteration, and the iterative process loops back.

In this section, a relatively simple electrostatic problem of a conductor in a homogeneous medium is treated. For this case, if the terminating surface S collapses onto the conductor surface S_o, a boundary element integral equation approach is obtained. Here, the quantities to be solved are only the surface charges on the conductor.

6.3.3 Numerical Results

6.3.3.1 *Circular Cylinder*

As the first example, let us consider a charged circular cylinder of radius 0.1 m. The conductor potential is assumed to be 10 V with respect to the reference point at 1 km from the cylinder axis. A two-layer triangular finite element mesh for the problem is shown in Figure 6.5. There are 40 boundary nodes: 20 of them are on the inside boundary (conductor surface, S_o) and the other 20 nodes are on the outside boundary (terminating surface, S). The nodes on the terminating surface are at a distance of 0.2 m from the axis (0.1 m from the conductor surface). The mesh also has a total of 20 free nodes, which are located in one layer, at a distance of 0.15 m from the cylinder axis (0.05 m from the conductor surface).

The error in successive iterations of the relaxation process is plotted in Figure 6.6. As it can be seen, the error reduces rapidly after 15 to 20 iterations. The potential at the outside boundary is found to be 6.81 V after the procedure converged. This value compares favorably with 6.96 V obtained by the MOM (with 20 pulses used to approximate the conductor surface charges and the point matching procedure) and 6.99 V obtained analytically. The total p.u.l charge of the cylinder obtained from the FEM solution was found to be 235 pC/m, the MOM solution provided 240 pC/m, whereas the analytical solution is 242 pC/m.

Keeping the terminating surface at a radius of 0.2 m, now one more layer of finite elements is introduced in between the object surface and the terminating surface. The potential at the outside boundary is found to be 6.89 V after the procedure converged. Similarly, after introducing another layer (four layers all together) in between, 6.92 V for the potential at the terminating boundary is obtained. The p.u.l. charge density of the cylinder for three layers of finite elements was found to be 238 pC/m, whereas the four-layer scheme gave 239 pC/m. As the number of layers is increased while keeping the outer boundary fixed, the results improve and compare very well with the MOM value.

For the next set of examples, compared with the mesh of Figure 6.5, other layers of finite elements are added, maintaining the same distance between the layers of nodes. Hence, the outer boundary is moved out by 0.05 m per added layer. For three layers of finite elements, when the outside boundary is at 0.25 m from the axis, the potential at 0.2 m from the axis was found to be 6.84 V. Similarly, with four layers of finite elements, with the boundary placed at 0.3 m, the potential at 0.2 m was found be 6.86 V. The improvement of the results is not significant as for the previous case, but still the values are getting closer to the MOM result.

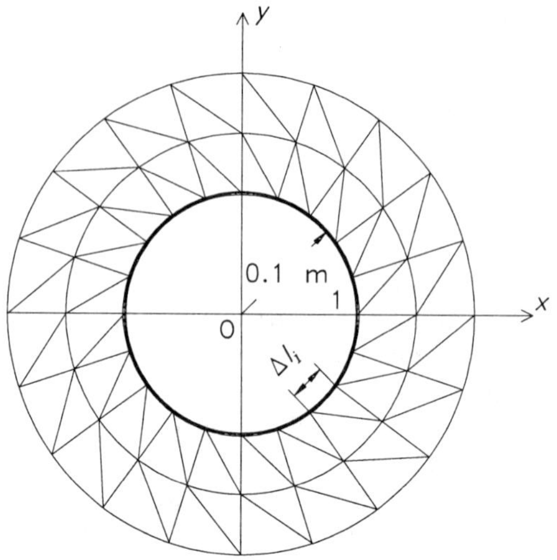

Figure 6.5 Finite element mesh for a circular cylinder.

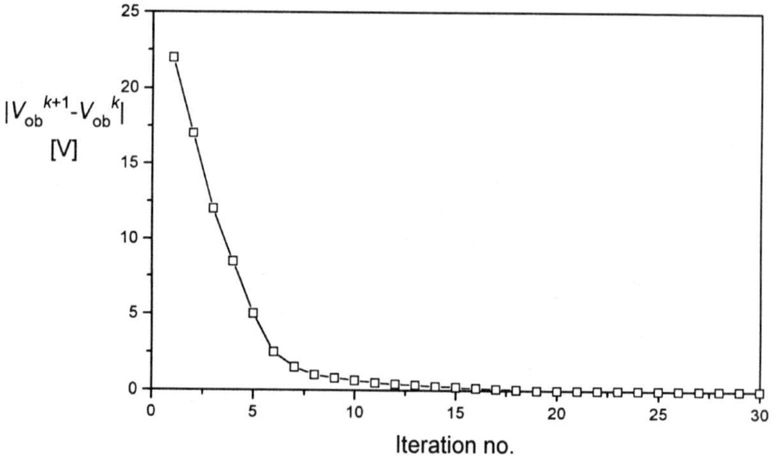

Figure 6.6 Maximum error for the potentials at nodes on the terminating surface, in Volts, versus the iteration count.

6.3.3.2 *Square Cylinder*

Consider a charged square cylinder of dimensions 0.1 m × 0.1 m, which is at a potential of 10 V with respect to the reference point 1 km away. The finite element mesh scheme is shown in Figure 6.7. As in the previous example, two layers of triangular elements are used to form the computational domain. There are 32 free nodes, 24 nodes on the conductor surface, and 40 nodes on the outer boundary. In this case, the terminating surface is 0.033 m away from the conducting surface. Now, because of the singularity of the electric field at the corners, 63 iterations were required to reduce the maximum error on the terminating surface to 10^{-6}.

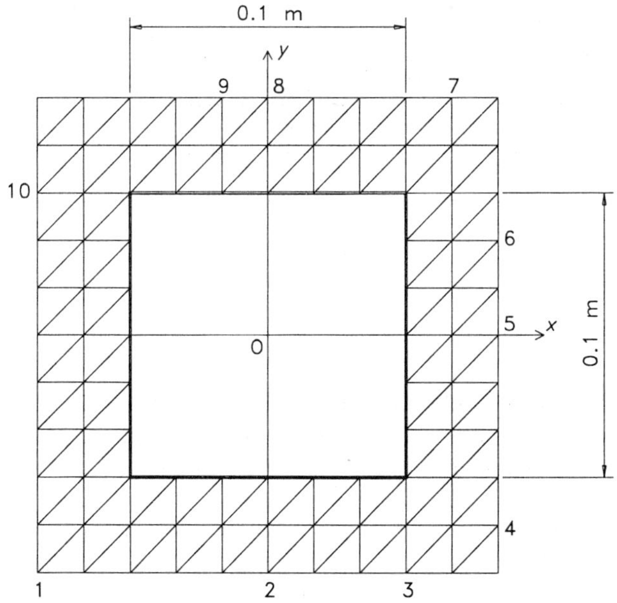

Figure 6.7 Finite element mesh for a square cylinder.

The computed potentials at various locations resulting from the present solution, along with MOM values, are given in Table 6.1. For MOM, 24 subsections were used with pulse basis and point matching testing procedure. The maximum error between the two sets of results is about 1%.

Table 6.1
Computed potentials for the square cylinder of Figure 6.7

Point	x (m)	y (m)	MOM (Volt)	Present method (Volt)
1	-0.0833	-0.0833	9.194	9.295
2	0.0000	-0.8333	9.598	9.606
3	0.0500	-0.8333	9.391	9.494
4	0.0833	-0.0666	9.291	9.401
5	0.0833	0.0000	9.518	9.606
6	0.0833	0.0333	9.462	9.560
7	0.0666	0.0833	9.297	9.401
8	0.0000	0.0833	9.518	9.606
9	-0.0333	0.0833	9.462	9.560
10	-0.0833	0.0500	9.390	9.949

6.3.3.3 *Semicircular Cylinder*

Consider a charged semicircular cylinder of radius 0.06 m, at a potential of 10 V with respect to the reference point 1 km away. The finite element grid is shown in Figure 6.8. The terminating surface of the mesh is not conformal to the body, just to illustrate it as a possibility.

There are 26 free nodes. 26 nodes are on the outer boundary, and 16 nodes on the conductor surface. The same iterative procedure as in the other cases was used here. This structure is a good example to show how the computation of the normal component of the field should be done in a general case. Here, both the x component and the y component of the electric field should be computed for each node at the surface of the structure, following expressions similar to (6.59) and (6.60); i.e., a smooth global expression for both components of the field should be obtained. Then the expression of the normal component at the node of interest should be taken into account and the scalar product between both vectors performed. This is easily done by performing all computations at the parent element (see Appendix C.1). The error criterion was satisfied after 76 iterations. Again, there is a nonsmooth behavior of the electric field. The computed potentials at various locations, along with MOM values, are presented in Table 6.2. For MOM calculations, 16 subsections were used to model the surface of the conductor. The maximum error is found to be around 2%.

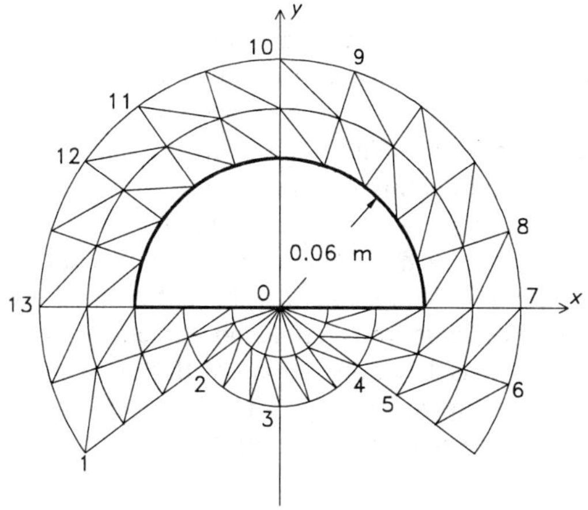

Figure 6.8 Finite element mesh for a semicircular cylinder.

Table 6.2
Computed potentials for the semicircular cylinder of Figure 6.8

Point	x (m)	y (m)	MOM (Volt)	Hybrid method (Volt)
1	-0.0809	-0.0587	9.075	9.127
2	-0.0323	-0.0235	9.664	9.776
3	0.0	-0.0400	9.506	9.671
4	0.0323	-0.0235	9.664	9.876
5	0.0485	-0.0352	9.458	9.596
6	0.0951	-0.0309	9.162	9.366
7	0.1	0.0	9.256	9.307
8	0.0951	0.0309	9.328	9.326
9	0.0309	0.0951	9.398	9.421
10	0.0	0.1	9.401	9.418
11	-0.0587	0.0809	9.388	9.394
12	-0.0809	0.0587	9.368	9.368
13	-0.1	0.0	9.256	9.263

6.3.3.4 *Bow-Tie Cylinder*

To illustrate the application of the method to more complicated geometries, a charged bow-tie cylinder is considered (Figure 6.9) of inner radius r_1 = 0.0667 m and outer radius r_2 = 0.2 m, at a potential of 10 V with respect to the reference point 1 km away. The wings subtend an angle of 72°. The finite element mesh is conformal to the body. There are 28 nodes on the terminating surface. The nearest nodes are located on an arc of radius 0.2 m, and the farthest on an arc of radius 0.333 m. There are 28 free nodes and 28 nodes on the conductor surface. The normal component of the electric field was evaluated as in Section 6.3.3.3. With standard values of α and β, the iterative procedure converged after 82 steps. Again, there is a nonsmooth behavior of the electric field.The computed potentials at various locations, along with MOM values, are presented in Table 6.3. For MOM calculations, 28 subsections to model the conductor were used. The maximum difference between the two techniques is in the second decimal place.

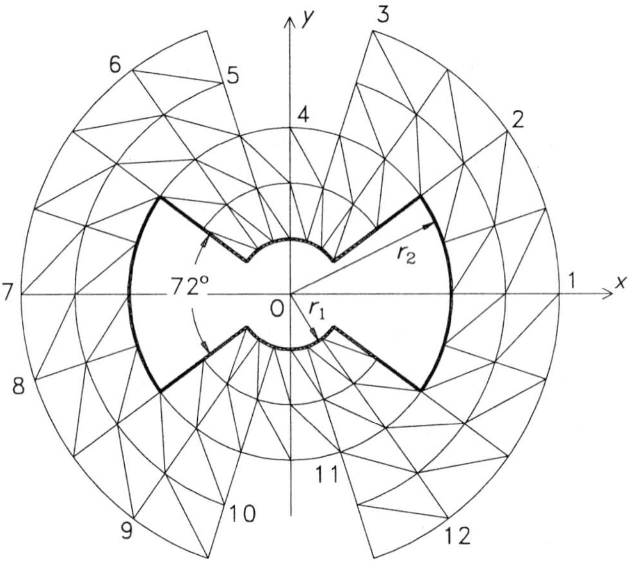

Figure 6.9 Finite element mesh for a bow-tie cylinder.

Table 6.3
Computed potentials for the bow-tie cylinder of Figure 6.9

Point	x (m)	y (m)	MOM (Volt)	Hybrid method (Volt)
1	0.0333	0.0	9.275	9.267
2	0.2696	0.1959	9.213	9.211
3	0.1029	0.3169	9.101	9.117
4	0.0	0.1999	9.515	9.556
5	-0.0823	0.2535	9.312	9.335
6	-0.2696	0.1959	9.213	9.216
7	-0.3333	0.0	9.275	9.267
8	-0.3169	-0.1029	9.260	9.253
9	-0.1959	-0.2696	9.150	9.158
10	-0.0823	-0.1901	9.312	9.335
11	0.0617	-0.1901	9.545	9.584
12	0.1959	-0.2696	9.150	9.158

6.3.4 Conclusion

In this section, the application of the hybrid FE-BI procedure to the solution of Laplace's equation for two-dimensional open-region electrostatic problems was demonstrated. The iterative technique is easy to implement and produces acceptable results. Although the number of iterations required for the convergence is already small, it can be further reduced by choosing appropriate values of the over-relaxation factors (α and β). The results are compared with the solution obtained using the MOM. The maximum difference in the results for the potentials and charges is of the order of 1%.

In the iterative solution, the convergence may be jeopardized when the terminating surface S is extremely close to the surface with field sources (S_o), i.e., the conductor surface, because in that case the potential on S is predominantly determined by the nearby charges. To assure the convergence, the relaxation process must be slowed down by using the under-relaxation for the nodes on S. However, the over-relaxation can still be used for the free nodes. For such cases, a direct solution of the resulting system of linear equations was found to be much more efficient.

Good results can be achieved for fairly arbitrary shapes of the terminating surface, which need not be conformal to the surface of the conducting body. The terminating surface can be placed at a relatively small distance from the conducting body, thus keeping small the domain of finite elements. It is sufficient to work with two layers of elements to achieve acceptable results. The combined FE-BI method yields a moderately sparse matrix, and it can be used efficiently for large and complex open-region problems.

6.4 TM SCATTERING

6.4.1 Introduction

The objective of this and the following sections is to demonstrate that the hybrid FE-BI method also works for dynamic open-region problems and investigate the nature of the solution. A simple two-dimensional problem of TM and TE scattering from a perfectly conducting infinitely long cylinder residing in vacuum is chosen for that purpose. In this section, the TM case is treated, while the TE case is analyzed in Section 6.5. The cylinder is assumed to be illuminated by a time-harmonic uniform plane wave of angular frequency ω, whose electric field is parallel to the cylinder axis (z). The direction of wave propagation is perpendicular to the z-axis. This incident field produces axially directed induced surface currents on the conductor. These currents produce an axially directed scattered electric field, while the magnetic field is purely transverse to the cylinder axis.

In this analysis, the finite element technique in conjunction with integral boundary conditions is used to solve the Helmholtz equation for two-dimensional TM problems in a homogeneous region, i.e.,

$$\nabla^2 E_z + k^2 E_z = \frac{\partial^2 E_z(x,y)}{\partial x^2} + \frac{\partial^2 E_z(x,y)}{\partial y^2} + k^2 E_z = 0 \tag{6.63}$$

where E_z is the axial component of electric field, which does not depend on the axial coordinate (z). The time dependence is assumed through the term $e^{j\omega t}$, which is suppressed in equations. The free-space wave number k is given by (6.12), with ε and μ substituted by ε_0 and μ_0, respectively.

Triangular finite elements are used for the discretization [Roy et al. 1996]. As in the electrostatic cases [Roy et al. 1995] presented in the previous section, to expedite computations the finite elements extend only two layers from the conducting structure. However, this is not necessary in the general case, as the mesh-terminating surface can have an arbitrary shape, which need not conform to the conductor surface.

6.4.2 Formulation

In this analysis, it is assumed a monochromatic plane wave incident on the conductor, as shown in Figure 6.10.

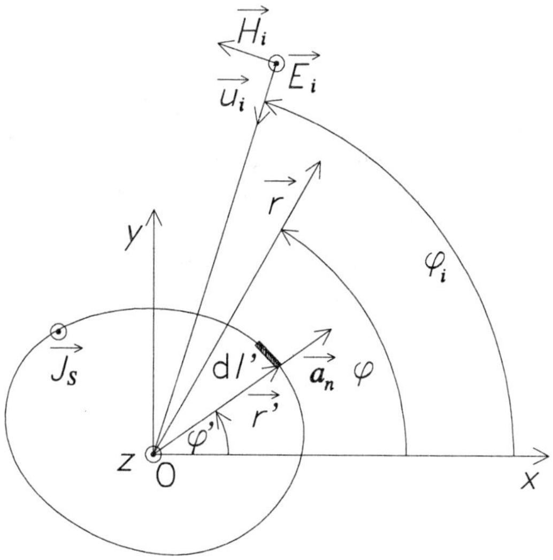

Figure 6.10 Cross section of a scattering cylinder (TM case) and the coordinate system.

The wave arrives from a direction defined by the cylindrical angle φ_i. The electric field of this wave is given by

$$\bar{E}_i = E_o \, e^{jk(x\cos\varphi_i + y\sin\varphi_i)} \, \bar{a}_z \tag{6.64}$$

where $\bar{a}_z$ is the unit vector of the z-axis. Equation (6.63) is satisfied by the incident electric field ($\bar{E}_i$), the scattered electric field ($\bar{E}_s$) produced by the currents induced on the conductor, as well as by the total electric field ($\bar{E}$), which is given by

$$\bar{E} = \bar{E}_i + \bar{E}_s \tag{6.65}$$

The scattered electric field is given by

$$\bar{E}_s = -j\omega \bar{A} = -j\omega \mu_0 \oint_C \bar{J}_s \, G(\bar{r};\bar{r}')\,dl' \qquad (6.66)$$

where $\bar{A}$ is the magnetic vector-potential, $\bar{J}_s$ is the density of the induced surface currents, which is given by

$$\bar{J}_s = J_{sz}\,\bar{a}_z \qquad (6.67)$$

Green's function for this problem is given by (6.14), and dl' is an element of the contour C that bounds the conductor cross section. Since J_{sz} is independent of z, there are no charges associated with this current. Hence, the electric scalar-potential, V, is zero everywhere. The transverse components of the electric and magnetic fields are, in vector form,

$$\bar{E}_t = 0 \qquad (6.68)$$

and

$$\bar{H}_t = \frac{1}{j\omega\mu_o}\,\bar{a}_z \times \bar{\nabla}_t E_z \qquad (6.69)$$

respectively. Denoting by $\bar{a}_x$ and $\bar{a}_y$ the unit vectors along the x- and y-axis, respectively, the transverse "del" operator is given by

$$\bar{\nabla}_t = \bar{a}_x \frac{\partial}{\partial x} + \bar{a}_y \frac{\partial}{\partial y} \qquad (6.70)$$

The transverse Cartesian components of the electric fields are, from (6.68),

$$E_x = E_y = 0 \qquad (6.71)$$

and from (6.69) and (6.70) one gets the transverse components of the magnetic field,

$$H_x = -\frac{1}{j\omega\mu_o}\,\frac{\partial E_z}{\partial y} \qquad (6.72)$$

$$H_y = \frac{1}{j\omega\mu_o}\frac{\partial E_z}{\partial x} \tag{6.73}$$

By the definition of the TM case, the axial component of the magnetic field is zero, i.e.,

$$H_z = 0 \tag{6.74}$$

Since the cylinder is perfectly conducting, the boundary conditions require that the tangential component of the total electric field be zero on the conductor surface. Hence,

$$E_z = 0 \tag{6.75}$$

on the conductor surface. The surface-current density on the conductor may be expressed in terms of the tangential magnetic field just outside the conductor from (6.1). In this case, one has

$$\bar{J}_s = \bar{a}_n \times \bar{H}_t \tag{6.76}$$

where $\bar{a}_n$ is the outward unit normal on the conductor surface. Hence, all the field and source components have been completely defined in terms of E_z.

In the FEM solution, the open region surrounding the body is subdivided into nonoverlapping finite elements up to a certain distance from the cylinder. Note that one is solving for a scalar function (let us denote it here by Φ), like in the electrostatic case, except that the scalar function is now a field component, E_z, instead of the electrostatic potential, V. More precisely, (6.63) may be formulated in terms of the scattered field, $\underline{E}_s$, or the total field, $\bar{E}$. These two formulations are theoretically equivalent but may yield different results in a numerical procedure due to the discretization. Also, the two procedures have different behavior regarding the erroneous solutions due to the internal resonances of the domain [Jin 1993; Ch. 9]. If one uses the formulation for the scattered field, then $\Phi = E_{sz}$. If one uses the formulation for the total field, then $\Phi = E_z$.

To find the scalar function Φ for the two-dimensional solution region, one seeks an approximation for Φ within the eth element and then interrelate, these approximations in adjacent elements to have a continuous function across the interelement boundaries. Using a polynomial approximation for $\Phi(x,y)$ over each element, the solution for the whole region is given by

$$\Phi(x, y) \approx \sum_{e=1}^{N_e} \Phi^e(x, y) \tag{6.77}$$

where N_e is the number of triangular elements into which the solution region is divided. The most simple form of approximation for Φ^e within an element is a linear polynomial. The analysis is carried out based on that.

The functional given by expression (3.71) is valid for this problem where all boundaries support Dirichlet conditions. The only difference is that k_c in (3.71) is now equal to k. Using the procedure as described earlier, the functional $I(\Phi^e)$, which is proportional to the energy p.u.l. associated with the eth element, is given by

$$I(\Phi^e) = \frac{1}{2} \int_{S^e} \left(|\bar{\nabla}_t \Phi^e|^2 - k^2 (\Phi^e)^2 \right) \mathrm{d}S \tag{6.78}$$

where S^e denotes the surface of the element. One further has

$$\bar{\nabla}_t \Phi^e(x,y) = \sum_{i=1}^{3} \Phi_i^e \bar{\nabla}_t N_i^e \tag{6.79}$$

where Φ_i^e are the values of the scalar function at the nodes of the eth element and the coefficients N_i^e are given by (6.31) to (6.33). Substituting (6.79) into (6.78), one gets

$$I(\Phi^e) = \frac{1}{2} \int_{S^e} \left(|\sum_{i=1}^{3} \Phi_i^e \bar{\nabla}_t N_i^e|^2 - k^2 \left(\sum_{i=1}^{3} \Phi_i^e N_i^e \right)^2 \right) \mathrm{d}S \tag{6.80}$$

Expanding (6.80), one gets

$$\begin{aligned} I(\Phi^e) = & \frac{1}{2} \sum_{i=1}^{3}\sum_{j=1}^{3} \Phi_i^e \Phi_j^e \int_{S^e} \bar{\nabla}_t N_i^e \cdot \bar{\nabla}_t N_j^e \, \mathrm{d}S \\ & - k^2 \frac{1}{2} \sum_{i=1}^{3}\sum_{j=1}^{3} \Phi_i^e \Phi_j^e \int_{S^e} N_i^e N_j^e \, \mathrm{d}S \end{aligned} \tag{6.81}$$

which transforms into the following compact form

$$I(\Phi^e) = \frac{1}{2} \sum_{i=1}^{3}\sum_{j=1}^{3} \Phi_i^e \left(k_{ij}^e - k^2 m_{ij}^e \right) \Phi_j^e \tag{6.82}$$

where the coefficients k_{ij}^e are defined in (6.40) and (6.44) to (6.48), while the coefficients m_{ij}^e are given by

$$m_{ij}^{e} = \int_{S^e} N_i^e N_j^e \, \mathrm{d}S \tag{6.83}$$

Equation (6.82) can now be written in matrix form as

$$I(\Phi^e) = \frac{1}{2}\{\Phi^e\}^T[k^e]\{\Phi^e\} - \frac{k^2}{2}\{\Phi^e\}^T[m^e]\{\Phi^e\} \tag{6.84}$$

where

$$m_{ij}^{e} = \begin{cases} \dfrac{A_e}{6}, & i=j \\[2ex] \dfrac{A_e}{12}, & i \neq j \end{cases} \tag{6.85}$$

and A_e is the area of the eth triangular element. Assembling all such elements in the solution region, the total energy of the assemblage is given by

$$I(\Phi) = \sum_{e=1}^{N_e} I(\Phi^e) = \frac{1}{2}\{\Phi\}^T[K]\{\Phi\} - \frac{k^2}{2}\{\Phi\}^T[M]\{\Phi\} \tag{6.86}$$

The matrices $[K]$ and $[M]$ are the assemblage of individual matrices $[k^e]$ and $[m^e]$, respectively. The column matrix $\{\Phi\}$ contains E_z at the corresponding nodes.

The nodes of the finite element mesh are divided into three groups (the nodes on the terminating surface, the nodes on the conductor surface, and the remaining nodes), as described in Section 6.3.2. Equation (6.86) can be rearranged to produce a form similar to (6.52), as

$$I(\Phi) = \frac{1}{2}\begin{Bmatrix}\{\Phi_b\} \\ \{\Phi_f\}\end{Bmatrix}^T \left(\begin{bmatrix}[K_{bb}] & [K_{bf}] \\ [K_{fb}] & [K_{ff}]\end{bmatrix} - k^2\begin{bmatrix}[M_{bb}] & [M_{bf}] \\ [M_{fb}] & [M_{ff}]\end{bmatrix}\right)\begin{Bmatrix}\{\Phi_b\} \\ \{\Phi_f\}\end{Bmatrix} \tag{6.87}$$

where the subscript "b" refers to the nodes on the boundary surfaces (the mesh terminating surface and the conductor surface) and "f" to the free nodes. To find the minimum of the functional $I(\Phi)$, it is set as

$$\frac{\partial I}{\partial \Phi_f} = 0 \tag{6.88}$$

which in general yields

$$\Big(\big[[K_{fb}]\ [K_{ff}]\big] - k^2\big[[M_{fb}]\ [M_{ff}]\big]\Big)\begin{Bmatrix}\{\Phi_b\}\\ \{\Phi_f\}\end{Bmatrix} = \{0\} \tag{6.89}$$

This equation can be written as

$$\big([K_{ff}] - k^2[M_{ff}]\big)\{\Phi_f\} = \big(k^2[M_{fb}] - [K_{fb}]\big)\{\Phi_b\} \tag{6.90}$$

According to (6.75), the tangential component of the total field on the conductor is zero. If one uses the formulation utilizing the total field, then $\Phi = 0$ at the nodes residing on the conductor surface. If one uses the formulation for the scattered field, then at the conductor surface, the tangential component of the scattered electric field must cancel the tangential component of the incident field. Hence, at the corresponding nodes, $\Phi = -E_{iz}$. In both cases, the values of Φ at the nodes on the conductor surface are thus fixed and correspond to a Dirichlet condition. Equations expressing this condition should be added to (6.90) and to equations expressing the boundary conditions on the mesh terminating surface, explained below.

The values of Φ on the outer boundary surface are evaluated in terms of the integral boundary conditions. The z-component of the scattered electric fields is expressed by (6.66) in terms of the surface currents induced on the conductor. Taking into account the segmentation of the conductor surface and substituting (6.14) into (6.66) yields

$$E_{sz} = -j\omega\mu_0 \sum_i \int_{\Delta l_i} J_{szi} \frac{1}{4j} H_0^{(2)}(k|\bar{r}-\bar{r}'|)\,dl' \tag{6.91}$$

where J_{szi} is the surface-current density over the ith segment on the conductor surface, i.e., Δl_i, and dl' is an element of that segment. The field sources (J_{szi}) can be evaluated from (6.76), (6.72), and (6.73) in terms of the z-component of the total field as

$$J_{sz} = \frac{1}{j\omega\mu_0}\frac{dE_z}{dn} \tag{6.92}$$

The normal derivative of the electric field may be computed as a postprocess once the degrees of freedom are obtained. This is analogous to the electrostatic case when computing the normal derivative of the potential using (6.58) to (6.60). Substituting (6.92) into (6.91) and using (6.65), one obtains an equation suitable for the formulation in terms of the scattered field:

$$E_{sz} = -\frac{1}{4\mathrm{j}} \sum_i \int_{\Delta l_i} \left(\frac{\mathrm{d}(E_{iz}+E_{sz})}{\mathrm{d}n} \right) H_0^{(2)}(k\,|\bar{r}-\bar{r}'|)\,\mathrm{d}l' \tag{6.93}$$

where the integral and the field components are evaluated on the outer terminating surface (S in Figure 6.3) and E_{iz} is the z-component of the electric field of the incident wave on the segment Δl_i. The normal derivative of the incident field can be found explicitly. This normal derivative turns out to be identical to the tangential component of the incident magnetic field on the conductor surface multiplied by $\mathrm{j}\omega\mu_0$. This becomes obvious when (6.76) and (6.92) are written for the incident field. Referring to (6.19), for this formulation, the excitation appears in all the elements of $\{E_1\}$ and in those elements of $\{E_2\}$ that correspond to the Dirichlet condition on the conductor surface.

For the formulation in terms of the total field, the z-component of the total electric field at the mesh termination is needed, which is given by

$$E_z = E_{iz} - \frac{1}{4\mathrm{j}} \sum_i \int_{\Delta l_i} \left(\frac{\mathrm{d}E_z}{\mathrm{d}n} \right) H_0^{(2)}(k\,|\bar{r}-\bar{r}'|)\,\mathrm{d}l' \tag{6.94}$$

where E_{iz} is the z-component of the electric field of the incident wave on the outer terminating surface. For this formulation, the excitation appears only in the elements of $\{E_1\}$.

At the beginning of the iterative process used to solve the resulting linear equations, Φ_{ob} is arbitrarily assumed to be zero for the scattered field formulation and E_{iz} for the total field formulation. (In the examples presented in Section 6.4.3, the formulation for the total field is used.) The values of Φ on the conductor surface (the inner boundary) are fixed, as explained before. Otherwise, the iterations are carried out as in the electrostatic case.

In particular, if the vector of nodal values of the scalar function Φ on the terminating surface is denoted by $\{\Phi_{ob}\}$ and a vector of values at the free nodes by $\{\Phi_f\}$, and if the iteration count (k) is denoted by a superscript k, then these vectors at the iteration (k+1) are given by

$$\{\Phi_{ob}^{k+1}\} \leftarrow \{\Phi_{ob}^{k}\} + \alpha(\{\Phi_{ob}^{k+1}\} - \{\Phi_{ob}^{k}\}) \tag{6.95}$$

and

$$\{\Phi_{f}^{k+1}\} \leftarrow \{\Phi_{f}^{k}\} + \beta(\{\Phi_{f}^{k+1}\} - \{\Phi_{f}^{k}\}) \tag{6.96}$$

respectively. As in the electrostatic case, the factors α and β are selected in the range 0.1 to 1.0 and 1 to 1.9, respectively. The iterative procedure is stopped when the variations of the electric field on the outer terminating surface become less than a specified value.

The most useful practical parameter evaluated for a scatterer is its cross section. To derive it, one has to find the scattered field in the radiation (far) zone. An asymptotic expression for the Green's function is required first. It is given by [Harrington 1961]

$$\mathscr{L}t_{x \to \infty} \; H_0^{(2)}(x) \simeq \sqrt{\frac{2}{\pi x}} \; e^{-j(x-\pi/4)} \tag{6.97}$$

Referring to Figure 6.10, $\bar{r}$ and $\bar{r}'$ are the position-vectors of the field and source point, respectively, so the distance between the two points is

$$|\bar{r}-\bar{r}'| = \sqrt{(x-x')^2+(y-y')^2} = \sqrt{r^2+r'^2-2r'r\cos(\varphi-\varphi')} \tag{6.98}$$

where r and r' are moduli of $\bar{r}$ and $\bar{r}'$, respectively. Since in the far-field region $r \gg r'$, from (6.98) one gets

$$|\bar{r}-\bar{r}'| \approx r - r'\cos(\varphi-\varphi') \tag{6.99}$$

Replacing (6.99) into (6.97) for x leads to

$$\begin{aligned} H_0^{(2)}(k|\bar{r}-\bar{r}'|) &\approx \sqrt{\frac{2}{\pi k r}} \; e^{-j\left(kr-kr'\cos(\varphi-\varphi')-\pi/4\right)} \\ &= \sqrt{\frac{2}{\pi k r}} \; e^{-j(kr-\pi/4)} \; e^{jkr'\cos(\varphi-\varphi')} \end{aligned} \tag{6.100}$$

To get the far-field, (6.100) is replaced into (6.91) to get

$$E_{sz}(\bar{r}) \approx -\frac{1}{4\mathrm{j}}\sqrt{\frac{2}{\pi k r}}\, \mathrm{e}^{-\mathrm{j}(kr-\pi/4)} \sum_i \left(\frac{\mathrm{d}E_z}{\mathrm{d}n}\right) \int_{\Delta l_i} \mathrm{e}^{\,\mathrm{j}k(x'\cos\varphi + y'\sin\varphi)}\,\mathrm{d}l'$$

$$= \sqrt{\frac{1}{8\pi k r}}\, \mathrm{e}^{-\mathrm{j}(kr-3\pi/4)} \sum_i \left(\frac{\mathrm{d}E_z}{\mathrm{d}n}\right) \mathrm{e}^{\,\mathrm{j}k(x_i\cos\varphi + y_i\sin\varphi)}\,\Delta l_i \qquad (6.101)$$

where (x_i,y_i) is the midpoint of the ith segment (Δl_i). Changing the notation from φ to φ_s for the angle in which the scattering is considered, the scattering cross section can be written as

$$\sigma(\varphi_i,\varphi_s) = \frac{2\pi r\,|\bar{E}_s|^2}{|\bar{E}_i|^2} = \frac{1}{4k}\,|\{V\}^T\{J\}|^2 \qquad (6.102)$$

where the electric field has only the z component in this case. The intensity of the electric field of the incident wave is assumed to be unity (i.e., E_o = 1 V/m), and

$$\{V\} = \{\Delta l\ \mathrm{e}^{\,\mathrm{j}k(x\cos\varphi_s + y\sin\varphi_s)}\} \qquad (6.103)$$

and

$$\{J\} = \left\{\frac{\mathrm{d}E_z}{\mathrm{d}n}\right\} \qquad (6.104)$$

are column matrices.

6.4.3 Numerical Results

6.4.3.1 *Elliptic Cylinder*

As the first example, let us consider an elliptic cylinder with the major axis $2\lambda/3$ and minor axis $8\lambda/15$, where λ is the wavelength at the operating frequency. The incident wave arrives from $\phi_i = 0$. In all the examples in Section 6.4.3 we assume the intensity of the electric field of the incident wave to be 1 V/m. A two-layer finite element mesh for the problem is shown in Figure 6.11. In this example, the terminating surface is 0.2λ away from the conductor surface. There are 40 nodes on the conductor surface, 40 nodes on the outer terminating surfaces, and 40 free nodes between the conductor and the terminating surface. Note that in all the examples in Section 6.4.3, the meshes are sketched only qualitatively; i.e., the number of nodes is shown smaller than actually used in computations.

For the iterative solution, the values of 0.15 and 1.9 were chosen for α and β, respectively. The process was terminated after 302 iterations when, in successive iterations, the field values at the outer boundary surface differed for less than 10^{-4}. The same error criterion was used in the other examples of TM and TE scattering.

The real and imaginary parts of the induced surface currents on the conductor are plotted in Figures 6.12 and 6.13, respectively, along with results obtained by the MOM. Currents are plotted from node 1 in the counterclockwise sense, as in all the other examples of Sections 6.4.3 and 6.5.3. In MOM, 40 subsections were chosen on the conductor surface and the solution was obtained using pulse expansion functions and the point matching testing procedure. The two sets of results agree very well.

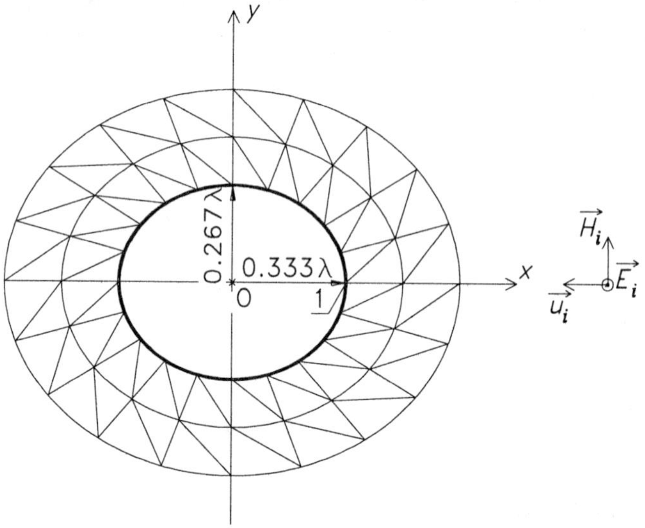

Figure 6.11 Finite element mesh for an elliptic cylinder.

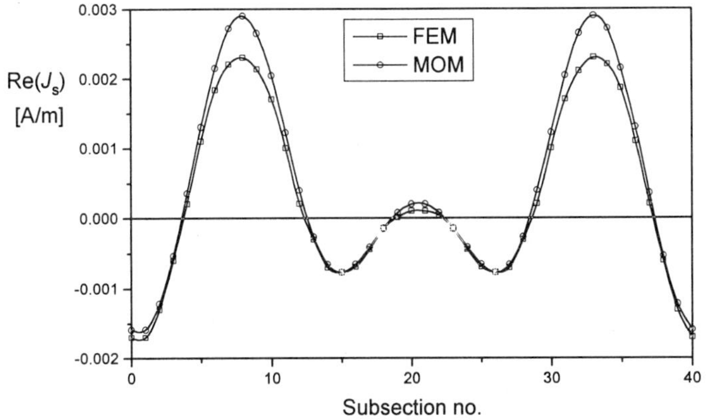

Figure 6.12 Real part of the induced surface currents on the elliptic cylinder of Figure 6.11.

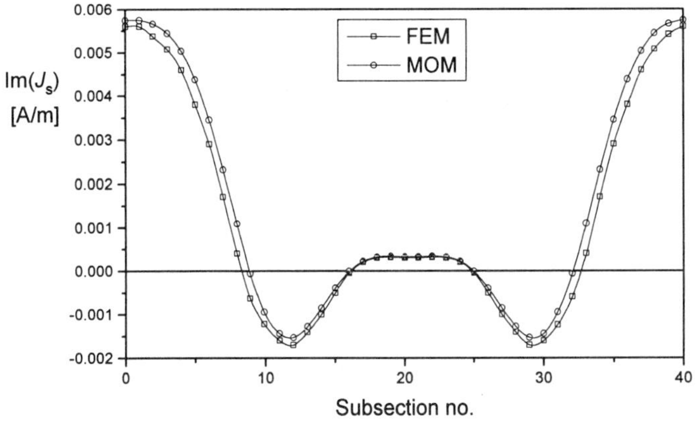

Figure 6.13 Imaginary part of the induced surface currents on the elliptic cylinder of Figure 6.11.

6.4.3.2 *Square Cylinder*

Consider a square cylinder of side $4\lambda/3$, illuminated by a uniform plane wave arriving from $\phi_i = 0$. The finite element grid for this problem is shown in Figure 6.14. The outer terminating surface is placed at a distance of $4\lambda/15$ away from the conductor surface. There are 65 nodes on the outer terminating surface, 48 nodes on the conductor surface, and 56 free nodes.

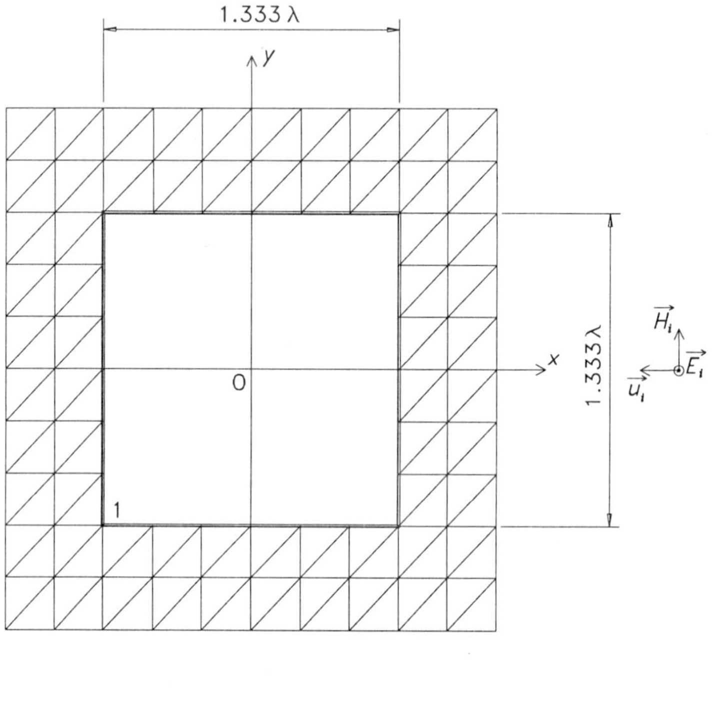

Figure 6.14 Finite element mesh for a square cylinder.

The iterative procedure, with $\alpha = 0.2$ and $\beta = 1.3$, terminated after 188 iterations. The real and the imaginary parts of the induced surface currents on the conductor are plotted in Figures 6.15 and 6.16, respectively, along with MOM values (evaluated using 48 subsections on the conductor surface). The two sets of results agree well except at cylinder edges. This is to be expected since it is the inherent nature of FEM using unknowns associated with nodes. Since the current density and the electric field are infinitely large at the edges, their modeling is difficult. One possible way to improve the results is to use edge-based elements or higher order polynomials for triangles near the edges. Another option is to apply a self-adaptive procedure, described elsewhere in this book.

Let us see how the error between the electric fields at the terminating surface in two successive iterations depends on the relaxation factors. As seen in Figure 6.17, increasing the value of α significantly hastens the convergence. As α increases, the refinement of the boundary condition on the terminating surface moves faster to the numerically exact value. Increasing β has a small effect on the rate of convergence because the unknowns pertinent to the free nodes are bounded on one side by the conductor boundary condition (which is fixed, not iterated) and on the other side by the integral boundary condition on the outer terminating surface.

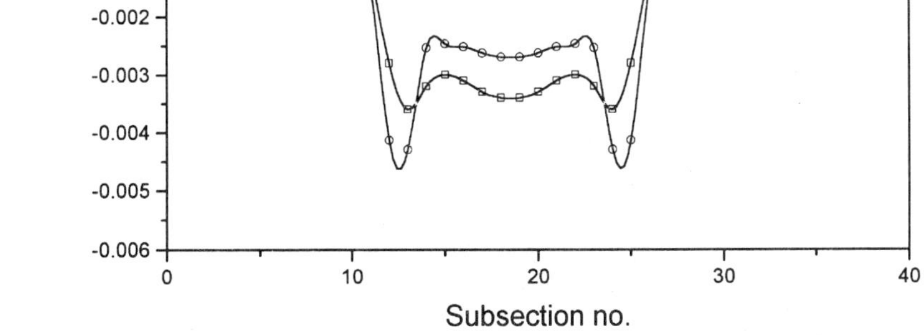

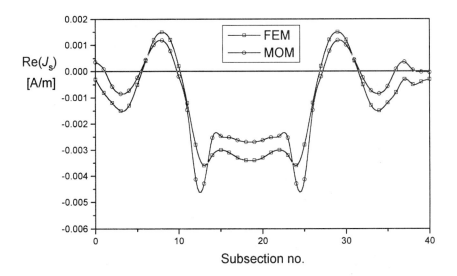

Figure 6.15 Real part of the induced currents on the square cylinder of Figure 6.14.

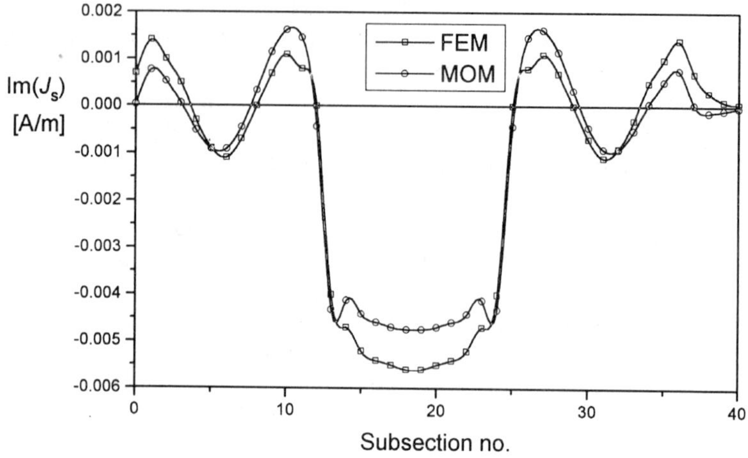

Figure 6.16 Imaginary part of the induced currents on the square cylinder of Figure 6.14.

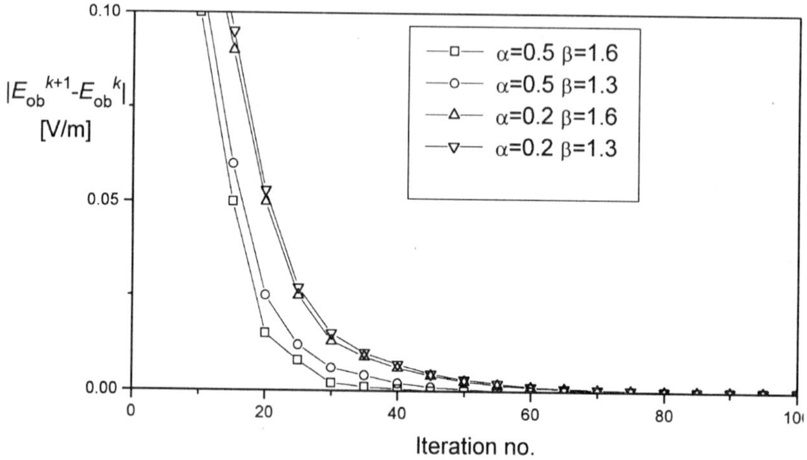

Figure 6.17 Maximum error for the electric field at nodes on the terminating surface, in V/m, versus the iteration count, with the relaxation factors α and β as parameters.

6.4.3.3 *Semicircular Cylinder*

In this example, the case of a semicircular cylinder is considered. It has a diameter of 1.73λ, as shown in Figure 6.18. The incident wave arrives from $\phi_i = 90°$. In the finite element meshing, the terminating surface is two layers away from the conductor (at a distance of 0.173λ). There are 44 free nodes, 40 nodes on the conductor surface, and 48 nodes on the outer terminating surface. The real and imaginary parts of the induced surface currents on the conductor are plotted in Figures 6.19 and 6.20, respectively, along with values obtained by MOM (with 40 pulses and point matching). Currents are plotted from node 1 in the counterclockwise sense. The FEM results agree well with the MOM values on the illuminated side of the cylinder (top). However, the agreement for the real part of the surface current in the shadow region of the conductor (bottom) is worse because the current density is very low in this region. Hence, the relative accuracy of computations is strongly affected by various numerical errors, particularly those introduced by the numerical differentiation to get the normal derivative in (6.92).

Compute the scattering cross section of the cylinder. Once the distribution of the induced surface currents on the conductor is found, the scattering cross section can be computed by (6.102). The results obtained by FEM and MOM are plotted in Figure 6.21. The agreement is reasonably good.

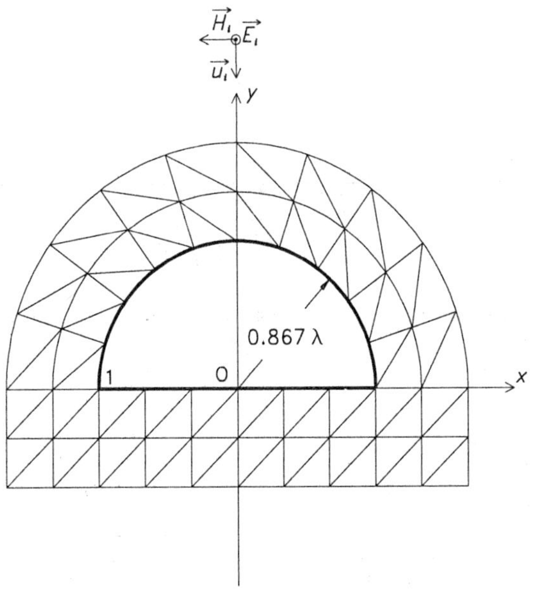

Figure 6.18 Finite element mesh for a semicircular cylinder.

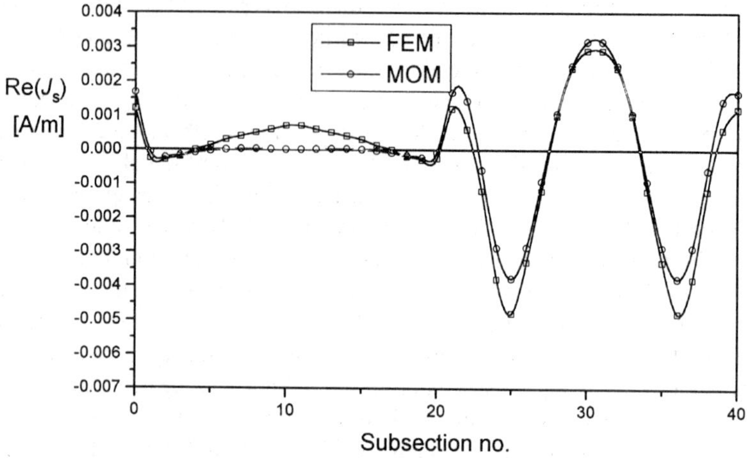

Figure 6.19 Real part of the induced currents on the semicircular cylinder of Figure 6.18.

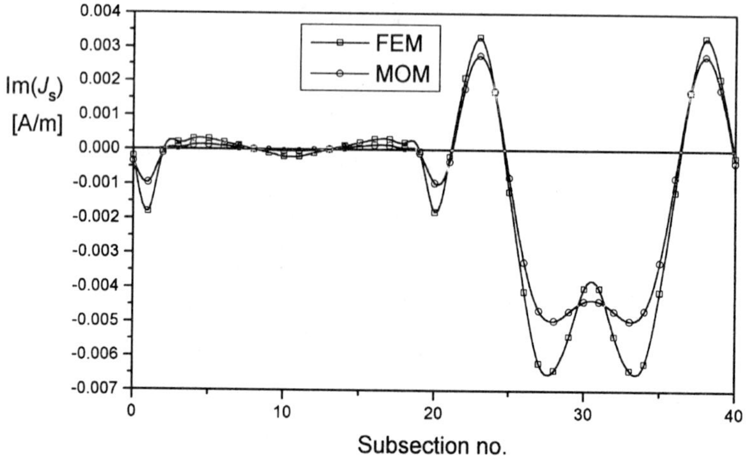

Figure 6.20 Imaginary part of the induced currents on the semicircular cylinder of Figure 6.18.

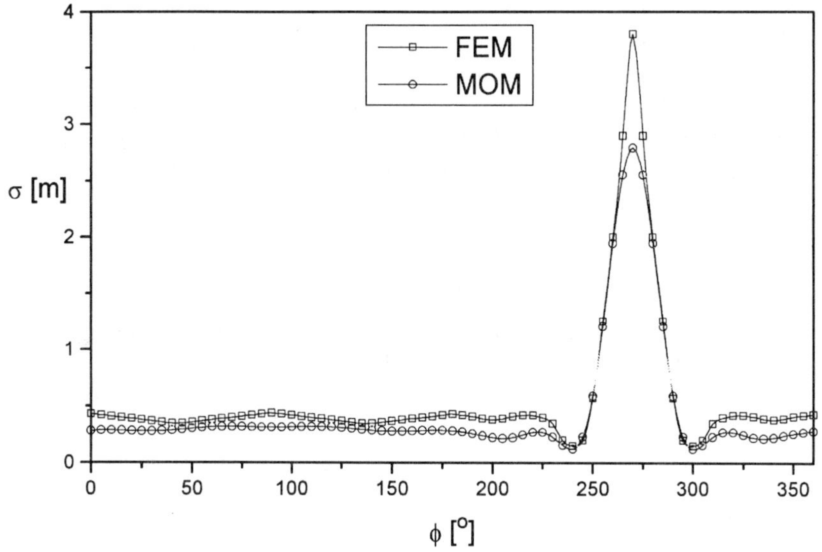

Figure 6.21 Scattering cross section of the semicircular cylinder of Figure 6.18.

6.4.4 Conclusion

A hybrid method is presented for the solution of the Helmholtz equation in two dimensions for open-region TM scattering problems. The results are in reasonable agreement with the solutions of MOM. It is seen that by changing the relaxation parameters (α and β), the rate of convergence is changed. In this context, it should be mentioned that placing the boundary too close to the body or keeping it too far away from it introduces problems. When placing it too close, a slow rate of convergence is obtained due to the local interaction of the fields. When keeping it very far away from the body, but using only two layers of finite elements, an error is introduced in computing the normal derivative of the electric field, which results in erroneous values of the induced currents. This, in turn, moves the radiation condition away from convergence. The field values at the sharp edges cannot be modeled with node-based elements. As the field goes to infinity on the edges, denser triangular grids are necessary to model those regions for accurate results, but this in turn increases the computational domain.

6.5 TE SCATTERING

6.5.1 Introduction

In this section, the two-dimensional problem of TE scattering from a perfectly conducting cylinder residing in a vacuum is considered. The cylinder is assumed to be illuminated by a time-harmonic uniform plane wave of angular frequency ω, whose magnetic field is parallel to the cylinder axis (z). This incident field produces transverse induced surface currents on the conductor. The scattered magnetic field produced by these currents is axially directed, while the electric field is purely transverse to the cylinder axis.

To analyze this scattering problem, the finite element technique is used in conjunction with the integral equation boundary conditions to solve the 2D Helmholtz equation. The approach parallels that of Section 6.4, but there are certain differences involved. Instead of E_z, now one deals with the axial component of the magnetic field, H_z, which does not depend on z. Hence, instead of (6.63), the Helmholtz equation now reads

$$\nabla^2 H_z + k^2 H_z \equiv \frac{\partial^2 H_z(x,y)}{\partial x^2} + \frac{\partial^2 H_z(x,y)}{\partial y^2} + k^2 H_z = 0 \tag{6.105}$$

As in previous sections, triangular finite elements are used to discretize the computational domain. In numerical examples presented in Section 6.5.3, these elements form two layers between the conductor boundary and the terminating surface. However, generally, the shape of the terminating surface need not be conformal to the structure and the number of layers need not be restricted to two.

6.5.2 Formulation

In this analysis, a monochromatic plane wave is assumed to be incident on the conductor (Figure 6.22). The magnetic field of this wave is given by

$$\bar{H}_i = H_0 \, \mathrm{e}^{jk(x\cos\varphi_i + y\sin\varphi_i)} \, \bar{a}_z \tag{6.106}$$

where $\bar{a}_z$ is the unit vector in the direction of the z-axis.

Equation (6.105) is satisfied by the incident magnetic field ($\bar{H}_i$), the scattered magnetic field (H_s) produced by the surface currents induced on the conductor, as well as by the total magnetic field (H) given by

$$\bar{H} = \bar{H}_i + \bar{H}_s \tag{6.107}$$

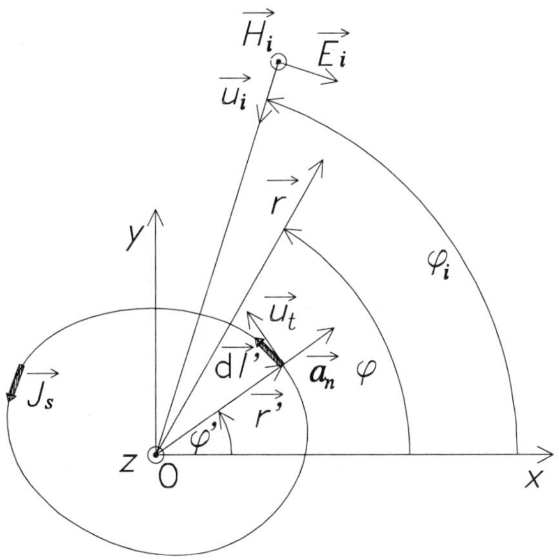

Figure 6.22 Cross section of a scattering cylinder (TE case) and the coordinate system.

The scattered field is given in terms of the surface current by

$$\bar{H}_s = \frac{1}{\mu}\bar{\nabla} \times \bar{A} = \bar{\nabla} \times \oint_C \bar{J}_s G(\bar{r}; \bar{r}')\, dl' \tag{6.108}$$

where $\bar{A}$ represents the magnetic vector-potential; C is the contour of the conductor cross section; $\bar{r}$ and $\bar{r}'$ are the position-vectors of the field and source point, respectively; $G(\bar{r}\,; \bar{r}')$ is the Green's function given by (6.14); and $\bar{J}_s$ is the density of the induced surface currents on the conductor, which is given as

$$\bar{J}_s = \bar{a}_\tau J_s \tag{6.109}$$

where $\bar{a}_\tau$ denotes the transverse unit vector, which is tangential to the conductor surface. The transverse components of the magnetic and electric fields are

$$\bar{H}_t = 0 \tag{6.110}$$

and

$$\bar{E}_t = -\frac{1}{j\omega\varepsilon_o}\,\bar{a}_z \times \bar{\nabla}_t H_z \tag{6.111}$$

respectively. The transverse "del" operator (∇_t) is given in (6.70). From (6.110), the Cartesian components of the transverse magnetic field are

$$H_x = H_y = 0 \tag{6.112}$$

Expanding the transverse "del" operator in (6.111), one gets the following transverse components of the electric field:

$$E_x = \frac{1}{j\omega\varepsilon_o}\frac{\partial H_z}{\partial y} \tag{6.113}$$

$$E_y = -\frac{1}{j\omega\varepsilon_o}\frac{\partial H_z}{\partial x} \tag{6.114}$$

From the definition of the TE case, one gets

$$E_z = 0 \tag{6.115}$$

Hence, all the field components have been expressed in terms of H_z.

Since the cylinder is perfectly conducting, the boundary conditions require the tangential component of the total electric field be zero on the conductor. Hence,

$$\bar{a}_n \times \bar{E}_t = \bar{0} \tag{6.116}$$

where $\bar{a}_n$ is the outward unit normal on the conductor surface. One can express the surface current density on the conductor in terms of the tangential magnetic field just outside the conductor as

$$\bar{J}_s = \bar{a}_n \times \bar{H}_t \tag{6.117}$$

This equation can be further written as

$$\bar{J}_s = \bar{a}_n \times \bar{H}_z = -H_z \bar{a}_\tau \tag{6.118}$$

where

$$\bar{a}_\tau = -\bar{a}_n \times \bar{a}_z \tag{6.119}$$

can now be interpreted more precisely as the unit vector tangential to the contour of the conductor cross section (C) and oriented in the anti-clockwise direction in Figure 6.22.

Rewriting (6.108) for the scattered magnetic field and replacing $\bar{J}_s$ from (6.118) yield

$$\begin{aligned}\bar{H}_s &= \bar{\nabla} \times \oint_C \bar{J}_s \frac{1}{4\mathrm{j}} H_0^{(2)}(k|\bar{r}-\bar{r}'|)\mathrm{d}l' \\ &= -\frac{1}{4\mathrm{j}} \bar{\nabla} \times \oint_C H_z H_0^{(2)}(k|\bar{r}-\bar{r}'|)\overline{\mathrm{d}l'}\end{aligned} \tag{6.120}$$

where

$$\overline{\mathrm{d}l}' = \bar{a}_\tau \mathrm{d}l' \tag{6.121}$$

is a vector element of the contour C. After differentiating the Hankel's function in (6.120) yields

$$\bar{H}_s = \frac{k}{4\mathrm{j}} \oint_C H_z H_1^{(2)}(k|\bar{r}-\bar{r}'|)\, \bar{a}_R \times \overline{\mathrm{d}l'} \tag{6.122}$$

where $\bar{a}_R$ is the unit vector of

$$\bar{R} = \bar{r} - \bar{r}' \tag{6.123}$$

and $H_1^{(2)}$ is Hankel's function of the second kind and order 1.

The region adjacent to the cylinder is subdivided into nonoverlapping finite elements. The solution region is kept small and the terminating surface is positioned only two layers from the structure. In this analysis, too, triangular finite elements with a linear polynomial approximation are used over the element surface.

As in the TM case, the equations can be formulated in terms of the total magnetic field ($\bar{H}$), in which case one has $\Phi = H_z$, or in terms of the scattered magnetic field ($\bar{H}_s$), when $\Phi = H_{sz}$. In any case, to solve for the magnetic field in the two-dimensional solution region, one seeks an approximation for Φ within the eth element and interrelate these approximations in adjacent elements so that the scalar function Φ is continuous across the boundaries between the elements. Formally, the same FEM procedure is followed as described by (6.77) to (6.90) in Section 6.4.2, which shall not be repeated here. However, an important difference between both formulations is that the inner boundary, i.e., the conductor surface, is no longer a Dirichlet type boundary but a Neumann type boundary. Thus, a term given by $-\int \varepsilon_0 \Phi \bar{a}_n \cdot \bar{\nabla}_t \Phi \, dS'$ extended to the inner boundary S_o should be added to the previous functional.

The integral boundary conditions on the mesh terminating surface can be derived following the general approach of this chapter. The field sources in the present case are the induced currents on the conductor surface, and their magnetic field (i.e., the scattered field) at an arbitrary point is given by (6.122). When specialized for a segmented conductor surface, this equation reads

$$\bar{H}_s = \frac{k}{4j} \sum_i \int_{\Delta l_i} H_{zi} \, H_1^{(2)}(k|\bar{r}-\bar{r}'|) \, \bar{a}_R \times \overline{dl}' \tag{6.124}$$

where H_{zi} is the total magnetic field just outside the ith segment on the conductor surface (Δl_i). For the scattering field formulation, from (6.124) the following integral boundary condition is obtained

$$\Phi_{ob} = \frac{k}{4j} \sum_i \int_{\Delta l_i} (\Phi_i + H_{izob}) \, H_1^{(2)}(k|\bar{r}-\bar{r}'|) \, \bar{a}_R \times \overline{dl}' \tag{6.125}$$

where the subscript "ob" denotes the outer terminating surface, and H_{izob} is the z-component of the magnetic field of the incident wave on the segment Δl_i. Similarly, for the total field formulation, one has

$$\Phi_{ob} = H_{iz} + \frac{k}{4j} \sum_i \int_{\Delta l_i} \Phi_i \, H_1^{(2)}(k|\bar{r}-\bar{r}'|) \, \bar{a}_R \times \overline{dl}' \tag{6.126}$$

where H_{iz} is the z-component of the magnetic field of the incident wave on the outer terminating surface.

On the conductor surface, the boundary conditions require that (6.116) be fulfilled. When the transverse component of the electric field is substituted from (6.111) into (6.116), one gets that the normal derivative of the z-component of the total magnetic field must be zero, i.e.,

$$\bar{a}_n \cdot \bar{\nabla}_t H_z \equiv \frac{\mathrm{d}H_z}{\mathrm{d}n} = 0 \tag{6.127}$$

Hence, for the case of the total field formulation, we have a homogeneous Neumann boundary condition over the conductor surface. Thus, the additional boundary term previously mentioned may be set equal to zero. In this way Neumann boundaries condition are imposed as natural ones.

For the formulation in terms of the scattered magnetic field, from (6.127) it is first obtained that

$$\bar{a}_n \cdot \bar{\nabla}_t H_{sz} = -\bar{a}_n \cdot \bar{\nabla}_t H_{iz} \tag{6.128}$$

Let us denote by $\bar{a}_i$ the unit vector of the direction of propagation of the incident wave. According to Figure 6.22, this vector is given by

$$\bar{a}_i = -\bar{a}_x \cos\varphi_i - \bar{a}_y \sin\varphi_i \tag{6.129}$$

From (6.106) it is deduced that

$$\bar{\nabla}_t H_{iz} = -\mathrm{j}kH_{iz}\bar{a}_i \tag{6.130}$$

so that

$$\bar{a}_n \cdot \bar{\nabla}_t H_{sz} = \mathrm{j}kH_{iz}\bar{a}_n \cdot \bar{a}_i \tag{6.131}$$

which is proportional to the tangential component of the incident electric field. Then this is the Neumann condition for the formulation in terms of the scattered magnetic field. Hence, the right hand term (6.131) is substituted on the boundary term mentioned before obtaining $-\mathrm{j}\int \varepsilon_0 \Phi k H_{iz}\bar{a}_n \cdot \bar{a}_i \, \mathrm{d}S$, extended to S_o. It is seen that this term involves the unknown function Φ to be discretized by the linear combination of basis functions while the rest are known quantities.

Referring to (6.19), the excitation appears in $\{E_1\}$ for both formulations. At the beginning of the iterative solution, the magnetic field at the outer terminating surface is arbitrarily assumed to be zero for the scattered field formulation, but $\bar{H}_{iz}$ for the total field formulation. The scalar function Φ at nodes on the conductor surface is not fixed, unlike the electrostatic case and TM scattering. Otherwise, an identical iterative technique as described in Section 6.4.2 is used, including (6.95) and (6.96), and the corresponding factors α and β.

To evaluate the cross section of the scatterer, a similar procedure as in Section 6.4.2 is followed. The asymptotic expression for $H_1^{(2)}$ for large arguments is given by [Harrington 1961]

$$\mathop{\mathcal{L}t}_{x \to \infty} \quad H_1^{(2)}(x) \;\to\; \sqrt{\frac{2}{\pi x}}\; e^{-j\left(x - \frac{3\pi}{4}\right)} \tag{6.132}$$

Substituting (6.99) into (6.132) for x, one gets

$$\begin{aligned} H_1^{(2)}(k\,|\bar{r}-\bar{r}'\,|) &\approx \sqrt{\frac{2}{\pi k r}}\; e^{-j\left(kr - kr'\cos(\varphi-\varphi') - \frac{3\pi}{4}\right)} \\ &= \sqrt{\frac{2}{\pi k r}}\; e^{-j\left(kr - \frac{3\pi}{4}\right)}\, e^{jkr'\cos(\varphi-\varphi')} \end{aligned} \tag{6.133}$$

To get the far magnetic field, (6.133) is replaced in (6.124), yielding

$$\begin{aligned} \bar{H}_s(\bar{r}) &= \frac{1}{j}\sqrt{\frac{k}{8\pi r}}\; e^{-j\left(kr - \frac{3\pi}{4}\right)} \sum_i H_{zi} \int_{\Delta l_i} e^{jk(x'\cos\varphi + y'\sin\varphi)}\; \bar{a}_R \times \overline{dl}' \\ &= \sqrt{\frac{k}{8\pi r}}\; e^{-j\left(kr - \frac{\pi}{4}\right)} \sum_i H_{zi}\; e^{jk(x_i\cos\varphi + y_i\sin\varphi)}\; \bar{a}_{Ri} \times \bar{a}_{ti}\,\Delta l_i \end{aligned} \tag{6.134}$$

The scattering cross section can now be written as

$$\sigma(\varphi_i, \varphi_s) = \frac{2\pi r\,|H_s|^2}{|H_i|^2} = \frac{k}{4}\,|\{\bar{V}\}^T\{J\}\,|^2 \tag{6.135}$$

where it is assumed that the intensity of the incident magnetic field is unity ($H_o = 1$ A/m), while the column matrices are given by

$$\{\bar{V}\} = \{e^{jk(x\cos\varphi_s + y\sin\varphi_s)}\ \bar{a}_R \times \bar{a}_\tau\ \Delta l\} \tag{6.136}$$

and

$$\{J\} = \{H_z\} \tag{6.137}$$

6.5.3 Numerical Results

6.5.3.1 *Square Cylinder*

Let us consider a square cylinder of side 0.9λ, illuminated by a uniform plane wave. In all the examples in Section 6.5.3, it is assumed that the intensity of the magnetic field of the incident wave is 1 A/m. The finite element grid of this problem is similar to that shown in Figure 6.14. Note, however, that the polarization of the incident wave is now different, as it corresponds to the TE case. In the present example, the terminating surface has 64 nodes, there are 48 nodes on the conductor surface, and 56 interior free nodes. The terminating surface is two layers away from the cylinder and placed at a distance of 0.09λ. For this particular structure, with $\alpha = 0.1$ and $\beta = 1.0$, 821 iterations were required for the field values at the terminating surface to achieve an error of 10^{-4}. The real and imaginary parts of the induced surface currents on the conductor are plotted in Figures 6.23 and 6.24, respectively, along with results obtained using MOM. The MOM solution is based on the magnetic field integral equation. The conductor surface was divided into with 48 subsections, each corresponding to one pulse basis function. For testing, the point matching procedure was used. The FEM and MOM results agree well.

6.5.3.2 *Circular Cylinder*

In this example, a circular cylinder of diameter $\lambda = 1$ m is considered. The incident wave arrives from the negative *x*-direction ($\phi_i = 180°$). A mesh that is conformal to the body, like that shown in Figure 6.5, is used. The terminating surface is 0.067λ away from the conductor surface. There are 40 nodes on the conductor surface, 40 nodes on the outer boundary, and 40 interior free nodes. The real and imaginary parts of the induced surface currents on the conductor are plotted in Figures 6.25 and 6.26, respectively, along with MOM results (with 40 subsections on the conducting surface). Again, a very good agreement is obtained between the FEM and MOM solutions.

For the same circular cylinder, the scattering cross section is plotted in Figure 6.27 along with MOM results. The agreement between the two sets of results is, again, very good.

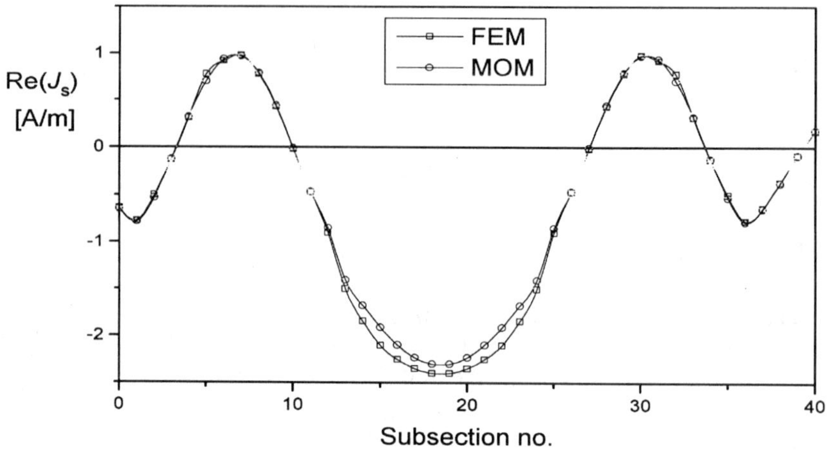

Figure 6.23 Real part of the induced surface currents on the square cylinder of Figure 6.14.

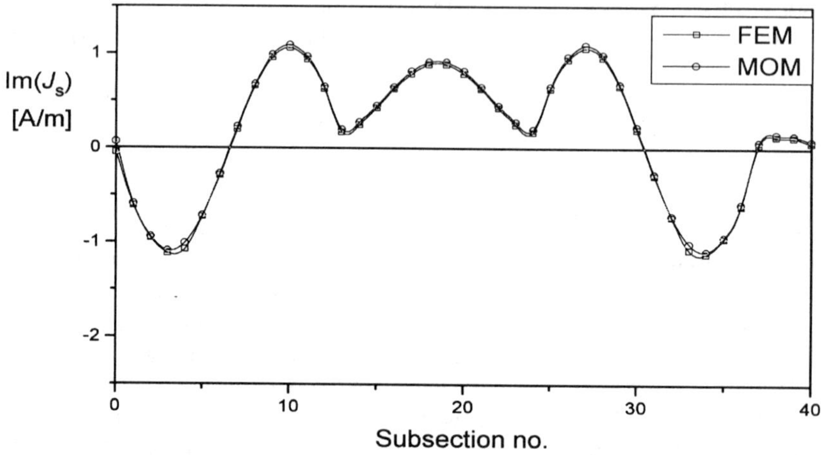

Figure 6.24 Imaginary part of the induced surface currents on the square cylinder of Figure 6.14

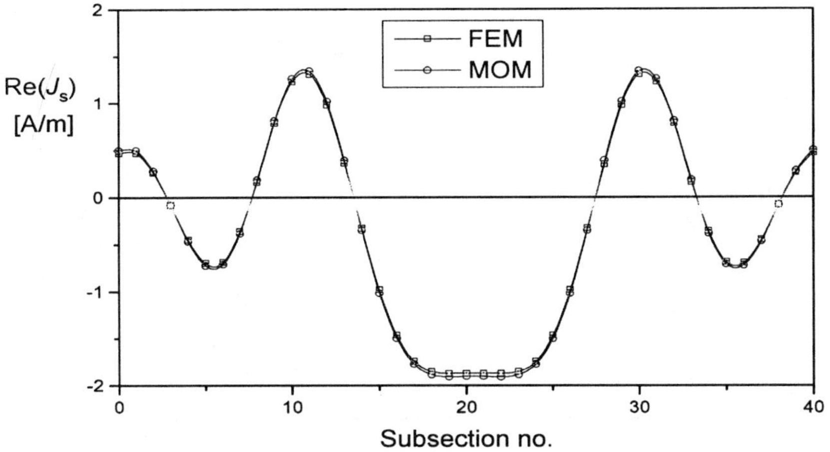

Figure 6.25 Real part of the induced surface currents on the circular cylinder of Figure 6.5.

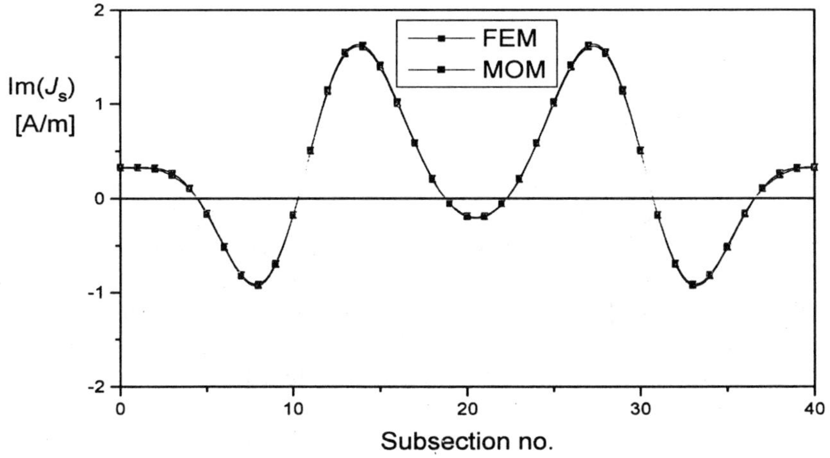

Figure 6.26 Imaginary part of the induced surface currents on the circular cylinder of Figure 6.5.

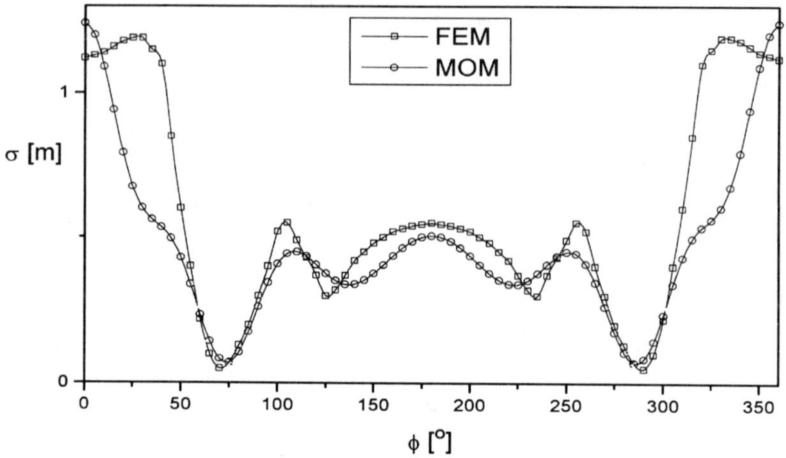

Figure 6.27 Scattering cross section of the circular cylinder of Figure 6.5.

6.5.3.3 *Semicircular Cylinder*

In this example, a semicircular cylinder of diameter 0.867λ, like that shown in Figure 6.18, is considered. The incident wave arrives from $\phi_i = 90°$. The meshing scheme has 44 interior free nodes, 40 nodes on the conductor, and 48 nodes on the terminating surface. The outer boundary surface is at a distance of 0.0867λ from the conductor surface. With $\alpha = 0.4$ and $\beta = 1.2$, 535 iterations were required to reduce the maximum error in successive values of the magnetic field below 10^{-4}. The real and imaginary parts of the induced surface currents on the conductor are plotted in Figures 6.28 and 6.29, respectively, along with MOM results (with 40 subsections). Currents are plotted from node 1 in the counterclockwise sense. The agreement between the two sets of results is worse than in the previous example but still within acceptable limits. This is primarily due to the existence of sharp edges in the present case and a smaller total number of nodes used in the computations. This example obviously calls for an adaptive meshing algorithm.

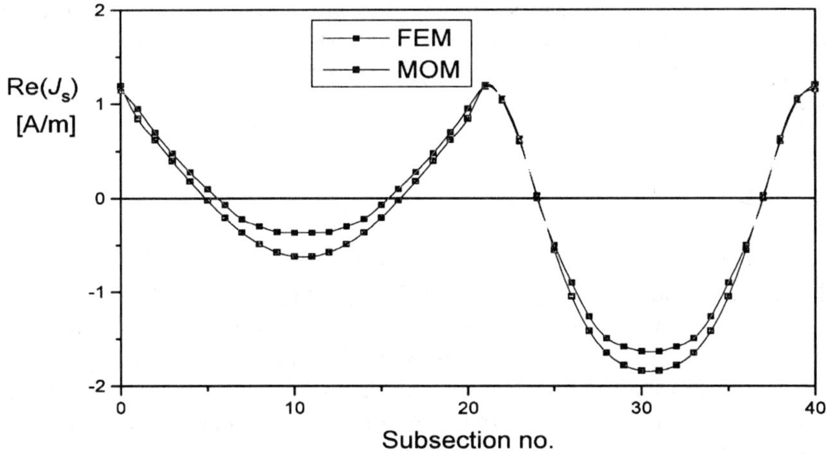

Figure 6.28 Real part of the induced surface currents on the semicircular cylinder of Figure 6.18.

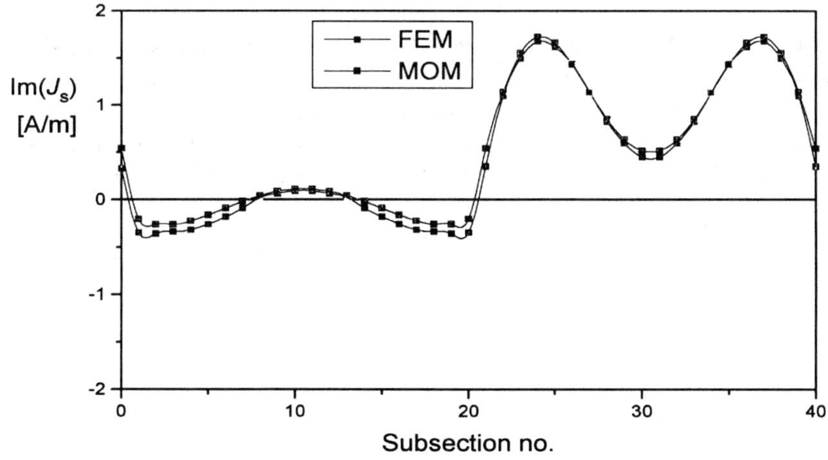

Figure 6.29 Imaginary part of the induced surface currents on the semicircular cylinder of Figure 6.18.

6.5.4 Conclusion

A hybrid method has been presented for the solution of the Helmholtz equation in two dimensions for open-region TE scattering problems. The results are in a better agreement with the solution of MOM than in the TM case. The primary reason is that in the TE case, the fields and currents have weaker singularities at sharp edges than in the TM case.

6.6 SUMMARY

In this chapter, it has been demonstrated how to utilize the FEM in combination with integral boundary conditions to solve open-region problems. The FEM discretization is terminated at a surface that tightly encloses all material inhomogeneities of the system. The integral conditions involve appropriate Green's functions that exactly reflect the influence of the surrounding homogeneous space on the fields inside the terminating surface. Hence, in dynamic cases, Sommerfeld's radiation condition is automatically taken into account.

The FE and BI equations have been arranged to form a simultaneous system of linear equations. The matrix for the FEM equations is sparse, as usual, while the matrix for the BI equations is denser. Yet, sparse matrix solvers, including iterative techniques, can be used efficiently. In all examples presented in this chapter, the coupled FE-BI equations have been solved by an iterative procedure (relaxation), although this may not be the only nor the fastest procedure.

The examples presented in this chapter are related to two-dimensional electrostatic problems and to TM and TE scattering. The results obtained by the hybrid FE-BI method have been compared with MOM solutions. In all cases, a moderate to very good agreement between the two sets of results is achieved. Still better results can be achieved by implementing self-adaptive FEM procedures. However, the purpose of this chapter is just to demonstrate the principles of the hybrid technique, not to develop an optimal algorithm. More efficient solutions, as well as examples of three-dimensional dynamic problems, are left for Chapter 7.

Chapter 7

Finite Element Analysis of Three-Dimensional Electromagnetic Problems

7.1 INTRODUCTION

In the previous chapters, various types of open region electromagnetic problems have been solved using the Finite Element Method (FEM). In all the cases, only two-dimensional (2D) problems have been considered. The present chapter applies FEM to three-dimensional (3D) problems.

Tetrahedral curl-conforming elements are used for the discretization process. A review of their features in connection with the existence of spurious modes is included in Section 7.2. The detailed description of these elements may be found in Section 2.5.1.2, and Appendix C.4.2.2 and C.4.3. An eigenvalue problem consisting of finding the resonances of arbitrarily shaped 3D cavities is introduced to illustrate the particular characteristics of these elements (Section 7.3). A deterministic problem is presented in Section 7.4 with the FEM analysis of discontinuities in 3D waveguiding structures. Section 7.5 describes an iterative technique for analyzing open-region problems using FEM, specifically, 3D scattering and radiation problems. The proposed approach makes use of the free-space Green's function, thus allowing the placement of the external boundary close to the sources of the problem. However, the sparsity of the FEM matrices is preserved. It represents an extension to 3D problems of the method presented in Chapter 6, although using a different implementation.

7.2 SPURIOUS MODES AND CURL-CONFORMING ELEMENTS

Curl-conforming elements (see Section 2.5.1.2, and Appendix C.4.2 and C.4.3) are known to overcome the problem of spurious modes. In this section, the question of the origin of the spurious modes in vector FEM formulations is addressed. A review of the various approaches published in the literature to solve the problem of spurious modes is included. The roles of both the formulation and the discretization processes are considered. It will be shown how the use of curl-conforming elements, also known as vector elements, in conjunction with the double-curl weak formulation, represents a consistent and suitable way for analyzing 3D electromagnetic problems.

7.2.1 Origin of Spurious Modes

First of all, let us define what it is meant by a *spurious mode* or a *spurious solution*. In the following, a spurious solution will be referred to as those *numerical solutions that are an approximation of a non-physical solution* [Fernandes and Sabbi 1992]. Non-physical solutions are those that do not satisfy Maxwell's equations and the boundary conditions of the problem. The definition is applicable in the context of an eigenvalue problem and that is why a *spurious solution* is also recognized in this case as a *spurious mode*. In a deterministic problem the solution is a linear combination of the eigenfunctions. Hence, the solution may be contaminated with spurious modes. In that case, the solution of the deterministic problem may have non-zero components of some of the spurious eigenmodes of the associated eigenvalue problem [Wong and Csendes 1989]. If in the computation of the eigenvalues for a deterministic problem, the presence of spurious modes is overcome, then the deterministic solution will not be contaminated by spurious solutions. Considering the definition of a spurious mode made previously, the problem of the spurious modes can be solved, as it will be shown later, in two ways: either by eliminating the appearance of spurious modes or, by just the opposite, approximating well the spurious solutions of the discretized problem. If the associated eigenvalue problem is not free of spurious modes, then it is possible that the deterministic solution may contain spurious modes, since it is not known *a priori* if a spurious solution will exist or not [Picon 1988], [Pinchuk et al. 1988].

Historically, the application of FEM to electromagnetic problems started with scalar formulations, i.e., formulations in which the unknown to be discretized is a scalar, as, for example, solution for the spectrum of the hollow waveguides [Ahmed 1968], [Ahmed and Daly 1969a], [Silvester 1969a,b]. The application of FEM to more general inhomogeneous and anisotropic waveguides leads to vector formulations,[1] i.e., formulations where the unknown or unknowns are vectors. Different combinations of the electric, $\overline{E}$, and the magnetic, $\overline{H}$, fields, have been employed, including combinations of the axial and transverse components of the fields. As an extension of the scalar FEM formulations, they are directly discretized with classical Lagrange elements (also known as nodal elements, see Section 2.5.1.2) by employing a separate scalar approximation for each one of the components (such as Cartesian and polar) of the vector fields. These approaches produce spurious modes that appear in the spectrum within the range of the values of the physical modes [Davies et al. 1982], [Konrad 1976], [Koshiba et al. 1985 a,b], [Rahman and Davies 1984a].

---

1 The formulation with E_z - H_z for inhomogeneous waveguides [Csendes and Silvester 1970] has been traditionally considered in the FEM literature as a vector formulation in the sense that the unknown is not just one scalar quantity; namely, in this case, the unknowns are two scalar quantities (see Section 2.5.1.2). However, in this chapter the term vector formulation applies to any formulation having one or more vector fields as unknowns. Therefore, the formulation with E_z - H_z is not considered to be a vector formulation. In fact, the cause of the nonphysical solutions in the E_z - H_z formulation has a different source of origin than those concerned in this section [Jin 1993; Ch. 7].

7.2.1.1 *Some Mathematical Concepts Related to the Spurious Modes*

Consider Maxwell's equations in a source-free region Ω. For the sake of clarity, let us assume that the inhomogeneous domain Ω is divided into different homogeneous media (i.e., a piecewise-homogeneous medium), with permeability $\underline{\underline{\mu}}_i$ and permittivity $\underline{\underline{\varepsilon}}_i$, i.e., $\Omega = \cup\, \Omega_i$ for $i = 1,\ldots, N_M$, where N_M is the total number of different homogeneous media. Hence, the constitutive relations can be written as

$$\bar{D} = \underline{\underline{\varepsilon}}_i \bar{E}; \qquad \bar{B} = \underline{\underline{\mu}}_i \bar{H}, \qquad \text{in} \quad \Omega_i \tag{7.1}$$

Maxwell's equations may be written in this case as:

$$\bar{\nabla} \times \bar{E} = -j\omega\underline{\underline{\mu}}_i \bar{H}, \qquad \text{in} \quad \Omega_i \tag{7.2}$$

$$\bar{\nabla} \cdot (\underline{\underline{\mu}}_i \bar{H}) = 0, \qquad \text{in} \quad \Omega_i \tag{7.3}$$

$$\bar{\nabla} \times \bar{H} = j\omega\underline{\underline{\varepsilon}}_i \bar{E}, \qquad \text{in} \quad \Omega_i \tag{7.4}$$

$$\bar{\nabla} \cdot (\underline{\underline{\varepsilon}}_i \bar{E}) = 0, \qquad \text{in} \quad \Omega_i \tag{7.5}$$

and the boundary conditions at the interfaces between the ith and jth media, i.e., at Γ_{ij}, are

$$\bar{a}_n \times (\bar{E}_i - \bar{E}_j) = 0, \qquad \text{on} \quad \Gamma_{ij} \tag{7.6}$$

$$\bar{a}_n \cdot (\underline{\underline{\mu}}_i \bar{H}_i - \underline{\underline{\mu}}_j \bar{H}_j) = 0, \qquad \text{on} \quad \Gamma_{ij} \tag{7.7}$$

$$\bar{a}_n \times (\bar{H}_i - \bar{H}_j) = 0, \qquad \text{on} \quad \Gamma_{ij} \tag{7.8}$$

$$\bar{a}_n \cdot (\underline{\underline{\varepsilon}}_i \bar{E}_i - \underline{\underline{\varepsilon}}_j \bar{E}_j) = 0, \qquad \text{on} \quad \Gamma_{ij} \tag{7.9}$$

For simplicity, the boundary conditions considered here are either those of a perfect electric wall, Γ_{pe}, or a perfect magnetic wall, Γ_{pm}, that can be written as

$$\bar{a}_n \times \bar{E} = 0, \qquad \text{on } \Gamma_{pe} \tag{7.10}$$

$$\bar{a}_n \cdot \bar{H} = 0, \qquad \text{on } \Gamma_{pe} \tag{7.11}$$

$$\bar{a}_n \times \bar{H} = 0, \qquad \text{on } \Gamma_{pm} \tag{7.12}$$

$$\bar{a}_n \cdot \bar{E} = 0, \qquad \text{on } \Gamma_{pm} \tag{7.13}$$

where $\bar{a}_n$ is the unit vector normal to surfaces Γ_{ij} , Γ_{pe} , and Γ_{pm}.

It is worth noting that the continuity of the tangential components is derived from the curl equations (7.2) and (7.4), while the continuity of the normal component comes from the divergence equations (7.3) and (7.5).

These equations formally constitute the strong formulation for the closed-region problems consisting of the analysis of the modes in a cavity resonator, which is an eigenvalue problem. This case has been chosen to illustrate the problem of spurious modes. However, as mentioned in the introduction to Section 7.2.1, the conclusions for other cases (as deterministic problems) may be easily deduced from this problem.

When $\omega = 0$, the $\bar{E}$ and $\bar{H}$ fields are decoupled. It is easily shown that the only solution of equations (7.2) to (7.13) (source free) is the trivial solution i.e., $\bar{E}$ and $\bar{H}$ equal to the null vector. When $\omega \neq 0$, Maxwell's equations (7.2) to (7.5) are dependent, and hence, the boundary conditions (7.6) to (7.13) are also dependent. Divergence equations (7.3) and (7.5) are deduced from the curl equations (7.2) and (7.4), respectively. The same situation occurs for the boundary conditions (7.6) to (7.13) since they are deduced from the curl and divergence form of Maxwell's equations. Therefore, for $\omega \neq 0$, some of the dependent equations of the set (7.2) to (7.13) do not need to be included in the formulation of the problem, since they are implicitly present in the other equations.

However, the dependence mentioned previously no longer applies when $\omega = 0$. Hence, if some of the equations dropped from the formulation of the problem are not implicit when $\omega = 0$, additional solutions may be obtained for $\omega = 0$ besides the trivial one. These additional solutions are not physical solutions because they do not satisfy all of Maxwell's equations. However, these nonphysical solutions are mathematically valid solutions of the formulation. They are the most common sources of the spurious modes.[2] In the following, spurious modes will be referred to as the numerical approximation of these zero frequency nonphysical solutions.

---

2 Some other types of nonphysical solutions may also appear in the discretized problem, for example, due to a dependent set of basis functions (see [Lee et al. 1991b]).

To illustrate the points mentioned earlier, an alternative differential formulation to the original set of equations (7.2) to (7.13) is considered. This alternative formulation has only one of the fields as unknown ($\bar{E}$ or $\bar{H}$).

In order to eliminate one of the unknowns ($\bar{H}$ or $\bar{E}$) of problem (7.2) to (7.13), the fields ($\bar{H}$ or $\bar{E}$) obtained from (7.2) or (7.4) are substituted into (7.4) or (7.2), resulting in an equation with ($\bar{E}$ or $\bar{H}$) as the only unknown. The following set of equations, equivalent[3] to the set (7.2) to (7.13), is obtained:

$$\bar{\nabla} \times (\underline{\underline{f}}_i^{-1} \bar{\nabla} \times \bar{V}) = \omega^2 \underline{\underline{p}}_i \bar{V}, \qquad \text{in } \Omega_i \tag{7.14}$$

$$\bar{\nabla} \cdot (\underline{\underline{p}}_i \bar{V}) = 0, \qquad \text{in } \Omega_i \tag{7.15}$$

$$\bar{a}_n \times \bar{V} = 0, \qquad \text{on } \Gamma_D \tag{7.16}$$

$$\bar{a}_n \times \underline{\underline{f}}^{-1} \bar{\nabla} \times \bar{V} = 0, \qquad \text{on } \Gamma_N \tag{7.17}$$

$$\bar{a}_n \cdot \underline{\underline{p}} \bar{V} = 0, \qquad \text{on } \Gamma_N \tag{7.18}$$

$$\bar{a}_n \times \left(\bar{V}_i - \bar{V}_j\right) = 0, \qquad \text{on } \Gamma_{ij} \tag{7.19}$$

$$\bar{a}_n \times \left(\underline{\underline{f}}_i^{-1} \bar{\nabla} \times \bar{V}_i - \underline{\underline{f}}_j^{-1} \bar{\nabla} \times \bar{V}_j\right) = 0, \qquad \text{on } \Gamma_{ij} \tag{7.20}$$

$$\bar{a}_n \cdot \left(\underline{\underline{p}}_i \bar{V}_i - \underline{\underline{p}}_j \bar{V}_j\right) = 0, \qquad \text{on } \Gamma_{ij} \tag{7.21}$$

Equation (7.14) is commonly known as the double-curl differential equation. In the following, formulations making use of the double-curl equation will be referred to as the double-curl formulation.

The vector field $\bar{V}$ stands for either the electric field $\bar{E}$ or the magnetic field $\bar{H}$. The correspondences between the different variables are shown in Table 7.1.

Note that the preceding substitution process is only valid for $\omega \neq 0$. However, (7.14) together with the rest of the equations (7.15) to (7.21) is equivalent to the original set of Maxwell's equations (7.2) to (7.13). This equivalence is clear for $\omega \neq 0$. For $\omega = 0$ it may be easily shown that the set of equations (7.14) to (7.21) has only the trivial solution: $\bar{E} = \bar{H} = \bar{0}$, and so does the set (7.2) to (7.13). Then, the numerical

[3] Two problems or sets of equations are said to be equivalent if they both have the same solutions.

22solution of equations (7.14) to (7.21) will not provide spurious modes or spurious solutions (see, for example, [Hayata et al. 1986], [Israel and Miniowitz 1990]). Of course, the same stands for the problem posed by (7.2) to (7.13) (see, e.g., [Svedin 1989], [Lebaric and Kajfez 1989]).

Table 7.1
Correspondences for the double-curl differential formulation of (7.14) to (7.21)

	$\bar{E}$ formulation	$\bar{H}$ formulation
$\bar{V}$	$\bar{E}$	$\bar{H}$
$\underline{\underline{f}}$	$\underline{\underline{\mu}}$	$\underline{\underline{\varepsilon}}$
$\underline{\underline{p}}$	$\underline{\underline{\varepsilon}}$	$\underline{\underline{\mu}}$
Γ_D	Γ_{pe}	Γ_{pm}
Γ_N	Γ_{pm}	Γ_{pe}

However, a weak formulation used in finite element discretizations do not include the entire set of equations (7.14) to (7.21). The set of equations (7.14) to (7.21) contains more equations (dependent for $\omega \neq 0$) than unknowns.

The usual weak formulation corresponding to the double-curl differential equation is

$$\int_{\Omega}\left(\bar{\nabla} \times \bar{V}'\right) \cdot \left(\underline{\underline{f}}^{-1} \bar{\nabla} \times \bar{V}\right) d\Omega = \omega^2 \int_{\Omega}\left(\bar{V}' \cdot \underline{\underline{p}}\,\bar{V}\right) d\Omega \tag{7.22}$$

where $\bar{V}'$ is a test function satisfying suitable boundary conditions. The weak formulation given by (7.22) constitutes an eigenvalue problem with ω^2 as the eigenvalue and $\bar{V}$ (electric or magnetic field) as the eigenmode.

In order to arrive at (7.22), equations (7.14), (7.17), and (7.20) have been used (see the details in Section 7.3.1). Thus, the boundary conditions of (7.17) and (7.20) are implicit in (7.22). The boundary conditions imposed in this way are known as the natural boundary conditions of the formulation (see Section 2.3.3.1 and Appendix A.2). They are satisfied in a weak, or distributional sense.

However, the divergence-free condition of (7.15) and its associated boundary conditions (7.18), (7.21) are not, either explicitly or implicitly, present in (7.22) when $\omega = 0$. Following the reasoning expressed previously, nonphysical solutions may exist corresponding to the eigenvalue $\omega = 0$. Therefore, these zero-frequency spurious solutions will not satisfy the divergence-free condition. For the case of $\omega \neq 0$, some weak enforcement of the divergence-free condition occurs (see Section 7.3.1.2).

It may be shown that the dimension of the vector space formed by these zero eigenvalue solutions is infinite, since it is satisfied

$$\bar{\nabla} \times \bar{V} = 0, \qquad \forall \bar{V} \ : \bar{V} = -\bar{\nabla}\phi \tag{7.23}$$

i.e., any $\bar{V}$, that can be derived as the gradient of a scalar function ϕ, i.e., $\bar{V} = -\bar{\nabla}\phi$, will satisfy (7.23) and hence will be a valid eigenvector of (7.22) with eigenvalue $\omega^2 = 0$.

Note that solutions of the form $\bar{\nabla}\phi$ correspond to static solutions of Maxwell's equations when free charges are present. However, this is not the case considered in sets (7.2) to (7.13) or (7.14) to (7.21) and the solutions of the form $\bar{V} = -\bar{\nabla}\phi$ are physically incorrect solutions of the problem analyzed here.

It is clear from (7.23) that the space spanned by the spurious modes is just the null space of the curl operator. This was mentioned in [Hara et al. 1983] and [Hano 1984] and later on in [Crowley et al. 1988] and [Wong and Csendes 1988].

Therefore, due to the infinite dimension of the null space of the curl operator, nonconsistent approximations of the solutions belonging to this space may provide spurious modes for the discretized problem, with numerical eigenvalues far from zero. In general, the spurious eigenvalues will be within the range of eigenvalues of the physical modes, hence polluting the spectrum. In general, a deterministic problem the solution will also be polluted by spurious components (depending on the excitation).

This actually happens when using conventional Lagrange C^0 elements[4] [Davies et al. 1982], [Konrad 1976], [Koshiba et al. 1985a,b], [Rahman and Davies 1984a]. Later in Section 7.2.2, several approaches to avoid the problem of spurious modes when using Lagrange elements will be presented. Their advantages and drawbacks will be briefly reviewed.

On the other hand, curl-conforming elements, i.e., those that only provide continuity of the tangential components, are known to overcome the problem of spurious modes [Bossavit and Verité 1982], [Hano 1984], [Mur and de Hoop 1985], [Welij 1985], [Barton and Csendes 1987], [Csendes and Wong 1987], [Crowley et al. 1988]. Employing these kinds of elements (in our cases in the weak formulation of (7.22)), nonphysical solutions corresponding to the zero eigenvalue of the differential formulation appear as spurious modes that are described by the eigenvectors corresponding to the numerically zero eigenvalues of the discretized problem. Hence, the spurious modes are easily distinguished from the physical solutions, which have strictly positive eigenvalues. Hence, the spurious modes do not pollute the spectrum. In addition, curl-conforming elements satisfy the continuity of the tangential components of the fields and, hence, are consistent with the behavior of the electromagnetic fields at the interface between different media. The perfect electric or magnetic wall Dirichlet boundary conditions are easily imposed directly even when the walls are not parallel to the coordinate axes of the problem. Also, the fact that the nodes are not at the

---

4 C^0 stands for continuous functions without any continuous derivatives. See Appendix A.1.3.

vertices of the mesh provides better convergence in problems with sharp corners. Advantages and disadvantages of the curl-conforming elements are discussed in more detail in Section 7.2.2.2.

The next question seems to be obvious: why do the curl-conforming elements approximate well the zero-frequency eigenvalues and Lagrange elements do not?

It has been shown earlier that spurious modes are static solutions satisfying

$$\bar{V}_{null} = -\bar{\nabla}\phi \tag{7.24}$$

where $\bar{V}_{null}$ stands for the spurious electric or magnetic field. The scalar function ϕ can be seen as a potential function that, although never computed, must exist.

If Lagrange C^0 elements are employed, the discretization of the vector field is made by approximating separately each one of its scalar components (e.g., its Cartesian components). Therefore, after the conventional FEM assembly procedure has taken place, the discretized vector field $\bar{V}_{null}$ is continuous. Now, in order for $\bar{V}_{null}$ to be a continuous function and according to (7.24), ϕ must have at least continuous first-order derivatives, i.e., ϕ must be a C^1 function.[5] But no such C^1 piecewise polynomial interpolation exists, in general, on arbitrary shaped meshes.[6] An inconsistency is then found in the discretization of $\bar{V}_{null}$ since the potential ϕ from which the field is derived may not exist. Thus, Lagrange C^0 elements are, in general, unable to properly model $\bar{V}_{null}$, i.e., fields derived from the gradient of a scalar function, as in (7.24). The inconsistency in the modeling of the null space of the curl operator leads to poor approximations of the field $\bar{V}_{null}$ for the discretized problem. Then, these approximations of the field $\bar{V}_{null}$, i.e., the spurious modes, appear with numerical eigenvalues far different from zero and, hence, polluting the spectrum of the operator.

On the other hand, if curl-conforming elements are used to discretize the field (and hence, $\bar{V}_{null}$), the potential ϕ is not required to be C^1, C^0 continuity is enough. The gradient of a C^0 finite element interpolation function has continuity of the tangential but not of the normal component. This is just the type of continuity imposed by curl-conforming elements. Then, in order to be consistent with (7.24), scalar C^0, not C^1, functions, are required to exist for the mesh. C^0 piecewise polynomial interpolation functions exist on arbitrary meshes and then the field $\bar{V}_{null}$ and the potential ϕ form a consistent pair. Thus, the spurious solutions are well approximated and the eigenvalues of the spurious modes appear as numerically zero values for the discretized problem. Hence, they do not pollute the physical solutions. The problem of the spurious modes is overcome.

---

5 C^1 stands mathematically for functions with continuity of their first-order derivatives. See Appendix A.1.3.

6 C^1 basis functions can exist on special type of meshes (for more details please refer to [Csendes and Wong 1987] or [Wong and Csendes 1988], [Sun et al. 1995]). There, globally C^1 polynomial interpolations over those meshes are built from conventional Lagrange C^0 elements.

2 The number of C^0 polynomial interpolation functions on the mesh, i.e., the number of zero eigenvalues in the discretized problem, will depend on the order and type of polynomial functions involved in the discretization (see Section 7.3). In particular, an example is shown in Section 7.3.2. There, first- and second-order curl-conforming Nédélec basis functions are employed and a different number of zero eigenvalues is obtained. The explanation is given in terms of the dimension of the space formed by the independent scalar potential functions supported by the mesh.

Note that in the case of curl-conforming elements, the spurious modes are not eliminated. It is just the opposite situation. The nonphysical modes are properly modeled and appear at $\omega = 0$, so they do not interfere with the physical modes of the discretized problem.

In summary, the origin of the spurious modes in FEM vector formulations has been addressed. It has been shown how the spurious modes, or spurious solutions, correspond to numerical approximations of nonphysical solutions, i.e., solutions that do not satisfy all of Maxwell's equations simultaneously. These nonphysical modes appear as mathematically valid solutions of the FEM formulation of the problem. This is due to the lack of enforcement of some of Maxwell's equations. The role of the finite element discretization for the problem of spurious modes (Lagrange elements versus curl-conforming elements) has also been explained. In the following, a brief presentation of some early ideas regarding spurious modes is introduced. The different approaches given in the literature to eliminate the spurious modes problem are also addressed.

7.2.1.2 *Some Early Ideas Regarding Spurious Modes*

In this section, a discussion is made of the various ideas about spurious modes in FEM held by the early researchers in this field, which were demonstrated later to be erroneous. However, some of these ideas may still be seen in relatively recent papers

- The origin of the spurious modes was first attributed to the lack of enforcement of the boundary condition $\bar{a}_n \cdot \bar{H}$ (or $\bar{a}_n \cdot \bar{E}$) on perfect electric (or magnetic) walls [Konrad 1976]. This idea was demonstrated to be erroneous as shown in [Davies et al. 1982].
- The spurious modes were thought to be a problem related only to eigenvalue problems (e.g., [Moyer and Schroeder 1991]). However, it had been shown that the deterministic solution could also be contaminated by nonphysical spurious mode components (depending on the excitation) whenever the associated discrete eigenvalue problem suffers from poor approximations of the spurious eigenfunctions [Crowley et al. 1988], [Pinchuk et al. 1988], [Picon 1988], [Ise et al. 1990].
- The origin of the spurious modes was suggested to be related to an inconsistent modeling of the fields at the media interfaces when using Lagrange elements. Thus, the spurious modes were supposed to appear only when applied to

problems with inhomogeneous media. Earlier, this idea was shown to be not true. Spurious solutions are also present in problems with homogeneous media [Rahman and Davies 1984b], [Webb 1985].

As it is well known, the direct discretization of a vector using Lagrange elements, i.e., an independent scalar approximation for each one of the components of the vector, provides an inconsistent modeling of the electric or magnetic fields at the interfaces between different materials. Lagrange elements impose the continuity between elements of all components of the field, while the electric (or magnetic) field is known to exhibit discontinuities of the normal component at the boundaries between media with different permittivity (or permeability). These inconsistencies may always be avoided by placing a pair of nodes at either side of the interfaces and writing explicit equations for these nodes in order to impose the right discontinuous behavior of the field (e.g., [Yuan et al. 1991]). However, this excessively complicates the FEM implementation. On the other hand, curl-conforming elements are consistent with the behavior of the electromagnetic field. They exhibit the continuity of the tangential component but are unable to approximate the jump in the normal components. This leads to the idea that there is a connection between inconsistent modeling of the fields at the media interfaces and spurious modes. Note, however, that the reasoning made in Section 7.2.1.1, relative to the origin of spurious modes, is still valid when the domain Ω is homogeneous, i.e., when no inconsistencies, in the previous sense, are present.

Moreover, if the connection between inconsistent modeling of the fields and the spurious mode would be true, then spurious modes would not be present in formulations with the magnetic field formulation when all the materials have the same permeability. It is well known that this is not the case (e.g., [Koshiba et al. 1985a], [Rahman and Davies 1984b]). An analogous conclusion can be reached for formulations with the electric field when the media are of the same permittivity.

- It was thought erroneously that first-order curl-conforming elements, i.e., the edge elements, eliminated the spurious modes because of the solenoidal nature of their approximation. This idea is still seen in [Wang and Ida 1991] and [Boyse et al. 1992]. As explained in Section 7.2.1.1, the divergence-free equation (7.15) is not implicit in the weak formulation of (7.22) for $\omega = 0$, i.e., for the spurious modes. Since the basis functions of the first-order curl-conforming elements, i.e., edge elements, are divergence-free, it gave rise to the thought that edge elements are eliminating the spurious modes because of their solenoidal nature.

First of all, it should be stressed that spurious modes are not eliminated when using curl-conforming elements. The spurious modes are present in the discretized problem but they correspond to proper approximations of the zero frequency nonphysical solutions. Thus, they appear with numerically zero eigenvalues and hence do not interfere with the physical solutions.

Secondly, it is worth pointing out that edge elements are not globally divergence-free. If they could provide zero divergence on the whole domain Ω, they should exhibit the continuity of the normal component. Edge elements are divergence-free inside each element but the lack of continuity of the normal component is translated into Dirac (delta) functions on the boundary elements and, hence, exhibit nonzero divergence.[7] Also, if edge elements would provide globally divergence-free approximations, they could not approximate fields with nonzero divergence (the spurious solutions), as in fact they do. In fact, higher order curl-conforming elements do not have solenoidal basis functions and still provide solutions not contaminated by spurious modes (see [Peterson 1994] for 2D and [Lee et al. 1991b], [Nédélec 1980] for 3D). Further information may also be found in Section 2.5.1.2 and Appendix C.4.2 and C.4.3.

7.2.2 Solution to the Problem of Spurious Modes

From the very early beginning, the spurious modes have been found to have a nonzero divergence, and this property has been used to distinguish them from the physical modes [Davies et al. 1982] since the divergence-free condition of (7.15) is not implicit in the FEM formulation when $\omega = 0$, as was shown earlier. This procedure may be considered as the first attempt at determining the solution of the problem of spurious modes in FEM. Obviously, this approach was far from the appropriate way to solve the spurious modes problem. In addition, it could not be used for deterministic problems.

As shown in Section 7.2.1.1, the problem of the spurious modes can be solved basically either by eliminating them at the formulation stage or by means of a correct approximation in the discretization process (using curl-conforming elements).

The first approach consists basically of the enforcement of the divergence-free equation. Thus, the zero-frequency spurious modes will not be present. Although it can be done either at the formulation stage or at the discretization level, more techniques are found in the literature on the first approach. The second approach is just the opposite. No attempt is made in this case to impose the solenoidal nature of the field. In this case, the nonphysical solutions are allowed to be present, and it depends on the discretization procedure to separate them. This is the case for curl-conforming elements, in which the zero-frequency nonphysical solutions appear as numerically zero eigenvalues for the spurious modes in the discretized problem. Hence, the spurious solutions can easily be handled. It will be shown that curl-conforming elements are the most suitable choice for 3D vector field problems.

In the following, the role of the FEM formulation is addressed first in Section 7.2.2.1. The role of the discretization is presented afterward in Section 7.2.2.2, paying special attention to the features of curl-conforming elements.

---

7 From a mathematical standpoint, the div operator is not defined as the normal components of the fields are not continuous. Thus, this statement must be understood in a distributional sense.

7.2.2.1 *At the Formulation Stage*

Various approaches regarding the solution of the problem of spurious modes in FEM utilizing vector formulations are reviewed. All of them have in common that the solenoidal character of the field (7.15) is imposed as a constraint in the formulation of the problem. They differ in the way this is achieved.

A rigorous method employed to impose the constraints for the divergence is the employment of *Lagrange multipliers* [Zienkiewicz 1977; Chs. 3 and 12]. However, this approach has not received much attention for the analysis of problems described here, probably due to its mathematical character and the complexity of the resulting formulation (more unknowns need to be added to the original unknowns). On the other hand, the *penalty function method*, which can be seen as a simplified version of the Lagrange multipliers method, has become a common approach in imposing the divergence-free condition in finite element vector formulations.

For the penalty function method, the divergence-free constraint is imposed by adding a penalty term multiplied by a positive real number, χ, in the weak formulation

$$\int_{\Omega}\left(\bar{\nabla}\times\bar{V}'\right)\cdot\left(\underline{\underline{f}}^{-1}\bar{\nabla}\times\bar{V}\right)\mathrm{d}\Omega + \chi\int_{\Omega}\left(\bar{\nabla}\cdot\underline{\underline{p}}\bar{V}'\right)\left(\bar{\nabla}\cdot\underline{\underline{p}}\bar{V}\right)\mathrm{d}\Omega = \omega^2\int_{\Omega}\left(\bar{V}'\cdot\underline{\underline{p}}\bar{V}\right)\mathrm{d}\Omega \qquad (7.25)$$

where $\underline{\underline{p}}$ is usually omitted in the penalty term when $\bar{V}$ is continuous. The vector field $\bar{V}$ is continuous for electrically homogeneous materials (ε = const.) when $\bar{V} = \bar{E}$ or for magnetically homogeneous materials (μ = const) when $\bar{V} = \bar{H}$. The multiplier χ, known as the penalty factor, is no longer a new unknown (as it is in the method of Lagrange multipliers) but an a priori fixed value.

It may be shown [Webb 1985] that the formulation of (7.25) possesses two sets of solutions. Solutions of the first set are divergence-free, and these are the physical modes. Solutions of the second set are curl-free (i.e., their curl is zero) and they are the spurious modes. The solutions of the first set are independent of the factor χ, while the solutions of the second set depend on χ. For the spurious modes, the resonant frequency, i.e., the eigenvalue, increases with the penalty parameter. Hence, the spurious modes are detected by observing modes whose eigenvalues vary with χ. Even more important, by scaling the value of χ the spurious eigenvalues can be forced up and out of the range of the physical modes of interest.

The penalty function method has been applied to several problems, mainly to eigenvalue problems such as waveguide problems and 3D cavity problems [Hara et al. 1983], [Webb 1985], [Koshiba et al. 1985a,b]. It has also been used in deterministic problems [Picon 1988]. The penalty function method has several advantages since there is no need for additional unknowns. They lead to sparse matrices and can be applied to 3D problems. However, the selection of a suitable penalty parameter χ is difficult. If the factor χ is large, then the spurious eigenvalues are shifted out of the range of physical eigenvalues of interest, but the accuracy of the desired physical eigenvalues is seriously affected. On the other hand, if the penalty parameter is very low, then the spurious modes appear between the physical eigenvalues of the spectrum. Furthermore, the selection of the optimum value of the penalty parameter depends on the mode. A cumbersome optimization procedure may be necessary. The choice of the penalty parameter in deterministic problems is also very critical.

The philosophy of the remaining methods is to formulate the problem in such a way that the spurious solutions are no longer a valid mathematical solution of the problem. These methods eliminate the spurious modes of the spectrum, in contrast to the penalty function method, which only pushes them toward the upper part of the spectrum. They eliminate the spurious solutions by imposing the divergence-free condition. Thus the only valid solution for $\omega = 0$ is the trivial one, $\bar{V} = 0$.

A family of methods, sometimes referred to as *reduction methods*, uses the divergence-free condition to reduce the number of degrees of freedom of the problem. In the following, some common approaches of this type are presented.

In some problems, it is possible to write one of the vector components as a function of the other variables, through (7.15). This is the case for waveguides and axisymmetric cavities in which the variation with one of the coordinates of the problem is known. For example, in a waveguide when all the materials are of the same magnetic permeability μ, the longitudinal component of the magnetic field may be written as

$$H_z = \frac{1}{j\beta}\left(\frac{\partial H_x}{\partial x} + \frac{\partial H_y}{\partial y}\right) \tag{7.26}$$

where β is the propagation constant in the longitudinal direction.

This type of approach is known to eliminate the spurious modes, but it has several serious drawbacks. First of all, it is not applicable, in general, to 3D problems where the variation of the field with respect to the coordinates of the problem is not known a priori. On the other hand, if the divergence-free constraint is imposed in the formulation after the discretization, then inversion of matrices is involved and hence the sparsity of the original system is lost [Hayata et al. 1989]. If the divergence-free condition is imposed before the discretization process, then a weak formulation in terms of the transverse components is obtained [Lu and Fernández 1993a,b]. Sparsity of the matrices is retained. However, the formulation in this case requires complicated C^1 elements, as was pointed out in [Wong and Csendes 1989]. As mentioned in Section 7.2.1.1, C^1 global basis functions can exist only for special meshes using conventional C^0 Lagrange finite elements. For more details, please refer to [Csendes and Wong 1987] or [Wong and Csendes 1988], [Sun et al. 1995]. On conventional meshes at least a fifth-order Hermitian polynomial is needed to achieve the required C^1 continuity (see [Israel and Miniowitz 1990]).

An alternate way to eliminate the spurious modes is to introduce a vector potential $\bar{\Psi}$ in such a way that $\bar{V} = \bar{\nabla}\times\bar{\Psi}$ [Wong and Csendes 1988]. The last expression is substituted in (7.22) and Lagrange elements are used for discretizing $\bar{\Psi}$. Thus, the solenoidal nature of the fields is automatically satisfied and, hence, no spurious modes appear. However, C^1 continuity is also required in this case.

Another way to impose the solenoidal constraint is the one proposed in [Kobelansky and Webb 1986]. There, globally divergence-free basis functions are constructed as a linear combination of the original finite element basis functions. This is done by finding the divergence-free eigenmodes of the original problem. The divergence-free eigenmodes are expressed as a function of the original finite element basis functions. The number of divergence-free eigenmodes will be less than the number of original basis functions. With these eigenmodes as the new basis functions, a new eigenvalue problem (of a reduced size compared with the original one) is solved. The spurious modes are not present, but at the expense of solving two eigenvalue problems. It should be noted that the idea of using divergence-free fields for finding the modes of three-dimensional cavities was first introduced by [Konrad 1985]. However, no further development on this technique seems to have been made thereafter.

Another technique, which is completely different from the previous ones, is the formulation with potentials instead of fields. However, this approach is beyond the scope of the book. This section deals with vector electric and/or magnetic field formulations. Only a few remarks are made. The use of potentials has several advantages. Potentials are smoother than fields, and they do not have singularities (although the potentials may have singular derivatives). They can be employed in 3D problems. They provide sparse systems. They can be chosen to be continuous. Lagrange elements are the appropriate basis for their discretization. However, a gauge must be defined in order to assure the uniqueness of the potentials. In addition, the fields are obtained from the potentials and, hence, have a lower accuracy than the potentials. Different types of gauges may be defined. The Coulomb gauge (e.g., [Bárdi and Biro 1991]) and the Lorentz gauge (e.g., [Boyse et al. 1992]) may be found in formulations employing potentials. The formulations with potentials overcome the problem of spurious modes through the fulfillment of the gauge equation. The interested reader may refer to [Bárdi and Biro 1991] and [Boyse et al. 1992a,b].

A comparison between two formulations employing potentials and the electric field may be found in [Bárdi et al. 1993] utilizing the case of loaded waveguides as an example. Lagrange elements (eight-node rectangular isoparametric elements) are used for the discretization of the potentials and rectangular edge elements for the electric field. In [Bárdi et al. 1993], it is shown that there are no significant differences in the accuracy obtained for the eigenvalues. However, a better convergence of the eigenvalues is reported when edge elements are used for structures with sharp corners. As mentioned before, the fields are computed with a lower accuracy in the formulations with potentials because they must be obtained by differentiating the potential.

In summary, there is no satisfactory approach at the formulation level, which is directly applicable to 3D problems, to solve the problem of the spurious modes. The family of methods known as reduction methods are not applicable to 3D problems, where the variation of the field is, in general, not known a priori. The penalty function methods are suitable for 3D problems, but they require the selection of a penalty parameter. This selection is very critical, and hence, the procedure cannot be automated. Other methods need to use complicated C^1 elements in the discretization procedure (e.g.,

[Wong and Csendes 1988]) or require the solution of two eigenvalue problems like the method in [Kobelansky and Webb 1986]. The formulation in terms of potentials is not considered because they do not involve the fields as unknowns. However, their disadvantages regarding the selection of a gauge and the computation of the fields from the potentials have also been outlined previously.

7.2.2.2 *At the Discretization Stage*

In the present section, the role of the discretization process regarding spurious modes is discussed. Two approaches are presented with special emphasis on the second one. The first approach maintains the philosophy of some of the methods of Section 7.2.2.1. Specifically, it consists of discretizing the fields in such a way that the nonphysical solutions are not allowed, i.e., they are eliminated. The second approach consists of using curl-conforming elements (already introduced in Section 7.2.1) to discretize the fields.

For this, the goal of *eliminating* the nonphysical solutions is achieved using a solenoidal basis at the element discretization stage. An attempt is made in [Konrad 1989], but no results have been obtained with these hypothetical divergence-free elements. No further development seems to have been made in this field. It is worth noting that the edge elements are not included here since they do not provide a divergence-free discretization of the fields.

The next choice is to employ a formulation that allows nonphysical solutions to exist and rely on the discretization procedure to avoid the contamination of the spectrum. This is the case when using curl-conforming elements. In the following, the features of the curl-conforming elements are reviewed. Nonphysical solutions corresponding to the zero eigenvalues of the differential formulation appear as a numerically zero eigenvalue of the discretized problem when curl-conforming elements are employed. Thus, the spurious modes may be easily distinguished and discarded from the physical solutions, which have strictly positive eigenvalues. Spurious modes do not pollute the spectrum. Thus, classical formulations, such as the double-curl formulation of (7.22), can be directly employed without any modification.

This second approach is that which has been followed in the FEM solution of the problems that appear later in this chapter. It will be shown that, besides solving elegantly the problem of spurious modes, curl-conforming elements have additional advantages that make them the most suitable choice for discretizing the electric or magnetic field for 3D vector field problems. A review of the features exhibited by this kind of elements in comparison with conventional Lagrange elements is included.

Moreover, the number of zero eigenvalues in the discrete domain can be known in advance and coincides with the dimension of the discrete vector space formed by the gradient of the potentials in the mesh (equation (7.24)). The eigenvectors corresponding to the zero eigenvalue are generated from the gradient of a potential. Thus, in the discrete domain, the number of numerically zero eigenvalues will coincide with the dimension of the finite vector space formed by the gradient of the potentials supported by the given mesh. The dimension of the gradient space will be the number of independent potentials allowed to exist on the mesh minus one. This is because the field solution corresponding to a constant potential on the mesh is the trivial solution, $\bar{V} = \bar{0}$. Therefore, one of the fields obtained from the gradient of the potential (the one that is derived from the constant potential solution) must be discarded. The number of discrete potentials for the mesh will depend on the polynomial interpolation. As a general rule, it may be said that the higher the order of the polynomial interpolation, the larger the number of potentials and, hence, the number of zero eigenvalues. A larger number of unknowns provide a larger number of zero eigenvalues. However, it is not possible to specify the exact number of eigenvalues without knowing the type of polynomial expressions employed in the approximation (e.g., mixed-order and complete). An example with first- and second-order Nédélec (mixed-order) interpolation functions based on tetrahedrons may be found in Section 7.3.2.

Besides solving the spurious modes problem, the use of curl-conforming elements presents additional advantages with respect to many of the approaches mentioned in Section 7.2.2.1. The commonality of the previous approaches has been the use of Lagrange elements to discretize each scalar component of the vector field. Then, the continuity of all the components of the vector field is explicitly enforced. This provides inconsistencies at material interfaces, where the electric or the magnetic field is known to exhibit discontinuities of the normal components. In the case of divergence-free elements, the continuity for the normal component would also be enforced in order to preserve the zero-divergence condition across elements. The inconsistencies at material discontinuities may be prevented by avoiding the conventional node-based FEM assembly related to elements that have part of their boundary at the interfaces. Explicit linear relations are written for these nodes in order to impose the right behavior of the field at the discontinuity (e.g., [Yuan et al. 1991]). However, this procedure excessively complicates the implementation of FEM.

The curl-conforming elements are consistent with the behavior of the electromagnetic field. They exhibit the continuity of the tangential components between elements but provide the jump in the normal component. Thus, the field at material interfaces is properly modeled, provided that different media correspond to different elements. The details about basis functions of the curl-conforming elements employed in this chapter may be found in Section 2.5.1.2, and Appendix C.4.2 and C.4.3. The continuity relationships across elements are shown to be consistent with the boundary conditions of the electromagnetic field. The field components $\bar{a}_n \times \bar{V}$ and $\bar{a}_n \times (\bar{\nabla} \times \bar{V})$ are continuous because the basis functions explicitly enforce this continuity in a strong sense.[8] Also, it enables the jump of the normal component of the field $\bar{V}$ and of the tangential component of the curl of $\bar{V}$. The continuity of $\bar{a}_n \times \underline{f}^{-1} \bar{\nabla} \times \bar{V}$, i.e., of the magnetic field when $\bar{V} = \bar{E}$ or of the electric field when $\bar{V} = \bar{H}$, is imposed by the formulation in a distributional sense. The continuity of the normal component of $\underline{p}\bar{V}$, i.e., the magnetic induction $\bar{B}$ when $\bar{V} = \bar{E}$ or the electric induction $\bar{D}$ when $\bar{V} = \bar{H}$, is

8 The continuity of the normal component of the curl is always enforced in a weak or distributional sense as a consequence of the tangential continuity of the field (Stokes theorem). See Section 7.3.1.2.

also enforced in a distributional sense. More details about the kind of continuity satisfied by the different electromagnetic quantities, in the context of the weak formulation of (7.22), may be found in Section 7.3.1.

Elements conforming in $H(\mathrm{div},\Omega)$ (see Appendix A.1.4, Section 2.5.1.2, and Appendix C.4.1 and C.4.3) or div-conforming elements have the continuity in the sense of the div operator, i.e., the continuity of the normal component as opposed to curl-conforming elements that have continuity of the tangential components. They are suitable for the discretization of the electric and magnetic inductions, $\overline{D}$ and $\overline{B}$. On the other hand, elements conforming in $H(\mathrm{curl},\Omega)$ are appropriate to approximate the electric and magnetic fields, $\overline{E}$ and $\overline{H}$. Therefore, elements in $H(\mathrm{curl},\Omega)$, in conjunction with elements in $H(\mathrm{div},\Omega)$, provide a consistent way of approximating the vector electromagnetic field quantities in opposition to Lagrange elements that are appropriate for scalar unknowns. The discussion will be focused on the $H(\mathrm{curl},\Omega)$ elements because the formulations in this chapter have the electric and the magnetic fields as unknowns.

$H(\mathrm{div},\Omega)$ elements were proposed by Raviart and Thomas for second-order 2D problems [Raviart and Thomas 1977a]. They were extended to 3D for cubes and tetrahedrons [Nédélec 1980], [Thomas and Joly 1981]. In [Nédélec 1980] 3D elements conforming in $H(\mathrm{curl},\Omega)$ were also described.[9] Different types of curl-conforming elements have appeared since then, including those of [Hano 1984], [Mur and de Hoop 1985], and [Brezzi et al. 1985]; a new family in 3D proposed by Nédélec based on this work of Brezzi et al. [Nédélec 1986]; and the covariant projection elements [Crowley et al. 1988], among others. There is a difference between complete polynomial elements (e.g., [Mur and de Hoop 1985], [Brezzi et al. 1985], [Nédélec 1986]) and the mixed-order elements (e.g., [Nédélec 1980], [Hano 1984]). Mixed-order elements have basis functions that provide a different order of polynomial approximation in one direction than in the others. For example, in the case of a first-order triangle, the variation of a component of the field in a given direction (e.g., an edge) is of zero order along that direction and linear in any other direction. On the other hand, polynomial complete elements provide the same order of approximation in any direction. There are other different curl-conforming elements in the literature, but most of them are equivalent to those mentioned previously.

Curl-conforming elements were applied first to solving eddy current problems [Bossavit and Verité 1982] and later on applied to a wide variety of applications like analysis of waveguiding structures ([Hano 1984], [García-Castillo 1992], [García-Castillo and Salazar-Palma 1992a,b], [Blanc-Castillo 1994], [Blanc-Castillo et al. 1995], [Miniowitz and Webb 1991], [Koshiba and Inoue 1992], [Lee et al. 1991 a], [Hirayama et al. 1996], [Wu 1996]), cavity resonances ([Chatterjee et al. 1992], and [Lee et al. 1993]), radiation and scattering problems ([Bossavit and Mayergoyz 1989], [D'Angelo and Mayergoyz 1991], [Jin and Volakis 1991], [Gong et al. 1994], [Peterson 1994]).

Before continuing with the description of the general features of the curl-conforming elements, a remark concerning edge elements is given. In many papers, especially in the early days, the term edge elements was used to designate the kind of

9 The tetrahedrons proposed in [Nédélec 1980] are the elements used in this chapter.

elements with the continuity of the tangential component but not of the normal component, i.e., what nowadays are known as curl-conforming elements, as opposed to the Lagrange elements that had full continuity (tangential and normal). Later, the term edge element was used only for the first-order curl-conforming elements. It is worth clarifying why this happened. The degrees of freedom of the first-order curl-conforming elements are defined as quantities proportional to the tangential components of the field at the edges of the element. Thus, it is logical in this case to relate the name of these elements to the term edge. However, the degrees of freedom of the higher order curl-conforming elements are not only associated with the edges of the mesh, but also with faces and volumes for example. Therefore, the term edge element does not seem to be appropriate to designate these higher order curl-conforming elements, but only those of first order, independent of the geometry of the element, i.e., triangle, cube, and tetrahedron. Note that the triangular edge element corresponds to the $H(\text{curl},\Omega)$ counterpart of the functions proposed by [Rao et al. 1982] to discretize the electric current over triangular patches in the MOM.

Edge elements belong to a wider family of geometric objects referred to as the Whitney forms [Whitney 1957]. Four finite-dimensional vector spaces, W^p, $p = 0,\ldots,3$, are defined as associated with nodes, edges, facets, and volumes of a mesh, respectively. W^0 and W^4 are composed of scalar functions. W^1 and W^2 are made of vector fields. Whitney forms of degree p ($p = 0,\ldots,3$) are associated with these p-simplexes (nodes, edges, facets, and volumes) and generate the spaces W^p. A property of Whitney p-forms is that the integral of a Whitney p-form on a p-simplex is 1 over the p-simplex to which it belongs and 0 elsewhere. Thus, the Lagrange elements are recognized as Whitney 0-forms and the edge elements as Whitney 1-forms. An interested reader may find additional information in [Bossavit 1988] and [Wong et al. 1993]. In the mentioned references, a detailed description of the connection between curl- and div-conforming elements within Whitney forms and its application to electromagnetics is given. Here, only an interesting relation between the W^0 and W^1 spaces in connection with the problem of spurious modes is pointed out:

$$W^0 \quad \overset{\text{grad}}{\rightarrow} \quad W^1 \tag{7.27}$$

i.e., the gradient of a finite element 0-form basis may be expanded in terms of 1-form basis. The 0-form basis, i.e., the basis based on the Lagrange element, is the natural basis of the scalar functions in the mesh, i.e., the potentials of the mesh. On the other hand, the 1-form basis corresponds to the edge element basis functions. Then, according to (7.27), the edge element basis (expanding W^1) is able to approximate the gradient of the C^0 piecewise potentials of the mesh (expanding W^0).

This property is mentioned in Section 7.2.1.1 regarding the feature of curl-conforming elements to approximate the gradients of the potentials of the mesh. In that section, it is shown that the gradient of the C^0 piecewise potentials of the finite element mesh have the continuity of the tangential components but not of the normal components. This is the kind of continuity across elements, exhibited by curl-conforming elements and, hence, in particular by edge elements. Thus, curl-conforming elements, and in particular edge elements, are able to approximate the fields derived as the gradient of a potential, i.e., the fields expanding the null space of the curl operator. Hence, the nonphysical solutions of the FEM formulation, recognized as static solutions derived from a potential, are properly modeled, hence not polluting the spectrum.

Curl-conforming elements, in addition, have other features that make them appropriate candidates for the approximation of the electromagnetic field. Namely, they can impose the Dirichlet boundary conditions on arbitrarily shaped boundaries in an orderly fashion. They have better convergence for the solution in structures with sharp corners. They may provide a direct computation from the degrees of freedom of some of the quantities of interest, such as currents and the Poynting vector.

These advantages are all a consequence of the definition of the degrees of freedom related to the element boundary. Namely, those degrees of freedom are proportional to the tangential component of the field. Equivalently, the component of the field tangential to the boundary of a curl-conforming element depends only on the degrees of freedom related to that boundary. Therefore, the discretized FEM formulation is expressed in terms of the connection across elements of the tangential components of the electric and/or the magnetic fields. The tangential components are, precisely, the components that guarantee the uniqueness theorem and satisfy boundary conditions at perfect electric or perfect magnetic conductors or conditions at the interfaces between different media, for example.

The boundary conditions on perfect electric walls (7.10) or perfect magnetic walls (7.12) only involve the tangential components of the field. Then, for curl-conforming elements, the boundary conditions may be easily imposed by nulling the corresponding degrees of freedom on the walls; i.e., those boundary conditions are homogeneous Dirichlet boundary conditions for those degrees of freedom. Analogous conclusions are derived for interelement continuity relations. They are also easily imposed in terms of the tangential components. On the other hand, the enforcement of boundary conditions related to tangential components with Lagrange elements may be cumbersome because it implies a linear combination of the degrees of freedom of each scalar component of the vector field.

For example, let us consider the case of boundary conditions on perfect electric conductors. Let the boundary be defined by the equation

$$a_{n_x} x + a_{n_y} y + a_{n_z} z = \text{const} \tag{7.28}$$

and let it be a perfect electric wall. The unit normal $\overline{a}_n$ in matrix form is

$$\{a_n\} = \begin{bmatrix} a_{n_x} & a_{n_y} & a_{n_z} \end{bmatrix}^T \tag{7.29}$$

where ^{T} stands for the transpose of a matrix. Then, the boundary condition (7.10) for this case will be

$$\bar{a}_n \times \bar{E} = \bar{0} \Rightarrow \begin{Bmatrix} a_{n_y} E_z - a_{n_z} E_y \\ -(a_{n_x} E_z - a_{n_z} E_x) \\ a_{n_x} E_y - a_{n_y} E_x \end{Bmatrix} = \begin{Bmatrix} 0 \\ 0 \\ 0 \end{Bmatrix} \tag{7.30}$$

Therefore, for the case of Lagrange elements, the degrees of freedom of $\bar{E}$ should satisfy the following relations at the perfect electric wall:

$$\begin{aligned} a_{n_y} d^g_{E_z} - a_{n_z} d^g_{E_y} &= 0 \\ a_{n_x} d^g_{E_z} - a_{n_z} d^g_{E_x} &= 0 \\ a_{n_x} d^g_{E_y} - a_{n_y} d^g_{E_x} &= 0 \end{aligned} \tag{7.31}$$

where ${d_{Ex}}^g$, ${d_{Ey}}^g$ and ${d_{Ez}}^g$ are the global degrees of freedom of the scalar approximations of the x, y, and z components, respectively, of the electric vector field $\bar{E}$.

Obviously, the implementation of (7.31) is not straightforward. It implies that one needs to obtain the unit normal $\bar{a}_n$ at each node of the boundary mesh and it is necessary to impose the linear relations of (7.31) in the FEM matrix. Furthermore, the unit normal $\bar{a}_n$ for the nodes placed at the corners is not uniquely defined.

The previous situation may be compared with the curl-conforming elements in which only homogeneous Dirichlet boundary conditions on the degrees of freedom, associated with nodes on the wall need to be applied, i.e.,:

$$d_i^{g\bar{E}} = 0, \qquad \forall\, i \mid \text{ node } i \in \Gamma_{pe} \tag{7.32}$$

where $d_i^{g\bar{E}}$ are the degrees of freedom of the vector approximation of the field $\bar{E}$.

In addition, the nodes of the curl-conforming elements are never at the vertices and, hence, are never placed at the corners of the structure. Therefore, there is no ambiguity in the definition of the unit normal $\bar{a}_n$ in this case.

Since the nodes of the curl-conforming elements are not at the vertices, a better convergence of the solution is achieved in problems with sharp corners as compared to the use of Lagrange elements (e.g., [Miniowitz and Webb 1991]). This is due to the fact that the field at the corners may be singular.

It is also worth mentioning that the curl-conforming elements can obtain directly, as a postprocess, various electromagnetic quantities of interest. Examples are the calculation of the Poynting vector, or the current sources, among others. For instance, the current on the conductors is directly related to the values of the degrees of freedom of the magnetic field associated with the nodes on those conductors. The correspondence is in such a way that the integral of the current over a surface may be obtained by basically adding the values of the degrees of freedom belonging to the nodes on the surface. Another example is the computation of the field tangential to the ports of a 3D structure as in problems of waveguide discontinuities (see Section 7.4). In this case, the excitation of a given port is done by associating the appropriate values of the degrees of freedom with that port.

One drawback of the curl-conforming elements is that they give rise to larger sized FEM matrices than conventional Lagrange elements. The number of nodes in a mesh for curl-conforming elements is larger than for Lagrange elements (for the same number of elements and order of the polynomial interpolation). Although the discretization of the vector field with Lagrange elements needs three (for 3D) or two (for 2D) separate scalar approximations, i.e., three or two degrees of freedom per node respectively, the previous statement regarding the size of the FEM matrices still holds. However, the connectivity of the nodes is lower with curl-conforming elements than with Lagrange elements. Therefore, very sparse matrices are obtained when using curl-conforming elements and it leads to savings in storage and CPU requirements.

In summary, the features of curl-conforming elements when utilized in vector field FEM formulations for the approximation of the electric or magnetic field have been presented. It has been shown that, in addition to the characterization of the spurious modes, curl-conforming elements exhibit other advantages that make them favorites over other approaches. Besides the fact that some of these approaches are not suitable for 3D problems (e.g., the reduction methods), the procedures described in Section 7.2.2.1 need to alter the original weak formulation in order to avoid the pollution by spurious modes. Thus, complicated C^1 elements, the selection of a critical penalty parameter, or dense matrices, for example, become necessary. Also, all the procedures have in common the use of Lagrange elements to discretize the vector field. Then, full continuity of the field is achieved across elements, leading mainly to inconsistencies at material interfaces and cumbersome procedures in imposing boundary conditions on arbitrarily shaped domains. These problems may be prevented by avoiding the conventional FEM assembly, but it excessively complicates the implementation of FEM. On the other hand, curl-conforming elements deal with the tangential components of the vector field across elements, as it happens in Maxwell's equations. Hence, it provides a natural and consistent way of approximating the electromagnetic field. They allow the use of the original weak formulations without any modification. They constitute the preferred choice for the FEM solution of 3D vector field problems.

Note that the features presented by the curl-conforming elements for the discretization of the electric or magnetic field have their analogous counterparts in the div-conforming elements when the latter are employed for the discretization of the electric induction $\overline{D}$ or the magnetic induction $\overline{B}$. They have not been discussed here because the formulations concerned in this chapter deal with the fields, not with inductions. The role played by the continuity of the tangential components with curl-conforming elements is now replaced by the continuity of the normal components with div-conforming elements. Thus, the homogeneous Dirichlet boundary conditions (7.11) and (7.13) may easily be imposed over arbitrarily shaped boundaries. Also, the normal (but not the tangential) continuity is imposed between elements. Hence, it is consistent with the physical behavior of the inductions. Finally, various quantities may be easily computed directly from the degrees of freedom, e.g., the charge on a conductor from the degrees of freedom associated with the electric induction $\underline{D}$.

7.3 ANALYSIS OF THREE-DIMENSIONAL CAVITY RESONANCES USING THE FINITE ELEMENT METHOD

This section describes the computation of the resonance modes of arbitrarily shaped 3D inhomogeneously loaded cavities. Authors have analyzed these structures using a variety of FEM formulations. In many cases, due to particular symmetry conditions of the structure and the modes of interest, a 2D problem is solved [Hernández-Gil and Pérez-Martínez 1985, 1987]. However, a 3D analysis is required in most cases, although for axisymmetric structures, cylindrical coordinates have been preferred by some authors. Formulations in terms of the electric or/and the magnetic field or scalar and vector potentials have been used. Lagrange elements were first utilized. Different curl-conforming elements are used nowadays. References related to FEM analysis of cavities are [Bárdi et al. 1991], [Bossavit 1990], [Chatterjee et al. 1992], [Davies et al. 1982], [Davies 1993], [Dibben and Metaxas 1997], [Konrad 1985, 1986], [Lebaric and Kajfez 1989], [Lee and Mittra 1992], [Lee et al. 1993], [Oldfield and Ida 1985], [Philippou et al. 1981], [Pichon and Bossavit 1993], [Wang and Ida 1991, 1992], [Wang et al. 1996], [Webb 1985, 1988a], among others.

In this section, FEM formulations with the electric or the magnetic field are employed together with tetrahedral curl-conforming elements. In fact, the cavity eigenvalue problem has been chosen first in order to illustrate some of the ideas presented in Section 7.2 regarding curl-conforming elements and spurious modes.

The FEM formulation is presented in Section 7.3.1. The role of the natural boundary conditions implicit in the weak FEM formulation of the problem is addressed. Features of the curl-conforming basis functions employed in the FEM discretization are also described. Emphasis is placed on satisfying the solution of the boundary conditions either in a strong or in a weak sense. The issue of the number of numerically zero eigenvalues, i.e., the number of spurious modes in the discretized problem, is discussed in Section 7.3.2. This is illustrated through an example utilizing two different sets of basis functions. Finally, some numerical results are presented in Section 7.3.3 showing the validity of the approach.

7.3.1 Finite Element Method Formulation

The FEM formulation used to compute the resonance modes of 3D microwave cavities is the double-curl weak formulation of (7.22) (Section 7.2.1.1). The procedure on how to obtain this formulation starts from Maxwell's equations and is explained in detail.

As shown in Section 7.2.1.1, a differential equation, in terms of the electric or the magnetic field may be obtained from Maxwell's curl equations, (7.2) and (7.4). The electric field $\underline{E}$ or the magnetic field $\underline{H}$ is obtained from one of the Maxwell's curl equations and then substituted into the other, resulting in a differential equation with $\bar{E}$ or $\bar{H}$ as the only unknown. The differential equation is the double-curl equation of (7.14) that is repeated here for convenience:

$$\bar{\nabla} \times \left(\underline{\underline{f}}_i^{-1} \, \bar{\nabla} \times \bar{V} \right) = \omega^2 \, \underline{\underline{p}}_i \, \bar{V}, \qquad \text{in } \Omega_i \tag{7.33}$$

The vector field $\bar{V}$ stands for either the electric field $\bar{E}$ or the magnetic field $\bar{H}$, similarly, with the tensors $\underline{\underline{f}}$ and $\underline{\underline{p}}$ that are used to denote the permittivity and the permeability of the anisotropic media. The correspondences are shown in Table 7.1.

The permittivity and permeability tensors are considered to be real and symmetric in order to obtain FEM matrices that are real and symmetric. With this kind of tensor for the permittivity and permeability, the field $\bar{V}$ may be considered to be real.

The differential equation (7.33) corresponds to the strong formulation (see Section 2.2) of an eigenvalue problem in which ω^2 is the eigenvalue and the vector field $\bar{V}$ is the eigenvector. According to the notation of (2.2), the eigenvalue problem of (7.33) may be written as

$$\mathcal{L}_1 \bar{V} - \lambda \mathcal{L}_2 \bar{V} = \bar{0} \tag{7.34}$$

or, equivalently

$$\mathcal{L} \bar{V} = \bar{0} \tag{7.35}$$

where λ is the eigenvalue, i.e., ω^2 and $\mathcal{L}$, $\mathcal{L}_1$, $\mathcal{L}_2$ are vector differential operators. Here:

$$\mathcal{L}_1 = \bar{\nabla} \times \underline{\underline{f}}^{-1} \, \bar{\nabla} \times \tag{7.36}$$

$$\mathcal{L}_2 = \underline{\underline{p}} \tag{7.37}$$

$$\mathcal{L} = \mathcal{L}_1 - \lambda \mathcal{L}_2 \tag{7.38}$$

The domain Ω is considered to be enclosed by the boundary Γ. Note that the domain does not need to be connected, e.g., a cavity formed by two concentric spheres. The boundary conditions on Γ are either due to a perfect electric wall or due to a perfect magnetic wall. The part of the boundary Γ that is a perfect electric wall is denoted by Γ_{pe}. The perfect magnetic wall of the boundary is referred to as Γ_{pm}. Therefore, it may be written that $\Gamma = \Gamma_{pe} \cup \Gamma_{pm}$.

A perfect electric wall stands for the boundary condition on a perfect conductor:

$$\Gamma_{pe} : \bar{a}_n \times \bar{E} = 0 \tag{7.39}$$

This is the condition used over the conducting surface of the cavity. Also, it models boundaries as symmetry planes that are of the short circuit type.

On the other hand, the perfect magnetic wall boundary condition is the dual of the perfect conductor boundary condition:

$$\Gamma_{pm} : \bar{a}_n \times \bar{H} = 0 \tag{7.40}$$

This is the condition used to model symmetry planes of open circuit type.

In some situations, the geometry of the microwave cavity is such that the problem may be analyzed over one-half or one-fourth, for example, of the structure. Symmetry planes (perfect electric and magnetic walls) are introduced in that case in order to reduce the domain Ω and, hence, the computational cost. The cavity is then analyzed several times with a combination of those two boundary conditions (7.39), (7.40) on the symmetry planes. For example, if the microwave cavity is such that one symmetry plane is introduced, then the problem domain will be reduced by one-half of the original cavity domain; i.e., the cavity problem is solved with half the number of unknowns than the original problem. The reduced problem will have to be solved twice, one with a perfect electric wall boundary condition on the symmetry plane and a second one with a perfect magnetic wall boundary condition. The whole spectrum of the original cavity is then obtained by adding the eigenvalues of the two reduced problems considered, i.e., the one with a perfect electric wall on the symmetry plane and the one with a perfect magnetic wall boundary condition on the symmetry plane. This procedure allows one to minimize the computational cost because the reduced problem has a considerably less number of unknowns. Considering that the computational cost of solving a linear system of equations with N_d unknowns is proportional to N_d^{α}, where α is always greater than 1, it is advantageous to solve a half-sized problem twice than to solve the original problem with double the number of unknowns N_d. This is also true when more than two symmetry planes are defined. Whenever the number of unknowns N_d is large, computational efficiency is achieved by introducing symmetry planes. For example, if two symmetry planes are found in the cavity, the reduced problem domain would be one-fourth the size of the original domain and it should be analyzed four times, with four different combinations of the perfect electric and magnetic boundary conditions on the two symmetry planes.

In summary, when the formulation uses the electric field, i.e., $\bar{V} = \bar{E}$, the boundary conditions are given by:

- Perfect electric wall (Γ_{pe}) $\Rightarrow \bar{a}_n \times \bar{E} = \bar{0}$;
- Perfect magnetic wall (Γ_{pm}) $\Rightarrow \bar{a}_n \times \bar{H} = \bar{0} \equiv \bar{a}_n \times \underline{\underline{\mu}}^{-1}\, \bar{\nabla} \times \bar{E} = \bar{0}$.

When the formulation utilizes the magnetic field, i.e., $\bar{V} = \bar{H}$, the boundary conditions may be written as:

- Perfect electric wall (Γ_{pe}) $\Rightarrow \bar{a}_n \times \bar{E} = \bar{0} \equiv \bar{a}_n \times \underline{\underline{\varepsilon}}^{-1}\, \bar{\nabla} \times \bar{H} = \bar{0}$;
- Perfect magnetic wall (Γ_{pm}) $\Rightarrow \bar{a}_n \times \bar{H} = \bar{0}$.

Equivalently the boundary conditions are given by (7.16) and (7.17)

$$\bar{a}_n \times \bar{V} = \bar{0}\ , \quad \text{on } \Gamma_D \tag{7.41}$$

$$\bar{a}_n \times \underline{\underline{f}}^{-1}\, \bar{\nabla} \times \bar{V} = \bar{0}\ , \quad \text{on } \Gamma_N \tag{7.42}$$

where $\Gamma = \Gamma_D \cup \Gamma_N$.

The symbol Γ_D denotes Γ_{pe} when $\bar{V} = \bar{E}$, i.e., when the formulation is with the electric field, and Γ_{pm} when $\bar{V} = \bar{H}$, i.e., when the formulation is with the magnetic field. It will be shown later that the boundary condition on Γ_D, i.e., (7.41), represents an essential boundary condition for the weak formulation with FEM.

On the other hand, the symbol Γ_N denotes Γ_{pm} when $\bar{V} = \bar{E}$ and Γ_{pe} when $\bar{V} = \bar{H}$. The boundary condition on Γ_N, i.e., (7.42), is imposed as a natural boundary condition for the FEM weak formulation, as it will be shown later in this section.

Now, in order to solve the problem defined by (7.33) on a computer, a variational method is applied to the differential equation of the problem. It may consist of the minimization of a suitable functional or the weighted residual technique. For the formulation given by (7.33) and making suitable assumptions, it will be shown that both procedures are equivalent. Further information about the different variational methods used with FEM may be found in Section 2.4.

7.3.1.1 *Variational Formulation*

The method of weighted residuals is introduced first. Consider the residual of the differential equation (7.33). It will be exactly zero when $\bar{V}$ is the solution of the problem. For an approximation to the solution V, denoted by $\bar{\upsilon}$, the residual will not be exactly zero but close to zero:

$$R(\bar{\upsilon}) = \bar{\nabla} \times \left(\underline{\underline{f}}^{-1}\, \bar{\nabla} \times \bar{\upsilon} \right) - \omega^2 \underline{\underline{p}}\, \bar{\upsilon} \approx 0 \tag{7.43}$$

where $\bar{\upsilon}$ consists of a linear combination of a given number, N_d, of vector functions:

$$\bar{\upsilon} = \sum_{i=1}^{N_d} \bar{N}_i\, d_i \tag{7.44}$$

and the functions $\bar{N}_i$ are usually referred to as the basis functions.

The Hilbert space to which the solution $\bar{V}$ belongs is referred as $H(\bar{V})$. The closed subspace of the Hilbert space $H(\bar{V})$ spanned by the approximations $\bar{\upsilon}$ will be denoted as $H(\bar{\upsilon})$. The Hilbert spaces $H(\bar{V})$ and $H(\bar{\upsilon})$ are mapped by the differential operator $\mathcal{L}$ into the Hilbert spaces referred to as $H(\bar{V}')$ and $H(\bar{\upsilon}')$, respectively. The vector functions of the Hilbert spaces $H(\bar{V}')$ and $H(\bar{\upsilon}')$ that satisfy the homogeneous form of the essential boundary conditions required on $\bar{V}$ and $\bar{\upsilon}$ are denoted as $H(\bar{V}_0')$ and $H(\bar{\upsilon}_0')$, respectively. This is illustrated by Figure 7.1.

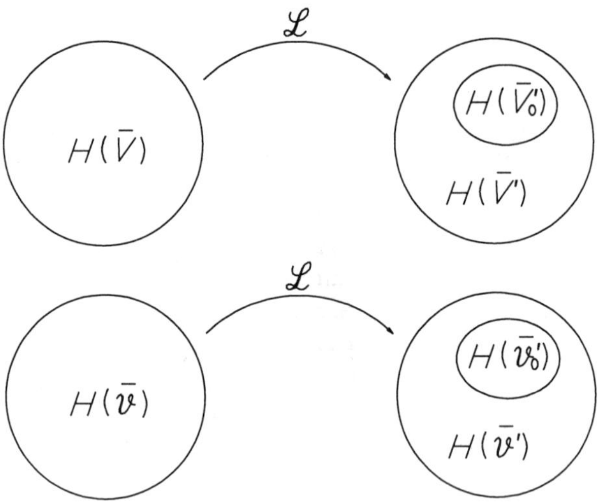

Figure 7.1 Relations between Hilbert spaces for basis and test functions.

Next, by using the Best Approximation Theorem (also known as Projection Theorem), the linear approximation is made orthogonal to the error [Debnath and Mikusiński 1990], i.e., to the residual. This is shown in Figure 7.2.

Then the best approximation $\bar{\upsilon}$ will satisfy

$$< \bar{\upsilon}' ; \; R(\bar{\upsilon}) > \; = 0 \tag{7.45}$$

where $<\cdot \, ; \, \cdot>$ denotes the inner product and $\bar{\upsilon}'$ is any function belonging to $H(\bar{\upsilon}_0')$ The functions $\bar{\upsilon}'$ are referred to as test functions.

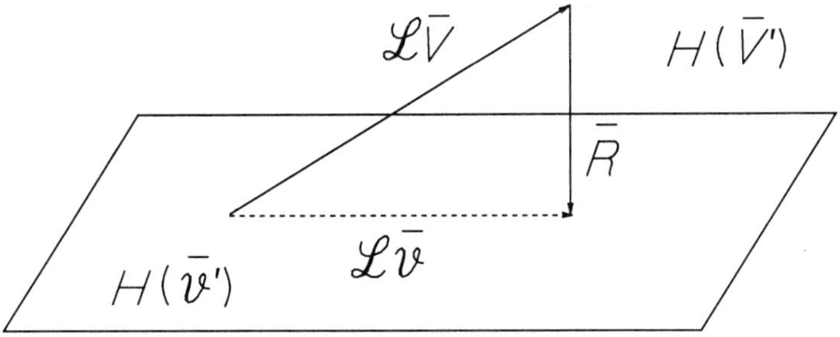

Figure 7.2 Projection Theorem.

The term *best approximation* is applied to the approximation obtained under the condition of minimum norm, i.e.,

$$\| \mathcal{L}\bar{\upsilon} - \mathcal{L}\bar{V} \| \leq \| \bar{\upsilon}' - \mathcal{L}\bar{V} \| \,, \quad \forall \; \bar{\upsilon}' \in H(\bar{\upsilon}') \tag{7.46}$$

where $\|\cdot\|$ is the norm induced by the inner product.

The inner product employed in this case is the conventional inner product for real-valued vector functions, i.e.,

$$<\bar{t}\,;\,\bar{q}> = \int_{\Omega} \bar{t} \cdot \bar{q} \; d\Omega \tag{7.47}$$

where the dot stands for the scalar product. Thus, (7.45) may be written as

$$\left\langle \bar{\upsilon}' ; R(\bar{\upsilon}) \right\rangle = \int_{\Omega} \left(\bar{\upsilon}' \cdot \left(\bar{\nabla} \times \underline{f}^{-1} \, \bar{\nabla} \times \bar{\upsilon} \right) - \omega^2 \, \bar{\upsilon}' \cdot \underline{p} \, \bar{\upsilon} \right) d\Omega = 0 \tag{7.48}$$

Equation (7.48) must be satisfied for all $\bar{\upsilon}'$ belonging to $H(\bar{\upsilon}_0')$. In practice, it is forced to be satisfied by a basis of $H(\bar{\upsilon}_0')$.

Equation (7.48) constitutes a weak formulation of the problem defined by (7.33). However, it may be observed that, in the formulation of (7.48), the vector function $\bar{\upsilon}$, and hence, each of the basis functions, should have second-order derivatives. In addition, they must satisfy all the boundary conditions of the problem (7.41), (7.42). Similarly, $\bar{\upsilon}'$ must satisfy the homogeneous form of the boundary conditions. Note that in this case, the boundary conditions of the problem (7.41), (7.42) are homogeneous, so no distinction needs to be made between the boundary conditions satisfied by $\bar{\upsilon}'$ and those satisfied by $\bar{\upsilon}$.

Usually, in FEM formulations, whenever possible, an integration by parts is performed in order to weaken the derivability conditions of $\bar{\upsilon}$. Then, the derivability conditions of $\bar{\upsilon}'$ turn out to be stronger. In this process, the different boundary conditions that $\bar{\upsilon}$ and $\bar{\upsilon}'$ should satisfy are reduced. Some boundary integral terms are obtained via this procedure. They determine the nature (essential or natural) of the boundary conditions of the problem.

For the formulation of (7.48), the first vector Green's theorem is employed, namely,

$$\iiint [(\bar{\nabla}\times\bar{F}_1)\cdot(\bar{\nabla}\times\bar{F}_2) - \bar{F}_1\cdot\bar{\nabla}\times\bar{\nabla}\times\bar{F}_2]\, d\Omega = \iint \bar{a}_n\cdot(\bar{F}_1\times\bar{\nabla}\times\bar{F}_2)\, d\Gamma \tag{7.49}$$

where $\bar{a}_n$ is the unit vector, normal to surface Γ.

Taking into account (7.49), (7.48) may be written as

$$\begin{aligned}\langle\bar{\upsilon}';R(\bar{\upsilon})\rangle &= \int_\Omega\left((\bar{\nabla}\times\bar{\upsilon}')\cdot\left(\underline{\underline{f}}^{-1}\bar{\nabla}\times\bar{\upsilon}\right) - \omega^2\,\bar{\upsilon}'\cdot\underline{\underline{p}}\,\bar{\upsilon}\right) d\Omega \\ &\quad - \int_\Gamma \bar{a}_n\cdot\left(\bar{\upsilon}'\times\underline{\underline{f}}^{-1}\,\bar{\nabla}\times\bar{\upsilon}\right)\, d\Gamma = 0\end{aligned} \tag{7.50}$$

Note that in (7.50), $\bar{\upsilon}$ should have only first-order derivatives and not second order derivatives as in (7.48). However, now $\bar{\upsilon}'$ also requires first-order derivatives. In addition, the boundary conditions on $\bar{\upsilon}'$ and $\bar{\upsilon}$ may be relaxed; i.e., both $\bar{\upsilon}'$ and $\bar{\upsilon}$ do not need to satisfy all the boundary conditions of the problem (7.41) and (7.42). It is shown below that $\bar{\upsilon}$ must satisfy only the essential boundary conditions of the problem (7.41). Similarly, $\bar{\upsilon}'$ (the test function) must satisfy the homogeneous form of the essential boundary conditions of the problem. However, in this case, the essential boundary conditions are homogeneous and, hence, there is no need to distinguish between the boundary conditions of $\bar{\upsilon}$ and $\bar{\upsilon}'$.

The concepts of essential and natural boundary conditions were already introduced in Section 2.3.3.1 and Appendix A.2. However, they are discussed again in the context for the particular case of the weak formulation of (7.50).

The key is the boundary integral that appears in (7.50). It is important to note that the only components that take part in the boundary integral are tangential components. The boundary integral may be written as

$$\int_\Gamma \bar{a}_n \cdot \left(\bar{\upsilon}' \times \underline{\underline{f}}^{-1}\bar{\nabla}\times\bar{\upsilon}\right) \mathrm{d}\Gamma = -\int_\Gamma \bar{\upsilon}' \cdot \left(\bar{a}_n \times \underline{\underline{f}}^{-1}\bar{\nabla}\times\bar{\upsilon}\right) \mathrm{d}\Gamma \tag{7.51}$$

or

$$\int_\Gamma \bar{a}_n \cdot \left(\bar{\upsilon}' \times \underline{\underline{f}}^{-1}\bar{\nabla}\times\bar{\upsilon}\right) \mathrm{d}\Gamma = \int_\Gamma \left(\bar{a}_n \times \bar{\upsilon}'\right) \cdot \left(\underline{\underline{f}}^{-1}\bar{\nabla}\times\bar{\upsilon}\right) \mathrm{d}\Gamma \tag{7.52}$$

Observing (7.51), it may be seen that the unknown of the problem, $\bar{\upsilon}$, appears in the term $\bar{a}_n \times \underline{\underline{f}}^{-1}\bar{\nabla}\times\bar{\upsilon}$ multiplying the test functions $\bar{\upsilon}'$. This term is related to the natural boundary conditions of the weak formulation. Thus, if a problem has only natural boundary conditions, then these conditions may be imposed by substituting their known value (i.e., the value of $\bar{a}_n \times \underline{\underline{f}}^{-1}\bar{\nabla}\times\bar{\upsilon}$) in (7.51) (i.e., in (7.50)). Because of this procedure, the natural boundary conditions are embedded in an implicit way in the formulation of the problem. Note that the natural boundary conditions will be satisfied by $\bar{\upsilon}$ only in a distributional sense due to the presence of the weighted integral. As a consequence, the basis functions are not required to fulfill any conditions over that part of the boundary.

The remaining boundary condition, i.e., $\bar{a}_n \times \bar{V} = \bar{0}$, constitutes the essential boundary condition of the formulation. As opposed to the natural boundary condition, the essential boundary condition is not implicit in the formulation. Thus, it must be explicitly satisfied by $\bar{\upsilon}$.

By considering that $\Gamma = \Gamma_D \cup \Gamma_N$, the boundary term of (7.50) may be further written as

$$\int_\Gamma \bar{a}_n \cdot \left(\bar{\upsilon}' \times \underline{\underline{f}}^{-1}\bar{\nabla}\times\bar{\upsilon}\right) \mathrm{d}\Gamma = \int_{\Gamma_D} \bar{a}_n \cdot \left(\bar{\upsilon}' \times \underline{\underline{f}}^{-1}\bar{\nabla}\times\bar{\upsilon}\right) \mathrm{d}\Gamma + \int_{\Gamma_N} \bar{a}_n \cdot \left(\bar{\upsilon}' \times \underline{\underline{f}}^{-1}\bar{\nabla}\times\bar{\upsilon}\right) \mathrm{d}\Gamma \tag{7.53}$$

Because of (7.52), the first term on the right-hand side of (7.53) is exactly zero if the test functions, $\bar{\upsilon}'$, are chosen so that they satisfy the homogeneous form of the essential boundary conditions over Γ_D, i.e., if they belong to $H(\bar{\upsilon}_0')$:

$$\begin{aligned} \text{On } \Gamma_D : (\bar{a}_n \times \bar{\upsilon}') = 0 \Rightarrow & \int_{\Gamma_D} \bar{a}_n \cdot \left(\bar{\upsilon}' \times \underline{\underline{f}}^{-1}\bar{\nabla}\times\bar{\upsilon}\right) \mathrm{d}\Gamma \\ &= \int_{\Gamma_D} \left(\bar{a}_n \times \bar{\upsilon}'\right) \cdot \left(\underline{\underline{f}}^{-1}\bar{\nabla}\times\bar{\upsilon}\right) \mathrm{d}\Gamma = 0 \end{aligned} \tag{7.54}$$

Because of (7.51) and (7.42), the second term is also zero:

$$\begin{aligned} \text{On } \Gamma_N : \bar{a}_n \times \underline{\underline{f}}^{-1}\bar{\nabla}\times\bar{\upsilon} = 0 \Rightarrow & \int_{\Gamma_N} \bar{a}_n \cdot \left(\bar{\upsilon}' \times \underline{\underline{f}}^{-1}\bar{\nabla}\times\bar{\upsilon}\right) \mathrm{d}\Gamma \\ &= -\int_{\Gamma_N} \bar{\upsilon}' \cdot \left(\bar{a}_n \times \underline{\underline{f}}^{-1}\bar{\nabla}\times\bar{\upsilon}\right) \mathrm{d}\Gamma = 0 \end{aligned} \tag{7.55}$$

Note that enforcing the boundary integral on Γ_N to zero, as shown in (7.55), is equivalent to imposing (7.42) in a distributional sense, i.e., as a natural boundary condition.

Taking into account (7.53), (7.54), and (7.55), (7.50) is written as

$$< \bar{\upsilon}' ; R(\bar{\upsilon})> = \int_{\Omega} \left(\left(\bar{\nabla} \times \bar{\upsilon}' \right) \cdot \left(\underline{\underline{f}}^{-1} \bar{\nabla} \times \bar{\upsilon} \right) - \omega^2 \, \bar{\upsilon}' \cdot \underline{\underline{p}} \, \bar{\upsilon} \right) d\Omega = 0 \tag{7.56}$$

Thus, (7.56) is the weak formulation employed in this section and, as explained above, already incorporates the boundary condition (7.42) in a distributional sense.

The next step is the discretization of (7.56) in order to obtain an algebraic system of equations. The approximation $\bar{\upsilon}$ of the exact solution consists of a linear combination of N_d basis functions. The vector function $\bar{\upsilon}$ must satisfy (7.56) for a finite number of M independent test functions $\bar{\upsilon}'$, together with the essential boundary conditions. Usually, $M = N_d$ and, hence, a determined system of equations is obtained. Depending on the choice of the test functions, the method of the weighted residuals leads to different procedures. Examples are the Galerkin method when the test functions are identical to the basis functions and the Collocation method (also known as point-matching method) where the test functions are Dirac delta functions. More information may be found in Section 2.4.2.

The Galerkin method has been chosen for the formulation of this section and in the remaining sections of the chapter. Therefore, the test functions $\bar{\upsilon}'$ are equal to the basis functions $\bar{N}_i$ of (7.44). Thus, the weak formulation of (7.56) leads to the following system of equations:

$$<\bar{N}_i ; R(\bar{\upsilon})> = 0 , \quad i = 1, \dots , N_d \tag{7.57}$$

that is,

$$\int_{\Omega} \left(\left(\bar{\nabla} \times \bar{N}_i \right) \cdot \left(\underline{\underline{f}}^{-1} \bar{\nabla} \times \bar{\upsilon} \right) - \omega^2 \bar{N}_i \cdot \underline{\underline{p}} \, \bar{\upsilon} \right) d\Omega = 0 , \quad i = 1, \dots , N_d \tag{7.58}$$

Note that the use of the basis functions as test functions is valid, from a mathematical point of view, when certain conditions are satisfied. In particular, if $H(\bar{V}')$ $\subseteq H(\bar{V})$ (or equivalently if $H(\bar{\upsilon}') \subseteq H(\bar{\upsilon})$), then the basis functions span the Hilbert space of the test functions and, hence, the basis functions may be used as test functions. This is the case when the differential operator is self-adjoint [Sarkar 1985].

It may be shown that the operator $\mathfrak{L}$ of (7.38), together with the boundary conditions (7.41) and (7.42), is self-adjoint. Therefore, the Galerkin method may be utilized.

In the procedure described previously, the test functions have been required to satisfy over Γ_D the homogeneous form of the essential boundary conditions of the problem. However, the basis functions $\overline{N}_i$ of (7.44), in general, will not satisfy those conditions. Moreover, $\overline{\upsilon}$ should satisfy the essential boundary conditions. This apparent contradiction is easily solved in the frame of FEM because it performs a discretization of the problem by subdomains. Taking into account that each basis function is identically zero all over the domain, except for a given subdomain, the approximation of the vector field, may be written as

$$\overline{\upsilon} = \overline{N}_0 + \sum_{i=1}^{N_k} \overline{N}_i \, d_i \qquad (7.59)$$

where $\overline{N}_0$ satisfies the essential boundary conditions and $\overline{N}_i$ satisfies the homogeneous form of the essential boundary conditions of the problem. Here, N_k stands for the number of degrees of freedom not related to the boundaries with essential conditions.

Until now, the way in which the unknown $\overline{V}$ is approximated through (7.44) has not been mentioned. FEM will now be used to discretize the problem. What FEM does is to provide an automatic procedure, independent of the geometry of the problem domain Ω, to build the approximation $\overline{\upsilon}$ utilized in the variational method. The variational formulation is applied to each one of the finite elements of the mesh, and a set of discrete integral forms is obtained for each finite element. The problem is then characterized by a global system of equations obtained after the assembly of the local discrete integral forms and the enforcement of the essential boundary conditions. Thus, FEM may be seen as an application by subdomains (in each finite element of the mesh) of the variational method, followed by the assembly procedure.

The application of the variational formulation of (7.50) to each finite element of the mesh leads to

$$\sum_{e=1}^{N_e} \left[\int_{\Omega^e} \left((\overline{\nabla} \times \overline{\upsilon}'^{\,e}) \cdot \left((\underline{f}^e)^{-1} \overline{\nabla} \times \overline{\upsilon}^e \right) - \omega^2 \, \overline{\upsilon}'^{\,e} \cdot \underline{p}^e \, \overline{\upsilon}^e \right) d\Omega \right. \qquad (7.60)$$
$$\left. - \int_{\Gamma^e} \overline{a}_n^e \cdot \left(\overline{\upsilon}'^{\,e} \times (\underline{f}^e)^{-1} \overline{\nabla} \times \overline{\upsilon}^e \right) d\Gamma \right] = 0$$

where N_e is the total number of elements. Here the superscript stands for the variables specialized to the eth element.

The tensors $\underline{f}^e$ and $\underline{g}^e$ denote the permittivity or permeability dyadics of the media in the eth element (see Table 7.1). Thus, by assigning different tensors to the elements, inhomogeneous media are easily handled. This corresponds to a piecewise-homogeneous medium in which the permittivity and permeability tensors are constant over the element. This is the case considered in the problems presented in this chapter. However, the situation can be more general. The permittivity and the permeability may be a function of the position inside the element.

Observing (7.60), it is seen that, as in (7.50), a boundary integral is present in the formulation but now is extended to the element boundary Γ^e. Then, the same procedure explained previously, with respect to the implicit enforcement of natural boundary conditions on the cavity boundary may be applied to the boundary conditions across elements. The procedure is completely analogous to that performed in (7.50) to (7.56).

Taking into account that the boundary integral of (7.60) extended to the cavity boundaries is zero, as shown in (7.54) and (7.55), (7.60) may be written as

$$\sum_{e=1}^{N_e} \left[\int_{\Omega^e} \left((\bar{\nabla} \times \bar{\upsilon}'^{\,e}) \cdot \left((\underline{\underline{f}}^e)^{-1} \, \bar{\nabla} \times \bar{\upsilon}^e \right) - \omega^2 \, \bar{\upsilon}'^{\,e} \cdot \underline{\underline{p}}^e \, \bar{\upsilon}^e \right) d\Omega \right. \\ \left. - \int_{\Gamma^e \not\subset \Gamma} \bar{a}_n^e \cdot \left(\bar{\upsilon}'^{\,e} \times (\underline{\underline{f}}^e)^{-1} \, \bar{\nabla} \times \bar{\upsilon}^e \right) d\Gamma \right] = 0 \tag{7.61}$$

If the boundary integral extended to the internal element boundaries (internal in the sense that they do not have any edge or face belonging to the boundary Γ) is equated to zero:

$$\sum_{e=1}^{N_e} \left(\int_{\Gamma^e \not\subset \Gamma} \bar{a}_n^e \cdot \left(\bar{\upsilon}'^{\,e} \times (\underline{\underline{f}}^e)^{-1} \, \bar{\nabla} \times \bar{\upsilon}^e \right) d\Gamma \right) = 0 \tag{7.62}$$

then the following equation is obtained:

$$\sum_{e=1}^{N_e} \left(\int_{\Omega^e} \left((\bar{\nabla} \times \bar{\upsilon}'^{\,e}) \cdot \left((\underline{\underline{f}}^e)^{-1} \, \bar{\nabla} \times \bar{\upsilon}^e \right) - \omega^2 \, \bar{\upsilon}'^{\,e} \cdot \underline{\underline{p}}^e \, \bar{\upsilon}^e \right) d\Omega \right) = 0 \tag{7.63}$$

and the FEM formulation will be a conforming one.

The weak formulation of (7.63) is the element-by-element counterpart of (7.56). However, it is important to remark that (7.63) implies (7.62). If a Galerkin procedure is used, i.e., if the test functions coincide with the basis functions used in the approximation of the unknown, (7.63) leads to the following system of equations:

$$\sum_{e=1}^{N_e} \left(\int_{\Omega^e} \left((\bar{\nabla} \times \bar{N}_i) \cdot \left((\underline{\underline{f}}^e)^{-1} \, \bar{\nabla} \times \bar{\upsilon}^e \right) - \omega^2 \bar{N}_i \cdot \underline{\underline{p}}^e \, \bar{\upsilon}^e \right) d\Omega \right) = 0 \,, \quad i = 1, \ldots, n_e \tag{7.64}$$

where n_e is the number of degrees of freedom of element Ω^e. In the following, it will be shown that, whenever the test functions $\bar{\upsilon}'^e$ are chosen to have continuous tangential components across element interfaces, then (7.62) is equivalent, in a distributional sense, to enforcing the continuity of $\bar{a}_n \times \underline{\underline{f}}^{-1} \, \bar{\nabla} \times \bar{\upsilon}$ across elements.

Consider the interface between elements Ω^i and Ω^j, Γ^{ij}. A unit vector normal to the interface $\bar{a}_n^{ij}$ is defined directed from element Ω^i to element Ω^j. The outward unit normal vectors for the element Ω^i, $\bar{a}_n^{\,i}$, and Ω^j, $\bar{a}_n^{\,j}$ satisfy

$$\bar{a}_n^{ij} = \bar{a}_n^{i} = -\bar{a}_n^{j} \tag{7.65}$$

and, therefore, if (7.62) is specialized to the interface between elements Ω^i and Ω^j, it may be written as

$$\begin{aligned}&\int_{\Gamma^{ij}} \left(\bar{a}_n^{\,i} \cdot \left(\bar{\upsilon}'^{\,i} \times (\underline{f}^{\,i})^{-1}\, \bar{\nabla} \times \bar{\upsilon}^i \right) + \bar{a}_n^{\,j} \cdot \left(\bar{\upsilon}'^{\,j} \times (\underline{f}^{\,j})^{-1}\, \bar{\nabla} \times \bar{\upsilon}^j \right) \right) \mathrm{d}\Gamma \\ &= \int_{\Gamma^{ij}} \bar{a}_n^{\,ij} \cdot \left(\bar{\upsilon}'^{\,i} \times (\underline{f}^{\,i})^{-1}\, \bar{\nabla} \times \bar{\upsilon}^i - \bar{\upsilon}'^{\,j} \times (\underline{f}^{\,i})^{-1}\, \bar{\nabla} \times \bar{\upsilon}^j \right) \mathrm{d}\Gamma\end{aligned} \tag{7.66}$$

If $\bar{t}$ and $\bar{q}$ are vectors with three components and $\bar{a}_n$ is the unit vector normal to a surface, the following identities hold:

$$\bar{a}_n \cdot (\bar{t} \times \bar{q}) = \bar{q} \cdot (\bar{a}_n \times \bar{t}) = \bar{q} \cdot (\bar{a}_n \times (\bar{t}_\tau + \bar{t}_n)) = \bar{q} \cdot (\bar{a}_n \times \bar{t}_\tau) = \bar{a}_n \cdot (\bar{t}_\tau \times \bar{q}) \tag{7.67}$$

where $\bar{t}_n$ is the normal component of $\bar{t}$ (in the direction of $\bar{a}_n$) and $\bar{t}_\tau$ is the tangential component of $\bar{t}$. Then, (7.66) leads to

$$\begin{aligned}&\int_{\Gamma^{ij}} \bar{a}_n^{\,ij} \cdot \left(\bar{\upsilon}'^{\,i} \times (\underline{f}^{\,i})^{-1}\, \bar{\nabla} \times \bar{\upsilon}^i - \bar{\upsilon}'^{\,j} \times (\underline{f}^{\,j})^{-1}\, \bar{\nabla} \times \bar{\upsilon}^j \right) \mathrm{d}\Gamma \\ &= \int_{\Gamma^{ij}} \bar{a}_n^{\,ij} \cdot \left(\bar{\upsilon}'^{\,i}_\tau \times (\underline{f}^{\,i})^{-1}\, \bar{\nabla} \times \bar{\upsilon}^i - \bar{\upsilon}'^{\,j}_\tau \times (\underline{f}^{\,j})^{-1}\, \bar{\nabla} \times \bar{\upsilon}^j \right) \mathrm{d}\Gamma\end{aligned} \tag{7.68}$$

where $\bar{\upsilon}'^{\,i}_\tau$ and $\bar{\upsilon}'^{\,j}_\tau$ are the tangential components of $\bar{\upsilon}'^i$ and $\bar{\upsilon}'^j$, respectively.

Hence, if the test functions $\bar{\upsilon}'$ have continuity of the tangential component across element interfaces, then the boundary integral between two elements given by expression (7.66) may be written as

$$\int_{\Gamma^{ij}} \bar{\upsilon}'^{\,ij}_\tau \cdot \left(\bar{a}_n^{ij} \times (\underline{f}^{\,i})^{-1}\, \bar{\nabla} \times \bar{\upsilon}^i - \bar{a}_n^{ij} \times (\underline{f}^{\,j})^{-1}\, \bar{\nabla} \times \bar{\upsilon}^j \right) \mathrm{d}\Gamma \tag{7.69}$$

where $\bar{\upsilon}'^{\,ij}_\tau = \bar{\upsilon}'^{\,i}_\tau = \bar{\upsilon}'^{\,j}_\tau$. Therefore, (7.62) is seen to be equivalent to imposing the continuity of $\bar{a}_n \times \underline{f}^{-1}\, \bar{\nabla} \times \bar{\upsilon}$ across element interfaces whenever the test functions $\bar{\upsilon}'$ have continuous tangential components across element interfaces. It is clear that, because of the weighted integral, that continuity is imposed in a weak sense.

Moreover, for $\omega \neq 0$, it is seen that

$$\bar{a}_n \times \underline{f}^{-1} \bar{\nabla} \times \bar{V} \propto \bar{a}_n \times \bar{U} \tag{7.70}$$

where the symbol $\propto$ means proportionality between the two quantities at both sides of the symbol. Here $\bar{U}$ is the dual of vector $\bar{V}$, i.e., $\bar{H}$ if $\bar{V} = \bar{E}$ or $\bar{E}$ if $\bar{V} = \bar{H}$. In other words, the continuity of $\bar{a}_n \times \underline{f}^{-1} \bar{\nabla} \times \bar{\upsilon}$ is equivalent to the continuity of the tangential component of the dual field ($\bar{H}$ when $\bar{V} = \bar{E}$ or $\bar{E}$ when $\bar{V} = \bar{H}$). Thus, the satisfaction of the weak formulation (7.63) implies the satisfaction, in a distributional sense, of the continuity of the tangential components of the dual field across elements. It is worth noting that the test functions, $\bar{\upsilon}'$, are required to be tangentially continuous across element interfaces. The weak FEM formulation has been obtained employing the method of weighted residuals. The essential and natural boundary conditions of the formulation have been identified. In the following, equivalent expressions are obtained by means of the stationarity of a functional. In the functional approach, the weak solution of the problem is obtained making stationary a suitable functional. Hence, the first variation of the functional equated to zero gives the weak formulation corresponding to the differential equation of the problem. This functional may be known *a priori* or may be obtained from the differential equation of the problem. However, the functional whose stationarity provides the weak solution of the differential equation may not exist.

The differential operator $\mathcal{L}$ of (7.36) to (7.38) together with the boundary condition of (7.41) and (7.42) is a self-adjoint operator. For the case of a self-adjoint operator, the functional exists and is given by

$$F(\bar{\upsilon}) = \frac{1}{2} < \mathcal{L}\bar{\upsilon} ; \bar{\upsilon} > \tag{7.71}$$

In order to prove that the stationarity of (7.71) is equivalent, in a weak sense, to the differential equation (7.35), the first variation of the functional is considered:

$$\delta F = \frac{1}{2} < \mathcal{L}\delta\bar{\upsilon} ; \bar{\upsilon} > + \frac{1}{2} < \mathcal{L}\bar{\upsilon} ; \delta\bar{\upsilon} > \tag{7.72}$$

Now, because the operator $\mathcal{L}$ is self-adjoint, the following expression holds:

$$< \mathcal{L}\delta\bar{\upsilon} ; \bar{\upsilon} > = < \delta\bar{\upsilon} ; \mathcal{L}\bar{\upsilon} > \tag{7.73}$$

Then, the first variation of the functional (7.72) may be written as

$$\delta F = \frac{1}{2} < \delta\bar{\upsilon} ; \mathcal{L}\bar{\upsilon} > + \frac{1}{2} < \mathcal{L}\bar{\upsilon} ; \delta\bar{\upsilon} > \tag{7.74}$$

From the symmetry of the inner product (real-valued functions), it is seen that

$$\delta F = <\delta \bar{\upsilon}; \mathcal{L}\bar{\upsilon}> \tag{7.75}$$

The stationarity of the functional (i.e., equating to zero its first variation) provides the weak formulation for (7.35):

$$<\delta \bar{\upsilon} \; ; \; \mathcal{L}\bar{\upsilon}> = 0 \; , \quad \forall \; \delta \bar{\upsilon} \tag{7.76}$$

The weak formulation given by (7.76) is equivalent to the weak formulation provided by the method of weighted residuals, i.e., expression (7.45). This may be shown by considering the variation of the unknown, $\delta\bar{\upsilon}$, in (7.76) as the test function $\bar{\upsilon}'$ in the formulation of (7.45). The residual $R(\bar{\upsilon})$ is $\mathcal{L}(\bar{\upsilon}')$ because the excitation term (the right-hand side in the differential formulation) is zero. By considering (7.36) to (7.38), the functional given by (7.71) may be written as

$$F(\bar{\upsilon}) = \frac{1}{2} \int_\Omega \left(\bar{\upsilon} \cdot \left(\bar{\nabla} \times \underline{\underline{f}}^{-1} \; \bar{\nabla} \times \bar{\upsilon} \right) - \omega^2 \; \bar{\upsilon} \cdot \underline{\underline{p}} \; \bar{\upsilon} \right) d\Omega \tag{7.77}$$

Applying the first vector Green's theorem (see (7.49)), expression (7.77) becomes

$$\begin{aligned} F(\bar{\upsilon}) = & \frac{1}{2} \int_\Omega \left((\bar{\nabla} \times \bar{\upsilon}) \cdot (\underline{\underline{f}}^{-1} \bar{\nabla} \times \bar{\upsilon}) - \omega^2 \bar{\upsilon} \cdot \underline{\underline{p}} \bar{\upsilon} \right) d\Omega \\ & - \frac{1}{2} \int_\Gamma \bar{a}_n \cdot (\bar{\upsilon} \times \underline{\underline{f}}^{-1} \bar{\nabla} \times \bar{\upsilon}) \; d\Gamma \end{aligned} \tag{7.78}$$

The boundary integral of (7.78) is again zero because of the boundary conditions of the problem (7.41) and (7.42). The boundary integral represents the flux of the Poynting vector across the boundary of the microwave cavity. From the physical point of view it is known that there is no flux of energy coming out of a microwave cavity enclosed by perfect conductors (perfect electric walls). The same happens with perfect magnetic conductors. Perfect magnetic conductors do not exist physically but they are used as boundary conditions (perfect magnetic walls) on symmetry planes. In summary, there is no flux through the boundary Γ for the type of boundary conditions used in this problem. Thus, it can be stated that the boundary integral of (7.78) is always zero, leading to the following functional:

$$F(\bar{\upsilon}) = \frac{1}{2} \int_\Omega \left((\bar{\nabla} \times \bar{\upsilon}) \cdot (\underline{\underline{f}}^{-1} \bar{\nabla} \times \bar{\upsilon}) - \omega^2 \bar{\upsilon} \cdot \underline{\underline{p}} \bar{\upsilon} \right) d\Omega \tag{7.79}$$

The variational formulation is applied to each finite element of the mesh:

$$F(\bar{\upsilon}) = \sum_{e=1}^{N_e} \left(\frac{1}{2} \int_{\Omega^e} \left((\bar{\nabla} \times \bar{\upsilon}^e) \cdot \left((\underline{f}^e)^{-1} \bar{\nabla} \times \bar{\upsilon}^e \right) - \omega^2 \bar{\upsilon}^e \cdot \underline{p}^e \bar{\upsilon}^e \right) d\Omega \right) \tag{7.80}$$

where the sum of the boundary integral terms at element interfaces has been equated to zero, i.e.:

$$\sum_{e=1}^{N_e} \left(\int_{\Gamma^e \not\subset \Gamma} \bar{a}_n^e \cdot (\bar{\upsilon}^e \times \underline{f}^{e-1} \bar{\nabla} \times \bar{\upsilon}^e) \, d\Gamma \right) = 0 \tag{7.81}$$

Equation (7.81) stands for the continuity of the flux of the Poynting vector across two elements of the mesh. By equating to zero the first variation of the functional (7.80), an equivalent formulation to that of (7.63) is obtained. In fact, the formulation provided by the first variation of (7.80) may be written as

$$\delta F = \sum_{e=1}^{N_e} \left(\int_{\Omega^e} \left((\bar{\nabla} \times \delta\bar{\upsilon}^e) \cdot \left((\underline{f}^e)^{-1} \bar{\nabla} \times \bar{\upsilon}^e \right) - \omega^2 \delta\bar{\upsilon}^e \cdot \underline{p}^e \bar{\upsilon}^e \right) d\Omega \right) = 0 \tag{7.82}$$

which is identical to (7.63), where $\delta\bar{\upsilon}$ may be identified with the test function of the weighted residual method.

However, for $\delta\bar{\upsilon}$ (or $\delta\bar{\upsilon}^e$ when specialized to the eth element) to be a valid test function it must satisfy the homogeneous form of the essential boundary conditions of the problem. This is easily shown to be accomplished by considering that $\bar{a}_n \times \bar{\upsilon}$ must be enforced to satisfy the essential boundary conditions on Γ_D. Thus, its first variation will be identically zero over Γ_D.

For the problem under study, the value of the tangential component on the boundary Γ_D is given by (7.41). Hence, one obtains

$$\delta(\bar{a}_n \times \bar{\upsilon}) = \bar{a}_n \times \delta\bar{\upsilon} = 0 \quad \text{and} \quad \delta(\bar{a}_n \times \bar{\upsilon}^e) = \bar{a}_n \times \delta\bar{\upsilon}^e = 0 \,, \quad \text{on } \Gamma_D \tag{7.83}$$

i.e., $\delta\bar{\upsilon}$ and $\delta\bar{\upsilon}_e$ satisfy the homogeneous form of the essential boundary conditions of the problem.

Since the approximate vector function $\bar{\upsilon}$ consists of a linear combination of basis functions, as shown in (7.44) or (7.59), its variation may then be written as

$$\delta\bar{\upsilon} = \sum_i \bar{N}_i \delta d_i \tag{7.84}$$

Considering (7.84) and taking into account that the variation must be zero, independently of the values of δd_i, one finally obtains

$$\sum_{e=1}^{N_e} \int_{\Omega^e} \left\{ \left(\nabla \times \bar{N}_i\right) \cdot \left((\underline{\underline{f}}^e)^{-1} \nabla \times \bar{\upsilon}^e \right) - \omega^2 \bar{N}_i \cdot \underline{\underline{p}}^e \bar{\upsilon}^e \right\} \mathrm{d}\Omega = \{0\} \,, \quad i = 1, \ldots, n_e \tag{7.85}$$

which is a system of equations identical to that obtained by the Galerkin method (7.64), where n_e is the number of degrees of freedom for the eth element.

Thus, the equivalence of the two variational approaches has been obtained, namely the Galerkin method (that leads to (7.64) and the stationarity of the functional (7.80), leading to (7.85).

In summary, the FEM formulation used to obtain the resonant modes of inhomogeneous, anisotropic, and arbitrarily shaped 3D microwave cavities has been presented. Different issues regarding the variational formulation have also been addressed: the role of the essential and natural boundary conditions, both at the boundary and across elements; the requirements to be satisfied by basis and test functions; and the equivalence between the Galerkin method and the stationarity of a functional, for example. Next the discretization process by FEM is discussed.

7.3.1.2 *Discretization by Curl-Conforming Elements*

The discretization of the FEM weak formulation presented in Section 7.3.1.1 is now addressed. The application of the method of weighted residuals or the stationarity of a functional has been shown to lead to an identical system of equations. In practice, for its computer implementation, a normalization procedure is applied to the double-curl weak formulation presented in the previous section. The details of this normalization process are shown below. Then, an algebraic system of equations with the degrees of freedom d_i as the unknowns is obtained. The expression for the coefficients of the matrices of the system is given.

The weak formulation to be discretized was presented in (7.56) of Section 7.3.1.1. First the field is normalized as

$$\bar{V}_n = \sqrt{p_0} \;\; \bar{V} \tag{7.86}$$

Taking into account that

$$\begin{aligned} \underline{\underline{f}} &= f_0 \, \underline{\underline{f}}_r \\ \underline{\underline{p}} &= p_0 \, \underline{\underline{p}}_r \end{aligned} \tag{7.87}$$

the normalized formulation can be written as

$$\int_{\Omega} \left((\bar{\nabla} \times \bar{\upsilon}_n') \cdot (\underline{\underline{f}}_r^{-1} \nabla \times \bar{\upsilon}_n) - k_0^2 (\bar{\upsilon}_n' \cdot \underline{\underline{p}}_r \bar{\upsilon}_n) \right) \mathrm{d}\Omega = 0 \tag{7.88}$$

where the correspondences are given in Table 7.2 and k_0 is the free-space wavenumber.

Table 7.2

Correspondences for the normalized double-curl formulation

	$\bar{E}$ formulation	$\bar{H}$ formulation
$\bar{V}$	$\bar{E}$	$\bar{H}$
f_0	μ_0	ε_0
p_0	ε_0	μ_0
$\underline{\underline{f}}_r$	$\underline{\underline{\mu}}_r$	$\underline{\underline{\varepsilon}}_r$
$\underline{\underline{p}}_r$	$\underline{\underline{\varepsilon}}_r$	$\underline{\underline{\mu}}_r$

The following set of discrete local integral forms is obtained for each element when the Galerkin procedure is applied to (7.88):

$$\{W^e\} = ([k^e] - k_0^2 [m^e]) \{d^e\} \tag{7.89}$$

where

$$k_{ij}^e = \int_{\Omega^e} \left(\bar{\nabla} \times \bar{N}_i^e\right) \cdot \left((\underline{\underline{f}}_r^{\,e})^{-1} \bar{\nabla} \times \bar{N}_j^e\right) \mathrm{d}\Omega \,, \quad i,j = 1, \cdots, n_e \tag{7.90}$$

$$m_{ij}^e = \int_{\Omega^e} \bar{N}_i^e \cdot \underline{\underline{p}}_r^{\,e} \bar{N}_j^e \, \mathrm{d}\Omega \,, \quad i,j = 1, \cdots, n_e \tag{7.91}$$

The FEM assembly procedure (see Chapter 2) leads to

$$([K] - k_0^2 [M]) \{D\} = \{0\} \tag{7.92}$$

where [K] and [M] are square $N_d \times N_d$ matrices. {D} is a $N_d \times 1$ column matrix. N_d is the total number of degrees of freedom. Matrices [K] and [M] have been obtained from

$$[K] = \sum_{e=1}^{N_e} [k^e] \; , \quad [M] = \sum_{e=1}^{N_e} [m^e] \tag{7.93}$$

where the sign Σ should be understood not as an ordinary summation but as the assembly of all the elements. The coefficients of the column matrix $\{D\} = \{d^g\}$ are the degrees of freedom d_i but following a global numbering. A + or - sign must be considered in order to uniquely define the global degree of freedom of nodes shared by different elements. This issue is explained in Appendix C.4.3.

Equation (7.92) represents an algebraic eigenvalue problem with the square of the free-space wavenumber as the eigenvalue and $\{D\}$ as the eigenvector. It is worth noticing that in (7.92) all the basis functions have been used as test functions. Hence, rigorously speaking, (7.92) is applicable only when no essential boundary conditions exist. If essential boundary conditions are present, the basis functions associated with nodes placed on the boundaries of the domain with essential conditions, Γ_D, should not be used as test functions. This is so because these basis functions do not satisfy homogeneous essential boundary conditions over Γ_D. However, in the following it will be shown that in practical FEM implementation (7.92) is also written. The enforcement of the essential boundary conditions over (7.92) leads to a global system of equations. This system of equations is equivalent to that obtained by using only basis functions satisfying homogeneous boundary conditions on Γ_D.

Let us consider the column matrix $\{D\}$ of the global degrees of freedom arranged as

$$\{D\} = \begin{Bmatrix} \{d^{gf}\} \\ \{d^{ge}\} \end{Bmatrix} \tag{7.94}$$

where $\{d^{ge}\}$ stands for the degrees of freedom associated with nodes on Γ_D (degrees of freedom with essential conditions) and the coefficients of $\{d^{gf}\}$ are the rest of the degrees of freedom (free ones). Note that the coefficients of $\{d^{ge}\}$ are not unknowns, while the rest should be solved for, i.e., they are obtained through the solution of the algebraic system of equations.

Taking into account expression (7.94), (7.92) may be written as

$$\left(\begin{bmatrix} [K_{11}] & [K_{12}] \\ [K_{21}] & [K_{22}] \end{bmatrix} - k_0^2 \begin{bmatrix} [M_{11}] & [M_{12}] \\ [M_{21}] & [M_{22}] \end{bmatrix} \right) \begin{Bmatrix} \{d^{gf}\} \\ \{d^{ge}\} \end{Bmatrix} = \{0\} \tag{7.95}$$

Now, if only the basis functions related with the degrees of freedom $\{d^{gf}\}$ are used as test functions, the following system of equations would be obtained:

$$([K_{11}] - k_0^2[M_{11}])\{d^{gf}\} = -([K_{12}] - k_0^2[M_{12}])\{d^{ge}\} \tag{7.96}$$

For the cavity problem, the essential boundary conditions of the problem are homogeneous. Hence, the degrees of freedom associated with that part of the boundary are zero, i.e., $\{d^{ge}\} = \{0\}$. Therefore, (7.96) leads to

$$([K_{11}] - k_0^2[M_{11}])\{d^{gf}\} = \{0\} \tag{7.97}$$

which constitutes an eigenvalue problem with the square of the free-space wavenumber as the eigenvalue. It is worth noting that (7.97) is identical to (7.92) when no essential boundary conditions are present. However, in practice, no distinction is made between the case of existence and the case of non-existence of essential boundary conditions. In both cases, the discretization is applied in each element of the mesh according to (7.89) to (7.91). The assembly procedure is performed and the global system of equations (7.92) is obtained. Then, the essential boundary conditions, if any, are imposed on the global system according to

$$\left(\begin{bmatrix} [K_{11}] & [K_{12}] \\ [0] & \chi[I] \end{bmatrix} - k_0^2 \begin{bmatrix} [M_{11}] & [M_{12}] \\ [0] & [I] \end{bmatrix}\right)\begin{Bmatrix} \{d^{gf}\} \\ \{d^{ge}\} \end{Bmatrix} = \{0\} \tag{7.98}$$

where χ is a very large number. Note that (7.98) is equivalent to (7.97) together with $\{d^{ge}\} = \{0\}$, i.e., the use of only those basis functions related with $\{d^{gf}\}$ is equivalent to the use of all the basis functions as test functions plus the enforcement of essential boundary conditions, as previously stated.

The basis functions utilized in this chapter are the mixed-order vector basis functions proposed in [Nédélec 1980] for tetrahedrons. The main features of these curl-conforming basis functions are shown in Section 7.2.2.2. In the following, a few remarks are given regarding the continuity properties of vector quantities approximated by means of such basis functions. The approximate vector solution $\overline{\upsilon}$ exhibits continuity of the tangential component across element interfaces in a strong sense because it is explicitly enforced by the basis functions. The basis functions utilized also enforce the continuity across element interfaces of the normal component of the curl of the vector solution.

However, it is worth noticing that the continuity of the normal component of the curl would have been enforced in a weak or distributional sense just by using basis functions that satisfy the continuity of the tangential components. This is easily seen from the application of the Stokes theorem to the faces of the element:

$$\iint_F \overline{a}_n \cdot (\overline{\nabla} \times \overline{V})\, dS = \oint_e \overline{V} \cdot \overline{dl} \tag{7.99}$$

where F refers to a face of an element, while e stands for the edges of that face. The right-hand term of (7.99) is continuous across elements because of the tangential continuity of the basis functions across edges. Then, the continuity of the left-hand side of (7.99) is inferred. However, as mentioned before, in the case of the basis functions utilized, the continuity of the normal component of the curl is strongly enforced, because the normal components of the curl of the basis functions are continuous across element interfaces.

The component of the solution vector, normal to element interfaces, and the tangential components of the curl are discontinuous across element interfaces.

Therefore, if $\bar{V} = \bar{E}$ (or $\bar{V} = \bar{H}$) the continuity of the tangential components of $\bar{E}$ (or $\bar{H}$) and the continuity of the normal component of $\bar{B}$ (or $\bar{D}$) are satisfied. Instead, the jump across element interfaces of the normal component of $\bar{E}$ (or $\bar{H}$) and the tangential component of $\bar{B}$ (or $\bar{D}$) is enabled. Thus, the continuity conditions of the vector solution are in agreement with the electromagnetic behavior of the electromagnetic field across media interfaces.

For $\bar{V} = \bar{E}$, the continuity of the tangential component of the magnetic field, $\bar{H}$, is imposed in a distributional sense as a natural boundary condition (see expressions (7.62), (7.66) to (7.70) of Section 7.3.1.1). Similarly, for $\bar{V} = \bar{H}$, the continuity of the tangential component of the electric field, $\bar{E}$, is imposed in a distributional sense.

The continuity of the normal component of the vector function $\underline{p}\,\bar{V}$, i.e., of the displacement vector $\bar{D}$ when $\bar{V} = \bar{E}$ or of the magnetic induction $\bar{B}$ when $\bar{V} = \bar{H}$, is also enforced in a distributional sense. Specifically, for $k_0 \neq 0$, one gets the following expression involving the divergence of $\underline{p}\,\bar{V}$:

$$\int_\Omega \Phi' \; \bar{\nabla} \cdot (\underline{p}\,\bar{V}) = 0 \tag{7.100}$$

where Φ' is a scalar test function that must be zero on Γ_D. Expression (7.100) is obtained considering the double-curl weak formulation of Section 7.3.1.1 with a test vector function given by

$$\bar{\upsilon}' = \bar{\nabla}\Phi' \tag{7.101}$$

which leads to

$$-k_0^2 \int_\Omega \bar{\nabla}\,\Phi' \cdot \underline{p}\,\bar{V}\; d\Omega = 0 \tag{7.102}$$

It may be shown that from (7.102), if $k_0 \neq 0$ and $\Phi' = 0$ on Γ_D, one may deduce expression (7.100). Therefore, under those conditions the continuity of the normal component of $\underline{p}\,\bar{V}$ will be satisfied in the distributional sense given by (7.100). However, this is true only for $k_0 \neq 0$. In conclusion, the divergence condition will not be satisfied for the spurious modes (solutions with $k_0 = 0$) even in the distributional sense of (7.100). On the other hand, to satisfy (7.100), the test functions must be able to approximate the gradient of a scalar function. As was mentioned in Section 7.2.1, curl-conforming basis functions fulfill that condition, in contrast with Lagrange basis functions. Thus, using Lagrange elements the continuity of the normal component of $\underline{\underline{p}}\,\overline{V}$ will not be satisfied even in the weak sense of (7.100).

In summary, it has been shown that curl-conforming elements, together with the natural boundary conditions of the double-curl weak formulation of Section 7.3.1.1, provide a field solution consistent with the boundary conditions of the electromagnetic field.

7.3.2 Dimension of the Vector Space Spanned by the Spurious Modes

As shown in Section 7.2, spurious modes are numerical approximations of nonphysical solutions. The nonphysical solutions are static solutions derived from the gradient of a scalar function or scalar potential function, ϕ (see (7.24)). The nonphysical solutions belong to the null space of the curl operator and, hence, appear as modes corresponding to a zero eigenvalue in double-curl formulations, as in (7.33).

The dimension of the vector space corresponding to the zero eigenvalue is the number of independent scalar functions supported by the mesh minus one. The number of different electromagnetic field configurations having zero eigenvalue is one less than the number of independent scalar functions because the field solution equal to the gradient of a constant potential function is the trivial solution, $\overline{V}_{null} = \overline{0}$. Hence, this trivial solution must not be considered as a spurious mode.

In order to estimate the number of spurious solutions of zero eigenvalue that will be obtained with the FEM double-curl formulation, it is necessary to know the number of independent potential functions that will support the corresponding mesh. The number of independent potential functions for a mesh will depend on the type of FEM basis functions used in the approximation of the field.

In order to illustrate this issue, a simple example is presented. It deals with the application of (7.92) to two different meshes. The first consists of only one tetrahedron, while the second considers two tetrahedrons. The tetrahedrons have unit dimensions in the three-coordinate axis (see Figure 7.3).

First- and second-order curl-conforming tetrahedral Nédélec elements are used. In 1980, Nédélec proposed a given space of vector polynomial functions for the approximation of vector quantities by means of these elements (see Appendix A.1.1). The space was derived through the imposition of a given set of conditions to be fulfilled by the polynomial of order p of each Cartesian component of the approximate vector functions. These conditions are called the Nédélec constraints [Nédélec 1980]. The idea behind the satisfaction of the Nédélec constraints will be clear later on in this section.

The first-order tetrahedral Nédélec element, i.e., the edge element, is considered first. The first-order polynomial basis functions satisfying the Nédélec constraints (see Section 2.5.1.2, Table 2.8 and Appendix C.4.2.2) are the following:

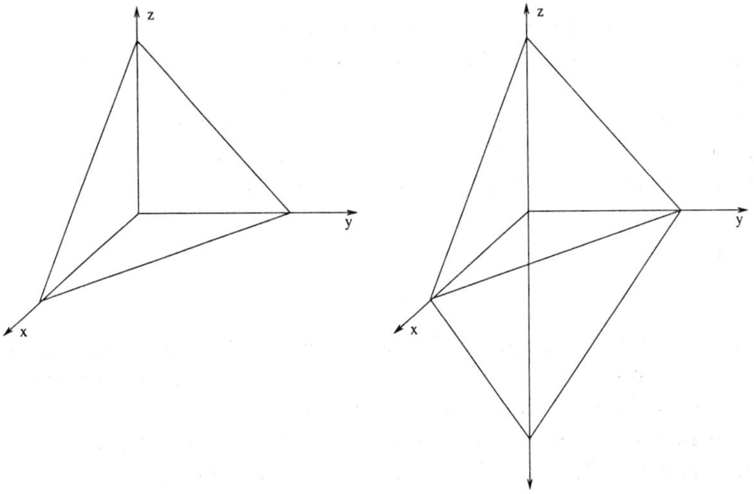

Figure 7.3 One-tetrahedron and two-tetrahedrons meshes.

$$\{N_i\} = \begin{Bmatrix} \gamma_1^{(i)} + \alpha_1^{(i)} y + \beta_1^{(i)} z \\ \gamma_2^{(i)} - \alpha_1^{(i)} x + \beta_2^{(i)} z \\ \gamma_3^{(i)} - \beta_1^{(i)} x - \beta_2^{(i)} y \end{Bmatrix} \tag{7.103}$$

where $\gamma_1^{(i)}$, $\gamma_2^{(i)}$, $\gamma_3^{(i)}$, $\alpha_1^{(i)}$, $\beta_1^{(i)}$, and $\beta_2^{(i)}$ are real constant coefficients and $i = 1, ..., 6$.

There are six coefficients, i.e., six unknowns to be determined for each basis function. Therefore, the degrees of freedom of the element are six. The specific values of the coefficients of the expansion functions or basis functions are obtained by imposing six conditions on each basis function, $\bar{N}_i$. In this case (first order), the six basis functions may be associated with points or nodes on the edges. More details may be found in Section 2.5.1.2 and Appendix C.4.2.2.

Now, which scalar potential function will provide a gradient having polynomial components with an x, y, z, dependence as the vector functions $\bar{N}_i$ of (7.103)? If first-order polynomial basis functions are used for the vector approximation, the potential scalar functions must be, at most, a complete polynomial of second order:

$$\phi = A_1 x^2 + A_2 y^2 + A_3 z^2 + A_4 xy + A_5 xz + A_6 yz + A_7 x + A_8 y + A_9 z + A_{10} \tag{7.104}$$

However, it may be observed that some of the monomial terms of (7.104) are unnecessary. An example is the term of x^2 that would provide along the x-axis a component of the field proportional to x. As seen from (7.103), there is no term in x along the x-component of the field. Analogous conclusions are obtained for the terms in y^2 and z^2. Also, the cross terms in xy, xz, and yz should not be present in the spurious field $\overline{V}_{null}$ because they can never provide a polynomial approximation of (7.103).

In summary, the potential, whose gradient is compatible with the field expanded by (7.103), has a dependence with x, y, and z of the form

$$\phi = A_7 x + A_8 y + A_9 z + A_{10} \tag{7.105}$$

It is worth noting that the polynomial function of (7.105) is not quadratic but linear, similar to the approximation of the vector function obtained through a linear combination of the basis functions given by (7.103).

There are four coefficients to be determined in the polynomial expression for the potential ϕ of (7.105). Therefore, four is the number of independent potential functions for the tetrahedral element of first order. Thus, the number of independent spurious fields, i.e., the number of zero eigenvalues, in the case of the one-tetrahedron mesh of Figure 7.3, will be three. This is shown in Table 7.3.

Table 7.3

Number of zero eigenvalues for the meshes of Figure 7.3

Mesh	First-order	Second-order
One tetrahedron	3	9
Two tetrahedrons	4	13

The reason behind the Nédélec constraints is to make the order of the polynomial of the scalar potential function from which the spurious fields are derived equal to the order of the polynomial basis functions employed for the approximation of the field. Then the number of degrees of freedom used to model the space of the spurious fields will be reduced with respect to the conventional polynomial functions, i.e., basis functions not satisfying the Nédélec constraints (see the comparison given in [Peterson and Wilton 1996]). In this way, a higher number of degrees of freedom with respect to the total one is used in the approximation of the physical fields and, hence, an improved accuracy on the computation of the physical modes corresponding to nonzero eigenvalues should be expected.

The reduction of the degrees of freedom leads to the appearance of mixed-order polynomial functions, i.e., the so-called mixed-order elements, which are the kind of curl-conforming elements utilized in this book [Nédélec 1980]. In mixed-order elements, the order of the polynomial approximation of the field is not the same in all directions as opposed to complete polynomial approximations. Specifically, the basis functions given by (7.103) provide a zero-order approximation along edges of the component of the field tangential to that edge and a linear approximation in any other direction. Also, the number of independent scalar potential functions and, hence, the number of zero eigenvalues, are different for mixed-order or polynomial complete elements. The procedure to estimate the number of zero eigenvalues has been explained here for the Nédélec mixed-order edge element. A similar procedure may be used for polynomial complete curl-conforming elements.

Up to now, only the case of a single-element mesh has been considered. The next question would be: how many independent potential functions exist for an arbitrary mesh, when in each element the potential is a linear polynomial function, as in (7.105)? To answer this question, it is worth noting that Lagrange elements constitute a natural way to approximate scalar functions over a FEM mesh. A piecewise polynomial interpolation of a scalar function corresponds to a Lagrange element approximation. In fact, the scalar function of (7.105) corresponds to the first-order Lagrange element. Note that a first-order Lagrange tetrahedron has four degrees of freedom as per (7.100), i.e., A_7, A_8, A_9, and A_{10}.

The number of independent potential functions in a mesh for the case of first-order elements coincides with the number of nodes that would be in the same mesh considering first-order Lagrange tetrahedral elements. As the nodes of a first-order Lagrange element are at the vertices of the element, the number of independent potentials in the mesh coincides with the number of vertices of the mesh. Thus, the number of spurious modes for a given mesh is the number of vertices of the mesh minus one.

This can be seen with the example of the two-tetrahedron mesh of Figure 7.3. The number of vertices of the mesh is 5. Then, the number of zero eigenvalues is 4, as shown in Table 7.3. The application of the FEM code with first-order Nédélec elements to the meshes shown in Figure 7.3 has verified the previously mentioned theory.

The conclusion reached (i.e., for first-order Nédélec elements, the number of zero eigenvalues equals the number of vertices minus one) is valid only when all the degrees of freedom of the polynomial interpolation are free (in the sense that there are no Dirichlet boundary conditions). When Dirichlet boundary conditions should be imposed in some parts of the boundary, the degrees of freedom associated with nodes on those parts of the boundary are fixed, i.e., they are no longer unknowns. Therefore, the number of degrees of freedom available for independent scalar potential functions and, hence, for the spurious modes is reduced.

It has been observed that when homogeneous Dirichlet boundary conditions are imposed, the number of zero eigenvalues with first-order elements coincides with the number of internal vertices plus the number of boundary vertices not placed on the Dirichlet boundaries of the mesh. This has also been reported in the literature for 2D elements (see, e.g., [Hano 1984] for rectangles and [Tanner and Peterson 1989] for triangles).

When some of the degrees of freedom are fixed to be zero, as is the case when homogeneous essential boundary conditions are present in the formulation (e.g., a perfect electric conductor in the case of the $\bar{E}$-field formulation), the potential must be constant on that conductor boundary; i.e., all the degrees of freedom of the potential associated with the nodes corresponding to Lagrange elements (vertices for the first-order element) on the conductor boundary must have the same value. Thus, the number of independent scalar potential functions in the mesh is equal to the number of free nodes of the Lagrange approximation (i.e., those not placed on the conductor boundary) plus one. Since one of the solutions for the potential is the constant one, which provides a null field, the number of nontrivial independent spurious fields (with zero eigenvalues) is, in this case, directly related to the number of free nodes of the Lagrange approximation. For the first-order case, the nodes of the Lagrange approximation are at the vertices, thus providing a number of zero eigenvalues equal to the free vertices of the problem, i.e., the number of internal vertices plus the number of boundary vertices not on the conductor surfaces. This conclusion has been verified for all the examples analyzed. In the following section, some of the structures that have been studied are presented.

The case of the second-order element is analogous to the case of the first-order element detailed previously. The second-order polynomial basis functions satisfying the Nédélec constraints are the following:

$$\tilde{N}_i = \begin{Bmatrix} a_1^{(i)} + a_2^{(i)}x + a_3^{(i)}y + a_4^{(i)}z \\ b_1^{(i)} + b_2^{(i)}x + b_3^{(i)}y + b_4^{(i)}z \\ c_1^{(i)} + c_2^{(i)}x + c_3^{(i)}y + c_4^{(i)}z \end{Bmatrix} + \begin{Bmatrix} D^{(i)}y^2 - F^{(i)}xy - G^{(i)}xz + H^{(i)}z^2 + J^{(i)}yz \\ -D^{(i)}xy - E^{(i)}yz + F^{(i)}x^2 + I^{(i)}z^2 - J^{(i)}xz + K^{(i)}xz \\ E^{(i)}y^2 + G^{(i)}x^2 - H^{(i)}xz - I^{(i)}yz - K^{(i)}xy \end{Bmatrix} \tag{7.106}$$

Similarly to the first-order case, it may be shown that the scalar potential function compatible with a vector function as in (7.106) is a second-order polynomial scalar function as given by (7.104). It is seen that the second-order scalar function has ten degrees of freedom. Then the number of independent scalar potential functions in one element is ten. Therefore, the number of spurious modes (with zero eigenvalue) is nine for the case of the one-tetrahedron mesh, as shown in Table 7.3.

For an arbitrary mesh, and considering the equivalence between piecewise scalar functions and Lagrange elements, the number of independent scalar potential functions coincides with the number of nodes of that mesh for second-order Lagrange tetrahedral elements.

Thus, the number of modes associated with the zero eigenvalue will be the number of nodes dealing with Lagrange elements in the mesh minus one. In the example of the two-tetrahedron mesh of Figure 7.3, the number of zero eigenvalues is 13, as is shown in Table 7.3. The second-order Lagrange tetrahedral element has ten nodes, placed at the vertices and at the midpoint of edges. The number of Lagrange nodes of the two-tetrahedron mesh is 14; hence, the number of zero-frequency spurious modes is one less. This is for the case of no essential boundary conditions.

The remarks made for the first-order case, with respect to the reduction of the number of zero eigenvalues when some degrees of freedom are fixed (Dirichlet boundary conditions), are also valid for the second-order element.

In summary, the question of the number of the zero-eigenvalue modes in the FEM discretization of the double-curl formulation has been addressed. An explanation in terms of the number of independent scalar functions in the mesh, given a type of piecewise polynomial approximation of the field, has been given. Specifically, first- and second- order Nédélec basis functions have been employed in a simple example in order to illustrate this issue.

An equivalent explanation regarding the number of zero eigenvalues may also be given in terms of the trees and cotrees of a mesh (graph theory). The interested reader is encouraged to refer to a review on spurious modes given in [Sun et al. 1995]. In that work, the number of zero eigenvalues for triangles is shown in detail. The case for tetrahedrons is outlined in [Webb 1993].

7.3.3 Numerical Results

Some numerical results are presented here to illustrate the validity of the 3D curl-conforming elements in double-curl FEM formulations and also to illustrate some of the conclusions stated in the previous sections. Specifically, mixed-order Nédélec tetrahedrons [Nédélec 1980] have been employed for the discretization procedure. First- and second-order elements have been used.

Results corresponding to both formulations, the $\bar{E}$-field formulation and the $\bar{H}$-field formulation, are shown. It is worth pointing out here that the number of zero eigenvalues obtained, i.e., the number of zero-frequency spurious modes, depends on the type of formulation utilized ($\bar{E}$-field or $\bar{H}$-field). This is due to the boundary conditions of the problem, which are of essential type for one formulation and of natural type for the other. This is better explained with an example.

Consider a closed-domain problem with only perfect electric walls as boundary conditions. This case corresponds to the analysis of cavities with perfect conductors without exploiting any possible symmetry of the structure.

In the case of the formulation with the magnetic field, the perfect electric wall is a natural boundary condition and, hence, no Dirichlet boundary conditions are present; i.e., all the degrees of freedom are free. As stated in Section 7.3.2, in this case the number of zero eigenvalues is the number of Lagrange nodes of the mesh minus one. The number of Lagrange nodes means the number of nodes of the mesh with Lagrange elements of the same order as the curl-conforming elements used. Specifically, for the first-order case, the nodes of the Lagrange approximation are at the vertices. Thus, for the first-order approximation, the number of zero eigenvalues is equal to the number of vertices of the mesh minus one. Consider an empty cavity whose dimensions in the x-, y-, and z-axes are 1, 0.5, and 0.75, respectively. The generation of the mesh has been done with the automatic mesher of MODULEF [Bernadou et al. 1985]. The domain is subdivided into a number of cubes and each cube into six tetrahedrons. Figure 7.4 shows the subdivision of a cube into six tetrahedral elements. A specific mesh is shown in Figure 7.5. For illustrative purposes, the elements presented are shrunk. Using first order elements, 63 zero eigenvalues were obtained with the $\overline{H}$-field formulation; i.e., there are 63 spurious modes. From Figure 7.5, it is seen that the number of vertices is 64, i.e., the number of spurious modes plus one.

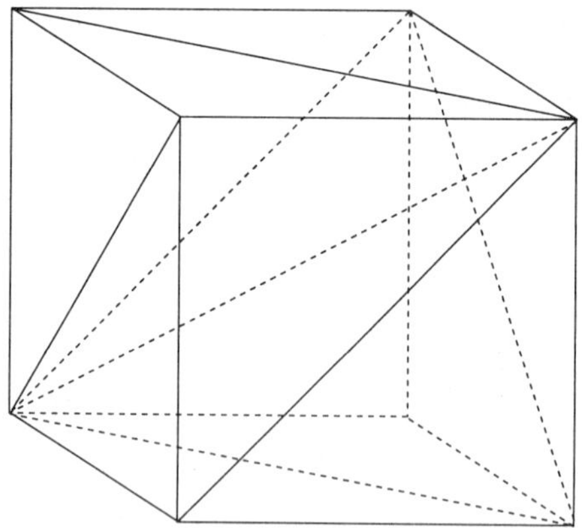

Figure 7.4 Subdivision of a cube into six tetrahedrons.

On the other hand, when the $\overline{E}$-field formulation is used, some of the degrees of freedom of the mesh are fixed to zero, specifically, those associated with the conductor boundaries. Thus, as stated in Section 7.3.2, the number of independent scalar potential functions on the mesh is reduced and so is the number of zero eigenvalues. In this case, the number of zero eigenvalues is equal to the number of free Lagrange nodes of the mesh. Then, when the whole boundary is a perfect conductor, the number of zero eigenvalues is found to be the number of internal Lagrange nodes (vertices for first-order elements) of the mesh. For the structure under study, using first-order elements, eight zero eigenvalues were computed utilizing the $\overline{E}$-field formulation and using the mesh of Figure 7.5. Again, this result coincides with the theoretical prediction.

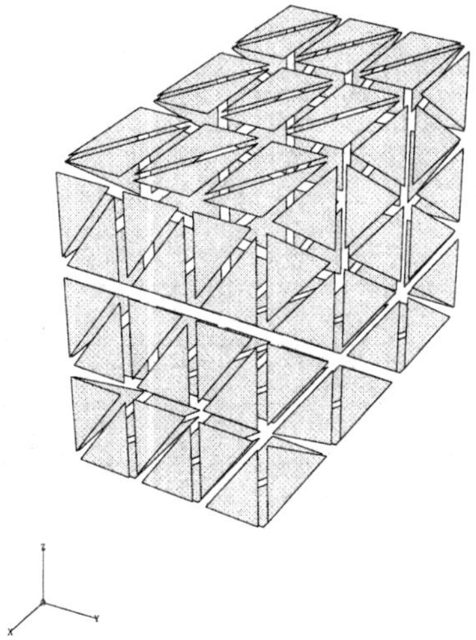

Figure 7.5 Empty cavity of dimensions 1×0.5×0.75. Mesh A (162 elements, 279 nodes, 64 vertices: 8 internal, 56 on the boundary).

In the following, numerical results corresponding to various cavities are given. The first structure to be analyzed is the empty rectangular cavity just mentioned. This simple structure has been chosen first because of the availability of analytic solutions.

First the dominant mode, called TE_{z101}, has been analyzed in order to study the convergence of the first- and second-order Nédélec tetrahedral elements. Table 7.4 shows the results for the first-order element, while Table 7.5 gives results for the second-order one. The first column of the tables gives the number of subdivisions in each direction for the generation of cubes. Figure 7.6 shows the rate of convergence of the eigenvalue (the square of the free-space wavenumber) for the first- and second-order elements and the theoretical slope when using expression (2.85) of Section 2.5.3. Taking into account that the study is done for the eigenvalues, the square of expression (2.85) should be considered; i.e., a slope of -2/3 should be expected for the first-order element, while -4/3 should be the expected slope for the second-order element, as shown in Figure 7.6.

For the higher order modes, the meshes shown in Figures 7.5 and 7.7 have been used. The square root of the first eigenvalues, i.e., the free-space wavenumber k_0, is shown in Table 7.6 and Table 7.7 for the cases of Mesh A (Figure 7.5) and of the Mesh B (Figure 7.7), respectively. As it can be seen, symmetries have not been exploited in this case. Results corresponding to both formulations, with the electric field ($\overline{V} = \overline{E}$) and with the magnetic field ($\overline{V} = \overline{H}$), are presented. In this case, the only difference between the two formulations comes from the way the boundary conditions are imposed. The conductors are assumed to be of infinite conductivity, and so the boundary condition

Table 7.4
Results for $\tilde{k}_0^2$ for the TE_{z101} resonant mode of the 1×0.5×0.75 empty cavity. First-order tetrahedral elements. Exact value k_0^2: 27.415696

Number of cubic cell divisions	Number of elements	Number of degrees of freedom	$\tilde{k}_0^2$	Relative error
1×1×1	6	19	30.8108	12.38436×10^{-2}
2×1×2	24	57	28.6489	4.49866×10^{-2}
4×2×4	192	330	27.6455	0.83880×10^{-2}
8×4×8	1536	2196	27.4552	0.14456×10^{-2}

Table 7.5
Results for $\tilde{k}_0^2$ for the TE_{z101} resonant mode of the 1×0.5×0.75 empty cavity. Second-order tetrahedral elements. Exact value k_0^2: 27.415696

Number of cubic cell divisions	Number of elements	Number of degrees of freedom	$\tilde{k}_0^2$	Relative error
2×1×2	24	242	27.5272	0.40722×10^{-2}
4×2×4	192	1556	27.4284	0.046952×10^{-2}
8×4×8	1536	11048	27.4165	0.0034368×10^{-2}

is a perfect electric wall. The perfect electric wall constitutes an essential boundary condition in the formulation with the electric field and a natural boundary condition in the formulation with the magnetic field. The analytic wavenumbers together with the errors associated with the computed wavenumbers are listed in the tables. Despite the coarseness of the meshes, reasonable accuracy is obtained for the resonant frequencies. It may be seen that degenerate modes pose numerical problems. It is also evident how the accuracy of the computation depends on the value of the exact eigenvalue, as mentioned in Section 2.5.3. Thus, higher order modes require more refined meshes in order to obtain the same accuracy than for the lower ones. To be more precise, what is required is a mesh well adapted to the electromagnetic field behavior of each mode.

For the next example, consider a ridge cavity. It consists of a rectangular cavity with a ridge along one of its faces. The dimensions of the rectangular cavity in the x, y, and z directions are 1, 0.5, and 0.75, respectively. The size of the ridge is 0.2 along

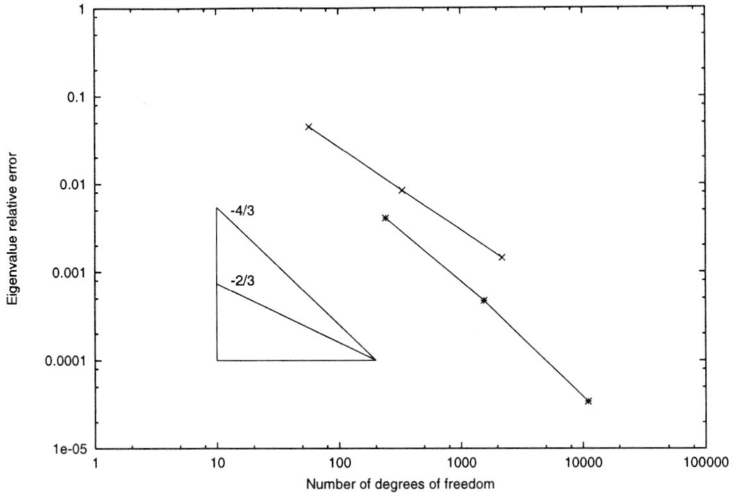

Figure 7.6 Rate of convergence for $\tilde{k}_0^2$ of the empty cavity of dimensions 1×0.5×0.75: –×–, first-order elements; –*–, second-order elements.

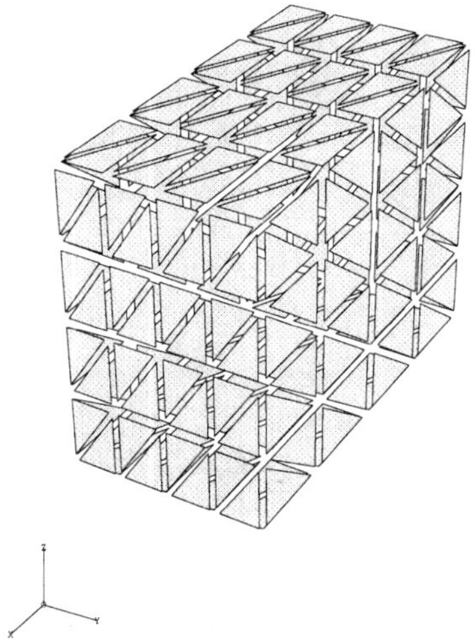

Figure 7.7 Empty cavity 1×0.5×0.75. Mesh B (384 elements, 604 nodes).

Table 7.6
Resonance wavenumbers for the empty cavity (Mesh A)

Mode	Analytic	$\bar{E}$ formulation	$\bar{H}$ formulation	$\bar{E}$ f. error (%)	$\bar{H}$ f. error (%)
TM_{z110}	7.025	6.534	6.784	-6.99	-3.43
TE_{z011}	7.531	6.789	7.542	-9.85	0.14
TE_{z201}	7.531	7.501	7.692	-0.40	2.14
TM_{z111}	8.179	7.547	7.853	-7.72	-3.99
TE_{z111}	8.179	8.059	8.249	-1.47	0.86
TM_{z210}	8.886	8.170	8.376	-8.06	-5.74

Table 7.7
Resonance wavenumbers for the empty cavity (Mesh B)

Mode	Analytic	$\bar{E}$ formulation	$\bar{H}$ formulation	$\bar{E}$ f. error (%)	$\bar{H}$ f. error (%)
TM_{z110}	7.025	6.724	6.881	-4.29	-2.05
TE_{z011}	7.531	7.119	7.550	-5.48	0.25
TE_{z201}	7.531	7.589	7.663	0.78	1.76
TM_{z111}	8.179	7.899	7.996	-3.42	-2.24
TE_{z111}	8.179	8.183	8.232	0.05	0.65
TM_{z210}	8.886	8.495	8.608	-4.40	-3.12

the x direction and 0.1 along the y direction. The meshes employed for this structure are shown in Figure 7.8 and Figure 7.9.

The first few wavenumbers k_0 for the ridge cavity are shown in Table 7.8. No symmetry has been taken into account. The eigenvalues have been obtained with the electric field formulation and have been compared with those shown in [Chatterjee et al. 1992] (an edge element FEM formulation with the electric field). A good agreement is observed. It has not been possible to reproduce the meshes used in the mentioned paper because the authors only provide the number of unknowns after deleting the degrees of freedom having essential boundary conditions. It is worth noting that (as can be seen in Figure 7.8 and Figure 7.9), the meshes employed in the analysis are almost uniform, i.e., no special care has been taken to refine the mesh near the corners of the

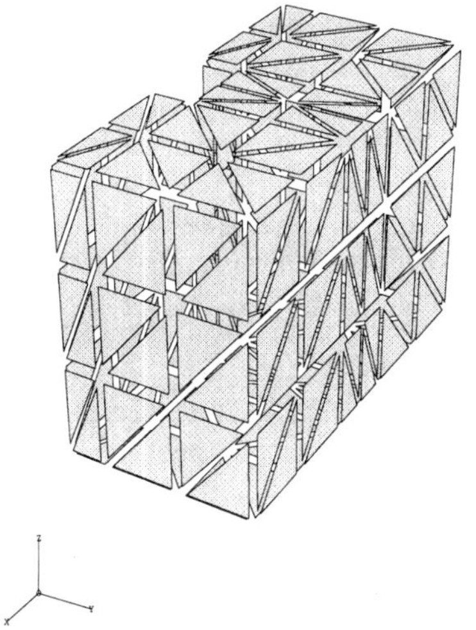

Figure 7.8 Ridge cavity. Mesh A (288 elements, 487 nodes).

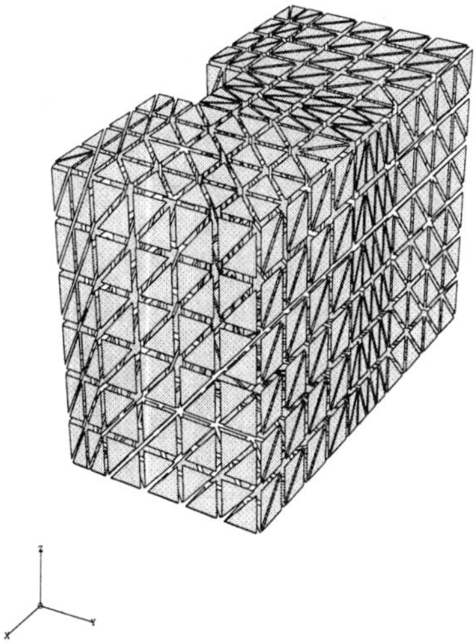

Figure 7.9 Ridge cavity. Mesh B (2304 elements, 3266 nodes).

ridge. The number of zero eigenvalues found with the electric field formulation is 14 for Mesh A and 225 for Mesh B. With the magnetic field formulation, the numbers of zero eigenvalues are 107 and 594, respectively. These numbers coincide with the number of zero eigenvalues expected according to the theory explained previously.

Table 7.8

Resonance wavenumbers for the ridge cavity ($\bar{E}$-field formulation)

Mode	[Chatterjee et al. 1992]	Mesh A	Mesh B
1	4.999	4.761	4.975
2	7.354	6.894	7.325
3	7.832	7.041	7.542
4	7.942	7.205	7.906
5	7.959	7.873	7.983
6	8.650	7.920	8.257
7	8.916	8.153	8.791

An example of an inhomogeneous cavity is the next structure. The dimensions of the cavity in the x, y, and z directions are 1, 0.1, and 1, respectively. The upper half of the cavity is filled with a dielectric of $\varepsilon_r = 2$. The wavenumbers of the resonant modes may be computed analytically. The dominant mode has been analyzed first in order to verify for an inhomogeneous structure the rate of convergence obtained for the empty cavity, with first- and second-order elements. Table 7.9 shows results for the first-order element, using uniform refinements in all directions, while Table 7.10 gives the same type of results for second-order elements. The results of Table 7.11 refer to a refinement of the mesh adapted to the electromagnetic field behavior. Namely, no refinement takes place in the y direction, in which the mode does not vary. First-order elements have been used for the adapted mesh. Figure 7.10 shows the corresponding rates of convergence for all three analyses. With uniform refinements the theoretical values for the slopes are achieved. A superconvergent rate of convergence is obtained for the adapted mesh.

The first few higher order wavenumbers have also been analyzed using the mesh of Figure 7.11. Results for the dominant and higher order modes are compared with the analytical ones in Table 7.12. No symmetry has been taken into account in this case. Good accuracy is achieved for both electric and magnetic field formulations. A superconvergent behavior for some of the higher order modes is obtained due to the fact

Table 7.9

Results for $\tilde{k}_0^2$ for the TE_{z101} resonant mode of the half-filled cavity.
First-order tetrahedral elements. Exact value k_0^2: 12.517444

Number of cubic cell divisions	Number of elements	Number of degrees of freedom	$\tilde{k}_0^2$	Relative error
2x1x2	24	57	14.0394	12.162978×10^{-2}
4x2x4	192	330	12.7523	1.879923×10^{-2}
8x4x8	1536	2196	12.5625	0.363266×10^{-2}

Table 7.10

Results for $\tilde{k}_0^2$ for the TE_{z101} resonant mode of the half-filled cavity.
Second-order tetrahedral elements. Exact value k_0^2: 12.517444

Number of cubic cell divisions	Number of elements	Number of degrees of freedom	$\tilde{k}_0^2$	Relative error
2x1x2	24	242	12.5736	0.45179×10^{-2}
4x2x4	192	1556	12.5307	0.10953×10^{-2}
6x3x6	648	4854	12.5195	0.020053×10^{-2}
8x4x8	1536	11048	12.5173	0.0026364×10^{-2}

Table 7.11

Results for $\tilde{k}_0^2$ for the TE_{z101} resonant mode of the half-filled cavity.
First-order tetrahedral elements. Adapted mesh. Exact value k_0^2: 12.517444

Number of cubic cell divisions	Number of elements	Number of degrees of freedom	$\tilde{k}_0^2$	Relative error
2x1x2	24	57	14.0394	12.162978×10^{-2}
4x1x4	96	193	12.7385	1.769593×10^{-2}
6x1x6	216	409	12.5876	0.564033×10^{-2}
8x1x8	384	705	12.5458	0.230087×10^{-2}
10x1x10	600	1081	12.5295	0.099864×10^{-2}

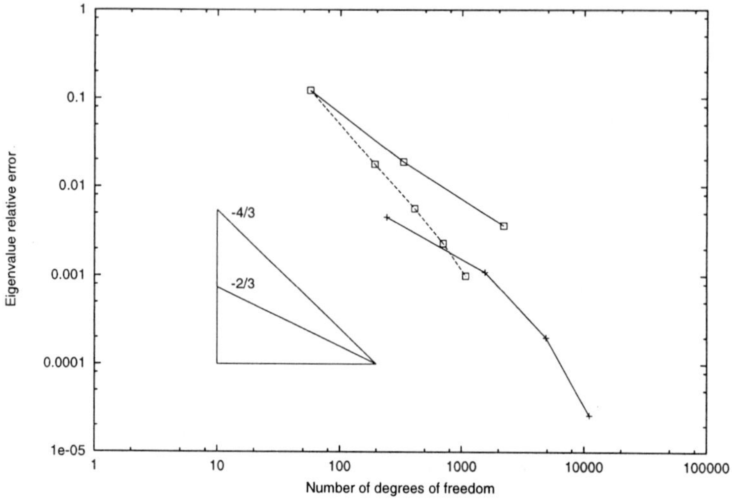

Figure 7.10 Rate of convergence of $\tilde{k}_0^2$ of the half-filled cavity: —□—, first-order, uniform refinement; —+—, second-order elements, uniform refinement; --□--, first-order, adapted mesh.

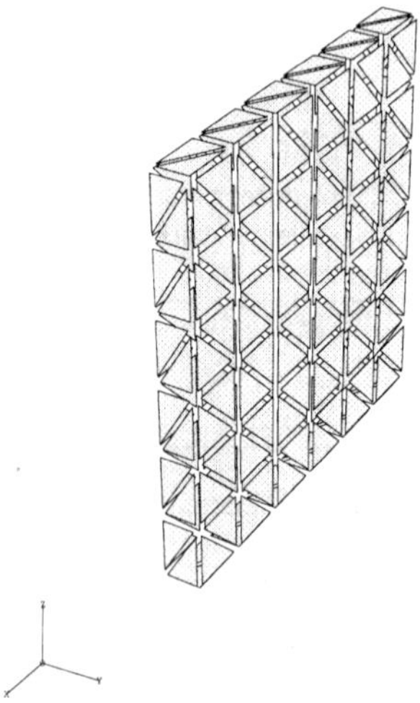

Figure 7.11 Half-filled cavity. Mesh (216 elements, 409 nodes).

Table 7.12
Resonance wavenumbers for the half-filled cavity

Mode	Analytic	$\bar{E}$ formulation	$\bar{H}$ formulation	$\bar{E}$ f. error (%)	$\bar{H}$ f. error (%)
TE_{z101}	3.538	3.547	3.548	0.24	0.28
TE_{z201}	5.445	5.431	5.445	-0.26	-0.01
TE_{z102}	5.935	5.980	5.945	0.75	0.16
TE_{z301}	7.503	7.378	7.371	-1.67	-1.76
TE_{z202}	7.633	7.614	7.443	-0.25	-2.49
TE_{z103}	8.096	8.121	7.784	0.31	-3.86

that the mesh is well adapted to the configuration of the electromagnetic fields. The number of zero eigenvalues found with the magnetic field formulation is 97. On the other hand, the electric field formulation did not provide any zero eigenvalues because this mesh has no internal vertices.

The next example is the dielectric loaded cavity shown in Figure 7.12. This problem was studied by several researchers, e.g., [Webb 1985], [Webb 1988a], [Bárdi et al. 1991], [Pichon and Bossavit 1993], and [Golias and Papagiannakis 1994]. It has been analyzed exploiting the symmetry of the structure. Specifically, only one-fourth of the cavity has been meshed as shown in Figure 7.13. The symmetry planes are the *x-z* plane and the *y-z* plane. Because of the two symmetry planes, it is necessary to run the problem four times in order to get all the possible modes of the structure, namely, the following (*x-z* PEC, *y-z* PMC), (*x-z* PMC, *y-z* PMC), (*x-z* PMC, *y-z* PEC), (*x-z* PEC, *y-z* PEC). Here PEC and PMC stands for perfect electric and magnetic conductors, respectively. This structure demands a large number of unknowns to obtain accurate results. It was long thought that the dominant mode had PMC symmetry in both symmetry planes, *x-z* and *y-z* (e.g., [Webb 1985, 1988a], [Bárdi et al. 1991]). Recently, using more refined meshes, this has been shown to be wrong [Golias and Papagiannakis 1994]. Utilizing an adaptive FEM mesh (meshes with nearly 20 thousands nodes are obtained), it has been shown that the dominant mode exhibits PEC symmetry in the *x-z* plane and PMC symmetry in the *y-z* plane. In Tables 7.13 and 7.14, the numerical results obtained with the mesh of Figure 7.13 are compared with those obtained in [Golias and Papagiannakis 1994]. As it can be seen from the tables, the mesh of Figure 7.13 is well adapted to the dominant mode when using the $\bar{H}$-formulation, depicting a correct approximation. However, a larger number of nodes or an adaptive mesh procedure is necessary to improve the accuracy of the resonance frequencies for the higher order modes when using an $\bar{E}$-formulation.

The number of spurious solutions in this example depends on the type of symmetries imposed. The number of zero eigenvalues obtained in each case is given in Table 7.15. They coincide in all cases with the theoretical predictions.

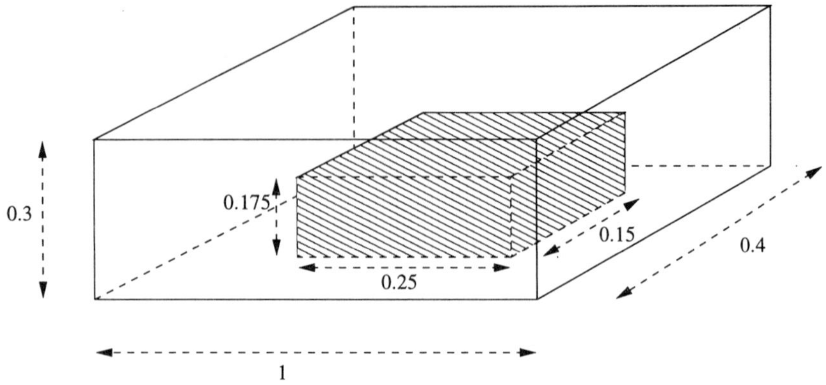

Figure 7.12 Dielectric loaded cavity, ε_r=16.

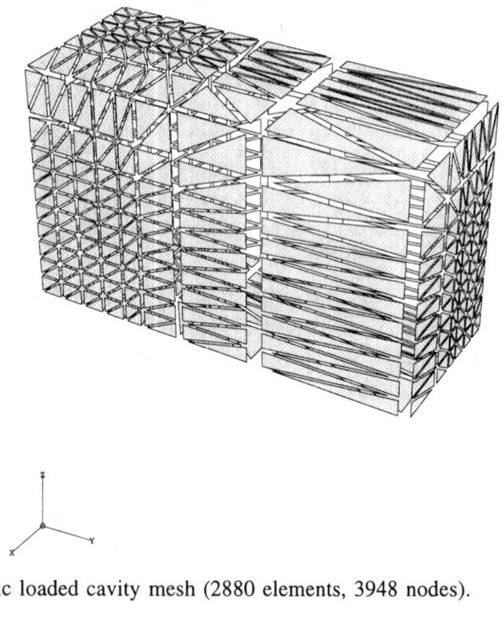

Figure 7.13 Dielectric loaded cavity mesh (2880 elements, 3948 nodes).

Table 7.13
Resonance wavenumbers for the dielectric loaded cavity ($\bar{H}$-field formulation)

Mode	[Golias and Papagiannakis 1994]	Present method	Error (%)
1	5.329	5.266	-1.182
2	5.428	5.363	-1.200
3	6.451	6.355	-1.490
4	7.574	7.564	-0.001
5	7.807	7.665	-1.820

Table 7.14
Resonance wavenumbers for the dielectric loaded cavity ($\bar{E}$-field formulation)

Mode	[Golias and Papagiannakis 1994]	Present method	Error (%)
1	5.354	5.387	0.61
2	5.394	5.315	-1.46
3	6.469	6.465	-0.06
4	7.659	7.664	0.06
5	7.838	7.5639	-3.50

Table 7.15
Number of zero eigenvalues for the mesh of Figure 7.13

Symmetry	$\bar{H}$ formulation	$\bar{E}$ formulation
(x-z PEC, y-z PMC)	595	387
(x-z PMC, y-z PMC)	530	432
(x-z PMC, y-z PEC)	617	360
(x-z PEC, y-z PEC)	692	315

It is worth pointing out that for certain structures both formulations (the $\bar{E}$-field formulation and the $\bar{H}$-field formulation) converge to the exact eigenvalue from opposite directions, i.e., one of the formulations gives an upper bound while the other one gives a lower bound of the exact solution. This behavior may be explained in terms of the duality of the formulations. However, it is not clear that a formulation will always provide an upper or a lower bound for the wavenumber k_0, especially if there are degenerate modes. It may also be observed that certain modes are obtained with different accuracy depending on the formulation ($\overline{E}$ or $\overline{H}$) used.

In summary, numerical results of the analysis of various 3D microwave cavities have been presented, showing the corresponding rate of convergence and the validity of the 3D tetrahedral curl-conforming elements in double-curl FEM formulations. Accurate results have been obtained. The spurious modes appear as eigenvectors corresponding to the numerically zero eigenvalue, but they do not pollute the spectrum. In addition, they are easily discarded. In addition, the number of zero eigenvalues may be known in advance, as was shown.

7.4 ANALYSIS OF DISCONTINUITIES IN WAVEGUIDES USING THE FINITE ELEMENT METHOD

In this section, the problem of characterization of discontinuities in waveguiding structures is considered. The first FEM approaches to discontinuity problems were 2D formulations (e.g., axisymmetric problems or E-plane and H-plane problems) using Lagrange elements, and later on, curl-conforming elements. In [Koshiba 1992; Ch. 7] a number of references may be found. However, the general discontinuity problem requires a 3D formulation. Initially Lagrange elements were also used in conjunction with different procedures to eliminate the spurious modes [Ise et al. 1990], [Picon 1988]. Later on, curl-conforming elements were utilized [Ise et al. 1991], [Foo and Silvester 1993], [Hirayama et al. 1994]. Examples of practical problems that may be analyzed as discontinuity problems are [Aregba et al. 1994], [Cousty et al. 1992], and [Wu 1996].

The problem, in this case, deals with the solution of Maxwell's equations in an arbitrarily shaped domain in which energy is injected or extracted through waveguiding structures (see Figure 7.14). Since the waveguiding structures are assumed to be infinitely long, the problem domain is then unbounded. However, for FEM analysis, the domain is truncated by placing fictitious planes perpendicular to the translational axis of the waveguiding structures. These planes are the ports of the problem. Then, the discontinuity is located inside a cavity in which energy is injected or extracted through the ports. The analysis of the discontinuity is a deterministic problem using the double-curl formulation discussed in Section 7.3 with an excitation term. Formulations based on scalar and vector potentials have also been used [Dyczij-Edlinger and Biro 1996]. Different approaches have been utilized for the formulation of the excitation term. Here, the length of the waveguiding structures up to the corresponding port is taken to be long enough to assume that only the fundamental mode exists at the ports, simplifying the boundary conditions to be imposed.

The wavenumber is assumed to be known. A field configuration is used to inject energy through one of the ports of the structure. The FEM analysis provides the field solution in all the problem domains, including the ports. Thus, the scattering parameters

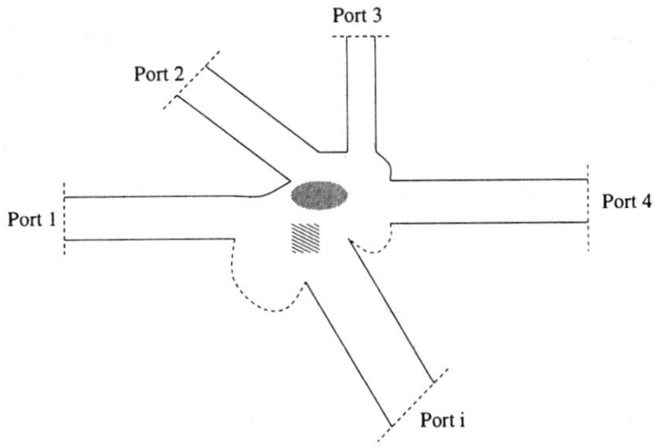

Figure 7.14 Geometry of the problem.

of the discontinuity for the excitation mode at a given frequency can easily be obtained by comparing the energy that is going outward from the cavity through each one of the ports with the energy that was injected inside. In order to characterize the discontinuity, this process needs to be done as many times as the number of ports that exist in the structure. However, in practice, the number of times the analysis has to be done may be reduced if reciprocity and symmetries are applied to the problem under study.

The FEM weak formulation of the problem is presented next in Section 7.4.1. Basically, it is the same double-curl formulation employed in Section 7.3 for the analysis of cavities. However, the formulation here does not lead to an eigenvalue problem, but to a deterministic problem, due to the inhomogeneous boundary conditions at the ports of the structure. As in Section 7.3, tetrahedral curl-conforming elements are used in the discretization of the field. Section 7.4.2 shows some numerical results for the analysis of discontinuities in rectangular waveguides.

7.4.1 Finite Element Formulation

7.4.1.1 *Variational Formulation*

A weak formulation in terms of the electric or the magnetic field is obtained from the double-curl differential formulation of (7.33). The solution of the problem is obtained by making stationary a suitable functional. For the analysis of discontinuities, the following functional is considered:

$$F(\bar{V}) = \frac{1}{2}\int_\Omega \left(\left(\bar{\nabla}\times\bar{V}^*\right)\cdot\left(\underline{\underline{f}}^{-1}\,\bar{\nabla}\times\bar{V}\right) - \omega^2\,\bar{V}^*\cdot\underline{\underline{p}}\bar{V}\right)\,d\Omega - \int_\Gamma \bar{a}_n\cdot\left(\bar{V}^*\times\underline{\underline{f}}^{-1}\,\bar{\nabla}\times\bar{V}\right)\,d\Gamma \qquad (7.107)$$

where the various variables are explained in Table 7.1.

The functional (7.107) is analogous to that of (7.78). However, it is worth noting the presence in (7.107) of the symbol *, which stands for "complex conjugate." In this case, and in contrast with the cavity problem of Section 7.3, the vector V, i.e., the FEM approximation of the electric or the magnetic field, cannot be assumed to be real. An inner product, valid for complex-valued vector functions (see Appendix A.1.3), is defined by

$$<\bar{f}\,;\,\bar{g}> = \int_{\Omega} \bar{f}^{*} \cdot \bar{g}\; d\Omega \tag{7.108}$$

Observing the boundary integral term of (7.107) it is easy to see that the boundary condition at the ports should relate the tangential components of the field and its curl. If it is assumed that only one mode is present at the ports, then the following boundary condition is applicable:

$$\bar{a}_n \times (\bar{\nabla} \times \bar{V}) + \gamma \bar{a}_n \times (\bar{a}_n \times \bar{V}) = \bar{U}^{inc} \tag{7.109}$$

where $\bar{a}_n$ stands for the outward unit vector normal to the port surface and γ is the complex propagation constant of the propagating mode that exists at the port. The vector $\bar{U}^{inc}$ constitutes the excitation of the problem and it is related to the value of the incident field, according to

$$\bar{U}^{inc} = 2\gamma \bar{a}_n \times (\bar{a}_n \times \bar{V}^{inc}) \tag{7.110}$$

$\bar{U}^{inc}$ will be different from zero only at the ports that are excited. The FEM analysis is performed to obtain the scattering parameters for the discontinuity. This is achieved by exciting one port at a time, hence

$$\bar{a}_n \times (\bar{\nabla} \times \bar{V}) + \gamma \bar{a}_n \times (\bar{a}_n \times \bar{V}) = \bar{U}^{inc} \qquad \text{on } \Gamma_{S_{inc}} \tag{7.111}$$

$$\bar{a}_n \times (\bar{\nabla} \times \bar{V}) + \gamma \bar{a}_n \times (\bar{a}_n \times \bar{V}) = \bar{0} \qquad \text{on } \Gamma_{S_k}\; ; \quad \forall\, k \mid S_k \neq S_{inc} \tag{7.112}$$

where S_{inc} denotes the port that is excited.

It is worth noting that the boundary condition (7.109) is the relation between the electric and the magnetic fields satisfied by the modes of a waveguiding structure, i.e., the tangential components of the vector electric and magnetic fields. To illustrate this, consider a local coordinate axis (ξ,η,ζ) for each port. The port is assumed to be contained in the ξ-η plane. The waveguiding structure of that port will have translational symmetry in the ζ direction. The ζ-axis is considered to be toward the discontinuity. Thus, at the excited port, there will be incident and reflected fields:

$$\bar{V}(\xi,\eta,\zeta) = \bar{V}^{inc}(\xi,\eta,\zeta) + \bar{V}^{ref}(\xi,\eta,\zeta) = \bar{V}^{inc}(\xi,\eta)\, e^{-\gamma\zeta} + \bar{V}^{ref}(\xi,\eta)\, e^{\gamma\zeta} \tag{7.113}$$

The curl of the field tangential to the port may be written as

$$\begin{aligned} -\bar{a}_\zeta \times (\bar{\nabla} \times \bar{V}) &= -\bar{a}_\zeta \times \Big((\bar{\nabla}_{\xi\eta} \times \bar{V}_{\xi\eta}) + (\bar{\nabla}_{\xi\eta} \times \bar{V}_\zeta) + (\bar{\nabla}_\zeta \times \bar{V}_{\xi\eta}) + (\bar{\nabla}_\zeta \times \bar{V}_\zeta) \Big) \\ &= -\bar{a}_\zeta \times (\bar{\nabla}_\zeta \times \bar{V}_{\xi\eta}) \end{aligned} \tag{7.114}$$

where $\bar{a}_\zeta$ is the unit vector in the ζ direction. Equation (7.114) is valid for TE modes, with $\bar{V} = \bar{E}$, or TM modes, with $\bar{V} = \bar{H}$. However, an expression analogous to (7.114) can be written for the general case. By using (7.113), expression (7.114) leads to

$$\begin{aligned} -\bar{a}_\zeta \times (\bar{\nabla} \times \bar{V}) &= -\gamma\Big(-\bar{a}_\zeta \times (\bar{a}_\zeta \times \bar{V}^{inc})\Big) + \gamma\Big(-\bar{a}_\zeta \times (\bar{a}_\zeta \times \bar{V}^{ref})\Big) \\ &= \gamma\Big(-\bar{a}_\zeta \times (\bar{a}_\zeta \times \bar{V})\Big) - 2\gamma\Big(-\bar{a}_\zeta \times (\bar{a}_\zeta \times \bar{V}^{inc})\Big) \end{aligned} \tag{7.115}$$

Thus, taking into account that $\bar{a}_n = -\bar{a}_\zeta$ and (7.110), (7.115) finally leads to (7.109).

The ports should be placed far enough from the discontinuity to ensure that only one mode exists at the ports. Thus, the distance from the discontinuity to the ports must be large enough so that the higher order evanescent modes reach the ports very attenuated and, hence, its effect may be discarded. Also, the range of frequencies of the analysis is limited to the interval in which only the fundamental mode is propagating. Then, the boundary condition of (7.109) is suitable only for the mono-mode case. However, a more elaborate boundary condition can be developed by considering the eigenfunction expansion of the field. In that case, the expansion of the field with a finite number of significant eigenmodes should be included in (7.110) and the orthogonality relations of the modes are employed. This leads to a boundary condition as the following:

$$\bar{a}_n \times (\bar{\nabla} \times \bar{V}) + P(\bar{V}) = \bar{U}^{inc} \tag{7.116}$$

in which $P(\bar{V})$ includes the information about the propagation constants and the field configuration of the different modes considered. Thus, the boundary condition of (7.116) allows the analysis in the frequency range where the waveguides permit multimode propagation. Also, it enables one to place the ports closer to the discontinuity, leading to FEM meshes with a lower number of unknowns.

In summary, the boundary condition at the ports is taken into account in the formulation. This is done through the boundary integral of the weak formulation. Thus, it is concluded that the boundary condition of the ports is a natural boundary condition of the problem.

The same normalization performed in Section 7.3 (equations (7.86) and (7.87) and Table 7.2) is applied here. Therefore, taking into account (7.109), the functional of (7.107) may be further written as

$$F = 1/2 \int_{\Omega} \left((\bar{\nabla} \times \bar{V}^{*}) \cdot (\underline{\underline{f}}_{r}^{-1} \bar{\nabla} \times \bar{V}) - k_0^2 \bar{V}^{*} \cdot \underline{\underline{p}}_{r} \bar{V} \right) \mathrm{d}\Omega + \sum_{k} \left(\int_{\Gamma_{S_k}} \gamma (\bar{a}_n \times \bar{V}^{*}) \cdot (\bar{a}_n \times \bar{V}) \, \mathrm{d}\Gamma \right) - \int_{\Gamma_{S_{inc}}} \bar{V}^{*} \cdot \bar{U}^{inc} \, \mathrm{d}\Gamma \tag{7.117}$$

with the vector field $\bar{V}$ referring here to the normalized field. It is worth noting that the sum of index k includes S_{inc}.

Applying the stationarity of the preceding functional over each element, a set of discrete integral forms is obtained for every element. Curl-conforming tetrahedrons are employed, namely, mixed-order Nédélec elements. The properties of these elements and of their basis functions have been explained in the previous sections of this chapter. Thus, considering vector basis functions of that type the following set of discrete integral forms at the element level is achieved:

$$\{W^e\} = [k^e]\{d^e\} - \{f^e\} \tag{7.118}$$

where

$$k_{ij}^e = \int_{\Omega^e} \left((\bar{\nabla} \times \bar{N}_i^e) \cdot \left((\underline{\underline{f}}_r^e)^{-1} \bar{\nabla} \times \bar{N}_j^e \right) - k_0^2 (\bar{N}_i^e \cdot \underline{\underline{p}}_r^e \bar{N}_j^e) \right) \mathrm{d}\Omega + \sum_{k} \int_{\Gamma_{S_k}^e} \gamma (\bar{a}_n \times \bar{N}_i^e) \cdot (\bar{a}_n \times \bar{N}_j^e) \, \mathrm{d}\Gamma \tag{7.119}$$

and

$$f_i^e = \int_{\Gamma_{S_{inc}}^e} \bar{N}_i^e \cdot \bar{U}^{inc} \, \mathrm{d}\Gamma \tag{7.120}$$

Then, the assembly of the local integral forms and the imposition of the essential boundary conditions lead to a global system of equations, which characterizes the electromagnetic problem:

$$[K]\{D\} = \{F\} \tag{7.121}$$

$$[K] = \sum_{e=1}^{N_e} [k^e]; \quad \{F\} = \sum_{e=1}^{N_e} \{f^e\} \tag{7.122}$$

Here (7.109) or (7.110) are used to match the ports to the fundamental mode or the various propagating modes. An alternate approach not requiring matched ports is to use together with FEM the Matrix Pencil Method that provides the amplitudes and propagation constants of the forward and backward waves of the dominant and higher order modes from which the scattering parameters are computed [Sarkar et al. 1992].

7.4.1.2 *Computation of the Scattering Parameters*

Once the preceding system of equations is solved, the field is known in the whole domain. To obtain the scattering parameters of the discontinuity, the FEM analysis is performed several times, each time exciting the structure through a different port. Thus, by analyzing the problem as many times as there are ports, all the coefficients of the scattering matrix are obtained. However, the number of analysis may be reduced exploiting the possible reciprocity and symmetry conditions of the structure.

In the following, the expressions for the scattering parameters in terms of the electric or magnetic field at the ports are presented. A local Cartesian system of coordinates is considered at each one of the ports. The cross section of the waveguiding structure defines the ξ-η plane. Then, the direction orthogonal to the plane in the counterclockwise sense defines the ζ axis. The orientation of the system of coordinates must be chosen so that the ζ-axis is on the outward direction. The transverse field component of the mth mode of the waveguide at the ith port may be written as

$$\overline{V}_m^{(i)}(\xi,\eta,\zeta) = V_m^{(i)}\,\overline{v}_m^{(i)}(\xi,\eta)\,e^{\pm\gamma\zeta} \tag{7.123}$$

The vector $\overline{v}_m^{(i)}(\xi,\eta)$ gives the spatial dependence of the mode. It may be obtained either analytically or numerically (e.g. by means of a 2D FEM analysis). The complex quantity $V_m^{(i)}$ gives the relative amplitude and phase of the mode. In what follows, for ease of notation, the ports are considered to be at the plane $\zeta = 0$ and the dependence with (ξ,η) will not be explicitly shown. From $\overline{v}_m^{(i)}$ and Maxwell's equation the dual vector $\overline{v}_{dm}^{(i)}$ may be obtained. It is easy to show that in the case of separable permittivity and permeability tensors those two vectors may be considered to be real (see Appendix D.3.2). They may be normalized in a suitable way. Here, the normalization is chosen so that, together with the orthogonality properties of the modes, the following relation is satisfied

$$\int_{S_i} (\overline{v}_m^{(i)} \times \overline{v}_{dn}^{(i)}) \cdot \overline{a}_\zeta \, dS = \delta_{mn} \tag{7.124}$$

Here S_i is the cross section of the ith port. With such normalization the complex quantity $V_m^{(i)}$ corresponds to the power wave of the mth mode at the ith port. Thus, the scattering parameter relating the power waves of the pth port (mode m) and qth port (mode n) is defined as $S_{pq} = V_m^{(p)}/V_n^{(q)}$.

For the computation of the S_{ii} parameter the problem is excited with the fundamental mode of ith port (considered to be the 0th mode) and solved by means of FEM. Then, the field at the ith port may be considered to be the superposition of the incident field and an infinite number of modes reflected at the discontinuity

$$\bar{V}_t^{(i)} = V_{inc}\,\bar{v}_0^{(i)} + \sum_{m=0}^{\infty} V_m^{(i)}\,\bar{v}_m^{(i)} \tag{7.125a}$$

Performing the vector product of (7.125a) with $\bar{v}_{d0}^{(i)}$, integrating over the ith port, and considering (7.124), the S_{ii} parameter is obtained as:

$$S_{ii} = \frac{\int_{S_i} (\bar{V}_t^{(i)} \times \bar{v}_{d0}^{(i)}) \cdot \bar{a}_\zeta \, dS}{V_{inc}} - 1 \tag{7.125b}$$

For the S_{ji} parameter ($j \neq i$), the field at the jth port may be written as

$$\bar{V}_t^{(j)} = \sum_{m=0}^{\infty} V_m^{(j)}\,\bar{v}_m^{(j)} \tag{7.126a}$$

which is identical to (7.125a) except that the incident field is not present. Let us denote the mode considered at the jth port for the computation of S_{ji} as the kth mode. Then, analogously to the procedure employed for S_{ii}, the following expression is obtained

$$S_{ji} = \frac{\int_{S_j} (\bar{V}_t^{(j)} \times \bar{v}_{dk}^{(j)}) \cdot \bar{a}_\zeta \, dS}{V_{inc}} \tag{7.126b}$$

It is worth noting the generality of the procedure that has been described. However, in the particular case of homogeneous waveguiding structures $\bar{v}_k^{(j)}$ and $\bar{v}_{dk}^{(j)}$ are related through the mode impedance. Therefore, the integrand in (7.125b) and (7.126b) may be substituted by the scalar product of the variables involved.

7.4.2 Numerical Results: Application to Rectangular Waveguides

The formulation in Section 7.4.1 does not assume a particular geometry of the discontinuity or of the waveguiding structures of the ports. However, the field configuration $\bar{v}(\zeta,\eta)$ and propagation constant of the fundamental modes of the waveguiding structures must be supplied. They may be known analytically (e.g., for rectangular waveguides), or they may be obtained numerically. In the latter case, a 2D FEM may be used (see Chapters 2 and 3). In the following, numerical results for several discontinuities in rectangular waveguides (i.e., mode TE_{10} assumed) are shown.

First a two-port network consisting of a waveguide section of length L is analyzed; i.e., no discontinuity is present (see Figure 7.15). Thus, $S_{11} = S_{22} = 0$, and $S_{21} = S_{12} = e^{-\gamma L}$. The structure could have been analyzed by means of a 2D formulation, taking advantage of its symmetry with respect to the x-z plane. However, the 3D formulation was applied to this benchmark example to validate it. Different meshes were used. The tetrahedral elements of the meshes were obtained, as in Section 7.3, by subdividing the cubes into six tetrahedrons. Values of S_{11} different from those of S_{22} were obtained. This was observed for all meshes, independently of the density of the mesh and of the direction of propagation through the tetrahedrons. It has been concluded that the reason for those results is the asymmetry of the meshes at the ports. Symmetric meshes were made by gluing two halves of the original domain, one being the mirror image of the other. Thus, the final mesh was symmetric with respect to $z = L/2$. As expected, with the symmetric mesh, S_{11} and S_{22} are equal. This results are summarized in Figure 7.16, which shows the magnitude of the reflection coefficient in the range of frequencies of mono-mode propagation in the waveguide. Also, it may be seen that the reflection level depends on the frequency when nonsymmetric meshes are used, while it remains constant when a symmetric mesh is utilized in the analysis.

Figure 7.17 shows the rate of convergence of the magnitude of S_{11} and S_{22} for the symmetric and nonsymmetric meshes. It is observed that the values of the reflection coefficient at the ports tend to zero as the meshes are refined. Also, it is seen that the rate of convergence is the same in all cases, as expected. However, independently of the density of the mesh, the values of S_{11} and S_{22} do not coincide when nonsymmetric meshes are used. On the other hand, the transmission coefficient behaves as expected, i.e., $S_{21} = S_{12}$. Figure 7.18 shows the numerical dispersion error of the propagation constant β ($\gamma = j\beta$) for the same three meshes considered in Figure 7.16. The error is obtained by comparing the phase of the parameter S_{21} (or S_{12}) with its analytical value.

In summary, it may be concluded that a more reliable behavior of the scattering parameters is obtained when symmetric meshes are employed in the FEM analysis.

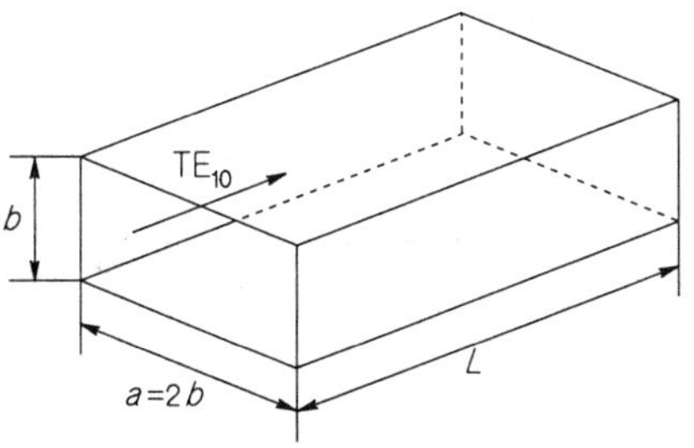

Figure 7.15 Section of a rectangular waveguide.

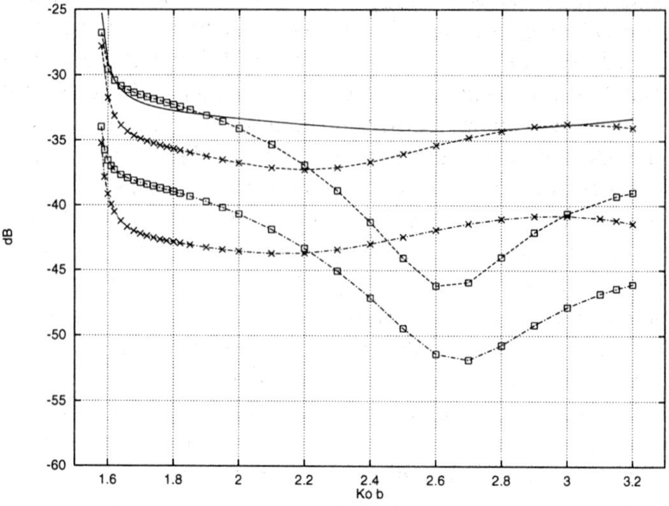

Figure 7.16 Rectangular waveguide. —, symmetric mesh (768 elements $S_{11} = S_{22}$); --, nonsymmetric mesh (384 elements): ×, S_{11}, □, S_{22}; -·-, nonsymmetric mesh (1296 elements): ×, S_{11}, □, S_{22}.

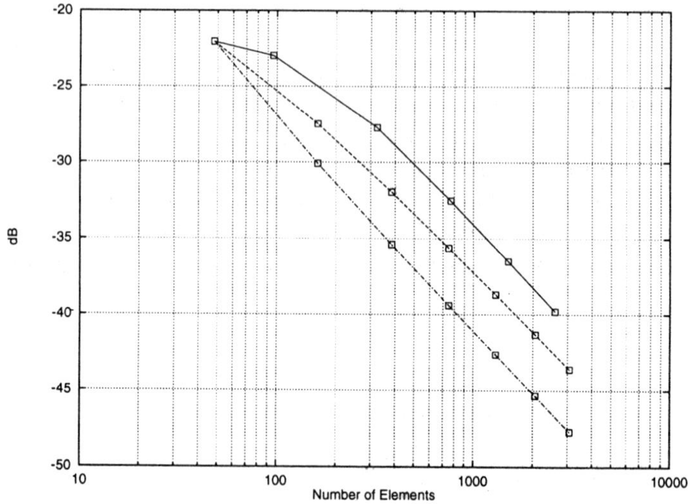

Figure 7.17 Convergence of S_{11} and S_{22}. —, symmetric mesh ($S_{11} = S_{22}$); --, nonsymmetric mesh (S_{11}); -·-, nonsymmetric mesh (S_{22}).

Symmetric meshes lead to identical numerical values of the reflection coefficient at both ports. Also, they provide a flat response along the whole range of frequencies. The rest of the results that will be presented here correspond to symmetric meshes.

In order to avoid the gluing of a mesh and its mirror image, to generate a symmetric mesh a different methodology for generating tetrahedral mesh from cubes is proposed. Figure 7.19 shows the subdivision of a cube into five tetrahedrons. It is seen that the subdivision is symmetric no matter from which direction of the cube the

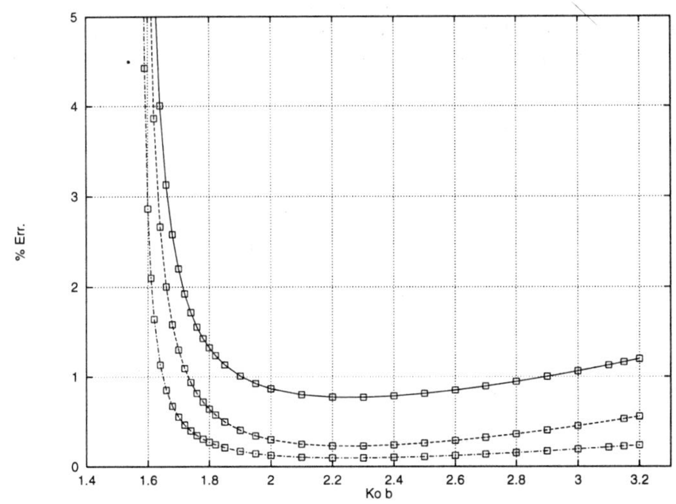

Figure 7.18 Relative error in β. —, symmetric mesh (768 elements); --, nonsymmetric mesh (384 elements); -·-, nonsymmetric mesh (1296 elements).

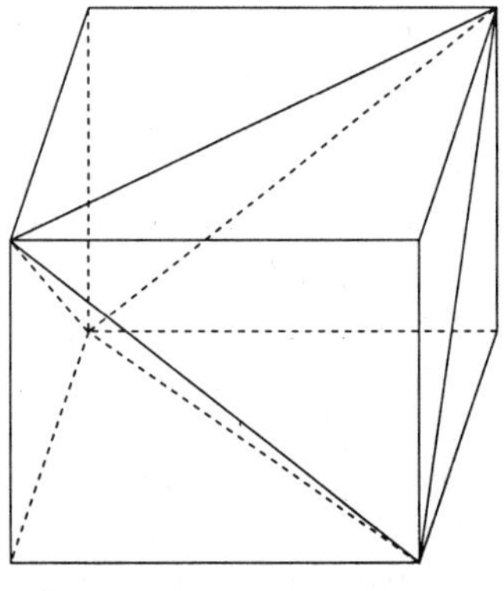

Figure 7.19 Subdivision of a cube into five tetrahedrons.

electromagnetic field is propagating. Table 7.16 shows results for the analysis of the previous waveguide using meshes generated from cubes subdivided into six tetrahedrons and cubes subdivided into five tetrahedrons. Propagation from the *x*, *y*, and *z* directions has been considered. A free-space wavenumber, k_0, equal to 1.76 has been selected, so $\beta = 0.794$. The mesh generated with subdivisions of cubes into five tetrahedrons is symmetric and leads to more accurate results.

Table 7.16

S parameters of a rectangular waveguide section using meshes generated by subdividing a cube into 6 tetrahedrons (nonsymmetric mesh) and into five tetrahedrons (symmetric mesh)

	$\|S_{11}\|$	Phase(S_{11})	$\|S_{21}\|$	Phase(S_{21})	$\|S_{22}\|$	Phase(S_{22})
6 Tetra. *x* direction	0.469	-0.535	0.739	-1.031	0.309	1.224
6 Tetra. *y* direction	0.406	0.716	0.747	-1.022	0.486	-0.05
6 Tetra. *z* direction	0.309	1.224	0.739	-1.031	0.459	-0.535
5 Tetra. Any direct.	0.206	0.067	0.811	-0.829	0.206	0.067

For the next example, consider a 90 degree bend. The conventional bend provides a poor reflection coefficient. Therefore, it is common to make a miter in the bend in order to improve the reflection performances of the bend. A mitered *E*-plane bend and a mitered *H*-plane bend have been analyzed. The mitered E-plane bend is considered first (see Figure 7.20). Only half of the structure has been meshed and a perfect magnetic symmetry wall has been employed in the analysis. The results shown here have been obtained employing the meshes that appear in Figure 7.21 (Mesh A) and Figure 7.22 (Mesh B), which correspond to a miter of $d/b = 0.707$ and a length of the ports equal to 1.5. The magnitude of S_{11} for both the meshes is shown in Figure 7.23 together with the mode-matching results of [Reiter and Arndt 1994]. A good agreement is observed except in the lower frequency band close to the cut-off frequency of the TE_{10} mode, in which, logically, the S_{11} parameter changes abruptly. On the other hand, Figure 7.24 shows the effective length of the bend (normalized to *b*) for Mesh A and Mesh B, which is obtained from the phase of S_{21}.

Next, consider the mitered *H*-plane bend (see Figure 7.25). No symmetry wall can be applied in the *H*-plane bend. Therefore the whole structure is meshed (Figures 7.26 and 7.27). Both the bends have a miter of $d/a = 0.707$ and the ports are located at a distance of $L = 1.5$. The magnitude of S_{11} for both meshes is shown in Figure 7.28 together with the results of [Reiter and Arndt 1994]. Figure 7.29 shows the comparison

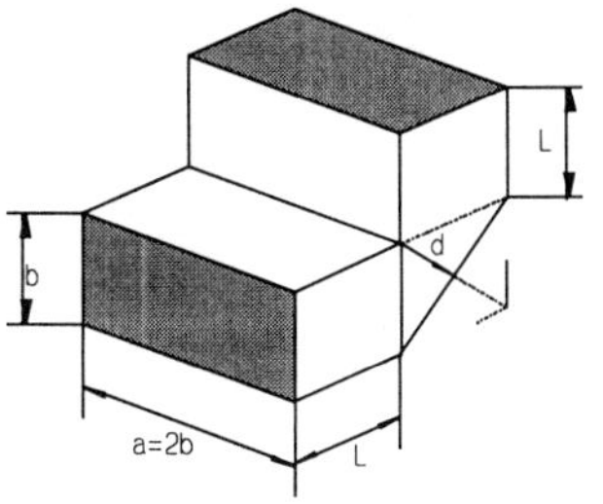

Figure 7.20 Mitered E-plane bend.

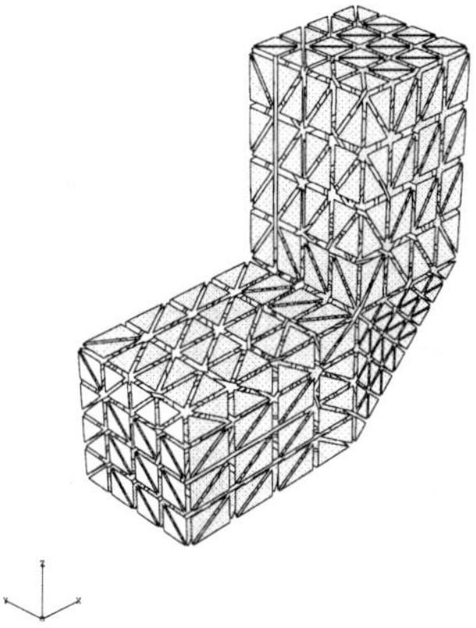

Figure 7.21 Mitered E-plane bend. d/b = 0.707, L = 1.5. Mesh A (1152 elements).

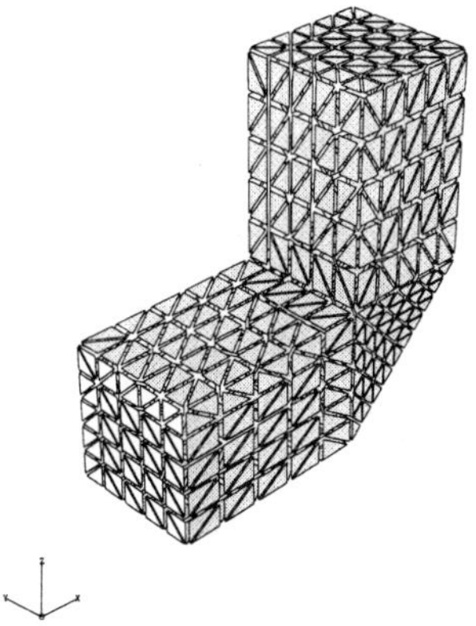

Figure 7.22 Mitered E-plane bend. d/b = 0.707, L = 1.5. Mesh B (2250 elements).

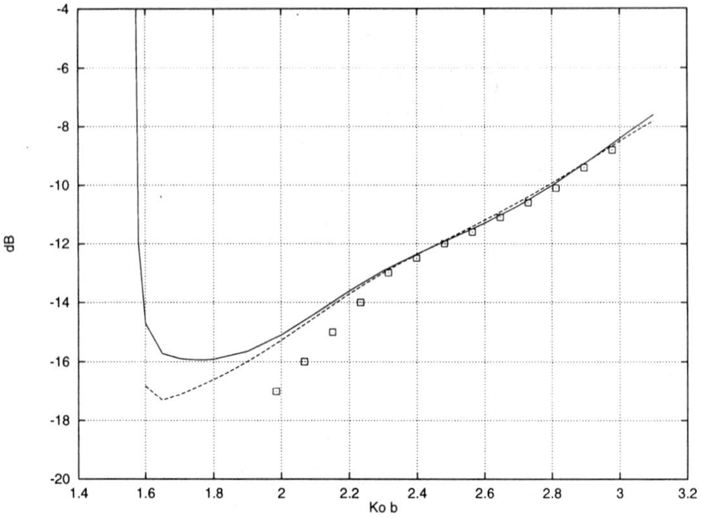

Figure 7.23 Mitered *E*-plane bend. $|S_{11}|$: —, mesh A; --, mesh B; □, [Reiter and Arndt 1994].

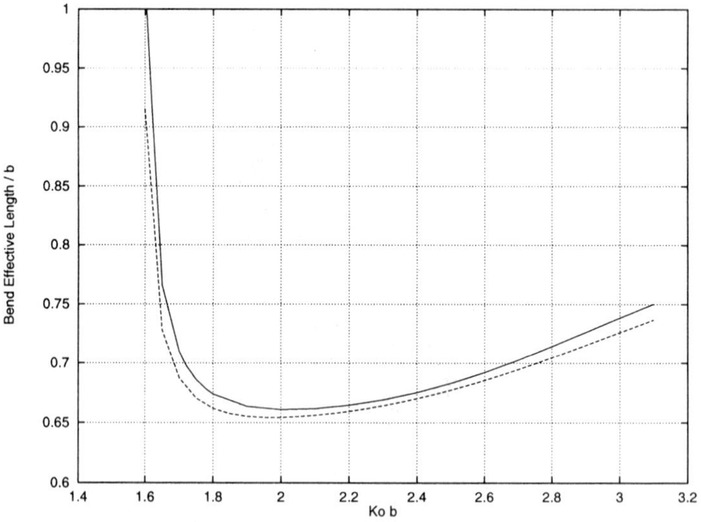

Figure 7.24 Mitered *E*-plane bend. Effective length / *b*. —, mesh A; --, mesh B.

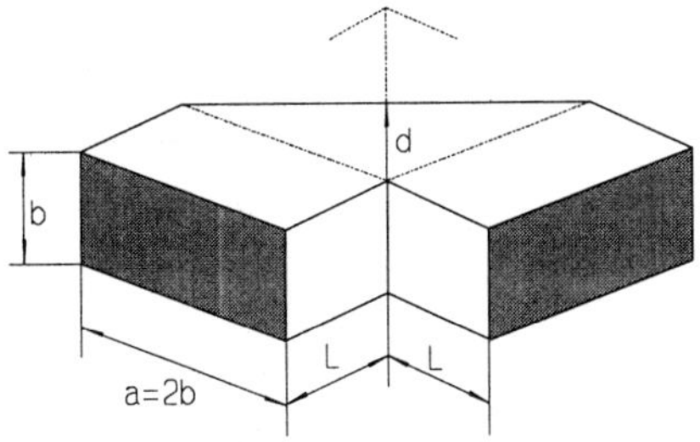

Figure 7.25 Mitered *H*-plane bend.

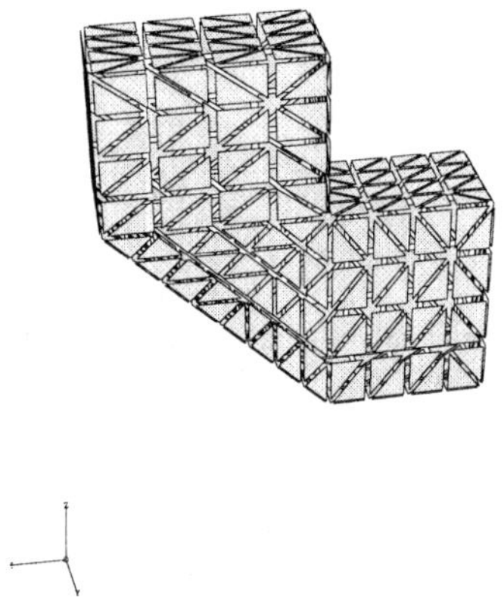

Figure 7.26 Mitered *H*-plane bend. $d/a = 0.707$, $L = 1.5$. Mesh A (1152 elements).

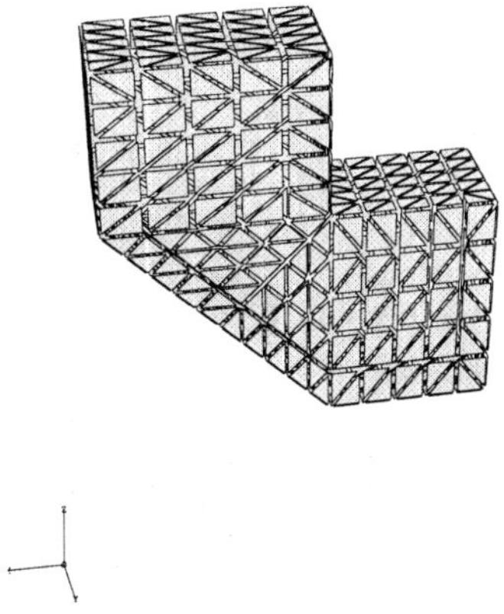

Figure 7.27 Mitered *H*-plane bend. $d/a = 0.707$, $L = 1.5$. Mesh B (2250 elements).

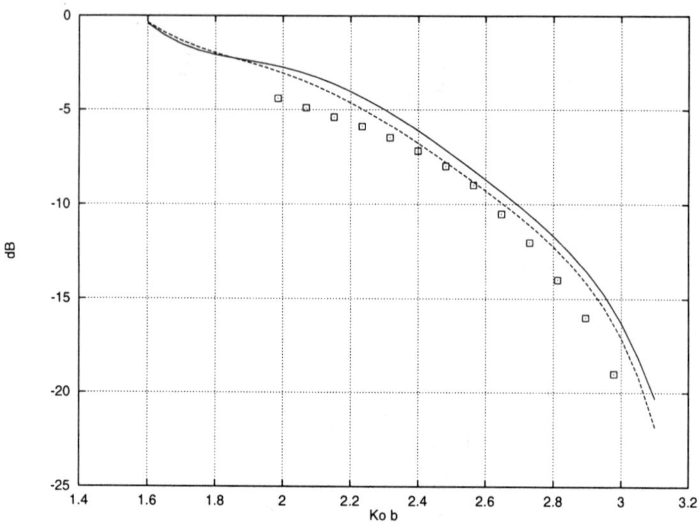

Figure 7.28 Mitered H-plane bend. $|S_{11}|$: —, mesh A; --, mesh B; □, [Reiter and Arndt 1994].

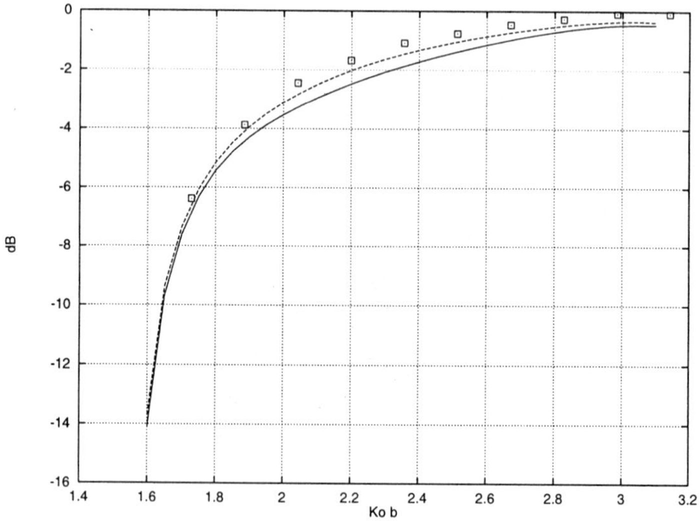

Figure 7.29 Mitered H-plane bend. $|S_{21}|$: —, mesh A; --, mesh B; □, [Koshiba et al. 1986].

between the results for S_{21} obtained with this method and those of [Koshiba and Suzuki 1986] (utilizing a boundary element method). A very good agreement is observed.

Next, consider a dielectric obstacle ($\varepsilon_r = 6$) which is placed inside a waveguide. The dimensions of the waveguide are $2b \times b$. The geometry of the problem appears in Figure 7.30. The values of the geometric parameters are $w = 0.888b$, $h = 0.399b$, and $d = 0.8b$. The ports are located at a distance of $L = 1$. Due to the symmetry of the problem, only one-half of the structure has been analyzed. A perfect magnetic wall has been used as the boundary condition in the symmetry plane. The meshes for one-half of the structure, referred to as Mesh A and Mesh B, are shown in Figure 7.31 and Figure 7.32, respectively. The reflection at the port is presented in Figure 7.33, which shows the absolute value of S_{11} in the range of frequencies corresponding to the mono-mode propagation of mode TE_{10}. Results are obtained with Mesh A (768 elements) and Mesh B (2592 elements). It may be observed that the results converge to that of [Ise et al. 1990]. A FEM approach is also employed in [Ise et al. 1990] but with the use of the penalty function method to avoid the effect of the spurious modes. There is no pollution by spurious modes in the results due to the use of curl-conforming elements (see Section 7.2).

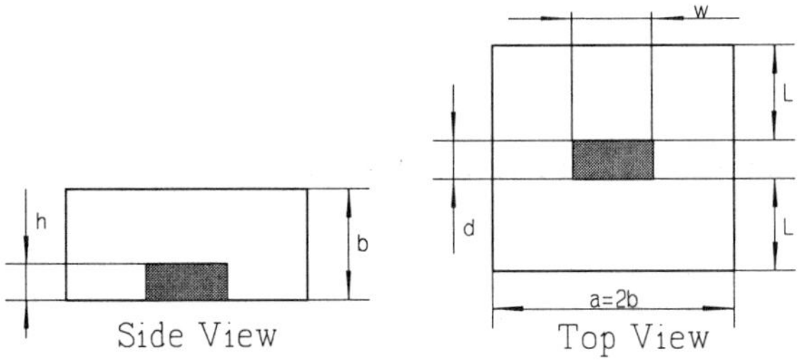

Figure 7.30 Dielectric obstacle in rectangular waveguide.

The last example consists of a dielectric slab in the middle of a rectangular waveguide as shown in Figure 7.34. This structure is usually employed as a phase shifter in waveguide technology (e.g., in array antennas). The ports are at a distance of $L = 1$. The length of the slab is $d = 0.2b$ and its dielectric constant is $\varepsilon_r = 2.03$. As in the case of the dielectric obstacle, only one-half of the structure has been meshed and a perfect magnetic symmetry wall has been employed in the analysis (see the meshes in Figures 7.35 and 7.36). The absolute value of the S_{11} parameter is shown in Figure 7.37. The results are compared with those of [Wang and Mittra 1994], which use a FEM technique to characterize the discontinuity. A good agreement is observed. On the other hand, the additional phase introduced by the slab is shown in Figure 7.38. This phase parameter is obtained by comparing the phase of S_{21} with the phase that would have been obtained if no dielectric slab were present. The constant phase shift introduced by the slab over a wide frequency range may be seen from Figure 7.38.

Hence, it can be characterized as a good phase shifter. It is illustrated in the figures that sufficient accuracy is achieved with Mesh A (1536 elements).

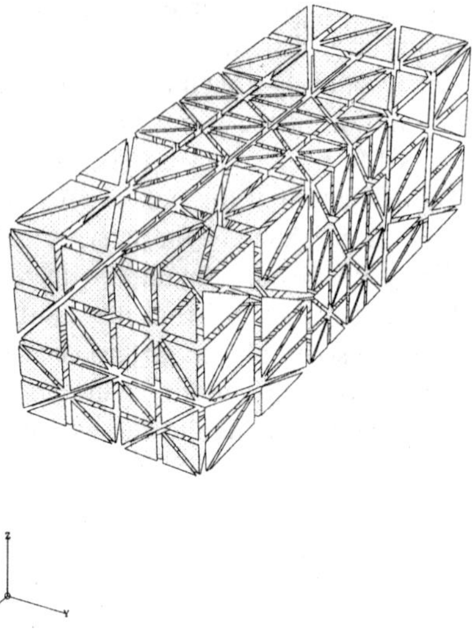

Figure 7.31 Dielectric obstacle. Mesh A (768 elements).

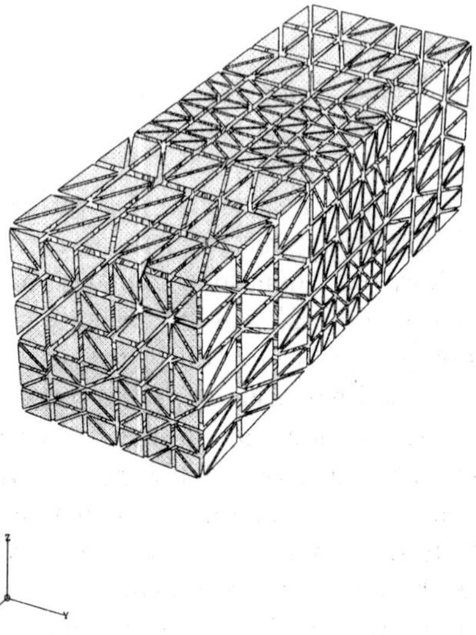

Figure 7.32 Dielectric obstacle. Mesh B (2592 elements).

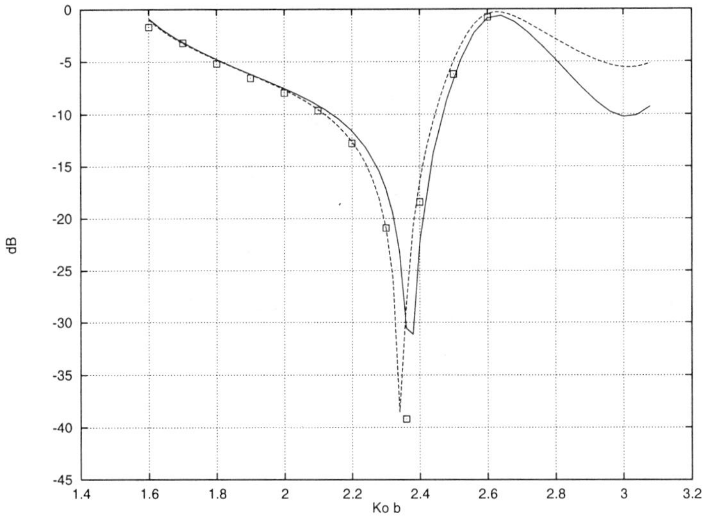

Figure 7.33 Dielectric obstacle. $|S_{11}|$: —, mesh A; --, mesh B; □, [Ise et al. 1990].

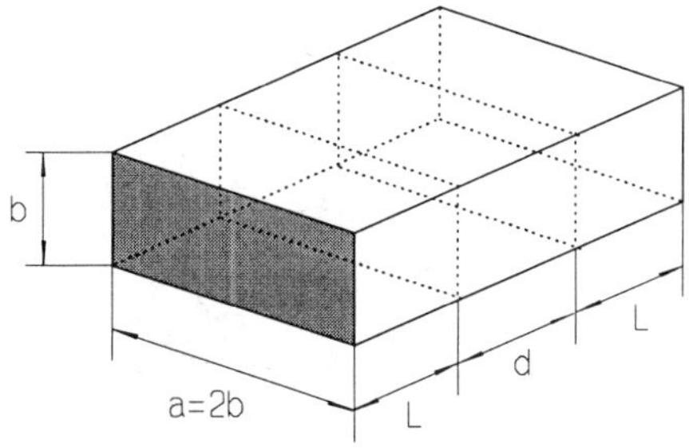

Figure 7.34 Dielectric slab discontinuity in a rectangular waveguide.

7.4.3 Conclusions

The analysis of the discontinuities in the waveguides have been solved by means of FEM. A double-curl formulation analogous to that used in the eigenvalue problem of Section 7.3 has been employed. The FEM formulation makes use of a simple boundary condition (mono-mode) for the ports. Tetrahedral mixed-order Nédélec elements have been used in the discretization process. Numerical results of its application to several discontinuities in rectangular waveguides have been included, showing a good agreement with those obtained by other methods. No pollution by spurious modes has been found in any of the structures analyzed.

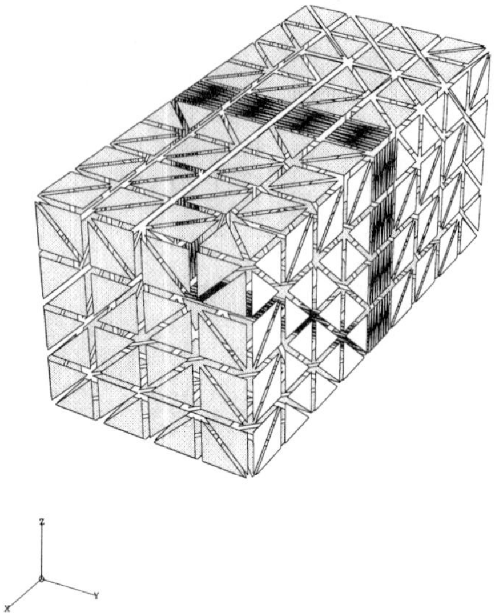

Figure 7.35 Dielectric slab discontinuity. Mesh A (1536 elements).

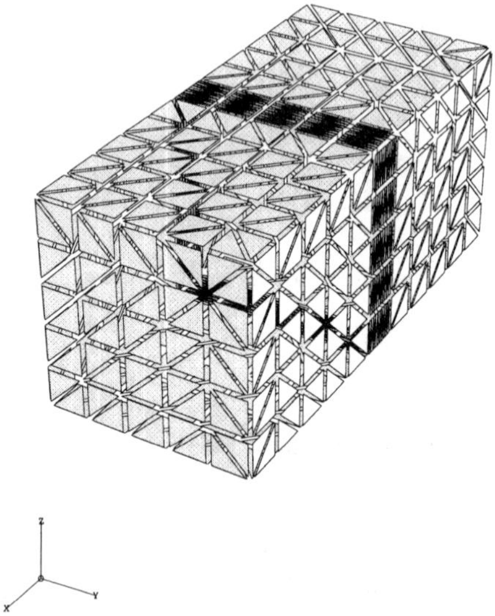

Figure 7.36 Dielectric slab discontinuity. Mesh B (3000 elements).

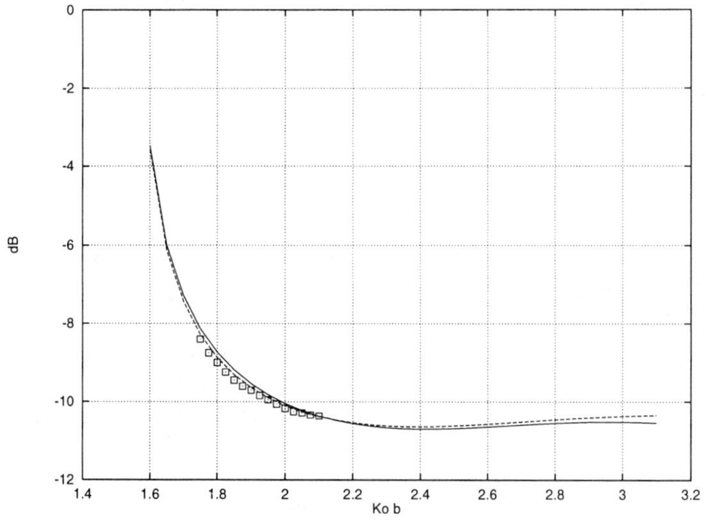

Figure 7.37 Dielectric slab discontinuity. $|S_{11}|$: —, mesh A; --, mesh B; □, [Wang and Mittra 1994].

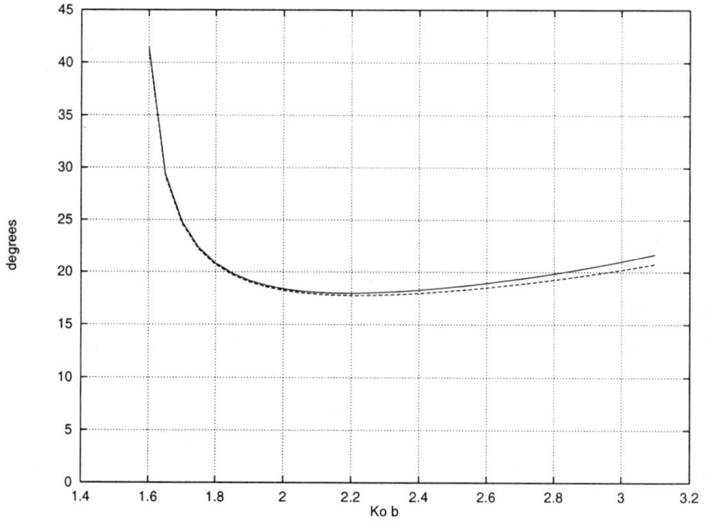

Figure 7.38 Dielectric slab discontinuity (phase shift introduced by the slab): —, mesh A; --, mesh B.

7.5 ANALYSIS OF SCATTERING AND RADIATION FROM THREE-DIMENSIONAL OPEN REGIONS USING THE FINITE ELEMENT METHOD

An iterative technique is presented for the FEM analysis of three-dimensional (3D) problems with open regions. It makes use of the boundary integral representation of the exterior field, allowing one to truncate the mesh close to the original boundary of the problem. Furthermore, and in contrast with the so-called hybrid FEM-BEM approaches, it does retain the original sparsity of the FEM matrices. It represents an extension of the method presented in Chapter 6 to three-dimensional problems. However, the approach presented in this chapter exhibits important differences with respect to that shown in the previous chapter.

In Chapter 6, the method has been applied to electrostatic problems and to the TM and TE scattering from perfectly conducting cylinders, i.e., to scalar formulations. In this chapter, the method is applied to vector formulations. A double-curl weak formulation in terms of the electric field or the magnetic field is employed. In fact, the weak FEM formulation for an open-region problem is formally identical to that presented in Section 7.4. The iterative method allows the analysis of general open-region problems. Here, its application to full-wave scattering and radiation (antenna) problems is considered. Tetrahedral curl-conforming elements are used for the discretization of the field.

The general basis and philosophy of the method has already been introduced in Chapter 6. In Section 7.5.1, the method is reviewed in the context of full-wave problems concerned in this chapter. Emphasis is made of its special features and differences with respect to the approach of Chapter 6. Finally, some numerical results of its application to scattering and radiation problems are shown in Section 7.5.2.

7.5.1 The Method

7.5.1.1 *Introduction*

When dealing with open-region problems, an artificial boundary S must be used to truncate the FEM mesh in order to keep the number of unknowns finite. A boundary condition must be enforced over the artificial boundary S. This boundary condition should account for the infinite extent of the original geometric region and, hence, should provide a good approximation of the solution of the original unbounded problem.

Several characteristics are desired of the boundary condition on S. The first one is accuracy. The FEM field solution obtained using the truncated mesh should be close to the true solution of the original open problem.

A second requirement is the capability of the artificial boundary to be placed close to the original sources of the problem (such as scatterer, inhomogeneities, and radiation sources) in order to obtain meshes with a lower number of nodes. Thus, the number of unknowns can be kept within reasonable limits, increasing the efficiency of the approach. This is especially important for 3D problems where the number of unknowns grows very fast with the size of the mesh.

A third desirable feature is the ability to retain the original structure of the FEM matrices. This has two important consequences. One is related to the efficiency of the approach due to the suitability of the use of sparse solvers. The other is the possibility of the employment of conventional FEM codes (in the sense of codes adapted to nonopen problems) for the analysis of open-region problems.

The method presented here makes use of the boundary integral representation of the exterior field to truncate the mesh, but at the same time it does not disturb the structure of the original FEM matrices. This is achieved at the expense of performing a number of iterations. The FEM system of equations obtained with this method will be shown later to be formally identical to that corresponding to the FEM analysis of a closed-problem (like the one in Section 7.4). Thus, the artificial boundary can be placed close to the sources of the problem (the Sommerfeld's radiation condition is satisfied through the use of the Green's function) and efficient sparse solvers may be employed. In addition, its implementation is very simple. The algorithm of the iterative method is explained next and is summarized in Table 7.17.

7.5.1.2 *Description of the Method*

Consider a problem with an open region. It may include different anisotropic materials, conducting objects, and symmetry walls, for example. An artificial boundary S is used to close the problem (see Figure 7.39). The closed domain, bounded by the surface S, is referred to as Ω. This is the domain employed in the FEM analysis. Therefore, the domain Ω is meshed and FEM equations are written inside. In this case, the double-curl weak formulation shown in the previous sections of this chapter has been employed. Details of the formulation are given later. However, the method is independent of the formulation. The only requirement of the method is that the FEM characterization of the interior field be carried out by one formulation or another.

The next step is the solution of the FEM system of equations. First, it is necessary to provide a boundary condition on S in order to uniquely define the problem. Here is the key of the method. The values of the degrees of freedom on S are supposed to be known,[10] i.e., the artificial boundary condition on S appears as a Dirichlet boundary condition of the formulation. At the first iteration, the degrees of freedom are fixed to some initial values. In this case, the degrees of freedom are related to the tangential components of the field, and they are made equal to zero for a radiation problem and are proportional to the values of the tangential components of the incident

---

10 The degrees of freedom do not have a meaningful geometric sense. They are related to geometric points of the domain through the nodes. Then, from a rigorous point of view, it should be said "the degrees of freedom associated with nodes on S" instead of "the degrees of freedom on S." However, for the sake of brevity, the expression "the degrees of freedom on S" will be used in the following to refer to the degrees of freedom associated with those nodes of the mesh placed on S.

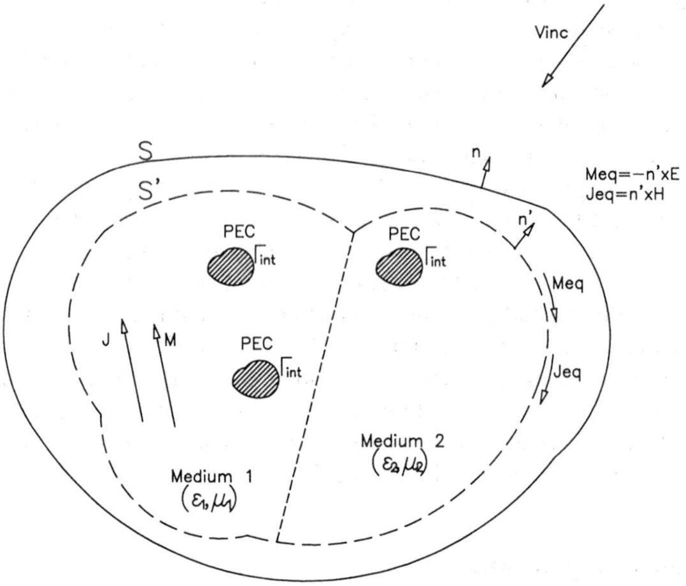

Figure 7.39 Illustration of an open-region problem. Scattering and radiation.

field for the case of the total field formulation of a scattering problem.

Then, after imposing in the FEM system the values of the degrees of freedom on S, the FEM system is solved. The solution of the system is performed through the employment of any direct (e.g., Gaussian elimination) or iterative (e.g., conjugate gradient) solver. From the FEM field solution, and making use of the Equivalence Principle, equivalent sources (electric and magnetic currents in this case) are computed over an auxiliary surface S' (see Figure 7.39). Outside S', these equivalent sources provide the same field as that of the original sources (within S'). Inside S', the total field of the equivalent problem is zero. The auxiliary surface S' is placed in such a way that it encloses all inhomogeneities and conducting objects, for example, and internal sources (in the case of a radiation problem). Thus, the free-space Green's function may be employed to compute the value of the field outside S'. In particular, the free-space Green's function is used to obtain the values of the degrees of freedom on the external boundary S, i.e., the values of the tangential field at the nodes placed on S.

These new values for the degrees of freedom of S computed through the equivalent sources on S' and the Green's function will be, in general, different from those values used previously for the FEM system. In that case, a new iteration takes place. The new values of the degrees of freedom of S are used as the new boundary condition on S. The FEM system with these new values for the boundary condition is again solved. Equivalent sources are computed from the field solution, for example. Note that the new iteration starts in step vi of Table 7.17; it is not necessary to fill the

Table 7.17
Summary of the iterative method for the analysis of 3D open-region problems

Step	Operation
i)	Mesh the domain Ω enclosed by S
ii)	Obtain the FEM discrete integral forms in each element
iii)	Assemble the element discrete forms $\Rightarrow$ Global FEM discrete form $\{W\} = [K]\{D\}\text{-}\{F\}$
iv)	Enforce the boundary conditions not related to $S \Rightarrow \{W'\} = [K']\{D\}\text{-}\ \{F'\}$
v)	Choose the initial values for the degrees of freedom on S
vi)	Enforce the values of the degrees of freedom on $S \Rightarrow \{W''\} = [K'']\{D\}\text{-}\{F''\}$
vii)	Solve the FEM system $[K'']\{D\} = \{F''\}$
viii)	Compute the equivalent sources on surface S' from the FEM field solution
ix)	Compute the new values of the degrees of freedom on S from the equivalent sources and the Green's function
x)	Check the error criteria between two consecutive sets of values of the degrees of freedom on S
xi)	If error criteria is satisfied, then stop. If not, continue from step vi) with the last set of values of degrees of freedom on S computed in ix)

FEM matrix again. The solution of the FEM system in step vii) can be done efficiently either with a direct or an iterative solver. Observe that if the method 2 of Section 2.5.2.3 is used to impose the Dirichlet conditions, the matrix $[K'']$ is the same for all iterations. Only $\{F''\}$ will be different. Thus, if a direct solver is utilized the factorization of $[K'']$ (which is the most computation intensive process) can be performed only once at the first iteration. The system is solved at subsequent iterations by simple back-substitution. For the iterative method the solution of the previous iteration may be used as an initial guess for the next iteration. The process continues until a certain error criteria is satisfied. The error criteria is based on the difference between the values of the degrees of freedom on S between two consecutive iterations.

7.5.1.3 *Finite Element Formulation and Features of the Iterative Method*

The full-wave formulation that has been used in the 3D implementation of the iterative method for open-region problems is presented. The main features of the iterative method are also shown here.

The problem with an open region has been reduced to the solution of Maxwell's equations in a closed domain Ω subject to Dirichlet and Neumann boundary conditions.

Thus, the structure of the FEM system corresponding to the open-region problem (at each iteration cycle) is identical to that of a closed-region problem (as in Section 7.4).

The FEM weak formulation used for open-region problems is derived from the stationarity of the following functional (see Table 7.18, where $Z_0 = (\mu_0/\varepsilon_0)^{1/2}$):

$$
\begin{aligned}
F = 1/2 \int_{\Omega} &\left(\left(\bar{\nabla} \times \bar{V}^*\right) \cdot \left(\underline{\underline{f}}_r^{-1} \bar{\nabla} \times \bar{V}\right) - k_0^2 \, \bar{V}^* \cdot \underline{\underline{p}}_r \bar{V} \right) d\Omega \\
&+ \sum_k \int_{\Gamma_{S_k}} \gamma (\bar{a}_n \times \bar{V}^*) \cdot (\bar{a}_n \times \bar{V}) \, d\Gamma \\
&+ \int_{\Omega} \left(j k_0 \sqrt{\frac{f_o}{p_o}} \left(\bar{V}^* \cdot \bar{Q}\right) + \bar{V}^* \cdot \left(\nabla \times \underline{\underline{f}}_r^{-1} \bar{T}\right) \right) d\Omega \\
&- \int_{\Gamma_{S_{inc}}} \bar{V}^* \cdot \bar{U}_{inc} \, d\Gamma
\end{aligned}
\qquad (7.127)
$$

Table 7.18
Correspondences for (7.127)

	$\bar{H}$ formulation	$\bar{E}$ formulation
$\bar{V}$	$\bar{H}$	$\bar{E}$
$\bar{T}$	$-\bar{J}$	$\bar{M}$
$\bar{Q}$	$\bar{M}$	$\bar{J}$
$\underline{\underline{f}}_r$	$\underline{\underline{\varepsilon}}_r$	$\underline{\underline{\mu}}_r$
$\underline{\underline{p}}_r$	$\underline{\underline{\mu}}_r$	$\underline{\underline{\varepsilon}}_r$
$(f_o/p_o)^{1/2}$	Z_0^{-1}	Z_0

The functional of (7.127) is recognized to be the same functional employed in Section 7.4 for the analysis of discontinuities in waveguides, with the exception of the two excitation terms with the electric and the magnetic currents according to Table 7.18. The reason for these two terms is that in this case, the domain of the problem is considered to contain sources. Thus, the procedure to get to the functional for the problem with an open region starts with the following differential equation:

$$
\bar{\nabla} \times (\underline{\underline{f}}_r^{-1} \bar{\nabla} \times \bar{V}) - k_0^2 \underline{\underline{p}}_r \bar{V} = -\left\{ j k_0 \sqrt{\frac{f_o}{p_o}} \, \bar{Q} + (\nabla \times \underline{\underline{f}}_r^{-1} \bar{T}) \right\} \qquad (7.128)
$$

which is the same double-curl differential equation of previous sections but including the current terms.

Another difference with the case of Section 7.4 is the existence of the artificial boundary S. However, it is worth noting that the surface integral of the weak formulation, which appears when applying the first vector Green's theorem, is zero over the external surface S. The field V on S is known; i.e., the boundary condition on S is an essential boundary condition. Therefore, as explained in Section 7.3.1.1, the surface integral on S may be omitted and, hence, does not appear in the expression of the functional.

It is important to point out that the particular formulation of (7.127) is not the only FEM formulation to be employed with the iterative method described in Section 7.5.1.2. A different formulation could have been used, and the steps of the method (summarized in Table 7.17) would have remained unchanged.

The discretization of (7.127) leads to sparse matrices, analogous to the case of Section 7.4. Tetrahedral curl-conforming elements are used in the discretization of the field. The features of the discretization with curl-conforming elements have been illustrated in previous sections.

Following the procedure shown in a previous section of this chapter, the coefficients of the element matrices can be evaluated as

$$k_{ij}^{e} = \int_{\Omega^e}\left(\left(\bar{\nabla}\times\bar{N}_i^{e}\right)\cdot\left((\underline{\underline{f}}_r^{e})^{-1}\bar{\nabla}\times\bar{N}_j^{e}\right) - k_0^2\bar{N}_i^{e}\cdot\underline{\underline{p}}_r^{e}\bar{N}_j^{e}\right)\mathrm{d}\Omega + \sum_k\int_{\Gamma_{S_k}}\gamma(\bar{a}_n\times\bar{N}_i^{e})\cdot(\bar{a}_n\times\bar{N}_j^{e})\,\mathrm{d}\Gamma \tag{7.129}$$

$$f_i^{e} = -\int_{\Omega^e}\left(\mathrm{j}k_o\sqrt{\frac{f_o}{p_o}}\,(\bar{N}_i^{e}\cdot\bar{Q}) + (\underline{\underline{f}}_r^{e})^{-1}\,\bar{T}\cdot(\nabla\times\bar{N}_i^{e})\right)\mathrm{d}\Omega + \int_{\Gamma^e_{S_{inc}}}\bar{N}_i^{e}\cdot\bar{U}_{inc}\,\mathrm{d}\Gamma \tag{7.130}$$

which are recognized to be the same as those of Section 7.4, but with two terms added to the excitation vector $\{f^e\}$.

As seen from (7.130), the internal excitation may come from a nonzero field impressed at a port aperture or by the existence of known currents (electric or magnetic) distributed over the problem domain. These cases correspond to a radiation problem. For scattering problems and total field formulation, $\{f^e\} = \{0\}$. The excitation comes from the incident field. The initial values of the degrees of freedom on S are chosen to be proportional to the values of the tangential component of the incident field. Then, after enforcing the boundary condition of S (step vi of Table 7.17), they constitute the excitation for the system.

The only change in the global FEM system is the imposition of the additional Dirichlet boundary condition on S. Thus, no difference exists in the structure of the FEM matrices between the formulation (7.127) and that shown in Section 7.4. This has some important advantages. First of all, it allows the employment of sparse solvers for the solution of the FEM system, with the corresponding saving in storage and CPU resources. On the other hand, existing FEM computer codes for nonopen-region problems may be easily incorporated into the analysis of open-region problems. The only part that needs to be added is that which corresponds to the computation of the field from the equivalent sources and the Green's function (see Figure 7.40).

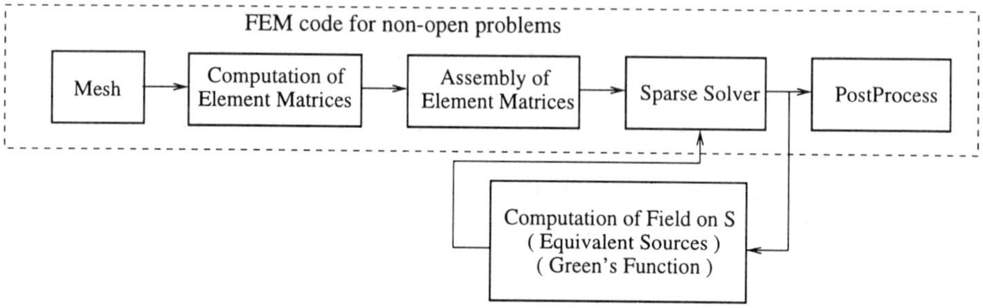

Figure 7.40 FEM implementation of the iterative method.

Thus, the method leads to sparse matrices and, at the same time, provides an accurate representation of the radiation boundary condition when the artificial surface S is placed close to the original sources of the problem. Indeed, the radiation boundary condition is satisfied exactly, in a numerical sense, through the employment of the Green's function. The approximation of the solution comes from the approximate nature of the equivalent sources because the equivalent sources are obtained from the FEM field solution. Also, the values for the degrees of freedom on S are approximate since they are computed from the equivalent sources.

Then, as the values of the degrees of freedom on S (the tangential component of the field on S) approach the true values, the equivalent sources tend also to the real values. When this happens, the computation of the tangential field on S via the Green's function and the equivalent sources, provides the same values (with a certain tolerance) as those used as the boundary condition for the FEM system. In this sense, the method presented here may be seen as an adaptive method, since the field solution is adapting to the true one as it does for the boundary condition on S.

The integral equation employed to compute the values of the degrees of freedom from the equivalent sources is now presented. It makes use of the free-space Green's function and provides the values of the field $\bar{V}$. The iteration cycle is given by

$$\begin{aligned}\bar{V}^{(i+1)}(\bar{r}) &= \bar{V}^{(i)}_{inc} + \int_{S'} \Big(((\bar{a}_{n'} \times \bar{V}^{(i)}(\bar{r}')) \times \bar{\nabla}' g_o(\bar{r};\bar{r}') \\ &\quad + (\bar{a}_{n'} \times \bar{\nabla}' \times \bar{V}^{(i)}(\bar{r}'))\, g_o(\bar{r};\bar{r}') \\ &\quad + 1/k_o^2\, (\bar{a}_{n'} \times \bar{\nabla}' \times \bar{V}^{(i)}(\bar{r}'))\, \underline{\underline{\bar{\nabla}\bar{\nabla}}}\, g_o(\bar{r};\bar{r}') \Big) dS' \end{aligned} \tag{7.131}$$

where (i) denotes the present iteration and $(i + 1)$ the next iteration cycle. Here $\underline{\underline{\bar{\nabla}\bar{\nabla}}}$ represents $\{\nabla\}\{\nabla\}^T$ in vector notation. The free-space Green's function, g_0, is

$$g_o(\bar{r};\bar{r}') = \frac{e^{-jk_0R}}{4\pi R}; \qquad R = |\bar{r} - \bar{r}'| \tag{7.132}$$

The vector $\bar{V}^{(i)}(\vec{r}')$ stands for the value of the field on the surface S'. It is obtained from the solution of the FEM system in the ith iteration. On the other hand, the vector $\bar{V}^{(i+1)}(\bar{r})$ refers to the value of the field on the surface S, which will be used as the Dirichlet boundary condition to be imposed for the FEM system in the next iteration, i+1. The vector $\bar{V}^{(i)}_{inc}(\vec{r})$ is the incident field. It is considered to be zero for a radiation problem. For scattering problems, a plane wave incidence is assumed. Hence

$$\bar{V}_{inc} = (V_\theta \bar{a}_\theta + V_\phi \bar{a}_\phi)\, e^{-j\bar{k}_i \cdot \bar{r}} \tag{7.133}$$

$$\bar{k}_i = -k_0 \,(\sin\theta_i \cos\phi_i\, \bar{a}_x + \sin\theta_i \sin\phi_i\, \bar{a}_y + \cos\theta_i\, \bar{a}_z) \tag{7.134}$$

where $(\theta_i\,;\,\phi_i)$ is the direction of arrival of the wave.

It is worth noting that the term iterative refers here to the method itself and not to the nature of the solver used in the solution of the algebraic system of equations. The approach presented here leads, at each iteration, to a system of equations that may be solved by an iterative (e.g., conjugate gradient) or a noniterative (e.g., Gaussian elimination or so) solver.

Another issue is the location of the auxiliary surface S'. The auxiliary surface S' and the artificial surface S have been assumed to be distinct, but they may also coincide. The former option is preferred because of the appearance of some numerical problems (see the comments of Chapter 6) when S and S' coincide.

As mentioned previously the auxiliary surface S' encloses all inhomogeneities and conducting objects. Then the free-space Green's function may be used (7.132). For a scattering problem, if only one scatterer is present, S' is usually chosen to be the surface of the scatterer. If two or more scatterers exist, the surface S' is chosen in such a way that it encompasses all the scatterers. On the other hand, in a radiation problem, surface S' encloses the radiating sources.

7.5.2 Numerical Results

In this section, numerical results utilizing the iterative method described in Section 7.5.1 are presented in order to show the validity of the approach for scattering and radiation problems. In all cases, the difference of the degrees of freedom on S between two consecutive iterations is made as a stopping criteria for the method. An error of 10^{-4} in the Euclidean norm of the two sets of values of the degrees of freedom at each iteration is used as a measure of accuracy.

The iterative approach allows the placement of the artificial boundary S close to the internal sources (radiation) or the scatterer (scattering). Specifically, the distance chosen between surfaces S and S' in the examples that follow has been approximately $0.1\lambda_0$ (one or two layers of elements). This helps to keep the number of unknowns of the discretized problem within computationally reasonable limits. On the other hand, sparse solvers have been employed, taking advantage of the high sparsity provided by the curl-conforming elements. Specifically, first-order tetrahedral elements are employed. The density of elements in all of the meshes is between 10 and 20 cubes per wavelength (one cube is divided into six tetrahedrons).

7.5.2.1 *Radiation*

As a first example, consider a dipole antenna. Figure 7.41 shows the schematic of the configuration of the meshes employed for its analysis. The rectangular artificial boundary S encloses the dipole. The auxiliary surface S', where the equivalent currents are calculated, is chosen to be rectangular. The distance S-S' is around $0.1\lambda_0$ and the distance between the dipole and S' varies between $0.05\lambda_0$ and $0.1\lambda_0$.

The excitation of the problem is a unit amplitude z-directed electric current J located at the dipole. Two cases are studied: a filament current (zero area cylinder) and a volumetric current (rectangular shaped cylinder). The latter is employed for both $\bar{E}$

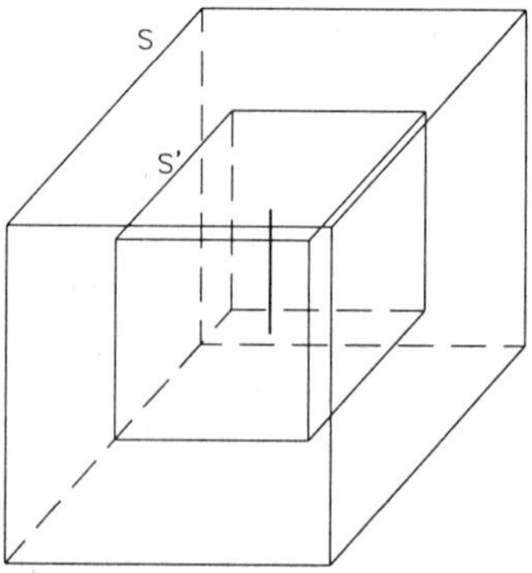

Figure 7.41 Configuration for the dipole problem.

and $\overline{H}$ formulations; the former only with the $\overline{E}$ formulation. This is because for the latter case, the expression of the excitation (7.130), has a curl over the region where the current is non-zero, and hence, it cannot be applied over a zero area cylindrical region.

Two meshes (referred to as Mesh A and Mesh B) have been employed for the analysis of the radiation from the filament current. The dipole is short and its length is $0.1\lambda_0$. Both meshes have identical dimensions but different densities: 878 nodes (Mesh A) and 4184 nodes (Mesh B). In addition, Mesh A has one layer of elements between S and S' while Mesh B has two layers, as shown in Figures 7.42 and 7.43.

The evolution of the error, i.e., the 2-norm of the difference of the values of the degrees of freedom on S between two consecutive iterations, is presented in Figure 7.44 for Mesh A. It may be observed that four iterations are enough to get an error less than 10^{-4}. The far-field is shown in Figure 7.45. It may be seen how the far field converges to the analytic solution when the denser mesh is used.

The far-field has been obtained from the solution at the end of the iteration number 4. However, it is worth noting that no significant improvement is found in the far-field solution after the second iteration. Therefore, a more refined mesh must be used in order to improve the accuracy of the far-field solution instead of requiring a more demanding error criteria to stop the iterative process.

The case of the dipole modeled as a volumetric current is addressed next. Three different lengths of the dipole are considered: $0.1\lambda_0$, $0.5\lambda_0$, and $1\lambda_0$. In all cases, a square cross-sectional cylinder has been employed as the region for the electric current. The three meshes, which have been used for the analysis of the $0.1\lambda_0$, $0.5\lambda_0$, and $1\lambda_0$ dipoles, lead to 1854, 2592, and 6144 nodes, respectively, as depicted in Figures 7.46 to 7.48.

Both formulations, with the electric and the magnetic fields, have been used. A comparison of the far-field solution corresponding to the preceding three cases is shown in Figure 7.49 ($\overline{H}$ formulation) and Figure 7.50 ($\overline{E}$ formulation). A good agreement with the analytic solutions is observed in all cases. The evolution of the error for the $0.1\lambda_0$ dipole is shown in Figure 7.51. The convergence is achieved in four and five iterations for the magnetic and electric field formulations, respectively.

7.5.2.2 Scattering

The problem of scattering by a homogeneous dielectric cube of length $\lambda_0/5$ is studied here. Both isotropic and anisotropic cubes have been considered. Figure 7.52 shows the configuration of the FEM meshes employed in its analysis. The artificial boundary S consists of a cube conforming to the scatterer. The auxiliary surface S' is chosen to be directly the scatterer surface. The distance between S and S', i.e., between the scatterer and the external boundary, is $0.1\lambda_0$. A unit amplitude plane wave is assumed to be incident from the $(\theta_i ; \phi_i)$ direction. At the first iteration cycle, the values of the degrees of freedom on S are chosen proportional to the tangential component of the incident field on S. Thus, by imposing these nonzero values of the degrees of freedom on S in the FEM system, the excitation of the problem is achieved.

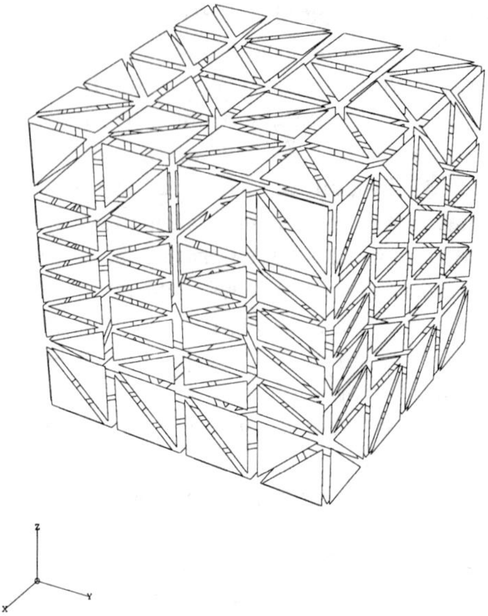

Figure 7.42 Dipole problem (filament current). Mesh A: 576 elements, 878 unknowns, 1 layer S-S'.

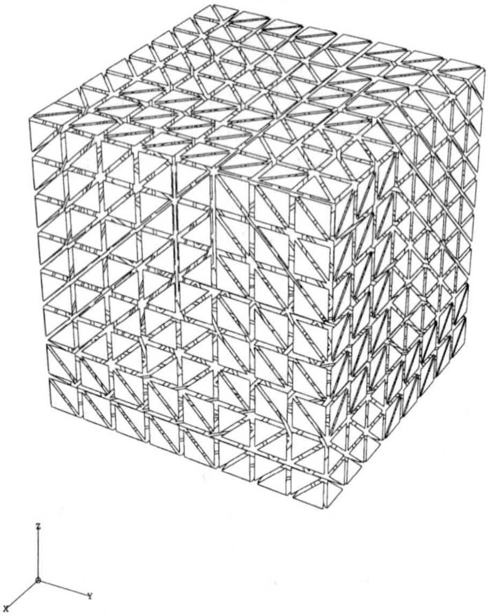

Figure 7.43 Dipole problem (filament current). Mesh B: 3072 elements, 4184 unknowns, 2 layers S-S'.

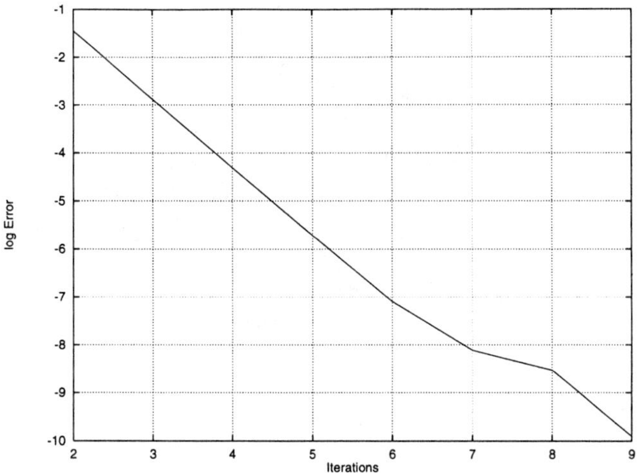

Figure 7.44 Relative error on S (Mesh A).

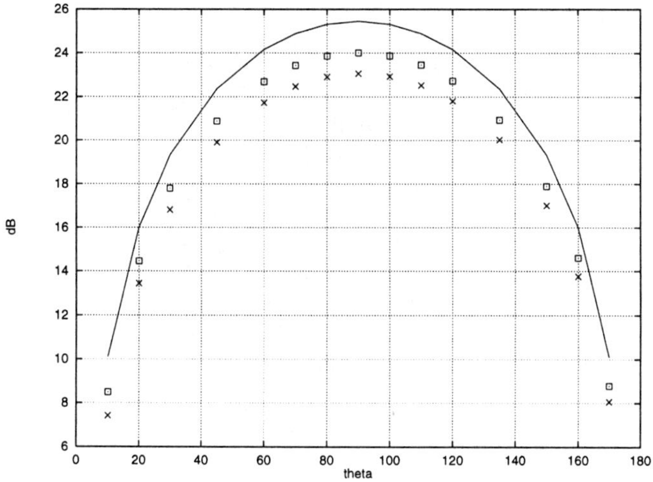

Figure 7.45 Far field (plane $\phi = 0°$): —, analytic; ×, mesh A; □, mesh B.

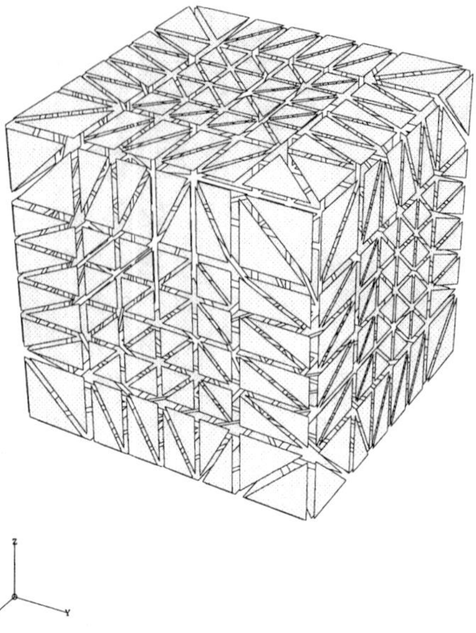

Figure 7.46 Dipole problem (volumetric current). Mesh $0.1\lambda_0$.

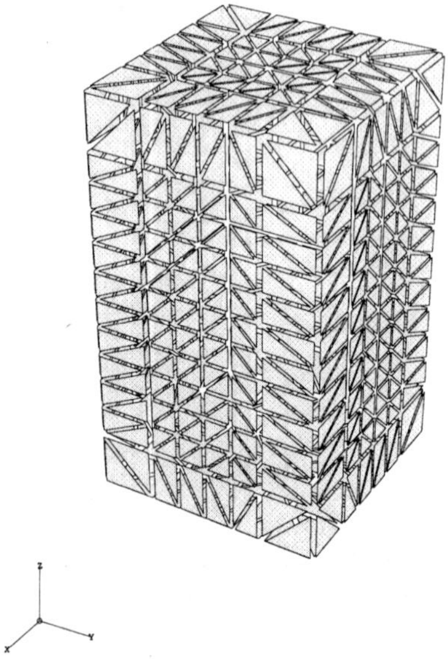

Figure 7.47 Dipole problem (volumetric current). Mesh $0.5\lambda_0$.

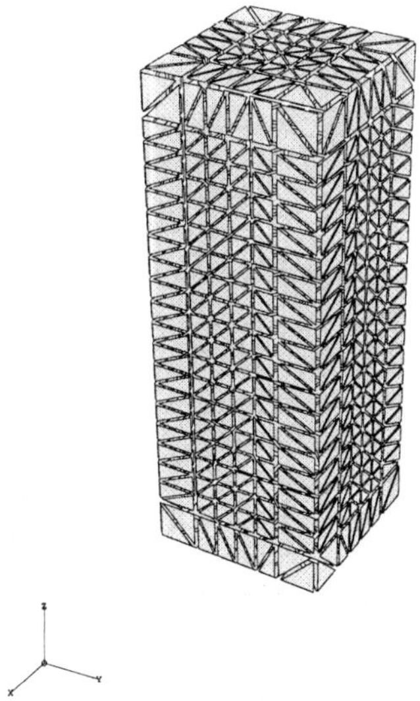

Figure 7.48 Dipole problem (volumetric current). Mesh $1\lambda_0$.

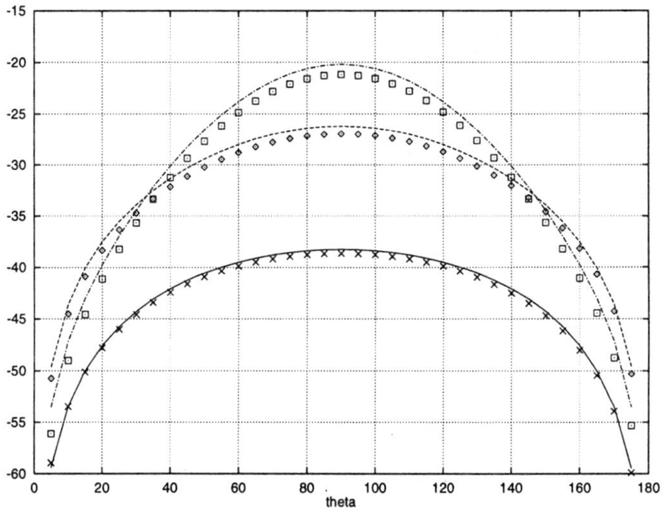

Figure 7.49 Far field (dB) ($\bar{H}$ formulation) (plane $\phi = 0°$): —, 0.1 λ_0; --, 0.5 λ_0; -·-, 1.0 λ_0. Lines: analytic results. Marks: this method.

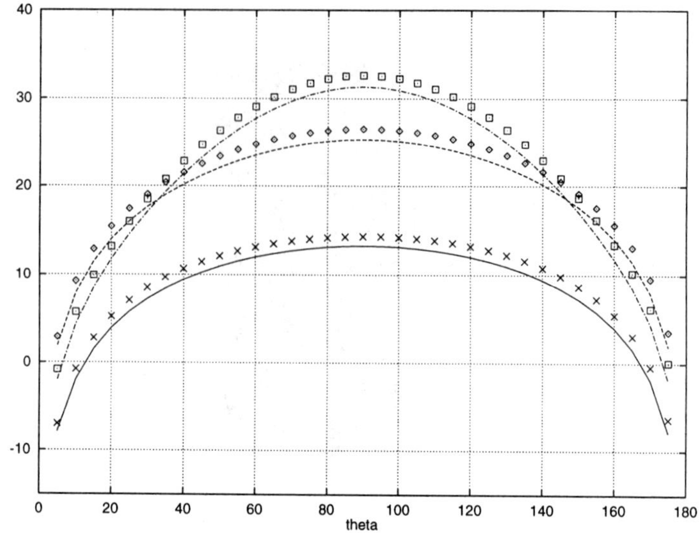

Figure 7.50 Far field (dB) ($\bar{E}$ formulation) (plane $\phi = 0°$): —, 0.1 λ_0; --, 0.5 λ_0; -·-, 1.0 λ_0. Lines: analytic results. Marks: this method.

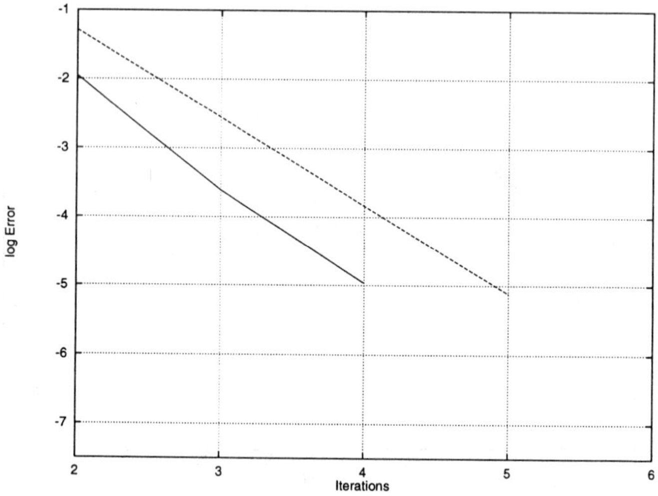

Figure 7.51 Relative error on S: —, $\bar{H}$ formulation; --, $\bar{E}$ formulation.

Three meshes - one with 1854 nodes (Mesh A, Figure 7.53), a second one with 4184 nodes (Mesh B, Figure 7.54), and a third utilizing a coarse mesh with 604 nodes (Mesh C, Figure 7.55) - have been employed in the analysis of the dielectric cube. Mesh A and Mesh C have one layer of elements between S and S' surfaces. On the other hand, Mesh B has two layers of elements.

First of all, a dielectric cube with $\varepsilon_r = 1$ is considered. This was done to study the level of reflection from the external boundary condition provided by the iterative method. The results of the scattered far-field for the case of the $\bar{H}$ formulation and meshes A, B, and C are shown in Figure 7.56. They correspond to a unit amplitude plane wave coming from $\theta_i = 0°$ and polarized in the y-direction. The results of Figure 7.56 are for the plane $\phi = 90°$. However, similar results are obtained in other planes. It may be observed that, despite the coarseness of meshes A and B, the level of reflection achieved is approximately better than 30 dB. It may also be seen that Mesh A and Mesh C (both with one layer of elements between S and S') get very similar levels of reflection. On the other hand, Mesh B leads to the more accurate results (35 dB). This is not only due to a more dense mesh over the scatterer but also due to a smaller size of the elements in the region S-S'. Thus, not only the field on S but also $\bar{V} = \bar{H}$ or the equivalent electric current for $\bar{V} = \bar{E}$ is computed with more accuracy in (7.126), providing an accurate field on S' and a better boundary condition on S.

Next, consider the case of an isotropic dielectric cube of $\varepsilon_r = 4$. A unit amplitude x-polarized plane wave is assumed to be incident from $\theta_i = 180°$. The evolution of the

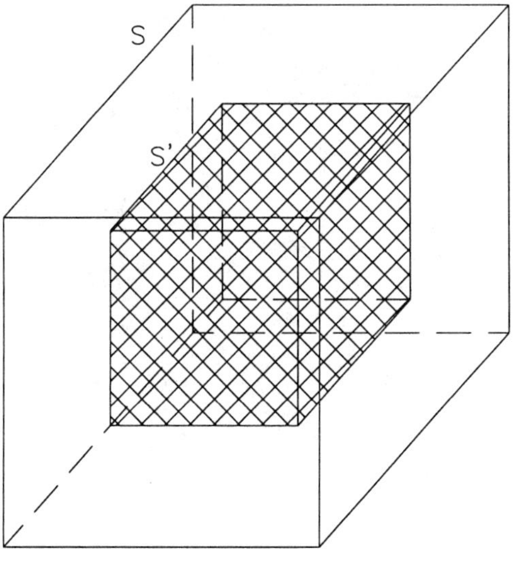

Figure 7.52 Configuration for the dielectric cube problem.

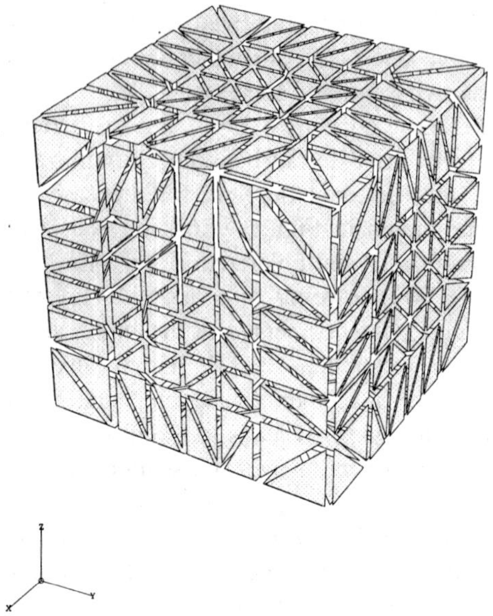

Figure 7.53 Dielectric cube. Mesh A (1296 elements, 1854 nodes, 1 layer S-S').

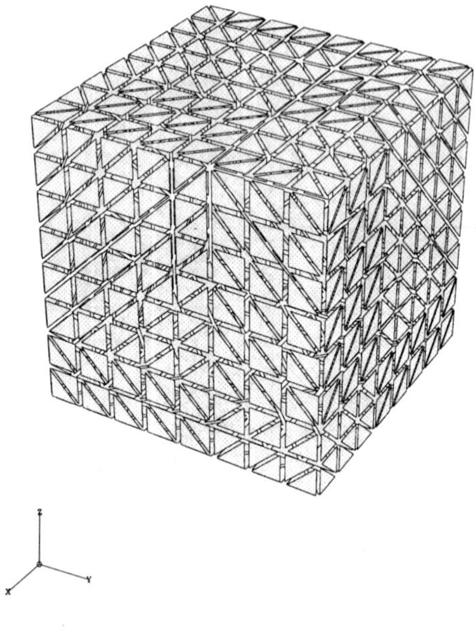

Figure 7.54 Dielectric cube. Mesh B (3072 elements, 4184 nodes, 2 layers S-S').

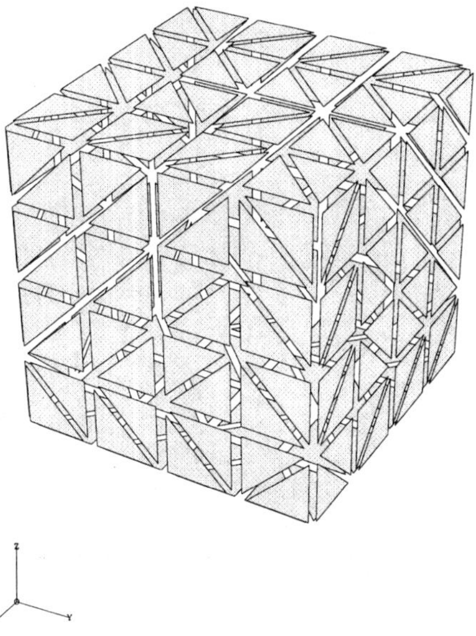

Figure 7.55 Dielectric cube. Mesh C (384 elements, 604 nodes, 1 layer *S*-*S'*).

error, for both formulations and Mesh A, is presented in Figure 7.57. The normalized scattered field obtained with the $\bar{H}$ and $\bar{E}$ formulations for Meshes A and B is shown in Figures 7.58 and 7.59. It is compared with the MOM results obtained in [Sarkar et al. 1989b] using surface and volumetric formulations in terms of the electric field. The smooth curve of the figures corresponds to the plane $\phi = 90°$, and the one with the dip for the plane $\phi = 0°$. It may be observed that the $\bar{E}$-field formulation provides better agreement with the MOM results than the $\bar{H}$-field formulation. However, the results approach the MOM results as the density of the mesh is increased.

Finally, consider the scattering from an anisotropic dielectric cube. This example has been treated in [Su 1989]. The anisotropic material is characterized by the permittivity tensor

$$\underline{\underline{\varepsilon}}_r = \begin{bmatrix} 4.0 & 0.5 & 0.5 \\ 0.5 & 3.5 & 0.5 \\ 0.5 & 0.5 & 4.5 \end{bmatrix} \tag{7.135}$$

A unit amplitude plane wave coming from ($\theta_i = 90°$, $\phi_i = 0°$) is assumed. Two different polarizations are considered in the y- and z-directions. The normalized scattered field obtained with the $\bar{H}$ formulation and Mesh B is shown in Figure 7.60.

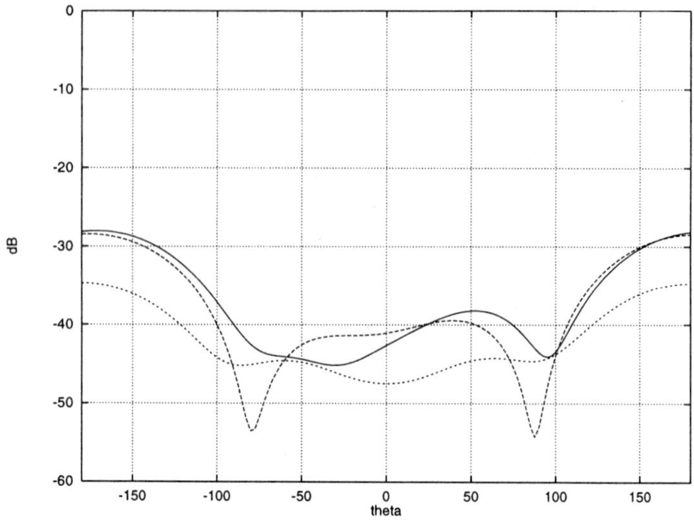

Figure 7.56 Level of reflection from S. Plane $\phi = 90°$. --: mesh A; ⋯: mesh B; —: mesh C.

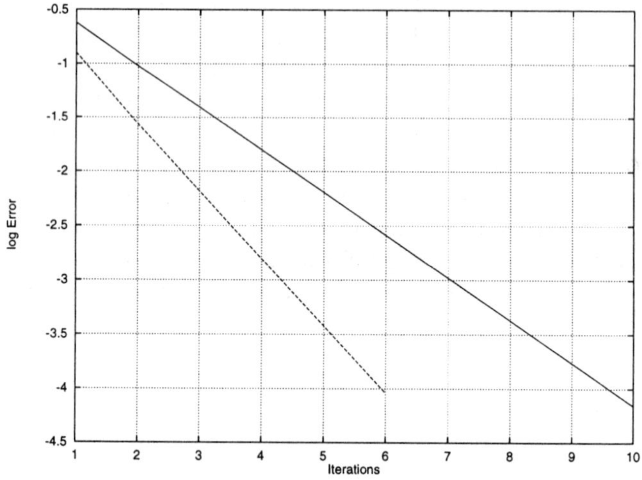

Figure 7.57 Relative error on S (mesh A): —, $\bar{E}$ formulation; --, $\bar{H}$ formulation.

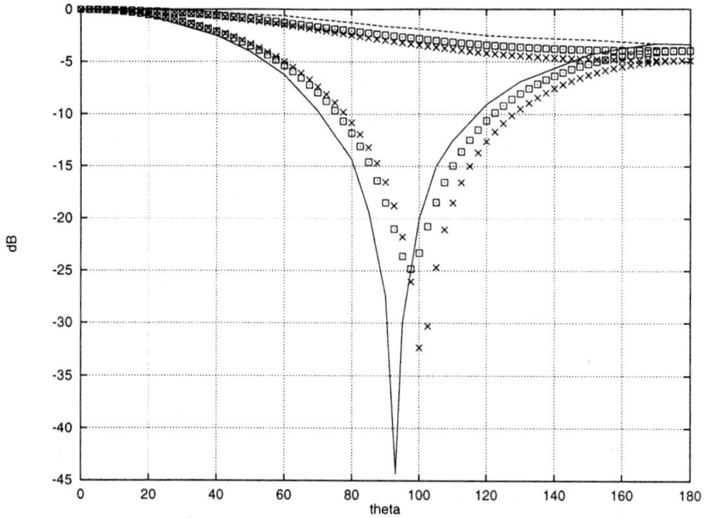

Figure 7.58 Normalized scattered field ($\bar{H}$ formulation): ×, mesh A; □, mesh B; —, --, [Sarkar et al. 1989b].

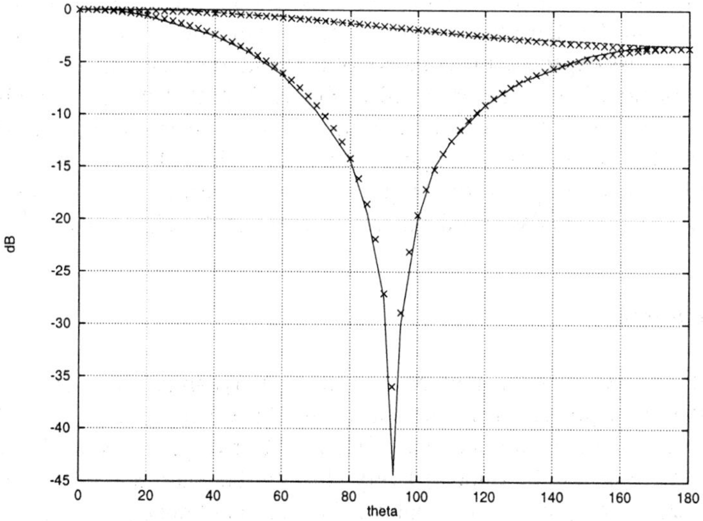

Figure 7.59 Normalized scattered field ($\bar{E}$ formulation): ×, mesh A; —, --, [Sarkar et al. 1989b].

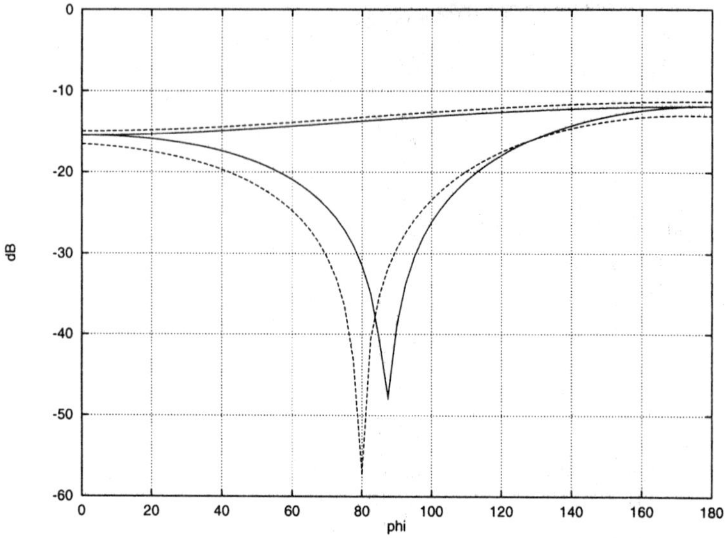

Figure 7.60 Scattered field (plane $\theta = 90°$): —, isotropic; --, anisotropic.

The anisotropic case (permittivity tensor according to (7.135)) is compared with the isotropic case ($\varepsilon_r = 4$). The smooth curve corresponds to the case of the polarization in the z-direction. On the other hand, the curve with the dip is obtained with the y-polarized wave. As expected, the behavior of the scatterer is slightly different depending on the polarization. It may be observed in Figure 7.60 that the curve with the dip (y-polarization) shifts downward when the anisotropic case is considered. The opposite happens with the smooth curve (z-polarization), which is shifted upward. This is due to the smaller value of ε_{yy} and the larger value of ε_{zz}, respectively, compared with the isotropic case ($\varepsilon_r = 4$). Also, a shift in the location of the dip is observed.

7.5.3 Conclusions

In this section, an iterative method for the 3D dynamic analysis of open-region problems has been presented. Numerical results have been presented to illustrate the validity of the method and its main features. Convergence is achieved in a few iterations. The mesh can be truncated very close to the sources of the original problem ($0.1\lambda_0$ to $0.2\lambda_0$ in the previous examples). The external boundary condition appears as a pointwise Dirichlet boundary condition for the FEM formulation. Hence, the structure of the FEM matrices remains banded and sparse. The radiation condition for the open-region FEM problems has been implemented in an essentially exact fashion and the problem is solved adaptively.

Appendix A

A Mathematical Overview

A.1 SOME CONCEPTS OF FUNCTIONAL ANALYSIS

A real linear space consists of a set, V, of mathematical elements in which two operations are defined:

Addition: $(v_1, v_2) \rightarrow v_1 + v_2$, from $V \times V$ into V (A.1a)

Multiplication by a scalar: $(\gamma, v) \rightarrow \gamma v$, from $\mathbb{R} \times V$ into V (A.1b)

where $\mathbb{R}$ is the set of real numbers. The first operation has the associative and the commutative properties and provides the existence of the zero element and the negative element $-v$ corresponding to an element v. The second operation has the associative and distributive properties and defines the unity scalar. The elements of V (v_1, v_2, and v in the previous definitions) are called vectors in a generalized sense. The elements of $\mathbb{R}$ ($\gamma \in \mathbb{R}$) are called scalars.

Examples of linear spaces are the aforementioned set of real numbers, $\mathbb{R}$, in which the elements of the space are scalar quantities, and the n-dimensional Euclidean space, $\mathbb{R}^n$, in which the elements $\bar{r}$ are vector quantities with n components; i.e., they are defined by the n-tuple (α_1, α_2, ..., α_n) [Celia and Gray 1992; Appendix A], [Kardestuncer and Norrie 1987; P. 1, Ch. 1].

In an analogous fashion, when γ belongs to the set of complex numbers, $\mathbb{C}$, linear complex spaces may be defined [Stakgold 1979; Ch. 4].

A.1.1 Dimension of a Space. Finite-Dimensional Spaces

Let V be a linear space and A be a nonempty subset of V. The elements of A are said to be linearly independent, if and only if, for any linear combination of a finite number of elements, v_i, of the set A, the expression $\Sigma_i \beta_i v_i = 0$ implies that $\beta_i = 0$, for any i. The elements of A are said to form a basis, if they generate V. The number of independent elements of A is called the dimension of V. It may be 0 (trivial case), finite or infinite.

Interesting examples of finite-dimensional spaces are those defined by means of polynomials of n variables over an open subset of $\mathbb{R}^n$, Ω:

1. The space of polynomials $P_p^{(n)}(\Omega)$: Its elements are scalar quantities. They consist of polynomials (of n variables) of total degree less than or equal to p, where $p \in \mathbb{N}$ ($\mathbb{N}$ is the set of natural numbers). Thus a polynomial containing all possible monomials will be of total degree p and it will be complete up to the degree p. The dimension of the space is

$$\dim P_p^{(n)} = \frac{1}{n!}(p+1)\cdots(p+n) \tag{A.2a}$$

   where $n!$ stands for the factorial of n. For example, the basis for the polynomial space $P_3^{(2)}$ is given by the set $(1;\ x,\ y;\ x^2,\ xy,\ y^2;\ x^3,\ x^2y,\ xy^2,\ y^3)$. The dimension of the space $P_3^{(2)}$ is 10 [Kardestuncer and Norrie 1987; P. 1, Ch. 1].
2. The space of polynomials $Q_p^{(n)}(\Omega)$: Its elements are scalar quantities. They consist of polynomials (of n variables) of degree less than or equal to p in each variable, where $p \in \mathbb{N}$. Thus, a polynomial containing all possible monomials will be of total degree np, but it will contain a complete polynomial of degree p only. Its dimension is given by

$$\dim Q_p^{(n)} = (p+1)^n \tag{A.2b}$$

   An example may be $Q_3^{(2)}$, whose basis is given by the set $(1;\ x,\ xy,\ y;\ x^2,\ xy^2,\ x^2y^2,\ x^2y,\ y^2;\ x^3,\ x^3y,\ x^3y^2,\ x^3y^3,\ x^2y^3,\ xy^3,\ y^3)$. Its dimension is 16 [Kardestuncer and Norrie 1987; P. 1, Ch. 1].
3. The space of vectors $D_p^{(n)}(\Omega)$: Its elements are vectors $\bar{d}$ (in matrix notation, $\{d\}$), whose ith component is a polynomial of n variables of degree p, as

$$D_p^{(n)} = \left[\{d\} = \begin{Bmatrix} d_1 \\ \vdots \\ d_n \end{Bmatrix};\quad p_0, p_1, \ldots, p_n \in P_{p-1}^{(n)}; \right.$$
$$d_1(x_1,\ldots,x_n) = p_1(x_1,\ldots,x_n) + x_1 p_0(x_1,\ldots,x_n), \tag{A.2c}$$
$$\cdots\cdots\cdots\cdots\cdots\cdots$$
$$\left. d_n(x_1,\ldots,x_n) = p_n(x_1,\ldots,x_n) + x_n p_0(x_1,\ldots,x_n) \right]$$

   where x_i stands for the ith coordinate of an n-dimensional Cartesian system.

Thus, the elements are vectors whose components are polynomials of maximum total degree p, being complete up to degree p-1. It is easy to see that the polynomial p_0 may be a homogeneous polynomial of degree p-1 (a polynomial of degree k is said to be homogeneous if it contains only monomials of degree k). That is, $p_0 \in P_{0,p-1}^{(n)}$, where $P_{0,p-1}^{(n)}$ stands for the space of homogeneous polynomials (of n variables) of degree p-1. The dimension of $D_p^{(n)}$ is given by

$$\dim D_p^{(n)} = n\binom{n+p-1}{n} + \binom{n+p-2}{n-1} = (n+p)\binom{n+p-2}{n-1} = \frac{(n+p)(n+p-2)!}{(n-1)!(p-1)!} \tag{A.2d}$$

Alternatively, the space $D_p^{(n)}$ may be defined as

$$D_p^{(n)} = \left(P_{p-1}^{(n)}\right)^n \oplus M_p^{(n)}, \quad M_p^{(n)} = \left\{p_0\,\bar{r} \mid p_0 \in P_{0,p-1}^{(n)}\right\} \tag{A.2e}$$

where $(P_{p-1}^{(n)})^n$ is the space of vectors with n-components, each component belonging to the space $P_{p-1}^{(n)}$; the symbol $\oplus$ stands for the direct addition of spaces, and $\bar{r}$ is the position vector of a point, i.e., a vector of two coefficients (the coordinates x and y of the point) for $n = 2$, i.e., for a two-dimensional (2D) domain, or three coefficients (coordinates x, y, and z) for $n = 3$, i.e., for a three-dimensional (3D) domain [Nédélec 1980], [Raviart and Thomas 1977a], [Thomas and Joly 1981; Ch. 3].

For the two-dimensional (2D) case, the space will be given by vectors having two components, $d_x(x,y)$, and $d_y(x,y)$, that are polynomials (in two variables) of degree p, according to

$$D_p^{(2)} = \left[\{d\} = \begin{Bmatrix} d_x \\ d_y \end{Bmatrix}; \quad p_1, p_2 \in P_{p-1}^{(2)}\ ; \ p_0 \in P_{0,p-1}^{(2)}\ ; \quad \begin{aligned} d_x(x,y) &= p_1(x,y) + x p_0(x,y) \\ d_y(x,y) &= p_2(x,y) + y p_0(x,y) \end{aligned} \right] \tag{A.2f}$$

From (A.2d), the dimension of $D_p^{(2)}$ is given by

$$\dim D_p^{(2)} = p\,(p+2) \tag{A.2g}$$

Thus, for $p = 1$, $\dim D_1^{(2)} = 3$, while for $p = 2$, $\dim D_2^{(2)} = 8$.

For the 3D case, the space will consist of vectors having three components, $d_x(x,y,z)$, $d_y(x,y,z)$, and $d_z(x,y,z)$, that are polynomials (in three variables) of degree p, according to

$$D_p^{(3)} = \left[\{d\} = \begin{Bmatrix} d_x \\ d_y \\ d_z \end{Bmatrix} ; \quad p_1, p_2, p_3 \in P_{p-1}^{(3)} \; ; \; p_0 \in P_{0,p-1}^{(3)} \; ; \right. \qquad \text{(A.2h)}$$

$$\left. \begin{aligned} d_x(x,y,z) &= p_1(x,y,z) + x p_0(x,y,z), \\ d_y(x,y,z) &= p_2(x,y,z) + y p_0(x,y,z), \\ d_z(x,y,z) &= p_3(x,y,z) + z p_0(x,y,z) \end{aligned} \right]$$

From (A.2d), the dimension of $D_p^{(3)}$ is given by

$$\dim D_p^{(3)} = \frac{p(p+1)(p+3)}{2} \qquad \text{(A.2i)}$$

Thus, for $p = 1$, $\dim D_1^{(3)} = 4$, while for $p = 2$, $\dim D_2^{(3)} = 15$.

4. The space of vectors $R_p^{(n)}(\Omega)$: Its elements are vectors, d, whose components are polynomials of degree p, complete up to degree p-1, defined by [Nédélec 1980]

$$R_p^{(n)} = \left(P_{p-1}^{(n)}\right)^n \oplus S_p^{(n)}, \quad S_p^{(n)} = \left\{ \bar{d}_s \in \left(P_{0,p}^{(n)}\right)^n \mid \bar{d}_s \cdot \bar{r} = 0 \right\} \qquad \text{(A.2j)}$$

where the dot stands for the scalar product between two vectors. The dimension of this space is given by

$$\dim R_p^{(n)} = n \binom{n+p}{n} - \binom{n+p}{n-1} = \frac{p(n+p)!}{(n-1)!(p+1)!} \qquad \text{(A.2k)}$$

For the two-dimensional case, the space $R_p^{(2)}(\Omega)$ may also be defined as

$$R_p^{(2)} = \left[\{d\} = \begin{Bmatrix} d_x \\ d_y \end{Bmatrix} ; \quad p_1, p_2 \in P_{p-1}^{(2)} \; ; \; p_0 \in P_{0,p-1}^{(2)} \; ; \right. \qquad \text{(A.2l)}$$

$$\left. \begin{aligned} d_x(x,y) &= p_1(x,y) - y p_0(x,y), \\ d_y(x,y) &= p_2(x,y) + x p_0(x,y) \end{aligned} \right]$$

From (A.2k), the dimension of $R_p^{(2)}$ is given by

$$\dim R_p^{(2)} = p\,(p+2) \tag{A.2m}$$

that turns out to be identical to (A.2g). Thus, for $p = 1$, $\dim R_1^{(2)} = 3$, while for $p = 2$, $\dim R_2^{(2)} = 8$.

It is easy to see that the $R_p^{(2)}$ space is related to the $D_p^{(2)}$ space. The elements of the former space may be obtained from the elements of the latter one by a shift of $\pi/2$ of the axes [Nédélec 1980]. In other words, $\bar{d}_R = \bar{a}_z \times \bar{d}_D$, where $\bar{d}_R \in R_p^{(2)}$, $\bar{a}_z$ is the unit vector in the z-direction, $\times$ stands for the vector product, and $\bar{d}_D \in D_p^{(2)}$.

For the three-dimensional case, it is not possible to arrive to an expression similar to (A.2l). There is no direct relation between the spaces $R_p^{(3)}$ and $D_p^{(3)}$. However, the definition of the space $R_p^{(3)}$ for a specific order p may be easily deduced from (A.2j). From (A.2k) the dimension of $R_p^{(3)}$ is given by [Nédélec 1980]

$$\dim R_p^{(3)} = \frac{(p+3)(p+2)p}{2} \tag{A.2n}$$

Thus, for $p = 1$, $\dim R_1^{(3)} = 6$, while for $p = 2$, $\dim R_2^{(3)} = 20$. In the following, the specific definitions of spaces $R_1^{(3)}$ and $R_2^{(3)}$ are given.

For $p = 1$, after obtaining the basis of the space $S_1^{(3)}$, it is easy to see that the space $R_1^{(3)}$ may be defined in a compact form as

$$R_1^{(3)} = \left\{ \bar{d} = \bar{d}_1 + \bar{d}_2 \times \bar{r}\ ;\ \bar{d}_1, \bar{d}_2 \in (P_0)^3 \right\} \tag{A.2o}$$

or, equivalently, as

$$R_1^{(3)} = \left[\{d\} = \begin{Bmatrix} d_x \\ d_y \\ d_z \end{Bmatrix} ;\quad a_i, b_i \in \mathbb{R},\ i = 1,2,3\ ; \right.$$
$$\left. \begin{aligned} d_x(y,z) &= a_1 - b_3 y + b_2 z\,, \\ d_y(x,z) &= a_2 + b_3 x - b_1 z\,, \\ d_z(x,y) &= a_3 - b_2 x + b_1 y \end{aligned} \right] \tag{A.2p}$$

For $p = 2$, after obtaining the basis of the space $S_2^{(3)}$, it is easy to see that the space $R_2^{(3)}$ may be defined as

$$R_2^{(3)} = \left[\{d\} = \begin{Bmatrix} d_x \\ d_y \\ d_z \end{Bmatrix} ; \quad p_1, p_2, p_3 \in P_1^{(3)} \ ; \ a_i \ , \ i = 1, \cdots, 8 \in \mathbb{R} \ ; \right.$$

$$\left. \begin{aligned} d_x &= p_1(x,y,z) + a_1 xy + a_2 xz + a_3 y^2 + a_4 yz + a_5 z^2, \\ d_y &= p_2(x,y,z) - a_1 x^2 - a_3 xy + a_6 xz + a_7 yz + a_8 z^2, \\ d_z &= p_3(x,y,z) - a_2 x^2 - (a_4 + a_6) xy - a_5 xz - a_7 y^2 - a_8 yz \end{aligned} \right] \tag{A.2q}$$

A.1.2 Functional Forms. Linear and Bilinear Operators

Let V and W be linear spaces. Let $\mathcal{L}$ be an operator that maps V into W. It is said that $\mathcal{L}$ is a linear operator from V into W, if and only if the following expressions hold:

$$\mathcal{L}(v_1 + v_2) = \mathcal{L}(v_1) + \mathcal{L}(v_2) ; \quad \forall \ v_1, v_2 \in V \tag{A.3a}$$

$$\mathcal{L}(\alpha v) = \alpha \mathcal{L}(v) ; \ \forall \alpha \in \mathbb{R} ; \ \forall v \in V \tag{A.3b}$$

in which case it may be written as

$$\mathcal{L}(v) = \mathcal{L} v \tag{A.4}$$

Let U, V, and W be linear spaces. It is said that $a(u,v)$ is a bilinear operator from $U \times V$ into W if and only if a is linear on U and on V.

A bilinear operator, $a(u,v)$, is said to be symmetric when

$$a(u,v) = a(v,u) \tag{A.5}$$

The problems to be analyzed in this book are described by means of partial differential equations where all the operators involved are linear [Kardestuncer and Norrie 1987; P.1, Ch. 1].

A linear operator from V into $\mathbb{R}$ is called a linear form on V (or a linear functional on V). Similarly, a bilinear operator from $U \times V$ into $\mathbb{R}$ is said to be a bilinear form (or a bilinear functional) on $U \times V$. A bilinear form, $B(u,v)$, is symmetric if

$$B(u,v) = B(v,u) \tag{A.6}$$

Let V be a linear space. G is said to be a quadratic form (or a quadratic functional) on V, if and only if, there exists a bilinear form, A, from $V \times V$ into $\mathbb{R}$, such that the following expression holds:

$$G(v) = A(v,v) \ ; \ \forall\, v \in V \tag{A.7}$$

If G is a quadratic form on V, there exists exactly one symmetric bilinear form, $B(v,v)$, from $V \times V$ into $\mathbb{R}$, such that

$$G(v) = B(v,v) \ ; \quad \forall \ v \in V \tag{A.8}$$

The forms to be handled in this book are integrals (so, they are also called integral forms). Therefore, they are scalar functions having as arguments one or more functions and their corresponding derivatives. Hence for this reason, they are also called functionals. They may be expressed in one of the two following forms.

1. A linear integral form, such as

$$\begin{aligned} L(v) &= L\left(v, \frac{\partial v}{\partial x}, \frac{\partial v}{\partial y}, \frac{\partial v}{\partial z}, \ldots \right) \\ &= \int_\Omega f_1 \left(v, \frac{\partial v}{\partial x}, \frac{\partial v}{\partial y}, \frac{\partial v}{\partial z}, \ldots \right) \mathrm{d}\Omega + \int_\Gamma g_1 (v, \ldots) \, \mathrm{d}\Gamma \end{aligned} \tag{A.9a}$$

   where Ω defines a domain and Γ is its boundary; and, in general, $v = v(x,y,z)$. Hence f_1 and g_1 are linear operators so that L is a linear integral form.
2. A bilinear integral form, such as:

$$\begin{aligned} A(u,v) &= A\left(u, v, \frac{\partial u}{\partial x}, \frac{\partial u}{\partial y}, \frac{\partial u}{\partial z}, \frac{\partial v}{\partial x}, \frac{\partial v}{\partial y}, \frac{\partial v}{\partial z}, \ldots \right) \\ &= \int_\Omega f_2 \left(u, v, \frac{\partial u}{\partial x}, \frac{\partial u}{\partial y}, \frac{\partial u}{\partial z}, \frac{\partial v}{\partial x}, \frac{\partial v}{\partial y}, \frac{\partial v}{\partial z}, \ldots \right) \mathrm{d}\Omega \\ &\quad + \int_\Gamma g_2 (u, v, \ldots) \, \mathrm{d}\Gamma \end{aligned} \tag{A.9b}$$

   where, in general, $u = u(x,y,z)$, and $v = v(x,y,z)$. Hence, f_2 and g_2 are bilinear operators so that A is a bilinear integral form.

In general, the integral forms would have an expression given by

$$W(u,v) = \int_\Omega f\left(u, v, \frac{\partial u}{\partial x}, \frac{\partial u}{\partial y}, \frac{\partial u}{\partial z}, \frac{\partial v}{\partial x}, \frac{\partial v}{\partial y}, \frac{\partial v}{\partial z}, \ldots\right) d\Omega + \int_\Gamma g(u, v, \ldots)\, d\Gamma \tag{A.10a}$$

where f and g are operators that have both a linear part and a bilinear part; so $W(u,v)$ may be written as

$$W(u,v) = A(u,v) - L(v) \tag{A.10b}$$

where $A(u,v)$ is a bilinear integral form and $L(v)$ is a linear one. In many problems an integral form may be reduced to an expression in which the bilinear form is symmetric

$$W(u,v) = B(u,v) - L(v) \tag{A.11}$$

where $B(u,v)$ fulfills (A.6). Sometimes in order to arrive at an expression of this type it may be required to perform an integration by parts (see Appendix A.2). This will lead to

$$A(u,v) = B(u,v) + \int_\Gamma m(v)\, d\Gamma \tag{A.12}$$

where $B(u,v)$ has a symmetric form and where the second term of the right-hand side is a boundary integral term. Hence, after taking into account the natural boundary conditions (see Appendix A.2) the boundary integral term may be expressed by a linear form.

This procedure allows one to broaden the definition of symmetry given by (A.6). If a problem is formulated according to the expressions of Chapter 2, (2.1) or (2.2), by an operator $\mathcal{L}$ or, in general, by a matrix of operators, $[\mathcal{L}]$ (in matrix notation), then

$$\langle\{v\};[\mathcal{L}]\{u\}\rangle = \int_\Omega \{v\}^T [\mathcal{L}]\{u\}\, d\Omega = \int_\Omega \{u\}^T [\mathcal{L}]\{v\}\, d\Omega + b.t. = \langle[\mathcal{L}]^a\{v\};\{u\}\rangle \tag{A.13}$$

where $<\cdot;\cdot>$ denotes the inner product defined by (A.18), ^{T} denotes the transpose of a matrix, and *b.t.* indicates a boundary term. Expression (A.13) defines the adjoint operator of $\mathcal{L}$, $\mathcal{L}^a$. The operator $\mathcal{L}$ is symmetric if $\mathcal{L} = \mathcal{L}^a$. In addition, the operator is self-adjoint if $\mathcal{L}$ and $\mathcal{L}^a$ have the same boundary conditions [Stakgold 1979; Ch. 5]. A self-adjoint operator $\mathcal{L}$ is positive definite when

$$\int_\Omega \{v\}^T [\mathcal{L}]\{v\}\, d\Omega > 0, \quad \{v\} \neq \{0\} \tag{A.14}$$

A.1.3 Hilbert and Sobolev Spaces

Consider a collection of scalar elements. Such a collection is said to be a metric space if to each pair of elements u,v there is associated a real number $d(u,v)$ called the distance between u and v. The distance function is known as the metric for that space [Stakgold 1979, Ch. 4]. Figure A.1 illustrates the properties of the metric.

The concept of metric is also related to the concept of sequences of scalar elements of the collection under consideration. The sequence of points u_k, $k = 1, \ldots$, is said to converge to a limit u if and only if the sequence of real numbers $d(u,u_k)$ converges to zero. A particular form of these sequences, namely, a Cauchy sequence, is of interest. A sequence is a Cauchy sequence if for each number greater than zero there exists a natural number N such $d(u_m,u_p) \leq \varepsilon$ for $m,p > N$. A metric space is said to be complete if every Cauchy sequence of points from that space converges to a limit in the same space.

Consider a set of elements of a linear vector space. Since a vector is endowed with the properties of direction and magnitude, one can associate an inner product space with the direction and a normed linear space with the magnitude. An inner product space is a linear space in which a real-valued function of a pair of vectors, called an inner product, is defined with properties again given by Figure A.1. A normed linear space is a linear space in which a real-valued function known as norm is defined. Its properties are also given in Figure A.1. In addition, a metric can also characterize linear spaces resulting in a linear metric space. These various spaces may or may not be complete. A complete normed linear metric space is called a Banach space and a normed linear inner product space complete with its natural metric is called a Hilbert space as illustrated by Figure A.1.

Let u be a scalar function defined in a domain Ω, $\Omega \in \mathbb{R}^n$, and the αth partial derivative be given by

$$D^{\alpha} u = \frac{\partial^{\alpha} u}{\partial^{\alpha_1} x_1 \, \partial^{\alpha_2} x_2 \, \cdots \, \partial^{\alpha_n} x_n} \, , \quad \alpha = \sum_{i=1}^{n} \alpha_i \tag{A.15}$$

where α_i for $i = 1, \ldots, n$ are natural numbers and x_i stands for the ith coordinate of a n-dimensional coordinate system. It is clear that for $\alpha > 0$ and $n > 1$ there are more than one αth partial derivatives. In the expressions to follow, $D^{\alpha}u$ should be understood as to involve all possible αth partial derivatives unless the contrary is explicitly stated.

The linear space of scalar functions u, having continuous partial derivatives in Ω up to order m ($D^{\alpha}u$, $0 \leq \alpha \leq m$), is represented by $C^m(\Omega)$. For example, $C^0(\Omega)$ stands for the space of continuous functions and $C^{\infty}(\Omega)$ represents the space of functions having infinite continuous derivatives.

Analogous definitions may be given for the case of vector functions, $\bar{u}$, with n components so that $(C^m(\Omega))^n$ represents the linear space of vector functions with n continuous components defined over Ω having continuous partial derivatives up to the order m. Similarly $(C^0(\Omega))^n$ stands for the space of vector functions with n continuous

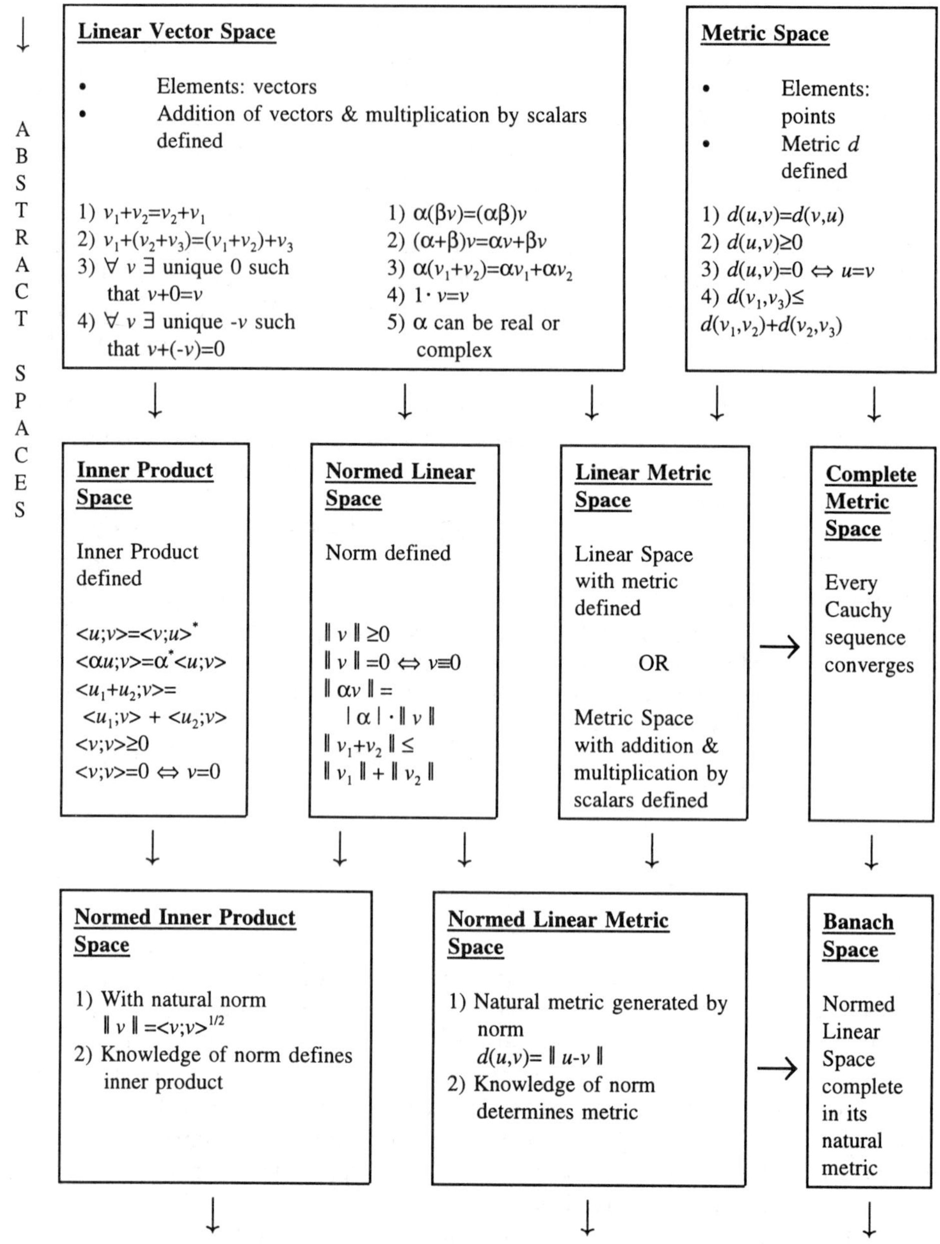

Figure A.1 A summary of various spaces.

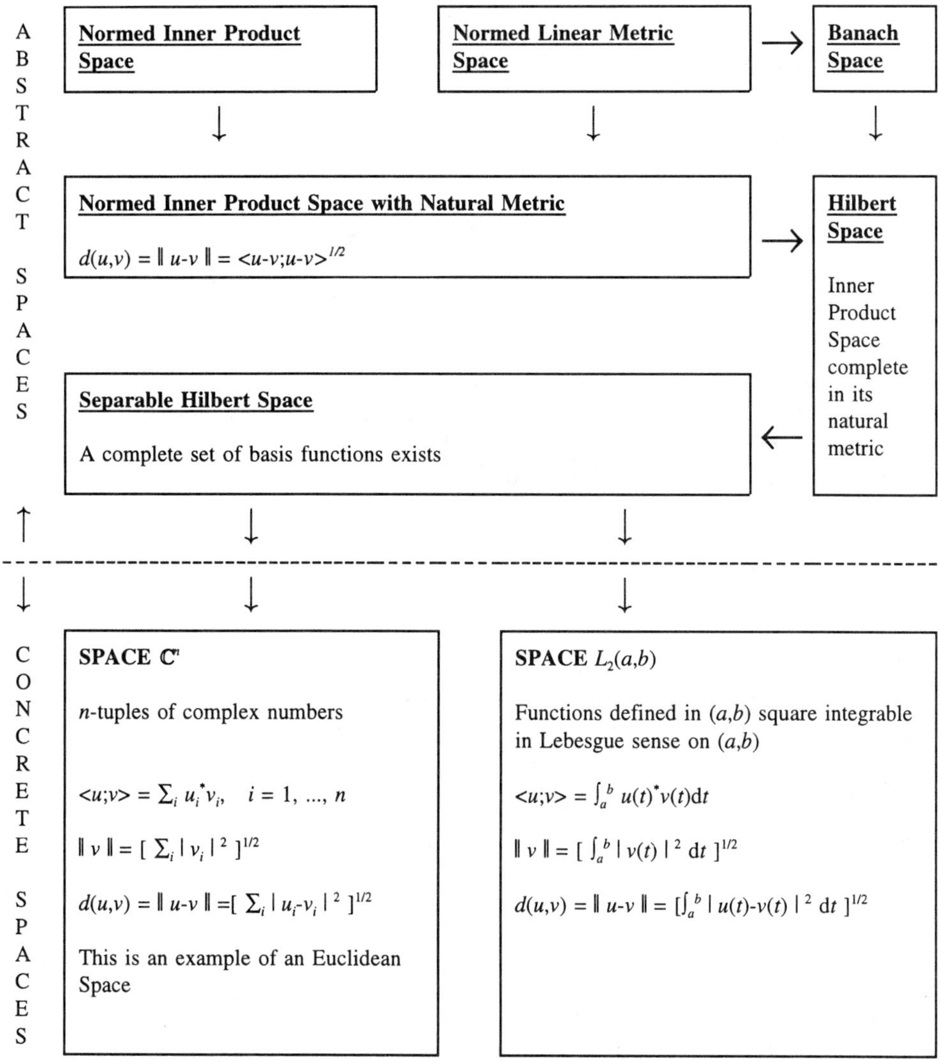

Note: Linear spaces may be finite or infinite dimensional.

Figure A.1 (Continued.)

components and $(C^{\infty}(\Omega))^n$ for that of vectors with n continuous components having infinite continuous derivatives.

The linear space of scalar functions, u, being p-integrable is called $L^p(\Omega)$, where $1 \leq p < \infty$. The definition of p-integrability of a scalar function u is given by

$$\int_{\Omega} |u|^p \, d\Omega < \infty \tag{A.16a}$$

where $|u|$ represents the absolute value; for the complex case

$$|u| = (u^* u)^{1/2} \tag{A.16b}$$

where u^* stands for the complex conjugate of u.

For the case of vector functions $\bar{u} \in \mathbb{R}^n$ with n components $u_1, \ldots, u_n$, the definition of p-integrability is given by (A.16a) and still holds with $|u| = |\bar{u}|$ where the Euclidean norm of the vector $\bar{u}$ is given by

$$|\bar{u}| = (\bar{u} \cdot \bar{u})^{1/2} = \left(\sum_{i=1}^{n} u_i^2 \right)^{1/2} \tag{A.16c}$$

Here the dot stands for the scalar product between two vectors. For the complex case the following definition holds [Stakgold 1979; Ch. 4], [Noble and Daniel 1988; Ch. 5]:

$$|\bar{u}| = (\bar{u}^* \cdot \bar{u})^{1/2} = \left(\sum_{i=1}^{n} u_i^* u_i \right)^{1/2} \tag{A.16d}$$

The corresponding spaces of p-integrable vector functions are represented by $(L^p(\Omega))^n$.

Important examples of these spaces are the space of square integrable scalar functions $L^2(\Omega)$ and that of square integrable vector functions with two components $(L^2(\Omega))^2$ or with three components, $(L^2(\Omega))^3$.

A measure or norm may be defined for p-integrable spaces, according to

$$\| u \|_{L^p(\Omega)} = \left[\int_{\Omega} | u |^p \, d\Omega \right]^{1/p} \tag{A.17a}$$

An alternate notation for this norm is

$$\| u \|_{L^p(\Omega)} = \| u \|_{0,p,\Omega} = \| u \|_{0,p} \tag{A.17b}$$

The justification of the second notation will be clear later on when defining Sobolev spaces because the space $L^p(\Omega)$ provided with the norm (A.17a) is a Sobolev space of order $(0,p)$. In the third notation the reference to Ω has been suppressed. That may be done whenever there is no ambiguity with respect to the domain of definition of the space.

If the function is a real or complex scalar, $|u|$ will be the absolute value of the function. If the function is a real or complex vector, $\bar{u}$, then

$$\|\bar{u}\|_{L^p(\Omega)} = \|\bar{u}\|_{0,p,\Omega} = \|\bar{u}\|_{0,p} = \left[\int_\Omega |\bar{u}|^p \, d\Omega\right]^{1/p} \tag{A.17c}$$

where $|\bar{u}|$ is defined in (A.16c) or in (A.16d). Spaces $L^p(\Omega)$ provided with the aforementioned norm are Banach spaces. For spaces of square integrable scalar functions, $L^2(\Omega)$, the norm is given by:

$$\|u\|_{L^2(\Omega)} = \|u\|_{0,2,\Omega} = \|u\|_{0,2} = \|u\|_0 = \left[\int_\Omega |u|^2 \, d\Omega\right]^{1/2} \tag{A.17d}$$

and in the case of spaces of square integrable vector functions by

$$\|\bar{u}\|_{L^2(\Omega)} = \|\bar{u}\|_{0,2,\Omega} = \|\bar{u}\|_{0,2} = \|\bar{u}\|_0 = \left[\int_\Omega |\bar{u}|^2 \, d\Omega\right]^{1/2} \tag{A.17e}$$

When the notation $\|u\|_0$ is used it should be understood that the norm refers to square integrable spaces, i.e., Sobolev spaces of order (0,2).

An inner product may be defined for square integrable spaces by

$$<u;v>_{L^2(\Omega)} = \int_\Omega uv \, d\Omega \tag{A.18a}$$

for real scalar functions, or

$$<u;v>_{L^2(\Omega)} = \int_\Omega u^* v \, d\Omega \tag{A.18b}$$

for complex scalar functions.

For real vector functions, the following expression holds:

$$<\bar{u};\bar{v}>_{L^2(\Omega)} = \int_\Omega \bar{u}\cdot\bar{v} \, d\Omega \tag{A.18c}$$

which in matrix notation may be written as

$$<\{u\};\{v\}>_{L^2(\Omega)} = \int_\Omega \{u\}^T \{v\}\, \mathrm{d}\Omega \tag{A.18d}$$

If $\bar{u}$ and $\bar{v}$ (or $\{u\}$ and $\{v\}$) are complex [Stakgold 1979; Ch. 4], [Noble and Daniel 1988; Ch. 5], the inner product is defined as

$$<\bar{u};\bar{v}>_{L^2(\Omega)} = \int_\Omega \bar{u}^* \cdot \bar{v}\, \mathrm{d}\Omega \tag{A.18e}$$

or, in matrix notation,

$$<\{u\};\{v\}>_{L^2(\Omega)} = \int_\Omega \{u\}^H \{v\}\, \mathrm{d}\Omega \tag{A.18f}$$

where the superscript ^{H} stands for the Hermitian transpose of a matrix, i.e., in this case, $\{u\}^H = \{u^*\}^T$.

A linear space of scalar functions u defined in Ω is called a Hilbert space of order m and is represented by $H^m(\Omega)$ if u are square integrable functions with square integrable derivatives, up to order m, that is,

$$\begin{aligned} H^m(\Omega) &= \{u \mid u \in L^2(\Omega) : D^\alpha u \in L^2(\Omega)\ ;\ 1 \le \alpha \le m\} \\ &= \{u \mid D^\alpha u \in L^2(\Omega)\ ;\ 0 \le \alpha \le m\} \end{aligned} \tag{A.19a}$$

and an inner product and a norm are provided.

For real scalar Hilbert spaces the inner product is defined by

$$\begin{aligned} <u;v>_{H^m(\Omega)} &= \sum_{0\le\alpha\le m} \left(\sum_\alpha < D^\alpha u\, , D^\alpha v >_{L^2(\Omega)} \right) \\ &= \sum_{0\le\alpha\le m} \left(\sum_\alpha \int_\Omega D^\alpha u\ D^\alpha v\ \mathrm{d}\Omega \right) \\ &= \int_\Omega \left[\sum_{0\le\alpha\le m} \left(\sum_\alpha D^\alpha u\ D^\alpha v \right) \mathrm{d}\Omega \right] \end{aligned} \tag{A.19b}$$

where the summation for α is extended to the product of all the αth partial derivatives of the same indices.

For complex scalar spaces the inner product is given by

$$<u;v>_{H^m(\Omega)} = \sum_{0\le\alpha\le m} \left(\sum_\alpha <D^\alpha u, D^\alpha v>_{L^2(\Omega)}\right)$$

$$= \sum_{0\le\alpha\le m} \left(\sum_\alpha \int_\Omega D^\alpha u^* \, D^\alpha v \; d\Omega\right) \tag{A.19c}$$

$$= \int_\Omega \left[\sum_{0\le\alpha\le m} \left(\sum_\alpha D^\alpha u^* \, D^\alpha v \; d\Omega\right)\right]$$

A norm may be defined according to

$$\| u \|_{H^m(\Omega)} = \| u \|_{m,2,\Omega} = \| u \|_m = \left[\sum_{0\le\alpha\le m} \left(\sum_\alpha \|D^\alpha u \|^2_{L^2(\Omega))}\right)\right]^{1/2}$$

$$= \left[\sum_{0\le\alpha\le m} \left(\sum_\alpha \int_\Omega |D^\alpha u|^2 \; d\Omega\right)\right]^{1/2} \tag{A.19d}$$

as well as a seminorm

$$|u|_{H^m(\Omega)} = |u|_{m,2,\Omega} = |u|_{m,\Omega} = |u|_m = \left[\sum_{\alpha=m} \|D^\alpha u\|^2_{L^2(\Omega)}\right]^{1/2}$$

$$= \left[\sum_{\alpha=m} \int_\Omega |D^\alpha u|^2 \; d\Omega\right]^{1/2} \tag{A.19e}$$

With the corresponding definitions for the inner product, norm, and seminorm, it may be observed that

$$H^0(\Omega) = L^2(\Omega) \tag{A.20a}$$

Some of the problems analyzed in this book have been formulated in terms of a weak formulation. They require for their solution, scalar functions belonging to the space $H^1(\Omega)$, that is, square integrable functions having also square integrable first derivatives.

However, for the solution of boundary value problems of mathematical physics it is convenient to introduce the Sobolev space [Sobolev 1963a,b], which has some interesting properties. First, one can deal with generalized derivatives with p-integrability instead of the ordinary ones. Secondly, when dealing with a Lipschitz boundary (most smooth and piecewise-smooth boundaries are Lipschitz boundaries), the norm defining the Sobolev space assumes a simplified form [Stakgold 1979, Ch. 4]. Thirdly, every function in the Sobolev space $W^{m,p}(\Omega)$ can be arbitrarily well approximated by functions having partial derivatives of order m that are uniformly continuous in the domain Ω.

A Sobolev space of order (m,p) in Ω, called $W^{m,p}(\Omega)$ is a linear space of functions u belonging to the space $L^p(\Omega)$, whose partial derivatives $D^\alpha u$ also belong to $L^p(\Omega)$, for all $\alpha \leq m$; that is,

$$W^{m,p}(\Omega) = \{u \,|D^\alpha u \in L^p(\Omega) \;;\; 0 \leq \alpha \leq m \;,\; 1 \leq p < \infty\} \tag{A.20b}$$

Sobolev spaces may be provided with the norm

$$\|u\|_{W^{m,p}(\Omega)} = \|u\|_{m,p,\Omega} = \|u\|_{m,p} = \left[\sum_{0\leq\alpha\leq m}\left(\sum_{\alpha}\int_\Omega |D^\alpha u|^p \mathrm{d}\Omega\right)\right]^{1/p} \;,\; 1 \leq p < \infty \tag{A.20c}$$

where $D^\alpha u$ is defined by (A.15). Since these derivatives may be generalized derivatives, the space is a normed linear space complete in its natural metric. Hence it is also a Banach space. It can be shown that

$$W^{0,2}(\Omega) = L^2(\Omega) \tag{A.20d}$$

Sobolev spaces of order $(m,2)$ in Ω are called $H^m(\Omega)$. These spaces, provided with the aforementioned inner product (A.19b) or (A.19c) and the norm (A.19d) (i.e., the norm (A.20c) for $p = 2$), are Hilbert spaces. This is evident from the pictorial representation of Figure A.1. It is seen that when an inner product is introduced in the characterization of a Banach space it results in a Hilbert space. The association of the inner product with the norm and the metric is possible in this case only for $p = 2$. For other values of p, it is a Banach space and not a Hilbert space.

Figure A.1 summarizes the various definitions of all the spaces. An interested reader may find further information in [Kardestuncer and Norrie 1987; P. 1, Chs. 1 , 2], [Mitchell 1973], [Noble and Daniel 1988; Ch. 5], [Stakgold 1979; Ch. 4], [Zlamal 1973].

Finally, for the sake of completeness, it may be pointed out that sometimes spaces are represented by $W^{-m,q}(\Omega)$. In this case, functions that are elements of $W^{-m,q}(\Omega)$ form the space adjoint to the Banach space $W^{m,p}(\Omega)$ with $1/p + 1/q = 1$.

A.1.4 $H(\text{div},\Omega)$ Spaces

Let $\bar{q}(x,y)$ be a vector variable with two components, $q_x(x,y)$ and $q_y(x,y)$, defined in $\Omega \in \mathbb{R}^2$, along the two orthogonal directions of a two-dimensional Cartesian system of coordinates. In matrix notation it may be written as

$$\{q\} = \begin{Bmatrix} q_x(x,y) \\ q_y(x,y) \end{Bmatrix} \tag{A.21}$$

Its divergence is defined as

$$\text{div}\,\bar{q} = \bar{\nabla}_t \cdot \bar{q} = \frac{\partial q_x}{\partial x} + \frac{\partial q_y}{\partial y} \tag{A.22a}$$

where the two-dimensional gradient operator defined in Chapter 2, (2.7a), has been used. In matrix notation it may be written as

$$\text{div}\,\bar{q} = \{\nabla_t\}^T\{q\} = \begin{Bmatrix} \dfrac{\partial}{\partial x} \\ \dfrac{\partial}{\partial y} \end{Bmatrix}^T \{q\} = \frac{\partial q_x}{\partial x} + \frac{\partial q_y}{\partial y} \tag{A.22b}$$

The linear space of functions $\bar{q}$ having square integrable components and square integrable divergence is called $H(\text{div},\Omega)$ and is

$$H(\text{div},\Omega) = \{\, \bar{q} \mid \bar{q} \in \left(L^2(\Omega)\right)^2 ;\ \text{div}\,\bar{q} \in L^2(\Omega)\,]\} \tag{A.23}$$

The following norm is associated with the space

$$\| \bar{q} \|_{H(\text{div},\Omega)} = \left[\, \| \bar{q} \|_0^2 + \| \text{div}\,\bar{q} \|_0^2 \,\right]^{1/2} \tag{A.24}$$

where

$$\| \bar{q} \|_0^2 = \| q_x \|_0^2 + \| q_y \|_0^2 \tag{A.25}$$

so that the space $H(\text{div},\Omega)$ is a Hilbert space [Raviart and Thomas 1977a], [Thomas and

Joly 1981; Ch. 1]. In an analogous way the $H(\mathrm{div},\Omega)$ space for vectors with three components over an open subset, $\Omega \in \mathbb{R}^3$, can be defined using the 3D operator defined in Chapter 2, (2.7b) [Nédélec 1980], [Thomas and Joly 1981; Ch. 1]. Some problems may require for its solution through a weak formation, vector functions belonging to spaces $H(\mathrm{div},\Omega)$, namely, vector functions whose components and divergence may not be continuous over Ω but should be square integrable.

A.1.5 *H*(curl,Ω) Spaces

Let $\bar{q}(x,y)$ be a vector variable with two components. Its curl may be defined as

$$\mathrm{curl}\ \bar{q} = \bar{\nabla} \times \bar{q} = \bar{a}_z \left(\frac{\partial q_y}{\partial x} - \frac{\partial q_x}{\partial y} \right) \tag{A.26a}$$

where × stands for the vector product between two vectors. In matrix notation

$$\{\nabla\} \times \begin{Bmatrix} q_x \\ q_y \\ 0 \end{Bmatrix} = \begin{Bmatrix} 0 \\ 0 \\ \dfrac{\partial q_y}{\partial x} - \dfrac{\partial q_x}{\partial y} \end{Bmatrix} \tag{A.26b}$$

Let $\bar{q}(x,y,z)$ be a vector variable with three components $q_x(x,y,z)$, $q_y(x,y,z)$, and $q_z(x,y,z)$ defined over an open subset $\Omega \in \mathbb{R}^3$. The curl of that vector is given by

$$\mathrm{curl}\ \bar{q} = \bar{\nabla} \times \bar{q} = \bar{a}_x \left(\frac{\partial q_z}{\partial y} - \frac{\partial q_y}{\partial z} \right) + \bar{a}_y \left(\frac{\partial q_x}{\partial z} - \frac{\partial q_z}{\partial x} \right) + \bar{a}_z \left(\frac{\partial q_y}{\partial x} - \frac{\partial q_x}{\partial y} \right) \tag{A.26c}$$

and in matrix notation it can be written as

$$\{\nabla\} \times \begin{pmatrix} q_x \\ q_y \\ q_z \end{pmatrix} = \begin{Bmatrix} \dfrac{\partial q_z}{\partial y} - \dfrac{\partial q_y}{\partial z} \\ \dfrac{\partial q_x}{\partial z} - \dfrac{\partial q_z}{\partial x} \\ \dfrac{\partial q_y}{\partial x} - \dfrac{\partial q_x}{\partial y} \end{Bmatrix} \tag{A.26d}$$

The linear space of vector functions $\bar{q}$ having square integrable components and square integrable curl is called $H(\text{curl},\Omega)$, according to

$$H(\text{curl},\Omega) = \{\bar{q} \mid \bar{q} \in \left(L^2(\Omega)\right)^2 ;\ \text{curl}\,\bar{q} \in L^2(\Omega)\} \tag{A.27a}$$

for the space of vectors with two components and

$$H(\text{curl},\Omega) = \{\bar{q} \mid \bar{q} \in \left(L^2(\Omega)\right)^3 , \text{curl}\,\bar{q} \in \left(L^2(\Omega)\right)^3\} \tag{A.27b}$$

for the space of vectors with three components.

For these spaces a norm can be defined by

$$\| \bar{q} \|_{H(\text{curl},\Omega)} = \left[\| \bar{q} \|_0^2 + \| \text{curl}\,\bar{q} \|_0^2 \right]^{1/2} \tag{A.28}$$

so that they are Hilbert spaces [Nédélec 1980].

Some problems defined through a weak formulation, may require for a solution, vector functions belonging to spaces $H(\text{curl},\Omega)$, namely, vector functions whose components and curl may not be continuous over Ω but should be square integrable.

A.1.6 $H^1(\Omega)\times H(\text{curl},\Omega)$ Space

Let $\bar{q}(x,y)$ be a vector variable with three components $q_x(x,y)$, $q_y(x,y)$, and $q_z(x,y)$ defined over a two-dimensional open subset $\Omega \in \mathbb{R}^2$, so that the vector variable $\bar{q}_t(x,y) = \bar{a}_x\, q_x(x,y) + \bar{a}_y\, q_y(x,y)$ belongs to the space $H(\text{curl},\Omega)$ in two dimensions and the scalar variable $q_z(x,y)$ belongs to the space $H^1(\Omega)$. Then it is said that the variable $\bar{q}(x,y)$ belongs to the product space $H^1(\Omega)\times H(\text{curl},\Omega)$:

$$\begin{aligned} H^1(\Omega)\times H(\text{curl},\Omega) = \{\bar{q} \mid D^\alpha q_z \in L^2(\Omega), \forall\ 0 \le \alpha \le 1 ; \\ \bar{q}_t \in \left(L^2(\Omega)\right)^2 ;\ \text{curl}\,\bar{q}_t \in L^2(\Omega)\} \end{aligned} \tag{A.29}$$

This space is provided with the corresponding norms according to the previous paragraphs. Some problems analyzed in this book through weak formulations require for their solutions vector functions belonging to the product space $H^1(\Omega)\times H(\text{curl},\Omega)$.

A.2 WEAK INTEGRAL FORMULATIONS BY THE WEIGHTED RESIDUAL METHOD: INTEGRATION BY PARTS. ESSENTIAL AND NATURAL BOUNDARY CONDITIONS

In Section 2.3.3.1, the residual $\{R(\{u\})\}$ in Ω corresponding to the boundary value problems defined by the partial differential expressions (2.1) for a deterministic problem or (2.2) for an eigenvalue problem is defined as

$$\{R(\{u\})\} = [\mathfrak{L}]\{u\} - \{f_\Omega\} \tag{A.30a}$$

or

$$\{R(\{u\})\} = [\mathfrak{L}]_1\{u\} - \lambda[\mathfrak{L}]_2\{u\} \tag{A.30b}$$

respectively. An integral form may be obtained weighting (A.30) with a vector of test functions, $\{v\}$, and integrating the product over the domain, Ω, to yield

$$W(\{u\},\{v\}) = \int_\Omega \{v\}^T\{R(\{u\})\}\, \mathrm{d}\Omega \tag{A.31a}$$

obtaining

$$\begin{aligned} W(\{u\},\{v\}) &= \int_\Omega \{v\}^T[\mathfrak{L}]\{u\}\, \mathrm{d}\Omega - \int_\Omega \{v\}^T\{f_\Omega\}\, \mathrm{d}\Omega \\ &= A(\{u\},\{v\}) - L(\{v\}) \end{aligned} \tag{A.31b}$$

for the deterministic problem (2.1) of Chapter 2, and

$$\begin{aligned} W(\{u\},\{v\}) &= \int_\Omega \{v\}^T[\mathfrak{L}_1]\{u\}\, \mathrm{d}\Omega - \lambda\int_\Omega \{v\}^T[\mathfrak{L}_2]\{u\}\, \mathrm{d}\Omega \\ &= A_1(\{u\},\{v\}) - \lambda A_2(\{u\},\{v\}) \end{aligned} \tag{A.31c}$$

for the eigenvalue problem (2.2) of Chapter 2. Here $A(\{u\},\{v\})$, $A_1(\{u\},\{v\})$, and $A_2(\{u\},\{v\})$ are bilinear forms and $L(\{v\})$ is a linear form (see Appendix A.1.2).

The weighted residuals method provides a weak formulation whose solution would be a set of functions $\{u\}$ such that the following expression will hold:

$$W(\{u\},\{v\}) = A(\{u\},\{v\}) - L\{v\} = 0 \tag{A.32a}$$

for the deterministic problem, (2.1) of Chapter 2, or

$$W(\{u\},\{v\}) = A_1(\{u\},\{v\}) - \lambda A_2(\{u\},\{v\}) = 0 \tag{A.32b}$$

for the case of the eigenvalue problem (2.2) of Chapter 2, for any weighting function $\{v\}$ belonging to a given set V. Functions $\{u\}$ should belong to the set of functions satisfying the boundary conditions (2.1b) or (2.2b) of Chapter 2 and having derivatives up to order m, where m is the order of the differential operators $[\mathcal{L}]$, $[\mathcal{L}_1]$, and $[\mathcal{L}_2]$ so that the integrability of (A.31b) and (A.31c) is ensured.

In Section 2.3.3.1 it is explained in which sense the expressions (A.32) constitute weak formulations for (2.1) or (2.2) of Chapter 2. In order to obtain even weaker formulations, the integration by parts may be employed. For scalar functions in a two-dimensional domain Ω defined by the boundary Γ, the following expressions are of interest:

$$\int_\Omega \psi(\bar{\nabla}_t \phi)\, d\Omega = -\int_\Omega (\bar{\nabla}_t \psi)\,\phi\, d\Omega + \oint_\Gamma \psi\,\phi\,\bar{a}_n\, d\Gamma \tag{A.33a}$$

$$\int_\Omega \psi(\bar{\nabla}_t \cdot \underline{\underline{a}}\bar{\nabla}_t \phi)\, d\Omega = -\int_\Omega (\bar{\nabla}_t \psi \cdot \underline{\underline{a}}\bar{\nabla}_t \phi)\, d\Omega + \oint_\Gamma \psi\,(\bar{a}_n \cdot \underline{\underline{a}}\bar{\nabla}_t \phi)\, d\Gamma \tag{A.33b}$$

where ψ is a scalar weighting (or test) function and ϕ is a scalar unknown function, both having first- order derivatives. Here $\bar{a}_n = a_{nx}\bar{a}_x + a_{ny}\bar{a}_y$ is the unit vector normal to $\Gamma(x,y)$ directed toward the outside of Ω and $\underline{\underline{a}}$ is a dyadic. Expressions (A.33a) and (A.33b) are also valid for scalar functions in a three-dimensional domain Ω, having a boundary Γ. For the three-dimensional case, the two-dimensional gradient operator $\bar{\nabla}_t$ is replaced by the three-dimensional operator, $\bar{\nabla}$ (see (2.7b)). Expression (A.33b) is a generalization of the first scalar Green's theorem. In its original formulation $\underline{\underline{a}}$ is a scalar equal to unity.

From (A.33a) and (A.33b) it may be concluded that an integration by parts provides a new integral form in which the order of the derivatives of $\{\phi\}$ will be reduced. Thus, the regularity conditions on $\{\phi\}$ will be weaker. At the same time, terms containing some of the boundary conditions will appear. Actually the conormal derivative (see (2.9a)) with respect to the gradient operator appears in the boundary integral term of (A.33b). That term may be split into two parts. The first one, extended to the boundaries having Dirichlet boundary conditions where the test function ψ is chosen to be zero. Hence it will be identically zero. The second one is extended to those parts of the boundary having Neumann boundary conditions. Here the value of the conormal derivative is known (see (2.11a)). Thus, that integral involves only known values and the known test functions. The unknown variable is not directly involved. Hence, it is not required for the unknown variable to exactly satisfy the boundary conditions of the Neumann type because they are taken into account in a weak way through the boundary integral form.

Another consequence of the integration by parts is that the integral over the domain includes derivatives of the weighting function, so its regularity conditions are now stronger. Finally, it should be noticed that the unknown u must be explicitly enforced to fulfill the prescribed Dirichlet conditions (see (2.10a)).

In 2D problems where vector and scalar variables are involved (as in the case of the mixed formulation for the quasi-static analysis of transmission lines, dealt with in Section 3.2.3.3 and Appendix E.3) the following expression is useful:

$$\int_\Omega (\bar{s}\cdot\bar{\nabla}_t\varphi)\,\mathrm{d}\Omega = -\int_\Omega \varphi\,(\bar{\nabla}_t\cdot\bar{s})\,\mathrm{d}\Omega + \oint_\Gamma \varphi\,(\bar{a}_n\cdot\bar{s})\,\mathrm{d}\Gamma \tag{A.33c}$$

where $\bar{s}(x,y)$ is a vector function with two components and $\varphi(x,y)$ is a scalar function. From (A.33c) it is clear that also in this case the regularity conditions of φ have been weakened, although not those of $\bar{s}$, which now requires integrable divergence.

For problems involving vectors with three components (as in the case of the formulation by means of the transverse and the longitudinal components of either the electric field or the magnetic field for the full wave analysis of wave-guiding structures, dealt with in Section 3.3.3.2 and Appendix F.3, or in the cases of cavity, discontinuity, or 3D scattering and antenna problems, dealt with in Sections 7.3, 7.4, and 7.5, respectively), the following expressions are of interest:

$$\int_\Omega (\bar{v}_1\cdot(\bar{\nabla}\times\bar{v}_2))\,\mathrm{d}\Omega = \int_\Omega ((\bar{\nabla}\times\bar{v}_1)\cdot\bar{v}_2)\,\mathrm{d}\Omega + \oint_\Gamma ((\bar{v}_1\times\bar{a}_n)\cdot\bar{v}_2)\,\mathrm{d}\Gamma \tag{A.33d}$$

$$\int_\Omega (\bar{v}_1\cdot\bar{\nabla}\times\underline{\underline{a}}\,\bar{\nabla}\times\bar{v}_2)\,\mathrm{d}\Omega = \int_\Omega (\bar{\nabla}\times\bar{v}_1\cdot\underline{\underline{a}}\,\bar{\nabla}\times\bar{v}_2)\,\mathrm{d}\Omega - \oint_\Gamma (\bar{a}_n\cdot\bar{v}_1\times\underline{\underline{a}}\,\bar{\nabla}\times v_2)\,\mathrm{d}\Gamma \tag{A.33e}$$

where $\bar{v}_1(x,y,z)$ and $\bar{v}_2(x,y,z)$ are vector functions with three components and $\underline{\underline{a}}$ is a dyadic. It is evident that those expressions lead to weaker derivative conditions for $\bar{v}_2$. Expression (A.33e) is a generalization of the first vector Green's theorem. In its original formulation $\underline{\underline{a}}$ is a scalar equal to unity.

As an example, consider a 2D problem formulated by means of second-order partial differential equations according to expressions (2.6a), (2.8), (2.10a), and (2.11a) of Chapter 2. After an integration by parts, a formulation is obtained in which the bilinear forms would have a general expression given by

$$A_{kl}(v_n,u_m) = \int_\Omega (\bar{\nabla}_t v_n\cdot\underline{\underline{a}}\,\bar{\nabla}_t u_m + v_n b u_m)\,\mathrm{d}\Omega - \int_\Gamma v_n\frac{\partial u_m}{\partial n_{\mathcal{L}_{kl}}}\,\mathrm{d}\Gamma \tag{A.34a}$$

and the linear forms are given by

$$L_{k_\Omega}(v_n) = \int_\Omega v_n\,f_m\,\mathrm{d}\Omega \tag{A.34b}$$

As a consequence:

1. Functions $\{u\}$ and $\{v\}$ should have first derivatives. Thus, polynomial functions (even of the first degree) are admissible as basis functions and test functions.
2. In order to perform the integration, it is sufficient if $\{u\}$ and $\{v\}$ are continuous functions with discontinuous first derivatives. Thus C^0 continuity may be sought. Actually the only requirement is that $\{u\}$ and $\{v\}$ as well as their first derivatives be square integrable. In summary, $\{u\}$ and $\{v\}$ may belong to the first order Hilbert space $H^1(\Omega)$ (see Appendix A.1.3).
3. Functions $\{u\}$ must exactly fulfill the Dirichlet conditions on the boundaries of this type. These conditions are called essential. They must be exactly enforced.
4. Functions $\{v\}$ may be selected in a way that fulfills the homogeneous Dirichlet conditions on the boundaries, where $\{u\}$ has essential conditions. Hence, the boundary integral that appears in (A.34a) can be extended only to the boundary Γ_N, on which $\{u\}$ has Neumann conditions. That boundary integral will be reduced to a linear form involving only known quantities and functions, namely, the natural conditions and the test functions, respectively,

$$\int_\Gamma \left(\frac{\partial u_m}{\partial n_{\mathfrak{L}_{kl}}} \right) v_n \, \mathrm{d}\Gamma = \int_{\Gamma_N} \left(\frac{\partial u_m}{\partial n_{\mathfrak{L}_{kl}}} \right) v_n \, \mathrm{d}\Gamma = \int_{\Gamma_N} g_n v_n \, \mathrm{d}\Gamma = L_{k_\Gamma}(v_n) \tag{A.34c}$$

5. It is not necessary for $\{u\}$ to explicitly fulfill the boundary conditions of the Neumann type because those conditions are taken into account in the weak integral form as shown in (A.34a) and (A.34c). Those conditions are called natural ones. They will be satisfied in a weak or distributional sense.

A.3 AN OVERVIEW OF VARIATIONAL CALCULUS

A.3.1 Variational Principles: Properties

Let a functional be given by the general expression

$$\begin{aligned} F(\{u\}) &= F\left(\{u\}, \left\{\frac{\partial u}{\partial x}\right\}, \left\{\frac{\partial u}{\partial y}\right\}, \left\{\frac{\partial u}{\partial z}\right\}, \ldots\right) \\ &= \int_\Omega f\left(\{u\}, \left\{\frac{\partial u}{\partial x}\right\}, \left\{\frac{\partial u}{\partial y}\right\}, \left\{\frac{\partial u}{\partial z}\right\}, \ldots\right) \mathrm{d}\Omega + \int_\Gamma g(\{u\}, \ldots)\, \mathrm{d}\Gamma \end{aligned} \tag{A.35}$$

where $\{u\}$ may be a set of variables in two, $\{u\}=\{u(x,y)\}$, or three dimensions, $\{u\}=\{u(x,y,z)\}$. The expression (A.35) implies that for a given value of the independent variables (x,y) or (x,y,z) the functional F is a function of $\{u\}$ and of its partial derivatives. Thus a functional is a function of functions.

A variation in $\{u\}$ is represented by

$$\delta\{u\} = \{\delta u\} = \{\gamma v\} \tag{A.36}$$

where the coefficients, γ, are constants. The variation δu of a function u represents an admissible modification of the function $u(x,y)$ or $u(x,y,z)$ for fixed values of (x,y) or (x,y,z). If u is specified at a given point (as in the case of points belonging to boundaries with conditions of the Dirichlet type), δu would be zero at that point because u must not change. For any other point the variation δu would be arbitrary, as shown in (A.36) [Hildebrand 1965; Ch. 2]. Associated with the variation $\{\delta u\}$ in $\{u\}$ is the variation in F, δF, that may be written as

$$\begin{aligned} \delta F &= \left\{\frac{\partial F}{\partial u}\right\}^T \{\delta u\} + \left\{\frac{\partial F}{\partial\left(\dfrac{\partial u}{\partial x}\right)}\right\}^T \left\{\delta\left(\frac{\partial u}{\partial x}\right)\right\} + \left\{\frac{\partial F}{\partial\left(\dfrac{\partial u}{\partial y}\right)}\right\}^T \left\{\delta\left(\frac{\partial u}{\partial y}\right)\right\} + \cdots \\ &= \{\delta u\}^T \left\{\frac{\partial F}{\partial u}\right\} + \left\{\delta\left(\frac{\partial u}{\partial x}\right)\right\}^T \left\{\frac{\partial F}{\partial\left(\dfrac{\partial u}{\partial x}\right)}\right\} + \left\{\delta\left(\frac{\partial u}{\partial y}\right)\right\}^T \left\{\frac{\partial F}{\partial\left(\dfrac{\partial u}{\partial y}\right)}\right\} + \cdots \end{aligned} \tag{A.37}$$

where

$$\left.\begin{aligned} \delta\left(\frac{\partial u}{\partial x}\right) &= \frac{\partial \delta u}{\partial x} \\ \delta\left(\frac{\partial u}{\partial y}\right) &= \frac{\partial \delta u}{\partial y} \\ \delta\left(\frac{\partial u}{\partial y}\right) &= \frac{\partial \delta u}{\partial z} \end{aligned}\right\} \tag{A.38a}$$

$$\left.\begin{aligned} \delta\int_\Omega u \, d\Omega &= \int_\Omega \delta u \, d\Omega \\ \delta\int_\Gamma u \, d\Gamma &= \int_\Gamma \delta u \, d\Gamma \end{aligned}\right\} \tag{A.38b}$$

So from (A.35) the following expression can be obtained:

$$\delta F = \int_\Omega \left\{ l\left(\{u\}, \left\{\frac{\partial u}{\partial x}\right\}, \left\{\frac{\partial u}{\partial y}\right\}, \cdots \right)\right\}^T \{\delta u\}\, \mathrm{d}\Omega$$
$$+ \int_\Gamma \left\{ m\left(\{u\}, \left\{\frac{\partial u}{\partial x}\right\}, \left\{\frac{\partial u}{\partial y}\right\}, \cdots \right)\right\}^T \{\delta u\}\, \mathrm{d}\Gamma \tag{A.39}$$

where $\{l\}$ and $\{m\}$ are differential operators of the type described in expressions (2.1a) or (2.2a) and (2.1b) or (2.2b) of Chapter 2, respectively. Expression (A.39) may be transformed if desired via integration by parts using (A.33).

The condition (2.16) of Chapter 2

$$\delta F(\{u\}) = 0 \tag{A.40}$$

refers to the stationarity principle formulated in Section 2.3.3.2, so it may be written as

$$\delta F = \int_\Omega \left\{ l\left(\{u\}, \left\{\frac{\partial u}{\partial x}\right\}, \left\{\frac{\partial u}{\partial y}\right\}, \ldots \right)\right\}^T \{\delta u\}\, \mathrm{d}\Omega$$
$$+ \int_\Gamma \left\{ m\left(\{u\}, \left\{\frac{\partial u}{\partial x}, \right\}\left\{\frac{\partial u}{\partial y}\right\}, \ldots \right)\right\}^T \{\delta u\}\, \mathrm{d}\Gamma = 0 \tag{A.41}$$

Since it must be fulfilled for any variation $\{\delta u\}$, it implies

$$\left\{ l\left(\{u\}, \left\{\frac{\partial u}{\partial x}\right\}, \left\{\frac{\partial u}{\partial y}\right\}, \ldots \right)\right\} = \{0\}, \quad \text{in } \Omega \tag{A.42a}$$

$$\left\{ m\left(\{u\}, \left\{\frac{\partial u}{\partial x}\right\}, \left\{\frac{\partial u}{\partial y}\right\}, \ldots \right)\right\} = \{0\}, \quad \text{on } \Gamma \tag{A.42b}$$

Thus, expressions (A.42) represent the partial differential equations or Euler equations corresponding to the functional (A.35). Because of this, the functional is also called the variational principle of the set of equations (A.42). If the set (A.42) coincides with (2.1) or (2.2) of Chapter 2 (i.e., with the expressions corresponding to the strong formulation of the problem), it is said that the variational principle is a natural one.

Given a functional, it is always possible to find the Euler equations that can be derived from it. The inverse is not always true. In fact, given a system of partial differential equations like (2.1) or (2.2) of Chapter 2, it is not always possible to find a functional from which it is derived.

For a linear system of partial differential linear equations, as those represented by (2.1) or (2.2) of Chapter 2, it may be demonstrated that a symmetric variational principle may be constructed if the operators $[\mathcal{L}]$ (or $[\mathcal{L}_1]$ and $[\mathcal{L}_2]$) are self-adjoint or symmetric [Dhatt and Touzot 1981; Ch. 3], [Jin 1993; Ch. 6], [Reddy 1984, 1993; Ch. 2], [Wexler 1969], [Zienkiewicz 1977; Ch. 3]. If that condition is fulfilled, like in the case of the problems that are considered in this book, a weak formulation of the weighted residual type as given in (2.15) of Chapter 2 or its equivalent (A.32), may be obtained. In those expressions the bilinear forms A, A_1, and A_2 are symmetric. Then it can be demonstrated that the corresponding variational principles are given by

$$F(\{u\}) = \frac{1}{2} A(\{u\},\{u\}) - L(\{u\}) \tag{A.43a}$$

for the deterministic problem (2.1) of Chapter 2 or

$$F(\{u\}) = \frac{1}{2}\left(A_1(\{u\},\{u\}) - \lambda A_2(\{u\},\{u\})\right) \tag{A.43b}$$

for the eigenvalue problem (2.2) of Chapter 2.

In many cases, the variational principle will be established through physical laws so that (2.1) or (2.2) of Chapter 2 defining the problem are the Euler equations corresponding to the stationarity of the natural variational principle [Berk 1956], [Jin 1993; Ch. 6], [Konrad 1976], [Morishita and Kumagai 1977].

In other cases, variational principles may be obtained through nonself-adjoint operators by modifying (2.1) or (2.2) of Chapter 2 without essentially transforming them [Zienkiewicz 1977; Ch. 3].

A functional is said to be quadratic, if it is quadratic in functions $\{u\}$ and their derivatives. Sometimes this nomenclature is used to refer to a functional with a linear part and a quadratic part. A pure quadratic functional may be expressed by

$$F = \frac{1}{2} \int_\Omega \begin{Bmatrix} \{u\} \\ \left\{\dfrac{\partial u}{\partial x}\right\} \\ \vdots \end{Bmatrix}^T [G] \begin{Bmatrix} \{u\} \\ \left\{\dfrac{\partial u}{\partial x}\right\} \\ \vdots \end{Bmatrix} \mathrm{d}\Omega \tag{A.44}$$

where matrix $[G]$ is symmetric and does not depend on $\{u\}$. This functional is positive definite when matrix $[G]$ is positive definite. This holds when all of its eigenvalues are positive or if the following condition holds: $\{v\}^T[G]\{v\} > 0$, $\{v\}\neq\{0\}$, for $\{v\}$ real. Once the stationarity of the functional is established ($\delta F = 0$), it immediately follows that the second variation $\delta^2 F$ will be positive. This means that the solution that makes stationary a positive definite quadratic functional gives a minimum of that functional,

F, and will give a maximum of the complementary functional [Hildebrand 1965; Ch.1], [Dhatt and Touzot 1981; Ch. 3].

A.3.2 Generalized, Complementary, and Mixed Variational Principles by Means of Lagrange Multipliers

The method of Lagrange multipliers is one of the tools that may be employed to obtain alternate variational principles. The method provides weak formulations with interesting features like, for example, the introduction of new unknowns, the establishment of less stringent regularity conditions for some of the unknowns, and the relaxation of the boundary conditions of given variables [Dhatt and Touzot 1981; Ch. 3], [Fremond 1973], [Zienkiewicz 1977; Ch. 3], [Zienkiewicz and Taylor 1989; Chs. 12 , 13].

Assume a problem for which a solution is obtained at the stationary point of a functional $F(\{u\},\{q\})$. Assume that $\{u\}$ is the set of primary variables and $\{q\}$ is the set of secondary variables (see Section 2.3.3.2). The functional depends on both of them. They are related by m differential expressions g_i, i=1,..., m, according to

$$\{g(\{u\},\{q\})\} = \{0\} \tag{A.45}$$

Expression (A.45) imposes m conditions on the solution $(\{u\},\{q\})$. A way to obtain a variational principle without any condition would be to eliminate the variable $\{q\}$, leading to a direct variational principle $F_u(\{u\})$. It is also called the direct functional, from which a direct formulation will be obtained. This functional depends only on the primary variables. The expression for its stationarity is given by (2.16) of Chapter 2.

The method of Lagrange multipliers introduces m parameters (or multipliers), γ_1,..., γ_m (in vector notation, $\{\gamma\}$), to obtain a generalized variational principle:

$$F^g(\{u\},\{q\},\{\gamma\}) = F(\{u\},\{q\}) + \int \{\gamma\}^T\{g(\{u\},\{q\})\}\,\mathrm{d}\Omega \tag{A.46}$$

The stationarity conditions would now be written as:

$$\delta F^g = \{\delta u\}^T\left\{\frac{\partial F^g}{\partial u}\right\} + \{\delta q\}^T\left\{\frac{\partial F^g}{\partial q}\right\} + \{\delta \gamma\}^T\left\{\frac{\partial F^g}{\partial \gamma}\right\} = 0 \tag{A.47a}$$

which implies the following sets of equations:

$$\left\{\frac{\partial F^g}{\partial u}\right\} = \{0\} \quad ; \quad \left\{\frac{\partial F^g}{\partial q}\right\} = \{0\} \tag{A.47b}$$

together with

$$\left\{ \frac{\partial F^{g}}{\partial \gamma} \right\} = \{0\} \tag{A.47c}$$

This last set of equations assumes that condition (A.45) is satisfied. By using this procedure the number of unknowns has been increased from $(\{u\},\{q\})$ to $(\{u\},\{q\},\{\gamma\})$. The Finite Element Methods that are derived from functionals obtained through these procedures are usually called hybrid methods. In these formulations primary variables and secondary variables along with one or more Lagrange multipliers are involved. Depending on the problem, the Lagrange multipliers would be identified with the value (or trace) of some of the variables at the interfaces between elements or at the boundaries. This method allows the relaxation of the continuity conditions of the unknown variables [Fremond 1973], [Henshell 1973], [Kardestuncer and Norrie 1987; P. 2, Ch. 9], [Raviart and Thomas 1977b], [Thomas and Joly 1981; Ch. 4].

If the Lagrange multipliers are eliminated from the functional (A.46) by means of the relations (A.47c), a new functional will be obtained in which only the variables $\{u\}$ and $\{q\}$ will be involved. This new functional is called a mixed one, $F_m(\{u\},\{q\})$, for which the stationarity condition may be written as [Kardestuncer and Norrie 1987; P. 2, Ch. 9]:

$$\delta F_m(\{u\},\{q\}) = 0 \tag{A.48a}$$

leading to the following sets of equations:

$$\left\{ \frac{\partial F_m}{\partial u} \right\} = \{0\} \quad ; \quad \left\{ \frac{\partial F_m}{\partial q} \right\} = \{0\} \tag{A.48b}$$

Finally, if $\{u\}$ is eliminated through the conditions imposed on $\{q\}$, a complementary functional of $F_u(\{u\})$ is obtained that may be called $F_q(\{q\})$. In the functional $F_q(\{q\})$ only the secondary variables, $\{q\}$, would appear. This is called the complementary formulation. Then the stationarity condition will be

$$\delta F_q(\{q\}) = 0 \tag{A.49a}$$

leading to

$$\left\{ \frac{\partial F_q}{\partial q} \right\} = \{0\} \tag{A.49b}$$

Appendix B

Definitions of Convergence

B.1 TYPES OF CONVERGENCE

Consider a continuous problem of the type (2.1) or (2.2) of Chapter 2 for the unknown u that may be a real or complex, scalar or vector function. Also assume that the problem is discretized by utilizing a variational method. For a number of degrees of freedom, N_d (which is synonymous with the number of unknowns), consider that the approximate solution is u_{Nd}. It is said that the approximate solution converges to the exact one, u, if for a given $\varepsilon > 0$ a number of degrees of freedom N_d may be found such that the magnitude of the error will be less than ε. Depending on the way in which the error is measured, the following types of convergence may be defined.

1. Uniform convergence:

$$|u - u_{N_d}| < \varepsilon \quad \text{for a fixed } \varepsilon \text{ independent of } N_d \tag{B.1}$$

   where the definition of the absolute value $|\cdot|$ depends on whether u is a real or complex, scalar or vector function (see Appendix A.1.3, (A.16b) to (A.16d)). This type of convergence implies that u_{Nd} approximates u at all points in the domain.

2. Convergence in the norm:

$$\|u - u_{N_d}\|_m < \varepsilon \tag{B.2}$$

   or in the seminorm:

$$|u - u_{N_d}|_m < \varepsilon \tag{B.3}$$

   for scalar variables belonging to the space $H^m(\Omega)$ (see Appendix A.1.3 and the definitions for the norm and seminorm in (A.19d) and (A.19e), respectively); or

$$\|\bar{u}-\bar{u}_{N_d}\|_0 < \varepsilon \tag{B.4}$$

together with

$$\|\bar{\nabla}_t \cdot \bar{u} - \bar{\nabla}_t \cdot \bar{u}_{N_d}\|_0 < \varepsilon \tag{B.5a}$$

for the case of two-dimensional (2D) vector variables belonging to the space $H(\text{div},\Omega)$ (see Appendix A.1.4); or with

$$\|\bar{\nabla}_t \times \bar{u} - \bar{\nabla}_t \times \bar{u}_{N_d}\|_0 < \varepsilon \tag{B.5b}$$

for the case of 2D vector variables belonging to the space $H(\text{curl},\Omega)$ (see Appendix A.1.5). If three-dimensional (3D) vector variables are considered the 3D vector operator given by (2.7b) of Chapter 2 should be used in (B.5).

In any case, the convergence is strong or the convergence is in the mean, if the following expression holds:

$$\|u-u_{N_d}\|_0 < \varepsilon \tag{B.6}$$

where u may be a real or complex, scalar or vector function. This implies that although u_{Nd} is not approximating u at every point of the domain, the region in which they differ can be made arbitrarily small when N_d is increased.

3. Convergence in energy:

$$\left\langle (u-u_{N_d})\,;\,\mathcal{L}(u-u_{N_d})\right\rangle^{1/2} < \varepsilon \tag{B.7}$$

where $<\cdot\,;\,\cdot>$ stands for the usual inner product defined in (A.13) and $\mathcal{L}$ is the differential operator of expression (2.1) or the differential operator $\mathcal{L}_1 - \lambda\mathcal{L}_2$ of expression (2.2) of Chapter 2. For the problems considered, the left-hand side of (B.7) represents the energy of the error. The convergence in energy is only a combination or/and a particular case of (B.2) and (B.3).

4. Weak convergence:

$$\left\langle v\,;\,u-u_{N_d}\right\rangle^{1/2} < \varepsilon \tag{B.8}$$

for any v belonging to the space of test functions.

5. Strong convergence of the residual:

$$\| R(u_{N_d}) \|_0 = \| R(u - u_{N_d}) \|_0 < \varepsilon \tag{B.9}$$

where $R(u)$ is the residual corresponding to the expressions (2.1) or (2.2) of Chapter 2 as given by (2.12). Expression (B.9) makes use of the linearity of the differential operators involved in (2.1) or (2.2) of Chapter 2. It also considers that by definition $R(u) = 0$.

6. Weak convergence of the residual:

$$\left\langle v \, ; \, R(u_{N_d}) \right\rangle^{1/2} = \left\langle v \, ; \, R(u - u_{N_d}) \right\rangle^{1/2} < \varepsilon \tag{B.10}$$

It should be mentioned that if, for example, the objective of a problem is to compute accurately the energy it would be enough to obtain a convergence of that type, namely, that defined by (B.7), and it would not be necessary to enforce a convergence of type (B.6) or (B.9) [Sarkar 1983].

B.2 SOME GENERAL CONCLUSIONS REGARDING CONVERGENCE

The weighted residual method, or the method of moments (MOM), only ensures the weak convergence of the residuals [Sarkar 1985]. If certain conditions are fulfilled, the Galerkin method may provide strong convergence of the residual and of the solution [Raviart and Thomas 1983; Ch. 3], [Sarkar 1983]. In general, this method offers better results than the method of point matching or collocation [Sarkar 1985], [Kardestuncer and Norrie 1987; P. 2, Ch. 6].

The Ritz method, applied to a functional from which the classical formulation is derived, provides weak convergence of the residual. Strong convergence of the solution is ensured if the operator $[\mathcal{L}]$ of expression (2.1) or $[\mathcal{L}_1]$ and $[\mathcal{L}_2]$ of expression (2.2) are definite (either positive or negative). Strong convergence of the residual is not ensured in general. Weak convergence of the residual will be obtained if $[\mathcal{L}]\{P_{Nd}\}$ is bounded, where $\{P_{Nd}\}$ stands for the basis functions, corresponding to a number of degrees of freedom N_d [Sarkar 1983]. Under given conditions, strong convergence of the residuals may also be obtained.

The least squares method always ensures strong convergence both of the residuals and of the solution if the operator is definite. If convergence in energy of this method is compared with the convergence in energy obtained by means of the Ritz method for the same set of basis functions, it may be observed that the last method converges more rapidly. It may also be pointed out that the least squares method requires more CPU time [Sarkar 1983].

The conjugate gradient method ensures strong convergence of the solution when the operator is definite. It also provides strong convergence of the residual. It is quite efficient. It requires the computation of the basis functions at each iteration [Sarkar 1985].

Appendix C

Topics Related to Finite Elements

C.1 MAPPING BETWEEN PARENT FINITE ELEMENTS AND REAL FINITE ELEMENTS: PROPERTIES

In Section 2.5.1.2 the concept of parent element (also called, in the literature, master element or reference element) is introduced. Some important properties of the geometric transformations that generate real elements from a master element are summarized.

A parent finite element is given by the triplet:

1. A compact region of $\mathbb{R}^n$, called $\hat{\Omega}$, where n stands for the dimension of the element: 1 for one-dimensional (1D) elements, 2 for two-dimensional (2D) elements, or 3 for three-dimensional (3D) elements. A point of $\hat{\Omega}$ will be given by its coordinates, $(\hat{x})$ for the 1D case, $(\hat{x},\hat{y})$ for the 2D case, or $(\hat{x},\hat{y},\hat{z})$ for the 3D case. In the following, to avoid the repetition of expressions for 1D, 2D, or 3D cases, the coordinates of a point of the parent element will be represented by its position vector, which in matrix notation is written as $\{\hat{r}\}$. The coefficients of the column matrix $\{\hat{r}\}$ are the coordinates of the point. Parent elements have simple geometries and do not have curved contours but straight edges and faces. The geometry of a parent element, $\hat{\Omega}$, is defined by a set of points (called geometric points, to distinguish them from nodes) that includes the vertices. In fact, being straight elements, the definition of parent elements only requires the specification of the coordinates of the vertices. However, in order to generate real elements with distorted geometries, additional geometric points are required. The position vector of the ith geometric point is called $\{\hat{r}_{pi}\}$. Tables 2.1 to 2.9 of Chapter 2 show the geometry of the parent elements utilized in this book.
2. A space of polynomial scalar or vector functions defined over $\hat{\Omega}$. Namely, for Lagrange elements, a space of polynomial scalar functions. This space will be called generically $\hat{P}$ and consists of the space $P_p^{(n)}$ for simplexes (triangular or tetrahedral elements) or the space $Q_p^{(n)}$ for rectangular or cubic elements (see Appendix A.1.1). Here p is the order of the space. The functions belonging to the space $\hat{P}$ will be called $\phi(\{\hat{r}\})$. For div-conforming and curl-conforming simplexes, a space of vector functions is the second element of the triplet

defining a master element. This space will be called generically $\hat{D}$. It consists of the space $D_p^{(n)}$ for div-conforming elements or the space $R_p^{(n)}$ for curl-conforming elements (see Appendix A.1.1). The vector functions belonging to the space $\hat{D}$ will be called, in matrix notation, $\{\hat{t}(\{\hat{r}\})\}$.

3. A set of degrees of freedom A and, eventually, a set of nodes, called $\hat{\Sigma}$, defined over $\hat{\Omega}$, to which the values of the degrees of freedom are associated. The position vector of the ith node is denoted by $\{\hat{r}_{ni}\}$. Namely, for Lagrange elements, a node is associated with each degree of freedom. For div-conforming or curl-conforming elements, sampling points (i.e., nodes in a broad sense) may be associated with some of the degrees of freedom while other degrees of freedom are related to the surface of the element for the 2D case or to the volume of the element for the 3D case. However, for practical reasons, points of $\hat{\Omega}$ are usually associated with these degrees of freedom.

A real finite element is defined by a triplet:

1. A compact region of $\mathbb{R}^n$, called Ω^e. A point of Ω^e will be given by its position vector $\{r\}$. Real elements may have shapes distorted from those of parent elements and curved contours, for example. Their geometry may be defined by a set of geometric points $\{r_{pi}\}$.
2. A space of polynomial scalar or vector functions defined over Ω^e. A space of polynomial scalar functions is required for Lagrange elements. It will be called P. It consists of one of the two spaces $P_p^{(n)}$ or $Q_p^{(n)}$. The functions belonging to the space P will be called $\varphi(\{r\})$. A space of vector functions is needed for div-conforming or curl-conforming elements. It will be called D. It consists of the mentioned spaces $D_p^{(n)}$ or $R_p^{(n)}$. The vector functions belonging to the space D will be called $\{t(\{r\})\}$.
3. A set of degrees of freedom, A, and a set of nodes, called Σ, defined over Ω^e, with which the degrees of freedom are associated. The position vector of the ith node is denoted by $\{r_{ni}\}$.

Let F^e be a bidirectional mapping from $\mathbb{R}^n$ into $\mathbb{R}^n$ so that a point $\{r\}$ of Ω^e will be mapped into a point $\{\hat{r}\}$ of $\hat{\Omega}$ and *vice versa*, according to

$$\{r\} = F^e(\{\hat{r}\}) \tag{C.1a}$$

and

$$\{\hat{r}\} = (F^e)^{-1}(\{r\}) \tag{C.1b}$$

Expression (C.1a) may be further written as

$$\{r\} = F^e(\{\hat{r}\}) = [J^e]\{\hat{r}\} + \{b\} \tag{C.2a}$$

where $[J^e]$ is the Jacobian matrix of the geometric transformation, given by

$$[J^e] = \begin{bmatrix} \dfrac{\partial x}{\partial \hat{x}} & \dfrac{\partial x}{\partial \hat{y}} \\[2ex] \dfrac{\partial y}{\partial \hat{x}} & \dfrac{\partial y}{\partial \hat{y}} \end{bmatrix} \quad \text{or} \quad [J^e] = \begin{bmatrix} \dfrac{\partial x}{\partial \hat{x}} & \dfrac{\partial x}{\partial \hat{y}} & \dfrac{\partial x}{\partial \hat{z}} \\[2ex] \dfrac{\partial y}{\partial \hat{x}} & \dfrac{\partial y}{\partial \hat{y}} & \dfrac{\partial y}{\partial \hat{z}} \\[2ex] \dfrac{\partial z}{\partial \hat{x}} & \dfrac{\partial z}{\partial \hat{y}} & \dfrac{\partial z}{\partial \hat{z}} \end{bmatrix} \tag{C.2b}$$

for 2D or 3D, respectively.[1] The coefficients of the column matrix $\{b\}$ are constants.

The elements ($\hat{\Omega}$, $\hat{P}$ or $\hat{D}$, $A(\hat{\Sigma})$) and (Ω^e, P or D, $A(\Sigma)$) are said to be equivalent if the following conditions are fulfilled:

1. $\Omega^e = F^e(\hat{\Omega})$, that is , $\{r_{pi}\} = F^e(\{\hat{r}_{pi}\})$; (C.3a)

2. The value of each degree of freedom is invariant;

3. $\Sigma = F^e(\hat{\Sigma})$, that is , $\{r_{ni}\} = F^e(\{\hat{r}_{ni}\})$; (C.3b)

4. $\varphi(r) = \hat{\varphi}(\hat{r})$; (C.3c)

for the transformation of scalar functions. In addition, in order to have invariant vector function spaces, the transformation of vectors must be done according to

$$\{t(\{r\})\} = \frac{1}{|J^e|}[J^e]\{\hat{t}(\{\hat{r}\})\} \tag{C.3d}$$

for div-conforming elements or

$$\{t(\{r\})\} = ([J^e]^T)^{-1}\{\hat{t}(\{\hat{r}\})\} \tag{C.3e}$$

---

1 Different authors give different definitions of $[J]$. Here we use the definition as used in [Johnson 1992; Ch. 12], [Kardestuncer and Norrie 1987; P. 2, Ch. 3], [Norrie and Vries 1978; Ch. 2], [Pearson 1974], [Raviart and Thomas 1983; Ch. 5]. Others define $[J]$ alternately, as the transpose of (C.2b) [Akin 1982; Ch. 5], [Bathe and Wilson 1976; Ch. 4], [Bathe 1982; Ch. 5], [Dhatt and Touzot 1981; Ch. 1], [Jin 1993; Ch. 8], [Reddy 1993; Ch. 9], [Szabó and Babuška 1991; Ch. 6].

for curl-conforming elements. Here $|J^e|$ is the determinant of the Jacobian matrix.

Expressions (C.1) to (C.3) imply:

1. For F^e to be invertible, $|J^e|$ must not be zero at any point belonging to the element Ω^e.
2. To each vertex or geometric point of the element Ω^e there should correspond a vertex or a geometric point of the element $\hat{\Omega}$ in a pre-established order and vice versa.
3. To each portion of the boundary of Ω^e, Γ^e, defined by its vertices and geometric points, there should correspond a portion of the boundary $\hat{\Gamma}$ of $\hat{\Omega}$ defined by the corresponding vertices and geometric points of $\hat{\Omega}$ and vice versa.
4. To each node of the set Σ there should correspond a node of the set $\hat{\Sigma}$ in a pre-established order and vice versa.
5. To each scalar function $\hat{\phi}(\hat{r})$ defined over $\hat{\Omega}$ there should correspond a scalar function $\phi(x)$ over Ω^e according to

$$\phi(\{r\}) = \hat{\phi}(\{\hat{r}\}) \tag{C.4}$$

with $\{\hat{r}\}$ given by (C.1b).

6. From the Lagrange basis function $\hat{N}_j(\{\hat{r}\})$ corresponding to the jth degree of freedom or node of $\hat{\Omega}$, the basis function $N_j(\{r\})$ corresponding to the jth degree of freedom or node of Ω^e may be obtained according to the following transformation:

$$N_j(\{r\}) = \hat{N}_j(\{\hat{r}\}) \ , \quad j=1,\ldots,n_e \tag{C.5}$$

where n_e is number of degrees of freedom for the element and $\{\hat{r}\}$ is given by (C.1b).

7. To each vector function $\{\hat{t}(\{\hat{r}\})\}$ defined over a master div-conforming element there corresponds a vector function $\{t(\{r\})\}$ over Ω^e so that

$$\{t(\{r\})\} = \frac{1}{|J^e|}[J^e]\{\hat{t}(\{\hat{r}\})\} \tag{C.6}$$

with $\{\hat{r}\}$ given by (C.1b).

8. From the vector basis function $\{\hat{N}_j(\{\hat{r}\})\}$ of a div-conforming parent element corresponding to the jth degree of freedom, the jth vector basis

function $\{N_j(\{r\})\}$ of Ω^e may be computed through the following transformation:

$$\{N_j(\{r\})\} = \frac{1}{|J^e|}[J^e]\{\hat{N}_j(\{\hat{r}\})\}\ , \quad j=1,\ldots,n_e \tag{C.7}$$

with $\{\hat{r}\}$ given by (C.1b).

9. To each vector function $\{\hat{t}(\{\hat{r}\})\}$ defined over a curl-conforming parent element there corresponds a vector function $\{t(\{r\})\}$ over Ω^e so that

$$\{t(\{r\})\} = ([J^e]^T)^{-1}\{\hat{t}(\{\hat{r}\})\} \tag{C.8}$$

with $\{\hat{r}\}$ given by (C.1b).

10. From the vector basis function $\{\hat{N}_j(\{\hat{r}\})\}$ of a curl-conforming parent element corresponding to the jth degree of freedom, the jth vector basis function $\{N_j(\{r\})\}$ of Ω^e may be computed through the following transformation:

$$\{N_j(\{r\})\} = ([J^e]^T)^{-1}\{\hat{N}_j(\{\hat{r}\})\}\ , \quad j=1,\ldots,n_e \tag{C.9}$$

with $\{\hat{r}\}$ given by (C.1b).

Once the basis functions of the master element are known (see Tables 2.1 to 2.9 of Chapter 2) the basis functions corresponding to each real element may be computed, according to (C.5), (C.7), and (C.9). Then the approximate expression for the unknown function in each real element will be given by the linear combination of its basis functions, where the coefficients of the linear combination are the invariant degrees of freedom (see (2.43), (2.49), (2.52a), for example). Thus, after substituting the approximation for the unknown in the selected variational formulation, the coefficients of matrices $[k^e]$, $[m^e]$ and vector $\{f^e\}$ for each element may be computed (see (2.4), (2.68), and (2.69)).

In practice, it is not necessary to explicitly calculate the basis functions for each real element. In fact, the coefficients of matrices $[k^e]$, $[m^e]$ and vector $\{f^e\}$ may be computed using the basis functions of the parent element and the expressions that transform scalar or vector functions from a real element to the parent element (see relations (C.4), (C.6), and (C.8)). This procedure allows one to program the computation of the local matrices in an efficient way because for each element one needs only to compute the Jacobian matrix and its determinant. The rest of the quantities involved will be identical for all the elements.

Differential and integral operators must also be transformed according to the following rules:

1. Differential operators:

- *First-order derivatives of a scalar function.* The first-order derivatives of a scalar function $f(\{r\})$ with respect to the coordinates of the real element may be computed in terms of the first-order derivatives of $\hat{f}(\{\hat{r}\})$ with respect to the coordinates of the parent element, according to

$$\begin{Bmatrix} \dfrac{\partial}{\partial x} \\ \\ \dfrac{\partial}{\partial y} \end{Bmatrix} f(x,y) = ([J^e]^T)^{-1} \begin{Bmatrix} \dfrac{\partial}{\partial \hat{x}} \\ \\ \dfrac{\partial}{\partial \hat{y}} \end{Bmatrix} \hat{f}(\hat{x},\hat{y}) \tag{C.10a}$$

for the 2D case and

$$\begin{Bmatrix} \dfrac{\partial}{\partial x} \\ \\ \dfrac{\partial}{\partial y} \\ \\ \dfrac{\partial}{\partial z} \end{Bmatrix} f(x,y,z) = ([J^e]^T)^{-1} \begin{Bmatrix} \dfrac{\partial}{\partial \hat{x}} \\ \\ \dfrac{\partial}{\partial \hat{y}} \\ \\ \dfrac{\partial}{\partial \hat{z}} \end{Bmatrix} \hat{f}(\hat{x},\hat{y},\hat{z}) \tag{C.10b}$$

for the 3D case. In order to avoid repetitions for the 2D and 3D cases, the column matrix of partial derivatives with respect to the coordinates of the parent element will be called $\{\partial_{\hat{r}}\}$, while the column matrix of partial derivatives with respect to the coordinates of the real element will be called $\{\partial_r\}$. It may be observed that the left-hand term of (C.10a) and (C.10b) is the matrix notation of the gradient of the scalar function $f(\{r\})$, while the column matrix multiplying the inverse of the transpose of the Jacobian matrix is the gradient of the scalar function $\hat{f}(\{\hat{r}\})$; that is, the following relation between the gradient of a scalar function computed over the real element and over the parent element is utilized:

$$\{\nabla\} f(\{r\}) = ([J^e]^T)^{-1} \{\hat{\nabla}\} \hat{f}(\{\hat{r}\}) \tag{C.10c}$$

The coordinate transformation is done usually by means of expression (2.46) of Chapter 2, that is repeated here

$$\{r\} = \left[\{r_1\}\ \{r_2\}\ \cdots\ \{r_m\}\right] \begin{Bmatrix} \hat{N}_1^{\,g}(\{\hat{r}\}) \\ \hat{N}_2^{\,g}(\{\hat{r}\}) \\ \vdots \\ \hat{N}_m^{\,g}(\{\hat{r}\}) \end{Bmatrix} \tag{C.11a}$$

The left-hand column matrix is the position vector of a generic point of the real element as defined previously. The jth column of the matrix of the right-hand term corresponds to the position vector of the jth geometric point of the real element. The number of points involved in the geometric transformation is assumed to be m. It may not be equal to n_e, the number of degrees of freedom. The coefficients of the column matrix of the right-hand term are the m geometric transformation basis functions (usually, the Lagrange basis functions of order m) over the parent element. This vector will be called, in the following, $\{\hat{N}^g(\{\hat{r}\})\}$ or simply $\{\hat{N}^g\}$, where the dependence on $\{\hat{r}\}$ is assumed. Matrix $[J^e]$ results in

$$[J^e] = \left[\{r_1\}\{r_2\}\ \cdots\ \{r_m\}\right]\left[\left\{\partial_{\hat{r}}\hat{N}_1^{\,g}\right\}\ \left\{\partial_{\hat{r}}\hat{N}_2^{\,g}\right\}\ \cdots\ \left\{\partial_{\hat{r}}\hat{N}_m^{\,g}\right\}\right]^T \tag{C.11b}$$

where the coefficients of the vector $\{\partial_{\hat{r}}\, \hat{N}_j^g\}$ are the partial derivatives with respect to $\hat{x}$ and $\hat{y}$ for the 2D case or with respect to $\hat{x}$, $\hat{y}$, and $\hat{z}$ for the 3D case of the jth geometrical basis function over the parent element.

- *Second-order derivatives of a scalar function.* The second-order derivatives with respect to the coordinates of the real element and those of the parent element for the 2D case are defined as

$$\begin{Bmatrix} \dfrac{\partial^2}{\partial x^2} \\ \dfrac{\partial^2}{\partial y^2} \\ \dfrac{\partial^2}{\partial x \partial y} \end{Bmatrix} f(\{r\}) = [T_1] \begin{Bmatrix} \dfrac{\partial}{\partial \hat{x}} \\ \dfrac{\partial}{\partial \hat{y}} \end{Bmatrix} \hat{f}(\{\hat{r}\}) + [T_2] \begin{Bmatrix} \dfrac{\partial^2}{\partial \hat{x}^2} \\ \dfrac{\partial^2}{\partial \hat{y}^2} \\ \dfrac{\partial^2}{\partial \hat{x} \partial \hat{y}} \end{Bmatrix} \hat{f}(\{\hat{r}\}) \tag{C.12a}$$

Matrices $[T_1]$ and $[T_2]$ are given by (C.12d) to (C.12f). For the 3D case, the relationship is given by

$$\begin{Bmatrix} \dfrac{\partial^2}{\partial x^2} \\ \dfrac{\partial^2}{\partial y^2} \\ \dfrac{\partial^2}{\partial z^2} \\ \dfrac{\partial^2}{\partial x \partial y} \\ \dfrac{\partial^2}{\partial y \partial z} \\ \dfrac{\partial^2}{\partial x \partial z} \end{Bmatrix} f(x,y) = [T_1] \begin{Bmatrix} \dfrac{\partial}{\partial \hat{x}} \\ \dfrac{\partial}{\partial \hat{y}} \\ \dfrac{\partial}{\partial \hat{z}} \end{Bmatrix} \hat{f}(\hat{x},\hat{y}) + [T_2] \begin{Bmatrix} \dfrac{\partial^2}{\partial \hat{x}^2} \\ \dfrac{\partial^2}{\partial \hat{y}^2} \\ \dfrac{\partial^2}{\partial \hat{z}^2} \\ \dfrac{\partial^2}{\partial \hat{x} \partial \hat{y}} \\ \dfrac{\partial^2}{\partial \hat{y} \partial \hat{z}} \\ \dfrac{\partial^2}{\partial \hat{x} \partial \hat{z}} \end{Bmatrix} \hat{f}(\hat{x},\hat{y}) \quad \text{(C.12b)}$$

where matrices $[T_1]$ and $[T_2]$ are given by (C.12d), (C.12g), and (C.12h). The previous relations may be written as

$$\{\partial_r^2\} = [T_1]\{\partial_{\hat{r}}\} + [T_2]\{\partial_{\hat{r}}^2\} \quad \text{(C.12c)}$$

where the definition for $\{\partial_r{}^2\}$ and $\{\partial_{\hat{r}}{}^2\}$ may be easily deduced. $[T_1]$ is

$$[T_1] = -[T_2][C_1]([J]^T)^{-1} \quad \text{(C.12d)}$$

For the 2D case, $[T_2]$ is given by

$$[T_2] = \begin{bmatrix} j_{11}^2 & j_{12}^2 & 2j_{11}j_{12} \\ j_{21}^2 & j_{22}^2 & 2j_{21}j_{22} \\ j_{11}j_{21} & j_{12}j_{22} & j_{11}j_{22} + j_{12}j_{21} \end{bmatrix} \quad \text{(C.12e)}$$

where j_{ij} is the ijth coefficient of the matrix $([J]^T)^{-1}$. $[C_1]$ is given by the following matrix, where J_{ij} is the ij-coefficient of $[J]^T$:

$$[C_1] = \begin{bmatrix} \dfrac{\partial J_{11}}{\partial \hat{x}} & \dfrac{\partial J_{12}}{\partial \hat{x}} \\ \dfrac{\partial J_{21}}{\partial \hat{y}} & \dfrac{\partial J_{22}}{\partial \hat{y}} \\ \dfrac{1}{2}\left(\dfrac{\partial J_{11}}{\partial \hat{y}} + \dfrac{\partial J_{21}}{\partial \hat{x}}\right) & \dfrac{1}{2}\left(\dfrac{\partial J_{12}}{\partial \hat{y}} + \dfrac{\partial J_{22}}{\partial \hat{x}}\right) \end{bmatrix} \tag{C.12f}$$

For the 3D case $[T_2]$ and $[C_1]$ are given by

$$[T_2] = \begin{bmatrix} j_{11}^2 & j_{12}^2 & j_{13}^2 & 2j_{11}j_{12} & 2j_{12}j_{13} & 2j_{13}j_{11} \\ j_{21}^2 & j_{22}^2 & j_{23}^2 & 2j_{21}j_{22} & 2j_{22}j_{23} & 2j_{23}j_{21} \\ j_{31}^2 & j_{32}^2 & j_{33}^2 & 2j_{31}j_{32} & 2j_{32}j_{33} & 2j_{33}j_{31} \\ j_{11}j_{21} & j_{12}j_{22} & j_{13}j_{23} & j_{11}j_{22}+j_{12}j_{21} & j_{12}j_{23}+j_{13}j_{22} & j_{11}j_{23}+j_{13}j_{21} \\ j_{21}j_{31} & j_{22}j_{32} & j_{23}j_{33} & j_{21}j_{32}+j_{22}j_{31} & j_{22}j_{33}+j_{23}j_{32} & j_{21}j_{33}+j_{23}j_{31} \\ j_{31}j_{11} & j_{32}j_{12} & j_{33}j_{13} & j_{31}j_{12}+j_{32}j_{11} & j_{32}j_{13}+j_{33}j_{12} & j_{31}j_{13}+j_{33}j_{11} \end{bmatrix} \tag{C.12g}$$

$$[C_1] = \begin{bmatrix} \dfrac{\partial J_{11}}{\partial \hat{x}} & \dfrac{\partial J_{12}}{\partial \hat{x}} & \dfrac{\partial J_{13}}{\partial \hat{x}} \\ \dfrac{\partial J_{21}}{\partial \hat{y}} & \dfrac{\partial J_{22}}{\partial \hat{y}} & \dfrac{\partial J_{23}}{\partial \hat{y}} \\ \dfrac{\partial J_{31}}{\partial \hat{z}} & \dfrac{\partial J_{32}}{\partial \hat{z}} & \dfrac{\partial J_{33}}{\partial \hat{z}} \\ \dfrac{1}{2}\left(\dfrac{\partial J_{11}}{\partial \hat{y}} + \dfrac{\partial J_{21}}{\partial \hat{x}}\right) & \dfrac{1}{2}\left(\dfrac{\partial J_{12}}{\partial \hat{y}} + \dfrac{\partial J_{22}}{\partial \hat{x}}\right) & \dfrac{1}{2}\left(\dfrac{\partial J_{13}}{\partial \hat{y}} + \dfrac{\partial J_{23}}{\partial \hat{x}}\right) \\ \dfrac{1}{2}\left(\dfrac{\partial J_{21}}{\partial \hat{z}} + \dfrac{\partial J_{31}}{\partial \hat{y}}\right) & \dfrac{1}{2}\left(\dfrac{\partial J_{22}}{\partial \hat{z}} + \dfrac{\partial J_{32}}{\partial \hat{y}}\right) & \dfrac{1}{2}\left(\dfrac{\partial J_{23}}{\partial \hat{z}} + \dfrac{\partial J_{33}}{\partial \hat{y}}\right) \\ \dfrac{1}{2}\left(\dfrac{\partial J_{31}}{\partial \hat{x}} + \dfrac{\partial J_{11}}{\partial \hat{z}}\right) & \dfrac{1}{2}\left(\dfrac{\partial J_{32}}{\partial \hat{x}} + \dfrac{\partial J_{12}}{\partial \hat{z}}\right) & \dfrac{1}{2}\left(\dfrac{\partial J_{33}}{\partial \hat{x}} + \dfrac{\partial J_{13}}{\partial \hat{z}}\right) \end{bmatrix} \tag{C.12h}$$

It may be observed that both $[C_1]$ and $[T_1]$ are equal to the null matrix whenever the coefficients of $[J^e]$ are constant. This is the case for triangular elements and tetrahedral elements when the mapping basis functions are the Lagrange linear basis functions, that is, for straight triangular and tetrahedral real elements. The previous relations are described in [Dhatt and Touzot 1981; Ch. 1].

2. Integral operators:

- *Integrand.* The integrand must be transformed according to (C.4), (C.6), and (C.8) and, in addition, (C.10) to (C.12) if derivatives are involved.
- *Domain of integration.* In order to perform the integration over the parent element $\hat{\Omega}$, besides transforming the integrand as mentioned before, the following expression holds:

$$\int_{\Omega^e} \cdots \, \mathrm{d}\Omega = \int_{\hat{\Omega}} \cdots \, |J^e| \, \mathrm{d}\hat{\Omega} \tag{C.13}$$

where $\mathrm{d}\Omega = \mathrm{d}x\,\mathrm{d}y$ for the 2D case, while $\mathrm{d}\Omega = \mathrm{d}x\,\mathrm{d}y\,\mathrm{d}z$ in the 3D case.

- *Boundary integrals.* Two types of boundary integrals may be involved:

  1. Curvilinear integrals in two or three dimensions: along part of the boundary of a 2D element, like, for example, along the ith edge, $\Gamma_i^e(x,y)$, or any of the edges of a 3D element, that is, the boundary of the 3D element, $\Gamma_i^e(x,y,z)$, but specialized to the ith edge. In any case, the analytical expression of the curve that defines the edge of interest, $\Gamma_i^e(\{r\}) = 0$, may be expressed as a function of a parameter ν (that will coincide with one of the coordinates $\hat{x}$, $\hat{y}$, or $\hat{z}$ of the parent element) by expressing

$$x = x(\nu)\,, \quad y = y(\nu)\,, \quad z = z(\nu) \tag{C.14a}$$

Then, after transforming the integrand according to (C.4), (C.6), (C.8), and (C.10) to (C.12), the result must be specialized for (C.14a). Additionally one has

$$\int_{\Gamma_i^e} \cdots \, \mathrm{d}\Gamma = \int_{\hat{\Gamma}_i} \cdots \, J_\Gamma^e \mathrm{d}\nu = \int_{\nu_1}^{\nu_2} \cdots \, J_\Gamma^e(\nu)\mathrm{d}\nu \tag{C.14b}$$

where J_Γ stands for the transformation coefficient from the real element to the parent element given by

$$J_\Gamma^e(\nu) = \left(\left(\frac{\partial x}{\partial \nu}\right)^2 + \left(\frac{\partial y}{\partial \nu}\right)^2 + \left(\frac{\partial z}{\partial \nu}\right)^2 \right)^{1/2} \tag{C.14c}$$

Here, ν_1, ν_2 are the values of ν for $\{r_1\}$ and $\{r_2\}$, the initial and final points of the edge Γ_i^e, respectively. For the 2D case, the term involving the partial derivative of z with respect to ν in (C.14c) does not exist. In the FEM formulations and for postprocesses of the problems analyzed in this book, one requires the analytical expression of the unit vector $\bar{a}_{\tau i}^{\,e}(\{r\})$ tangential to the edge Γ_i^e. The unit vector is given by

$$\bar{a}_{\tau_i}^{\,e} = \left(\left(\frac{\partial x}{\partial \nu}\right)^2 + \left(\frac{\partial y}{\partial \nu}\right)^2 + \left(\frac{\partial z}{\partial \nu}\right)^2 \right)^{-1/2} \begin{Bmatrix} \dfrac{\partial x}{\partial \nu} \\ \dfrac{\partial y}{\partial \nu} \\ \dfrac{\partial z}{\partial \nu} \end{Bmatrix} \tag{C.15a}$$

where, again, for the 2D case, the term involving the partial derivative of z with respect to ν does not exist. Similarly, one requires the expression of the unit vector normal to the edge, Γ_i^e, of a 2D element, $\bar{a}_{ni}^{\,e}(x,y)$. It is given by

$$\bar{a}_{n_i}^{\,e} = \left(\left(\frac{\partial x}{\partial \nu}\right)^2 + \left(\frac{\partial y}{\partial \nu}\right)^2 \right)^{-1/2} \begin{Bmatrix} \dfrac{\partial y}{\partial \nu} \\ -\dfrac{\partial x}{\partial \nu} \end{Bmatrix} \tag{C.15b}$$

2. Surface integrals in three dimensions: over the ith face of a 3D element $\Gamma_i^e(x,y,z)$. The analytical expression of the surface that defines the face of interest, $\Gamma_i^e(\{r\}) = 0$, may be expressed as a function of two parameters μ, ν (that will coincide with two of the parent element coordinates, $\hat{x},\hat{y}$ or $\hat{x},\hat{z}$ or $\hat{y},\hat{z}$) by writing

$$x = x(\mu,\nu)\ ,\quad y = y(\mu,\nu)\ ,\quad z = z(\mu,\nu) \tag{C.16a}$$

Then, after transforming the integrand according to (C.4), (C.6), (C.8), and (C.10) to (C.12), the result must be specialized for (C.16a). Additionally, one has

$$\int_{\Gamma_i^e} \cdots \, d\Gamma = \int_{\hat{\Gamma}_i} \cdots \, J_\Gamma^e \, d\mu d\nu = \int_{\mu_1}^{\mu_2}\int_{\nu_1}^{\nu_2} \cdots \, J_\Gamma^e(\mu,\nu)\, d\mu d\nu \tag{C.16b}$$

where J_Γ stands for the transformation coefficient from the real element to the parent element given by

$$J_\Gamma^e(\mu,\nu) = \left(\alpha_x^2+\alpha_y^2+\alpha_z^2\right)^{1/2} \tag{C.16c}$$

where α_x, α_y, α_z are given by

$$\begin{aligned}
\alpha_x &= \frac{\partial y}{\partial \mu}\frac{\partial z}{\partial \nu} - \frac{\partial z}{\partial \mu}\frac{\partial y}{\partial \nu}\\
\alpha_y &= \frac{\partial z}{\partial \mu}\frac{\partial x}{\partial \nu} - \frac{\partial x}{\partial \mu}\frac{\partial z}{\partial \nu}\\
\alpha_z &= \frac{\partial x}{\partial \mu}\frac{\partial y}{\partial \nu} - \frac{\partial y}{\partial \mu}\frac{\partial x}{\partial \nu}
\end{aligned} \tag{C.16d}$$

In (C.16b), μ_1,ν_1 and μ_2,ν_2 are the values of μ,ν for $\{r_1\}$ and $\{r_2\}$, the initial and final parameters of the face Γ_i^e, respectively. Expressions of vectors tangential to the face at a generic point are given by $\overline{t}_\mu$ and $\overline{t}_\nu$, which are functions of the variables μ,ν. In matrix notation it may be written as

$$\{t_\mu\} = \begin{Bmatrix} \dfrac{\partial x}{\partial \mu} \\ \dfrac{\partial y}{\partial \mu} \\ \dfrac{\partial z}{\partial \mu} \end{Bmatrix}, \quad \{t_\nu\} = \begin{Bmatrix} \dfrac{\partial x}{\partial \nu} \\ \dfrac{\partial y}{\partial \nu} \\ \dfrac{\partial z}{\partial \nu} \end{Bmatrix} \tag{C.16e}$$

In the FEM formulations and postprocesses of the problems analyzed in this book, one requires the analytical expression of the unit vector, normal to the face Γ_i^e of a 3D element, $\overline{a}_{ni}^{\,e}(\{r\})$. At a generic point, given by the parameters μ,ν, the expression may be written in terms of the previous vectors as

$$\overline{a}_{n_i}^{\,e} = \left(\alpha_x^2+\alpha_y^2+\alpha_z^2\right)^{-1/2}\,\overline{t}_\mu \times \overline{t}_\nu \tag{C.16f}$$

where a_x, a_y, a_z are given by (C.16d).

- *Additional expressions.* The following identities are also relevant for 2D div-conforming elements, where expression (C.6) has been taken into account and a matrix notation is used:

$$\int_{\Omega^e} \{\nabla_t\}^T \{t(\{r\})\}\, \varphi(\{r\})\, d\Omega = \int_{\hat{\Omega}} \{\hat{\nabla}_t\}^T \{\hat{t}(\{\hat{r}\})\}\, \hat{\varphi}(\{\hat{r}\})\, d\Omega \tag{C.17a}$$

$$\int_{\Gamma_i^e} \{t(\{r\})\}^T \{a_{n_i}^e(\{r\})\}\, \varphi(\{r\})\, d\Gamma = \int_{\hat{\Gamma}_i} \{\hat{t}(\{\hat{r}\})\}^T \{\hat{a}_{n_i}(\{\hat{r}\})\}\, \hat{\varphi}(\{r\})\, d\Gamma \tag{C.17b}$$

For 2D curl-conforming elements, the following expressions are also relevant, where expression (C.8) has been taken into account and the matrix notation is used:

$$\int_{\Omega^e} \{\nabla\} \times \{t(\{r\})\}\, \varphi(\{r\})\, d\Omega = \int_{\hat{\Omega}} \{\hat{\nabla}\} \times \{\hat{t}(\{\hat{r}\})\}\, \hat{\varphi}(\{\hat{r}\})\, d\Omega \tag{C.17c}$$

$$\int_{\Gamma_i^e} \{t(\{r\})\}^T \{a_{\tau_i}^e(\{r\})\}\, \varphi(\{r\})\, d\Gamma = \int_{\hat{\Gamma}_i} \{\hat{t}(\{\hat{r}\})\}^T \{\hat{a}_{\tau_i}(\{\hat{r}\})\}\, \hat{\varphi}(\{\hat{r}\})\, d\Gamma \tag{C.17d}$$

Expressions (C.17) state the invariance of those particular integrals when evaluated either over the real or the parent element.

C.2 LAGRANGE ORDINARY ELEMENTS

Next, the expressions for the basis functions of the parent element utilizing the ordinary Lagrange elements are derived.

C.2.1 Rectangular Parent Elements

Several families of rectangular elements of degree m in the x variable and of degree s in the y variable may be generated just by a transformation from a master element with nodes at the intersection of m subdivisions in the x direction and s subdivisions in the y direction, respectively. The basis functions will be obtained according to

$$\hat{N}_i(\hat{x},\hat{y}) = \hat{N}_{jk}(\hat{x},\hat{y}) = L_j^m(\hat{x})\, L_k^s(\hat{y}) \tag{C.18}$$

for $i = 1, ..., (m+1)(s+1)$; $j = 1, ..., m+1$; $k = 1, ..., s+1$, where $L_j^m(x)$ and $L_k^s(y)$ would be given by expression (2.44).

Table 2.1 of Chapter 2 shows the parent elements and the basis functions for rectangular elements, $Q(p)$, of first $Q(1)$, second $Q(2)$ and third order $Q(3)$, in both directions. These elements provide an interpolation of that consist on polynomials of maximum degree $2p$ complete in each variable up to order p.

Elements $Q(2)$ and $Q(3)$ have internal nodes. These nodes will not be shared by contiguous elements. They are said to support internal degrees of freedom, i.e., the associated degrees of freedom that are not shared by other elements. For these degrees of freedom there is no need to impose any continuity conditions and they will not be subjected to any boundary conditions. Thus, if possible, it would be convenient to eliminate them before proceeding to the matrix assembly process that will lead to the global system of equations. In that way the total number of unknowns will be reduced. There are two ways to eliminate them. The first procedure is to obtain the discrete integral form for each element as usual and then eliminate them. This may be done because the rows of the column matrix of the local integral form corresponding to those internal degrees of freedom may be equated to zero. Thus, the internal degrees of freedom may be expressed as a linear combination of the degrees of freedom associated with the nodes placed on the boundary of the element. Then, the discrete integral form of each element may be rewritten with a number of rows equal to the number of boundary nodes. A different approach is to generate elements in which those nodes are not present. This last choice leads to formulating new basis functions involving only the rest of the nodes, giving rise to elements termed of serendipity type (they are so called from the Prince of Serendip), $Q(p')$, [Dhatt and Touzot 1981; Ch. 2], [Zienkiewicz 1977; Ch. 7]. Table 2.1 of Chapter 2 shows the elements $Q(2')$ and $Q(3')$. It may be demonstrated that the interpolation polynomials are still complete up to order p.

C.2.2 Simplex Parent Elements

In order to generate the basis functions of the triangular and tetrahedral elements, it is practical to use the natural coordinates also called area or volume (in 2D or 3D, respectively) coordinates. The natural coordinates (λ,ξ,η) of a point of Cartesian coordinates (x,y) belonging to a triangular element of vertices (x_i,y_i), for $i = 1, 2, 3$, and the natural coordinates (λ,ξ,η,ζ) of a point of Cartesian coordinates (x,y,z) belonging to a tetrahedral element of vertices (x_i,y_i,z_i), for $i = 1, 2, 3, 4$, are given by

$$\begin{Bmatrix} x \\ y \\ 1 \end{Bmatrix} = \begin{bmatrix} x_1 & x_2 & x_3 \\ y_1 & y_2 & y_3 \\ 1 & 1 & 1 \end{bmatrix} \begin{Bmatrix} \lambda \\ \xi \\ \eta \end{Bmatrix}, \quad \begin{Bmatrix} x \\ y \\ z \\ 1 \end{Bmatrix} = \begin{bmatrix} x_1 & x_2 & x_3 & x_4 \\ y_1 & y_2 & y_3 & y_4 \\ z_1 & z_2 & z_3 & z_4 \\ 1 & 1 & 1 & 1 \end{bmatrix} \begin{Bmatrix} \lambda \\ \xi \\ \eta \\ \zeta \end{Bmatrix} \tag{C.19a}$$

respectively. For the simplex parent elements defined as shown in Figure 2.15 of

Chapter 2, the natural coordinates of a generic point $\{\hat{r}\}$ may be written as

$$\begin{Bmatrix} \lambda \\ \xi \\ \eta \end{Bmatrix} = \begin{Bmatrix} 1-\hat{x}-\hat{y} \\ \hat{x} \\ \hat{y} \end{Bmatrix} , \quad \begin{Bmatrix} \lambda \\ \xi \\ \eta \\ \zeta \end{Bmatrix} = \begin{Bmatrix} 1-\hat{x}-\hat{y}-\hat{z} \\ \hat{x} \\ \hat{y} \\ \hat{z} \end{Bmatrix} \tag{C.19b}$$

for the 2D and 3D simplexes, respectively. These coordinates are termed *natural* because they are invariant for a point $\{r\}$ of a real element and the corresponding point $\{\hat{r}\}$ of the parent element. Interior points of a simplex have positive natural coordinates [Kardestuncer and Norrie 1987; P. 2, Ch. 3], [Raviart and Thomas 1983; Ch. 4].

The basis functions of Lagrange triangular parent elements of order p may be generated according to the following expression:

$$\hat{N}_i(\hat{x},\hat{y}) = \hat{N}_i(\lambda,\xi,\eta) = \hat{N}_{jkl}(\lambda,\xi,\eta) = L_j^j(\lambda)\,L_k^k(\xi)\,L_l^l(\eta) \tag{C.20a}$$

where $j+k+l = p$; j, k, and l are non-negative integers, with $i = 1, \ldots, \frac{1}{2}(p+1)(p+2)$; and the expressions $L_j^j(\lambda)$, $L_k^k(\xi)$, $L_l^l(\eta)$ are given by expression (2.44).

Table 2.2 of Chapter 2 shows the geometry, node location, and basis functions of elements $P(1)$, $P(2)$, and $P(3)$ of first, second, and third orders. These basis functions generate interpolation polynomials of maximum order p containing a polynomial complete up to order p. The interior node of element $P(3)$ may be eliminated obtaining element $P(3')$, which is also included in Table 2.2 of Chapter 2. However, this element provides an interpolation that is complete only up to second-order.

Similarly, Lagrange tetrahedral parent elements of order p are generated from

$$\hat{N}_i(\hat{x},\hat{y},\hat{z}) = \hat{N}_i(\lambda,\xi,\eta,\zeta) = \hat{N}_{jklm}(\lambda,\xi,\eta,\zeta) = L_j^j(\lambda)L_k^k(\xi)L_l^l(\eta)L_m^m(\zeta) \tag{C.20b}$$

where $j+k+l+m = p$; and j, k, l, and m are non-negative integers, with $i = 1, \ldots, 1/6(p+1)(p+2)(p+3)$; and $L_j^j(\lambda)$, $L_k^k(\xi)$, $L_l^l(\eta)$, $L_m^m(\zeta)$ are given by expression (2.44).

Table 2.3 of Chapter 2 shows the geometry, node location, and basis functions of elements $T(1)$ and $T(2)$ of first and second orders. The interpolation polynomials will have a maximum order given by p and will be complete up to order p.

C.3 GENERATION OF ONE-DIMENSIONAL INFINITE ELEMENTS FOR ASYMPTOTIC APPROXIMATION OF THE UNKNOWN OF TYPE (1/r)

Consider a one-dimensional finite parent element defined by three geometric points of coordinates $\hat{x}_1 = -1$, $\hat{x}_2 = 1$, $\hat{x}_3 = 0$, respectively. Assume that such an element is transformed using a mapping function into a one-dimensional infinite real element defined by three geometric points x_1, x_2, x_3, where x_2 would be at infinity, according to Figure 2.19.

A mapping function that performs such a transformation is given by the following expression

$$x = \{\hat{N}_1^{\prime I}(\hat{x}), \hat{N}_2^{\prime I}(\hat{x}), \hat{N}_3^{\prime I}(\hat{x})\}^T \begin{Bmatrix} x_1 \\ x_2 \\ x_3 \end{Bmatrix} \tag{C.21}$$

where the following mapping basis functions have been utilized

$$\hat{N}_1^{\prime I}(\hat{x}) = -\frac{2\hat{x}}{1-\hat{x}} \tag{C.22a}$$

$$\hat{N}_2^{\prime I}(\hat{x}) = 0 \tag{C.22b}$$

$$\hat{N}_3^{\prime I}(\hat{x}) = \frac{1+\hat{x}}{1-\hat{x}} \tag{C.22c}$$

with the understanding that the product of $\hat{N}_2^{\prime I} x_2$ is equated to zero.

The expressions corresponding to the inverse transformation may be obtained from (C.21) and (C.22) as

$$\hat{x} = \frac{x-x_3}{x+x_3-2x_1} = \frac{x-x_3}{x+x_{31}-x_{10}-x_0} = \frac{x-x_3}{x-x_0} \tag{C.23}$$

where

$$x_{ij} = x_i - x_j \tag{C.24}$$

In addition, a point, called pole, has been introduced. It is located outside the real infinite element (i.e., it is located at the near-field region). It is defined by its coordinate x_0 so that

$$x_{10} = x_{31} \tag{C.25}$$

where x_{31} is called the size of the infinite element and x_{10} defines the position of the pole with respect to the interface of the near-field region and the far-field region, i.e., the first point of the real infinite element, x_1. The expression (C.23) may be written in terms of x_{10} and r, the distance of any point of the infinite element to the pole:

$$r = x - x_0 \tag{C.26}$$

giving

$$\hat{x} = 1 - \frac{2x_{10}}{r} \tag{C.27}$$

Now assume a function w that is going to be approximated by a first-order Lagrange interpolation in one dimension, according to

$$\Pi w = \{\hat{N}_1(\hat{x}), \hat{N}_2(\hat{x})\}^T \begin{Bmatrix} w_1 \\ w_2 \end{Bmatrix} \tag{C.28}$$

where the basis functions of the parent element are given by the ordinary first-order Lagrange basis function

$$\hat{N}_1(\hat{x}) = \frac{1}{2}(1 - \hat{x}) \tag{C.29a}$$

$$\hat{N}_2(\hat{x}) = \frac{1}{2}(1 + \hat{x}) \tag{C.29b}$$

It can be seen that those basis functions correspond to nodes placed at the geometric points of coordinates $\hat{x}_1$ and $\hat{x}_2$ of the parent element. It is required that the function w should go to zero at infinity, i.e., $w_2 = 0$. Substituting (C.27) into (C.28) and taking into account the condition $w_2 = 0$, the following expression is obtained:

$$\Pi w = \frac{x_{10}}{r} w_1 \tag{C.30}$$

Thus, an asymptotic approximation of type $1/r$, which goes to zero when r tends to infinity, has been obtained for the interpolated function.

In this way, an infinite one-dimensional element has been generated. It may be stressed that ordinary Lagrange basis functions of first order have been used for the interpolation, while special functions have been employed to map the finite parent element into the infinite real element. In this case, the infinite element may be considered to be superparametric because, although three points have been used for its geometric definition, only two nodes have been utilized for the interpolation functions.

Expressions (C.23) and (C.27) explain why the external geometric point of coordinate x_0 has been called a pole. For $x = x_0$, r would be equal to zero and the value of (C.30) would be infinite. This result is only theoretical since the approximation in (C.30) is valid only for points inside the infinite element. The pole is exterior to it. However, the position of the pole with respect to the interface between the near-field and the far-field region should be properly chosen, i.e., the *size* of the infinite element, x_{31}, which is related to x_{10}, will determine the quality of the approximation.

If for the approximation of w second-order Lagrange basis functions are used, the interpolated function would be

$$\Pi w = \{\hat{N}_1(\hat{x}), \hat{N}_2(\hat{x}), \hat{N}_3(\hat{x})\}^T \begin{Bmatrix} w_1 \\ w_2 \\ w_3 \end{Bmatrix} \tag{C.31}$$

with

$$\hat{N}_1(\hat{x}) = -\frac{1}{2}\hat{x}(1-\hat{x}) \tag{C.32a}$$

$$\hat{N}_2(\hat{x}) = \frac{1}{2}\hat{x}(1+\hat{x}) \tag{C.32b}$$

$$\hat{N}_3(\hat{x}) = 1-\hat{x}^2 \tag{C.32c}$$

where nodes are located at the geometric points of coordinates $\hat{x}_1$, $\hat{x}_2$, and $\hat{x}_3$ of the master element. Substituting (C.27) into (C.31) and taking into account the condition that $w_2 = 0$, a new expression for the interpolated function is obtained. It is given by

$$\Pi w = \frac{x_{10}}{r}(4w_3 - w_1) + \frac{2x_{10}}{r^2}(w_1 - 2w_3) \tag{C.33}$$

Thus, a more accurate approximation has been obtained by adding a second-order term to the dominant one of type $1/r$. In this case, the real infinite element would be isoparametric in the sense that the same number of nodes are used for both the geometric transformation and the interpolation.

The use of higher order interpolation functions would add terms of type $(1/r^3)$, and subsequent higher order which may improve the quality of the approximation but maintaining the dominant term of type $1/r$. Subparametric infinite elements may be generated in this way.

C.4 SOME TOPICS RELATED TO DIV-CONFORMING AND CURL-CONFORMING ELEMENTS

In the following, the expressions for the basis functions for div-conforming and curl-conforming parent elements used in this book are presented. Some important peculiarities of the assembly process when using these type of elements are also mentioned.

C.4.1 Div-Conforming Triangular Parent Elements

For the geometric description and the generation of the basis functions of these elements, the natural system of coordinates is used (see Figure 2.15 of Chapter 2 and expression (C.19b)). These elements have been proposed by [Raviart and Thomas 1977a] and [Nédélec 1980].

C.4.1.1 *First-Order Elements*

In this case, the space of vector functions is given by $D_1^{(2)}$ (see Appendix A.1.1). For the parent element, it is composed of vectors $\{\hat{t}(\xi,\eta)\}$ that can be expressed in matrix notation as

$$\{\hat{t}(\xi,\eta)\} = \begin{Bmatrix} \hat{t}_x \\ \hat{t}_y \end{Bmatrix} = \begin{Bmatrix} \alpha_1 + \alpha_0 \xi \\ \alpha_2 + \alpha_0 \eta \end{Bmatrix} \tag{C.34}$$

It can be seen that the number of coefficients that are required to specify (C.34), i.e., the number of degrees of freedom, is three. The definition of the degree of freedom is given by expression (2.51e) of Chapter 2, sampled at the midpoint of each edge. The normalization constant is chosen at the length of the corresponding edge. This choice provides invariant degrees of freedom between the parent and the real element. To the ith degree of freedom there corresponds a node (ξ_i,η_i) placed at the midpoint of the ith edge of the parent element (see Table 2.6 of Chapter 2).

Let $\{\hat{t}\}$ be the vector variable to be approximated over the parent element. The

ith degree of freedom may be defined as

$$d_i = \hat{L}_i \{\hat{a}_{n_i}\}^T \{\hat{t}(\xi_i,\eta_i)\} , \quad i=1,2,3 \tag{C.35}$$

where $\hat{L}_i$ is the length of the ith edge. $\{\hat{a}_{ni}\}$ is the unit vector, normal to the ith edge and directed outward from the parent element.

The expression for the interpolation is

$$\Pi \{\hat{t}(\xi,\eta)\} = \sum_{j=1}^{3} \{\hat{N}_j\} d_j \tag{C.36a}$$

where the vector basis functions are given by

$$\{\hat{N}_j\} = \begin{Bmatrix} \alpha_1^{(j)} + \alpha_0^{(j)}\xi \\ \alpha_2^{(j)} + \alpha_0^{(j)}\eta \end{Bmatrix} , \quad j=1,2,3 \tag{C.36b}$$

Substituting (C.36a) into (C.35) the following expression holds:

$$d_i = \hat{L}_i \left(\{a_{n_i}\}^T \sum_{j=1}^{3} \{\hat{N}_j(\xi_i,\eta_i)\} d_j \right) , \quad i=1,2,3 \tag{C.37}$$

In order to obtain the coefficients of the polynomials for each component of the vector basis functions (C.36b), the following conditions should be fulfilled:

$$\left.\begin{aligned} \hat{L}_i \{\hat{a}_{n_i}\}^T \{\hat{N}_j(\xi_i,\eta_i)\} &= 1, \quad i=j \\ \hat{L}_i \{\hat{a}_{n_i}\}^T \{\hat{N}_j(\xi_i,\eta_i)\} &= 0, \quad i\neq j \end{aligned}\right\} \begin{aligned} i&=1,2,3 \\ j&=1,2,3 \end{aligned} \tag{C.38}$$

These expressions are analogous to those given by (2.41) for Lagrange elements.The value of $\hat{L}_i$ and the expression of $\{\hat{a}_{ni}\}$, for i = 1, 2, 3, are easily obtained from the parent element. Substituting them into (C.38) as well as into (C.36b), a system of nine equations with nine unknowns, $\alpha_0^{(i)}$, $\alpha_1^{(i)}$, $\alpha_2^{(i)}$, i = 1, 2, 3, will be obtained. Its solution leads to the expression of the vector basis functions given in Table 2.6 of Chapter 2.

C.4.1.2 *Second-Order Elements*

In this case, the elements belong to the space $D_2^{(2)}$ (see Appendix A.1.1). For the parent element, the space is composed of vectors $\{\hat{t}(\xi,\eta)\}$ that may be expressed in matrix notation as

$$\{\hat{t}(\xi,\eta)\} = \begin{Bmatrix} a_1 + a_2\xi + a_3\eta + \xi(\alpha\xi + \beta\eta) \\ b_1 + b_2\xi + b_3\eta + \eta(\alpha\xi + \beta\eta) \end{Bmatrix} \tag{C.39}$$

It can be seen that the number of coefficients that are required to specify (C.39), i.e., the number of degrees of freedom, is eight. There are two types of degrees of freedom.

The definition of the first one is given again by expression (2.51e) of Chapter 2 sampled at two points (the Gauss-Legendre points) for each edge of the triangle. Again the normalization constant is the length of the edge. There are six degrees of freedom of this first type to which there correspond six nodes (ξ_{ij},η_{ij}) located at the jth Gauss integration point (j = 1, 2) of the ith edge of the parent element (see Table 2.6 of Chapter 2).

Let $\{\hat{t}\}$ be the vector function to be approximated. Those degrees of freedom may be defined as

$$d_k = \hat{L}_i \{\hat{a}_{n_i}\}^T \{\hat{t}(\xi_{ij},\eta_{ij})\}\ , \quad k = 2(i-1)+j\ , \quad i = 1,2,3\ , \quad j = 1,2 \tag{C.40a}$$

In order to simplify the notation, the node (ξ_{ij},η_{ij}) may be called the kth node, (ξ_k,η_k), to which $\hat{L}_k$ and $\{\hat{a}_{nk}\}$ (where $\hat{L}_1 = \hat{L}_2$, $\hat{L}_3 = \hat{L}_4$, $\hat{L}_5 = \hat{L}_6$, $\{\hat{a}_{n1}\} = \{\hat{a}_{n2}\}$, $\{\hat{a}_{n3}\} = \{\hat{a}_{n4}\}$, $\{\hat{a}_{n5}\} = \{\hat{a}_{n6}\}$) are associated, i.e., the length of the edges and the unit normal vectors corresponding to the kth node. Then, one obtains

$$d_k = \hat{L}_k \{\hat{a}_{n_k}\}^T \{\hat{t}(\xi_k,\eta_k)\}\ , \quad k = 1,\ldots,6 \tag{C.40b}$$

The other two degrees of freedom are internal to the element and are given by expressions (2.51f) of Chapter 2, leading to

$$d_7 = \int_\Omega t_x(\xi,\eta)\ d\Omega \tag{C.40c}$$

and

$$d_8 = \int_\Omega t_y(\xi,\eta)\ d\Omega \tag{C.40d}$$

The expression of the interpolation is given by

$$\Pi\{\hat{t}(\xi,\eta)\} = \sum_{r=1}^{8} \{\hat{N}_r(\xi,\eta)\} d_r \,, \quad r=1,\ldots,8 \tag{C.41a}$$

where the vector basis functions, in matrix notation, are given by

$$\{\hat{N}_r\} = \begin{Bmatrix} a_1^{(r)} + a_2^{(r)}\xi + a_3^{(r)}\eta + \xi(\alpha^{(r)}\xi + \beta^{(r)}\eta) \\ b_1^{(r)} + b_2^{(r)}\xi + b_3^{(r)}\eta + \eta(\alpha^{(r)}\xi + \beta^{(r)}\eta) \end{Bmatrix} \tag{C.41b}$$

Substituting (C.41a) into (C.40b) to (C.40d), one obtains:

$$d_k = \hat{L}_k\left(\{\hat{a}_{n_k}\}^T \sum_{r=1}^{8} \{\hat{N}_r(\xi_k,\eta_k)\} d_r \right), \quad k=1,\ldots,6 \tag{C.42a}$$

$$d_7 = \int_{\hat{\Omega}} \left(\sum_{r=1}^{8} \hat{N}_{xr}(\xi,\eta)\, d_r \right) \mathrm{d}\Omega \tag{C.42b}$$

$$d_8 = \int_{\hat{\Omega}} \left(\sum_{r=1}^{8} \hat{N}_{yr}(\xi,\eta) \right) d_r \, \mathrm{d}\Omega \tag{C.42c}$$

For those expressions to hold, the following conditions should be fulfilled:

$$\left.\begin{aligned} \hat{L}_k\{\hat{a}_{n_k}\}^T\{\hat{N}_r(\xi_k,\eta_k)\} &= 1 \,, \quad k=r \\ \hat{L}_k\{\hat{a}_{n_k}\}^T\{\hat{N}_r(\xi_k,\eta_k)\} &= 0 \,, \quad k\neq r \end{aligned}\right\} \begin{aligned} k&=1,\ldots,6 \\ r&=1,\ldots,8 \end{aligned} \tag{C.43a}$$

$$\left.\begin{aligned} \int_{\hat{\Omega}} \hat{N}_{xr}(\xi,\eta)\,\mathrm{d}\Omega &= 1 \,, \quad r=7 \\ \int_{\hat{\Omega}} \hat{N}_{xr}(\xi,\eta)\,\mathrm{d}\Omega &= 0 \,, \quad r\neq 7 \end{aligned}\right\} \quad r=1,\ldots,8 \tag{C.43b}$$

$$\left.\begin{aligned}\int_{\hat{\Omega}} \hat{N}_{yr}(\xi,\eta)\,\mathrm{d}\Omega &= 1\ , \quad r=8 \\ \int_{\hat{\Omega}} \hat{N}_{yr}(\xi,\eta)\,\mathrm{d}\Omega &= 0\ , \quad r\neq 8\end{aligned}\right\}\ r=1,\ldots,8 \tag{C.43c}$$

Expressions (C.43a) give 48 equations once the values of $\hat{L}_k$, $\{\hat{a}_{nk}\}$, and the expression (C.41b) are substituted. Additional 16 equations coming from (C.43b) and (C.43c) give a total of 64 equations with 64 unknowns. The values obtained for $a_1^{(r)}$, $a_2^{(r)}$, $a_3^{(r)}$, $b_1^{(r)}$, $b_2^{(r)}$, $b_3^{(r)}$, $\alpha^{(r)}$, $\beta^{(r)}$ for r = 1, ..., 8 are shown in Table 2.6 of Chapter 2.

The last two degrees of freedom have only local support, that is, they are not shared by other elements. Thus they are not necessary for the assembly process. Furthermore, when writing the set of eight local discrete integral forms for each element, the two last rows may be set equal to zero (see (2.68) of Chapter 2). Then, at least for deterministic problems, the two last degrees of freedom of each element may be written in terms of the other six. The corresponding expressions may be substituted in the first six rows so that the original set of eight local discrete integral forms will be reduced to six, written in terms of the six degrees of freedom related to the edges. For a better understanding of this issue, see Appendix E.3.2. The assembly of the reduced local integral forms and the enforcement of the essential boundary conditions lead to a global system of equations that will be of reduced size. Once this system solved the value of the last two degrees of freedom of any element may be obtained if desired from the corresponding expressions.

C.4.2 Curl-Conforming Simplex Parent Elements

In the following, the geometric description and the generation of the basis functions of 2D triangular and 3D tetrahedral curl-conforming parent elements of first and second order is given. The natural system of coordinates is used (see Figure 2.15 and (C.19)). These elements have been proposed in [Nédélec 1980].

C.4.2.1 *Triangular Elements*

First-Order Elements

The space of vector functions to be used is the space $R_1^{(2)}$ (see Appendix A.1.1). For the parent element it is composed of vectors $\{\hat{t}(\xi,\eta)\}$ that can be expressed in matrix notation as

$$\{\hat{t}(\xi,\eta)\} = \begin{Bmatrix} \hat{t}_x \\ \hat{t}_y \end{Bmatrix} = \begin{Bmatrix} \alpha_1 - \alpha_0\eta \\ \alpha_2 + \alpha_0\xi \end{Bmatrix} \tag{C.44}$$

It can be seen that the number of coefficients that are required to specify (C.44), i.e., the number of degrees of freedom, is three. The definition of the degree of freedom is given by expression (2.55p) of Chapter 2 sampled at the midpoint of each edge. Here the function q of (2.55p) is chosen as the length of the edge. This choice provides invariant degrees of freedom. To the ith degree of freedom there corresponds a node (ξ_i,η_i) placed at the midpoint of the ith edge of the parent element (see Table 2.7 of Chapter 2).

Let $\{\hat{t}\}$ be the vector variable to be approximated over the parent element. The ith degree of freedom may be defined as

$$d_i = \hat{L}_i\{\hat{a}_{\tau_i}\}^T\{\hat{t}(\xi_i,\eta_i)\}\ , \quad i=1,2,3 \tag{C.45}$$

where $\hat{L}_i$ is the length of the ith edge and $\{\hat{a}_{\tau i}\}$ is the unit vector, tangential to the ith edge of the parent element in a counterclockwise direction.

The expression for the interpolation is given by (C.36a) where the vector basis functions are given by

$$\{\hat{N}_j\} = \begin{Bmatrix} \alpha_1^{(j)} - \alpha_0^{(j)}\eta \\ \alpha_2^{(j)} + \alpha_0^{(j)}\xi \end{Bmatrix}, \quad j = 1,2,3 \tag{C.46}$$

Substituting (C.36a) into (C.45) the following expression holds:

$$d_i = \hat{L}_i\left(\{a_{\tau_i}\}^T\sum_{j=1}^{3}\{\hat{N}_j(\xi_i,\eta_i)\}\,d_j\right), \quad i=1,2,3 \tag{C.47}$$

In order to obtain the coefficients of the polynomials for each component of the vector basis functions (C.46), the following conditions should be fulfilled:

$$\left.\begin{aligned} \hat{L}_i\{\hat{a}_{\tau_i}\}^T\{\hat{N}_j(\xi_i,\eta_i)\} &= 1, \quad i=j \\ \hat{L}_i\{\hat{a}_{\tau_i}\}^T\{\hat{N}_j(\xi_i,\eta_i)\} &= 0, \quad i\neq j \end{aligned}\right\} \quad \begin{aligned} i&=1,2,3 \\ j&=1,2,3 \end{aligned} \tag{C.48}$$

These expressions are analogous to those given by (2.41) for Lagrange elements.The value of $\hat{L}_i$ and the expression of $\{\hat{a}_{\tau_i}\}$, i = 1, 2, 3, are easily obtained from the parent element. Substituting them into (C.48) and utilizing (C.36), a system of nine equations with nine unknowns, $\alpha_0^{(i)}$, $\alpha_1^{(i)}$, $\alpha_2^{(i)}$, i = 1, 2, 3, will be obtained. Its solution leads to the expression of the vector basis functions given in Table 2.7 of Chapter 2.

Second-Order Elements

The space of vector functions to be used is the space $R_2^{(2)}$ (see Appendix A.1.1) that for the parent element is composed of vectors $\{\hat{t}(\xi,\eta)\}$ that can be expressed in matrix notation as

$$\{\hat{t}(\xi,\eta)\} = \begin{Bmatrix} a_1 + a_2\xi + a_3\eta + \eta(\alpha\eta - \beta\xi) \\ b_1 + b_2\xi + b_3\eta + \xi(\beta\xi - \alpha\eta) \end{Bmatrix} \tag{C.49}$$

It can be seen that the number of coefficients that are required to specify (C.49), i.e., the number of degrees of freedom, is eight. There are two types of degrees of freedom.

The definition of the first one is given again by expression (2.55p) of Chapter 2 sampled at two points (the Gauss-Legendre points) for each edge. The function q is also chosen as the length of the edge. There are six degrees of freedom of the first type to which there correspond six nodes (ξ_{ij},η_{ij}) located at the jth Gauss integration point (j = 1, 2) of the ith edge of the parent element (see Table 2.7 of Chapter 2).

Let $\{\hat{t}\}$ be the vector function to be approximated. These degrees of freedom may be defined as

$$d_k = \hat{L}_i\{\hat{a}_{\tau_i}\}^T\{\hat{t}(\xi_{ij},\eta_{ij})\}\ , \quad k=2(i-1)+j\ , \quad i=1,2,3\ , \quad j=1,2 \tag{C.50a}$$

In order to simplify the notation, the node (ξ_{ij},η_{ij}) may be called the kth node, (ξ_k,η_k), to which $\hat{L}_k$ and $\{\hat{a}_{\tau k}\}$ (where $\hat{L}_1 = \hat{L}_2$, $\hat{L}_3 = \hat{L}_4$, $\hat{L}_5 = \hat{L}_6$, $\{\hat{a}_{\tau 1}\} = \{\hat{a}_{\tau 2}\}$, $\{\hat{a}_{\tau 3}\} = \{\hat{a}_{\tau 4}\}$, $\{\hat{a}_{\tau 5}\} = \{\hat{a}_{\tau 6}\}$) are associated, i.e., the length of the edges and the unit tangential vectors corresponding to the kth node. Then, one obtains

$$d_k = \hat{L}_k\{\hat{a}_{\tau_k}\}^T\{\hat{t}(\xi_k,\eta_k)\}\ , \quad k=1,\ldots,6 \tag{C.50b}$$

The other two degrees of freedom are internal to the element and they are given by expressions (2.55q) of Chapter 2. Choosing $\overline{q}$ as the unit vectors on the x- and y-directions one gets

$$d_7 = \int_{\hat{\Omega}} t_y(\xi,\eta)\, d\Omega \tag{C.50c}$$

and

$$d_8 = -\int_{\hat{\Omega}} t_x(\xi,\eta)\, d\Omega \tag{C.50d}$$

The expression of the interpolation is given by (C.41a), where the vector basis functions, in matrix notation, are given by

$$\{\hat{N}_r\} = \begin{Bmatrix} a_1^{(r)} + a_2^{(r)}\xi + a_3^{(r)}\eta + \eta(\alpha^{(r)}\eta - \beta^{(r)}\xi) \\ b_1^{(r)} + b_2^{(r)}\xi + b_3^{(r)}\eta + \xi(\beta^{(r)}\xi - \alpha^{(r)}\eta) \end{Bmatrix} \tag{C.51}$$

Substituting (C.41a) into (C.50b) to (C.50d), one obtains

$$d_k = \hat{L}_k\left(\{\hat{a}_{\tau_k}\}^T \sum_{r=1}^{8} \{\hat{N}_r(\xi_k,\eta_k)\}\, d_r \right), \quad k=1,\ldots,6 \tag{C.52a}$$

$$d_7 = \int_{\hat{\Omega}} \left(\sum_{r=1}^{8} \hat{N}_{yr}(\xi,\eta)\, d_r \right) d\Omega \tag{C.52b}$$

$$d_8 = -\int_{\hat{\Omega}} \left(\sum_{r=1}^{8} \hat{N}_{xr}(\xi,\eta)\, d_r \right) d\Omega \tag{C.52c}$$

For those expressions to hold, the following conditions should be fulfilled:

$$\left.\begin{aligned} \hat{L}_k\{\hat{a}_{\tau_k}\}^T\{\hat{N}_r(\xi_k,\eta_k)\} &= 1\,, \quad k=r \\ \hat{L}_k\{\hat{a}_{\tau_k}\}^T\{\hat{N}_r(\xi_k,\eta_k)\} &= 0\,, \quad k\neq r \end{aligned}\right\} \begin{aligned} k&=1,\ldots,6 \\ r&=1,\ldots,8 \end{aligned} \tag{C.53a}$$

$$\left.\begin{array}{l}\int_{\hat{\Omega}} \hat{N}_{yr}(\xi,\eta)\,\mathrm{d}\Omega = 1\ , \quad r=7 \\ \int_{\hat{\Omega}} \hat{N}_{yr}(\xi,\eta)\,\mathrm{d}\Omega = 0\ , \quad r\neq 7\end{array}\right\} \quad r=1,\ldots,8 \tag{C.53b}$$

$$\left.\begin{array}{l}-\int_{\hat{\Omega}} \hat{N}_{xr}(\xi,\eta)\,\mathrm{d}\Omega = 1\ , \quad r=8 \\ -\int_{\hat{\Omega}} \hat{N}_{xr}(\xi,\eta)\,\mathrm{d}\Omega = 0\ , \quad r\neq 8\end{array}\right\} \quad r=1,\ldots,8 \tag{C.53c}$$

Expressions (C.53a) would give 48 equations once the values of $\hat{L}_k$, $\{\hat{a}_{\tau k}\}$, and the expressions (C.51) are substituted. Additional 16 equations coming from (C.53b) and (C.53c) lead to 64 equations with 64 unknowns. The values obtained for the coefficients $a_1^{(r)}$, $a_2^{(r)}$, $a_3^{(r)}$, $b_1^{(r)}$, $b_2^{(r)}$, $b_3^{(r)}$, $\alpha^{(r)}$, $\beta^{(r)}$ for r = 1, ..., 8 are shown in Table 2.7 of Chapter 2. From that table it may be seen that all eight basis functions are of second order, i.e., none of the α and β coefficients are zero. Figure C.1 represents the eight basis functions. It may be noticed that along different directions their magnitude shows a different behavior (of linear or second order type). In fact, they perform an approximation of a vector variable with different order depending on the direction. For this reason, this element is categorized as a mixed-order element.

This element is different, as well as the first-order one, from those proposed in [Brezzi et al. 1985] and [Nédélec 1986]. These are called complete polynomial curl-conforming elements. They perform an approximation of a vector variable with the same order in all directions. From the discussion of Section 7.2, it is seen that when the latter elements are used for the full-wave analysis of waveguiding problems, the extra number of degrees of freedom does not contribute to modelling the real physical solutions, thus the number of zero eigenvalues (i.e., the number of spurious solutions) will be greater with the latter elements than with the former ones.

The second-order element presented here [Nédélec 1980] differs also from other mixed-order curl-conforming elements that have become popular. In Figure C.2, the basis functions of the second-order curl-conforming element of [Lee et al. 1991a,b] are represented. It may be noticed that the first six are linear functions. Only the last two are second-order functions. The definition of the first six degrees of freedom is identical to the elements of [Nédélec 1980]. However, they are associated with nodes at the vertices of the edges. The definition of the last two degrees of freedom also differs from those proposed by Nédélec. They are associated with the normal components of two of the three edges. Also, this element does not fulfill the Nédélec constraints (see Sections 2.5.1.2 and 7.3.2). In [Peterson 1994], other mixed type second-order curl-conforming elements have been presented. The first six are the same as in [Lee et al. 1991a,b],

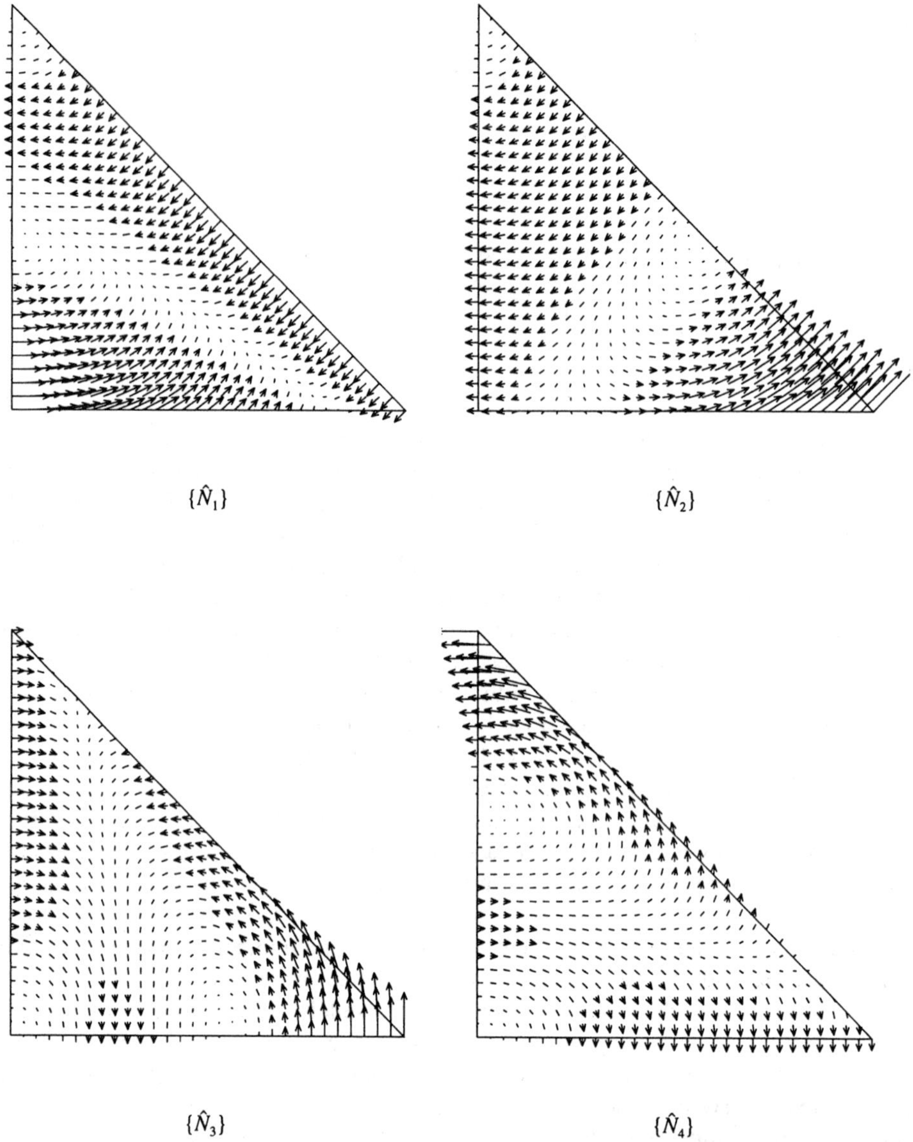

Figure C.1 Basis functions of the second-order curl-conforming triangular parent element of [Nédélec 1980].

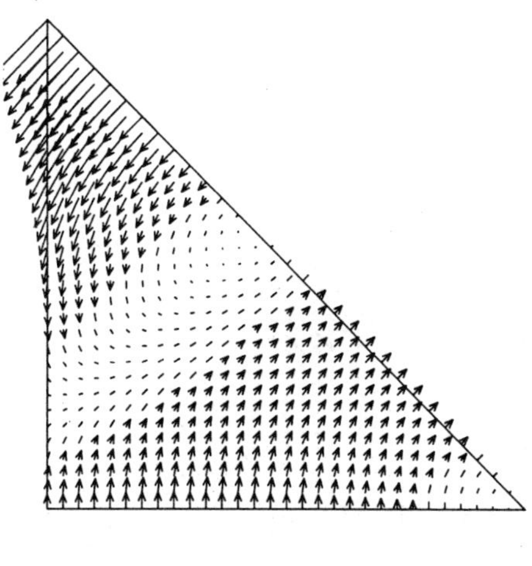

$\{\hat{N}_5\}$

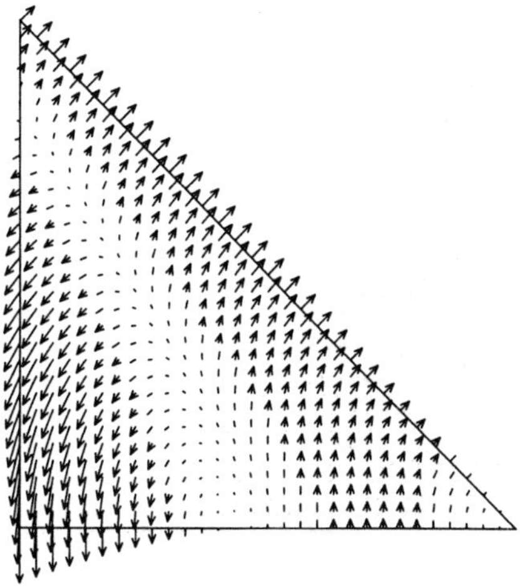

$\{\hat{N}_6\}$

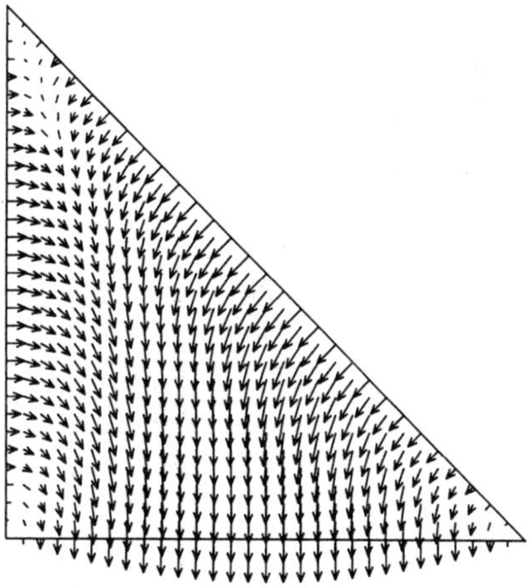

$\{\hat{N}_7\}$

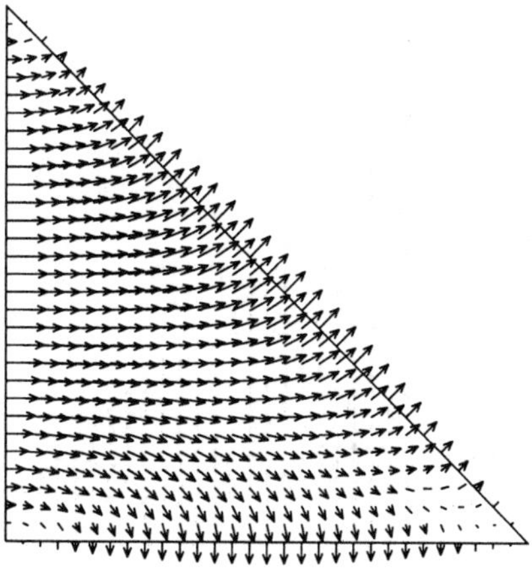

$\{\hat{N}_8\}$

Figure C.1 (Continued.)

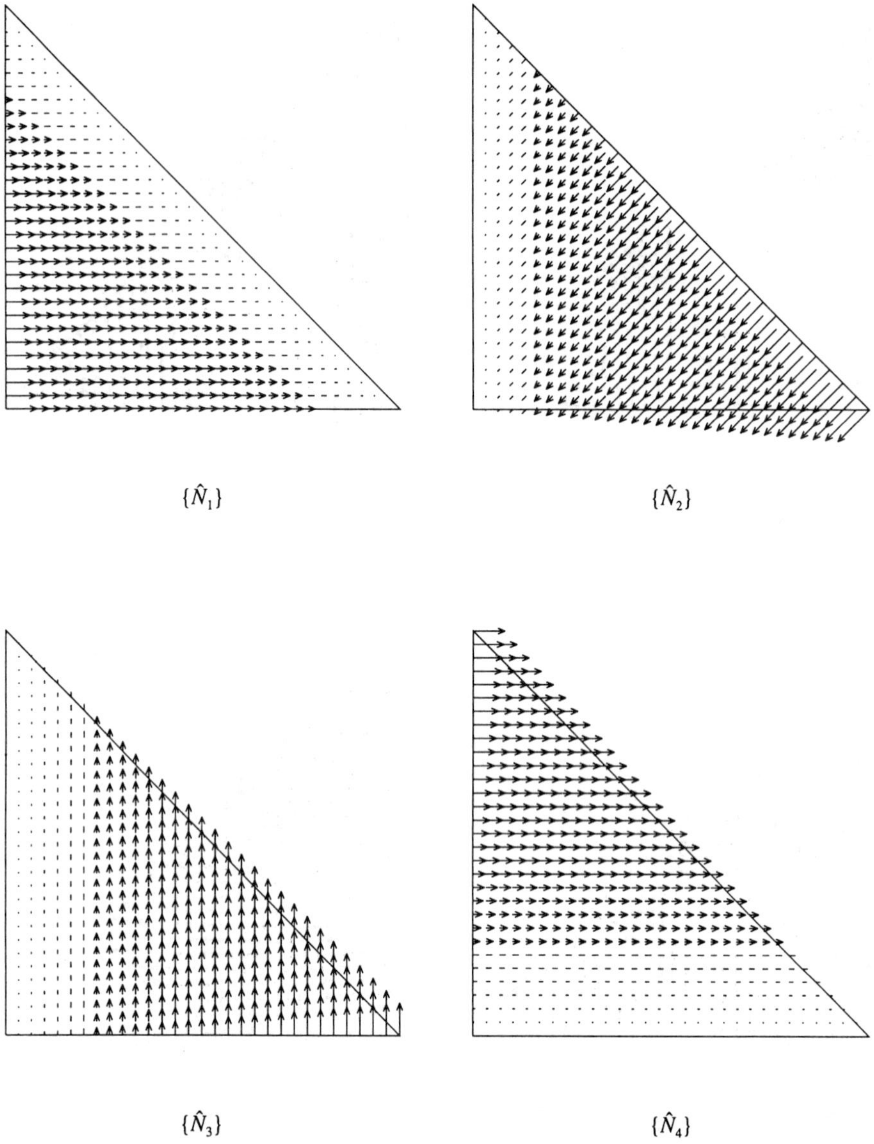

Figure C.2 Basis functions of the second-order curl-conforming triangular parent element of [Lee et al. 1991a,b].

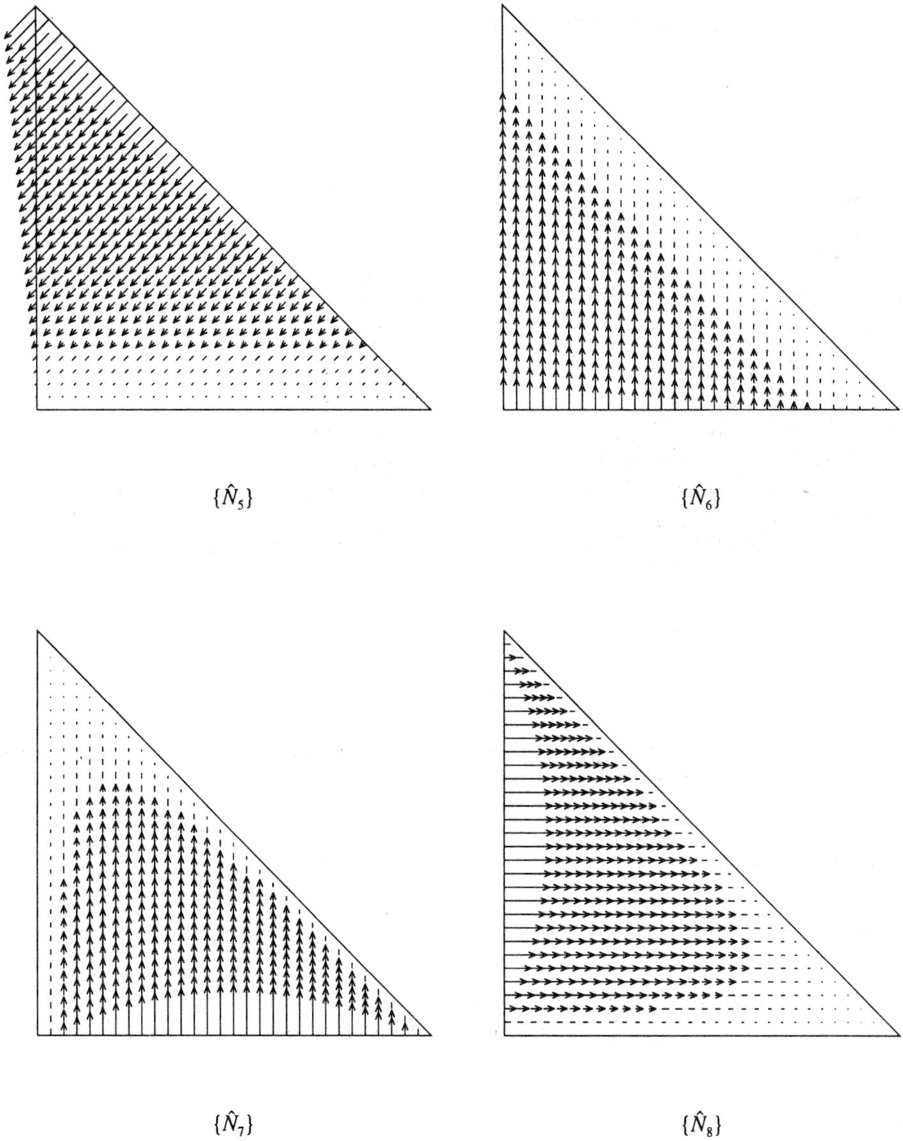

Figure C.2 (Continued.)

while the last two as shown in Figure C.3 differ from those of [Lee et al. 1991a,b]. The definition of the degrees of freedom and the location of the nodes are the same as in [Lee et al. 1991a,b]. These elements fulfill the Nédélec constraints.

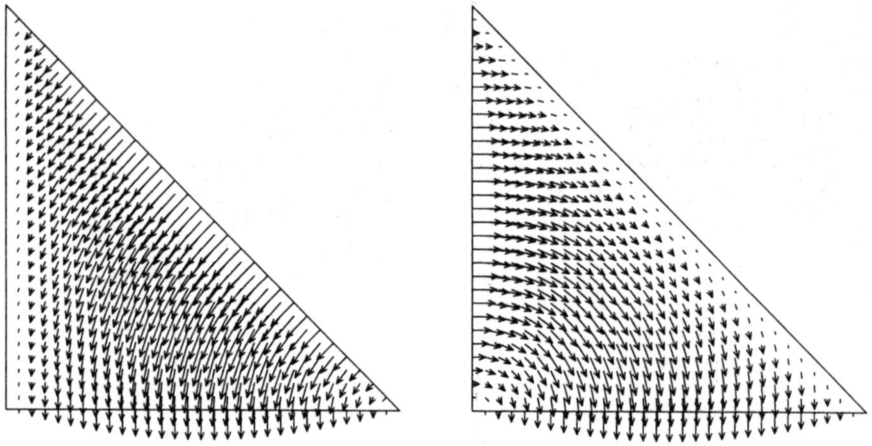

$\{\hat{N}_7\}$ $\{\hat{N}_8\}$

Figure C.3 Last two basis functions of the second-order curl-conforming triangular parent element of [Peterson 1994].

C.4.2.2 *Tetrahedral elements*

First-Order Elements

The space of vector functions to be used is the space $R_1^{(3)}$ (see Appendix A.1.1) that is composed of vectors $\{\hat{t}(\xi,\eta,\zeta)\}$ that can be expressed in matrix notation as

$$\{\hat{t}\} = \begin{Bmatrix} \hat{t}_x \\ \hat{t}_y \\ \hat{t}_z \end{Bmatrix} = \begin{Bmatrix} a_1 - b_3\eta + b_2\zeta \\ a_2 + b_3\xi - b_1\zeta \\ a_3 - b_2\xi + b_1\eta \end{Bmatrix} \tag{C.54}$$

It can be seen that the number of coefficients that are required to specify (C.54), i.e.

the number of degrees of freedom, is six. The definition of the degree of freedom is given by expression (2.55r) of Chapter 2 sampled at the midpoint of each edge. The function q is chosen as the length of the edge. This choice provides invariant degrees of freedom. Hence, to the ith degree of freedom there corresponds a node (ξ_i,η_i,ζ_i) placed at the midpoint of the ith edge of the parent element (see Table 2.8 of Chapter 2).

Let $\{\hat{t}\}$ be the vector variable to be approximated over the parent element. The ith degree of freedom may be defined as

$$d_i = \hat{L}_i \{\hat{a}_{\tau_i}\}^T \{\hat{t}(\xi_i,\eta_i,\zeta_i)\} \ , \quad i=1,\ldots,6 \tag{C.55}$$

where $\hat{L}_i$ is the length of the ith edge and $\{\hat{a}_{\tau i}\}$ is the unit vector, tangential to the ith edge of the parent element.

The expression for the interpolation is given by

$$\Pi\{\hat{t}(\xi,\eta\zeta)\} = \sum_{j=1}^{6} \{\hat{N}_j\} d_j \tag{C.56a}$$

where the vector basis functions are given by

$$\{\hat{N}\} = \begin{Bmatrix} a_1^{(i)} - b_3^{(i)}\eta + b_2^{(i)}\zeta \\ a_2^{(i)} + b_3^{(i)}\xi - b_1^{(i)}\zeta \\ a_3^{(i)} - b_2^{(i)}\xi + b_1^{(i)}\eta \end{Bmatrix} \tag{C.56b}$$

Substituting (C.56a) into (C.55) the following expression holds:

$$d_i = \hat{L}_i \left(\{a_{\tau_i}\}^T \sum_{j=1}^{6} \{\hat{N}_j(\xi_i,\eta_i,\zeta_i)\} d_j \right), \quad i=1,\ldots,6 \tag{C.57}$$

In order to obtain the coefficients of the polynomials for each component of the vector basis functions (C.56b), the following conditions should be fulfilled:

$$\left.\begin{aligned} \hat{L}_i \{\hat{a}_{\tau_i}\}^T \{\hat{N}_j(\xi_i,\eta_i,\zeta_i)\} &= 1, \quad i=j \\ \hat{L}_i \{\hat{a}_{\tau_i}\}^T \{\hat{N}_j(\xi_i,\eta_i,\zeta_i)\} &= 0, \quad i\neq j \end{aligned}\right\} \quad \begin{aligned} i&=1,\ldots,6 \\ j&=1,\ldots,6 \end{aligned} \tag{C.58}$$

These expressions are analogous to those given by (2.41) for Lagrange elements. The value of $\hat{L}_i$ and the expression of $\{\hat{a}_{\tau i}\}$, $i = 1, ..., 6$, are easily obtained from the parent element. Substituting them into (C.58) as well as into (C.56b), a system of 36 equations with 36 unknowns, $a_j^{(i)}$, $b_j^{(i)}$, $i, j = 1, ..., 6$, will be obtained. Its solution leads to the expression of the vector basis functions given in Table 2.8 of Chapter 2.

Second-Order Elements

The space of vector functions to be used is the space $R_2^{(3)}$ (see Appendix A.1.1) that is composed of vectors $\{\hat{t}(\xi,\eta,\zeta)\}$ that can be expressed in matrix notation as:

$$\{\hat{t}\} = \begin{Bmatrix} \hat{t}_x \\ \hat{t}_y \\ \hat{t}_z \end{Bmatrix} = \begin{Bmatrix} a_1 + a_2\xi + a_3\eta + a_4\zeta + D\eta^2 - F\xi\eta - G\xi\zeta + H\zeta^2 + J\eta\zeta \\ b_1 + b_2\xi + b_3\eta + b_4\zeta - D\xi\eta - E\eta\zeta + F\xi^2 + I\zeta^2 - J\xi\zeta + K\xi\zeta \\ c_1 + c_2\xi + c_3\eta + c_4\eta + E\eta^2 + G\xi^2 - H\xi\zeta - I\eta\zeta - K\xi\eta \end{Bmatrix} \qquad \text{(C.59)}$$

The number of coefficients that are required to specify (C.59), i.e., the number of degrees of freedom, is 20. There are two types of degrees of freedom. The definition of the first one is given again by expression (2.55r) of Chapter 2 sampled at two points (the two Gauss-Legendre points of integration) of each edge. The function q is chosen as the length of the edge. Hence, there are 12 degrees of freedom of the first type to which there correspond 12 nodes $(\xi_{ij},\eta_{ij},\zeta_{ij})$ located at the jth Gauss integration point (j = 1, 2) of the ith edge of the parent element (see Table 2.8 of Chapter 2).

Let $\{\hat{t}\}$ be the vector function to be approximated. The first 12 degrees of freedom are defined as

$$d_k = \hat{L}_i \{\hat{a}_{\tau_i}\}^T \{\hat{t}(\xi_{ij},\eta_{ij},\zeta_{ij})\}\ , \quad k = 2(i-1)+j\ , \quad i = 1,\dots,6\ , \quad j = 1,2 \qquad \text{(C.60a)}$$

In order to simplify the notation, the node $(\xi_{ij},\eta_{ij},\zeta_{ij})$ may be called the kth node, (ξ_k,η_k), to which $\hat{L}_k$ and $\{\hat{a}_{\tau k}\}$ (where $\hat{L}_1 = \hat{L}_2$, $\hat{L}_3 = \hat{L}_4$, $\hat{L}_5 = \hat{L}_6$, $\hat{L}_7 = \hat{L}_8$, $\hat{L}_9 = \hat{L}_{10}$, $\hat{L}_{11} = \hat{L}_{12}$, $\{\hat{a}_{\tau 1}\} = \{\hat{a}_{\tau 2}\}$, $\{\hat{a}_{\tau 3}\} = \{\hat{a}_{\tau 4}\}$, $\{\hat{a}_{\tau 5}\} = \{\hat{a}_{\tau 6}\}$, $\{\hat{a}_{\tau 7}\} = \{\hat{a}_{\tau 8}\}$, $\{\hat{a}_{\tau 9}\} = \{\hat{a}_{\tau 10}\}$, $\{\hat{a}_{\tau 11}\} = \{\hat{a}_{\tau 12}\}$) are associated, i.e., the length of the edges and the unit normal vectors corresponding to the kth node. Then, one obtains

$$d_k = \hat{L}_k \{\hat{a}_{\tau_k}\}^T \{\hat{t}(\xi_k,\eta_k,\zeta_k)\}\ , \quad k = 1,\dots,12 \qquad \text{(C.60b)}$$

The other eight degrees of freedom are related to the faces of the element. They are defined by expression (2.55s) of Chapter 2. There are two degrees of freedom per face.

Their definition requires the selection of two vectors tangential to the face. In theory, those vectors could be any two defining the plane of the given face of the tetrahedral parent element. However, the invariance of the degrees of freedom in the mapping from the parent element to the real element must be preserved. The expression (2.56), relevant for the transformation of vectors between the parent and the real element, does not guarantee that all vectors tangential to a given face of the parent element will be transformed into vectors tangential to the corresponding face of the real element. However, it is easy to demonstrate that vectors on the direction of the three edges of a face of the parent element will be mapped into vectors tangential to the corresponding face of the real element on the direction of the corresponding edges. In summary, in order to obtain invariant degrees of freedom, the vectors involved in the definition of this type of degrees of freedom should be chosen in the direction of the edges of the face of the parent element. In Table 2.8 of Chapter 2, the expressions for the three vectors $\{\hat{q}_i\}$, $i = 1, 2, 3$, for each face of the parent element are given. Then one has

$$d_k = \int_{\hat{\Gamma}_f} \left\{ \{\hat{t}(\xi,\eta,\zeta)\} \times \{\hat{a}_{n_f}\} \right\}^T \{\hat{q}_\alpha\} d\Gamma\,, \quad \alpha = 1, \text{or } 2, \text{or } 3\,,\ f = 1,\ldots,4\,,\ k = 12 + f \tag{C.60c}$$

$$d_k = \int_{\hat{\Gamma}_f} \left\{ \{\hat{t}(\xi,\eta,\zeta)\} \times \{\hat{a}_{n_f}\} \right\}^T \{\hat{q}_\beta\} d\Gamma\,, \quad \beta = 2, \text{or } 3, \text{or } 1\,,\ f = 1,\ldots,4\,,\ k = 16 + f \tag{C.60d}$$

where the unit vector normal to each face should be substituted. The interpolation is

$$\Pi\{\hat{t}(\xi,\eta,\zeta)\} = \sum_{j=1}^{20} \{\hat{N}_j\} d_j \tag{C.61}$$

and the expression of each vector basis functions will be given by (C.59). Thus, for each of the 20 basis functions, 20 coefficients should be computed. Proceeding in an analogous fashion as for the other elements, a system of 400 equations with 400 unknowns is obtained that will provide the coefficients of each basis function. As it will be clear from the discussion of Appendix C.4.3, for the correct assemblage of the last 8 degrees of freedom coming from contiguous elements, the basis functions for all three possible combinations of vectors $\{\hat{q}_\alpha\}$ and $\{\hat{q}_\beta\}$ are required (i.e., $\{\hat{q}_1\}$ and $\{\hat{q}_2\}$; $\{\hat{q}_1\}$ and $\{\hat{q}_3\}$; and $\{\hat{q}_2\}$ and $\{\hat{q}_3\}$). It is interesting to observe that the first 12 basis functions remain the same, while the rest depend on the choice, as shown in Table 2.8.

C.4.3 ON THE ASSEMBLY OF DIV-CONFORMING AND CURL-CONFORMING ELEMENTS

The definition of the first type of degrees of freedom of the div-conforming elements (see (2.51g) of Chapter 2), as well as the first and second types of degrees of freedom of the curl-conforming elements (see (2.55p), (2.55r), and (2.55s) of Chapter 2) involves the product of the vector unknown by the unit vector normal or tangential to the edges or the faces of the element. The unit vector normal to an element interface has been defined as directed toward the outside of the element. The definition for the unit vector tangential to an edge assumes the trigonometric direction (see, for the 2D case, Figure 2.22 of Chapter 2). Thus, the degrees of freedom associated with nodes located on an edge or face common to two different elements do not have a unique definition, because in each element the unit vector normal or tangential to that edge or face will have opposite signs, as shown in Figure C.4 for the case of two first-order div-conforming or curl-conforming triangles. In Figure C.4, the local numbering assigned to each node of the elements is shown. Bold numbers refer to the global numbering of nodes. Then it is seen that for the node that has been numbered globally as 5 (i.e., node number 2 for element Ω^1, that is also node number 3 for element Ω^2) the following relations hold:

$$\left\{a_{n_2}^1\right\} = -\left\{a_{n_3}^2\right\} \tag{C.62a}$$

$$\left\{a_{\tau_2}^1\right\} = -\left\{a_{\tau_3}^2\right\} \tag{C.62b}$$

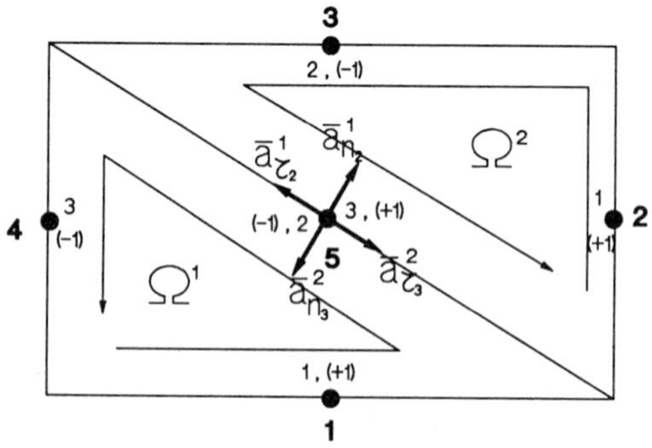

Figure C.4 Definition of the normal and tangential unit vectors. Assignment of sign to the edges.

Thus, the degrees of freedom associated with a node, common to two elements, will have a local definition that differs in sign according to one or the other element. The assembly process is based on the identity of the degrees of freedom associated with the nodes that are common to several elements. Thus, it is evident that for div-conforming and curl-conforming elements, it is required to adopt a unique definition of those degrees of freedom. This implies that one should select between one or the other local degree of freedom in order to define a unique global degree of freedom, e.g.,

$$d_5^g = -d_5^1 = d_5^2 \tag{C.63}$$

where d_i^g is the ith global degree of freedom and d_i^e is the corresponding local degree of freedom of the eth element. In general, it may be written as

$$d_i^e = d_i^g \, sign(i_{\Omega^e}) \qquad \text{(C.64)}$$

where $sign(i_{\Omega^e})$ is a function that assigns locally, in the element Ω^e, the value +1 or -1 to the variable called *sign*, so the value of the ith global degree of freedom should be multiplied by that variable in order to obtain the local degree of freedom. This assignment is done according to the sign that has been specified locally for the edge on which the node associated with that degree of freedom is located, following the algorithm:

- if $x_{ji} > 0$, or ($x_{ji} = 0$, and $y_{ji} > 0$), or ($x_{ji} = 0$, and $y_{ji} = 0$, and $z_{ji} > 0$), a positive sign is assigned;
- in all other cases, a negative sign is assigned,

where $x_{ji} = x_j - x_i$, $y_{ji} = y_j - y_i$, and $z_{ji} = z_j - z_i$. Here, (x_i, y_i, z_i), (x_j, y_j, z_j) are the coordinates of the initial and final vertices, respectively, of the edge, when the element is defined in the trigonometric sense. Figure C.4 gives an example of the sign allocations for the unit vectors normal and tangential to the edges of a 2D element.

In the case of 3D elements, the allocation of signs to the edges (i.e., to the unit tangential vectors) follows the same procedure. Here the faces of the element are run in the trigonometric sense when the element is observed from the outside. Figure C.5 shows this issue. For the case of the unit vector normal to the faces, ones need to compute it on the outward direction of the element. Once the initial and final points of the normal vector are computed, the preceding algorithm will give the corresponding sign.

In that way the proper sign may be allocated to the local degree of freedom. In conclusion, the definition of those degrees of freedom must be completed with the inclusion of the corresponding sign, so a definition of the local degrees of freedom $\{d^e\}$ as well as a unique definition of the global degrees of freedom $\{d^g\}$ will be obtained according to (C.64). The value +1 is assigned to the internal degrees of freedom.

In order to automatically assemble the matrices of the elements so that the global discrete integral form can be obtained the following procedure is adopted.

Let us assume a problem in which div-conforming or curl-conforming elements have been employed. The variational expression for a given element would be

$$W^e = \{\delta d^e\}^T [k^e] \{d^e\} - \{\delta d^e\}^T \{f^e\} \qquad \text{(C.65a)}$$

for a deterministic problem, or

$$W^e = \{\delta d^e\}^T ([k^e]\{d^e\} - \lambda [m^e]\{d^e\}) \qquad \text{(C.65b)}$$

for an eigenvalue problem.

If in the previous expressions the product that involves matrix $[k^e]$ is analyzed, then expressions of the following type will be obtained

$$\delta d_i^e k_{i1}^e d_1^e + \delta d_i^e k_{i2}^e d_2^e + \delta d_i^e k_{i3}^e d_3^e + \cdots \tag{C.66}$$

where $i, j = 1, 2, 3$, refer to the degrees of freedom corresponding to the eth element. That expression, with the help of (C.64), may be written as

$$\begin{aligned}
&\left(\delta d_i^g \, sign(i_{\Omega^e})\right) k_{i1}^e \left(sign(1_{\Omega^e})\, d_1^g\right) \\
&\quad + \left(\delta d_i^g \, sign(i_{\Omega^e})\right) k_{i2}^e \left(sign(2_{\Omega^e})\, d_2^g\right) \\
&\quad + \left(\delta d_i^g \, sign(i_{\Omega^e})\right) k_{i3}^e \left(sign(3_{\Omega^e})\, d_3^g\right) + \cdots \\
&\quad = (\delta d_i^g)\, d_1^g \left(k_{i1}^e \; sign(i_{\Omega^e}) \; sign(1_{\Omega^e})\right) \\
&\qquad + (\delta d_i^g)\, d_2^g \left(k_{i2}^e \; sign(i_{\Omega^e}) \; sign(2_{\Omega^e})\right) \\
&\qquad + (\delta d_i^g)\, d_3^g \left(k_{i3}^e \; sign(i_{\Omega^e}) \; sign(3_{\Omega^e})\right) + \cdots
\end{aligned} \tag{C.67}$$

leading to expressions of the type

$$W^e = \{\delta d^{ge}\}^T [k^{le}] \{d^{ge}\} - \{\delta d^{ge}\}^T \{f^{le}\} \tag{C.68a}$$

for deterministic problems, or

$$W^e = \{\delta d^{ge}\}^T ([k^{le}] - \lambda [m^{le}]) \{d^{ge}\} \tag{C.68b}$$

for eigenvalue problems. The variational formulation is given now in terms of the global expression of the degrees of freedom, so the usual formulation for each element is obtained

$$\{W^e\} = [k^{le}]\{d^{ge}\} - \{f^{le}\} \tag{C.69a}$$

for deterministic problems, or

$$\{W^e\} = ([k^{le}] - \lambda [m^{le}])\{d^{ge}\} \tag{C.69b}$$

for eigenvalue problems.

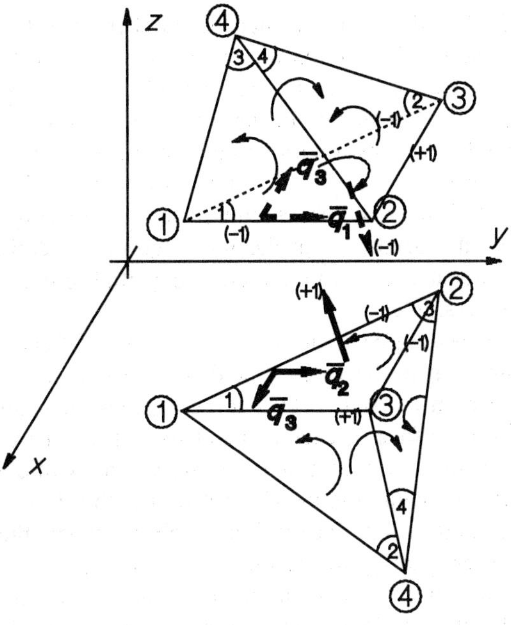

Figure C.5 Tetrahedrons showing the local numbering.

In conclusion, from the matrices of each element $[k^e]$, $[m^e]$, and $\{f^e\}$, the matrices $[k^{le}]$, $[m^{le}]$, and $\{f^{le}\}$ have been obtained, whose coefficients are given by

$$
\begin{aligned}
k_{ij}^{le} &= k_{ij}\ sign(i_{\Omega^e})\ sign(j_{\Omega^e}) \\
m_{ij}^{le} &= m_{ij}\ sign(i_{\Omega^e})\ sign(j_{\Omega^e}) \\
f_i^{le} &= f_i\ sign(i_{\Omega^e})
\end{aligned}
\tag{C.70}
$$

Once the matrices for each element of $[k^{le}]$, $[m^{le}]$, and $\{f^{le}\}$ are computed, their assemblage can be done as usual in order to obtain the global matrices $[K]$, $[M]$, and $\{F\}$, respectively. After imposing the boundary conditions, the solution of the corresponding system of equations will give the column matrix of global degrees of freedom $\{D\}$, whose values are the coefficients d_i^g.

For a particular postprocess, like the computation of the electric or the magnetic field, it will be necessary to know the values of the local degrees of freedom $\{d^e\}$ for the eth element. They can be obtained through expression (C.64).

In the case of 3D elements, a similar reasoning is required for the unit vector tangential to edges or the unit vector normal to faces. Figure C.5 shows two contiguous tetrahedrons. They have been shrunk for better visibility. The local numbering of vertices is shown. It should be noticed that the numbering of the vertices is obtained running the base of the tetrahedron on the trigonometric sense from the inside of the tetrahedron so that vertex 4 will be obtained according to the usual positive criterion (i.e., positive value of the determinant of the Jacobian matrix). The local numbering of the faces is determined by the local numbering of vertices (face 1: vertices 1, 3, 2; face 2: vertices 1, 4, 3; face 3: vertices 1, 2, 4; face 4: vertices 2, 3, 4). As mentioned before, faces are run following the trigonometric sense observing the element from the outside. These criterions together with the *sign* algorithm mentioned before give the sign that should be allocated to the unit vector tangential to edges or normal to faces. Figure C.5 shows an example.

There is an interesting peculiarity for the degrees of freedom related to faces in which vectors $\overline{q}$ tangential to the faces should be selected (see (2.55s) of Chapter 2). Those degrees of freedom are shared by two contiguous elements. Thus, one need to select the vectors $\overline{q}$ on the face of each element so that the degree of freedom will be invariant. This implies that for each element those vectors will correspond to different vectors of the parent element. Figure C.5 illustrates this issue. For the upper part tetrahedron, the face being an interface is face number 1, where the vectors selected are $\overline{q}_1$ and $\overline{q}_3$, (i.e., corresponding to vectors $\{\hat{q}_1\}$ and $\{\hat{q}_3\}$ of the parent element, respectively). Those forming an angle closest to 90° are chosen in order to optimize the condition number of matrices. For the lower part tetrahedron, the face being an interface is face number 1, where the corresponding vector should be $\overline{q}_2$ and $\overline{q}_3$, (i.e., corresponding to vectors $\{\hat{q}_2\}$ and $\{\hat{q}_3\}$ of the parent element, respectively). This means that for the upper part element, one needs to use the basis functions computed with $\{\hat{q}_1\}$ and $\{\hat{q}_3\}$ for face 1, while for the lower element, one needs to use the basis functions computed with $\{\hat{q}_2\}$ and $\{\hat{q}_3\}$ for face 1 (see Table 2.8 of Chapter 2). A criterion to distinguish which is the first and the second degree of freedom is also needed.

C.5 SOME GENERAL CONCLUSIONS REGARDING THE USE OF LAGRANGE ELEMENTS AND DIV-CONFORMING OR CURL-CONFORMING ELEMENTS

In the following, a comparison from the point of view of the type of element employed, between the different formulations mentioned in this book for 2D and 3D deterministic and eigenvalue problems is presented.

C.5.1 Two-Dimensional Deterministic Problems: Quasi-Static Analysis of Transmission Lines. Lagrange Elements Versus Div-Conforming Elements

Consider a second-order elliptic deterministic problem that is formulated according to (2.1) with (2.6) to (2.11). Let u be the primary variable that constitutes the scalar unknown of the problem. A vector variable $\overline{q}$ (or $\{q\}$) is introduced according to

$$\{q\} = [a]\{\nabla_t\}u \tag{C.71}$$

where $[a]$ is given by (2.6), the problem may be formulated in terms of $\{q\}$ (complementary formulation) or in terms of both u and $\{q\}$ (mixed formulation). The quasi-static analysis of transmission lines follows the direct formulation, where the electrostatic potential function $\phi(x,y)$ is the unknown u, and the mixed formulation, where the displacement vector $\bar{D}(x,y) = \underline{\underline{\varepsilon}}_r\bar{\nabla}_t\phi$ is the unknown $\{q\}$.

Both of the weak formulations employed in this book, the direct one and the mixed one, lead to positive-definite formulations. The number of unknowns of the continuous problem is also the same, because in the mixed formulation a penalty factor is employed in order to eliminate the scalar unknown (see Appendix E.3.2).

The weak formulation of the standard problem will provide an approximate solution, $\tilde{u}$, that should belong to the space $H^1(\Omega)$ (see Appendix A.1.1.3). The use of the Lagrange element of order p, means that the approximation of u, $\tilde{u}$, would be a polynomial of order p which could be a continuous scalar function. Its gradient, $\bar{\nabla}_t\tilde{u}$, would be a vector function whose components, in general, would be of p-1 order and is a discontinuous polynomial function, with discontinuous normal component and continuous tangential component at the interfaces between elements. This discretization is, thus, consistent with the mathematical formulation of the electrostatic problem that makes use of a potential function from which the electric field is derived since the potential function must be continuous over the domain. The condition of the continuity of the tangential component of the electric field across the dielectric interfaces is also fulfilled and the jump of the normal component of the electric field at those interfaces is also allowed. However, the continuity of the normal component of the electric field at a homogeneous media is not exactly fulfilled (because it may be discontinuous at the interfaces between elements) as well as the continuity of the normal component of the displacement vector at both interfaces between elements and dielectric interfaces.

If the mixed formulation is employed, the approximate solution, $\{\tilde{q}\}$, should belong to the space $H(\text{div},\Omega)$ (see Appendix A.1.1.4). The use of div-conforming elements of order p means that the approximation of $\{q\}$ will be a vector function whose Cartesian components would be a p degree polynomial that can be a discontinuous function. The normal component at the interfaces between elements will be a continuous polynomial of order p-1. Instead, the tangential component would be discontinuous. This formulation is, thus, consistent with the electromagnetic formulation of the electrostatic problem in terms of the displacement vector because the normal component of this vector at dielectric interfaces should be continuous and a jump of the tangential component of the displacement vector at interfaces is allowed. Nevertheless, it does not offer a continuous solution for the electric field in a homogenous medium and the continuity of the tangential components of the electric field at dielectric interfaces is not exactly satisfied. Given the relation between $\{q\}$ and u in (C.71), it is easy to demonstrate that the interpolation error and the convergence of functions $\{\tilde{q}\}$ are of the same order in both the cases if Lagrange elements are employed or if div-conforming elements of the same degree are used (see Section 2.5.1.2).

The matrix coefficients of the circuital primary parameters of a multiconductor transmission line are computed by evaluating the energy (through the derivative of the scalar unknown in the standard formulation) or by computing the charge on the conductors (through the normal component of the vector unknowns in the mixed formulation). The rate of convergence of such a computation for both formulations (the first one with Lagrange elements and the second one with div-conforming elements) is the same if elements of identical order are employed. So, in this sense, there is no reason to choose between one formulation or the other except for the fact that the postprocessing in the second formulation is much simpler (see Appendix E.3.2). A clear advantage of the formulation in terms of the displacement vector (i.e., using div-conforming elements) is that for the computation of the parameters of a multiconductor line the number of problems that should be solved is less than for the other case. It is interesting to note that each formulation offers a different type of bound for the exact solution. Hence, it may be useful to simultaneously apply both the formulations.

For the problems that are handled in this book, frequently there are singularities in the $\{q\}$ variables at the corners of the structures. Those corners coincide with the vertices of the elements of the mesh. If the Lagrange elements are employed, the singularity appears not in the u unknown but in its derivatives. This may be seen as an advantage. However, for the corners in which the singularities are present, a node is always located, which may cause some problems. On the other hand, if div-conforming elements are used, the unknown will be singular at corners but these will not coincide with the location nodes. Thus, good convergence will be obtained also in these cases.

For both Lagrange elements and div-conforming elements of the same order, the number of unknowns or the degrees of freedom of each element is identical. For example, they are three for the first order and six for the second order. This is due to the div-conforming element in which the last two degrees of freedom, i.e., the internal degrees of freedom, are eliminated at the local level (see Appendix E.3.2). Nevertheless, it is interesting to observe that the analysis of the problem by means of triangular div-conforming elements will provide a discrete system of equations with a larger number of unknowns than the case in which Lagrange triangular elements of the same order are used, as in the former case each node is common only to a maximum of two elements while in the later case, each node that is located at the vertex of a triangle may be common to more than two elements. In other words, Lagrange elements have a greater connectivity than div-conforming elements. Conversely, div-conforming elements will provide highly sparse matrices [Bossavit 1989], so memory, storage and CPU time requirements would be similar or even less than those for the Lagrange formulation.

Thus, no general rules may be derived regarding a clear preference for one formulation over the other because it highly depends on the objectives of the user.

C.5.2 Two-Dimensional Eigenvalue Problems. Full-Wave Analysis of Waveguiding Structures. Lagrange Elements Versus Lagrange/Curl-Conforming Elements

The full-wave formulations for the analysis of waveguiding structures employed in this book do not make use of scalar or vector potentials. They are written in terms of the electric field or the magnetic field (or some components of them). In this book, three different formulations have been used.

The first formulation employs the longitudinal components of the electric field and the magnetic field as scalar unknowns. The transverse components may be obtained from the longitudinal ones. For the case of homogenous structures, the number of unknowns may be reduced to one, although that implies that a separate analysis has to be carried out for the TM modes (in which case the unknown is the longitudinal component of the electric field) and for the TE modes (where the unknown is the longitudinal component of the magnetic field). This formulation leads to a positive-definite problem for the case of homogeneous structures. However, it is not definite for inhomogeneous structures. Depending on the structure, there may be problems with the numerical computations since there may be terms with nearly zero values in the denominator of the coefficients of the matrices of the discrete problem (see Appendix F.2).

The other two formulations employ as unknowns the longitudinal component and the transverse component of either the electric field or the magnetic field, respectively. The magnetic field or the electric field may be computed from the primary unknowns, respectively. Thus, the number of variational unknowns is also two: one of them is scalar (the magnitude of the longitudinal component) and the other one is a vector (the transverse component). A proper manipulation of the expressions will lead to formulations that are always positive definite and without numerical problems, so there is a clear advantage, in contrast to the formulation mentioned previously (see Appendix F.3).

The first formulation employs Lagrange elements for its discretization. Thus it is a vector formulation of the type mentioned in Section 2.5.1.2 because, for the case of inhomogeneous structures, two scalar unknowns for each node are employed. This discretization has been selected because it will lead to solutions that are consistent with the behavior of the electromagnetic field. The reason is that the longitudinal components of both the electric and the magnetic fields should be continuous over the waveguiding structure. Nevertheless, the transverse components for a waveguide problem are usually more important than the longitudinal ones and have a rate of convergence that is lower than for the longitudinal component; and the continuity conditions of the tangential components of the electric field and the magnetic field at dielectric interfaces will not be exactly fulfilled. A problem with this formulation arises for the case of inhomogeneous structures, where for given values of the propagation constant, spurious modes appear that can only be distinguished from the real ones by means of several postprocesses.

The other two formulations use the Lagrange/curl-conforming element described in Section 2.5.1.2. This discretization has been selected because it is consistent with the electromagnetic field behavior of the unknown, the reason being that the scalar unknown (the magnitude of the longitudinal component) will be continuous over the waveguiding structure because it is approximated by means of Lagrange elements and the transverse component will have continuous tangential components at dielectric interfaces because it is discretized by means of curl-conforming elements. It is evident that the continuity of the corresponding magnitude of the unknown that is acting as a secondary variable is not fulfilled. The formulation has no numerical problems and computes the spurious modes as eigenvectors corresponding to the zero eigenvalues.

Regarding the possible singularities that may be associated with the unknown or its derivatives, the same discussions made before still apply. This holds true also for the total number of degrees of freedom and the CPU time. The problem is with the application of the Lagrange elements or Lagrange/curl-conforming elements in the FEM formulation. In the formulation using longitudinal components, if the Lagrange elements are utilized, the unknown will not be singular but their derivatives will be. In the case of the Lagrange/curl-conforming element, the variables that would be singular at the corners (the transverse components) would be discretized employing nodes that are not located at those corners. Hence, the convergence problem is minimized.

Regarding the total number of degrees of freedom for the same spatial discretization (i.e., the same mesh) and for elements of same order, the number of unknowns will be greater in the case of the Lagrange/curl-conforming elements because nodes on the edges have a lower connectivity. But this disadvantage is compensated by the reduction in storage requirements and CPU time for the solution of the global system of equations since the matrix has a higher sparsity.

It may be concluded that for the case of full-wave analysis, it is clearly advantageous to use the formulation employing Lagrange/curl-conforming elements for the study of the inhomogeneous structures. In the case of homogeneous structures, the formulation employing the longitudinal components and Lagrange elements would be preferable, in principle, because the method reduces by more than half the number of unknowns. This is in spite of the fact that the structure should be analyzed twice in order to obtain the TE and TM modes.

C.5.3 Three-Dimensional Problems

In the case of three-dimensional problems, the advantages of using curl-conforming elements instead of Lagrange elements is evident. The former basis computes the spurious modes as eigenvectors corresponding to the zero eigenvalue. For the latter case, the spurious modes pollute the spectrum. A complete explanation of this issue is given in Section 7.2.

Appendix D

Maxwell's Equations in a Source-Free Region Specialized to Waveguiding Structures

D.1 INTRODUCTION

This appendix presents an overview of Maxwell's equations for a source-free medium and their specialization to waveguiding structures. The first section describes the electromagnetic properties of the medium and their mathematical properties. The second section deals with the mathematical characterization of the electric and the magnetic fields.

D.2 ELECTROMAGNETIC CHARACTERIZATION OF MEDIA IN ELECTROMAGNETIC STRUCTURES

The electromagnetic structures considered in this book and, in particular, waveguiding structures (structures with translational symmetry) are composed of one or more media in which a certain number of conductors are embedded. Guiding structures in which there are no conductors are also possible, provided there are at least two media.

The electromagnetic properties of a medium are determined by its conductivity σ (mhos/m), electric permittivity ε (Farads/m), and magnetic permeability μ (Henrys/m), which will generally vary with frequency f (Hertz). This implies that the medium is dispersive. The conductivity is generally taken into account through the imaginary part of a complex permittivity $\varepsilon = \varepsilon' - j\varepsilon''$ as $\sigma = \omega\varepsilon''$, where ω is the angular frequency (rad/s), corresponding to frequency f.

For an ideal conductor, the conductivity σ can be considered infinite for all frequencies of interest. In that case, the electromagnetic field inside the conductor is zero, whereas for an imperfect conductor of finite conductivity the field penetrates inside the material. The depth of penetration is characterized by δ_s for ordinary conductors or δ_{ssc} for superconductors (m^{-1}). The latter does not vary with frequency, while in the case of ordinary conductors the parameter δ_s is given by

$$\delta_s = (\pi f \mu_c \sigma)^{-1/2} \tag{D.1}$$

where μ_c is the magnetic permeability of the conductor.

A dielectric medium is characterized by a complex permittivity $\varepsilon = \varepsilon' - j\varepsilon''$, whereas for good dielectrics $\varepsilon'' \ll \varepsilon'$. A perfect dielectric is that for which $\varepsilon'' = 0$. The term $\varepsilon_r = \varepsilon/\varepsilon_0$ is called the relative permittivity of the medium. In general, dielectric media can have ohmic losses that are represented by means of a complex ε_r:

$$\varepsilon_r = \varepsilon_r' - j\varepsilon_r'' = |\varepsilon_r| e^{-j\delta_d} \tag{D.2a}$$

or equivalently by

$$\varepsilon_r = \varepsilon_r' \left(1 - j\frac{\sigma_d}{\omega \varepsilon_0 \varepsilon_r'} \right) \tag{D.2b}$$

where the equivalent conductivity of the dielectric is defined as

$$\sigma_d = \omega \varepsilon'' = \omega \varepsilon_0 \varepsilon_r'' \tag{D.2c}$$

The relative permittivity can also be written as

$$\varepsilon_r = \varepsilon_r' (1 - j \tan \delta_d) = \varepsilon_r' \left(1 - j\frac{\varepsilon_r''}{\varepsilon_r'} \right) \tag{D.2d}$$

where

$$\tan \delta_d = \frac{\sigma_d}{\omega \varepsilon_0 \varepsilon_r'} = \frac{\varepsilon_r''}{\varepsilon_r'} \tag{D.2e}$$

is the dielectric loss tangent of the material. It is important to note that the real and the imaginary parts of ε_r are mutually related by the fact that ε_r must be an analytical function in the right halfplane of the complex frequency. The relationships between them are the so-called Kronig-Kramer relations [Gardiol 1987; Ch. 4], [Ramo et al. 1965; Ch. 6]. The immediate consequence is that if a dielectric medium is dissipative ($\varepsilon_r'' \neq 0$), it is also dispersive, so $\varepsilon_r' = \varepsilon_r'(\omega)$ and $\varepsilon_r'' = \varepsilon_r''(\omega)$. However, at lower angular frequencies, ω, ε_r', and ε_r'' do not practically vary with frequency. In the microwave band, beyond a certain frequency, which is different for each material, most of the dielectrics are dispersive, having higher dispersion for ε_r'' than for ε_r'. However, dispersion is usually considered zero within a specific working band (from L to W band) [Elliot 1993; Ch. 7].

If the medium is non-magnetic, its permeability will be that of vacuum, which is μ_0. A magnetic medium will have, in general, a complex permeability $\mu = \mu' - j\mu''$. A perfect magnetic medium will be that in which $\mu'' = 0$ since it is not dispersive. The relative magnetic permeability is defined by $\mu_r = \mu/\mu_0$. In general, the magnetic media can have ohmic losses that are represented by means of a complex μ_r so that

$$\mu_r = \mu_r' - j\mu_r'' = |\mu_r| e^{-j\delta_m} \tag{D.3a}$$

or equivalently by

$$\mu_r = \mu_r'(1 - j\tan\delta_m) = \mu_r'\left(1 - j\frac{\mu_r''}{\mu_r'}\right) \tag{D.3b}$$

where

$$\tan\delta_m = \frac{\mu_r''}{\mu_r'} \tag{D.3c}$$

is the magnetic loss tangent [Harrington 1961; Ch. 1]. In general, the magnetic loss tangent of the materials most commonly used in microwave is much lower than the dielectric loss tangent. Generally, one would be dealing with materials where the losses are small, that is,

$$\tan\delta_d << 1 \tag{D.4a}$$

and

$$\tan\delta_m << 1 \tag{D.4b}$$

So far the material properties have been considered to be uniform (or isotropic) in all three Cartesian directions. However, in general, this may not be the case. A medium can be dielectrically and/or magnetically anisotropic, so its permittivity and/or permeability are given by three-dimensional (3D) second-order tensors with $\underline{\underline{\varepsilon}} = \varepsilon_0\, \underline{\underline{\varepsilon}}_r$ and $\underline{\underline{\mu}} = \mu_0\, \underline{\underline{\mu}}_r$. The tensors can be represented by the following 3×3 matrix:

$$[\xi_r] = \begin{bmatrix} \xi_{rxx} & \xi_{rxy} & \xi_{rxz} \\ \xi_{ryx} & \xi_{ryy} & \xi_{ryz} \\ \xi_{rzx} & \xi_{rzy} & \xi_{rzz} \end{bmatrix} \tag{D.5}$$

where $\xi_{r\alpha\beta} = \varepsilon_{r\alpha\beta}$ or $\mu_{r\alpha\beta}$ for $\alpha,\beta = x,y,z$. In this book, the previous matrix is assumed to be symmetric. In general, coefficients of (D.5) would be complex and dependent on the position, that is $\xi_{r\alpha\beta} = \xi_{r\alpha\beta}(x,y,z)$ (i.e., the medium will be in general lossy and inhomogeneous).

In this book, the media involved in waveguiding structures are assumed to have physical characteristics that may be described by means of tensors separable in a transverse and a longitudinal part. That is, (D.5) may be written as

$$[\xi_r] = \begin{bmatrix} \xi_{rxx} & \xi_{rxy} & 0 \\ \xi_{rxy} & \xi_{ryy} & 0 \\ 0 & 0 & \xi_{rzz} \end{bmatrix} \tag{D.6a}$$

with

$$[\xi_r] = \begin{bmatrix} [\xi_{rt}] & \{0\} \\ \{0\}^T & \xi_{rzz} \end{bmatrix} \tag{D.6b}$$

where $[\xi_{rt}]$ is a 2×2 symmetric matrix of relative transverse permittivity or permeability that represents the corresponding transverse tensor $\underline{\underline{\varepsilon}}_{rt}$ or $\underline{\underline{\mu}}_{rt}$ and ^{T} denotes the transpose of a matrix. The z-direction is assumed to be the direction of propagation of the electromagnetic energy in the waveguiding structure (i.e., the axis of translational symmetry). For the isotropic case

$$[\xi_r] = \xi_r[U] \tag{D.7a}$$

where $\xi_r = \varepsilon_r$ or μ_r and $[U]$ is the identity matrix defined as

$$[U] = \begin{bmatrix} 1 & 0 & 0 \\ 0 & 1 & 0 \\ 0 & 0 & 1 \end{bmatrix} \tag{D.7b}$$

If the material is perfect, then the matrix in (D.5) is real as well as symmetric and can be diagonalized with respect to a system of orthogonal axes. In dielectric materials those axes are called the main axes of the material. The coefficients of the diagonal matrix are the eigenvalues of (D.5). When the eigenvalues are different from each other, then the material is said to be biaxial. For these materials, the main axes are oriented along the coordinate system of the waveguiding structure so that $[\varepsilon_r]$ is diagonal, that is,

$$[\varepsilon_r] = \begin{bmatrix} \varepsilon_{rxx} & 0 & 0 \\ 0 & \varepsilon_{ryy} & 0 \\ 0 & 0 & \varepsilon_{rzz} \end{bmatrix} \tag{D.8}$$

In materials with one symmetry axis, called the optic axis, two of the eigenvalues are equal and the material is said to be uniaxial. The eigenvalue in the direction of the optic axis is usually represented by $\varepsilon_{r\|}$ and that in the perpendicular direction by $\varepsilon_{r\perp}$. The anisotropic substrates most commonly used in microwave circuits are uniaxial. They are normally used in planar structures in such a way that its optic axis is perpendicular to the plane of the substrate so that the permittivity matrix is diagonal with respect to the coordinates system of the structure. Therefore, one has

$$[\varepsilon_r] = \begin{bmatrix} \varepsilon_{r\|} & 0 & 0 \\ 0 & \varepsilon_{r\|} & 0 \\ 0 & 0 & \varepsilon_{r\perp} \end{bmatrix} \tag{D.9a}$$

In some cases it is preferable to rotate this axis so that

$$[\varepsilon_r] = \begin{bmatrix} \varepsilon_{r\perp}\cos^2\theta + \varepsilon_{r\|}\sin^2\theta & (\varepsilon_{r\perp} - \varepsilon_{r\|})\sin\theta\cos\theta & 0 \\ (\varepsilon_{r\perp} - \varepsilon_{r\|})\sin\theta\cos\theta & \varepsilon_{r\perp}\sin^2\theta + \varepsilon_{r\|}\cos^2\theta & 0 \\ 0 & 0 & \varepsilon_{r\perp} \end{bmatrix} \tag{D.9b}$$

where θ is the angle between the optical axis and the line perpendicular to the planar substrate axis.

In the problems considered in this book, the computation of the electromagnetic field is done under the assumption that all media are perfect (including conductors). Any possible losses will be taken into account by perturbational methods. This implies the assumption of low losses. Additionally, for waveguiding problems any inhomogeneity must obey the translational symmetry. Therefore, the most general expression for the relative electric permittivity and/or magnetic permeability for the calculation of the electromagnetic field in waveguiding structures will be given by

$$[\xi_r(x,y)] = \begin{bmatrix} \xi_{rxx}(x,y) & \xi_{rxy}(x,y) & 0 \\ \xi_{rxy}(x,y) & \xi_{ryy}(x,y) & 0 \\ 0 & 0 & \xi_{rzz}(x,y) \end{bmatrix} = \begin{bmatrix} [\xi_{rt}(x,y)]_t & \{0\} \\ \{0\}^T & \xi_{rzz}(x,y) \end{bmatrix} \tag{D.10}$$

where $[\xi_r]$ stands for $[\varepsilon_r]$ or $[\mu_r]$, and the coefficients of (D.10) are real.

D.3 STEADY-STATE MAXWELL'S EQUATIONS IN A SOURCE-FREE WAVEGUIDING STRUCTURE

D.3.1 Steady-State Maxwell's Equations in a Source-Free Structure

Maxwell's equations in a source-free medium (a cavity, or a waveguiding structure, for example) are given by

$$\bar{\nabla} \times \bar{e} = -\frac{\partial \bar{b}}{\partial t} \tag{D.11a}$$

$$\bar{\nabla} \times \bar{h} = \frac{\partial \bar{d}}{\partial t} \tag{D.11b}$$

$$\bar{\nabla} \cdot \bar{d} = 0 \tag{D.11c}$$

$$\bar{\nabla} \cdot \bar{b} = 0 \tag{D.11d}$$

where $\bar{e}(x,y,z,t)$ is the electric field intensity vector (volt/m), $\bar{h}(x,y,z,t)$ is the magnetic field intensity vector (ampere/m), $\bar{d}(x,y,z,t)$ is the electric flux density vector or electric displacement (coulomb/m^2), and $\bar{b}(x,y,z,t)$ is the magnetic flux density vector (weber/m^2). All of them have three components, in general, along each of the Cartesian coordinate directions [Ramo et al. 1965; Ch. 4]. $\bar{\nabla}$ is the 3D vector operator given by

$$\bar{\nabla} = \bar{a}_x \frac{\partial}{\partial x} + \bar{a}_y \frac{\partial}{\partial y} + \bar{a}_z \frac{\partial}{\partial z} \tag{D.11e}$$

The symbol $\times$ represents the vector product, while the symbol $\cdot$ denotes the scalar product between vectors, and $\bar{a}_x$, $\bar{a}_y$, $\bar{a}_z$ represent the unit vectors along the three Cartesian coordinates. The transverse operator $\bar{\nabla}_t$ is defined by

$$\bar{\nabla}_t = \bar{a}_x \frac{\partial}{\partial x} + \bar{a}_y \frac{\partial}{\partial y} \tag{D.11f}$$

For the sinusoidal steady-state, at angular frequency ω (rad/sec), the fields may be given by

$$\bar{e} = \mathrm{Re}[\underline{\bar{E}}(x,y,z)e^{j\omega t}] \tag{D.12a}$$

$$\bar{h} = \mathrm{Re}[\underline{\bar{H}}(x,y,z)e^{j\omega t}] \tag{D.12b}$$

$$\bar{d} = \mathrm{Re}[\underline{\bar{D}}(x,y,z)e^{j\omega t}] \tag{D.12c}$$

$$\bar{b} = \mathrm{Re}[\underline{\bar{B}}(x,y,z)e^{j\omega t}] \tag{D.12d}$$

where $\underline{\bar{E}}$, $\underline{\bar{H}}$, $\underline{\bar{D}}$, and $\underline{\bar{B}}$ are phasors that represent the vectors $\bar{e}$, $\bar{h}$, $\bar{d}$, and $\bar{b}$, respectively. Re[$\cdot$] denotes the real part of the complex expression between brackets. The nomenclature $\underline{\bar{V}}(x,y,z)$ has been used to refer to a vector with three components in the x, y, z directions, each of which is a function of the (x,y,z) coordinates

$$\underline{\bar{V}}(x,y,z) = \bar{a}_x \underline{V}_x(x,y,z) + \bar{a}_y \underline{V}_y(x,y,z) + \bar{a}_z \underline{V}_z(x,y,z) \tag{D.13}$$

The underline on a variable is used to differentiate it from the vectors that will later be represented by $\bar{V}(x,y)$, not underlined. These variables have three components and each one is a function of the (x,y) coordinates only:

$$\bar{V}(x,y) = \bar{a}_x V_x(x,y) + \bar{a}_y V_y(x,y) + \bar{a}_z V_z(x,y) \tag{D.14}$$

Using the phasor notation (in which factor $e^{j\omega t}$ is suppressed), one obtains

$$\bar{\nabla} \times \underline{\bar{E}} = -j\omega \underline{\bar{B}} \tag{D.15a}$$

$$\bar{\nabla} \times \underline{\bar{H}} = j\omega \underline{\bar{D}} \tag{D.15b}$$

$$\bar{\nabla} \cdot \underline{\bar{D}} = 0 \tag{D.15c}$$

$$\bar{\nabla} \cdot \underline{\bar{B}} = 0 \tag{D.15d}$$

Vectors $\underline{\bar{D}}$ and $\underline{\bar{B}}$ are related to $\underline{\bar{E}}$ and $\underline{\bar{H}}$ through the constitutive relations of the medium

$$\underline{\bar{D}} = \underline{\underline{\varepsilon}}\, \underline{\bar{E}} \tag{D.16a}$$

$$\underline{\bar{B}} = \underline{\underline{\mu}}\, \underline{\bar{H}} \tag{D.16b}$$

where $\underline{\underline{\varepsilon}}$ and $\underline{\underline{\mu}}$ are, in general, tensors that allow a matrix representation of type (D.5).

Equations (D.15) are, in general, represented as

$$\bar{\nabla} \times \underline{\bar{E}} = -\, j\omega\, \underline{\underline{\mu}}\, \underline{\bar{H}} \tag{D.17a}$$

$$\bar{\nabla} \times \underline{\bar{H}} = j\omega\, \underline{\underline{\varepsilon}}\, \underline{\bar{E}} \tag{D.17b}$$

$$\bar{\nabla} \cdot \underline{\underline{\varepsilon}}\, \underline{\bar{E}} = 0 \tag{D.17c}$$

$$\bar{\nabla} \cdot \underline{\underline{\mu}}\, \underline{\bar{H}} = 0 \tag{D.17d}$$

For $\omega \neq 0$, the last two equations may be obtained from the first two, since it is evident through the divergence of (D.17a,b), that

$$\bar{\nabla} \cdot (\bar{\nabla} \times \underline{\bar{E}}) \equiv 0 = -\, j\omega\, \bar{\nabla} \cdot \underline{\underline{\mu}}\, \underline{\bar{H}} \tag{D.18a}$$

$$\bar{\nabla} \cdot (\bar{\nabla} \times \underline{\bar{H}}) \equiv 0 = j\omega\, \bar{\nabla} \cdot \underline{\underline{\varepsilon}}\, \underline{\bar{E}} \quad \text{(D.18b)}$$

Hence, in principle, (D.17c,d) may be neglected in the problem formulation.

It is necessary to add to (D.17) the boundary conditions that must be fulfilled at the interfaces between different media, namely:

1. At an interface between two media:

$$\bar{a}_{n_1} \times \underline{\bar{E}}_1 + \bar{a}_{n_2} \times \underline{\bar{E}}_2 = 0 \quad \text{(D.19a)}$$

$$\bar{a}_{n_1} \times \underline{\bar{H}}_1 + \bar{a}_{n_2} \times \underline{\bar{H}}_2 = 0 \quad \text{(D.19b)}$$

$$\bar{a}_{n_1} \cdot \underline{\underline{\varepsilon}}_1\, \underline{\bar{E}}_1 + \bar{a}_{n_2} \cdot \underline{\underline{\varepsilon}}_2\, \underline{\bar{E}}_2 = 0 \quad \text{(D.19c)}$$

$$\bar{a}_{n_1} \cdot \underline{\underline{\mu}}_1\, \underline{\bar{H}}_1 + \bar{a}_{n_2} \cdot \underline{\underline{\mu}}_2\, \underline{\bar{H}}_2 = 0 \quad \text{(D.19d)}$$

where the subscript $i = 1$ or 2 refers to the ith medium and $\bar{a}_{n_i}$ is the unit vector normal to the interface and pointing out from the ith medium. Since $\bar{a}_{n_1} = -\bar{a}_{n_2}$, the first two equations imply the continuity of the tangential components of vectors $\underline{E}$ and $\underline{H}$. The last two equations represent the continuity of the normal components of vectors $\underline{D}$ and $\underline{B}$ at the boundary interface.

2. At a perfect electric wall, Γ_{pe}:

$$\bar{a}_n \times \underline{\bar{E}} = 0 \quad \text{(D.20a)}$$

$$\bar{a}_n \cdot \underline{\underline{\mu}}\, \underline{\bar{H}} = 0 \quad \text{(D.20b)}$$

where $\bar{a}_n$ is the unit vector normal to the electric wall, pointing toward the interior of the perfectly conducting medium. The first condition implies the vanishing of the tangential component of the electric field $\underline{E}$. The second equation shows that the normal component of the magnetic flux vector $\underline{B}$ is zero.

The surface charge density, ρ_s (coulomb/m^2), and current density, $\bar{J}_s$ (amp/m), on the conductor can be obtained through

$$\rho_s = -\bar{a}_n \cdot \underline{\underline{\varepsilon}}\, \underline{\bar{E}} \quad \text{(D.21a)}$$

$$\bar{J}_s = -\bar{a}_n \times \underline{\bar{H}} \tag{D.21b}$$

3. At a perfectly magnetic wall, Γ_{pm}:

$$\bar{a}_n \times \underline{\bar{H}} = 0 \tag{D.22a}$$

$$\bar{a}_n \cdot \underline{\underline{\varepsilon}}\, \underline{\bar{E}} = 0 \tag{D.22b}$$

where $\bar{a}_n$ is the unit vector normal to the magnetic wall.

The conditions concerning the normal components of the fields are dependent on the relationships of the tangential components, since the former are derived from the divergence equations of (D.15c,d). Thus, it would be enough to consider (D.19a,b), (D.20a), and (D.22a).

The sinusoidal steady-state fields in a source-free region are given by (D.17) and (D.19) to (D.22) for three-dimensional electromagnetic problems. Since there are more equations available than are necessary, the problems can be formulated in more ways than one.

A possible formulation may be to use (D.17a,b) along with the boundary conditions (D.19a,b), (D.20a), and (D.22a). It has been utilized in [Svedin 1989, 1990, 1991] for the analysis of waveguiding structures.

Another possibility is to use (D.17a,b) and solve for the electric field $\underline{\bar{E}}$ or the magnetic field $\underline{H}$. Multiplying the left-hand side of (D.17a) by $\underline{\underline{\mu}}^{-1}$ and taking the curl one obtains for $\omega \neq 0$

$$\bar{\nabla} \times (\underline{\underline{\mu}}^{-1} \bar{\nabla} \times \underline{\bar{E}}) = \omega^2 \, \underline{\underline{\varepsilon}}\, \underline{\bar{E}} \tag{D.23a}$$

By using an analogous procedure on (D.17b), one can obtain, for $\omega \neq 0$

$$\bar{\nabla} \times (\underline{\underline{\varepsilon}}^{-1} \bar{\nabla} \times \underline{\bar{H}}) = \omega^2 \, \underline{\underline{\mu}}\, \underline{\bar{H}} \tag{D.23b}$$

Expressions (D.23) constitute the generalized Helmholtz equations for the electric and the magnetic fields, respectively. They are called the double-curl equations

Equation (D.23a) is the double-curl formulation in terms of the $\underline{\bar{E}}$ field. For $\omega \neq 0$ the magnetic field could be obtained from the electric field using (D.17a):

$$\underline{\bar{H}} = \frac{j}{\omega} \underline{\underline{\mu}}^{-1} \bar{\nabla} \times \underline{\bar{E}} \tag{D.24a}$$

The boundary conditions in terms of $\underline{\bar{E}}$ are obtained by replacing (D.24a) in (D.19b) and (D.22a) together with (D.19a) and (D.20a). This formulation is used in [Chatterjee et al. 1992], [Konrad 1985], [Lee and Mittra 1992], [Wang and Ida 1991, 1992], and [Webb 1985], for example, for cavity problems, and in [Alam et al. 1994a, 1995], [Gil and Webb 1997], [Hayata et al. 1989], [Hano 1984], and [Koshiba et al. 1985b, 1994], among others, for waveguiding problems.

Expression (D.23b) provides the dual double-curl formulation in terms of the $\underline{\bar{H}}$ field vector. The electric field can be obtained from the magnetic field using the expression (D.17b), for $\omega \neq 0$, as

$$\underline{\bar{E}} = -\frac{j}{\omega}\,\underline{\underline{\varepsilon}}^{-1}\bar{\nabla} \times \underline{\bar{H}} \qquad \text{(D.24b)}$$

The boundary conditions in terms of $\underline{\bar{H}}$ are obtained by replacing (D.24b) in (D.19a) and (D.20a) together with (D.19b) and (D.22a). This formulation has been used in [Konrad 1985], [Wang and Ida 1992], and [Webb 1985], among others, for cavity analysis. It has been used for waveguiding problems in [Alam et al. 1994b], [Chaves et al. 1994], [Gil and Webb 1997], [Hano 1984], [Hayata et al. 1986], [Israel and Miniowitz 1987, 1990], [Koshiba et al. 1984a, 1985a, 1986, 1994], [Miniowitz and Webb 1991], [Valor and Zapata 1995, 1996], and [Rahman and Davies 1984a,b], among others.

Another possible formulation is to use both expressions in (D.23) with the boundary conditions (D.19a,b), (D.20a), and (D.22a), as in [Bossavit 1990], [Pichon and Bossavit 1993].

D.3.2 Specialization to Waveguiding Structures

For waveguiding problems it would be convenient to introduce the dependence of the fields along the longitudinal direction (i.e., along the direction of the wave propagation) in an explicit fashion. Hence

$$\underline{\bar{E}}(x,y,z) = \bar{E}(x,y)\,e^{\mp\gamma z} \qquad \text{(D.25a)}$$

$$\underline{\bar{H}}(x,y,z) = \bar{H}(x,y)\,e^{\mp\gamma z} \qquad \text{(D.25b)}$$

The factor $e^{\mp\gamma z}$ corresponds to the propagation in the positive (sign -) or negative (sign +) direction of the z-axis. Unless otherwise indicated, the propagation is assumed to be in the positive z direction. The propagation constant γ depends on frequency in general. It can be purely imaginary, $\gamma(\omega) = j\beta$. This implies that the mode is propagating with a phase velocity $v_f = \omega/\beta$. For the frequency range in which $\gamma(\omega)$ is real, $\gamma(\omega) = \alpha$, the mode is at cutoff, decaying according to $e^{-\alpha z}$. The cutoff angular frequency, ω_c, corresponds to the condition $\beta(\omega_c) = 0$. So below this value there would not be any propagation. In problems with losses, the propagation constant is given by $\gamma(\omega) = \alpha(\omega)+j\beta(\omega)$ and is a complex quantity.

In expressions (D.25a,b) vectors $\bar{E}$ and $\bar{H}$ are of the form (D.14), with three Cartesian components that are functions of the transverse coordinates. Taking the z-dependence into account and suppressing the factor $e^{-\gamma z}$, expressions (D.17a,b) become

$$\bar{\nabla}_t \times \bar{E} = \gamma\, \bar{a}_z \times \bar{E} - j\omega\, \underline{\underline{\mu}}\, \bar{H} \tag{D.26a}$$

$$\bar{\nabla}_t \times \bar{H} = \gamma\, \bar{a}_z \times \bar{H} + j\omega\, \underline{\underline{\varepsilon}}\, \bar{E} \tag{D.26b}$$

When the material parameters $\underline{\underline{\varepsilon}}$ and $\underline{\underline{\mu}}$ are tensors that may be represented by matrices of the form given by (D.6) it is convenient to split vectors $\bar{E}$ and $\bar{H}$ into their transverse and longitudinal components, according to

$$\bar{E}(x,y) = \bar{E}_t(x,y) + j\, E_z(x,y)\, \bar{a}_z \tag{D.27a}$$

$$\bar{H}(x,y) = \bar{H}_t(x,y) + j\, H_z(x,y)\, \bar{a}_z \tag{D.27b}$$

This results in

$$\bar{\nabla}_t \times \bar{E}_t = \omega\, \mu_0\, \mu_{rzz}\, H_z\, \bar{a}_z \tag{D.28a}$$

$$\bar{\nabla}_t \times \bar{H}_t = -\omega\, \varepsilon_0\, \varepsilon_{rzz}\, E_z\, \bar{a}_z \tag{D.28b}$$

$$\bar{\nabla}_t \times E_z\, \bar{a}_z = -j\gamma\, \bar{a}_z \times \bar{E}_t - \omega\, \mu_0\, \underline{\underline{\mu}}_{rt}\, \bar{H}_t \tag{D.28c}$$

$$\bar{\nabla}_t \times H_z\, \bar{a}_z = -j\gamma\, \bar{a}_z \times \bar{H}_t + \omega\, \varepsilon_0\, \underline{\underline{\varepsilon}}_{rt}\, \bar{E}_t \tag{D.28d}$$

It can be observed that for a lossless media and with propagating modes $\gamma = j\beta$ the last two equations become

$$\bar{\nabla}_t \times E_z\, \bar{a}_z = \beta\, \bar{a}_z \times \bar{E}_t - \omega\, \mu_0\, \underline{\underline{\mu}}_{rt}\, \bar{H}_t \tag{D.29a}$$

$$\bar{\nabla}_t \times H_z \bar{a}_z = \beta \bar{a}_z \times \bar{H}_t + \omega \varepsilon_0 \underline{\underline{\varepsilon}}_{rt} \bar{E}_t \qquad \text{(D.29b)}$$

which indicates that if $\bar{E}_t$ and $\bar{H}_t$ are real vectors, then E_z and H_z will also be real. Therefore according to (D.27), the longitudinal component of the fields would be out of phase $\pi/2$ with respect to the transverse components. For the lossless case, it is seen that (D.27) leads to a formulation in which only real quantities are involved.

If (D.27) is replaced in (D.17c,d), this will result in

$$\bar{\nabla}_t \cdot \varepsilon_0 \underline{\underline{\varepsilon}}_{rt} \bar{E}_t - j\,\gamma\,\varepsilon_0\,\varepsilon_{rzz} E_z = 0 \qquad \text{(D.30a)}$$

$$\bar{\nabla}_t \cdot \mu_0 \underline{\underline{\mu}}_{rt} \bar{H}_t - j\,\gamma\,\mu_0\,\mu_{rzz} H_z = 0 \qquad \text{(D.30b)}$$

which for the lossless case would be

$$\bar{\nabla}_t \cdot \varepsilon_0 \underline{\underline{\varepsilon}}_{rt} \bar{E}_t + \beta\,\varepsilon_0\,\varepsilon_{rzz} E_z = 0 \qquad \text{(D.31a)}$$

$$\bar{\nabla}_t \cdot \mu_0 \underline{\underline{\mu}}_{rt} \bar{H}_t + \beta\,\mu_0\,\mu_{rzz} H_z = 0 \qquad \text{(D.31b)}$$

The set of equations (D.28) and (D.30) is the starting point of many formulations. To this we need to add the boundary conditions.

1. At the boundary interface between two media

$$\bar{a}_{n_1} \times \bar{E}_{t_1} + \bar{a}_{n_2} \times \bar{E}_{t_2} = 0 \qquad \text{(D.32a)}$$

$$\bar{a}_{n_1} \times \bar{H}_{t_1} + \bar{a}_{n_2} \times \bar{H}_{t_2} = 0 \qquad \text{(D.32b)}$$

$$E_{z_1} = E_{z_2} \qquad \text{(D.32c)}$$

$$H_{z_1} = H_{z_2} \qquad \text{(D.32d)}$$

$$\bar{a}_{n_1} \cdot \underline{\underline{\varepsilon}}_{rt_1} \bar{E}_{t_1} + \bar{a}_{n_2} \cdot \underline{\underline{\varepsilon}}_{rt_2} \bar{E}_{t_2} = 0 \qquad \text{(D.32e)}$$

$$\bar{a}_{n_1} \cdot \underline{\underline{\mu}}_{rt_1} \bar{H}_{t_1} + \bar{a}_{n_2} \cdot \underline{\underline{\mu}}_{rt_2} \bar{H}_{t_2} = 0 \tag{D.32f}$$

2. At an electric wall

$$\bar{a}_n \times \bar{E}_t = 0 \tag{D.33a}$$

$$E_z = 0 \tag{D.33b}$$

$$\bar{a}_n \cdot \underline{\underline{\mu}}_{rt} \bar{H}_t = 0 \tag{D.33c}$$

and the following expressions

$$\rho_s = - \bar{a}_n \cdot \varepsilon_0 \underline{\underline{\varepsilon}}_{rt} \bar{E}_t \tag{D.34}$$

$$\bar{J}_s = \mathrm{j} \bar{J}_{st} + J_{sz} \bar{a}_z \tag{D.35a}$$

$$\bar{J}_{st} = - \bar{a}_n \times H_z \bar{a}_z \tag{D.35b}$$

$$\bar{J}_{sz} = - \bar{a}_n \times \bar{H}_t \tag{D.35c}$$

provide the surface charge density and the current density per unit width of the cross-section of the reference conductor.

3. At a magnetic wall

$$\bar{a}_n \times \bar{H}_t = 0 \tag{D.36a}$$

$$H_z = 0 \tag{D.36b}$$

$$\bar{a}_n \cdot \underline{\underline{\varepsilon}}_{rt} \bar{E}_t = 0 \tag{D.36c}$$

A possible mathematical formulation for the waveguide problem is to use the expression (D.28) together with conditions (D.32a-d), (D.33a,b), and (D.36a,b). This excludes the divergence equations and the related boundary conditions. However, the formulation involves all the field quantities $\bar{E}_t$, $\bar{H}_t$, E_z, and H_z as presented in (D.17a,b).

Another possibility is to replace the longitudinal components H_z and E_z by

$$H_z\, \bar{a}_z = \frac{1}{\omega\, \mu_0\, \mu_{rzz}} \bar{\nabla}_t \times \bar{E}_t \tag{D.37a}$$

$$E_z\, \bar{a}_z = -\frac{1}{\omega\, \varepsilon_0\, \varepsilon_{rzz}} \bar{\nabla}_t \times \bar{E}_t \tag{D.37b}$$

in (D.28c,d) and solve for only the transverse components $\bar{E}_t$ and $\bar{H}_t$, resulting in

$$\bar{\nabla}_t \times \frac{1}{\varepsilon_0\, \varepsilon_{rzz}} \bar{\nabla}_t \times \bar{H}_t = j\, \omega\, \gamma\, \bar{a}_z \times \bar{E}_t + \omega^2\, \mu_0\, \underline{\underline{\mu}}_{rt}\, \bar{H}_t \tag{D.38a}$$

$$\bar{\nabla}_t \times \frac{1}{\mu_0\, \mu_{rzz}} \bar{\nabla}_t \times \bar{E}_t = -\, j\, \omega\, \gamma\, \bar{a}_z \times \bar{H}_t + \omega^2\, \varepsilon_0\, \underline{\underline{\varepsilon}}_{rt}\, \bar{E}_t \tag{D.38b}$$

which for the lossless case reduces to

$$\bar{\nabla}_t \times \frac{1}{\varepsilon_0\, \varepsilon_{rzz}} \bar{\nabla}_t \times \bar{H}_t = -\omega\, \beta\, \bar{a}_z \times \bar{E}_t + \omega^2\, \mu_0\, \underline{\underline{\mu}}_{rt}\, \bar{H}_t \tag{D.39a}$$

$$\bar{\nabla}_t \times \frac{1}{\mu_0\, \mu_{rzz}} \bar{\nabla}_t \times \bar{E}_t = \omega\, \beta\, \bar{a}_z \times \bar{H}_t + \omega^2\, \mu_0\, \underline{\underline{\mu}}_{rt}\, \bar{E}_t \tag{D.39b}$$

The boundary conditions are given by (D.32a-d), (D.33a,b), and (D.36a,b) replacing (D.37). This formulation has been used in [Angkaew et al. 1987], [García-Castillo 1992], [García-Castillo and Salazar-Palma 1992a].

It is possible to obtain various formulations in terms of $\bar{E}_t$ and E_z, or $\bar{H}_t$ and H_z (or in terms of all the field components) by substituting (D.25) and (D.27) in (D.23) and (D.24), and isolating the equations relating the transverse components from those involving the longitudinal components. Then from (D.23a) results

$$\begin{aligned} &\mu_0^{-1}\, \bar{\nabla}_t \times \mu_{rzz}^{-1}\, \bar{\nabla}_t \times \bar{E}_t - j\, \gamma\, \mu_0^{-1}\, \bar{a}_z \times \underline{\underline{\mu}}_{rt}^{-1}\, \bar{\nabla}_t \times E_z\, \bar{a}_z \\ &\quad + \gamma^2\, \mu_0^{-1}\, \bar{a}_z \times \underline{\underline{\mu}}_{rt}^{-1}\, \bar{a}_z \times \bar{E}_t - \omega^2\, \varepsilon_0\, \underline{\underline{\varepsilon}}_{rt}\, \bar{E}_t = 0 \end{aligned} \tag{D.40a}$$

$$
\begin{aligned}
&j\,\mu_0^{-1}\,\bar{\nabla}_t \times \underline{\underline{\mu}}_{rt}^{-1}\,\bar{\nabla}_t \times E_z\,\bar{a}_z - \gamma\,\mu_0^{-1}\,\bar{\nabla}_t \times \underline{\underline{\mu}}_{rt}^{-1}\,\bar{a}_z \times \bar{E}_t \\
&\quad - j\,\omega^2\,\varepsilon_0\,\varepsilon_{rzz}\,E_z\,\bar{a}_z = 0
\end{aligned}
\tag{D.40b}
$$

while from (D.23b) one obtains

$$
\begin{aligned}
&\varepsilon_0^{-1}\,\bar{\nabla}_t \times \varepsilon_{rzz}^{-1}\,\bar{\nabla}_t \times \bar{H}_t - j\,\gamma\,\varepsilon_0^{-1}\,\bar{a}_z \times \underline{\underline{\varepsilon}}_{rt}^{-1}\,\bar{\nabla}_t \times H_z\,\bar{a}_z \\
&\quad + \gamma^2\,\varepsilon_0^{-1}\,\bar{a}_z \times \underline{\underline{\varepsilon}}_{rt}^{-1}\,\bar{a}_z \times \bar{H}_t - \omega^2\,\mu_0\,\underline{\underline{\mu}}_{rt}\,\bar{H}_t = 0
\end{aligned}
\tag{D.41a}
$$

$$
\begin{aligned}
&j\,\varepsilon_0^{-1}\,\bar{\nabla}_t \times \underline{\underline{\varepsilon}}_{rt}^{-1}\,\bar{\nabla}_t \times H_z\bar{a}_z - \gamma\,\varepsilon_0^{-1}\,\bar{\nabla}_t \times \underline{\underline{\varepsilon}}_{rt}^{-1}\,\bar{a}_z \times \bar{H}_t \\
&\quad - j\,\omega^2\,\mu_o\,\mu_{rzz}\,H_z\,\bar{a}_z = 0
\end{aligned}
\tag{D.41b}
$$

For the lossless case $\gamma = j\beta$. Multiplying (D.40) by μ_0 and (D.41) by ε_0 yields

$$
\begin{aligned}
&\bar{\nabla}_t \times \mu_{rzz}^{-1}\,\bar{\nabla}_t \times \bar{E}_t + \beta\,\bar{a}_z \times \underline{\underline{\mu}}_{rt}^{-1}\,\bar{\nabla}_t \times E_z\,\bar{a}_z \\
&\quad - \beta^2\,\bar{a}_z \times \underline{\underline{\mu}}_{rt}^{-1}\,\bar{a}_z \times \bar{E}_t - \omega^2\,\varepsilon_0\,\mu_0\,\underline{\underline{\varepsilon}}_{rt}\,\bar{E}_t = 0
\end{aligned}
\tag{D.42a}
$$

$$
\begin{aligned}
&\bar{\nabla}_t \times \underline{\underline{\mu}}_{rt}^{-1}\,\bar{\nabla}_t \times E_z\,\bar{a}_z - \beta\,\bar{\nabla}_t \times \underline{\underline{\mu}}_{rt}^{-1}\,\bar{a}_z \times \bar{E}_t \\
&\quad - \omega^2\,\varepsilon_0\,\mu_0\,\varepsilon_{rzz}\,E_z\,\bar{a}_z = 0
\end{aligned}
\tag{D.42b}
$$

as well as

$$
\begin{aligned}
&\bar{\nabla}_t \times \varepsilon_{rzz}^{-1}\,\bar{\nabla}_t \times \bar{H}_t + \beta\,\bar{a}_z \times \underline{\underline{\varepsilon}}_{rt}^{-1}\bar{\nabla}_t \times H_z\,\bar{a}_z \\
&\quad - \beta^2\,\bar{a}_z \times \underline{\underline{\varepsilon}}_{rt}^{-1}\,\bar{a}_z \times \bar{H}_t - \omega^2\,\varepsilon_0\,\mu_0\,\underline{\underline{\mu}}_{rt}\,\bar{H}_t = 0
\end{aligned}
\tag{D.43a}
$$

$$
\begin{aligned}
&\bar{\nabla}_t \times \underline{\underline{\varepsilon}}_{rt}^{-1}\,\bar{\nabla}_t \times H_z\,\bar{a}_z - \beta\,\bar{\nabla}_t \times \underline{\underline{\varepsilon}}_{rt}^{-1}\,\bar{a}_z \times \bar{H}_t \\
&\quad - \omega^2\,\varepsilon_0\,\mu_0\,\mu_{rzz}\,H_z\,\bar{a}_z = 0
\end{aligned}
\tag{D.43b}
$$

From the preceding equations, it is clear that the system of partial differential equations with real coefficients (D.42) and (D.43) defines an eigenvalue problem where the propagation constant β is an input data. The eigenvalues of the problem are the angular frequencies ω_i of modes propagating with such value of β, and the eigenvectors are the corresponding transverse and longitudinal components of both the electric and the magnetic fields. The problem may also be solved by using only two of those equations, either (D.42), in terms of the electric field components or (D.43), in terms of the magnetic field components. This is the formulation used in [Alam et al. 1997], [Bárdi et al. 1993], [Blanc-Castillo 1994], [Blanc-Castillo et al. 1995], [Bermúdez de Castro and Gómez-Pedreira 1992], [García-Castillo 1992], [García-Castillo and Salazar-Palma 1992a,b], [Helal et al. 1994], [Jin 1993; Ch. 8], [Koshiba and Inoue 1992], and [Salazar-Palma et al. 1996]. In particular equations (D.42) and (D.43) constitute the strong formulation used in Section 3.3.3.2 for the analysis of waveguiding structures in terms of either the electric or the magnetic field. From them a weak formulation has been developed and discretized by means of a finite element method that uses Lagrange/curl-conforming elements (see Appendix F.3.1).

Expressions (D.42) and (D.43) may be manipulated to obtain an eigenvalue problem where the input data is ω and the eigenvalue is β [Lee et al. 1991a,b], [Lee 1994] (see Section 3.3.3.2 and Appendix F.3.2).

By using the divergence equations it is possible to rewrite the longitudinal components in terms of the transverse ones, which would allow one to obtain formulations in terms of the transverse components of the electric or the magnetic field or both as shown in [Chew and Nasir 1989], [Fernández et al. 1993], [Fernández and Lu 1991], [Lu and Fernández 1993a,b, 1994], among others.

When the media in which the wave is propagating are characterized by diagonal tensors $\underline{\underline{\varepsilon}}$ and $\underline{\underline{\mu}}$, a formulation in terms of the longitudinal components E_z and H_z is obtained. In order to accomplish this the following decomposition for the fields is used:

$$\bar{E}_t = E_x\,\bar{a}_x + E_y\,\bar{a}_y \tag{D.44a}$$

$$\bar{H}_t = H_x\,\bar{a}_x + H_y\,\bar{a}_y \tag{D.44b}$$

By replacing (D.44) in (D.28) and taking into account the diagonal form of the permittivity and permeability tensors, the expressions for the transverse components E_x, E_y, H_x, and H_y as functions of the longitudinal components E_z and H_z are given by

$$E_x = \frac{1}{\omega^2\,\varepsilon_0\,\mu_0\,\varepsilon_{rxx}\,\mu_{ryy} + \gamma^2}\left[-\,j\,\gamma\,\frac{\partial E_z}{\partial x} + \omega\,\mu_0\,\mu_{ryy}\,\frac{\partial H_z}{\partial y}\right] \tag{D.45a}$$

$$E_y = \frac{1}{\omega^2 \varepsilon_0 \mu_0 \varepsilon_{ryy} \mu_{rxx} + \gamma^2} \left[-j \gamma \frac{\partial E_z}{\partial y} - \omega \mu_0 \mu_{rxx} \frac{\partial H_z}{\partial x} \right] \tag{D.45b}$$

$$H_x = \frac{1}{\omega^2 \varepsilon_0 \mu_0 \varepsilon_{ryy} \mu_{rxx} + \gamma^2} \left[-\omega \varepsilon_0 \varepsilon_{ryy} \frac{\partial E_z}{\partial y} - j \gamma \frac{\partial H_z}{\partial x} \right] \tag{D.45c}$$

$$H_y = \frac{1}{\omega^2 \varepsilon_0 \mu_0 \varepsilon_{rxx} \mu_{ryy} + \gamma^2} \left[\omega \varepsilon_0 \varepsilon_{rxx} \frac{\partial E_z}{\partial x} - j \gamma \frac{\partial H_z}{\partial y} \right] \tag{D.45d}$$

The partial differential equations in terms of E_z and H_z are obtained as

$$\begin{aligned} &- \left[\frac{\partial}{\partial x} \left(\frac{\varepsilon_{rxx}}{T_{xy}} \frac{\partial E_z}{\partial x} \right) + \frac{\partial}{\partial y} \left(\frac{\varepsilon_{ryy}}{T_{yx}} \frac{\partial E_z}{\partial y} \right) \right] \\ &+ \frac{j\gamma}{\omega \varepsilon_0} \left[\frac{\partial}{\partial x} \left(\frac{1}{T_{xy}} \frac{\partial H_z}{\partial y} \right) - \frac{\partial}{\partial y} \left(\frac{1}{T_{yx}} \frac{\partial H_z}{\partial x} \right) \right] - \omega^2 \varepsilon_0 \mu_0 \varepsilon_{rzz} E_z = 0 \end{aligned} \tag{D.46a}$$

$$\begin{aligned} &- \left[\frac{\partial}{\partial x} \left(\frac{\mu_{rxx}}{T_{yx}} \frac{\partial H_z}{\partial x} \right) + \frac{\partial}{\partial y} \left(\frac{\mu_{ryy}}{T_{xy}} \frac{\partial H_z}{\partial y} \right) \right] \\ &+ \frac{j\gamma}{\omega \mu_0} \left[-\frac{\partial}{\partial x} \left(\frac{1}{T_{yx}} \frac{\partial E_z}{\partial y} \right) + \frac{\partial}{\partial y} \left(\frac{1}{T_{xy}} \frac{\partial E_z}{\partial x} \right) \right] - \omega^2 \varepsilon_0 \mu_0 \mu_{rzz} H_z = 0 \end{aligned} \tag{D.46b}$$

where

$$T_{xy} = \varepsilon_{rxx} \mu_{ryy} + \frac{\gamma^2}{\omega^2 \varepsilon_0 \mu_0} \tag{D.47a}$$

$$T_{yx} = \varepsilon_{ryy} \mu_{rxx} + \frac{\gamma^2}{\omega^2 \varepsilon_0 \mu_0} \tag{D.47b}$$

For the lossless case $\gamma = j\beta$. Using the normalization

$$\beta_n = \frac{\beta}{\omega\,(\varepsilon_0\mu_0)^{1/2}} \tag{D.48}$$

of the phase constant with respect to that of the vacuum $k_0 = \omega(\varepsilon_0\mu_0)^{1/2}$, then one obtains

$$-\left[\frac{\partial}{\partial x}\left(\frac{\varepsilon_{rxx}}{T_{xy}}\frac{\partial E_z}{\partial x}\right) + \frac{\partial}{\partial y}\left(\frac{\varepsilon_{ryy}}{T_{yx}}\frac{\partial E_z}{\partial y}\right)\right]$$
$$- \beta_n\left(\frac{\mu_0}{\varepsilon_0}\right)^{1/2}\left[\frac{\partial}{\partial x}\left(\frac{1}{T_{xy}}\frac{\partial H_z}{\partial y}\right) - \frac{\partial}{\partial y}\left(\frac{1}{T_{yx}}\frac{\partial H_z}{\partial x}\right)\right] - k_0^2\,\varepsilon_{rzz}E_z = 0 \tag{D.49a}$$

$$-\left[\frac{\partial}{\partial x}\left(\frac{\mu_{rxx}}{T_{yx}}\frac{\partial H_z}{\partial x}\right) + \frac{\partial}{\partial y}\left(\frac{\mu_{ryy}}{T_{xy}}\frac{\partial H_z}{\partial y}\right)\right]$$
$$- \beta_n\left(\frac{\varepsilon_0}{\mu_0}\right)^{1/2}\left[-\frac{\partial}{\partial x}\left(\frac{1}{T_{yx}}\frac{\partial E_z}{\partial y}\right) + \frac{\partial}{\partial y}\left(\frac{1}{T_{xy}}\frac{\partial E_z}{\partial x}\right)\right] - k_0^2\,\mu_{rzz}H_z = 0 \tag{D.49b}$$

where

$$T_{xy} = \varepsilon_{rxx}\mu_{ryy} - \beta_n^2 \tag{D.50a}$$

$$T_{yx} = \varepsilon_{ryy}\mu_{rxx} - \beta_n^2 \tag{D.50b}$$

Again, expressions (D.49) form a system of partial differential equations with real coefficients and define an eigenvalue problem where for each value of β_n the solution of the problem will give the possible values of ω_i^2 (or, to be more precise, $k_{0i}^2 = \omega_i^2\varepsilon_0\mu_0$). The propagation constant is then given by $\beta = \beta_n\omega_i(\varepsilon_0\mu_0)^{1/2}$. The boundary conditions are enforced by replacing (D.45) with the relevant component of the fields [Ahmed 1968], [Ahmed and Daly 1969a,b], [Álvarez-Melcón et al. 1994], [Aubourg et al. 1983], [Chang et al. 1990, 1991], [Csendes and Silvester 1970], [Daly 1971], [Eswarappa et al. 1989], [García-Castillo et al. 1994, 1995], [Gil and Zapata 1994, 1995, 1997], [Hernández-Figueroa and Pagiatakis 1993], [Israel and Miniowitz 1987],

[Ikeuchi et al. 1981], [Jin 1993; Ch. 7], [Lagasse and Van Bladel 1972], [Mabaya et al. 1981], [McAulay 1977], [Okamoto and Okoshi 1978], [Sarkar et al. 1994], [Silvester 1969a,b], [Silvester and Ferrari 1983; Chs. 2,3], [Tzuang and Itoh 1986], [Vandenbulcke and Lagasse 1976], [Welt and Webb 1985], [Wu and Chen 1985a,b]. The system (D.49) has also been used in this book for the full-wave analysis of waveguiding structures using Lagrange finite elements (see Section 3.3.3.1 and Appendix F.2).

In addition, there exist other formulations specifically adapted to problems in which certain approximations can be carried out, such as in [Chiang 1985], [Hayata et al. 1988], [Koshiba et al. 1982b, 1984b, 1985c], [Koshiba and Suzuki 1982], and [Vandenbulcke and Lagasse 1976].

The use of scalar or vector potential functions from which the electromagnetic fields may be derived provides alternate sets of formulations, such as those used in [Bárdi and Biro 1991], [Bárdi et al. 1994], [Boyse et al. 1992, 1993], and [Paulsen et al. 1992].

Review papers where formulations of the waveguiding problem (and the cavity problem) based in the previous expressions may be found are [Dillon and Webb 1994], [Davies 1993], [Moyer and Schroeder 1991], [Rahman et al. 1991], [Selleri and Zoboli 1996]. In [Koshiba 1992; Chs. 2 to 4], a number of references using the different formulations may be found.

Appendix E

Weak Formulations for the Quasi-Static Analysis of Waveguiding Structures and Their Finite Element Discretization

E.1 INTRODUCTION

This appendix deals with the development of the weak formulation for the quasi-static analysis of waveguiding structures utilizing the Finite Element Method (FEM). The direct formulation is applied using the scalar potential function as the unknown from which the electric field is derived. Triangular and rectangular Lagrange elements of first-, second-, and third-order (serendipity and complete) have been used as basis functions in their straight and curved versions. The latter elements improve the quality of the solution particularly for structures with curved contours since they approximate the domain with high accuracy. Also, infinite Lagrange elements of the inverse type have been used for open structures.

A mixed formulation where the unknowns are the electric displacement vector and the scalar potential has also been used. A FEM with triangular first- and second-order div-conforming elements has been used for this formulation.

E.2 DIRECT FORMULATION. LAGRANGE ELEMENTS

E.2.1 Weak Formulation

As explained in Section 3.2.2.2 and 3.2.3, the quasi-static analysis of waveguiding structures with two or more conductors reduces to solving one or more electrostatic problems that may be formulated according to the following expressions over the two-dimensional (2D) cross section of the structure Ω, whose boundary is Γ:

$$\bar{E}_t = -\bar{\nabla}_t \phi , \qquad \text{in} \quad \Omega \tag{E.1a}$$

$$\bar{D}_t = \varepsilon_0 \underline{\underline{\varepsilon}}_{rt} \bar{E}_t , \qquad \text{in} \quad \Omega \tag{E.1b}$$

$$\bar{\nabla}_t \cdot \bar{D}_t = 0 \,, \qquad \text{in} \quad \Omega \tag{E.1c}$$

where $\bar{E}_t$ and $\bar{\nabla}_t$ have been defined in Appendix D, (D.27a), and (D.11f), respectively; $\phi(x,y)$ is a scalar function; $\cdot$ stands for the scalar product between vectors; and $\underline{\underline{\varepsilon}}_{rt}$ is the relative electric permittivity tensor of the different media, which is considered to be symmetric and real, i.e., corresponding to lossless media. Since in the quasi-static approach, the longitudinal component of the electric field E_z is assumed to be zero, the transverse electric field may be written as the gradient of a scalar electrostatic potential function defined over Ω.

The boundary conditions that the preceding problem must satisfy are:

1. At the electric walls or perfect conductors, i.e., on $\Gamma_{pe} = \cup \Gamma_{ci}$

$$\phi = \phi_{ci} \,, \quad \text{on} \ \Gamma_{ci} \,, \qquad i = 0, \dots, N_c \tag{E.2}$$

where Γ_{ci} is the ith conductor that is at a constant potential of value ϕ_{ci}. There are N_c+1 conductors. The conductors are considered to be perfect, i.e., lossless.

2. At the magnetic walls, i.e., on $\Gamma_{pm} = \cup \Gamma_{pmj}$

$$\bar{a}_n \cdot \bar{D}_t = 0 \,, \qquad \text{on} \ \Gamma_{pm} \tag{E.3}$$

where $\bar{a}_n$ is the unit vector normal to each magnetic wall.

3. At the interface boundary between two different media:

$$\bar{a}_{n1} \cdot \bar{D}_{t1} + \bar{a}_{n2} \cdot \bar{D}_{t2} = 0 \tag{E.4a}$$

$$\bar{a}_{n1} \times \bar{E}_{t1} + \bar{a}_{n2} \times \bar{E}_{t2} = 0 \tag{E.4b}$$

where the subscript i refers to the ith medium and $\bar{a}_{ni}$ is the unit vector normal to the interface and directed outward from the ith medium.

If equations (E.1) to (E.4) are expressed as functions of the potential $\phi(x,y)$, the strong formulation of the problem is obtained as follows:

Given $\underline{\underline{\varepsilon}}_{rt}$ in Ω, find ϕ so that:

$$-\bar{\nabla}_t \cdot \varepsilon_0 \underline{\underline{\varepsilon}}_{rt} \bar{\nabla}_t \phi = 0 \,, \qquad in \quad \Omega \tag{E.5}$$

under conditions:

a) *At the electric walls*

$$\phi = \phi_{ci} , \quad on \ \Gamma_{ci} , \ i = 0, \dots , N_c \tag{E.6}$$

b) *At the magnetic walls*

$$\bar{a}_n \cdot \varepsilon_0 \underline{\underline{\varepsilon}}_{rt} \bar{\nabla}_t \phi = 0 , \quad on \ \Gamma_{pm} \tag{E.7}$$

c) *At the interface between two media*

$$\bar{a}_{n1} \cdot \varepsilon_0 \underline{\underline{\varepsilon}}_{rt1} \bar{\nabla}_t \phi |_1 + \bar{a}_{n2} \cdot \varepsilon_0 \underline{\underline{\varepsilon}}_{rt2} \bar{\nabla}_t \phi |_2 = 0 \tag{E.8a}$$

$$\bar{a}_{n1} \times \bar{\nabla}_t \phi |_1 + \bar{a}_{n2} \times \bar{\nabla}_t \phi |_2 = 0 \tag{E.8b}$$

The operator defined in (E.5) with homogeneous Dirichlet boundary conditions on $\Gamma_{pe} = \cup \Gamma_{ci}$ and homogeneous Neumann boundary conditions on Γ_{pm}, with $\Gamma = \Gamma_{pe} \cup \Gamma_{pm}$ is positive definite and self-adjoint. However, the Dirichlet boundary conditions are not always homogeneous (see (E.6)). Hence, the operator in (E.5) to (E.7) is not self-adjoint. Nevertheless, a self-adjoint operator can be constructed via a change of variables $\phi' = \phi - u$, where u is any function satisfying the inhomogeneous Dirichlet boundary conditions. The operator acting on ϕ' would be self-adjoint. Thus, a modified variational principle can be constructed from which (E.5) to (E.7) are derived [Jin 1993; Ch. 6]. It is given by

$$F(\phi) = \frac{1}{2} \int_\Omega \bar{\nabla}_t \phi \cdot \varepsilon_0 \underline{\underline{\varepsilon}}_{rt} \bar{\nabla}_t \phi \, d\Omega - \int_{\Gamma_{pm}} \left(\phi \bar{a}_n \cdot \varepsilon_0 \underline{\underline{\varepsilon}}_{rt} \bar{\nabla}_t \phi \right) d\Gamma \tag{E.9}$$

where ϕ must satisfy (E.6). The condition (E.7) is imposed in a weak sense. In fact, taking into account that the right-hand side of (E.7) is zero, the functional becomes

$$F(\phi) = \frac{1}{2} \int_\Omega \bar{\nabla}_t \phi \cdot \varepsilon_0 \underline{\underline{\varepsilon}}_{rt} \bar{\nabla}_t \phi \, d\Omega \tag{E.10}$$

and the weak formulation of (E.5) to (E.8) can be stated as

Given $\underline{\underline{\varepsilon}}_{rt}$ *in* Ω*, find the function* ϕ *with* $\phi \in H^1(\Omega)$, $\phi = \phi_{ci}$ *on* Γ_{ci}, $i = 0, \dots, N_c$, *that makes the functional* (E.10) *stationary.*

When the operator in (E.10) is positive definite, the function ϕ obtained will give a minimum of the functional.

In order to obtain a weak formulation, the method of weighted residuals can also be used. In that case, one needs to multiply (E.5) by a weighting function ψ (that should satisfy the same continuity conditions as ϕ and should have homogeneous boundary conditions on the Dirichlet boundaries) and weigh the residual to zero

$$W = -\int_{\Omega} \psi \, \bar{\nabla}_t \cdot \varepsilon_0 \underline{\underline{\varepsilon}}_{rt} \bar{\nabla}_t \phi \, d\Omega = 0 \tag{E.11a}$$

Integrating by parts yields

$$\begin{aligned} W &= \int_{\Omega} \bar{\nabla}_t \psi \cdot \varepsilon_0 \underline{\underline{\varepsilon}}_{rt} \bar{\nabla}_t \phi \, d\Omega - \int_{\Gamma} \psi \, \bar{a}_n \cdot \varepsilon_0 \underline{\underline{\varepsilon}}_{rt} \bar{\nabla}_t \phi \, d\Gamma \\ &= \int_{\Omega} \bar{\nabla}_t \psi \cdot \varepsilon_o \underline{\underline{\varepsilon}}_{rt} \bar{\nabla}_t \phi \, d\Omega - \sum_i \int_{\Gamma_{ci}} \psi \, \bar{a}_n \cdot \varepsilon_0 \underline{\underline{\varepsilon}}_{rt} \bar{\nabla}_t \phi \, d\Gamma \\ &\quad - \int_{\Gamma_{pm}} \psi \, \bar{a}_n \cdot \varepsilon_0 \underline{\underline{\varepsilon}}_{rt} \bar{\nabla}_t \phi \, d\Gamma = 0 \end{aligned} \tag{E.11b}$$

Integrals along the boundaries Γ_{ci} are eliminated by choosing $\psi = 0$ on them. At the same time, the integral along Γ_{pm} is eliminated since (E.7) should be satisfied. This yields

$$W = \int_{\Omega} \bar{\nabla}_t \psi \cdot \varepsilon_0 \underline{\underline{\varepsilon}}_{rt} \bar{\nabla}_t \phi \, d\Omega = 0 \tag{E.12}$$

Hence, the problem of (E.5) to (E.8) can be formulated as

Given $\underline{\underline{\varepsilon}}_{rt}$ *in* Ω, *find the function* ϕ, $\phi \in H^1(\Omega)$, *with* $\phi = \phi_{ci}$ *on* Γ_{ci} , $i = 0, ..., N_c$, *so that* (E.12) *is satisfied for any function* ψ, $\psi \in H^1(\Omega)$, $\psi = 0$, *on* Γ_{ci}, $i = 0, ..., N_c$.

It can be observed that the integral form W, when $\underline{\underline{\varepsilon}}_{rt}$ is real and symmetric, becomes a real symmetric functional. This formulation has been applied together with FEM in [Costache 1987], [Decreton 1974], [Daly 1973a, 1984, 1985], [Khebir et al. 1990a,b], [Lee and Csendes 1987], [McDonald and Wexler 1972], [Medina et al. 1994], [Nortier and McNamara 1984], [Pantic and Mittra 1986], [Pantic-Tanner and Mittra 1988], [Richards and Wexler 1972], [Sadiku 1989], [Salazar-Palma and Hernández-Gil 1989, 1995], [Salazar-Palma et al. 1992], [Salazar-Palma and Recio-Peláez 1996], [Salazar-Palma and García-Castillo 1996], [Thatcher 1982], [Young and Smith 1987], [Waldow 1989], and [Wu and Davidovitz 1992].

E.2.2 Discretization by Means of Lagrange Finite Elements

The continuous problem can be discretized using any of the preceding weak formulations and applying a FEM that uses Lagrange elements. Triangular and rectangular elements as well as rectangular infinite elements have been used for closed and open structures. The triangular elements used are of first order (three nodes), second order (six nodes), straight (subparametric) and curved (isoparametric), third-order serendipity (nine nodes), straight and curved, and third-order complete (ten nodes), straight and curved. The rectangular elements used are of first-order (also called bilinear: four nodes), second-order serendipity (eight nodes), second-order complete (nine nodes), third-order serendipity (twelve nodes), and third-order complete (sixteen nodes), straight and curved. The infinite elements used are of first-order (four nodes, two of which are at infinity), second-order serendipity (eight nodes, three at infinity), and second-order complete (nine nodes, three at infinity).

The problem domain Ω is subdivided, if required, into a near field region where ordinary finite elements are used, and a far field region where infinite elements are utilized. The ordinary elements and the infinite elements are chosen so that they are mutually compatible. The finite region can be subdivided into triangular and/or rectangular elements.

In any case, the unknown ϕ in the eth element Ω^e is approximated by

$$\tilde{\phi}^e = \sum_{i=1}^{n_e} N_i^e d_i^e = \{N^e\}^T\{d^e\} \tag{E.13}$$

where N_i^e is the ith Lagrange basis function for the element Ω^e with n_e nodes and d_i^e is the ith local degree of freedom, which is defined as the value of the unknown ϕ at the ith node.

Introducing (E.13) and applying Ritz or Galerkin method to (E.10) or (E.12), normalized with respect to ε_0, lead to the following set of discrete integral forms in each element Ω^e:

$$W_i^e = \sum_{j=1}^{n_e} \left(\int_{\Omega^e} \bar{\nabla}_t N_i \cdot \underline{\underline{\varepsilon}}_{rt} \, \bar{\nabla}_t N_j \, \mathrm{d}\Omega \right) d_j^e \ , \quad i = 1, \ldots, n_e \tag{E.14}$$

which can be written in matrix form as

$$\{W^e\} = [k^e]\{d^e\} - \{f^e\} \tag{E.15a}$$

where

$$[k^e] = \int_{\Omega^e} [\nabla_t N]^T [\varepsilon_{rt}] [\nabla_t N] \, \mathrm{d}\Omega \tag{E.15b}$$

is the $n_e \times n_e$ stiffness matrix of element Ω^e,

$$\{d^e\} = \begin{Bmatrix} d_1^e \\ \vdots \\ d_{n_e}^e \end{Bmatrix} \tag{E.15c}$$

is the column matrix of local degrees of freedom for element Ω^e, and the n_e coefficients of the local column matrix of excitation $\{f_e\}$ are zero. The following notation has been used:

$$[\nabla_t N] = [\partial N] = \begin{bmatrix} \dfrac{\partial N_1}{\partial x} & \cdots & \dfrac{\partial N_{n_e}}{\partial x} \\ \\ \dfrac{\partial N_1}{\partial y} & \cdots & \dfrac{\partial N_{n_e}}{\partial y} \end{bmatrix} \tag{E.15d}$$

where N_i is the ith shape function for the triangular, ordinary rectangular, or infinite rectangular Lagrange element Ω^e.

To calculate the coefficients of $[k^e]$, it is more efficient to transform it to the parent (reference or master) element which results in

$$[k^e] = \int_{\hat{\Omega}} \left[\hat{\nabla}_t \hat{N}\right]^T [J^e]^{-1} [\varepsilon_{rt}] \left([J^e]^T\right)^{-1} \left[\hat{\nabla}_t \hat{N}\right] |J^e| \, \mathrm{d}\Omega \tag{E.16}$$

where $[\hat{\nabla}_t \hat{N}]$ (or $[\partial \hat{N}]$) is given by (E.15d) when the derivatives are taken with respect to the coordinates of the parent element, i.e., with respect to $\hat{x}$ and $\hat{y}$ (see Appendix C.1). The basis functions of the triangular parent elements and the rectangular elements are given in Tables 2.1 and 2.2 of Chapter 2. $[J^e]$ represents the Jacobian matrix of the geometric transformation of each element and $|J^e|$ is the determinant of the Jacobian matrix. They are computed from (C.11). When dealing with ordinary straight elements, mapping functions of (C.11) are the shape functions of the first-order triangular element $P(1)$, or rectangular element $Q(1)$. For the case of curved isoparametric elements, transformation functions are the shape functions of elements $P(2)$, $P(3')$, or $P(3)$, or $Q(2')$, $Q(2)$, $Q(3')$, or $Q(3)$, respectively (see Tables 2.1 and 2.2). If dealing with infinite elements, then the mapping functions of $I(6)$ given in Table 2.5, should be used.

It can be observed that by replacing the functions $[J^e]^{-1}$ and $([J^e]^T)^{-1}$ in (E.16) the coefficients of the matrix $[k^e]$ can be evaluated analytically only for the case of triangular straight elements (the subparametric elements of first, second, and third order) because $|J^e|$ is constant in each element and therefore the integrands are polynomials in $\hat{x}$ and $\hat{y}$ to which (2.71) of Chapter 2 can be applied. For the rest of the elements (triangular curved elements, rectangular straight and curved elements, and infinite elements), $|J^e|$ is either a polynomial in $\hat{x}$ and $\hat{y}$ (in the case of ordinary finite elements) or has a functional dependence of the form $\hat{x}^{a/m}$ and $\hat{y}^{b/m}$ (in the case of infinite elements), where m is the coefficient of the type of decay $(1/r)^m$ of the infinite elements used (see Section 2.5.1.2), and a and b are integer numbers. Hence, in those cases, the integrands are ratios of polynomials in $\hat{x}$ and $\hat{y}$ (or ratios of polynomials and irrational functions) and a numerical integration is required. Table 3.1 of Chapter 3 summarizes the order of the numerical integration carried out in each case. It can be observed that for certain elements, different number of integration points are used.

After calculating the elements of the matrix $[k^e]$ for each element Ω^e, one can obtain the global discrete integral form as explained in Section 2.5.2.2. Then, the Dirichlet boundary conditions are imposed following the second procedure outlined in Section 2.5.2.3, resulting in a global system of equations given by

$$[K]\{D\}=\{F\} \tag{E.17}$$

where $[K]$, $\{D\}$, and $\{F\}$ are, respectively, the global stiffness matrix, the global column matrix of degrees of freedom, and the global column matrix of excitations. Due to the non homogeneous Dirichlet boundary conditions of the problem a nonzero column matrix of excitations is obtained. It may be observed that the boundary conditions (E.3) and (E.4), which are of natural type, are only approximately fulfilled.

Matrix $[K]$ is sparse, real, symmetric, and positive definite. The method used for the solution of system (E.17) is the Choleski method with an algorithm adapted to the case of sparse matrices with sky-line storage of the upper triangular part. Also, the conjugate gradient method (without preconditioning, or with preconditioning, either through relaxation or an incomplete Choleski factorization) storing only the nonzero coefficients of the upper triangular part (see Section 2.5.2.4) has been used to solve the matrix equation. For the case of open structures using infinite elements, it is always convenient to choose the conjugate gradient method (and to use preconditioning through relaxation) since the matrix is not usually well conditioned.

Once the solution of (E.17) obtained, the approximate solution to the problem will be given by (E.13). The function $\tilde{\phi}$ will be continuous in Ω while its gradient will have continuous tangential component and discontinuous normal component across element interfaces.

As $[K]$ is positive definite, $\tilde{\phi}$ provides a minimum of the functional $F(\phi)$ given by (E.10) and, therefore, provides an upper bound of the exact value of $F(\phi)$. This is valid as long as the discretization error in the subdivision of the domain does not exist.

E.3 MIXED FORMULATION. DIV-CONFORMING ELEMENTS

E.3.1 Weak Formulation

If equations (E.1) to (E.4) are expressed as functions of ϕ and $\bar{D}_t$, the mixed formulation for the problem is obtained as:

Given $\underline{\underline{\varepsilon}}_{rt}$ in Ω, find $\bar{D}_t$ and ϕ so that

$$\bar{D}_t + \varepsilon_0 \underline{\underline{\varepsilon}}_{rt} \bar{\nabla}_t \phi = 0, \ in\ \Omega \tag{E.18a}$$

and

$$\bar{\nabla}_t \cdot \bar{D}_t = 0, \ in\ \Omega \tag{E.18b}$$

where ϕ and $\bar{D}_t$ have to satisfy the following boundary conditions:

a) At the electric walls

$$\phi = \phi_{ci}, \quad on\ \ \Gamma_{ci},\ i = 0, \ldots, N_c \tag{E.19a}$$

b) At the magnetic walls

$$\bar{a}_n \cdot \bar{D}_t = 0, \quad on\ \Gamma_{pm} \tag{E.19b}$$

c) At the interfaces between two media

$$\bar{a}_{n1} \cdot \bar{D}_{t1} + \bar{a}_{n2} \cdot \bar{D}_{t2} = 0 \tag{E.19c}$$

$$\bar{a}_{n1} \times \bar{\nabla}_t \phi \Big|_1 + \bar{a}_{n2} \times \bar{\nabla}_t \phi \Big|_2 = 0 \tag{E.19d}$$

In order to obtain a weak formulation, the method of weighted residuals is applied to (E.18). To this effect, a vector weighting function $\bar{q}_t$ and a scalar weighting function ψ are chosen. They must fulfill homogeneous boundary conditions on Γ_{pm} and Γ_{pe}, respectively, that is, $\bar{a}_n \cdot \bar{q}_t = 0$ on Γ_{pm} and $\psi = 0$ on Γ_{ci}, $i = 0, \ldots, N_c$. The vector equation (E.18a) is weighted with the vector function $\bar{q}_t$ and the scalar equation (E.18b) with the scalar function ψ, resulting in

$$W_1 = \int_\Omega \bar{q}_t \cdot (\varepsilon_0^{-1} \underline{\underline{\varepsilon}}_{rt}^{-1} \bar{D}_t + \bar{\nabla}_t \phi)\ d\Omega = 0 \tag{E.20a}$$

$$W_2 = \int_\Omega \psi \bar{\nabla}_t \cdot \bar{D}_t\ d\Omega = 0 \tag{E.20b}$$

Integrating W_1 by parts yields

$$\begin{aligned} W_1 &= \int_\Omega \bar{q}_t \cdot \varepsilon_0^{-1} \underline{\underline{\varepsilon}}_{rt}^{-1} \bar{D}_t\ d\Omega - \int_\Omega \bar{\nabla}_t \cdot \bar{q}_t \phi\ d\Omega + \int_\Gamma \phi\ \bar{a}_n \cdot \bar{q}_t\ d\Gamma \\ &= \int_\Omega \bar{q}_t \cdot \varepsilon_0^{-1} \underline{\underline{\varepsilon}}_{rt}^{-1} \bar{D}_t\ d\Omega - \int_\Omega \bar{\nabla}_t \cdot \bar{q}_t \phi\ d\Omega \\ &\quad + \sum_i \int_{\Gamma_{ci}} \phi\ \bar{a}_n \cdot \bar{q}_t\ d\Gamma + \int_{\Gamma_{pm}} \phi\ \bar{a}_n \cdot \bar{q}_t\ d\Gamma = 0 \end{aligned} \tag{E.21}$$

The integral on the contour Γ_{pm} is zero in a strong sense since the vector function $\bar{q}_t$ is chosen so that $\bar{a}_n \cdot \bar{q}_t = 0$ on Γ_{pm}. The integrals along Γ_{ci}, $i = 0,...,N_c$, do not vanish in general, since ϕ assumes the values ϕ_{ci} corresponding to each conductor Γ_{ci}. This yields

$$\begin{aligned} W_1 &= \int_\Omega \bar{q}_t \cdot \varepsilon_0^{-1} \underline{\underline{\varepsilon}}_{rt}^{-1} \bar{D}_t\ d\Omega - \int_\Omega \bar{\nabla}_t \cdot \bar{q}_t\ \phi\ d\Omega \\ &\quad + \sum_i \int_{\Gamma_{ci}} \phi_{ci}\ \bar{a}_n \cdot \bar{q}_t\ d\Gamma = 0 \end{aligned} \tag{E.22a}$$

$$W_2 = \int_\Omega \psi\ \bar{\nabla}_t \cdot \bar{D}_t\ d\Omega = 0 \tag{E.22b}$$

Thus, the problem (E.18) to (E.19) can be reformulated as:

Given the invertible tensor $\underline{\underline{\varepsilon}}_{rt}$ *in* Ω, *let* X^0 *be the space of functions* $\bar{q}_t = q_x(x,y)\bar{a}_x + q_y(x,y)\bar{a}_y$ *so that* $\bar{q}_t \in H(\text{div},\Omega)$ *and* $\bar{a}_n \cdot \bar{q}_t = 0$ *on* Γ_{pm}. *Then find* $\bar{D}_t \in X^0$ *and* $\phi \in L^2(\Omega)$ *so that* (E.22a) *is fulfilled for all functions* $\bar{q}_t \in X^0$ *and* (E.22b) *is satisfied for all functions* $\psi \in L^2(\Omega)$.

This formulation is used in [Arnold 1990], [Bossavit 1988], [Brezzi and Marini 1994], [Raviart and Thomas 1977a], and [Salazar-Palma et al. 1991].

The number of variational unknowns is two, ϕ and D_t. This would imply that this method has a disadvantage compared to the standard formulation. However, one of the unknowns, the variable ϕ, can be eliminated. To accomplish this, a penalty function method is applied to (E.22b) that may be written as

$$W_{2p} = \int_\Omega \psi \bar{\nabla}_t \cdot \bar{D}_t \, \mathrm{d}\Omega + \lambda \int_\Omega \psi \, \phi \, \mathrm{d}\Omega = 0 \tag{E.23}$$

If λ is small enough, expression (E.23) is a good representation of (E.22b). It can be observed that W_{2p} is the integral obtained from the application of the weighted residual method to

$$\bar{\nabla}_t \cdot \bar{D}_t = -\lambda\phi \tag{E.24}$$

that will be approximately zero for λ sufficiently small.

Substituting ϕ given by (E.24) in W_1 as given in (E.22a) and substituting $\alpha = 1/\lambda$, the penalty function method results in

$$W_p = \int_\Omega \bar{q}_t \cdot \varepsilon_0^{-1} \underline{\underline{\varepsilon}}_{rt}^{-1} \bar{D}_t \, \mathrm{d}\Omega + \alpha \int_\Omega \bar{\nabla}_t \cdot \bar{q}_t \, \bar{\nabla}_t \cdot \bar{D}_t \, \mathrm{d}\Omega + \sum_i \int_{\Gamma_{ci}} \phi_{ci} \, \bar{a}_n \cdot \bar{q}_t \, \mathrm{d}\Gamma = 0 \tag{E.25}$$

Moreover, if $\bar{D}_t$ is normalized according to

$$\bar{D}_{tn} = \frac{\bar{D}_t}{\varepsilon_0} \tag{E.26}$$

then the normalized integral form is obtained as:

$$W_{pn} = \int_\Omega \bar{q}_{tn} \cdot \underline{\underline{\varepsilon}}_{rt}^{-1} \bar{D}_{tn} \, \mathrm{d}\Omega + \alpha_n \int_\Omega (\bar{\nabla}_t \cdot \bar{q}_{tn}) (\bar{\nabla}_t \cdot \bar{D}_{tn}) \, \mathrm{d}\Omega + \sum_i \int_{\Gamma_{ci}} \phi_{ci} \, \bar{a}_n \cdot \bar{q}_{tn} \, \mathrm{d}\Gamma = 0 \tag{E.27}$$

The weak formulation would then be:

Given the invertible tensor $\underline{\underline{\varepsilon}}_{rt}$ in Ω, find $\bar{D}_{tn} \in X^0$ so that (E.27) *is fulfilled for all $\bar{q}_{tn} \in X^0$ and for a sufficiently large value of α_n.*

This penalty function method also eliminates certain problems associated with the original mixed formulation [Mathew 1993a,b].

The weak formulation (E.27) can also be obtained from the complementary variational principle of (E.9) (see Section 2.3.3.2).

E.3.2 Discretization by Means of Div-Conforming Finite Elements

In order to solve the previous problem by FEM, the Galerkin method is applied to the expression (E.27), i.e., the same expansion and weighting functions are chosen, belonging to the space X^0. This implies that one should choose a finite element providing solutions belonging to $H(\text{div},\Omega)$. Also the boundaries Γ_{pm} (magnetic walls or symmetry walls) must be considered as Dirichlet boundaries on which $\bar{a}_n \cdot \bar{D}_{tn} = 0$ must be imposed. The boundaries where the potential is prescribed (i.e., the conductors Γ_{ci}, $i = 0, ..., N_c$) contribute to the boundary excitation term as shown in (E.27). Div-conforming triangular elements of first- and second-order are chosen so that they fulfill the first requirement (i.e., the approximate solution will belong to $H(\text{div},\Omega)$). The definition of the degrees of freedom that are associated with the edges of the elements is proportional to $\bar{a}_n \cdot \bar{D}_{tn}$, (see the description of div-conforming elements in Section 2.5.1.2). Thus, the essential boundary conditions are easily imposed as zero values of the degrees of freedom on the edges placed at symmetry walls.

For the first-order element, three degrees of freedom are required. Each of them is related to an edge. The degree of freedom is defined as the length of the edge multiplied by the normal component of the vector unknown at the midpoint of the edge. It is necessary to attribute a sign to the local value of the degree of freedom since the definition of the unit vector normal to an edge being the interface between two contiguous elements differs in the sign depending on the element that is being dealt with. This subject is covered in Appendix C.4.3. It should be mentioned that the value of the degrees of freedom is the value of the flux of the unknown vector along the edge of the element computed through numerical integration by means of one sampling point. Thus, for edges located on conductors the value of the degrees of freedom will give the electric charge on the edge.

For the second-order element, eight degrees of freedom are required, six of which are associated with the edges and the other two with the element surface. Two nodes are defined at each edge, located at the two Gauss integration points of the edge. The corresponding degrees of freedom are defined as the length of the edge multiplied by the normal component of the vector unknown at the node. As when dealing with the first-order element, a sign must be attributed to the local value of this type of degree of freedom. The internal degrees of freedom are defined as the surface integral of the x and y components of the vector unknown, respectively. A node at the centroid of the element may be assigned to these degrees of freedom (see Section 2.5.1). For second-order elements, the value of the flux of the vector unknown along the edge (i.e., of the electric charge, in the case of an edge located on a conductor) may be obtained averaging the values of the two degrees of freedom associated to the edge.

Once the domain Ω has been discretized with triangular elements and the nodes have been defined at their edges, the unknown $\bar{D}_{tn}$ at the eth element is approximated by $\breve{D}_{tn}^{\,e}$ given by

$$\breve{D}_{tn}^{\,e} = \sum_{i=1}^{3 \text{ or } 8} \bar{N}_i^{\,e} d_i^{\,e} = \{\bar{N}^{\,e}\}^T \{d^{\,e}\} \qquad \text{(E.28)}$$

The upper limit of the summation is 3 for the first-order element and 8 for the second-order element. Here $\bar{N}_i^e$ is the ith vector basis function and d_i^e is the ith degree of freedom of the element Ω^e. Replacing (E.28) in (E.27) and applying the Galerkin method the following integral form at each element Ω^e is obtained:

$$W_i^e = \sum_{j=1}^{3\ or\ 8} \left[\int_{\Omega^e} \bar{N}_i \cdot \underline{\underline{\varepsilon}}_{rt}^{-1} \bar{N}_j \, d\Omega + \alpha_n \int_{\Omega^e} \left(\bar{\nabla}_t \cdot \bar{N}_i\right) \left(\bar{\nabla}_t \cdot \bar{N}_j\right) d\Omega \right] d_j^e$$
$$+ \sum_{k=1}^{3} \int_{\Gamma_k \subset \Gamma_{pe}} \phi_{ck} \bar{a}_n \cdot \bar{N}_i \, d\Gamma , \qquad i = 1, \ldots, 3 \text{ or } 8 \tag{E.29}$$

which can be expressed in matrix form as

$$\{W^e\} = [k^e]\{d^e\} - \{f^e\} \tag{E.30a}$$

where $[k^e]$ is the stiffness matrix, $\{d^e\}$ is the column matrix that contains the degrees of freedom for the element, and $\{f^e\}$ is the excitation column matrix such that

$$[k^e] = \int_{\Omega^e} [\bar{N}]^T [\varepsilon_{rt}]^{-1} [\bar{N}] \, d\Omega + \alpha_n \int_{\Omega^e} \left[\{\nabla_t\}^T [\bar{N}]\right] \left[\{\nabla_t\}^T [\bar{N}]\right]^T d\Omega \tag{E.30b}$$

$$\{f^e\} = -\sum_{k=1}^{3} \int_{\Gamma_k \subset \Gamma_{pe}} \phi_{ck} \, [\bar{N}]^T \begin{Bmatrix} a_{nx} \\ a_{ny} \end{Bmatrix} d\Gamma \tag{E.30c}$$

where

$$[\bar{N}] = \{\bar{N}\}^T = \begin{bmatrix} N_{x1} & \cdots & N_{x3} \text{ or } N_{x8} \\ N_{y1} & \cdots & N_{y3} \text{ or } N_{y8} \end{bmatrix} \tag{E.30d}$$

$$\{\nabla_t\} = \begin{Bmatrix} \dfrac{\partial}{\partial x} \\ \dfrac{\partial}{\partial y} \end{Bmatrix} \tag{E.30e}$$

and

$$\{a_n\} = \begin{Bmatrix} a_{nx} \\ a_{ny} \end{Bmatrix} \tag{E.30f}$$

is the unit vector normal to the edge of the triangle. In (E.30c) $[\bar{N}]$ and $\{\bar{a}_n\}$ must be specialized to the kth side of the triangle provided that it coincides with a conducting boundary. Finally, ϕ_{ck} is the voltage of the kth edge of the triangle provided that it is a conducting edge raised to a voltage ϕ_{ck}.

The coefficients $[k^e]$ and $[f^e]$ are easily computed by transforming each element Ω^e to the parent element. To this effect, utilizing expressions from Appendix C, (C.3d), (C.10a), and (C.16b), one obtains

$$[k^e] = \int_{\hat{\Omega}} \frac{1}{|J^e|} [\breve{N}]^T [J^e]^T [\varepsilon_{rt}]^{-1} [J^e] [\breve{N}] \, d\Omega + \alpha_n \int_{\hat{\Omega}} \frac{1}{|J^e|} \left[\{\hat{\nabla}_t\}^T [\breve{N}] \right] \left[\{\hat{\nabla}_t\}^T [\breve{N}] \right]^T d\Omega \tag{E.31a}$$

and

$$\{f^e\} = -\sum_{k=1}^{3} \int_{\hat{\Gamma}_k | \Gamma_k \subset \Gamma_{pe}} \phi_{ck} [\breve{N}]^T \begin{Bmatrix} a_{nx} \\ a_{ny} \end{Bmatrix}_k d\Gamma \tag{E.31b}$$

where $[\breve{N}]$ refers to the vector shape functions of the parent element, and $\{\hat{\nabla}_t\}$ denotes the gradient operator in the parent element. According to (C.17b), the computation of (E.30c) may be done directly over the parent element as expressed by (E.31b). The shape functions of the reference elements are given in Table 2.6 of Chapter 2.

When straight elements are used, $|J^e|$ is constant on each element and so are the coefficients of $[J^e]$. Thus the integrands of (E.31a) are polynomials and integration of the elements of $[k^e]$ can be carried out analytically. For curved elements numerical integration is necessary. Since $[\varepsilon_{rt}]$ is symmetric, $[k^e]$ will also be symmetric. The coefficients in (E.31b) are calculated at one point of numerical integration for the case of the first-order element and at two points for the case of the second-order. Since the integration points coincide with the edge nodes, the calculation of (E.31b) becomes straightforward. The parameter α_n has been assumed to be $\alpha_n = 10^6$.

Before the final system of equations is formed, a sign must be assigned to the local degrees of freedom according to Appendix C.4.3. For the case of the second order element, a positive sign is assigned to the last two degrees of freedom.

Therefore (E.30a) becomes

$$\{W^e\} = [k^{le}]\{d^{ge}\} - \{f^{le}\} \tag{E.32a}$$

where

$$k_{ij}^{le} = sign(i)\ sign(j)\ k_{ij}^{e}\ , \quad i,j=1,\ldots,3 \text{ or } 8 \tag{E.32b}$$

$$f_i^{le} = sign(i)\, f_i^{e} \tag{E.32c}$$

Here the variable *sign* takes the values +1 or -1 as explained in Appendix C.4.4, and $\{d^{ge}\}$ represents the column matrix of the global degrees of freedom in the *e*th element.

For the second-order elements, the number of unknowns may be reduced from eight per element to six. It can be observed that the last two degrees of freedom are not shared by contiguous elements. Hence, the idea of eliminating them at a local level becomes feasible since the independent terms f_7^e and f_8^e are always zero as the following identities hold

$$\breve{N}_7(\hat{x}_j, \hat{y}_j) \cdot \bar{a}_{nj} \equiv 0, \ j=1,\ldots,6 \tag{E.33a}$$

$$\breve{N}_8(\hat{x}_j, \hat{y}_j) \cdot \bar{a}_{nj} \equiv 0, \ j=1,\ldots,6 \tag{E.33b}$$

where $(\hat{x}_j, \hat{y}_j)$ are the coordinates of the *j*th node, for j = 1, ..., 6, and $\bar{a}_{nj}$ is the unit vector normal to the edge at the *j*th node of the parent element in the outward direction. Expression (E.30a) can therefore be written as

$$\{W^e\} = \begin{bmatrix} [k_a^{le}] & [k_b^{le}] \\ [k_b^{le}]^T & [k_c^{le}] \end{bmatrix} \begin{Bmatrix} \{d_a^{ge}\} \\ \{d_c^{ge}\} \end{Bmatrix} - \begin{Bmatrix} \{f_a^{le}\} \\ 0 \\ 0 \end{Bmatrix} \tag{E.34a}$$

where

$$[k_a^{le}] = \begin{bmatrix} k_{11}^{le} & \ldots & k_{16}^{le} \\ \ldots & \ldots & \ldots \\ k_{61}^{le} & \ldots & k_{66}^{le} \end{bmatrix} \tag{E.34b}$$

with $k_{ij}^{le} = k_{ji}^{le}$, for i, j = 1, ..., 6;

$$[k_b^{le}] = \begin{bmatrix} k_{17}^{le} & k_{18}^{le} \\ \cdots & \cdots \\ k_{67}^{le} & k_{68}^{le} \end{bmatrix} \tag{E.34c}$$

$$[k_c^{le}] = \begin{bmatrix} k_{77}^{le} & k_{78}^{le} \\ k_{87}^{le} & k_{88}^{le} \end{bmatrix} \tag{E.34d}$$

with $k_{78}{}^{le} = k_{87}{}^{le}$; and

$$\{d_c^{ge}\} = \begin{Bmatrix} d_7^{ge} \\ d_8^{ge} \end{Bmatrix}, \quad \{d_a^{ge}\} = \begin{Bmatrix} d_1^{ge} \\ \vdots \\ d_6^{ge} \end{Bmatrix} \tag{E.34e}$$

$$\{f_a^{le}\} = \begin{Bmatrix} f_1^{le} \\ \vdots \\ \vdots \\ f_6^{le} \end{Bmatrix} \tag{E.34f}$$

Because the degrees of freedom $\{d_c^{ge}\}$ are not shared by any other element, one gets

$$[k_b^{le}]^T\{d_a^{ge}\} + [k_c^{le}]\{d_c^{ge}\} = \begin{Bmatrix} 0 \\ 0 \end{Bmatrix} \tag{E.35}$$

Hence, $\{d_c^{ge}\}$ may be written in terms of $\{d_a^{ge}\}$ as

$$\{d_c^{ge}\} = -[k_c^{le}]^{-1}[k_b^{le}]^T\{d_a^{ge}\} \tag{E.36}$$

Substituting (E.36) into the first six rows of $\{W^e\}$ in (E.34a) one obtains a new reduced set of integral forms $\{W^{le}\}$

$$\{W^{le}\} = [k_a^{le}]\{d_a^{ge}\} - [k_b^{le}][k_c^{le}]^{-1}[k_b^{le}]^T\{d_a^{ge}\} - \{f_a^{le}\} \tag{E.37}$$

from which the reduced local integral form for each element Ω^e is generated as

$$\{W^{le}\} = [k^{lle}]\{d_a^{ge}\} - \{f_a^{le}\} \tag{E.38a}$$

where

$$[k^{lle}] = [k_a^{le}] - [k_b^{le}][k_c^{le}]^{-1}[k_b^{le}]^T \tag{E.38b}$$

Once the discrete local integral form (E.32) (for first-order elements) or (E.38) (for second-order elements) has been generated, the essential boundary conditions at the magnetic walls (or symmetry walls) are imposed. Hence, a global system of equations is generated as

$$[K]\{D\} = \{F\} \tag{E.39}$$

in which the matrix $[K]$ is sparse, real, symmetric, and positive definite. The methods used for the solution of (E.39) are either the Choleski method with sky-line storage of the upper triangular part of the matrix $[K]$, or the conjugate gradient method with any type of preconditioning, storing only the nonzero coefficients of the upper triangular part of $[K]$.

After (E.39) has been solved, the approximate solution of the unknown as per (E.28) can be obtained. The solution will satisfy exactly (E.19b,c) but conditions (E.19a,d) will be satisfied only in a distributional sense. Since the solution obtained through this method minimizes the complementary functional of the functional (E.10), one obtains a lower bound for the exact value of the functional (E.10).

Appendix F

Weak Formulations for the Full-Wave Analysis of Waveguiding Structures and Their Finite Element Discretization

F.1 INTRODUCTION

This appendix deals with the development of several weak formulations for the full-wave analysis of waveguide structures and their discretization utilizing the Finite Element Method (FEM). The first formulation uses as unknowns the longitudinal components of the electric and the magnetic fields. The Lagrange element is used for its discretization. A formulation that uses the longitudinal and the transverse components of the electric or the magnetic field is also presented. A combination of the curl-conforming element and the Lagrange element is utilized.

F.2 FORMULATION UTILIZING THE LONGITUDINAL COMPONENTS OF THE ELECTRIC AND THE MAGNETIC FIELDS

This section is divided into two parts. The first part deals with the general case of inhomogeneous and anisotropic structures, in which the Lagrange elements with two degrees of freedom per node are used for the discretization of the electromagnetic problem. The second part deals with homogeneous and isotropic structures. The analysis is performed in terms of one of the longitudinal components of the fields. Lagrange elements with one degree of freedom per node are utilized.

F.2.1 Inhomogeneous and Anisotropic Structures

F.2.1.1 *Weak Formulation*

As explained in Appendix D.3, when the matrices describing the tensors $\underline{\underline{\varepsilon}}$ and $\underline{\underline{\mu}}$ are diagonal, a classical or strong formulation of the guided wave problem may be obtained in terms of the longitudinal components of the electric and the magnetic fields. More specifically, the unknowns would be E_z and H_z, according to (D.27). The transverse components of the electric and the magnetic fields can be written as functions of the longitudinal components according to (D.44) and (D.45), yielding the strong formulation given by (D.46) with the corresponding boundary conditions (D.33a,b) and (D.36a,b).

If the different material media of the problem are considered to be perfect and the conductors are ideal, the propagation constant will be purely imaginary, $\gamma = j\beta$, and the expressions of (D.46) are reduced to (D.49), which are

$$
\begin{aligned}
&-\left[\frac{\partial}{\partial x}\left(\frac{\varepsilon_{rxx}}{T_{xy}}\frac{\partial E_z}{\partial x}\right)+\frac{\partial}{\partial y}\left(\frac{\varepsilon_{ryy}}{T_{yx}}\frac{\partial E_z}{\partial y}\right)\right] \\
&\quad -\beta_n\left(\frac{\mu_o}{\varepsilon_o}\right)^{1/2}\left[\frac{\partial}{\partial x}\left(\frac{1}{T_{xy}}\frac{\partial H_z}{\partial y}\right)-\frac{\partial}{\partial y}\left(\frac{1}{T_{yx}}\frac{\partial H_z}{\partial x}\right)\right]-k_0^2\varepsilon_{rzz}E_z = 0
\end{aligned}
\tag{F.1a}
$$

$$
\begin{aligned}
&-\left[\frac{\partial}{\partial x}\left(\frac{\mu_{rxx}}{T_{yx}}\frac{\partial H_z}{\partial x}\right)+\frac{\partial}{\partial y}\left(\frac{\mu_{ryy}}{T_{xy}}\frac{\partial H_z}{\partial y}\right)\right] \\
&\quad -\beta_n\left(\frac{\varepsilon_0}{\mu_0}\right)^{1/2}\left[-\frac{\partial}{\partial x}\left(\frac{1}{T_{yx}}\frac{\partial E_z}{\partial y}\right)+\frac{\partial}{\partial y}\left(\frac{1}{T_{xy}}\frac{\partial E_z}{\partial x}\right)\right]-k_0^2\mu_{rzz}H_z = 0
\end{aligned}
\tag{F.1b}
$$

where

$$T_{xy} = \varepsilon_{rxx}\mu_{ryy} - \beta_n^2 \tag{F.2a}$$

$$T_{yx} = \varepsilon_{ryy}\mu_{rxx} - \beta_n^2 \tag{F.2b}$$

k_0 is the wave number in the vacuum and is defined as

$$k_0 = \omega(\varepsilon_0\mu_0)^{1/2} \tag{F.3}$$

and β_n is the normalized phase constant with respect to k_0 and is defined as

$$\beta_n = \frac{\beta}{\omega(\varepsilon_0\mu_0)^{1/2}} \tag{F.4}$$

If the following normalization is used:

$$\bar{E}_n = \left(\frac{\varepsilon_0}{\mu_0}\right)^{1/2} \bar{E} \tag{F.5a}$$

$$\bar{H}_n = \bar{H} \tag{F.5b}$$

where $\bar{E}$ and $\bar{H}$ are of the form (D.27) with (D.44), then the classical formulation of the problem results:

$$-\left[\frac{\partial}{\partial x}\left(\frac{\varepsilon_{rxx}}{T_{xy}}\frac{\partial E_{zn}}{\partial x}\right) + \frac{\partial}{\partial y}\left(\frac{\varepsilon_{ryy}}{T_{yx}}\frac{\partial E_{zn}}{\partial y}\right)\right] - \left[\frac{\partial}{\partial x}\left(\frac{\beta_n}{T_{xy}}\frac{\partial H_{zn}}{\partial y}\right) - \frac{\partial}{\partial y}\left(\frac{\beta_n}{T_{yx}}\frac{\partial H_{zn}}{\partial x}\right)\right]$$
$$- k_0^2 \varepsilon_{rzz} E_{zn} = 0 \tag{F.6a}$$

$$-\left[\frac{\partial}{\partial x}\left(\frac{\mu_{rxx}}{T_{yx}}\frac{\partial H_{zn}}{\partial x}\right) + \frac{\partial}{\partial y}\left(\frac{\mu_{ryy}}{T_{xy}}\frac{\partial H_{zn}}{\partial y}\right)\right] - \left[-\frac{\partial}{\partial x}\left(\frac{\beta_n}{T_{yx}}\frac{\partial E_{zn}}{\partial y}\right) + \frac{\partial}{\partial y}\left(\frac{\beta_n}{T_{xy}}\frac{\partial E_{zn}}{\partial x}\right)\right]$$
$$- k_0^2 \mu_{rzz} H_{zn} = 0 \tag{F.6b}$$

At the electric walls the following conditions must hold:

$$\bar{a}_n \times \bar{E}_{tn} = 0 \tag{F.7a}$$

$$E_{zn} = 0 \tag{F.7b}$$

and at the magnetic walls

$$\bar{a}_n \times \bar{H}_{tn} = 0 \tag{F.8a}$$

$$H_{zn} = 0 \tag{F.8b}$$

where

$$\bar{E}_{tn} = E_{xn}\bar{a}_x + E_{yn}\bar{a}_y \tag{F.9a}$$

$$\bar{H}_{tn} = H_{xn}\bar{a}_x + H_{yn}\bar{a}_y \tag{F.9b}$$

and the various transverse components of the fields are given by

$$E_{xn} = \frac{1}{k_0}\left[\frac{\beta_n}{T_{xy}}\frac{\partial E_{zn}}{\partial x} + \frac{\mu_{ryy}}{T_{xy}}\frac{\partial H_{zn}}{\partial y}\right] \tag{F.10a}$$

$$E_{yn} = \frac{1}{k_0}\left[\frac{\beta_n}{T_{yx}}\frac{\partial E_{zn}}{\partial y} - \frac{\mu_{rxx}}{T_{yx}}\frac{\partial H_{zn}}{\partial x}\right] \tag{F.10b}$$

$$H_{xn} = \frac{1}{k_0}\left[\frac{\beta_n}{T_{yx}}\frac{\partial H_{zn}}{\partial x} - \frac{\varepsilon_{ryy}}{T_{yx}}\frac{\partial E_{zn}}{\partial y}\right] \tag{F.10c}$$

$$H_{yn} = \frac{1}{k_0}\left[\frac{\beta_n}{T_{xy}}\frac{\partial H_{zn}}{\partial y} + \frac{\varepsilon_{rxx}}{T_{xy}}\frac{\partial E_{zn}}{\partial x}\right] \tag{F.10d}$$

$\bar{a}_n$ is the unit vector normal to the electric or the magnetic wall, so

$$\bar{a}_n = a_{nx}\bar{a}_x + a_{ny}\bar{a}_y \tag{F.11}$$

Omitting the factor $1/k_0$, the boundary conditions at the electric walls are given by

$$a_{nx}\left[\frac{\beta_n}{T_{yx}}\frac{\partial E_{zn}}{\partial y} - \frac{\mu_{rxx}}{T_{yx}}\frac{\partial H_{zn}}{\partial x}\right] - a_{ny}\left[\frac{\beta_n}{T_{xy}}\frac{\partial E_{zn}}{\partial x} + \frac{\mu_{ryy}}{T_{xy}}\frac{\partial H_{zn}}{\partial y}\right] = 0 \tag{F.12a}$$

$$E_{zn} = 0 \tag{F.12b}$$

and at the magnetic walls

$$a_{nx}\left[\frac{\beta_n}{T_{xy}}\frac{\partial H_{zn}}{\partial y} + \frac{\varepsilon_{rxx}}{T_{xy}}\frac{\partial E_{zn}}{\partial x}\right] - a_{ny}\left[\frac{\beta_n}{T_{yx}}\frac{\partial H_{zn}}{\partial x} - \frac{\varepsilon_{ryy}}{T_{yx}}\frac{\partial E_{zn}}{\partial y}\right] = 0 \tag{F.13a}$$

$$H_{zn} = 0 \tag{F.13b}$$

Equations (F.6) can be written in a matrix form as

$$-[G]^T [f] [G] \begin{Bmatrix} E_{zn} \\ H_{zn} \end{Bmatrix} - k_0^2 [p] \begin{Bmatrix} E_{zn} \\ H_{zn} \end{Bmatrix} = \begin{Bmatrix} 0 \\ 0 \end{Bmatrix} \tag{F.14a}$$

where ^{T} stands for the transverse of a matrix and matrices $[f]$, $[p]$, and $[G]$ are given by

$$[f] = \begin{bmatrix} \frac{\varepsilon_{rxx}}{T_{xy}} & 0 & 0 & \frac{\beta_n}{T_{xy}} \\ 0 & \frac{\varepsilon_{ryy}}{T_{yx}} & -\frac{\beta_n}{T_{yx}} & 0 \\ 0 & -\frac{\beta_n}{T_{yx}} & \frac{\mu_{rxx}}{T_{yx}} & 0 \\ \frac{\beta_n}{T_{xy}} & 0 & 0 & \frac{\mu_{ryy}}{T_{xy}} \end{bmatrix} \tag{F.14b}$$

$$[p] = \begin{bmatrix} \varepsilon_{rzz} & 0 \\ 0 & \mu_{rzz} \end{bmatrix} \tag{F.14c}$$

$$[G] = \begin{bmatrix} \frac{\partial}{\partial x} & 0 \\ \frac{\partial}{\partial y} & 0 \\ 0 & \frac{\partial}{\partial x} \\ 0 & \frac{\partial}{\partial y} \end{bmatrix} = \begin{bmatrix} \{\nabla_t\} & \{0\} \\ \{0\} & \{\nabla_t\} \end{bmatrix} \tag{F.14d}$$

It may be observed that matrix $[f]$ is symmetric while $[p]$ is diagonal.

Boundary conditions at the electric walls are

$$\begin{bmatrix} 0 \\ 0 \\ a_{nx} \\ a_{ny} \end{bmatrix}^T [f]\ [G] \begin{Bmatrix} E_{zn} \\ H_{zn} \end{Bmatrix} = 0 \tag{F.15a}$$

$$E_{zn} = 0 \tag{F.15b}$$

and at the magnetic walls

$$\begin{Bmatrix} a_{nx} \\ a_{ny} \\ 0 \\ 0 \end{Bmatrix}^T [f]\ [G] \begin{Bmatrix} E_{zn} \\ H_{zn} \end{Bmatrix} = 0 \tag{F.16a}$$

$$H_{zn} = 0 \tag{F.16b}$$

Equation (F.14), with conditions (F.15) and (F.16), corresponds to an eigenvalue problem of the type (2.2a) of Chapter 2

$$\left([\mathfrak{L}_1] - \lambda[\mathfrak{L}_2]\right)\{u\} = \{0\} \tag{F.17a}$$

where

$$[\mathfrak{L}_1] = -[G]^T\ [f]\ [G] \tag{F.17b}$$

$$[\mathfrak{L}_2] = [p] \tag{F.17c}$$

$$\{u\} = \begin{Bmatrix} E_{zn} \\ H_{zn} \end{Bmatrix} \tag{F.17d}$$

and $\lambda = k_0^2$ along with the boundary conditions of the form (2.2.b)

$$[\zeta_1]\{u\} - \lambda[\zeta_2]\{u\} = \{0\} \tag{F.17e}$$

The matrix operators $[\zeta_1]$ and $[\zeta_2]$ are given by (F.15) and (F.16).

The problem can be formulated using the classical or strong formulation as

Given the tensors $\underline{\underline{\varepsilon}}_r$ *and* $\underline{\underline{\mu}}_r$ *in* Ω, *in the form of a diagonal matrix and* β_n , *find the eigenvalues* λ *and the corresponding eigenvectors* $\{u\}$ *so that* (F.17) *is satisfied.*

It can be demonstrated that the operator in (F.17a) is self-adjoint. It is necessary to choose functions ϕ and ψ with the same properties and boundary conditions as those of E_{zn} and H_{zn}, respectively. The operator will be self-adjoint using the inner product of (A.18c) if it satisfies

$$\int_\Omega \begin{Bmatrix}\phi\\ \psi\end{Bmatrix}^T \left([\mathcal{L}_1] - k_0^2[\mathcal{L}_2]\right) \begin{Bmatrix}E_{zn}\\ H_{zn}\end{Bmatrix} d\Omega = \int_\Omega \begin{Bmatrix}E_{zn}\\ H_{zn}\end{Bmatrix}^T \left([\mathcal{L}_1] - k_0^2[\mathcal{L}_2]\right) \begin{Bmatrix}\phi\\ \psi\end{Bmatrix} d\Omega = 0 \tag{F.18}$$

where Ω is the domain of (F.14), with boundary conditions (F.15) and (F.16).

Integrating by parts the first term of (F.18) yields

$$- \int_\Omega \begin{Bmatrix}\phi\\ \psi\end{Bmatrix}^T [G]^T [f] [G] \begin{Bmatrix}E_{zn}\\ H_{zn}\end{Bmatrix} d\Omega - k_0^2 \int_\Omega \begin{Bmatrix}\phi\\ \psi\end{Bmatrix}^T [p] \begin{Bmatrix}E_{zn}\\ H_{zn}\end{Bmatrix} d\Omega$$

$$= \int_\Omega \left([G]\begin{Bmatrix}\phi\\ \psi\end{Bmatrix}\right)^T [f] [G] \begin{Bmatrix}E_{zn}\\ H_{zn}\end{Bmatrix} d\Omega - k_0^2 \int_\Omega \begin{Bmatrix}\phi\\ \psi\end{Bmatrix}^T [p] \begin{Bmatrix}E_{zn}\\ H_{zn}\end{Bmatrix} d\Omega \tag{F.19}$$

$$+ \int_\Gamma \begin{Bmatrix}\phi\\ \psi\end{Bmatrix}^T \begin{bmatrix} a_{nx} & 0\\ a_{ny} & 0\\ 0 & a_{nx}\\ 0 & a_{ny}\end{bmatrix}^T [f] [G] \begin{Bmatrix}E_{zn}\\ H_{zn}\end{Bmatrix} d\Gamma = 0$$

The third integral vanishes along the contour of the domain since it is equal to

$$\int_{\Gamma}(\phi k_0 \bar{a}_n \times \bar{H}_{tn} + \psi k_0 \bar{a}_n \times \bar{E}_{tn})\, d\Gamma$$

$$= \int_{\Gamma_{pe}} \phi k_0 \bar{a}_n \times \bar{H}_{tn}\, d\Gamma + \int_{\Gamma_{pm}} \phi k_0 \bar{a}_n \times \bar{H}_{tn}\, d\Gamma \tag{F.20}$$

$$+ \int_{\Gamma_{pe}} \psi k_0 \bar{a}_n \times \bar{E}_{tn}\, d\Gamma + \int_{\Gamma_{pm}} \psi k_0 \bar{a}_n \times \bar{E}_{tn}\, d\Gamma = 0$$

Note that ϕ and $\bar{a}_n \times \bar{E}_{tn}$ are zero at the electric walls and ψ and $\bar{a}_n \times \bar{H}_{tn}$ are zero at the magnetic walls. If the same procedure is carried out with the second term of (F.18), an analogous result is obtained regarding the boundary terms. Therefore, to prove that the operator is self-adjoint it will be enough to observe that due to the symmetry of [*f*] and [*p*] the following equality is satisfied:

$$\int_{\Omega} \left([G] \begin{Bmatrix} \phi \\ \psi \end{Bmatrix} \right)^T [f]\, [G] \begin{Bmatrix} E_{zn} \\ H_{zn} \end{Bmatrix} d\Omega - \int_{\Omega} k_0^2 \begin{Bmatrix} \phi \\ \psi \end{Bmatrix}^T [p] \begin{Bmatrix} E_{zn} \\ H_{zn} \end{Bmatrix} d\Omega$$

$$= \int_{\Omega} \left([G] \begin{Bmatrix} E_{zn} \\ H_{zn} \end{Bmatrix} \right)^T [f]\, [G] \begin{Bmatrix} \phi \\ \psi \end{Bmatrix} d\Omega - \int_{\Omega} k_0^2 \begin{Bmatrix} E_{zn} \\ H_{zn} \end{Bmatrix}^T [p] \begin{Bmatrix} \phi \\ \psi \end{Bmatrix} d\Omega = 0 \tag{F.21}$$

The fact that the operator is self-adjoint allows one to look for a functional from which equations (F.14) to (F.16) are derived. This functional can be constructed using the standard procedure, obtaining

$$F(E_{zn}, H_{zn}) = \frac{1}{2} \int_{\Omega} \left([G] \begin{Bmatrix} E_{zn} \\ H_{zn} \end{Bmatrix} \right)^T [f] [G] \begin{Bmatrix} E_{zn} \\ H_{zn} \end{Bmatrix} d\Omega - k_0^2 \frac{1}{2} \int_{\Omega} \begin{Bmatrix} E_{zn} \\ H_{zn} \end{Bmatrix}^T [p] \begin{Bmatrix} E_{zn} \\ H_{zn} \end{Bmatrix} d\Omega \tag{F.22}$$

However, the operator in (F.14) to (F.16) is not positive definite. Even though the matrix [*p*] is positive definite, the matrix [*f*] is not positive definite for any point in the domain Ω or for any value of β_n. In fact, the following expressions for the eigenvalues of [*f*] are obtained:

$$\lambda_{1,2} = \frac{1}{2} \frac{\varepsilon_{rxx} + \mu_{ryy}}{\varepsilon_{rxx}\mu_{ryy} - \beta_n^2} \pm \frac{1}{2} \left[\left(\frac{\varepsilon_{rxx} + \mu_{ryy}}{\varepsilon_{rxx}\mu_{ryy} - \beta_n^2} \right)^2 - \frac{4}{\varepsilon_{rxx}\mu_{ryy} - \beta_n^2} \right]^{1/2} \tag{F.23a}$$

$$\lambda_{3,4} = \frac{1}{2}\frac{\varepsilon_{ryy} + \mu_{rxx}}{\varepsilon_{ryy}\mu_{rxx} - \beta_n^2} \pm \frac{1}{2}\left[\left(\frac{\varepsilon_{ryy} + \mu_{rxx}}{\varepsilon_{ryy}\mu_{rxx} - \beta_n^2}\right)^2 - \frac{4}{\varepsilon_{ryy}\mu_{rxx} - \beta_n^2}\right]^{1/2} \tag{F.23b}$$

For $\beta_n = 0$, the eigenvalues are positive and given by $1/\mu_{ryy}$, $1/\varepsilon_{rxx}$, $1/\mu_{rxx}$, and $1/\varepsilon_{ryy}$. For $\beta_n^2 < \varepsilon_{rxx}\mu_{ryy}$, the first two values are positive, whereas for $\beta_n^2 > \varepsilon_{rxx}\mu_{ryy}$, one of them is positive and the other is negative. For $\beta_n^2 = \varepsilon_{rxx}\mu_{ryy}$, the matrix is singular. An analogous result is obtained for the third and fourth eigenvalues. They are positive only for values of $\beta_n^2 < \varepsilon_{ryy}\mu_{rxx}$. Matrix $[f]$ is singular also for $\beta_n^2 = \varepsilon_{ryy}\mu_{rxx}$.

Taking into account that the products $\varepsilon_{rxx}\mu_{ryy}$, $\varepsilon_{ryy}\mu_{rxx}$ will be different in each region of the problem domain, it can be concluded that the operator in (F.14) to (F.16) is positive definite only for the range of values of β_n^2 comprised between zero (the cut-off value of the mode) and the minimum value of the products of $\varepsilon_{rxx}(x,y)\mu_{ryy}(x,y)$ and $\varepsilon_{ryy}(x,y)\mu_{rxx}(x,y)$.

Even though a weak formulation given by (F.14) to (F.16) can be obtained for the problem by making the functional (F.22) stationary, this formulation has meaning only for values of β_n^2 that make the operator of (F.14) positive definite. Outside this range, the operator is not defined and finding a stationary point of the functional has no meaning. Therefore, it seems more suitable to obtain the weak formulation of the problem by the weighted residuals method and, specifically, by the Galerkin method. Obviously this formulation will formally agree with that obtained by the variational method.

The application of the weighted residuals method to (F.14) leads to the expression

$$-\int_\Omega \begin{Bmatrix}\phi\\ \psi\end{Bmatrix}^T [G]^T [f]\,[G] \begin{Bmatrix}E_{zn}\\ H_{zn}\end{Bmatrix} d\Omega - k_0^2 \int_\Omega \begin{Bmatrix}\phi\\ \psi\end{Bmatrix}^T [p] \begin{Bmatrix}E_{zn}\\ H_{zn}\end{Bmatrix} d\Omega = 0 \tag{F.24}$$

Integrating by parts yields

$$\int_\Omega \left([G]\begin{Bmatrix}\phi\\ \psi\end{Bmatrix}\right)^T [f]\,[G] \begin{Bmatrix}E_{zn}\\ H_{zn}\end{Bmatrix} d\Omega - k_0^2 \int_\Omega \begin{Bmatrix}\phi\\ \psi\end{Bmatrix}^T [p] \begin{Bmatrix}E_{zn}\\ H_{zn}\end{Bmatrix} d\Omega = 0 \tag{F.25}$$

where the cancellation of the integrals along the contour in (F.20) has already been taken into account.

Expression (F.25) with conditions (F.15b) and (F.16b) constitutes the weak formulation for the problem (F.14) to (F.16). It is identical to that which would be obtained from the functional (F.22) via the expression

$$\delta F(\delta E_{zn}, \delta H_{zn}) = 0 \tag{F.26}$$

provided that δE_{zn} is replaced by ϕ and δH_{zn} by ψ.

The weak formulation of problem (F.14) to (F.16) is

Let $H_{pe}^{1}(\Omega)$ be the space of functions containing φ so that $\varphi \in H^1(\Omega)$ and φ = 0 on Γ_{pe}. Let $H_{pm}^{1}(\Omega)$ be the space of functions containing γ so that $\gamma \in H^1(\Omega)$ and γ = 0 on Γ_{pm}. Here $\Gamma = \Gamma_{pe} \cup \Gamma_{pm}$ is the boundary of region Ω. Given two diagonal matrices $[\varepsilon_r]$, $[\mu_r]$ in Ω and β_n, find k_0^2 with $E_{zn} \in H_{pe}^{1}(\Omega)$ and $H_{zn} \in H_{pm}^{1}(\Omega)$ so that (F.25) *is satisfied for all $\phi \in H_{pe}^{1}(\Omega)$ and $\psi \in H_{pm}^{1}(\Omega)$.*

However, there are some numerical problems associated with (F.25) and (F.26) because for some values of β_n^2 the matrix $[f]$ is singular and it is positive definite only for values of β_n^2 between zero and the first value that makes the matrix $[f]$ singular. The singularity of matrix $[f]$ can be eliminated by multiplying (F.25) by the product $T_{xy}T_{yx}$. However, matrix $[p]$ would have to be multiplied by $T_{xy}T_{yx}$ and then it would no longer be positive-definite for all possible values of β_n^2. As will be shown later, in order for the methodology to be meaningful, only one of the matrices $[f]$ or $[p]$ has to be positive-definite. Then the formulation can be transformed to a positive-definite form by means of a translation (see Section 2.5.2.4). This is the formulation originally used in a FEM analysis of waveguiding structures. It can be seen in the works of [Ahmed and Daly 1969b], [Aubourg et al. 1983], [Chang et al. 1990, 1991], [Csendes and Silvester 1970], [Daly 1971], [Eswarappa et al. 1989], [Hernández-Figueroa and Pagiatakis 1993], [Ikeuchi et al. 1981], [Jin 1993; Ch. 7], [Mabaya et al. 1981], [McAulay 1977], [Okamoto and Okoshi 1978], [Tzuang and Itoh 1986], [Vandenbulcke and Lagasse 1976], [Welt and Webb 1985], and [Wu and Chen 1985a,b] for example.

F.2.1.2 *Discretization by Means of Lagrange Finite Elements*

Discretization of (F.25) is carried out via the Galerkin method and a FEM utilizing the Lagrange triangular elements. The choice of these elements is justified by the fact that the variational unknowns E_{zn} and H_{zn} should be continuous in the entire domain of the problem. In each element, each variational unknown is approximated by a linear combination of Lagrange triangular elements. In this work, linear elements and second-order straight and curved elements have been used for the solution of these classes of problems. Shape functions for the parent elements are shown in Table 2.2 of Chapter 2. Two degrees of freedom are therefore defined at each node. They correspond to the values of E_{zn} and H_{zn} at that point, respectively. Hence, vector Lagrange elements (see Section 2.5.1.2) have been used.

For the case of a linear element the unknown at each element is approximated by

$$\begin{Bmatrix} \tilde{E}^{\,e}_{zn} \\ \tilde{H}^{\,e}_{zn} \end{Bmatrix} = \begin{bmatrix} N_1^e & 0 & N_2^e & 0 & N_3^e & 0 \\ 0 & N_1^e & 0 & N_2^e & 0 & N_3^e \end{bmatrix} \begin{Bmatrix} d_{1e}^e \\ d_{1h}^e \\ d_{2e}^e \\ d_{2h}^e \\ d_{3e}^e \\ d_{3h}^e \end{Bmatrix} = [N^e]\,\{d^e\} \tag{F.27a}$$

whereas for the second-order elements

$$\begin{Bmatrix} \tilde{E}^{\,e}_{zn} \\ \tilde{H}^{\,e}_{zn} \end{Bmatrix} = \begin{bmatrix} N_1^e & 0 \\ 0 & N_1^e \\ N_2^e & 0 \\ 0 & N_2^e \\ N_3^e & 0 \\ 0 & N_3^e \\ N_4^e & 0 \\ 0 & N_4^e \\ N_5^e & 0 \\ 0 & N_5^e \\ N_6^e & 0 \\ 0 & N_6^e \end{bmatrix}^T \begin{Bmatrix} d_{1e}^e \\ d_{1h}^e \\ \cdot \\ \cdot \\ \cdot \\ d_{6e}^e \\ d_{6h}^e \end{Bmatrix} = [N^e]\,\{d^e\} \tag{F.27b}$$

where N_i^e refers to the ith shape function and $d_{ie}^{\,e}$, $d_{ih}^{\,e}$ correspond to the ith degree of freedom for the longitudinal component of the electric or the magnetic field, respectively, for the eth element. The values for the index i for the first- and second-order elements are $i = 1, ..., 3$ or 6, respectively.

Applying the Galerkin method to (F.25) leads to the following discrete integral form for each element:

$$\{W^e\} = \left[\int_{\Omega^e} \left([G]\,[N]\right)^T [f]\,[G]\,[N]\;\mathrm{d}\Omega \;-\; k_0^2 \int_{\Omega^e} [N]^T [p]\,[N]\;\mathrm{d}\Omega \right] \{d^e\} \tag{F.28}$$

where $[N^e]$ and $\{d^e\}$ are given by (F.27).

Expression (F.28) can be rewritten as

$$\{W^e\} = \left([k^e] - k_0^2\,[m^e]\right)\,\{d^e\} \tag{F.29}$$

Let the elements of the stiffness matrix be represented by $[k^e]$. Then, performing the computations over the parent element (see Appendix C.1), it can be shown that

$$[k^e] = \int_{\hat{\Omega}} [\partial\hat{N}]^T \begin{bmatrix} [J^e]^{-1} & \begin{matrix} 0 & 0 \\ 0 & 0 \end{matrix} \\ \begin{matrix} 0 & 0 \\ 0 & 0 \end{matrix} & [J^e]^{-1} \end{bmatrix} [f] \begin{bmatrix} \left([J^e]^{-1}\right)^T & \begin{matrix} 0 & 0 \\ 0 & 0 \end{matrix} \\ \begin{matrix} 0 & 0 \\ 0 & 0 \end{matrix} & \left([J^e]^{-1}\right)^T \end{bmatrix} [\partial\hat{N}] \; |J^e| \; d\Omega \tag{F.30}$$

where for the first-order element

$$[\partial\hat{N}] = \begin{bmatrix} \frac{\partial\hat{N}_1}{\partial\xi} & 0 & \frac{\partial\hat{N}_2}{\partial\xi} & 0 & \frac{\partial\hat{N}_3}{\partial\xi} & 0 \\ \frac{\partial\hat{N}_1}{\partial\eta} & 0 & \frac{\partial\hat{N}_2}{\partial\eta} & 0 & \frac{\partial\hat{N}_3}{\partial\eta} & 0 \\ 0 & \frac{\partial\hat{N}_1}{\partial\xi} & 0 & \frac{\partial\hat{N}_2}{\partial\xi} & 0 & \frac{\partial\hat{N}_3}{\partial\xi} \\ 0 & \frac{\partial\hat{N}_1}{\partial\eta} & 0 & \frac{\partial\hat{N}_2}{\partial\eta} & 0 & \frac{\partial\hat{N}_3}{\partial\eta} \end{bmatrix} \tag{F.31a}$$

while for the second-order reference element one has

$$[\partial\hat{N}] = \begin{bmatrix} \frac{\partial\hat{N}_1}{\partial\xi} & 0 & \frac{\partial\hat{N}_2}{\partial\xi} & 0 & \frac{\partial\hat{N}_3}{\partial\xi} & 0 & \frac{\partial\hat{N}_4}{\partial\xi} & 0 & \frac{\partial\hat{N}_5}{\partial\xi} & 0 & \frac{\partial\hat{N}_6}{\partial\xi} & 0 \\ \frac{\partial\hat{N}_1}{\partial\eta} & 0 & \frac{\partial\hat{N}_2}{\partial\eta} & 0 & \frac{\partial\hat{N}_3}{\partial\eta} & 0 & \frac{\partial\hat{N}_4}{\partial\eta} & 0 & \frac{\partial\hat{N}_5}{\partial\eta} & 0 & \frac{\partial\hat{N}_6}{\partial\eta} & 0 \\ 0 & \frac{\partial\hat{N}_1}{\partial\xi} & 0 & \frac{\partial\hat{N}_2}{\partial\xi} & 0 & \frac{\partial\hat{N}_3}{\partial\xi} & 0 & \frac{\partial\hat{N}_4}{\partial\xi} & 0 & \frac{\partial\hat{N}_5}{\partial\xi} & 0 & \frac{\partial\hat{N}_6}{\partial\xi} \\ 0 & \frac{\partial\hat{N}_1}{\partial\eta} & 0 & \frac{\partial\hat{N}_2}{\partial\eta} & 0 & \frac{\partial\hat{N}_3}{\partial\eta} & 0 & \frac{\partial\hat{N}_4}{\partial\eta} & 0 & \frac{\partial\hat{N}_5}{\partial\eta} & 0 & \frac{\partial\hat{N}_6}{\partial\eta} \end{bmatrix}$$

(F.31b)

In (F.29) $[m^e]$ is the element mass matrix. When computed over the parent element it is given by

$$[m^e] = \int_{\hat{\Omega}} [\hat{N}]^T [p] [\hat{N}] \, |J^e| \, \mathrm{d}\Omega \tag{F.32}$$

where $[\hat{N}]$ is defined by (F.27) with respect to the coordinates of $\hat{x}$ and $\hat{y}$ of the parent element, i.e., the natural coordinates ξ,η of the reference element, respectively. After utilizing a geometrical normalization constant one obtains from (F.29)

$$\{W^e\} = \left([k^e] - s^2[m^e]\right)\{d^e\} \tag{F.33}$$

where

$$s = k_0 a \tag{F.34}$$

Here a is a scale factor that allows the normalization of the dimensions of the structure with respect to this scale factor. This would not change the value of β_n since $\beta_n = \beta/k_0$. Therefore one would be multiplying and dividing by the same variable a.

For the case of the first-order element, the calculation of the coefficients of the matrices $[k^e]$ and $[m^e]$ are carried out in an analytical manner using (2.71). For the second-order elements, the elements of the matrix $[k^e]$ are calculated using a three point numerical integration and those of the matrix $[m^e]$ by a six-point numerical integration (see Table 2.11 of Chapter 2). These integrations are exact for straight elements (subparametric case) because of the number of integration points chosen and only approximate for curved isoparametric elements.

After calculating the element matrices $[k^e]$ and $[m^e]$, they are assembled and the Dirichlet type boundary conditions are applied. This results in equating to zero the degrees of freedom for E_{zn} located at the electric walls and the degrees of freedom for H_{zn} located at the magnetic walls. Computationally this operation is performed as explained in Section 2.5.2.4. Conditions (F.15a) are the natural boundary conditions for the unknown H_{zn} at the electric walls, and (F.15b) are the natural boundary conditions for E_{zn} at the magnetic walls. The natural boundary conditions will therefore be satisfied only approximately. Continuity of E_{zn} and H_{zn} in the entire domain is assured, but this does not guarantee the continuity of the derivatives. Specifically, conditions at the interfaces between two nonconducting media given by (D.32a,b) will only be fulfilled approximately.

Once the essential boundary conditions are imposed, the global system of equations is obtained:

$$\left([K] - s^2[M]\right)\{D\} = \{0\} \tag{F.35}$$

in which the matrices $[K]$ and $[M]$ are sparse, real, and symmetric. This system is solved for each value of β_n^2, resulting in the corresponding eigenvalues s_i^2 and the eigenvectors $\{F_i\}$. If the algorithms used for the solution of (F.35) cannot deal with the case when the matrix $[K]$ is singular, those values of β_n^2 that make $[f]$ singular must be avoided. Also in order to obtain correct numerical values for s^2 close to the singularities, the results obtained for values of β_n^2 close to those making $[f]$ singular should be interpolated.

Algorithms for the solution of the eigenvalue system of equations with sparse, real, and symmetric positive-definite matrices are used for the solution of (F.35). Consequently, because for some values of β_n^2, one is not dealing with positive-definite systems (recall that $[K]$ will not always be positive definite), a translation of the original eigenvalue problem is necessary, yielding

$$\left([K'] - \lambda[M]\right)\{D\} = \{0\} \tag{F.36}$$

where

$$[K'] = [K] + \theta[M] \tag{F.37a}$$

so that $[K']$ is positive definite. Here θ is a scalar quantity. The eigenvalues of the modified problem, λ, are related to the eigenvalues of the original problem, s^2, via

$$s^2 = \lambda - \theta \tag{F.37b}$$

This method assumes that $[M]$ is positive definite, as it is in this case. In this book, the technique used to solve for the eigenvalues of the modified systems is that of inverse iteration by subspaces utilizing the Rayleigh-Ritz method. For some values of β_n^2, spurious modes are generated [Jin 1993; Ch. 7]. The number of the spurious modes is equal to the number of nodes on the boundaries between dielectrics. It had been suggested that these modes may be eliminated by imposing the boundary conditions at the dielectric interfaces. This method has been applied in [Mabaya et al. 1981] with some success, considering all of the spurious modes were not eliminated.

In any case, these spurious modes can be detected in a variety of ways. For instance, they can be identified by looking at the eigenvectors since the spurious modes display a chaotic behavior for the field pattern.

F.2.2 Homogeneous and Isotropic Structures

F.2.2.1 *Weak Formulation*

For the case of a homogeneous and isotropic structure (F.1) leads to two decoupled problems, one for each longitudinal component

$$-\left[\frac{\partial}{\partial x}\left(\frac{\partial E_z}{\partial x}\right) + \frac{\partial}{\partial y}\left(\frac{\partial E_z}{\partial y}\right)\right] - k_0^2(\varepsilon_r\mu_r - \beta_n^2)E_z = 0 \tag{F.38a}$$

$$-\left[\frac{\partial}{\partial x}\left(\frac{\partial H_z}{\partial x}\right) + \frac{\partial}{\partial y}\left(\frac{\partial H_z}{\partial y}\right)\right] - k_0^2(\varepsilon_r\mu_r - \beta_n^2)H_z = 0 \tag{F.38b}$$

where

$$k_c^2 = k_0^2(\varepsilon_r\mu_r - \beta_n^2) \tag{F.39}$$

is the cutoff wave number. Alternately, (F.38) can be written as

$$-\Delta_t\phi - k_c^2\phi = 0 \tag{F.40}$$

where operator Δ_t is

$$\Delta_t = \bar{\nabla}_t \cdot \bar{\nabla}_t = \frac{\partial^2}{\partial x^2} + \frac{\partial^2}{\partial y^2} \tag{F.41a}$$

with $\bar{\nabla}_t$ given by (D.11f), or in a matrix notation

$$\Delta_t = \{\nabla_t\}^T\{\nabla_t\} = \begin{Bmatrix}\frac{\partial}{\partial x} & \frac{\partial}{\partial y}\end{Bmatrix}^T \begin{Bmatrix}\frac{\partial}{\partial x} \\ \frac{\partial}{\partial y}\end{Bmatrix} = \frac{\partial^2}{\partial x^2} + \frac{\partial^2}{\partial y^2} \tag{F.41b}$$

In (F.40) ϕ means E_z or H_z depending on whether the problem to be solved is (F.38a) or (F.38b), respectively.

Since in an homogeneous and isotropic guiding structure, the propagating modes are either TM modes (in which $H_z(x,y) = 0$ in Ω) or TE modes (in which $E_z(x,y) = 0$ in Ω), the analysis of such structures involves solving (F.40) under different boundary conditions, depending on whether the unknown is E_z (TM modes) or H_z (TE modes). Table 3.24 of Chapter 3 summarizes this formulation.

It is evident that k_c^2 neither depends on ω nor on β_n. It is, in fact, a constant and an eigenvalue of (F.40). The ith eigenvalue, k_{ci}^2, characterizes the ith mode that is propagating. Once the eigenvalue has been calculated, the corresponding value of β_n for each value of ω (of k_0) can be obtained from (F.39). The decoupled problems are:

a) TM modes: The unknown ϕ in (F.40) is E_z. The other electric and magnetic field components can be written as functions of ϕ according to the following expressions:

$$\bar{E}_t = \frac{\beta}{k_c^2} \bar{\nabla}_t \phi \tag{F.42a}$$

$$\bar{H}_t = \frac{\omega \varepsilon_0 \varepsilon_r}{k_c^2} \bar{a}_z \times \bar{\nabla}_t \phi \tag{F.42b}$$

$$H_z = 0 \tag{F.42c}$$

The electric walls in this problem are boundaries in which homogeneous Dirichlet constraints are imposed:

$$\phi = 0 \tag{F.43a}$$

whereas magnetic walls are under homogeneous Neumann type constraints since expression (D.36a) yields

$$\bar{a}_n \cdot \bar{\nabla}_t \phi = 0 \tag{F.43b}$$

b) TE modes: The unknown ϕ in (F.40) is H_z. The other magnetic and electric field components are given by

$$\bar{H}_t = \frac{\beta}{k_c^2} \bar{\nabla}_t \phi \tag{F.44a}$$

$$\bar{E}_t = -\frac{\omega \mu_0 \mu_r}{k_c^2} \bar{a}_z \times \bar{\nabla}_t \phi \tag{F.44b}$$

$$E_z = 0 \tag{F.44c}$$

Magnetic walls are now Dirichlet boundaries on which

$$\phi = 0 \tag{F.45a}$$

and electric walls are under Neumann type constraints since (D.33a) yields

$$\bar{a}_n \cdot \bar{\nabla}_t \phi = 0 \tag{F.45b}$$

Therefore, in both cases (F.40) must be solved under boundary conditions (F.43) identical to those in (F.45). The only difference is that for TM modes, the Dirichlet type boundaries are the electric walls and for the Neumann type they are the magnetic walls, whereas for TE modes it is the reverse situation.

The classical or strong formulation of the problem would then be posed as

Find k_c and ϕ that satisfy (F.40) *in Ω and with boundary conditions $\phi = 0$ on Γ_D and $\bar{a}_n \cdot \bar{\nabla}_t \phi = 0$ on Γ_N, where $\Gamma = \Gamma_D \cup \Gamma_N$ is the boundary of Ω.*

It is easy to demonstrate that the operator of (F.40) with conditions (F.43) or (F.44) is self-adjoint and positive definite. The functional for the classical formulation thus becomes

$$F = \frac{1}{2}\left(\int_\Omega \bar{\nabla}_t \phi \cdot \bar{\nabla}_t \phi \, \mathrm{d}\Omega - k_c^2 \int_\Omega \phi^2 \mathrm{d}\Omega\right) \tag{F.46}$$

where k_c^2 and ϕ are real. The terms containing the contour integrals cancel out since the boundary conditions have been taken into account.

The weak formulation of the problem can therefore be obtained by minimizing the functional according to

$$\delta F = 0 \tag{F.47}$$

Hence it can be formulated as

Let $H_0^1(\Omega)$ be the space of functions φ so that $\varphi \in H^1(\Omega)$ and $\varphi = 0$ on Γ_D, where Γ_D is a part of the contour of Ω (the Dirichlet one). Find $\phi \in H_0^1(\Omega)$ that makes the functional (F.46) *stationary.*

This formulation is typical in FEM and can be seen in the works of [Ahmed 1968], [Ahmed and Daly 1969a], [Álvarez-Melcón et al. 1994], [García-Castillo et al. 1994, 1995], [Gil and Zapata 1994, 1995, 1997], [Israel and Miniowitz 1987], [Jin

1993; Ch. 7], [Lagasse and Van Bladel 1972], [Sarkar et al. 1994], [Silvester 1969a,b], and [Silvester and Ferrari 1983; Chs. 3], among others.

F.2.2.2 *Discretization by Means of Lagrange Finite Elements*

Discretization of the previous problem is carried out by means of the Ritz method and a FEM that uses Lagrange triangular elements. This is justified since ϕ is continuous in the problem domain. The degree of freedom is defined at each node as the value of ϕ at the node. Linear triangular elements and second-order straight and curved elements have been used. Shape functions of the corresponding reference elements are shown in Table 2.2 of Chapter 2. In each element the unknown is approximated by

$$\tilde{\phi}^e = \{N^e\}^T\{d^e\} \tag{F.48}$$

Hence the functional (F.46) for each element results in

$$F^e(d^e) = \frac{1}{2}\left\{\int_{\Omega^e}\left[\{\nabla_t\}\left(\{N\}^T\{d^e\}\right)\right]^T\left[\{\nabla_t\}\left(\{N\}^T\{d^e\}\right)\right]\mathrm{d}\Omega - k_c^2\int_{\Omega^e}\left(\{N\}^T\{d^e\}\right)^2\mathrm{d}\Omega\right\} \tag{F.49}$$

In addition, the first variation will give the local discrete integral form

$$\delta F^e(d^e) = \delta F\left(\{N^e\}^T\{d^e\}\right) = \left\{\frac{\partial F^e}{\partial d^e}\right\}^T\{\delta d^e\} = \{W^e\}\{\delta d^e\} \tag{F.50}$$

where

$$\{W^e\} = \left(\int_{\Omega^e}\left[\{\nabla_t\}\{N\}^T\right]^T\left[\{\nabla_t\}\{N\}^T\right]\mathrm{d}\Omega - k_c^2\int_{\Omega^e}\{N\}\{N\}^T\,\mathrm{d}\Omega\right)\{d^e\} \tag{F.51}$$

resulting in

$$\{W^e\} = \left([k^e] - k_c^2[m^e]\right)\{d^e\} \tag{F.52a}$$

where

$$[k^e] = \int_{\Omega^e}[\partial N]^T[\partial N]\mathrm{d}\Omega \tag{F.52b}$$

$$[m^e] = \int_{\Omega^e} \{N\}\{N\}^T \, \mathrm{d}\Omega \tag{F.52c}$$

Converting the expressions back to the reference element it is seen that

$$[k^e] = \int_{\hat{\Omega}} [\partial\hat{N}]^T \, [J^e]^{-1}([J^e]^T)^{-1} \, [\partial\hat{N}] \, |J^e| \, \mathrm{d}\Omega \tag{F.53a}$$

where $[J^e]$ is the Jacobian matrix of the transformation of the eth element and $|J^e|$ is the determinant. Here $[\partial\hat{N}]$ is given by

$$[\partial\hat{N}] = \begin{bmatrix} \left\{\dfrac{\partial N}{\partial \hat{x}}\right\}^T \\ \left\{\dfrac{\partial N}{\partial \hat{y}}\right\}^T \end{bmatrix} \tag{F.53b}$$

and

$$[m^e] = \int_{\Omega^e} \{\hat{N}\}\{\hat{N}\}^T \, |J^e| \, \mathrm{d}\Omega \tag{F.53c}$$

For the calculation of $[m^e]$ and $[k^e]$, an analytical integration is used for the case of the first-order element. For the second-order elements, a numerical integration is used with three points for the elements of the matrix $[k^e]$ and six points for the elements of matrix $[m^e]$. These integrations are exact for straight elements because of the number of integration points chosen and approximate for curved elements.

After assembling the elements and imposing the Dirichlet boundary conditions, the following system is obtained:

$$\left([K] - k_c^2 [M]\right)\{D\} = \{0\} \tag{F.54a}$$

whose solution provides the different eigenvalues k_{ci}^2 and the corresponding eigenvectors $\{D_i\}$.

A geometric normalization may be used in this case as well, so the eigenvalue becomes

$$s^2 = (a k_c)^2 \tag{F.54b}$$

and the problem to be solved is

$$\left([K]-s^2[M]\right)\{D\} = \{0\} \tag{F.54c}$$

Matrices $[K]$ and $[M]$ are real, symmetric, sparse and positive definite. Thus, (F.54) can be solved efficiently utilizing the appropriate algorithms. The solution of (F.48), i.e., the longitudinal component of the electric or the magnetic field, will be a continuous function with discontinuous derivatives. Spurious modes do not appear in this formulation.

F.3 FORMULATION UTILIZING THE LONGITUDINAL AND TRANSVERSE COMPONENTS OF THE ELECTRIC OR MAGNETIC FIELD

This section deals with the development of the weak formulation for the full-wave analysis of waveguide problems that uses the transverse and the longitudinal components of the electric or the magnetic field. The discretization is done by a FEM that utilizes a combination of the Lagrange and the curl-conforming triangular elements.

The standard formulation is described first. This formulation leads to an eigenvalue problem where for a specified value of the propagation constant the eigenvalue related to the frequency is solved. This formulation has the advantage that the integral forms involved are real. The resultant matrices are real symmetric and positive semidefinite.

A variant of the previous formulation is shown next. This formulation is called nonstandard because it is obtained from the previous one after some manipulations. However, it gives a canonical form because the eigenvalue is the propagation constant for a given known frequency. The integral forms involved are also real and the resultant matrices of the discrete problem although symmetric and sparse are not positive definite.

F.3.1 Standard Formulation

F.3.1.1 *Weak Formulation*

The formulation developed in Section F.2.1 can only be applied to structures whose tensors $\underline{\underline{\varepsilon}}$ and $\underline{\underline{\mu}}$ can be represented by diagonal matrices. The formulation given in this section is more general and can be applied to almost all types of waveguide structures used in the microwaves and millimeter wave frequencies. The only requirement here is that the tensors $\underline{\underline{\varepsilon}}$ and $\underline{\underline{\mu}}$ can be separated along their cross and longitudinal sections as discussed in Appendix D.2. The formulation starts with the Helmholtz equations in a medium without any electric or magnetic losses for the electric or the magnetic field. This is characterized by (D.23) with conditions (D.19a,b), (D.20a), and (D.22a). Due to the duality between the electric and the magnetic fields the generic nomenclature $\underline{\bar{V}}$ has been used that refers to either $\underline{\bar{E}}(x,y,z)$ or $\underline{\bar{H}}(x,y,z)$ and $\underline{\underline{f}}$ and $\underline{\underline{p}}$ correspond to the tensors $\underline{\underline{\varepsilon}}$ and $\underline{\underline{\mu}}$ as shown in Table 3.26 of Chapter 3.

The operator (D.23) with its appropriate boundary conditions is self-adjoint and positive definite. Hence, the functional from which it is derived can be expressed as

$$F = \int_{\Omega} \left(\overline{\nabla}\times\overline{V}^{*}\right) \cdot \left(\underline{\underline{f}}^{-1}\overline{\nabla}\times\overline{V}\right) \mathrm{d}\Omega - \omega^{2}\int_{\Omega} \overline{V}^{*} \cdot \underline{\underline{p}}\,\overline{V}\,\mathrm{d}\Omega \tag{F.55}$$

with $\bar{a}_n \times \overline{V} = 0$ at magnetic or electric walls depending on whether $\overline{V}$ is the electric or the magnetic field and * denotes complex conjugate [Jin 1993; Ch. 6]. A boundary integral term has been equated to zero in (F.55), due to homogeneous natural boundary conditions.

Using the duality between the formulations that employ the electric and the magnetic fields, (D.23) and (F.55) can be expressed as functions of the unknown vector, $\overline{V}$, that is,

$$\overline{V} = \overline{V}_t + \mathrm{j}\bar{a}_z V_z \tag{F.56a}$$

In addition, the vector $\overline{V}$ can be normalized according to

$$\overline{V}_n = \overline{V}_{tn} + \mathrm{j}\bar{a}_z V_{zn} \tag{F.56b}$$

where

$$\overline{V}_n = \sqrt{p_0}\;\overline{V} \tag{F.56c}$$

$$\overline{V}_{tn} = \sqrt{p_0}\;\overline{V}_t \tag{F.56d}$$

$$V_{zn} = \sqrt{p_0}\;V_z \tag{F.56e}$$

When $\overline{V}$ is the electric field, $\overline{E}$, p_0 is the dielectric permittivity of vacuum, ε_0. When it is the magnetic field, $\overline{H}$, then p_0 is the magnetic permeability of vacuum, μ_0.

Let the dual variable be $\overline{U}$, where

$$\overline{U} = \overline{U}_t + \mathrm{j}\bar{a}_z U_z \tag{F.57a}$$

and introduce the following normalization:

$$\bar{U}_n = \bar{U}_{tn} + j\bar{a}_z U_{zn} \tag{F.57b}$$

where

$$\bar{U}_n = \sqrt{f_0}\,\bar{U} \tag{F.57c}$$

$$\bar{U}_{tn} = \sqrt{f_0}\,\bar{U}_t \tag{F.57d}$$

$$U_{zn} = \sqrt{f_0}\,U_z \tag{F.57e}$$

If $\bar{U}$ is the magnetic field, $\bar{H}$, then f_0 is the vacuum magnetic permeability, μ_0. When it is the electric field, $\bar{E}$, then f_0 is the vacuum dielectric permittivity, ε_0.

When the formulation is used in terms of the electric field, the magnetic field components are obtained from (D.28a) and (D.29a). For the dual situation, the electric field components can be obtained from (D.28b) and (D.29b), providing the general expressions

$$\bar{U}_{tn} = sign\left[\frac{1}{k_0}\underline{\underline{f}}_{rt}^{-1}\bar{a}_z \times \bar{V}_{tn} - \frac{1}{k_0}\underline{\underline{f}}_{rt}^{-1}\bar{\nabla}_t \times V_{zn}\bar{a}_z\right] \tag{F.58a}$$

$$U_{zn}\bar{a}_z = sign\left[\frac{1}{k_0}f_{rzz}^{-1}\nabla_t \times \bar{V}_{tn}\right] \tag{F.58b}$$

where $\underline{\underline{f}}_{rt}$ and f_{rzz} imply $\underline{\underline{\mu}}_{rt}$ and μ_{rzz}, respectively, when the formulation is in terms of the electric field and $\underline{\underline{\varepsilon}}_{rt}$ and ε_{rzz} when the unknown is the magnetic field. At the same time, the variable *sign* has a value of +1 for the formulation with the electric field and -1 for the formulation with the magnetic field.

With this notation, the strong formulation of the problem is given by

$$\bar{\nabla}_t \times f_{rzz}^{-1}\bar{\nabla}_t \times \bar{V}_{tn} + \beta\bar{a}_z \times \underline{\underline{f}}_{rt}^{-1}\bar{\nabla}_t \times V_{zn}\bar{a}_z - \beta^2\bar{a}_z \times \underline{\underline{f}}_{rt}^{-1}\bar{a}_z \times \bar{V}_{tn} - k_0^2\underline{\underline{p}}_{rt}\bar{V}_{tn} = 0 \tag{F.59a}$$

$$\bar{\nabla}_t \times \underline{\underline{f}}_{rt}^{-1}\bar{\nabla}_t \times V_{zn}\bar{a}_z - \beta\bar{\nabla}_t \times \underline{\underline{f}}_{rt}^{-1}\bar{a}_z \times \bar{V}_{tn} - k_0^2 p_{rzz}V_{zn}\bar{a}_z = 0 \tag{F.59b}$$

over the cross section, Ω, of the structure. Here $\underline{\underline{p}}_{rt}$ and p_{rzz} imply $\underline{\underline{\varepsilon}}_{rt}$ and ε_{rzz} in the case of a formulation with $\bar{E}_{tn}$ and E_{zn} or $\underline{\underline{\mu}}_{rt}$ and μ_{rzz} in the case of a formulation with $\bar{H}_{tn}$ and H_{zn}.

The Dirichlet boundary conditions are

$$\bar{a}_n \times \bar{V}_{tn} = 0 \tag{F.60a}$$

$$V_{zn} = 0 \tag{F.60b}$$

and the Neumann boundary conditions are given by

$$\bar{a}_n \times \beta \underline{\underline{f}}_{rt}^{-1} \bar{a}_z \times \bar{V}_{tn} - \bar{a}_n \times \underline{\underline{f}}_{rt}^{-1} \bar{\nabla}_t \times V_{zn} \bar{a}_z = 0 \tag{F.61a}$$

$$\bar{a}_z \cdot \bar{\nabla}_t \times \bar{V}_{tn} = \bar{a}_n \times \bar{\nabla}_t \times \bar{V}_{tn} = 0 \tag{F.61b}$$

Table 3.27 summarizes the definition of the different variables in (F.56) to (F.61) depending on whether the formulation is applied to the electric or the magnetic field.

The weak formulation can be obtained by starting from the functional

$$F_n = \frac{c_0^2}{2}\left[A_n - k_0^2 B_n\right] \tag{F.62a}$$

where

$$\begin{aligned} A_n = {} & \int_\Omega \bar{\nabla}_t \times V_{zn}\bar{a}_z \cdot \underline{\underline{f}}_{rt}^{-1} \bar{\nabla}_t \times V_{zn}\bar{a}_z \, d\Omega + \beta^2 \int_\Omega \bar{a}_z \times \bar{V}_{tn} \cdot \underline{\underline{f}}_{rt}^{-1} \bar{a}_z \times \bar{V}_{tn} \, d\Omega \\ & - 2\beta \int_\Omega \bar{\nabla}_t \times V_{zn}\bar{a}_z \cdot \underline{\underline{f}}_{rt}^{-1} \bar{a}_z \times \bar{V}_{tn} \, d\Omega + \int_\Omega \bar{\nabla}_t \times \bar{V}_{tn} \cdot f_{rzz}^{-1} \bar{\nabla}_t \times \bar{V}_{tn} \, d\Omega \end{aligned} \tag{F.62b}$$

and

$$B_n = \int_\Omega \bar{V}_{tn} \cdot \underline{\underline{p}}_{rt} \bar{V}_{tn} + V_{zn} p_{rzz} V_{zn} \, d\Omega \tag{F.62c}$$

The functional (F.62) can be obtained from (F.55) by specializing $\underline{\bar{V}}$ according to (D.25) and (D.27). It can be seen that (F.62) is real [García-Castillo 1992].

Therefore the problem can be formulated in a weak form as

Let $X^0(\Omega)$ be the space of the vector functions $\bar{q}$ in Ω so that $\bar{q} \in H(\mathrm{curl},\Omega)$ and $\bar{a}_n \times \bar{q} = 0$ on Γ_D. Let $H_0^1(\Omega)$ be the space of scalar functions ψ so that $\psi \in H^1(\Omega)$ and $\psi = 0$ on Γ_D, where $\Gamma = \Gamma_D \cup \Gamma_N$ is the boundary of Ω. Given β, $\underline{f}_r$ and $\underline{p}_r$ in Ω, separable along its cross and longitudinal parts, and assuming that the inverse of $\underline{f}_{rt}$ exists, find the three unknowns k_0^2, $\bar{V}_{tn}$, and V_{zn} with $\bar{V}_{tn} \in X^0$ and $V_{zn} \in H_0^1(\Omega)$ that make the functional (F.62) *stationary.*

The weak formulation of the problem can also be obtained using a weighted residual method. Let the residual in Ω be given by the complex vector

$$\bar{r}_n = \bar{r}_{tn} + j r_{zn} \bar{a}_z \tag{F.63a}$$

where

$$\bar{r}_{tn} = \bar{\nabla}_t \times f_{rzz}^{-1} \bar{\nabla}_t \times \bar{V}_{tn} + \beta \bar{a}_z \times \underline{f}_{rt}^{-1} \bar{\nabla}_t \times V_{zn} \bar{a}_z - \beta^2 \bar{a}_z \times \underline{f}_{rt}^{-1} \bar{a}_z \times \bar{V}_{tn} - k_0^2 \underline{p}_{rt} \bar{V}_{tn} \tag{F.63b}$$

and

$$r_{zn} \bar{a}_z = \bar{\nabla}_t \times \underline{f}_{rt}^{-1} \bar{\nabla}_t \times V_{zn} \bar{a}_z - \beta \bar{\nabla}_t \times \underline{f}_{rt}^{-1} \bar{a}_z \times \bar{V}_{tn} - k_0^2 p_{rzz} V_{zn} \bar{a}_z \tag{F.63c}$$

Hence the objective is to find a solution that will make the residual orthogonal to the conjugate of a weighted complex vector function (see Section 2.3.3.1)

$$\bar{\phi} = \bar{\phi}_t + j \phi_z \bar{a}_z \tag{F.64}$$

according to

$$W = \int_\Omega \bar{\phi}^* \cdot \bar{r}_n \, d\Omega = 0 \tag{F.65}$$

Therefore,

$$W = \int_\Omega \bar{\phi}_t \cdot \bar{r}_{tn} \, d\Omega + \int_\Omega \phi_z \bar{a}_z \cdot r_{zn} \bar{a}_z \, d\Omega = 0 \tag{F.66}$$

and this makes

$$W = W_1 - \beta(W_2 + W_3) + \beta^2 W_4 - k_0^2 W_5 = 0 \tag{F.67a}$$

where

$$W_1 = \int_\Omega \bar{\phi}_t \cdot (\bar{\nabla}_t \times f_{rzz}^{-1}\ \bar{\nabla}_t \times \bar{V}_{tn})\ d\Omega + \int_\Omega \phi_z \bar{a}_z \cdot (\bar{\nabla}_t \times \underline{\underline{f}}_{rt}^{-1}\ \bar{\nabla}_t \times V_{zn} \bar{a}_z)\ d\Omega \tag{F.67b}$$

$$W_2 = -\int_\Omega \bar{\phi}_t \cdot (\bar{a}_z \times \underline{\underline{f}}_{rt}^{-1}\ \bar{\nabla}_t \times V_{zn} \bar{a}_z)\ d\Omega \tag{F.67c}$$

$$W_3 = \int_\Omega \phi_z \bar{a}_z \cdot (\nabla_t \times \underline{\underline{f}}_{rt}^{-1}\ \bar{a}_z \times \bar{V}_{tn})\ d\Omega \tag{F.67d}$$

$$W_4 = -\int_\Omega \bar{\phi}_t \cdot (\bar{a}_z \times \underline{\underline{f}}_{rt}^{-1}\ \bar{a}_z \times \bar{V}_{tn})\ d\Omega \tag{F.67e}$$

$$W_5 = \int_\Omega \bar{\phi}_t \cdot \underline{\underline{p}}_{rt}\ \bar{V}_{tn}\ d\Omega + \int_\Omega \phi_z p_{rzz} V_{zn}\ d\Omega \tag{F.67f}$$

Integrating W_1 by parts and using the vectorial identities [Tai 1992] one obtains

$$\begin{aligned} W_1 = & \int_\Omega (\bar{\nabla}_t \times \bar{\phi}_t) \cdot (f_{rzz}^{-1} \bar{\nabla}_t \times \bar{V}_{tn})\ d\Omega + \int_\Omega (\bar{\nabla}_t \times \phi_z \bar{a}_z) \cdot (\underline{\underline{f}}_{rt}^{-1}\ \bar{\nabla}_t \times V_{zn} \bar{a}_z)\ d\Omega \\ & + \int_{\Gamma_N} \bar{\phi}_t \cdot (\bar{a}_n \times f_{rzz}^{-1} \bar{\nabla}_t \times \bar{V}_{tn})\ d\Gamma - \int_{\Gamma_D} (\bar{a}_n \times \bar{\phi}_t) \cdot (f_{rzz}^{-1}\ \bar{\nabla}_t \times \bar{V}_{tn})\ d\Gamma \\ & + \int_{\Gamma_N} \phi_z \bar{a}_z \cdot (\bar{a}_n \times \underline{\underline{f}}_{rt}^{-1}\ \bar{\nabla}_t \times V_{zn} \bar{a}_z)\ d\Gamma + \int_{\Gamma_D} \phi_z \bar{a}_z \cdot (\bar{a}_n \times \underline{\underline{f}}_{rt}^{-1} \times V_{zn} \bar{a}_z)\ d\Gamma \end{aligned} \tag{F.68a}$$

The first boundary integral is zero due to the Neumann boundary conditions. Since $\bar{\phi}_t$ and ϕ_z must satisfy (F.60a) and (F.60b) on the Dirichlet boundaries, respectively, the second and the fourth contour integrals are zero as well. Thus W_1 becomes

$$\begin{aligned} W_1 = & \int_\Omega (\bar{\nabla}_t \times \bar{\phi}_t) \cdot (f_{rzz}^{-1} \bar{\nabla}_t \times \bar{V}_{tn})\ d\Omega + \int_\Omega (\bar{\nabla}_t \times \phi_z \bar{a}_z) \cdot (\underline{\underline{f}}_{rt}^{-1}\ \bar{\nabla}_t \times V_{zn} \bar{a}_z)\ d\Omega \\ & + \int_{\Gamma_N} \phi_z \bar{a}_z \cdot (\bar{a}_n \times \underline{\underline{f}}_{rt}^{-1}\ \bar{\nabla}_t \times V_{zn} \bar{a}_z)\ d\Gamma \end{aligned} \tag{F.68b}$$

With regard to W_2, when one takes into account that $\underline{\underline{f}}_{rt}$ is symmetric, one obtains

$$W_2 = \int_\Omega (\bar{\nabla}_t \times V_{zn} \bar{a}_z) \cdot (\underline{\underline{f}}_{rt}^{-1}\ \bar{a}_z \times \bar{\phi}_t)\ d\Omega \tag{F.69}$$

Integrating by parts W_3 and taking into account (A.33d) yield

$$W_3 = \int_\Omega (\bar{\nabla}_t \times \phi_z \bar{a}_z) \cdot (\underline{\underline{f}}_{rt}^{-1}\, \bar{a}_z \times \bar{V}_{tn})\, \mathrm{d}\Omega$$
$$+ \int_{\Gamma_N} \phi_z \bar{a}_z \cdot (\bar{a}_n \times \underline{\underline{f}}_{rt}^{-1}\, \bar{a}_z \times \bar{V}_{tn})\, \mathrm{d}\Gamma + \int_{\Gamma_D} \phi_z \bar{a}_z \cdot (\bar{a}_n \times \underline{\underline{f}}_{rt}^{-1}\, \bar{a}_z \times \bar{V}_{tn})\, \mathrm{d}\Gamma \tag{F.70a}$$

The second contour integral is zero since ϕ_z must satisfy (F.60b) along Dirichlet boundaries, resulting in

$$W_3 = \int_\Omega (\bar{\nabla}_t \times \phi_z \bar{a}_z) \cdot (\underline{\underline{f}}_{rt}^{-1}\, \bar{a}_z \times \bar{V}_{tn})\, \mathrm{d}\Omega + \int_{\Gamma_N} \phi_z \bar{a}_z \cdot (\bar{a}_n \times \underline{\underline{f}}_{rt}^{-1}\, \bar{a}_z \times \bar{V}_{tn})\, \mathrm{d}\Gamma \tag{F.70b}$$

Applying vectorial identities, W_4 becomes

$$W_4 = \int_\Omega (\bar{a}_z \times \bar{\phi}_t) \cdot (\underline{\underline{f}}_{rt}^{-1}\, \bar{a}_z \times \bar{V}_{tn})\, \mathrm{d}\Omega \tag{F.71}$$

Hence (F.67a) can be written as

$$W = \int_\Omega (\bar{\nabla}_t \times \bar{\phi}_t) \cdot (f_{rzz}^{-1}\, \bar{\nabla}_t \times \bar{V}_{tn})\mathrm{d}\Omega + \int_\Omega (\bar{\nabla}_t \times \phi_z \bar{a}_z) \cdot (\underline{\underline{f}}_{rt}^{-1} \bar{\nabla}_t \times V_{zn} \bar{a}_z)\, \mathrm{d}\Omega$$
$$- \beta\left[\int_\Omega (\bar{\nabla}_t \times V_{zn} \bar{a}_z) \cdot (\underline{\underline{f}}_{rt}^{-1}\, \bar{a}_z \times \bar{\phi}_t)\, \mathrm{d}\Omega + \int_\Omega (\bar{\nabla}_t \times \phi_z \bar{a}_z) \cdot (\underline{\underline{f}}_{rt}^{-1}\, \bar{a}_z \times \bar{V}_{tn})\mathrm{d}\Omega\right]$$
$$+ \beta^2 \int_\Omega (\bar{a}_z \times \bar{\phi}_t) \cdot (\underline{\underline{f}}_{rt}^{-1} \bar{a}_z \times \bar{V}_{tn})\, \mathrm{d}\Omega - k_0^2\left[\int_\Omega \bar{\phi}_t \cdot \underline{\underline{p}}_{rt}\, \bar{V}_{tn}\, \mathrm{d}\Omega + \int_\Omega \phi_z p_{rzz} V_{zn}\, \mathrm{d}\Omega\right]$$
$$+ \int_{\Gamma_N} \phi_z \bar{a}_z \cdot (\bar{a}_n \times \underline{\underline{f}}_{rt}^{-1}\, \bar{\nabla}_t \times V_{zn} \bar{a}_z - \beta \bar{a}_n \times \underline{\underline{f}}_{rt}^{-1} \bar{a}_z \times \bar{V}_{tn})\, \mathrm{d}\Gamma = 0 \tag{F.72}$$

The contour integral is zero since (F.61a) is satisfied along Neumann boundaries. Therefore,

$$W = \int_\Omega (\bar{\nabla}_t \times \bar{\phi}_t) \cdot (f_{rzz}^{-1}\, \bar{\nabla}_t \times \bar{V}_{tn})\, \mathrm{d}\Omega + \int_\Omega (\bar{\nabla}_t \times \phi_z \bar{a}_z) \cdot (\underline{\underline{f}}_{rt}^{-1}\, \bar{\nabla}_t \times V_{zn} \bar{a}_z)\, \mathrm{d}\Omega$$
$$- \beta\left[\int_\Omega (\bar{\nabla}_t \times V_{zn} \bar{a}_z) \cdot (\underline{\underline{f}}_{rt}^{-1}\, \bar{a}_z \times \bar{\phi}_t)\mathrm{d}\Omega + \int_\Omega (\bar{\nabla}_t \times \phi_z \bar{a}_z) \cdot (\underline{\underline{f}}_{rt}^{-1}\, \bar{a}_x \times \bar{V}_{tn})\mathrm{d}\Omega\right] \tag{F.73}$$
$$+ \beta^2 \int_\Omega (\bar{a}_z \times \bar{\phi}_t) \cdot (\underline{\underline{f}}_{rt}^{-1} \bar{a}_z \times \bar{V}_{tn})\mathrm{d}\Omega - k_0^2\left[\int_\Omega \bar{\phi}_t \cdot \underline{\underline{p}}_{rt}\, \bar{V}_{tn}\mathrm{d}\Omega + \int_\Omega \phi_z p_{rzz} V_{zn}\, \mathrm{d}\Omega\right] = 0$$

is the weak formulation. It is easy to verify that if in (F.73) $\bar{\phi}_t$ is replaced by $\delta \bar{V}_{tn}$ and ϕ_z by δV_{zn}, the expression agrees with that obtained by minimizing (F.62)

$$\delta F_n = W\Big|_{\substack{\bar{\phi}_t = \delta \bar{V}_{tn} \\ \phi_z = \delta V_{zn}}} = 0 \qquad \text{(F.74)}$$

as expected. The weak formulation of the problem can therefore be expressed as

Let $X^0(\Omega)$ be the space of vectorial functions with two components $\bar{q}$ in Ω so that $\bar{q} \in H(\text{curl},\Omega)$ and $\bar{a}_n \times \bar{q} = 0$ on Γ_D and $H_0^1(\Omega)$ be the space of scalar functions ψ in Ω so that $\psi \in H^1(\Omega)$ and $\psi = 0$ on Γ_D, where $\Gamma = \Gamma_D \cup \Gamma_N$ is the boundary of Ω. Given β, $\underline{f}_r$ and $\underline{p}_r$ in Ω, separable along its transverse and longitudinal parts ($\underline{f}_{rt}$, f_{rzz}, $\underline{p}_{rt}$, p_{rzz}, and assuming the inverse of $\underline{f}_{rt}$ exists), find k_0^2, $\bar{V}_{tn}$ and V_{zn} [$\bar{V}_{tn} \in X^0$ and $V_{zn} \in H_0^1(\Omega)$], so that (F.73) is satisfied for all $\bar{\phi}_t \in X^0$ and $\phi_z \in H_0^1(\Omega)$.

Then the integral forms (F.73) and (F.74) are real, symmetric, and positive definite [Jin 1993; Chs. 6,8]. This forms an eigenvalue problem using (F.73) and (F.74), where the eigenvalue is k_0^2 for a given value of β.

Discretization of this formulation by means of FEM is carried out in [Alam et al. 1997], [Bárdi et al. 1993], [Blanc-Castillo 1994], [Blanc-Castillo et al. 1995], [Bermúdez de Castro and García-Pedreira 1992], [García-Castillo 1992], [García-Castillo and Salazar-Palma 1992a,b], [Helal et al. 1994], [Jin 1993; Ch. 8], [Koshiba and Inoue 1992], and [Salazar- Palma et al. 1996] among others, which use a variety of different finite elements.

F.3.1.2 *Discretization by Means of Lagrange/Curl-Conforming Finite Elements*

A FEM utilizing first- and second-order Lagrange/curl-conforming triangular element, whose parent elements are described in Section 2.5.1.2, is used to discretize the problem. These elements are a combination of the Lagrange and curl-conforming elements. For first-order, besides the straight isoparametric version (i.e., using first-order Lagrange basis functions in order to perform the transformation of each real element into the parent element), a curved superparametric version (i.e., using second-order Lagrange basis functions for the geometric transformation) has also been applied. Second-order curl-conforming/Lagrange elements are also of the straight type (subparametric) when using a linear transformation and curved type (isoparametric) when using a second-order transformation.

Lagrange/curl-conforming elements provide continuity of the scalar unknown in the whole domain because it is discretized through the Lagrange interpolation. In this case, the longitudinal component V_z is continuous along the boundaries between elements. The transverse component V_t is discretized by the curl-conforming element, thus the tangential component will be continuous across media interfaces. These characteristics fully justify the choice of this element since it naturally fits the problem of wave propagation in guided structures. The discretization process involves separating the problem domain into triangular elements, in each of which 6 (or 14, for second-order) nodes are defined. The first 3 (or 6, for second-order) nodes are located at the vertices (and midpoints of edges, for second-order) of the triangles and are used for the approximation of the longitudinal component V_z using first- (or second-) order Lagrange shape functions, N_i^e, according to:

$$\upsilon_z^e = \{N^e\}^T\{d_z^e\} = \sum_{i=1}^{3 \text{ or } 6} N_i^e d_{zi}^e \tag{F.75}$$

Here, υ_z^e stands for the approximation of V_z over the eth element, d_{zi}^e is the ith degree of freedom associated with these nodes, and N_i^e is the i-th Lagrange basis function in each element Ω^e. The degree of freedom is defined as the value of V_z at each node.

For the first-order element, the other three nodes are located at the midpoints of the edge of each triangle and are used for the approximation of the transverse component V_t. For the second-order element, the other eight nodes are located as follows: the first six at the edges, at the Gauss points of integration (two per edge). The last two nodes may be located at the same point, the centroid of the element. The transverse component is approximated by the vectorial shape functions of the corresponding elements, $\bar{N}_i$, according to

$$\bar{\upsilon}_t^e = \{\bar{N}^e\}^T\{d_t^e\} = \sum_{i=1}^{3 \text{ or } 8} \bar{N}_i^e d_{ti}^e \tag{F.76}$$

Here $\bar{\upsilon}_t^e$ stands for the approximation of $\bar{V}_t$ over the eth element, d_{ti}^e represents the ith degree of freedom associated with these nodes, and $\bar{N}_i^e$ is the ith vector shape function in each element Ω^e. The definition for the degrees of freedom (at edges) is the length of the edge multiplied by the component of V_t tangential to the edge at the node. This definition also involves the selection of a sign for the local degree of freedom as explained in Appendix C.4.3. The definition of the two internal degrees of freedom for the second-order element is the integral of the x and y components of the vector unknown over the element surface.

By using (F.75) and (F.76) together with the application of the Ritz method to the functional (F.62) or the Galerkin method to the integral form (F.73) leads to the discrete local integral form

$$\{W_z^e\} = \sum_{j=1}^{3 \text{ or } 6}\left[\int_{\Omega^e}\left(\bar{\nabla}_t \times N_i\bar{a}_z\right)\cdot\left(\underline{\underline{f}}_{rt}^{-1}\bar{\nabla}_t \times N_j\bar{a}_z\right)\mathrm{d}\Omega\right]d_{zj}^e - \beta\sum_{j=1}^{3 \text{ or } 8}\left[\int_{\Omega^e}\left(\bar{\nabla}_t \times N_i\bar{a}_z\right)\cdot\left(\underline{\underline{f}}_{rt}^{-1}\bar{a}_z \times \bar{N}_j\right)\mathrm{d}\Omega\right]d_{tj}^e$$

$$- k_0^2\sum_{j=1}^{3 \text{ or } 6}\left[\int_{\Omega^e} N_i p_{rzz} N_j \mathrm{d}\Omega\right]d_{zj}^e = 0, \quad \text{for } i = 1,\ldots,3 \text{ or } 6 \tag{F.77a}$$

$$\left\{W_t^e\right\} = -\beta \sum_{j=1}^{3 \text{ or } 8} \left[\int_{\Omega^e} \left(\bar{\nabla}_t \times N_j \bar{a}_z\right) \cdot \left(\underline{\underline{f}}_{rt}^{-1} \bar{a}_z \times \bar{N}_i\right) \mathrm{d}\Omega \;+\; \right] d_{zj}^e$$

$$+ \sum_{j=1}^{3 \text{ or } 8} \left[\beta^2 \int_{\Omega^e} \left(\bar{a}_z \times \bar{N}_i\right) \cdot \left(\underline{\underline{f}}_{rt}^{-1} \bar{a}_z \times \bar{N}_j\right) \mathrm{d}\Omega + \int_{\Omega^e} \left(\bar{\nabla}_t \times \bar{N}_i\right) \cdot \left(f_{rzz}^{-1} \bar{\nabla}_t \times \bar{N}_j\right) \mathrm{d}\Omega \right] d_{tj}^e$$

$$- k_0^2 \sum_{j=1}^{3 \text{ or } 8} \left[\int_{\Omega^e} \bar{N}_i \cdot \underline{\underline{p}}_{rt} \bar{N}_j \mathrm{d}\Omega \right] d_{tj}^e \quad \text{for } i = 1, \ldots, 3 \text{ or } 8 \tag{F.77b}$$

for each element Ω^e. Expression (F.77) is of the form

$$\left\{W^e\right\} = \left([k^e] - k_0^2 [m^e]\right)\{d^e\} \tag{F.78}$$

where $\{d^e\}$ is the vector of the six or fourteen local degrees of freedom

$$\{d^e\} = \begin{Bmatrix} \{d_z^e\} \\ \{d_t^e\} \end{Bmatrix} \tag{F.79}$$

k_0^2 is the wave number for free space, $[k^e]$ is a 6×6 or 14×14 element stiffness matrix, and $[m^e]$ is also a 6×6 or 14×14 element mass matrix. The expression for the coefficients of $[k^e]$ involves the phase constant β, and hence it is assumed to be known in this case since the goal is to find the free-space wave number.

To obtain the coefficients of matrices $[k^e]$ and $[m^e]$ the following equivalences between vector notation and matrix can be taken into account:

$$\bar{\nabla}_t \times \psi \bar{a}_z \rightarrow \begin{bmatrix} 0 & 1 \\ -1 & 0 \end{bmatrix} \{\nabla_t\} \psi = \begin{bmatrix} 0 & 1 \\ -1 & 0 \end{bmatrix} \{\nabla_t \psi\} \tag{F.80a}$$

$$\left(\bar{\nabla}_t \times \bar{q}\right) \cdot \bar{a}_z \rightarrow \{\nabla_t\}^T \begin{bmatrix} 0 & 1 \\ -1 & 0 \end{bmatrix} \begin{Bmatrix} q_x \\ q_y \end{Bmatrix} \tag{F.80b}$$

$$\bar{a}_z \times \bar{q} \rightarrow \begin{bmatrix} 0 & -1 \\ 1 & 0 \end{bmatrix} \begin{Bmatrix} q_x \\ q_y \end{Bmatrix} \tag{F.80c}$$

$$(\bar{q}_1 \times \bar{q}_2) \cdot \bar{a}_z \rightarrow \begin{Bmatrix} q_1 \\ q_{2y} \end{Bmatrix}^T \begin{bmatrix} 0 & -1 \\ 1 & 0 \end{bmatrix} \begin{Bmatrix} q_{2x} \\ q_{2y} \end{Bmatrix} \tag{F.80d}$$

where ψ represents a scalar and $\bar{q}, \bar{q}_1$, and $\bar{q}_2$ are the transverse vectors, i.e., vectors with components in the x and y directions.

Matrix $[k^e]$ can be written as

$$[k^e] = \begin{bmatrix} [k_a^e] & [k_b^e] \\ [k_b^e]^T & [k_c^e] \end{bmatrix} \tag{F.81a}$$

where

$$[k_a^e] = \int_{\Omega^e} [\nabla_t N]^T \begin{bmatrix} 0 & -1 \\ 1 & 0 \end{bmatrix} [f_{rt}]^{-1} \begin{bmatrix} 0 & 1 \\ -1 & 0 \end{bmatrix} [\nabla_t N] \, d\Omega \tag{F.81b}$$

$$[k_b^e] = -\beta \int_{\Omega^e} [\nabla_t N]^T \begin{bmatrix} 0 & -1 \\ 1 & 0 \end{bmatrix} [f_{rt}]^{-1} \begin{bmatrix} 0 & -1 \\ 1 & 0 \end{bmatrix} [\bar{N}] \, d\Omega \tag{F.81c}$$

$$[k_c^e] = \beta^2 \int_{\Omega^e} [\bar{N}]^T \begin{bmatrix} 0 & 1 \\ -1 & 0 \end{bmatrix} [f_{rt}]^{-1} \begin{bmatrix} 0 & -1 \\ 1 & 0 \end{bmatrix} [\bar{N}] \, d\Omega$$
$$+ \int_{\Omega^e} \left\{ \{\nabla_t\}^T \begin{bmatrix} 0 & 1 \\ -1 & 0 \end{bmatrix} [\bar{N}] \right\} f_{rzz}^{-1} \left\{ \{\nabla_t\}^T \begin{bmatrix} 0 & 1 \\ -1 & 0 \end{bmatrix} [\bar{N}] \right\}^T d\Omega \tag{F.81d}$$

where

$$[\nabla_t N] = [\partial N] = \begin{bmatrix} \begin{Bmatrix} \dfrac{\partial N}{\partial x} \end{Bmatrix}^T \\ \begin{Bmatrix} \dfrac{\partial N}{\partial y} \end{Bmatrix}^T \end{bmatrix} = \begin{bmatrix} \dfrac{\partial N_1}{\partial x} & \cdots & \dfrac{\partial N_3}{\partial x} & \text{or} & \dfrac{\partial N_6}{\partial x} \\ \dfrac{\partial N_1}{\partial y} & \cdots & \dfrac{\partial N_3}{\partial y} & \text{or} & \dfrac{\partial N_6}{\partial y} \end{bmatrix} \tag{F.82a}$$

and $[\bar{N}]$ is a matrix which columns are the components of the vector basis functions of the curl-conforming eth element

$$[\bar{N}] = \begin{bmatrix} N_{x1} & \dots & N_{x3} \text{ or } N_{x8} \\ N_{y1} & \dots & N_{y3} \text{ or } N_{y8} \end{bmatrix} \tag{F.82b}$$

For the matrix $[m^e]$ one obtains

$$[m^e] = \begin{bmatrix} [m_a^e] & [0] \\ [0] & [m_c^e] \end{bmatrix} \tag{F.83a}$$

where

$$[m_a^e] = \int_{\Omega^e} \{N\} p_{rzz} \{N\}^T \, d\Omega \tag{F.83b}$$

$$[m_c^e] = \int_{\Omega^e} [\bar{N}]^T [p_{rt}] [\bar{N}] d\Omega \tag{F.83c}$$

and

$$\{N\}^T = [N_1 \quad N_2 \quad N_3]^T \text{ or } [N_1 \cdots N_6]^T \tag{F.84}$$

The calculation of the coefficients of matrices $[k^e]$ and $[m^e]$ is done by transforming each element to the parent element. The corresponding expressions are:

$$[k_a^e] = \int_{\hat{\Omega}} \left[\hat{\nabla}_t \hat{N}\right]^T [J^e]^{-1} \left[f_{rt}\right]^{-1} \left([J^e]^T\right)^{-1} \left[\hat{\nabla}_t \hat{N}\right] |J^e| \, d\Omega \tag{F.85a}$$

$$[k_b^e] = \beta \int_{\hat{\Omega}} \left[\hat{\nabla}_t \hat{N}\right]^T [J^e]^{-1} [f_{rt}]^{-1} \left([J^e]^T\right)^{-1} [\breve{N}] \, |J^e| \, d\Omega \tag{F.85b}$$

$$[k_c^e] = \beta^2 \int_{\hat\Omega} [\breve N]^T \, [J^e]^{-1} [f_{rt}]^{-1} \left([J^e]^T\right)^{-1} \, [\breve N] \, |J^e| \, \mathrm{d}\Omega$$

$$+ \int_{\hat\Omega} \frac{1}{|J^e|} \left\{ \{\hat\nabla_t\}^T \begin{bmatrix} 0 & 1 \\ -1 & 0 \end{bmatrix} [\breve N] \right\} f_{rzz}^{-1} \left\{ \{\hat\nabla_t\}^T \begin{bmatrix} 0 & 1 \\ -1 & 0 \end{bmatrix} [\breve N] \right\}^T \mathrm{d}\Omega \tag{F.85c}$$

$$[m_a^e] = \int_{\hat\Omega} \{\hat N\} p_{rzz} \{\hat N\}^T \, |J^e| \, \mathrm{d}\Omega \tag{F.86a}$$

$$[m_c^e] = \int_{\hat\Omega} [\breve N]^T [J^e]^{-1} [p_{rt}] \left([J^e]^T\right)^{-1} [\breve N] \, |J^e| \, \mathrm{d}\Omega \tag{F.86b}$$

where $[\hat\nabla_t \hat N]$, $\{\hat N\}^T$ and $[\breve N]$ are expressions in (F.82a), (F.84), and (F.82b) for the reference element, i.e., in terms of $\hat x$ and $\hat y$. For matrix $[k_c^e]$, the second integral can be reduced to a vector whose coefficients are the curl of the shape functions of the reference element:

$$\left\{ \{\hat\nabla_t\}^T \begin{bmatrix} 0 & 1 \\ -1 & 0 \end{bmatrix} [\breve N] \right\}^T = \left\{ \hat\nabla_t \times [\breve N] \right\} \tag{F.87}$$

For the first-order element, the curl of the shape function is a constant equal to 2. For the second-order element, it is a first-order polynomial.

In [García-Castillo 1992], the calculation of the elements of the matrices $[k^e]$ and $[m^e]$ are shown in detail for the case of the first-order Lagrange/curl-conforming straight or isoparametric element, in which an analytical integration can be carried out. This is so because the geometric transformation obeys the expression

$$\{x, y\}^T = \{\hat N(\hat x, \hat y)\}^T \begin{bmatrix} x_1 & y_1 \\ x_2 & y_2 \\ x_3 & y_3 \end{bmatrix} \tag{F.88}$$

where the three coefficients of $\{\hat N\}$ are the first-order Lagrange shape functions and (x_i, y_i), $i = 1, \ldots, 3$, are the coordinates of the ith vertex of each element Ω^e. According to (F.88), the determinant of the Jacobian matrix of the transformation is constant over each element and therefore the integrands of the different coefficients of matrices $[k^e]$ or $[m^e]$ are polynomials in $\hat x$ and $\hat y$ (the parent element coordinates).

For the case of the Lagrange/curl-conforming first-order curved element, the geometric transformation obeys

$$\{x,\ y\}^T = \{\hat{N}(\hat{x},\hat{y})\}^T \begin{bmatrix} x_1 & y_1 \\ \vdots & \vdots \\ \vdots & \vdots \\ x_6 & y_6 \end{bmatrix} \tag{F.89}$$

where the six coefficients of $\hat{N}$ are the second-order Lagrange shape functions and (x_i,y_i), $i = 1, \ldots, 6$, are the coordinates for the ith node of the real triangular element. Therefore, the determinant of the Jacobian matrix is a second-order polynomial. This makes the integrands of the coefficients of $[k_a^e]$, $[k_b^e]$, $[k_c^e]$, and $[m_c^e]$ to be rational expressions in which the denominator is a second-order polynomial in $\hat{x}$ and $\hat{y}$. Hence, numerical integration is necessary. A three-point numerical integration is chosen as indicated in Table 2.9 of Chapter 2, which would be exact for second-order polynomials. The choice of this integration formula is based on a reduced integration rule that determines the choice of the number of integration points that provides an exact integration of the element area (i.e., of the determinant of the Jacobian matrix) [Blanc-Castillo 1994].

Details for the computation of matrices for the second-order straight and curved elements may be found in [Blanc-Castillo 1994].

As indicated in Appendix C.4.3, before proceeding with the assembly of $[k^e]$ and $[m^e]$, a sign must be allocated to the degrees of freedom associated with the nodes along the edges so that the definition associated with the corresponding global degrees of freedom is unique. A +1 sign is assigned to the degrees of freedom corresponding to the scalar unknown as well as to the two inner degrees of freedom. The coefficients of matrices $[k^{le}]$ and $[m^{le}]$ become

$$k_{ij}^{le} = k_{ij}^{e}\ sign(i)\ sign(j) \tag{F.90a}$$

and

$$m_{ij}^{le} = m_{ij}^{e}\ sign(i)\ sign(j) \tag{F.90b}$$

where the sign variable will have a +1 or -1 value as explained in Appendix C.4.3.

Assembling matrices $[k^{le}]$ and $[m^{le}]$ and imposing the boundary conditions, one obtains the global system

$$\left([K] - k_0^2 [M]\right)\{D\} = \{0\} \tag{F.91}$$

where $[K]$ and $[M]$ are the global stiffness and mass matrices, respectively, and $\{D\}$ is the vector containing the global degrees of freedom.

A geometrical normalization identical to that described in (F.34) may be used so that the eigenvalue will be

$$s^2 = a^2 k_0^2 \tag{F.92a}$$

and the phase constant β will be normalized according to

$$\beta_a = a\beta \tag{F.92b}$$

Therefore, for the calculation of the elements of matrices $[k^e]$ and $[m^e]$, β_a will have to be used instead of β. Then the following system of equations is obtained

$$\left([K(\beta_a)] - s^2[M]\right)\{D\} = \{0\} \tag{F.93}$$

From (F.85) and (F.86) it can be concluded that matrices $[K]$ and $[M]$ are sparse, real, symmetric, and positive definite. This allows the use of very efficient algorithms for a quick solution of the eigenvalue problem. The algorithm employed is the inverse iteration by subspaces. The formulation described has the added advantage that the nonzero cut-off frequencies in a guided structure may be obtained by simply setting $\beta_a = 0$. As pointed out in Section 7.2, in this formulation all the spurious modes appear as having zero eigenvalues. The discussion regarding the number of zero eigenvalues to be obtained is also given in Section 7.2.

This formulation has been used in [Blanc-Castillo 1994], [Blanc-Castillo et al. 1995], [García-Castillo 1992], [García-Castillo and Salazar-Palma 1992a,b], [Salazar-Palma et al. 1996] exploiting all the advantages of the structure of the matrices involved. The same formulation has been used by [Koshiba and Inoue 1992], but the authors eliminate the degrees of freedom corresponding to the field longitudinal component after the assemblage of the matrices in (F.91). Thus, the resulting matrices are no longer sparse.

F.3.2 Nonstandard Formulation

F.3.2.1 *A Variant of the Previous Weak Formulation*

A variant of the previous formulation is described next. The objective is to obtain an eigenvalue problem where the eigenvalue would be the phase constant, β, or, more specifically, the normalized phase constant, β_a, while k_0^2, or equivalently s^2, would be an input data.

Performing a change of variable

$$\bar{V}'_{tn} = \beta \bar{V}_{tn} \tag{F.94}$$

where $\bar{V}_{tn}$ is given by (F.56d), so the functional (F.62a) becomes

$$F'_n = A'_n - k_0^2 B'_n \tag{F.95a}$$

The integrals A_n' and B_n' are given by

$$\begin{aligned} A'_n = & \int_\Omega \left(\bar{\nabla}_t \times V_{zn}\bar{a}_z\right) \cdot \left(\underline{\underline{f}}_{rt}^{-1}\bar{\nabla}_t \times V_{zn}\bar{a}_z\right) d\Omega + \int_\Omega \left(\bar{a}_z \times \bar{V}'_{tn}\right) \cdot \left(\underline{\underline{f}}_{rt}^{-1}\bar{a}_z \times \bar{V}'_{tn}\right) d\Omega \\ & - 2\int_\Omega \left(\bar{\nabla}_t \times V_{zn}\bar{a}_z\right) \cdot \left(\underline{\underline{f}}_{rt}^{-1}\bar{a}_z \times \bar{V}'_{tn}\right) d\Omega + \frac{1}{\beta^2}\int_\Omega \left(\bar{\nabla}_t \times \bar{V}'_{tn}\right) \cdot \left(f_{rzz}^{-1}\bar{\nabla}_t \times \bar{V}'_{tn}\right) d\Omega \end{aligned} \tag{F.95b}$$

$$B'_n = \frac{1}{\beta^2}\int_\Omega \left(\bar{V}'_{tn} \cdot \underline{\underline{p}}_{rt}\bar{V}'_{tn}\right) d\Omega + \int_\Omega V_{zn} p_{rzz} V_{zn} d\Omega \tag{F.95c}$$

After multiplying both sides by β^2 and rearranging the terms one obtains another integral expression, F_n'', which is defined as

$$F''_n = A''_n(k_0^2) - \beta^2 B''_n(k_0^2) \tag{F.96a}$$

where

$$A''_n = \int_\Omega \left(\bar{\nabla}_t \times \bar{V}'_{tn}\right) \cdot \left(f_{rzz}^{-1}\bar{\nabla}_t \times \bar{V}'_{tn}\right) d\Omega - k_0^2\int_\Omega \left(\bar{V}'_{tn} \cdot \underline{\underline{p}}_{rt}\bar{V}'_{tn}\right) d\Omega \tag{F.96b}$$

and

$$\begin{aligned} B''_n = & -\int_\Omega \left(\bar{\nabla}_t \times V_{zn}\bar{a}_z\right) \cdot \left(\underline{\underline{f}}_{rt}^{-1}\bar{\nabla}_t \times V_{zn}\bar{a}_z\right) d\Omega - \int_\Omega \left(\bar{a}_z \times \bar{V}'_{tn}\right) \cdot \left(\underline{\underline{f}}_{rt}^{-1}\bar{a}_z \times \bar{V}'_{tn}\right) d\Omega \\ & + 2\int_\Omega \left(\bar{\nabla}_t \times V_{zn}\bar{a}_z\right) \cdot \left(\underline{\underline{f}}_{rt}^{-1}\bar{a}_z \times \bar{V}'_{tn}\right) d\Omega + k_0^2\int_\Omega V_{zn} p_{rzz} V_{zn} d\Omega \end{aligned} \tag{F.96c}$$

By forcing the functional (F.96) to be stationary for a specific value of k_0^2, one needs to solve for the square of the phase constant β^2 and the corresponding solutions for the field components V_{zn} and $\bar{V}_{tn}'$. However, by inspecting (F.96b) and (F.96c), it becomes clear that the integral forms A_n'' and B_n'', although real and symmetric, they do not form a semidefinite system.

The weak formulation could also be obtained using the weighted residuals method utilizing the Galerkin method and this would lead to a formally identical formulation to the one obtained from the application of the Ritz method to (F.96). The integral forms obtained are still nondefinite.

The integral form corresponding to the weighted residuals method can be derived from (F.73), and it results in

$$
\begin{aligned}
W'' = {} & \int_\Omega \left(\bar{\nabla}_t \times \bar{\phi}_t'\right) \cdot \left(f_{rzz}^{-1} \bar{\nabla}_t \times \bar{V}_{tn}'\right) \mathrm{d}\Omega - k_0^2 \int \bar{\phi}_t' \cdot \underline{\underline{p}}_{rt} \bar{V}_{tn}' \, \mathrm{d}\Omega \\
& - \beta^2 \Big[k_0^2 \int_\Omega \phi_z p_{rzz} V_{zn} \, \mathrm{d}\Omega + \int_\Omega \left(\bar{\nabla}_t \times V_{zn} \bar{a}_z\right) \cdot \left(\underline{\underline{f}}_{rt}^{-1} \bar{a}_z \times \bar{\phi}_t'\right) \mathrm{d}\Omega \\
& + \int_\Omega \left(\bar{\nabla}_t \times \phi_z \bar{a}_z\right) \cdot \left(\underline{\underline{f}}_{rt}^{-1} \bar{a}_z \times \bar{V}_{tn}'\right) \mathrm{d}\Omega - \int_\Omega \left(\bar{\nabla}_t \times \phi_z \bar{a}_z\right) \cdot \left(\underline{\underline{f}}_{rt}^{-1} \bar{\nabla}_t \times V_{zn} \bar{a}_z\right) \mathrm{d}\Omega \\
& - \int_\Omega \left(\bar{a}_z \times \bar{\phi}_t'\right) \cdot \left(\underline{\underline{f}}_{rt}^{-1} \bar{a}_z \times \bar{V}_{tn}'\right) \mathrm{d}\Omega \Big] = 0
\end{aligned}
\tag{F.97}
$$

The previous expression is the weak formulation for a waveguiding problem, where ϕ_z and $\bar{\phi}'_t$ must fulfill the same conditions as V_{zn} and $\bar{V}_{tn}'$. It may be expressed as:

Let $X^0(\Omega)$ be the space of vectorial functions with two components $\bar{q}$ in Ω so that $\bar{q} \in H(\mathrm{curl},\Omega)$ and $\bar{a}_n \times \bar{q} = 0$ on Γ_D and $H_0^1(\Omega)$ be the space of scalar functions ψ in Ω so that $\psi \in H^1(\Omega)$ and $\psi = 0$ on Γ_D where $\Gamma = \Gamma_D \cup \Gamma_N$ is the boundary of Ω. Given k_0, $\underline{\underline{f}}_r$ and $\underline{\underline{p}}_r$ in Ω, separable along their transverse and longitudinal parts ($\underline{\underline{f}}_{rt}$, f_{rzz}, $\underline{\underline{p}}_{rt}$, p_{rzz}, assuming that the inverse of $\underline{\underline{f}}_{rt}$ exists), find all the three unknowns β^2, $\bar{V}_{tn}'$, and V_{zn} with $\bar{V}_{tn}' \in X^0$ and $V_{zn} \in H_0^1(\Omega)$ so that (F.97) is fulfilled for all $\bar{\phi}_t' \in X^0$ and $\phi_z \in H_0^1(\Omega)$.

This formulation has been used in [Lee et al. 1991a,b], [Lee 1994], and [Salazar-Palma et al. 1994], among others, using different curl-conforming elements.

F.3.2.2 *Discretization by Means of Lagrange/Curl-Conforming Elements*

The same elements as in the previous formulation have been used to discretize the variables V_{zn} and $\bar{V}_{tn}'$. After applying the Galerkin method to (F.97) for each element Ω^e, the following local integral form is obtained:

$$\{W^e\} = \left([k^e] - \beta^2[m^e]\right) \tag{F.98}$$

where

$$[k^e] = \begin{bmatrix} 0 & 0 & 0 & 0 & 0 & 0 \\ 0 & 0 & 0 & 0 & 0 & 0 \\ 0 & 0 & 0 & 0 & 0 & 0 \\ 0 & 0 & 0 & & & \\ 0 & 0 & 0 & & [k_c^e] & \\ 0 & 0 & 0 & & & \end{bmatrix} \tag{F.99a}$$

and

$$[k_c^e] = \int_{\Omega^e} \left\{ \{\bar{\nabla}_t\}^T \begin{bmatrix} 0 & 1 \\ -1 & 0 \end{bmatrix} [\bar{N}] \right\}^T f_{rzz}^{-1} \left\{ \{\bar{\nabla}_t\}^T \begin{bmatrix} 0 & 1 \\ -1 & 0 \end{bmatrix} [\bar{N}] \right\} \mathrm{d}\Omega$$

$$- k_0^2 \int_{\Omega^e} [\bar{N}]^T [p_{rt}] [\bar{N}] \ \mathrm{d}\Omega \tag{F.99b}$$

and

$$[m^e] = \begin{bmatrix} [m_a^e] & [m_b^e] \\ [m_b^e]^T & [m_c^e] \end{bmatrix} \tag{F.100a}$$

with

$$[m_a^e] = -\int_{\Omega^e} \left[\bar{\nabla}_t\ N\right]^T \begin{bmatrix} 0 & 1 \\ -1 & 0 \end{bmatrix} [f_{rt}]^{-1} \begin{bmatrix} 0 & -1 \\ 1 & 0 \end{bmatrix} [\bar{\nabla}_t\ N] \ \mathrm{d}\Omega$$

$$+ k_0^2 \int_{\Omega^e} \{N\} p_{rzz} \{N\}^T \mathrm{d}\Omega \tag{F.100b}$$

$$[m_b^e] = \int_{\Omega^e} \left[\bar{\nabla}_t N\right]^T \begin{bmatrix} 0 & -1 \\ 1 & 0 \end{bmatrix} [f_{rt}]^{-1} \begin{bmatrix} 0 & -1 \\ 1 & 0 \end{bmatrix} [\bar{N}] \ \mathrm{d}\Omega \tag{F.100c}$$

$$[m_c^e] = -\int_{\Omega^e} [\bar{N}]^T \begin{bmatrix} 0 & 1 \\ -1 & 0 \end{bmatrix} [f_{rt}]^{-1} \begin{bmatrix} 0 & -1 \\ 1 & 0 \end{bmatrix} [\bar{N}] \, d\Omega \tag{F.100d}$$

After conversion to the reference element one obtains

$$[k_c^e] = \int_{\hat{\Omega}} \frac{1}{|J^e|} \left\{ \{\hat{\nabla}_t\}^T \begin{bmatrix} 0 & 1 \\ -1 & 0 \end{bmatrix} [\breve{N}] \right\} f_{rzz}^{-1} \left\{ \{\hat{\nabla}_t\}^T \begin{bmatrix} 0 & 1 \\ -1 & 0 \end{bmatrix} [\breve{N}] \right\}^T d\Omega$$

$$- k_0^2 \int_{\hat{\Omega}} [\breve{N}]^T \, [J^e]^{-1} [p_{rt}] \left([J^e]^T\right)^{-1} [\breve{N}] \, |J^e| \, d\Omega \tag{F.101}$$

$$[m_a^e] = -\int_{\hat{\Omega}} \left[\hat{\nabla}\hat{N}\right]^T [J^e]^{-1} [f_{rt}]^{-1} \left([J^e]^T\right)^{-1} \left[\hat{\nabla}_t \hat{N}\right] |J^e| \, d\Omega +$$

$$+ k_0^2 \int_{\hat{\Omega}} \{\hat{N}\} p_{rzz} \{\hat{N}\}^T \, |J^e| \, d\Omega \tag{F.102a}$$

$$[m_b^e] = -\int_{\hat{\Omega}} \left[\hat{\nabla}_t \hat{N}\right]^T [J^e]^{-1} [f_{rt}]^{-1} \left([J^e]^T\right)^{-1} [\breve{N}] \, |J^e| \, d\Omega \tag{F.102b}$$

$$[m_c^e] = -\int_{\hat{\Omega}} [\breve{N}]^T [J^e]^{-1} [f_{rt}]^{-1} \left([J^e]^T\right)^{-1} [\breve{N}] \, |J^e| \, d\Omega \tag{F.102c}$$

The calculation of the elements of $[k^e]$ and $[m^e]$ can be carried out as before, either analytically for straight elements (since $|J^e|$ is a constant in each element and the different integrands are polynomials in terms of the coordinates $\hat{x}$ and $\hat{y}$ of the reference element) or through numerical integration for the curved elements. After allocating the sign and applying the boundary conditions, the following global system is obtained

$$\left([K(s^2)] - \beta_a^2 [M(s^2)]\right)\{D\} = \{0\} \tag{F.103}$$

where the geometric normalization has been used again, according to (F.92), and the elements of matrices of $[K]$ and $[M]$ are now functions of s^2. The matrices $[K]$ and $[M]$ are nondefinite.

In [Lee et al. 1991a,b] and [Lee 1994] a curl-conforming vector element has been used that differs from the one used here.

Appendix G

Computation of Error Estimates and Indicators

G.1 INTRODUCTION

In Section 4.3.1, a general form of the local error estimate or, when applicable, the local error indicator, is presented. That error estimate corresponds to the problems considered in Section 2.3.2. Let $\in_e$ be the estimator of the error made in the element Ω^e:

$$\in_e^2 = f_s \frac{h_e^2}{a_{e\,min} p} \int_{\Omega^e} \{r_s^*\}^T \{r_s\} \, d\Omega + f_{ed} \frac{h_e}{a_{e\,min} p} \sum_{k=1}^{n_{ed}^e} \int_{\Gamma_k \not\subset \Gamma_D} \{r_{ed}^*\}^T \{r_{ed}\} \, d\Gamma \tag{G.1}$$

where f_s and f_{ed} are weighting factors for the terms due to the residue on the domain and along the boundaries of the element, respectively. Because error estimates have been computed in this book only for two-dimensional (2D) problems, the term surface is used in the following to refer to the domain of the element. The term edge is used to refer to the boundaries of the element. Here, h_e is the diameter of the element Ω^e (see Section 2.5.1.1); a_{emin} is the lowest eigenvalue of matrix $[a]$ in the element Ω^e according to the terminology used in Section 2.3.2; and p is the order of the approximation used. The vector $\{r_s\}$ is the vector of the surface residue. The quantity n_{ed}^e is the number of edges of the element Ω^e. Γ_k represents the kth edge of the element Ω^e. Γ_D represents the boundaries or contour of the domain Ω with Dirichlet conditions, and finally, $\{r_{ed}\}$ represents the vector of the residue at the edges.

For element Ω^e, the surface residue $\{r_s\}$ is a continuous quantity defined at points belonging to the element as the value of the residue at that point. The residue at edges $\{r_{ed}\}$ is a singular quantity. It is equal to zero all over the element except at the edges that are not boundaries with essential boundary conditions (Dirichlet boundaries conditions). The residue at the edges measures the discontinuity of a given component of the secondary variables (the variables involved with the natural boundary conditions) across element interfaces. For edges that are boundaries with natural boundary conditions (Neumann boundary conditions), it measures the difference between the value of the component of interest of the secondary variable at the edge and the value of the natural condition. Thus, the surface residue measures the error in the domain of the element, while the residue at the edges measures the error at its boundaries.

Expression (G.1) can only be computed once the solution of the problem has been obtained by FEM. In general, one may write

$$\int_{\Omega^e} \{r_s\}^T \{r_s\} \mathrm{d}\Omega = \{d^e\}^T [r_s] \{d^e\} \tag{G.2a}$$

$$\{r_{ed}\} = \{r_{edk}\}^T \{d^e\} + \{r_{edj}\}^T \{d^m\} \tag{G.2b}$$

where $\{d^e\}$ is the vector of degrees of freedom of element Ω^e and $\{d^m\}$ is the vector of the degrees of freedom of element Ω^m whose jth edge is at the same time the kth edge of element Ω^e. Hence, $\{d^e\}$ and $\{d^m\}$ can only be calculated after obtaining the column matrix $\{D\}$ of the global degrees of freedom according to expression (2.77). However, the matrices $[r_s]$, $\{r_{edk}\}$, and $\{r_{edj}\}$ depend only on the geometry and physical constant of the element Ω^e and Ω^m and on the geometric transformation utilized to convert real elements into the parent element. Thus, the coefficients of those matrices can be calculated at the same time as matrices $[k^e]$ and $[m^e]$ are evaluated. In summary, local matrices $[k^e]$, $[m^e]$, $[r_s]$, and $[r_{edk}]$ are filled up at the same time. Basically, for each element the only computation required is that of the Jacobian matrix because for a given element the other quantities may be precomputed. This procedure allows one to save CPU time. Once $\{D\}$ is computed, a postprocess formulated at each element according to (G.2) will give the estimate (G.1).

For eigenvalue problems, expression (G.2a) involves the computed eigenvalue $\tilde{\lambda}$ according to

$$\int_{\Omega^e} \{r_s\}^T \{r_s\} \mathrm{d}\Omega = \sum_{i=0}^{\alpha} \tilde{\lambda}^i \{d^e\}^T [r_{si}] \{d^e\} \tag{G.3}$$

where $\alpha = 1$ or 2 depending on the formulation used. The previous expression shows that, for eigenvalue problems, the surface residue depends on the computed eigenvalue; i.e., in the analysis of waveguiding structures the surface residue depends on the mode under study.

This appendix is devoted to the evaluation of the surface residue $\{r_s\}$ and the singular term $\{r_{ed}\}$. It provides the details of the calculation process of expressions (G.1) to (G.3) for each of the problems analyzed by FEM utilizing a self-adaptive mesh procedure (see Chapter 4).

The first section deals with the computations in the case of the analysis of guiding structures with a quasi-static approach using the direct formulation and Lagrange elements. It is followed by the calculations for the case of the full-wave analysis of guiding structures using the longitudinal components of the electric or magnetic fields and Lagrange elements. The computations for the case of the full-wave formulation that uses transverse and longitudinal components either from the electric or magnetic fields and uses the Lagrange/curl-conforming element are finally presented.

G.2 COMPUTATION OF LOCAL ERROR ESTIMATES FOR THE QUASI-STATIC ANALYSIS OF TRANSMISSION LINES USING THE DIRECT FORMULATION AND LAGRANGE ELEMENTS

This section is based on the strong formulation given in Appendix E.2.1. Assume that the problem has been discretized as indicated in Appendix E.2.2 so that the approximate solution of ϕ is $\tilde{\phi}$. According to Section 4.3.1, $\tilde{\phi}$ is an exact solution of the original problem with perturbed right-hand terms. This perturbed problem can be formulated as

Given $\underline{\underline{\varepsilon}}_{rt}(x,y)$ *in* Ω, *find* $\phi(x,y)$ *so that* $\phi = \phi_{ci}$ *on the ith conductor,* Γ_{ci}, $i = 0, \ldots, N_c$, *and*

$$-\bar{\nabla}_t \cdot \underline{\underline{\varepsilon}}_{rt} \bar{\nabla}_t \phi = r \ , \quad in \ \Omega = \cup \Omega^e \ , \quad e = 1, \ldots, N_e \tag{G.4a}$$

$$-\bar{a}_n \cdot \underline{\underline{\varepsilon}}_{rt} \bar{\nabla}_t \phi = \xi \ , \quad on \ \Gamma_{pm} \tag{G.4b}$$

where N^e *is the total number of elements;* r *is the residue in* Ω, *which can be written as*

$$r = r_s + \rho \tag{G.4c}$$

$$r_s = -\bar{\nabla}_t \cdot \underline{\underline{\varepsilon}}_{rt} \bar{\nabla}_t \tilde{\phi} \ , \quad in \ \Omega^e \ , \quad e = 1, \ldots, N_e \tag{G.4d}$$

$$\rho = \rho_k \delta_k = -\left[\bar{a}_n \cdot \underline{\underline{\varepsilon}}_{rt} \bar{\nabla}_t \tilde{\phi}\right]_e - \left[\bar{a}_n \cdot \underline{\underline{\varepsilon}}_{rt} \bar{\nabla}_t \tilde{\phi}\right]_m \ , \quad \text{on } \Gamma_k \ , \quad k = 1, \ldots, n_{edi} \tag{G.4e}$$

where r_s *is the surface residue in each element* Ω^e, ρ *is the residue on each internal edge which has a singularity.* n_{edi} *is the total number of internal edges. The residue* ξ *at the edges located on the Neumann boundary* Γ_N *is given by*

$$\xi = -\bar{a}_n \cdot \underline{\underline{\varepsilon}}_{rt} \bar{\nabla}_t \tilde{\phi}, \quad \text{on} \quad \Gamma_N \equiv \Gamma_{pm} \tag{G.4f}$$

where Γ_{pm} *is the perfect magnetic wall or symmetry wall.*

The definition of the variables is given in Appendices E.2.1 and E.2.2. Note that r_s is the residue corresponding to (E.5). The residue corresponding to (E.8a) is given by ρ_k. This latter quantity represents the jump of the normal component of the secondary variable, $\underline{\underline{\varepsilon}}_{rt} \bar{\nabla}_t \phi$, between two elements, Ω^e and Ω^m, sharing a common edge,

Γ_k. Thus, ρ_k measures the nonexact fulfillment of the continuity condition (E.8a) across element interfaces due to the fact that the above condition has not been exactly enforced. In an analogous way, ξ is the residue corresponding to equation (E.7) at the edges that have Neumann boundary conditions. Thus, ξ measures the nonexact fulfillment of Neumann boundary conditions because those conditions have not been exactly enforced.

Therefore, for this problem, in the element Ω^e, the surface residue r_s is a scalar given by

$$r_s = -\bar{\nabla}_t \cdot \underline{\underline{\varepsilon}}_{rt} \bar{\nabla}_t \tilde{\phi} \quad , \qquad \text{in} \quad \Omega^e \tag{G.5a}$$

At the kth edge, the residue r_{ed} is also scalar and given by

$$r_{ed} = -\left[\!\left[\bar{a}_n \cdot \underline{\underline{\varepsilon}}_{rt} \bar{\nabla}_t \tilde{\phi} \right]\!\right]_k , \quad \text{on} \quad \Gamma_k \subset \Gamma_{\Omega^e}, \quad \Gamma_k \not\subset \Gamma \tag{G.5b}$$

where $[\![\cdot]\!]_k$ represents the jump across an internal edge, Γ_k (i.e., an edge not belonging to the boundary of the domain of the problem, Γ), or by

$$r_{ed} = -\left[\bar{a}_n \cdot \underline{\underline{\varepsilon}}_{rt} \bar{\nabla}_t \tilde{\phi} \right]_k , \quad \text{on} \quad \Gamma_k \subset \Gamma_{\Omega^e}, \quad \Gamma_k \subset \Gamma_N \equiv \Gamma_{pm} \tag{G.5c}$$

for the cases where Γ_k is an external edge which is part of a Neumann type boundary. If Γ_k is part of the Dirichlet boundary, Γ_D, the value of r_{ed} is zero. For the quasi-static problem, the Neumann boundaries, Γ_N, are symmetry walls, also called magnetic walls, Γ_{pm}. The Dirichlet boundaries, Γ_D, are the electric walls, Γ_{pe}, or perfect conductors, Γ_{ci}, $i = 0, \ldots, N_c$, $\Gamma_{pe} = \cup\Gamma_{ci}$. Finally, $\Gamma_{\Omega e}$ means the contour of the element Ω^e.

The error estimator of element Ω^e can be written as

$$\epsilon_e^2 = f_s \frac{h_e^2}{\varepsilon_{rtmin}\, p} \int_{\Omega^e} r_s^2 \, d\Omega + f_{ed} \frac{h_e}{\varepsilon_{rtmin}\, p} \sum_{\substack{\Gamma_k \subset \Gamma_{\Omega^e} \\ \Gamma_k \not\subset \Gamma_D}} \int_{\Gamma_k} r_{ed}^2 \, d\Gamma \tag{G.6}$$

Ordinary Lagrange elements and, eventually, infinite elements are used for the discretization of the problem. The self-adaptive mesh procedure has been developed for first- and second-order straight and curved triangular elements along with compatible infinite elements. For the case of infinite elements, the error term is ignored, i.e., the error estimator is considered to be zero. Therefore, expression (G.6) can be written for triangular elements as

$$\epsilon_e^2 = f_s \frac{h_e^2}{\varepsilon_{rtmin} p} I_{\Omega^e} + f_{ed} \frac{h_e}{\varepsilon_{rtmin} p} \sum_{k=1}^{3} I_{\Gamma_k} \tag{G.7a}$$

where

$$I_{\Omega^e} = \int_{\Omega^e} \left(\bar{\nabla}_t \cdot \underline{\underline{\varepsilon}}_{rt} \bar{\nabla}_t \tilde{\phi} \right)^2 \, \mathrm{d}\Omega \tag{G.7b}$$

and $I_{\Gamma k}$ is

$$I_{\Gamma_k} = I_{\Gamma_{kN}} = \int_{\Gamma_k} \left[\bar{a}_n \cdot \underline{\underline{\varepsilon}}_{rt} \bar{\nabla}_t \tilde{\phi} \right]_k^2 \, \mathrm{d}\Gamma \ , \quad \text{for } \Gamma_k \subset \Gamma_N \equiv \Gamma_{pm} \tag{G.7c}$$

or

$$I_{\Gamma_k} = I_{\Gamma_{ki}} = \int_{\Gamma_k} [\![\bar{a}_n \cdot \underline{\underline{\varepsilon}}_{rt} \bar{\nabla}_t \tilde{\phi}]\!]_k^2 \, \mathrm{d}\Gamma \ , \quad \text{for } \Gamma_k \not\subset \Gamma = \Gamma_N \cup \Gamma_D = \Gamma_{pm} \cup \Gamma_{pe} \tag{G.7d}$$

or

$$I_{\Gamma_k} = I_{\Gamma_{kD}} = 0 \ , \quad \text{for } \Gamma_k \subset \Gamma_D \equiv \Gamma_{pe} = \cup \, \Gamma_{ci} \ , \quad i = 0, \ldots, N_c \tag{G.7e}$$

Since in the eth Lagrange element, Ω^e, we have

$$\tilde{\phi} = \{N\}^T \{d^e\} \tag{G.8}$$

where $\{N\}$ is the column matrix of the corresponding basis functions, one can write (G.7b) as

$$\begin{aligned} I_{\Omega^e} &= \int_{\Omega^e} \left[\{\nabla_t\}^T [\varepsilon_{rt}] \{\nabla_t\} \{N\}^T \{d^e\} \right]^T \left[\{\nabla_t\}^T [\varepsilon_{rt}] \{\nabla_t\} \{N\}^T \{d^e\} \right] \mathrm{d}\Omega \\ &= \{d^e\}^T \left[\int_{\Omega^e} \left\{ \{\nabla_t\}^T [\varepsilon_{rt}] [\nabla_t N] \right\} \left\{ \{\nabla_t\}^T [\varepsilon_{rt}] [\nabla_t N] \right\}^T \mathrm{d}\Omega \right] \{d^e\} \end{aligned} \tag{G.9}$$

where $[\nabla_t N]$ (also defined as $[\partial N]$) is given by (E.15d). Performing the computations in the parent element, $\hat{\Omega}$, yields

$$I_{\Omega^e} = \{d^e\}^T \left[\int_{\hat{\Omega}} \{\hat{r}_{se}\}\{\hat{r}_{se}\}^T \,|J|\, \mathrm{d}\Omega \right] \{d^e\} \tag{G.10a}$$

where $|J|$ is the determinant of the Jacobian matrix $[J]$ and $\{\hat{r}_{se}\}$ is a column matrix that stands for

$$\{\hat{r}_{se}\}^T = \{\nabla_t\}^T [\varepsilon_{rt}][\nabla_t N] \tag{G.10b}$$

with the understanding that expression (G.10b) should be evaluated for the parent element. The following subsections perform such an evaluation for the different parent elements used. Recalling

$$[r_s] = \int_{\hat{\Omega}} \{\hat{r}_{se}\}\{\hat{r}_{se}\}^T \,|J|\, \mathrm{d}\Omega \tag{G.11a}$$

(G.10a) may be written as

$$I_{\Omega^e} = \{d^e\}^T [r_s]\{d^e\} \tag{G.11b}$$

as previously stated in (G.2a).

Matrix $[r_s]$ is the postprocess matrix for the surface residue that will be computed for the eth element. As it can be seen, it depends on the Jacobian of the element and on a column matrix $\{r_{se}\}$ whose coefficients are evaluated from precomputed quantities (identical for all real elements generated by the same parent element) and the matrix $[\varepsilon_{rt}]$ of each element.

For the computation of I_{Γ_k}:

(1) If the kth side belongs to a Neumann boundary, then expression (G.7c) holds. For the quasi-static problem, the Neumann boundary are symmetry walls. Thus in order to obtain the same estimate for a problem solved in the whole domain, without making use of symmetry, the value of the residue has to be multiplied by the factor 2. Thus, one gets

$$I_{\Gamma_{kN}} = \int_{\Gamma_k} \left(2\{a_n\}^T [\varepsilon_{rt}]\{\nabla_t\}\{N\}^T\{d^e\} \right)^2 \mathrm{d}\Gamma = 4\int_{\Gamma_k} \left(\{a_n\}^T [\varepsilon_{rt}][\nabla_t N]\{d^e\} \right)^2 \mathrm{d}\Gamma \tag{G.12}$$

where $\{a_n\}$ stands for the matrix notation of the unit vector normal to the edge. By converting to the parent element this leads to

$$I_{\Gamma_{kN}} = 4\int_{v_1}^{v_2} \left(\{a_n\}^T[\varepsilon_{rt}]([J]^T)^{-1}\left[\hat{\nabla}_t\hat{N}\right]\{d^e\}\right)^2 J_\Gamma \mathrm{d}v \tag{G.13}$$

according to Appendix C.1. $\{a_n\}$ and J_Γ are given by (C.15b) and (C.14c), respectively. The previous expression can be rewritten as

$$I_{\Gamma_{kN}} = 4\int_{v_1}^{v_2} \left(\{a_n\}^T[\varepsilon_{rt}]([J]^T)^{-1}\left[\hat{\nabla}_t\hat{N}\right]J_\Gamma^{1/2}\{d^e\}\right)^2 \mathrm{d}v \tag{G.14a}$$

or

$$I_{\Gamma_{kN}} = 4\int_{v_1}^{v_2} \left(\{r_{edk}^e\}^T\{d^e\}\right)^2 \mathrm{d}v \tag{G.14b}$$

where

$$\{r_{edk}\}^T = \{a_n\}^T[\varepsilon_{rt}]\left([J]^T\right)^{-1}\left[\hat{\nabla}_t\hat{N}\right]J_\Gamma^{1/2} \tag{G.14c}$$

In (G.14c), an expression of the form (G.2b) may be recognized. One may evaluate the integral of (G.14c) by means of numerical integration for the corresponding edge of the parent element by

$$I_{\Gamma_{kN}} = \sum_{i=1}^{n_p} \left(\left\{r_{edk}^e(\hat{x}_{pi},\hat{y}_{pi})\right\}^T\{d^e\}\right)^2 w(i) \tag{G.15}$$

where $(\hat{x}_{pi},\hat{y}_{pi})$ are the coordinates of the ith integration point, $i = 1, ..., n^p$, and $w(i)$ are the corresponding weights.

(2) If the kth edge of element Ω^e is an internal edge, i.e., it is the same as the jth edge of element Ω^m, then

$$I_{\Gamma_{kS}} = \int_{\Gamma_k} \left(\left[\{a_n\}^T[\varepsilon_{rt}]\{\bar{\nabla}_t\}\{N\}^T\right]_k\{d^e\} + \left[\{a_n\}^T[\varepsilon_{rt}]\{\bar{\nabla}_t\}\{N\}^T\right]_j\{d^m\}\right)^2 \mathrm{d}\Gamma \tag{G.16}$$

By converting it into the parent element yields

$$I_{\Gamma_{kS}} = \int_{v_1}^{v_2}\left[\{r_{edk}\}^T\{d^e\} + \{r_{edj}\}^T\{d^m\}\right]^2 \mathrm{d}v \tag{G.17}$$

so

$$I_{\Gamma_{ks}} = \sum_{i=1}^{n_p} \left[\{r_{edk}(\hat{x}_{pi}, \hat{y}_{pi})\}^T \{d^e\} + \{r_{edj}(\hat{x}_{pi}, \hat{y}_{pi})\}^T \{d^m\} \right]^2 w(i) \tag{G.18}$$

where $\{r_{edj}\}$ refers to (G.14c) specialized for the jth edge of element Ω^m, which coincides with the kth edge of element Ω^e. Again, an expression of the form (G.2b) is obtained.

The calculation of the coefficients of $[r_s]$ and $\{r_{edk}\}$ depends on the type of element used. The first-order triangular element as well as the second-order straight (subparametric) and curved (isoparametric) element will be considered next. It is also explained how to calculate $I_{\Gamma_{ks}}$ in those cases for which Γ_k is a boundary between an ordinary element and an infinite one.

G.2.1 First-Order Straight Lagrange Triangular Element

Since the shape functions are of first order, the coefficients for $\{\hat{r}_{se}\}$ vanish because they involve second-order derivatives (see (G.10b)). Hence, in this case,

$$I_{\Omega^e} = 0 \tag{G.19}$$

Thus, only the computation of $I_{\Gamma k}$ is required. For a straight element, the mapping functions from the real elements to the parent element are the first-order Lagrange basis functions. Thus, the coefficients of the Jacobian matrix, $[J]$, are polynomials of zero order, i.e., constants. Its determinant, $|J|$, is also constant. And so are the coefficients of the first-order derivative matrix $[\hat{\nabla}_t \hat{N}]$ or $[\partial \hat{N}]$. The quantity J_Γ is also constant as well as the components of the unit normal vector. Therefore, the coefficients of $\{r_{edk}\}$ are zero-order polynomials. Hence, the integral $I_{\Gamma k}$ can be evaluated through a numerical integration at one point, the midpoint of the kth edge of the parent element, with unity value for the weighting factor.

Table G.1 provides the values of $[\partial N]$ for this case. Table G.2 summarizes the coefficients of matrix $[J]$ and of $|J|$. Lastly, expressions for the unit normal vector and of J_Γ for each side of the parent element can be found in Table G.3.

According to (G.14c), the column vectors $\{r_{edk}\}$ for k = 1, 2, 3 are given by

$$\{r_{ed1}\}^T = \frac{1}{\sqrt{L_1}} \begin{Bmatrix} y_{21} \\ -x_{21} \end{Bmatrix}^T [G] \tag{G.20a}$$

Table G.1
Coefficients of matrix $[\partial \hat{N}]$ for first-order Lagrange triangular elements

i	1	2	3
$\dfrac{\partial \hat{N}_i}{\partial \hat{x}}$	-1	1	0
$\dfrac{\partial \hat{N}_i}{\partial \hat{y}}$	-1	0	1

Table G.2
Coefficients of matrix $[J]^T$ and value of $|J|$ for linear mapping of a triangle of vertices (x_i,y_i), i = 1,2,3

$[J]^T = \begin{bmatrix} x_{21} & y_{21} \\ x_{31} & y_{31} \end{bmatrix}; \quad |J| = x_{21}y_{31} - x_{31}y_{21}$
$x_{ij} = x_i - x_j, \quad y_{ij} = y_i - y_j, \quad i, j = 1,2,3$

Table G.3
Expression for $\bar{a}_{nk}$, $J_{\Gamma k}$, and the length of each side L_k, k = 1,2,3, for a straight triangle with vertices at (x_i,y_i), i = 1,2,3

Side k	L_k	$J_{\Gamma k}$	$\{a_{nk}\}$
k = 1	$L_1 = \sqrt{x_{21}^2 + y_{21}^2}$	L_1	$\frac{1}{L_1}\begin{Bmatrix} y_{21} \\ -x_{21} \end{Bmatrix}$
k = 2	$L_2 = \sqrt{x_{32}^2 + y_{32}^2}$	L_2	$\frac{1}{L_2}\begin{Bmatrix} y_{32} \\ -x_{32} \end{Bmatrix}$
k = 3	$L_3 = \sqrt{x_{13}^2 + y_{13}^2}$	L_3	$\frac{1}{L_3}\begin{Bmatrix} y_{13} \\ -x_{13} \end{Bmatrix}$
$x_{ij} = x_i - x_j$; $y_{ij} = y_i - y_j$; i, j = 1, 2, 3			

$$\{r_{ed2}\}^T = \frac{1}{\sqrt{L_2}} \begin{Bmatrix} y_{32} \\ -x_{32} \end{Bmatrix}^T [G] \tag{G.20b}$$

$$\{r_{ed3}\}^T = \frac{1}{\sqrt{L_3}} \begin{Bmatrix} y_{13} \\ -x_{13} \end{Bmatrix}^T [G] \tag{G.20c}$$

where

$$[G] = [\varepsilon_{rt}]\left([J]^T\right)^{-1}\left[\hat{\nabla}_t\hat{N}\right] = \frac{1}{|J|}[\varepsilon_{rt}]\begin{bmatrix} y_{31} & -y_{21} \\ -x_{31} & x_{21} \end{bmatrix}\begin{bmatrix} -1 & 1 & 0 \\ -1 & 0 & 1 \end{bmatrix} \tag{G.20d}$$

As it can be seen $\{r_{edk}\}$ is a column matrix with three rows. Hence, I_{Γ_k} is given by

$$I_{\Gamma_k} = I_{\Gamma_{kN}} = 4\left(\{r_{edk}\}^T\{d^e\}\right)^2 \tag{G.21}$$

if the kth edge of element Ω^e is a Neumann boundary (symmetry wall) or

$$I_{\Gamma_k} = I_{\Gamma_{ki}} = \left(\{r_{edk}\}^T\{d^e\} + \{r_{edj}\}^T\{d^m\}\right)^2 \tag{G.22}$$

if the kth edge of element Ω^e is the same as the jth edge of element Ω^m. If element Ω^m contiguous to Ω^e is triangular, the expression for $\{r_{edj}\}$ will be given by (G.20). But if it is an infinite element (hence, in this case, a first order infinite element), it will be given by (G.38), which is defined later. Finally, if the kth side is part of a Dirichlet boundary, I_{Γ_k} vanishes as stated by (G.7e).

Thus, for this case ($p = 1$) the error estimator in element Ω^e reduces to

$$\epsilon_e^2 = f_{ed}\,\frac{h_e}{\varepsilon_{rtmin}}\sum_{k=1}^{3} I_{\Gamma_k} \tag{G.23}$$

where $h_e = \max(L_1, L_2, L_3)$.

G.2.2 Second-Order Lagrange Triangular Element

G.2.2.1 *Straight or Subparametric Element*

Here real elements have straight edges. Thus, as in the previous case, the Jacobian matrix $[J]$ and its determinant $|J|$ are given by Table G.2. $|J|$ and the coefficients of $[J]$ are zero-order polynomials, i.e., constants, within element Ω_e. Thus, $\{\hat{r}_{se}\}$ (see (G.10b)) can be expressed in the parent element as

$$\{\hat{r}_{se}\}^T = \{\hat{\nabla}_t\}^T\,[J]^{-1}[\varepsilon_{rt}]\left([J]^{-1}\right)^T\left[\hat{\nabla}_t\hat{N}\right] \tag{G.24}$$

where coefficients of $[\hat{\nabla}_t\hat{N}] = [\partial\hat{N}]$ are the first-order polynomials given by Table G.4.

Table G.4
Coefficients of $[\partial\hat{N}]$ for the second-order Lagrange triangular finite element

i	1	2	3	4	5	6
$\dfrac{\partial \hat{N}_i}{\partial \hat{x}}$	$-3+4(\hat{x}+\hat{y})$	$-1+4\hat{x}$	0	$4(1-2\hat{x}-\hat{y})$	$4\hat{y}$	$-4\hat{y}$
$\dfrac{\partial \hat{N}_i}{\partial \hat{y}}$	$-3+4(\hat{x}+\hat{y})$	0	$-1+4\hat{y}$	$-4\hat{x}$	$4\hat{x}$	$4(1-\hat{x}-2\hat{y})$

Defining the symmetric matrix $[T]$ by

$$[T] = [J]^{-1}[\varepsilon_{rt}]\left([J]^{-1}\right)^T = \begin{bmatrix} T_{11} & T_{12} \\ T_{12} & T_{22} \end{bmatrix} \tag{G.25a}$$

yields

$$\{\hat{r}_{se}\}^T = \begin{Bmatrix} \dfrac{\partial}{\partial \hat{x}} \\ \dfrac{\partial}{\partial \hat{y}} \end{Bmatrix}^T [T] \begin{bmatrix} \begin{Bmatrix} \dfrac{\partial \hat{N}}{\partial \hat{x}} \end{Bmatrix}^T \\ \begin{Bmatrix} \dfrac{\partial \hat{N}}{\partial \hat{y}} \end{Bmatrix}^T \end{bmatrix} = \begin{Bmatrix} T_{11}\dfrac{\partial^2 \hat{N}}{\partial \hat{x}^2} + T_{22}\dfrac{\partial^2 \hat{N}}{\partial \hat{y}^2} + 2T_{12}\dfrac{\partial^2 \hat{N}}{\partial \hat{x}\partial \hat{y}} \end{Bmatrix}^T \tag{G.25b}$$

The second derivatives of the Lagrange second-order basis functions are polynomials of zero order and are shown in Table G.5. Hence, $\{\hat{r}_{se}\}^T$ is a row matrix of constant coefficients:

$$\{\hat{r}_{se}\}^T = [\,4(T_{11}+2T_{12}+T_{22})\,;4T_{11}\,;4T_{22}\,;\,-8(T_{11}+T_{12})\,;8T_{12}\,;\,-8(T_{12}+T_{22})\,] \tag{G.26}$$

This allows an analytical integration, resulting in

$$[r_s] = \int_{\hat{\Omega}} \{\hat{r}_{se}\}^T \{\hat{r}_{se}\}\,|J|\,\mathrm{d}\Omega = \frac{|J|}{2}\{\hat{r}_{se}\}\{\hat{r}_{se}\}^T \tag{G.27}$$

Table G.5
Second derivatives for the basis functions of the second-order Lagrange triangular parent element

i	1	2	3	4	5	6
$\dfrac{\partial^2 \hat{N}_i}{\partial \hat{x}^2}$	4	4	0	-8	0	0
$\dfrac{\partial^2 \hat{N}_i}{\partial \hat{y}^2}$	4	0	4	0	0	-8
$\dfrac{\partial^2 \hat{N}_i}{\partial \hat{x} \partial \hat{y}}$	4	0	0	-4	4	-4

It is seen that $[r_s]$ is a 6×6 symmetric matrix. Finally, I_{Ω_e} is obtained by applying expression (G.11b).

For the calculation of I_{Γ_k} and thus of $\{r_{edk}\}$ according to (G.14c), it can be observed that the coefficients of $[\partial N]$ (see Table G.4) are first-order polynomials. Because the element is subparametric (straight), the rest of the variables involved in the computation of $\{r_{edk}\}$ are constant. They are given by Tables G.2 and G.3. Consequently, the integrands of $I_{\Gamma kN}$ and $I_{\Gamma ki}$ (see (G.14) and (G.17), respectively) are second-order polynomials. Thus, numerical integration at two points with a weight of 1/2 for each point may be used. In order to obtain the expression for $\{r_{edk}\}$ at the integration points corresponding to the kth side, the coefficients of $[\partial \hat{N}]$ (Table G.4) must be evaluated for the points of integration corresponding to each side of the parent element as given by Table G.6. Therefore $\{r_{edk}\}^T$, a row matrix with 6 columns for the kth side and at the ith integration point in the parent element, is given by

Table G.6
Coordinates of the points of integration for each side of the parent triangular element

Side k	Point 1 $(\hat{x}_{p1}, \hat{y}_{p1})$	Point 2 $(\hat{x}_{p2}, \hat{y}_{p2})$
$k = 1$	$(\alpha,0)$	$(\beta,0)$
$k = 2$	(β,α)	(α,β)
$k = 3$	$(0,\beta)$	$(0,\alpha)$
$\alpha = 0.5(1 - \sqrt{3})$, $\beta = 0.5(1 + \sqrt{3})$		

$$\{r_{edk}(\hat{x}_{pi},\hat{y}_{pi})\}^T = L_k^{1/2}\{a_{nk}\}^T[\varepsilon_{rt}]([J]^T)^{-1}\left[\hat{\nabla}_t\hat{N}\right]\Big|_{\substack{\hat{x}=\hat{x}_{pi}\\ \hat{y}=\hat{y}_{pi}}} \tag{G.28}$$

where L_k, $\{a_{nk}\}$, and $[J]$ are given by Tables G.3 and G.2. Thus, I_{Γ_k} will be given by

$$I_{\Gamma_k} = I_{\Gamma_{kN}} = 2\left(\left(\{r_{edk}(\hat{x}_{p1},\hat{y}_{p1})\}^T\{d^e\}\right)^2 + \left(\{r_{edk}(\hat{x}_{p2},\hat{y}_{p2})\}^T\{d^e\}\right)^2\right) \tag{G.29}$$

if the kth side is a Neumann boundary or

$$\begin{aligned} I_{\Gamma_k} = I_{\Gamma_{ki}} = &\frac{1}{2}\left(\{r_{edk}(\hat{x}_{p1},\hat{y}_{p1})\}^T\{d^e\} + \{r_{edj}(\hat{x}_{p2},\hat{y}_{p2})\}^T\{d^m\}\right)^2 \\ &+ \frac{1}{2}\left(\{r_{edk}(\hat{x}_{p2},\hat{y}_{p2})\}^T\{d^e\} + \{r_{edj}(\hat{x}_{p1},\hat{y}_{p1})\}^T\{d^m\}\right)^2 \end{aligned} \tag{G.30}$$

if the kth side of the element Ω^e is the same as the jth side of element Ω^m. Therefore, the first integration point in element Ω^e corresponds with the second in element Ω^m and vice versa, as has been reflected in (G.30). It should also be noticed that if Ω^m is a triangular element, then $\{r_{edj}\}^T$ is given by (G.28); but if it is a second-order serendipity or complete infinite element, then it is given by (G.38) (defined later). If the kth edge is part of a Dirichlet boundary, then $I_{\Gamma_k} = 0$.

For this case $p = 2$. Hence, the error estimate results in

$$\epsilon_e^2 = f_s\frac{h_e^2}{2\varepsilon_{rtmin}}I_{\Omega^e} + f_{ed}\frac{h_e}{2\varepsilon_{rtmin}}\sum_{k=1}^{3} I_{\Gamma_k} \tag{G.31}$$

with $h_e = \max(L_1, L_2, L_3)$ (Table G.3) and $I_{\Omega e}$ and $I_{\Gamma k}$ given by the corresponding expressions.

G.2.2.2 *Curved or Isoparametric Element*

In this case the mapping functions are the Lagrange second-order basis functions. Thus, expressions (C.12) must be taken into account to calculate $\{\hat{r}_{se}\}_T$ (see (G.10b)). The following expression is obtained, where matrices $[T_1]$ and $[T_2]$ are given by (C.12d) to (C.12f). They are given by Table G.7. The coefficients of the matrix that multiplies $[T_1]$, $[\partial\hat{N}]$, are given in Table G.4; and those of the matrix that multiplies $[T_2]$ are in Table G.5. Therefore,

$$\{\hat{r}_{se}\}^T = \{\nabla_t\}^T[\varepsilon_{rt}]\left[\nabla_t N\right] = \begin{Bmatrix}\varepsilon_{11}\\ \varepsilon_{22}\\ 2\varepsilon_{12}\end{Bmatrix}^T \begin{bmatrix} \left\{\dfrac{\partial^2 N}{\partial x^2}\right\}^T \\ \left\{\dfrac{\partial^2 N}{\partial y^2}\right\}^T \\ \left\{\dfrac{\partial^2 N}{\partial x \partial y}\right\}^T \end{bmatrix}$$

$$= \begin{Bmatrix}\varepsilon_{11}\\ \varepsilon_{22}\\ 2\varepsilon_{12}\end{Bmatrix}^T \left([T_1] \begin{bmatrix} \left\{\dfrac{\partial \hat{N}}{\partial \hat{x}}\right\}^T \\ \left\{\dfrac{\partial \hat{N}}{\partial \hat{y}}\right\}^T \end{bmatrix} + [T_2] \begin{bmatrix} \left\{\dfrac{\partial^2 \hat{N}}{\partial \hat{x}^2}\right\}^T \\ \left\{\dfrac{\partial^2 \hat{N}}{\partial \hat{y}^2}\right\}^T \\ \left\{\dfrac{\partial^2 \hat{N}}{\partial \hat{x} \partial \hat{y}}\right\}^T \end{bmatrix} \right) \tag{G.32}$$

According to (G.32) the coefficients of matrix $[r_s]$ (see (G.11a)) are surface integrals in which the integrand in the numerator is a polynomial of tenth order, whereas the denominator is a polynomial of order 12. To evaluate them, the reduced integration rule using three points located at the midpoints of the parent triangle edges with a weight of 1/6 for each of them is used. Thus, one needs to sample $[J]^T$, $|J|$, $[j]$, $[T_2]$, and $[T_1]$ at these points. Thus, the value of the first derivatives of the second-order basis functions at those sampling points are required. Table G.8 provides these values.

Matrix $[r_s]$ is a 6×6 matrix and is given by

$$[r_s] = \frac{1}{6}\sum_{i=1}^{3} \{\hat{r}_{se}(\hat{x}_{pi}\, ,\, \hat{y}_{pi})\}\, \{\hat{r}_{se}(\hat{x}_{pi}\, ,\, \hat{y}_{pi})\}^T \, |J(x_{pi}\, ,\, y_{pi})| \tag{G.33}$$

where $(\hat{x}_{pi},\hat{y}_{pi})$ are the coordinates of the ith integration point, i = 1, 2, 3. Then, (G.11b) gives I_{Ω_e}.

Table G.9 shows dΓ, J_Γ, and $\bar{a}_n$ for each side of the triangle as functions of the natural coordinates (λ,ξ,η) of the parent triangle. These expressions are deduced from (C.14) and (C.15b) according to

Table G.7

Matrices $[T_1]$ and $[T_2]$ for second-order mapping functions for triangles defined by (x_i, y_i), $i = 1,\ldots,6$

$[T_1]$	$[T_1] = -[T_2]\,[C_1]\,[j]$
$[J]^T$	$[J]^T = \begin{bmatrix} J_{11} & J_{12} \\ J_{21} & J_{22} \end{bmatrix} = \begin{bmatrix} \left\{ \dfrac{\partial \hat{N}}{\partial \hat{x}} \right\}^T \\ \left\{ \dfrac{\partial \hat{N}}{\partial \hat{y}} \right\}^T \end{bmatrix}_{(1)} \begin{bmatrix} x_1 & y_1 \\ \vdots & \vdots \\ x_6 & y_6 \end{bmatrix}$
$\lvert J \rvert$	$\lvert J \rvert = J_{11} J_{22} - J_{12} J_{21}$
$[j]$	$[j] = \begin{bmatrix} j_{11} & j_{12} \\ j_{21} & j_{22} \end{bmatrix} = \left([J]^T \right)^{-1} = \dfrac{1}{\lvert J \rvert} \begin{bmatrix} J_{22} & -J_{12} \\ -J_{21} & J_{11} \end{bmatrix}$
$[T_2]$	$[T_2] = \begin{bmatrix} j_{11}^2 & j_{12}^2 & 2 j_{11}\, j_{12} \\ j_{21}^2 & j_{22}^2 & 2 j_{21}\, j_{22} \\ j_{11}\, j_{21} & j_{12}\, j_{22} & j_{11}\, j_{22} + j_{12}\, j_{21} \end{bmatrix}$
$[C_1]$	$[C_1] = \begin{bmatrix} -4(x_{41} + x_{42}) & -4(y_{41} + y_{42}) \\ -4(x_{61} + x_{43}) & -4(y_{61} + y_{63}) \\ -4(x_{65} + x_{41}) & -4(y_{65} + y_{41}) \end{bmatrix}$

$x_{ij} = x_i - x_j; \quad y_{ij} = y_i - y_j; \quad i, j = 1, \ldots, 6$

(1) See table G.4

$$\begin{Bmatrix} x \\ y \end{Bmatrix} = \{ \hat{N}(\xi, \eta) \}^T \begin{bmatrix} x_1 & y_1 \\ \vdots & \vdots \\ x_6 & y_6 \end{bmatrix} \tag{G.34}$$

Table G.8
First derivatives of the second-order Lagrange triangular basis functions sampled at three points of integration of the parent element

	Point 1 (0.5,0)						Point 2 (0.5,0.5)						Point 3 (0,0.5)					
i	1	2	3	4	5	6	1	2	3	4	5	6	1	2	3	4	5	6
$\frac{\partial \hat{N}_i}{\partial \hat{x}}$	-1	1	0	0	0	0	1	1	0	-2	2	-2	-1	-1	0	2	2	-2
$\frac{\partial \hat{N}_i}{\partial \hat{y}}$	-1	0	-1	-2	2	2	1	0	1	-2	2	-2	-1	0	1	0	0	0

Table G.9
Expressions for $d\Gamma$, J_Γ, and $\overline{a}_n$ for the sides of a curved triangle for second-order geometric transformation

Side k	$k = 1$	$k = 2$	$k = 3$
	$\lambda = 1-\xi$ $\eta = 0$	$\lambda = 0$ $\xi = 1-\eta$	$\lambda = 1-\eta$ $\xi = 0$
$d\Gamma_k =$ $J_{\Gamma k} dv$	$\left[\left(\frac{\partial x}{\partial \xi}\right)^2 + \left(\frac{\partial y}{\partial \xi}\right)^2\right]^{1/2} d\xi$	$\left[\left(\frac{\partial x}{\partial \eta}\right)^2 + \left(\frac{\partial y}{\partial \eta}\right)^2\right]^{1/2} d\eta$	$\left[\left(\frac{\partial x}{\partial \eta}\right)^2 + \left(\frac{\partial y}{\partial \eta}\right)^2\right]^{1/2} d\eta$
$J_{\Gamma k}{}^2$	$[3x_{41}+x_{42}-4(x_{41}+x_{42})\xi]^2$ $+ [3y_{41}+y_{42}-4(y_{41}+y_{42})\xi]^2$	$[3x_{52}+x_{53}-4(x_{52}+x_{53})\eta]^2$ $+ [3y_{52}+y_{53}-4(y_{52}+y_{53})\eta]^2$	$[3x_{61}+x_{63}-4(x_{61}+x_{63})\eta]^2$ $+ [3y_{61}+y_{63}-4(y_{61}+y_{63})\eta]^2$
$\{a_{nk}\}$	$\frac{1}{J_{s1}}\begin{Bmatrix} \frac{\partial y}{\partial \xi} \\ -\frac{\partial x}{\partial \xi} \end{Bmatrix} =$	$\frac{1}{J_{s2}}\begin{Bmatrix} \frac{\partial y}{\partial \eta} \\ -\frac{\partial x}{\partial \eta} \end{Bmatrix} =$	$\frac{1}{J_{s3}}\begin{Bmatrix} -\frac{\partial y}{\partial \eta} \\ \frac{\partial x}{\partial \eta} \end{Bmatrix} =$
	$\frac{1}{J_{s1}}\begin{Bmatrix} 3y_{41}+y_{42}-4(y_{41}+y_{42})\xi \\ -3x_{41}-x_{42}+4(x_{41}+x_{42})\xi \end{Bmatrix}$	$\frac{1}{J_{s2}}\begin{Bmatrix} 3y_{52}+y_{53}-4(y_{52}+y_{53})\eta \\ -3x_{52}-x_{53}+4(x_{52}+x_{53})\eta \end{Bmatrix}$	$\frac{1}{J_{s3}}\begin{Bmatrix} -3y_{61}-y_{63}+4(y_{61}-y_{63})\eta \\ 3x_{61}+x_{63}-4(x_{61}+x_{63})\eta \end{Bmatrix}$

$x_{ij} = x_i - x_j$; $y_{ij} = y_i - y_j$; $i, j = 1,...,6$; $\lambda = 1 - \hat{x} - \hat{y}$; $\xi = \hat{x}$; $\eta = \hat{y}$

Expression (G.34) transforms a point of a real element Ω^e given by (x,y) into a point of the parent element (ξ,η), where $\xi = \hat{x}$, $\eta = \hat{y}$. Here, N are the Lagrange second-order basis functions (see Table 2.2, element P(2)); (x_i,y_i), $i = 1, ..., 6$, are the coordinates of the geometric points defining the real element Ω^e. In order to evaluate the quantities of Table G.9, expression (G.34) must be specialized to the kth side, as indicated in Table G.9.

For the calculation of $I_{\Gamma k}$, numerical integration is used to evaluate (G.29) and (G.30). Two points for each edge are used. Their coordinates in the parent triangle are given by Table G.6. Hence,

$$\{r_{edk}(\hat{x}_{pi}, \hat{y}_{pi})\}^T = \{a_{nk}(\hat{x}_{pi}, \hat{y}_{pi})\}^T[\varepsilon_{rt}]\left([J(\hat{x}_{pi}, \hat{y}_{pi})]^T\right)^{-1}\left[\hat{\nabla}_t\hat{N}\right]\Big|_{\substack{\hat{x}=\hat{x}_{pi}\\ \hat{y}=\hat{y}_{pi}}} J_{\Gamma k}(\hat{x}_{pi}, \hat{y}_{pi})^{1/2} \tag{G.35}$$

For the calculation of $I_{\Gamma k}$, it is enough to use the expressions (G.29) and (G.30), employing (G.35). The parameter h_e (see (G.1)) is also calculated through a numerical integration. As it is related to the length of the edge Γ_k, one obtains

$$L_k = \int_{\Gamma_k} d\Gamma_k = \int_{v_1}^{v_2} J_{\Gamma k}\, dv \tag{G.36}$$

Two points quadrature are used to evaluate (G.36), with 1/2 as the weighting factor. The coordinates of the integration points for each side of the parent triangle are given by Table G.6. Thus, $J_{\Gamma k}$ must be sampled at those points (see Table G.9), yielding

$$L_k = \frac{1}{2}\sum_{i=1}^{2} J_{sk}(\hat{x}_{pi}\ ;\ \hat{y}_{pi}), \quad k=1,2,3 \tag{G.37}$$

Finally, $h_e = \max(L_1, L_2, L_3)$. Then (G.31) provides the error estimator.

G.2.3 Infinite Elements. Computation of the Residue at an Interface with an Ordinary Element

As it was pointed out in Section G.1, the error estimator for infinite elements has been considered to be zero. This is done in order to avoid in the self-adaptive meshing procedure an infinite element which turns out to be a primary element, requiring its subdivision into more than two elements (see Chapter 4). Moreover, due to the nature of the infinite elements, located in the far-field region, the surface residue must be reasonably small. However, in order to obtain the error estimate of an ordinary element contiguous to an infinite one, one needs to evaluate the residue at the interface. In fact, this residue is involved in the computation of expression $I_{\Gamma ki}$ of the contiguous ordinary element through $\{r_{edj}\}$ as shown by (G.17), (G.18), (G.22), and (G.30).

Therefore, the coefficients of $\{r_{edj}\}^T$ along the fourth edge of the infinite elements must be computed. This is accomplished for the first order infinite element (with six points for the geometric transformation) compatible with the linear triangular element and for the second-order serendipity and complete infinite elements (with six geometric points for the mapping functions) compatible with second-order triangular elements. Their description is given in Section 2.5.1.2. The Γ_k edge at the interface between the triangular element Ω^e and the infinite element Ω^m will always be straight.

It should be noted that infinite elements are not symmetric like ordinary elements. This implies that a local numbering scheme has to be used, where the vertices associated with the edge of the parent element that is transformed into infinity should always be the second and the third vertices (it is recalled that vertices and edges are locally numbered following the trigonometric direction). According to that numbering scheme, the interface with an ordinary element will always be the fourth edge of the infinite element. In order to program this local numbering scheme, it is only necessary to identify the vertices of the edge transformed into infinity and to store the correspondence between the original numbering and the required one. Let $\{r_{ed4}\}$ be the expression of $\{r_{edj}\}$ which corresponds to the fourth edge. Then one obtains

$$\{r_{ed4}\}^T = \{\bar{a}_n\}^T \, [\varepsilon_{rt}] \left([J]^T\right)^{-1} [\hat{\nabla}_t \hat{N}] \, J_\Gamma^{1/2} \tag{G.38}$$

where $\{\bar{a}_n\}$ and J_Γ are constants in the element as shown in Table G.10. Expressions for the coefficients of $[\hat{\nabla}_t \hat{N}]$ depend on the order of the interpolation used. Instead, the expressions for the coefficients of $[J]$ are obtained from the mapping functions. In order to have compatibility between the definitions of the parent triangular elements and the parent infinite elements, one should define the latter parent element as a quadrangle with a unit area instead of the quadrangle of area equal to four given in Table 2.5. In other words, the infinite parent element is defined by the quadrangle of coordinates (0,0), (1,0), (1,1), (0,1) plus the two geometric points (0.5,0), (0.5,1). Thus, the following transformation between the parent element of Table 2.5 and the parent element of unit area holds:

$$\hat{x} = \frac{\xi + 1}{2}, \quad \hat{y} = \frac{\eta + 1}{2} \tag{G.39}$$

where (ξ,η) refers to the coordinates of the parent element $I(6)$, of Table 2.5, while $(\hat{x},\hat{y})$ refers to the new parent element. The mapping functions corresponding to the new parent element are given in Table G.11. Table G.12 provides their derivatives.

The coefficients of $[J]^T$ are obtained by making use of

$$[J]^T = \begin{bmatrix} \left\{ \dfrac{\partial \hat{N}^I}{\partial \hat{x}} \right\}^T \\ \left\{ \dfrac{\partial \hat{N}^I}{\partial \hat{y}} \right\}^T \end{bmatrix} \begin{bmatrix} x_1 & y_1 \\ \vdots & \vdots \\ x_6 & y_6 \end{bmatrix} \tag{G.40}$$

Table G.10
Expressions for $\bar{a}_n$ and J_Γ in the fourth edge of an infinite element of six geometric points of coordinates (x_i, y_i), $i = 1,\ldots,6$

J_s	$\sqrt{x_{41}^2 + y_{41}^2}$
$\bar{a}_n$	$\dfrac{1}{J_s} \begin{Bmatrix} -y_{41} \\ x_{41} \end{Bmatrix}$

$x_{41} = x_4 - x_1 ; \quad y_{41} = y_4 - y_1$

Table G.11
Mapping functions for the parent infinite element of unit area and six geometric points. Decay given by $(1/r)^n$ in the x direction

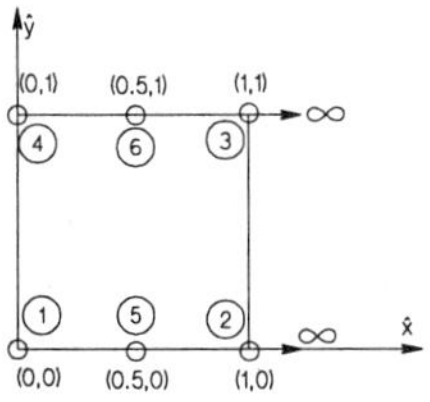

$$\hat{N}_1^I = \hat{N}_1^{In}(1-\hat{y})$$
$$\hat{N}_4^I = \hat{N}_1^{In}\hat{y} \; ; \qquad \hat{N}_2^I = 0$$
$$\hat{N}_5^I = \hat{N}_3^{In}(1-\hat{y}) \; ; \quad \hat{N}_3^I = 0$$
$$\hat{N}_6^I = \hat{N}_3^{In}\hat{y}$$

$$\hat{N}_1^{In} = k_2 \left[\frac{1}{(1-\hat{x})^{1/n}} - k_1 \right] ; \; \hat{N}_3^{In} = 1 - \hat{N}_1^{In} \; ; \; k_1 = 2^{1/n} \; ; \; k_2 = \frac{1}{1-k_1}$$

Table G.12
Expressions for the first derivatives of the functions of Table G.11

i	1	2	3	4	5	6
$\dfrac{\partial \hat{N}_i^I}{\partial \hat{x}}$	D_1^{In}	0	0	D_3^{In}	$-D_1^{In}$	$-D_3^{In}$
$\dfrac{\partial \hat{N}_i^I}{\partial \hat{y}}$	$-\hat{N}_1^{In}$	0	0	$\hat{N}_1^{In}$	$-\hat{N}_3^{In}$	$\hat{N}_3^{In}$

$$D_1^{In} = \frac{k}{n} \, \frac{1-\hat{y}}{(1-\hat{x})(1-\hat{x})^{1/n}} ; \qquad D_3^{In} = \frac{k}{n} \, \frac{\hat{y}}{(1-\hat{x})(1-\hat{x})^{1/n}} ; \qquad k = \frac{1}{1-2^{1/n}}$$

where (x_i,y_i) are the coordinates of the geometric points of the infinite element Ω^m with the appropriate local numbering. Hence the coefficients of (G.38) depend on $\hat{x}$ and $\hat{y}$ and it will be necessary to use numerical integration, whose order will depend on the order of the interpolation chosen. The following subsections provide details for first- and second-order infinite elements.

G.2.3.1 *First-Order Infinite Element*

In this case the integration is performed at one point, given by (0,0.5). Table G.13 gives the value of the derivatives of the mapping functions at that point. Then the value of $([J]^T)^{-1}$ at that point may be obtained. Table G.14 shows the value for the coefficients of $[\hat{\nabla}_t\hat{N}]$ for the first-order Lagrange rectangular element at the mentioned sampling point. Then the expression (G.38) may be evaluated at that point, leading to $\{r_{ed4}\}^T$. Finally, the term $\{r_{ed4}\}^T\,\{d^m\}$ required by (G.22) can be calculated, where $\{d^m\}$ is the vector of the four degrees of freedom of the first-order infinite element Ω^m.

Table G.13

Values of the expressions of Table G.12 at the point (0,0.5) of the parent element

i	1	2	3	4	5	6
$\frac{\partial \hat{N}_i^I}{\partial \hat{x}}$	$\frac{1}{2n(1-2^{1/n})}$	0	0	$\frac{1}{2n(1-2^{1/n})}$	$-\frac{1}{2n(1-2^{1/n})}$	$-\frac{1}{2n(1-2^{1/n})}$
$\frac{\partial \hat{N}_i^I}{\partial \hat{y}}$	-1	0	0	1	0	0

Table G.14

First derivatives of the basis functions of the first-order Lagrange rectangular element at the point (0,0.5) of the parent element of unit area

i	1	2	3	4
$\frac{\partial \hat{N}_i}{\partial \hat{x}}$	-0.5	0.5	0.5	-0.5
$\frac{\partial \hat{N}_i}{\partial \hat{y}}$	-1	0	0	1

G.2.3.2 *Serendipity and Complete Second-Order Infinite Elements*

The integration is performed at two points along the fourth side of the parent element of coordinates (0,β) and (0,α), respectively, where $\alpha = 0.5(1-\sqrt{3})$ and $\beta = 0.5(1+\sqrt{3})$. Thus, Table G.12 has to be evaluated for these coordinates. To compute $[\hat{\nabla}_t \hat{N}]$, the first derivatives of the unit area parent elements $Q(2')$ or $Q(2)$ (see Table 2.1) are specialized for the points mentioned previously. Then $\{\mathrm{r}_{ed4}(\hat{x}_{pi},\hat{y}_{pi})\}^T$ may be computed (see (G.38)). Finally, the terms $\{r_{ed4}(\hat{x}_{pi},\hat{y}_{pi})\}^T\{d^m\}$ are obtained using (G.30).

G.3 COMPUTATION OF LOCAL ERROR INDICATORS FOR THE FULL-WAVE ANALYSIS OF WAVEGUIDING STRUCTURES

G.3.1 Formulation by Means of the Longitudinal Components of the Electric or the Magnetic Field and a Lagrange Element-Based Finite Element Method

The problem addressed in this section uses the strong formulation that is given in Appendix F.2 for homogeneous and isotropic structures. Suppose that the problem is discretized as indicated in Appendix F.2 so that an approximate solution $\tilde{s}_l$ is obtained for the eigenvalue s_l corresponding to an approximate solution $\tilde{\phi}_l$ of the eigenvector ϕ_l. Here s_l and, in general, s are the cutoff wave numbers, k_c, normalized with respect to the geometric scale factor a, that is, $s=k_c a$.

According to Section 4.3.1, $\tilde{s}_l$ and $\tilde{\phi}_l$ are exact solutions of the original problem with perturbed second terms. This problem could be written as

Find s and ϕ in Ω so that $\phi = 0$ on Γ_{pe} (TM modes) or Γ_{pm} (TE modes), where $\Gamma = \Gamma_{pe} \cup \Gamma_{pm}$ is the boundary of Ω and

$$- \bar{\nabla}_t \cdot \bar{\nabla}_t \phi - s^2 \phi = r, \qquad in \;\; \Omega = \bigcup \Omega^e \;, \quad e=1,\ldots,N_e \tag{G.41}$$

$$- \bar{a}_n \cdot \bar{\nabla}_t \phi = \xi, \qquad on \; \Gamma_{pm} \qquad (for\ TM\ modes) \tag{G.42}$$

or

$$- \bar{a}_n \cdot \bar{\nabla}_t \phi = \xi, \qquad on \; \Gamma_{pe} \qquad (for\ TE\ modes) \tag{G.43}$$

where r is the residue in Ω, which can be written as

$$r = r_s + \rho \tag{G.44}$$

where r_s is the surface residue in each element Ω^e

$$r_s = -\bar{\nabla}_t \cdot \bar{\nabla}_t \tilde{\phi} - \tilde{s}^2 \tilde{\phi} \quad , \quad in \ \Omega^e \ , \ e=1,\ldots,N_e \tag{G.45}$$

and ρ *is the singular function*

$$\rho = \rho_k \delta_k = - [\![\bar{a}_n \cdot \bar{\nabla}_t \tilde{\phi}]\!]_k \ , \qquad on \ \ \Gamma_k \ , \qquad k=1,\ldots,n_{edi} \tag{G.46}$$

that represents the residue on the kth internal edge ($[\![\cdot]\!]$ *has been defined in Appendix G.2). Finally,* ξ *is the residue on the magnetic boundaries for TM modes or on the electric boundaries for TE modes:*

$$\xi = -\bar{a}_n \cdot \bar{\nabla}_t \tilde{\phi} \ , \qquad on \ \Gamma_{pm} \ \ (for\ TM\ modes) \ \ or \ \ \Gamma_{pe} \ \ (for\ TE\ modes) \tag{G.47}$$

Here N_e is the number of elements and n_{edi} is the number of internal edges.

The surface residue r_s is a scalar given by (G.45); for the kth edge of Ω^e. The residue r_{ed} is also a scalar given by

$$r_{ed} = \rho, \qquad \text{for} \ \ \Gamma_k \not\subset \Gamma \tag{G.48a}$$

or by

$$r_{ed} = \xi, \qquad \text{for} \ \ \Gamma_k \subset \Gamma_{pm} \ \text{(TM) or} \ \Gamma_{pe} \ \text{(TE)} \tag{G.48b}$$

as well as

$$r_{ed} = 0, \qquad \text{for} \ \Gamma_k \subset \Gamma_{pe} \ \text{(TM) or} \ \Gamma_{pm} \ \text{(TE)} \tag{G.48c}$$

The error indicator in element Ω^e can be written as

$$\epsilon_e^2 = f_s \frac{h_e^2}{p} \int_{\Omega^e} r_s^2 \, d\Omega + f_{ed} \frac{h_e}{p} \sum_{k=1}^{3} \int_{\Gamma_k} r_{ed}^2 \, d\Gamma \tag{G.49}$$

where for this problem only triangular elements have been used. Thus,

$$\epsilon_e^2 = f_s \frac{h_e^2}{p} I_{\Omega^e} + f_{ed} \frac{h_e}{p} I_{\Gamma_k} \tag{G.50a}$$

where

$$I_{\Omega^e} = \int_{\Omega^e} r_s^2 \, d\Omega \tag{G.50b}$$

and $I_{\Gamma k}$ is one of the following expressions:

$$I_{\Gamma_k} = I_{\Gamma_{ki}} = \int_{\Gamma_k} \rho^2 \, d\Gamma, \qquad \text{for} \quad \Gamma_k \not\subset \Gamma \tag{G.50c}$$

$$I_{\Gamma_k} = I_{\Gamma_{kN}} = \int_{\Gamma_k} \xi^2 \, d\Gamma, \qquad \text{for} \quad \Gamma_k \subset \Gamma_{pm}(\text{TM}) \;\; \text{or} \;\; \Gamma_{pe}(\text{TE}) \tag{G.50d}$$

$$I_{\Gamma_k} = I_{\Gamma_{kD}} = 0, \qquad \text{for} \quad \Gamma_k \subset \Gamma_{pe}(\text{TM}) \;\; \text{or} \;\; \Gamma_{pm}(\text{TE}) \tag{G.50e}$$

First- and second-order straight Lagrange triangular elements have been used. The calculation of $I_{\Omega e}$ and $I_{\Gamma k}$ is carried out within the parent element. Since (G.8) is valid for this case, (G.10a) is also valid, where

$$\{\hat{r}_{se}\}^T = -\{\hat{\nabla}_t\}^T [J]^{-1} \left([J]^T\right)^{-1} [\hat{\nabla}_t \hat{N}] - \tilde{s}^2 \{\hat{N}\}^T \tag{G.51}$$

for both the first-order and the second-order elements, since in both cases the mapping functions are linear. (G.11) is also applicable in this case. $I_{\Gamma k}$ is given by

$$I_{\Gamma_{kN}} = \sum_{i=1}^{n_p} \left(\{r_{edk}(\hat{x}_{pi}, \hat{y}_{pi})\}^T \{d^e\} \right)^2 w(i), \quad \text{for } \Gamma_k \subset \Gamma_N \tag{G.52a}$$

$$\{r_{edk}\}^T = \{a_n\}^T \left([J]^T\right)^{-1} [\hat{\nabla}_t \hat{N}] \, J_\Gamma^{1/2} \tag{G.52b}$$

or

$$I_{\Gamma_{ki}} = \sum_{i=1}^{n_p} \left(\{r_{edk}(\hat{x}_{pi}, \hat{y}_{pi})\}^T \{d^e\} + \{r_{edj}(\hat{x}_{pi}, \hat{y}_{pi})\}^T \{d^m\} \right)^2 w(i), \quad \text{for } \Gamma_k \not\subset \Gamma \tag{G.53}$$

where the different variables are given in Section G.2.

G.3.1.1 *First-Order Lagrange Straight Triangular Element*

For this case $\{\hat{r}_{se}\}^T$ becomes

$$\{\hat{r}_{se}\}^T = -\ \tilde{s}^2\{\hat{N}\}^T \tag{G.54}$$

and $[r_s]$ is a 3×3 matrix given by

$$[r_s] = \int_{\hat{\Omega}} \tilde{s}^4 \{\hat{N}\}\{\hat{N}\}^T \,|J|\, \mathrm{d}\,\Omega \tag{G.55a}$$

where $\{N\}^T$ is the 1×3 row matrix whose coefficients are the basis functions of the first-order element (see Table 2.2). It can be observed that

$$[r_s] = \tilde{s}^4 [m^e] \tag{G.55b}$$

where $[m^e]$ is the mass matrix of element Ω^e, given by (F.53c). The calculation of I_{Ω_e} is done according to (G.11b).

The coefficients of (G.52b) are constants. Thus, only one integration point is used, so one obtains

$$I_{\Gamma_{kN}} = \left(\{r_{edk}\}^T \{d^e\}\right)^2 \tag{G.56a}$$

and

$$I_{\Gamma_{ki}} = \left(\{r_{edk}\}^T \{d^e\} + \{r_{edj}\}^T \{d^m\}\right)^2 \tag{G.56b}$$

and

$$\{r_{edk}\}^T \doteq \{a_{nk}\}\left([J]^T\right)^{-1} [\hat{\nabla}_t \hat{N}]\, J_{\Gamma k}^{1/2} \tag{G.56c}$$

where $[\hat{\nabla}_t\hat{N}]$, $[J]^T$, $\bar{a}_{nk}$, and $J_{\Gamma k}$ are given by Tables G.1, G.2, and G.3. The diameter of the element h_e is calculated from Appendix G.2.1. For the evaluation of (G.50a) one uses $p = 1$.

G.3.1.2 *Second-Order Lagrange Straight Triangular Element*

For this case, $\{\hat{r}_{se}\}^T$ is given by expression (G.51), where $[J]^T$ is obtained from Table G.2. $\{\hat{N}\}^T$ are the shape functions of the second-order element. The coefficients of matrix $[\hat{\nabla}_t\hat{N}]$ are given in Table G.4. Let us define $[T]$ as the following symmetric matrix:

$$[T] = [J]^{-1}\left([J]^T\right)^{-1} = \begin{bmatrix} T_{11} & T_{12} \\ T_{12} & T_{22} \end{bmatrix} \tag{G.57}$$

then, one obtains

$$\{\hat{r}_{se}\}^T = -\begin{Bmatrix} 4(T_{11}+2T_{12}+T_{22}) \\ 4T_{11} \\ 4T_{22} \\ -8(T_{11}+T_{12}) \\ 8T_{12} \\ -8(T_{12}+T_{22}) \end{Bmatrix}^T - \tilde{s}^2\{\hat{N}\}^T \tag{G.58a}$$

and hence

$$\{r_{se}\}^T = -\{r_{se0}\}^T - \tilde{s}^2\{r_{se1}\}^T \tag{G.58b}$$

with

$$\{r_{se0}\}^T = \begin{Bmatrix} 4(T_{11}+2T_{12}+T_{22}) \\ 4T_{11} \\ 4T_{22} \\ -8(T_{11}+T_{12}) \\ 8T_{12} \\ -8(T_{12}+T_{22}) \end{Bmatrix}^T \tag{G.58c}$$

$$\{r_{se1}\}^T = \{\hat{N}\}^T \tag{G.58d}$$

From (G.11a) one obtains

$$[r_s] = [r_{s0}] + \tilde{s}^2\left([r_{s1}] + [r_{s1}]^T\right) + \tilde{s}^4[r_{s2}] \tag{G.59a}$$

where

$$[r_{s0}] = \int_{\hat{\Omega}} \{r_{se0}\}\{r_{se0}\}^T \, |J| \, d\Omega = \frac{|J|}{2}\{r_{se0}\}\{r_{se0}\}^T \tag{G.59b}$$

$$[r_{s1}] = \int_{\hat{\Omega}} \{r_{se0}\}\{\hat{N}\}^T \, |J| \, d\Omega \tag{G.59c}$$

$$[r_{s2}] = \int_{\hat{\Omega}} \{\hat{N}\}\{\hat{N}\}^T \, |J| \, d\Omega = [m^e] \tag{G.59d}$$

The computation of (G.59c) and (G.59d) can be carried out in an analytical way. Then I_{Ω_e} is obtained by applying expression (G.11b).

To calculate $I_{\Gamma k}$, two points of integration for each side of the element according to expressions (G.52a) and (G.53) are used, with $\{r_{edk}\}^T$ given by (G.52b). $\bar{a}_{nk}$, $[J]^T$, and $J_{\Gamma k}$ are constant on each side. They are given by Tables G.2 and G.3. The coefficients of $[\hat{\nabla}_t\hat{N}]$ (see Table G.4) must be sampled at each integration point, according to Table G.6. Summarizing, one gets

$$I_{\Gamma_{kN}} = \frac{1}{2} \sum_{i=1}^{2} \left(\{r_{edk}(\hat{x}_{pi}, \hat{y}_{pi})\}^T \{d^e\}\right)^2 \tag{G.60a}$$

$$\begin{aligned} I_{\Gamma_{ki}} = {} & \frac{1}{2}\left(\{r_{edk}(\hat{x}_{p1}, \hat{y}_{p1})\}^T\{d^e\} + \{r_{edj}(\hat{x}_{p2}, \hat{y}_{p2})\}^T\{d^m\}\right)^2 \\ & + \frac{1}{2}\left(\{r_{edk}(\hat{x}_{p2}, \hat{y}_{p2})\}^T\{d^e\} + \{r_{edj}(\hat{x}_{p1}, \hat{y}_{p1})\}^T\{d^m\}\right)^2 \end{aligned} \tag{G.60b}$$

with

$$\{r_{edk}(\hat{x}_{pi}, \hat{y}_{pi})\}^T = \{a_{nk}\}\left([J]^T\right)^{-1} [\hat{\nabla}_t\hat{N}]\Big|_{\substack{\hat{x}=\hat{x}_{pi} \\ \hat{y}=\hat{y}_{pi}}} J_{\Gamma k}^{1/2} \tag{G.60c}$$

To complete the computation of the indicator, the value of $h_e = \max(L_1, L_2, L_3)$ must be obtained.

G.3.2 Formulation by Means of the Transverse and Longitudinal Components of the Electric or Magnetic Field and a Lagrange/Curl-Conforming Element-Based Finite Element Method

The strong formulation of the problem is done according to Appendix F.3.1.1. Suppose that the problem is discretized as in Appendix F.3.1.2. Then, an approximate solution $\tilde{s}_l$ is obtained for the lth eigenvalue s_l, to which there corresponds the approximate solution $[\bar{\upsilon}_{tn}, j\upsilon_{zn}\bar{a}_z]_l^T$ of the lth exact eigenvector $[\bar{V}_{tn}, jV_{zn}\bar{a}_z]_l^T$. The meaning of s_l, $\bar{V}_{tn}$, and V_{zn} are explained in Appendix F.3.1.1. More exactly, the approximate normalized eigenvector can be written as $(\bar{\upsilon}_{tn}+j\upsilon_{zn}\bar{a}_z)e^{-j\beta_a z}$ for the normalized phase constant β_a utilized as input data. The approximate eigenvalue and eigenvector are an exact solution of the original problem with perturbed right-hand terms that could be written as

Given the invertible tensor $\underline{\underline{f}}_r(x,y)$ (or in matrix form, $[f_r(x,y)]$) and tensor $\underline{\underline{p}}_r(x,y)$ (or in matrix form, $[p_r(x,y)]$) in Ω, find for each value of β_a the values of s and $\underline{\bar{V}}_n(x,y,z)$, where

$$\underline{\bar{V}}_n(x,y,z) = \left(\bar{V}_{tn}(x,y) + jV_{zn}(x,y)\bar{a}_z\right) e^{-j\beta_a z} \tag{G.61a}$$

and for a null vector $\bar{0}$;

$$\bar{a}_n \times \underline{\bar{V}}_n(x,y,z) = \bar{0} \quad on \ \Gamma_D \tag{G.61b}$$

so that

$$\bar{\nabla} \times \underline{\underline{f}}_r^{-1} \bar{\nabla} \times \underline{\bar{V}}_n - s^2 \underline{\underline{p}}_r \underline{\bar{V}}_n = \bar{r}, \quad in \ \Omega \tag{G.61c}$$

and

$$\bar{a}_n \times \underline{\underline{f}}_r^{-1} \bar{\nabla} \times \bar{V}_n = \bar{\xi}, \quad on \ \Gamma_N \tag{G.61d}$$

where $\Omega = \cup\, \Omega^e$, $e = 1, \ldots, N_e$, N_e is the number of elements. Γ is the boundary of Ω, and

$$\bar{r} = \bar{r}_s + \bar{\rho} \tag{G.61e}$$

Here, $\bar{r}_s$ represents the surface residue for each element Ω^e:

$$\bar{r}_s = \bar{\nabla} \times \underline{\underline{f}}_r^{-1} \bar{\nabla} \times \underline{\bar{\upsilon}}_n - \tilde{s} \underline{\underline{p}}_r \underline{\bar{\upsilon}}_n \quad in \quad \Omega^e \ , \ e = 1, \ldots, N_e \tag{G.61f}$$

and $\bar{\rho}$ is the singular vector function that represents the residue at the internal edges:

$$\bar{\rho} = \rho_k \delta_k = [\![\bar{a}_n \times \underline{\underline{f}}_r^{-1} \bar{\nabla} \times \underline{\bar{\upsilon}}_n]\!]_k \quad on \quad \Gamma_k \subset \Gamma_{\Omega^e} \ ; \ \Gamma_k \not\subset \Gamma \tag{G.61g}$$

and finally, $\bar{\xi}$ is the vector function that represents the residue at the Neumann boundaries:

$$\bar{\xi} = \bar{a}_n \times \underline{\underline{f}}_r^{-1} \bar{\nabla} \times \underline{\bar{\upsilon}}_n , \quad on \quad \Gamma_k \subset \Gamma_{\Omega^e} \ ; \ \Gamma_k \subset \Gamma_N \tag{G.61h}$$

Therefore, the surface residue $\bar{r}_s$, in the element Ω^e is a vector and is given by

$$\bar{r}_s = \bar{\nabla} \times \underline{\underline{f}}_r^{-1} \bar{\nabla} \times \underline{\bar{\upsilon}}_n - \tilde{s}^2 \underline{\underline{p}}_r \underline{\bar{\upsilon}}_n \tag{G.62a}$$

The residue at the edges, $\bar{r}_{ed}$, is also a vector that is given by

$$\bar{r}_{ed} = \bar{\rho}, \qquad \Gamma_k \subset \Gamma_{\Omega^e}, \ \ \Gamma_k \not\subset \Gamma \tag{G.62b}$$

or

$$\bar{r}_{ed} = \bar{\xi}, \qquad \Gamma_k \subset \Gamma_{\Omega^e}, \ \ \Gamma_k \subset \Gamma_N \tag{G.62c}$$

as well as

$$\bar{r}_{ed} = 0, \qquad \Gamma_k \subset \Gamma_{\Omega^e}, \ \ \Gamma_k \subset \Gamma_D \tag{G.62d}$$

The error indicator in the element Ω^e can be written as

$$\epsilon_e^2 = f_s \frac{h_e^2}{I_{r\min} p} \int_{\Omega^e} \bar{r}_s^* \cdot \bar{r}_s \, d\Omega + f_{ed} \frac{h_e^2}{I_{r\min} p} \sum_{\substack{k=1 \\ \Gamma_k \subset \Gamma_{\Omega^e}}}^{3} \int_{\Gamma_k} \bar{r}_{ed}^* \cdot \bar{r}_{ed} \, d\Gamma \tag{G.63a}$$

resulting in (G.7a), which in this case would be

$$\epsilon_e^2 = f_s \frac{h_e^2}{I_{r\min} p} I_{\Omega^e} + f_{ed} \frac{h_e}{I_{r\min} p} \sum_{k=1}^{3} I_{\Gamma_k} \tag{G.63b}$$

with

$$I_{\Omega^e} = \int_{\Omega^e} \bar{r}_s^* \cdot \bar{r}_s \, \mathrm{d}\Omega \tag{G.63c}$$

$$I_{\Gamma_k} = \int_{\Gamma_k} \bar{r}_{ed}^* \cdot \bar{r}_{ed} \, \mathrm{d}\Gamma \tag{G.63d}$$

I_{rmin} is the lowest eigenvalue of $[I_r] = [f_r]^{-1}$, h_e is the diameter of Ω^e, and p is the order of the interpolation used.

To calculate I_{Ω_e} and $I_{\Gamma k}$, the form of $[f_r]$ and $[p_r]$ should be taken into account according to (D.10) as well as $\bar{\upsilon}$ according to (D.27), and $\bar{\upsilon}_n$ according to (G.61a).

The residual $\bar{r}_s$ results in

$$\bar{r}_s = (\bar{r}_{st} + j r_{sz} \bar{a}_z) e^{-j\beta_a z} \tag{G.64a}$$

where

$$\bar{r}_{st} = \bar{\nabla}_t \times f_{rzz}^{-1} \bar{\nabla}_t \times \bar{\upsilon}_{tn} + \beta_a \bar{a}_z \times \underline{\underline{f}}_{rt}^{-1} \bar{\nabla}_t \times \upsilon_{zn} \bar{a}_z - \beta_a^2 \bar{a}_z \times f_{rt}^{-1} \bar{a}_z \times \bar{\upsilon}_{tn} - \tilde{s}^2 \underline{\underline{p}}_{rt} \bar{\upsilon}_{tn} \tag{G.64b}$$

$$r_{sz} \bar{a}_z = \bar{\nabla}_t \times \underline{\underline{f}}_{rt}^{-1} \bar{\nabla}_t \times \upsilon_{zn} \bar{a}_z - \beta_a \bar{\nabla}_t \times \underline{\underline{f}}_{rt}^{-1} \bar{a}_z \times \bar{\upsilon}_{tn} - \tilde{s}^2 p_{rzz} \upsilon_{zn} \bar{a}_z \tag{G.64c}$$

It can be observed that $\bar{r}_{st}$ and r_{sz} are real; hence,

$$\bar{r}_s^* \cdot \bar{r}_s = \bar{r}_{st} \cdot \bar{r}_{st} + r_{sz} \bar{a}_z \cdot r_{sz} \bar{a}_z \tag{G.64d}$$

and therefore I_{Ω^e} is

$$I_{\Omega^e} = I_{t\Omega^e} + I_{z\Omega^e} \tag{G.65a}$$

where

$$I_{t\Omega^e} = \int_{\Omega^e} \bar{r}_{st} \cdot \bar{r}_{st} \, \mathrm{d}\Omega \tag{G.65b}$$

$$I_{z\Omega^e} = \int_{\Omega^e} r_{sz}\bar{a}_z \cdot r_{sz}\bar{a}_z \, d\Omega \tag{G.65c}$$

For the Lagrange/curl-conforming elements, the following approximation holds:

$$\upsilon_{zn} = \sum_{i=1}^{n_L} N_i d_i \tag{G.66a}$$

$$\bar{\upsilon}_{tn} = \sum_{i=n_L+1}^{n} \bar{N}_i d_i \tag{G.66b}$$

where N_i is the ith Lagrange basis function, n_L is the number of Lagrange degrees of freedom, $\bar{N}_i$ is the ith vector basis functions of curl-conforming element type, n_e-n_L is the number of degrees of freedom of that type, d_i is the ith local degree of freedom, and n_e is the total number of degrees of freedom. Hence, $\bar{r}_{st}$ is given by

$$\begin{aligned} \bar{r}_{st} = & \sum_{i=1}^{n_L} \left(\beta_a \bar{a}_z \times \underline{\underline{f}}_{rt}^{-1} \bar{\nabla}_t \times N_i \bar{a}_z \right) d_i \\ & + \sum_{i=n_L+1}^{n_e} \left(\bar{\nabla}_t \times f_{rzz}^{-1} \bar{\nabla}_t \times \bar{N}_i - \beta_a^2 \bar{a}_z \times \underline{\underline{f}}_{rt}^{-1} \bar{a}_z \times \bar{N}_i - \tilde{s}^2 \underline{\underline{p}}_{rt} \bar{N}_i \right) d_i \end{aligned} \tag{G.67a}$$

which can be written in a general way as

$$\bar{r}_{st} = \{\bar{r}_{ste}\}^T \{d^e\} \tag{G.67b}$$

where $\{\bar{r}_{ste}\}$ is a column matrix whose coefficients are the vectors between parenthesis in (G.67a). Similarly,

$$r_{sz}\bar{a}_z = \sum_{i=1}^{n_L} \left(\bar{\nabla}_t \times \underline{\underline{f}}_{rt}^{-1} \bar{\nabla}_t \times N_i \bar{a}_z - \tilde{s}^2 p_{rzz} N_i \bar{a}_z \right) d_i - \sum_{i=n_L+1}^{n_e} \left(\beta \bar{\nabla}_t \times \underline{\underline{f}}_{rt}^{-1} \bar{a}_z \times \bar{N}_i \right) d_i \tag{G.68a}$$

that may be rewritten in an analogous fashion as

$$r_{sz}\bar{a}_z = \{\bar{r}_{sze}\}^T \{d^e\} \tag{G.68b}$$

where $\{\bar{r}_{ste}\}$ is a column vector whose coefficients are the vectors between parenthesis in (G.68a). In conclusion (G.65b), may be written as

$$I_{t\Omega^e} = \{d^e\}^T \left[r_{st}\right] \{d^e\} \tag{G.69a}$$

with

$$\left[r_{st}\right] = \int_{\hat{\Omega}} \{\check{r}_{ste}\} \{\check{r}_{ste}\}^T |J| \; \mathrm{d}\Omega \tag{G.69b}$$

where $\check{r}$ stands for vector $\bar{r}$ transformed to the parent element. Equation (G.65c) may also be written as

$$I_{z\Omega^e} = \{d^e\}^T \left[r_{sz}\right] \{d^e\} \tag{G.70a}$$

with

$$\left[r_{sz}\right] = \int_{\hat{\Omega}} \{\check{r}_{sze}\} \{\check{r}_{sze}\}^T \; |J| \; \mathrm{d}\Omega \tag{G.70b}$$

where it has been assumed that coefficients of $\{\check{r}_{ste}\}^{\mathrm{T}}$ and $\{\check{r}_{sze}\}^{\mathrm{T}}$ are calculated in the reference element.

Regarding $I_{\Gamma k}$, one can write

$$\bar{\xi} = \bar{a}_n \times \left(j\bar{u}_{tn} + u_{zn}\bar{a}_z \right) e^{-j\beta_a z} \quad , \quad \Gamma_k \subset \Gamma_n \tag{G.71a}$$

where

$$\bar{u}_{tn} = \underline{\underline{f}}_{rt}^{-1} \bar{\nabla}_t \times \bar{\upsilon}_{zn} \bar{a}_z - \beta_a \underline{\underline{f}}_{rt}^{-1} \bar{a}_z \times \bar{\upsilon}_{tn} \tag{G.71b}$$

and

$$u_{zn}\bar{a}_z = f_{rzz}^{-1} \bar{\nabla}_t \times \bar{\upsilon}_{tn} \tag{G.71c}$$

It is seen that both $\bar{u}_{tn}$ and $u_{zn}\bar{a}_z$ are proportional to the approximate values of the transverse component $\underline{U}_{tn}$ and the longitudinal component $U_{zn}\bar{a}_z$ of the dual variable, respectively (see (F.58)). Therefore, the components of the vector ξ are proportional to the values of the tangential components at the edge Γ_k of the dual variable $\underline{U}_\tau$.

For the internal edges one obtains

$$\bar{\rho} = [\![\bar{a}_n \times (j\bar{u}_{tn} + u_{zn}\bar{a}_z) \, e^{-j\beta_o z}]\!]_k \quad , \text{ for } \Gamma_k \not\subset \Gamma \tag{G.72}$$

from which it is deduced that the components of the vector $\bar{\rho}$ provide a measure of the discontinuity in the tangential components of the dual variable across the kth internal edge. Thus one obtains

$$I_{\Gamma_k} = I_{\Gamma_{kN}} = \int_{\Gamma_k} \bar{\xi}^* \cdot \bar{\xi} \, d\Gamma = \int_{\Gamma_k} (\bar{a}_n \times u_{zn}\bar{a}_z) \cdot (\bar{a}_n \times u_{zn}\bar{a}_z) \, d\Gamma + \int_{\Gamma_k} (\bar{a}_n \times \bar{u}_{tn}) \cdot (\bar{a}_n \times \bar{u}_{tn}) \, d\Gamma \quad , \text{ for } \Gamma_k \subset \Gamma_N \tag{G.73a}$$

$$I_{\Gamma_k} = I_{\Gamma_{ki}} = \int_{\Gamma_k} \bar{\rho}^* \cdot \bar{\rho} \, d\Gamma = \int_{\Gamma_k} [\![\bar{a}_n \times \tilde{u}_{zn}\bar{a}_z]\!]_k \cdot [\![\bar{a}_n \times \tilde{u}_{zn}\bar{a}_z]\!]_k \, d\Gamma + \int_{\Gamma_k} [\![\bar{a}_n \times \bar{u}_{tn}]\!]_k \cdot [\![\bar{a}_n \times \bar{u}_{tn}]\!]_k \, d\Gamma \, , \text{ for } \Gamma_k \not\subset \Gamma \tag{G.73b}$$

$$I_{\Gamma_k} = I_{\Gamma_{kD}} = 0, \quad \textit{for} \quad \Gamma_k \subset \Gamma_D \tag{G.73c}$$

Taking into account (G.66), before converting into the reference element and using numerical integration, the previous expressions can be written as

$$I_{\Gamma_{kN}} = \sum_{i=1}^{n_{p\tau}} \{\breve{r}_{ed\tau k}(\hat{x}_{pi} \, , \, \hat{y}_{pi})\}^T \{d^e\} \cdot \{\breve{r}_{ed\tau k}(\hat{x}_{pi} \, , \, \hat{y}_{pi})\}^T \{d^e\} w_\tau(i) + \sum_{i=1}^{n_{pz}} \{\breve{r}_{edzk}(\hat{x}_{pi} \, , \, \hat{y}_{pi})\}^T \{d^e\} \cdot \{\breve{r}_{edzk}(\hat{x}_{pi} \, , \, \hat{y}_{pi}\}^T \{d^e\} w_z(i) \tag{G.74a}$$

$$\begin{aligned} I_{\Gamma_{ks}} = & \sum_{i=1}^{n_{p\tau}} \left[\{\breve{r}_{ed\tau k}(\hat{x}_{pi} \, , \, \hat{y}_{pi})\}^T \{d^e\} + \{\breve{r}_{ed\tau j}(\hat{x}_{pi} \, , \, \hat{y}_{pi})\}^T \{d^m\} \right] \\ & \cdot \left[\{\breve{r}_{ed\tau k}(\hat{x}_{pi} \, , \, \hat{y}_{pi})\}^T \{d^e\} + \breve{r}_{ed\tau j}(\hat{x}_{pi} \, , \, \hat{y}_{pi})\}^T \{d^m\} \right] w_\tau(i) \\ & + \sum_{i=1}^{n_{pz}} \left[\{\breve{r}_{edzk}(\hat{x}_{pi} \, , \, \hat{y}_{pi})\}^T \{d^e\} + \{\breve{r}_{edzj}(\hat{x}_{pi} \, , \, \hat{y}_{pi})\}^T \{d^m\} \right] \\ & \cdot \left[\{\breve{r}_{edzk}(\hat{x}_{pi} \, , \, \hat{y}_{pi})\}^T \{d^e\} + \{\breve{r}_{edzj}(\hat{x}_{pi} \, , \, \hat{y}_{pi})\}^T \{d^m\} \right] w_z(i) \end{aligned} \tag{G.74b}$$

where

$$\{\bar{r}_{ed\tau k}\}^T\{d^e\} = \left(\bar{a}_{nk}\times f_{rzz}^{-1}\bar{\nabla}_t\times\sum_{i=n_L+1}^{n_e}\bar{N}_i d_i^e\right)J_{\Gamma k}^{1/2} \tag{G.74c}$$

$$\{\bar{r}_{edzk}\}^T\{d^e\} = \left(\bar{a}_{nk}\times\underline{\underline{f}}_{rt}^{-1}\bar{\nabla}_t\times\sum_{i=1}^{n_L}N_i\bar{a}_z d_i^e - \beta_a\bar{a}_{nk}\times\underline{\underline{f}}_{rt}^{-1}\bar{a}_z\times\sum_{i=n_L+1}^{n_e}\bar{N}_i d_i^e\right)J_{\Gamma k}^{1/2} \tag{G.74d}$$

where the vector in the parentheses must be transformed to the parent element. To calculate (G.74a,b) expressions (G.74c,d) must be evaluated at each point of integration corresponding to the kth side of element Ω^e, whose vector of local degrees of freedom is $\{d^e\}$. To calculate (G.74b) it must be taken into account that the kth edge of element Ω^e is the jth edge of element Ω^m of local degrees of freedom $\{d^m\}$. The distinction between the weights w_τ and w_z reflects the fact that the different expressions between brackets may require a different number of points of integration, $n_{p\tau}$ and n_{pz}. From (G.74c,d) the following expressions are deduced:

$$\{\breve{r}_{ed\tau k}\} = \begin{Bmatrix} \{0\} \\ \{J_{\Gamma k}^{1/2}\,\bar{a}_{nk}\times f_{rzz}^{-1}\,\bar{\nabla}_t\times\bar{N}\} \end{Bmatrix} \tag{G.74e}$$

whose coefficients are vectors along the kth edge direction and

$$\{\breve{r}_{edzk}\} = \begin{Bmatrix} \{J_{\Gamma k}^{1/2}\,\bar{a}_{nk}\times\underline{\underline{f}}_{rt}^{-1}\bar{\nabla}_t\times N\,\bar{a}_z\} \\ \{-\beta_a J_{\Gamma k}^{1/2}\,\bar{a}_{nk}\times\underline{\underline{f}}_{rt}^{-1}\bar{a}_z\times\bar{N}\} \end{Bmatrix} \tag{G.74f}$$

whose coefficients are vectors in the longitudinal direction.

Expressions (G.69), (G.70), and (G.74) are valid for any curl-conforming Lagrange element, of any order, curved or straight. The self-adaptive mesh procedure has been developed for the straight first-order element. Thus these expressions are specialized for this case.

The coefficients of matrices $[r_{st}]$ in (G.69b) and $[r_{sz}]$ in (G.70b) can be calculated through analytical integration in the reference element, taking into account that, in this case, the coefficients of the Jacobian matrix and its determinant are polynomials of zero order, i.e., constants in each element Ω^e.

Matrix $[r_{st}]$ is a 6×6 symmetric matrix and has the form

$$\left[r_{st}\right] = \begin{bmatrix} [r_{stA}] & [r_{stB}] \\ [r_{stB}]^T & [r_{stC}] \end{bmatrix} \tag{G.75a}$$

where each submatrix is a 3×3 matrix whose coefficients are given by

$$\begin{aligned} r_{stAi,j} = \int_{\Omega^e} & \left(\beta_a \bar{a}_z \times \underline{\underline{f}}_{rt}^{-1} \bar{\nabla}_t \times N_i \bar{a}_z \right) \\ & \cdot \left(\beta_a \bar{a}_z \times f_{rt}^{-1} \bar{\nabla}_t \times N_j \bar{a}_z \right) \, d\Omega \qquad \text{with} \quad i = 1,2,3 \ ; \quad j = 1,2,3 \end{aligned} \tag{G.75b}$$

$$\begin{aligned} r_{stBi,j} = \int_{\Omega^e} & \left(\beta_a \bar{a}_z \times \underline{\underline{f}}_{rt}^{-1} \bar{\nabla}_t \times N_i \bar{a}_z \right) \cdot \left(\nabla_t \times f_{rzz}^{-1} \bar{\nabla}_t \times \bar{N}_j \right. \\ & \left. - \beta_a^2 \bar{a}_z \times \underline{\underline{f}}_{rt}^{-1} \bar{a}_z \times \bar{N}_j - \tilde{s}^2 \underline{\underline{p}}_{rt} \bar{N}_j \right) \, d\Omega, \quad \text{with} \quad i = 1,2,3 \ ; \ j = 1,2,3 \end{aligned} \tag{G.75c}$$

$$\begin{aligned} r_{stCi,j} = \int_{\Omega^e} & \left(\bar{\nabla}_t \times f_{rzz}^{-1} \bar{\nabla}_t \times \bar{N}_i - \beta_a^2 \bar{a}_z \times \underline{\underline{f}}_{rt}^{-1} \bar{a}_z \times \bar{N}_i - \tilde{s}^2 \underline{\underline{p}}_{rt} \bar{N}_i \right) \\ & \cdot \left(\bar{\nabla}_t \times f_{rzz}^{-1} \bar{\nabla}_t \times \bar{N}_j - \beta_a^2 \bar{a}_z \times \underline{\underline{f}}_{rt}^{-1} \bar{a}_z \times \bar{N}_j - \tilde{s}^2 \underline{\underline{p}}_{rt} \bar{N}_j \right) d\Omega, \\ & \text{for} \quad i = 4,5,6 \ ; \quad j = 4,5,6 \end{aligned} \tag{G.75d}$$

Transforming to the parent element,

$$\begin{aligned} [r_{stA}] = \beta_a^2 \int_{\hat{\Omega}} & [\hat{\nabla}_t \hat{N}]^T \, [J]^{-1} \begin{bmatrix} 0 & -1 \\ 1 & 0 \end{bmatrix} [f_{rt}]^{-1} \\ & \cdot [f_{rt}]^{-1} \begin{bmatrix} 0 & 1 \\ -1 & 0 \end{bmatrix} \left([J]^T\right)^{-1} [\hat{\nabla}_t \hat{N}] \ |J| \ d\Omega \end{aligned} \tag{G.76a}$$

where the symmetry of $[f_{rt}]$ has been taken into account as well as the equality

$$\begin{bmatrix} 0 & 1 \\ -1 & 0 \end{bmatrix} \begin{bmatrix} 0 & -1 \\ 1 & 0 \end{bmatrix} = \begin{bmatrix} 1 & 0 \\ 0 & 1 \end{bmatrix} \tag{G.76b}$$

Values for $[\hat{\nabla}_t \hat{N}]$, $[J]$, and $|J|$ can be found in Tables G.1 or G.2. $[A]$ is the symmetric 2×2 matrix

$$[A] = [J]^{-1}\begin{bmatrix}0 & -1\\1 & 0\end{bmatrix}[f_{rt}]^{-1}[f_{rt}]^{-1}\begin{bmatrix}0 & 1\\-1 & 0\end{bmatrix}\left([J]^{T}\right)^{-1} \tag{G.76c}$$

one obtains

$$[r_{stA}] = \frac{\beta_a^2}{2}\,|J|\begin{bmatrix}A_{11}+2A_{12}+A_{22} & -(A_{11}+A_{12}) & -(A_{12}+A_{22})\\-(A_{11}+A_{12}) & A_{11} & A_{12}\\-(A_{12}+A_{22}) & A_{12} & A_{22}\end{bmatrix} \tag{G.76d}$$

With respect to $[r_{stB}]$, given that the first and the second factors of the integrand from (G.75c) vanish since it involves second derivatives of first-order polynomials, one obtains

$$\begin{aligned}[r_{stB}] = &- \beta_a^3\int_{\hat{\Omega}}[\hat{\nabla}_t\hat{N}]^T[J]^{-1}\begin{bmatrix}0 & -1\\1 & 0\end{bmatrix}[f_{rt}]^{-1}[f_{rt}]^{-1}\begin{bmatrix}0 & -1\\1 & 0\end{bmatrix}\left([J]^{T}\right)^{-1}[\breve{N}]\;|J|\;d\Omega\\&- \beta_a\tilde{s}^2\int_{\hat{\Omega}}[\hat{\nabla}_t\hat{N}]^T[J]^{-1}\begin{bmatrix}0 & -1\\1 & 0\end{bmatrix}[f_{rt}]^{-1}\begin{bmatrix}0 & 1\\-1 & 0\end{bmatrix}[p_{rt}]\left([J]^{T}\right)^{-1}[\breve{N}]\;|J|\;d\Omega\end{aligned} \tag{G.77a}$$

Taking into account that

$$-[A] = [J]^{-1}\begin{bmatrix}0 & -1\\1 & 0\end{bmatrix}[f_{rt}]^{-1}[f_{rt}]^{-1}\begin{bmatrix}0 & -1\\1 & 0\end{bmatrix}\left([J]^{T}\right)^{-1} \tag{G.77b}$$

and defining $[B]$ as the 2x2 matrix

$$[B] = \begin{bmatrix}B_{11} & B_{12}\\B_{21} & B_{22}\end{bmatrix} = [J]^{-1}\begin{bmatrix}0 & -1\\1 & 0\end{bmatrix}[f_{rt}]^{-1}\begin{bmatrix}0 & 1\\-1 & 0\end{bmatrix}[p_{rt}]\left([J]^{T}\right)^{-1} \tag{G.77c}$$

yields

$$[r_{stB}] = \beta_a^3\int_{\hat{\Omega}}[\hat{\nabla}_t\hat{N}]^T[A][\breve{N}]\,|J|\;d\Omega - \beta_a\tilde{s}^2\int_{\hat{\Omega}}[\hat{\nabla}_t\hat{N}]^T[B][\breve{N}]\,|J|\;d\Omega \tag{G.77d}$$

where $[\breve{N}]$ is the 2×3 matrix whose coefficients are the basis functions of the first-order triangular curl-conforming parent element given by Table 2.7 of Chapter 2. The analytical integration of the coefficients from (G.77d) results in

$$[r_{stB}] = \frac{\beta_a^3}{6}|J|\begin{bmatrix} -2A_{11}-3A_{12}-A_{22} & A_{11}-A_{22} & A_{11}+3A_{12}+2A_{22} \\ 2A_{11}+A_{12} & -A_{11}+A_{12} & -A_{11}-2A_{12} \\ 2A_{12}+A_{22} & -A_{12}+A_{22} & -A_{12}-2A_{22} \end{bmatrix} \tag{G.77e}$$

$$+ \quad \tilde{s}^2\frac{\beta_a}{6}|J|\begin{bmatrix} 2B_{11}+B_{12}+2B_{21}+B_{22} & -B_{11}+B_{12}-B_{21}+B_{22} & -(B_{11}+2B_{12}+B_{21}+2B_{22}) \\ -(2B_{11}+B_{12}) & B_{11}-B_{12} & B_{11}+2B_{12} \\ -(2B_{21}+B_{22}) & B_{21}-B_{22} & B_{21}+2B_{22} \end{bmatrix}$$

Going to the computation of $[r_{stC}]$, it is seen that the first term of each of the two factors involved in the product of (G.75d) is zero. Then one obtains:

$$\begin{aligned}
[r_{stC}] = {} & \beta_a^4 \int_{\hat{\Omega}} [\breve{N}][J]^{-1}\begin{bmatrix} 0 & 1 \\ -1 & 0 \end{bmatrix}[f_{rt}]^{-1}[f_{rt}]^{-1}\begin{bmatrix} 0 & -1 \\ 1 & 0 \end{bmatrix}\left([J]^T\right)^{-1}[\breve{N}]\,|J|\;d\Omega \\
& + \beta_a^2\tilde{s}^2\int_{\hat{\Omega}}[\breve{N}]^T[J]^{-1}\begin{bmatrix} 0 & 1 \\ -1 & 0 \end{bmatrix}[f_{rt}]^{-1}\begin{bmatrix} 0 & 1 \\ -1 & 0 \end{bmatrix}[p_{rt}]\left([J]^T\right)^{-1}[\breve{N}]\,|J|\;d\Omega \\
& + \beta_a^2\tilde{s}^2\int_{\hat{\Omega}}[\breve{N}]^T[J]^{-1}[p_{rt}]\begin{bmatrix} 0 & -1 \\ 1 & 0 \end{bmatrix}[f_{rt}]^{-1}\begin{bmatrix} 0 & -1 \\ 1 & 0 \end{bmatrix}\left([J]^T\right)^{-1}[\breve{N}]\,|J|\;d\Omega \\
& + \tilde{s}^4\int_{\hat{\Omega}}[\breve{N}]^T[J]^{-1}[p_{rt}][p_{rt}]\left([J]^T\right)^{-1}[\breve{N}]\,|J|\;d\Omega
\end{aligned} \tag{G.78a}$$

Given that

$$[J]^{-1}\begin{bmatrix} 0 & 1 \\ -1 & 0 \end{bmatrix}[f_{rt}]^{-1}[f_{rt}]^{-1}\begin{bmatrix} 0 & -1 \\ 1 & 0 \end{bmatrix}\left([J]^T\right)^{-1} = [A] \tag{G.78b}$$

$$[J]^{-1}\begin{bmatrix}0 & 1\\ -1 & 0\end{bmatrix}[f_{rt}]^{-1}\begin{bmatrix}0 & 1\\ -1 & 0\end{bmatrix}[p_{rt}]\left([J]^T\right)^{-1} = -[B] \tag{G.78c}$$

and defining $[C]$ as the 2×2 symmetric matrix

$$[C] = \begin{bmatrix}C_{11} & C_{12}\\ C_{12} & C_{22}\end{bmatrix} = [J]^{-1}\,[p_{rt}]\,[p_{rt}]\left([J]^T\right)^{-1} \tag{G.78d}$$

one obtains:

$$\begin{aligned}[r_{stC}] \;=\;& \beta_a^4 \int_{\hat\Omega} [\breve N]^T\,[A]\,[\breve N]\,|J|\,d\Omega \\ &- \beta_a^2\,\tilde s^2 \int_{\hat\Omega} [\breve N]^T\,([B]+[B]^T)\,[\breve N]\,|J|\,d\Omega \\ &+ \tilde s^4 \int_{\hat\Omega} [\breve N]^T\,[C]\,[\breve N]\,|J|\,d\Omega\end{aligned} \tag{G.78e}$$

yielding

$$\begin{aligned}[r_{stC}] \;=\;& \frac{\beta_a^4}{12}\,|J| \begin{bmatrix} 3A_{11}+3A_{12}+A_{22} & -A_{11}+A_{12}+A_{22} & -A_{11}-3A_{12}-A_{22}\\ -A_{11}+A_{12}+A_{22} & A_{11}-A_{12}+A_{22} & A_{11}+A_{12}-A_{22}\\ -A_{11}-3A_{12}-A_{22} & A_{11}+A_{12}-A_{22} & A_{11}+3A_{12}+3A_{22}\end{bmatrix}\\
&+ \tilde s^2 \frac{\beta_a^2}{12}\,|J| \begin{bmatrix} -6B_{11}-3(B_{12}+B_{21})-2B_{22} & 2B_{11}-B_{12}-B_{21}-2B_{22} & 2B_{11}+3(B_{12}+B_{21})+2B_{22}\\ 2B_{11}-B_{12}-B_{21}-2B_{22} & -2B_{11}+B_{12}+B_{21}-2B_{22} & -2B_{11}-B_{12}-B_{21}+2B_{22}\\ 2B_{11}+3(B_{12}+B_{21})+2B_{22} & -2B_{11}-B_{12}-B_{21}+2B_{22} & -2B_{11}-3(B_{12}+B_{21})-6B_{22}\end{bmatrix}\\
&+ \tilde s^4 \frac{1}{12}\,|J| \begin{bmatrix} 3C_{11}+3C_{12}+C_{22} & -C_{11}+C_{12}+C_{22} & -C_{11}-3C_{12}-C_{22}\\ -C_{11}+C_{12}+C_{22} & C_{11}-C_{12}+C_{22} & C_{11}+C_{12}-C_{22}\\ -C_{11}-3C_{12}-C_{22} & C_{11}+C_{12}-C_{22} & C_{11}+3C_{12}+3C_{22}\end{bmatrix}\end{aligned} \tag{G.78f}$$

In summary

$$[r_{st}] = [r_{st0}] + \tilde{s}^2[r_{st1}] + \tilde{s}^4[r_{st2}] \tag{G.79}$$

where $[r_{st\alpha}]$, for $\alpha = 0,1,2$, are 6×6 symmetric matrices. Hence, once the solution $\tilde{s}$ and the local degrees of freedom vector $\{d^e\}$ corresponding to element Ω^e have been obtained, $I_{t\Omega^e}$ can be obtained by applying (G.69a). In order to carry out this operation from the global degrees of freedom $\{d^{ge}\}$ of element Ω^e, it is necessary to assign the appropriate sign (see Appendix F.3.1.2) to the terms of (G.69a), resulting in

$$I_{t\Omega^e} = \{d^{ge}\}^T\,[r^l_{st}]\,\{d^{ge}\} \tag{G.80a}$$

where

$$[r^l_{st}] = [r_{st0}] + \tilde{s}^2[r_{st1}] + \tilde{s}^4[r_{st2}] \tag{G.80b}$$

where $[r_{st\alpha}]$, for $\alpha = 0, 1, 2$, are 6×6 symmetric matrices whose coefficients can be calculated from Table G.15.

Matrix $[r_{sz}]$ from (G.70b) is a symmetric matrix with dimensions 6×6 as well and is of the form

$$[r_{sz}] = \begin{bmatrix} [r_{szA}] & [r_{szB}] \\ [r_{szB}]^T & [r_{szC}] \end{bmatrix} \tag{G.81a}$$

where the submatrices are matrices of dimensions 3×3, whose coefficients are:

$$\begin{aligned} r_{szAi,j} = \int_{\Omega^e} &\left(\bar{\nabla}_t\times\underline{\underline{f}}_{rt}^{-1}\bar{\nabla}_t\times N_i\bar{a}_z - \tilde{s}^2 p_{rzz}N_i\bar{a}_z\right) \\ &\cdot\left(\bar{\nabla}_t\times\underline{\underline{f}}_{rt}^{-1}\bar{\nabla}_t\times N_j\bar{a}_z - \tilde{s}^2 p_{rzz}N_j\bar{a}_z\right)\,d\Omega \end{aligned} \tag{G.81b}$$

$$r_{szBi,j} = \int_{\Omega^e}\left(\bar{\nabla}_t\times\underline{\underline{f}}_{rt}^{-1}\bar{\nabla}_t\times N_i\bar{a}_z - \tilde{s}^2 p_{rzz}N_i\bar{a}_z\right)\cdot\left(-\beta_a\bar{\nabla}_t\times\underline{\underline{f}}_{rt}^{-1}\bar{a}_z\times\bar{N}_j\right)\,d\Omega \tag{G.81c}$$

$$r_{szCi,j} = \int_{\Omega^e}\left(-\beta_a\bar{\nabla}_t\times\underline{\underline{f}}_{rt}^{-1}\bar{a}_z\times\bar{N}_i\right)\cdot\left(-\beta_a\bar{\nabla}_t\times\underline{\underline{f}}_{rt}^{-1}\bar{a}_z\times\bar{N}_j\right)\,d\Omega \tag{G.81d}$$

When transforming to the master element one obtains

Table G.15

Matrices for the calculation of the surface error estimate due to the transverse residue. The upper triangular part is shown

$$[r_{st0}^{///}]=\begin{bmatrix} 6(A_{11}+2A_{12}+A_{22}) & -6(A_{11}+A_{12}) & -6(A_{12}+A_{22}) & -2\beta_a(2A_{11}+3A_{12}+A_{22}) & 2\beta_a(A_{11}-A_{22}) & 2\beta_a(A_{11}+3A_{12}+2A_{22}) \\ & 6A_{11} & 6A_{12} & 2\beta_a(2A_{11}+A_{12}) & 2\beta_a(-A_{11}+A_{12}) & -2\beta_a(A_{11}+2A_{12}) \\ & & 6A_{22} & 2\beta_a(2A_{12}+A_{22}) & 2\beta_a(-A_{12}+A_{22}) & -2\beta_a(A_{12}+2A_{22}) \\ & & & \beta_a^2(3A_{11}+3A_{12}+A_{22}) & \beta_a^2(-A_{11}+A_{12}+A_{22}) & -\beta_a^2(A_{11}+3A_{12}+A_{22}) \\ & & & & \beta_a^2(A_{11}-A_{12}+A_{22}) & \beta_a^2(A_{11}+A_{12}-A_{22}) \\ & & & & & \beta_a^2(A_{11}+3A_{12}+3A_{22}) \end{bmatrix}$$

$$[r_{st1}^{///}]=\begin{bmatrix} 0 & 0 & 0 & 2(2B_{11}+B_{12}+2B_{21}+B_{22}) & 2(-B_{11}+B_{12}-B_{21}+B_{22}) & -2(B_{11}+2B_{12}+B_{21}+2B_{22}) \\ & 0 & 0 & -2(2B_{11}+B_{12}) & 2(B_{11}-B_{12}) & 2(B_{11}+2B_{12}) \\ & & 0 & -2(2B_{21}+B_{22}) & 2(B_{21}-B_{22}) & 2(B_{21}+2B_{22}) \\ & & & -\beta_a(6B_{11}+3B_{12}+3B_{21}+2B_{22}) & \beta_a(2B_{11}-B_{12}-B_{21}-2B_{22}) & \beta_a(2B_{11}+3B_{12}+3B_{21}+2B_{22}) \\ & & & & \beta_a(-2B_{11}+B_{12}+B_{21}-2B_{22}) & \beta_a(-2B_{11}-B_{12}-B_{21}+2B_{22}) \\ & & & & & -\beta_a(2B_{11}+3B_{12}+3B_{21}+6B_{22}) \end{bmatrix}$$

Table G.15 (continued)

Matrices for the calculation of the surface error estimate due to the transverse residue. The upper triangular part is shown

$$[r''_{st2}] = \begin{bmatrix} 0 & 0 & 0 & 0 & 0 & 0 \\ & 0 & 0 & 0 & 0 & 0 \\ & & 0 & 0 & 0 & 0 \\ & & & 3C_{11}+3C_{12}+C_{22} & -C_{11}+C_{12}+C_{22} & -(C_{11}+3C_{12}+C_{22}) \\ & & & & C_{11}-C_{12}+C_{22} & C_{11}+C_{12}-C_{22} \\ & & & & & C_{11}+3C_{12}+3C_{22} \end{bmatrix}$$

$$[r''_{st0}] = \beta_a^2 [r'''_{st0}]; \quad [r''_{st1}] = \beta_a [r'''_{st1}];$$

$$r_{st\alpha i,j} = \frac{|J|}{12} \, sign(i) \, sign(j) \, r''_{st\alpha i,j}; \qquad \alpha = 0,1,2; \quad j = 1,\ldots,6; \quad i = j,..,6$$

$$[r_{szA}] = \tilde{s}^4 p_{rzz}^2 \int_{\hat{\Omega}} [\hat{N}]^T [\hat{N}] \, |J| \, d\Omega = \tilde{s}^4 p_{rzz}^2 \, |J| \, \frac{1}{24} \begin{bmatrix} 2 & 1 & 1 \\ 1 & 2 & 1 \\ 1 & 1 & 2 \end{bmatrix} \tag{G.82}$$

$$[r_{szB}] = \tilde{s}^2 \beta_a p_{rzz} \int_{\hat{\Omega}} [\hat{N}]^T \begin{Bmatrix} \dfrac{\partial}{\partial \hat{x}} \\ \dfrac{\partial}{\partial \hat{y}} \end{Bmatrix}^T [J]^{-1} \begin{bmatrix} 0 & 1 \\ -1 & 0 \end{bmatrix} [f_{rt}]^{-1} \begin{bmatrix} 0 & -1 \\ 1 & 0 \end{bmatrix} \left([J]^T\right)^{-1} [\breve{N}] \; |J| \; \mathrm{d}\Omega \tag{G.83a}$$

Defining $[T]$ as the 2×2 symmetric matrix

$$[T] = \begin{bmatrix} T_{11} & T_{12} \\ T_{12} & T_{22} \end{bmatrix} = [J]^{-1} \begin{bmatrix} 0 & 1 \\ -1 & 0 \end{bmatrix} [f_{rt}]^{-1} \begin{bmatrix} 0 & -1 \\ 1 & 0 \end{bmatrix} \left([J]^T\right)^{-1} \tag{G.83b}$$

one obtains

$$[r_{szB}] = \tilde{s}^2 \beta_a \, p_{rzz} \int_{\hat{\Omega}} [\hat{N}]^T \begin{Bmatrix} \dfrac{\partial}{\partial \hat{x}} \\ \dfrac{\partial}{\partial \hat{y}} \end{Bmatrix}^T [T] \, [\breve{N}] \, |J| \, \mathrm{d}\Omega$$

$$= \tilde{s}^2 \beta_a p_{rzz} \int_{\hat{\Omega}} [\hat{N}]^T \begin{Bmatrix} 0 \\ 0 \\ 0 \end{Bmatrix} |J| \quad \mathrm{d}\Omega \tag{G.83c}$$

thus

$$[r_{szB}] = [0] \tag{G.83d}$$

and similarly

$$[r_{szC}] = [0] \tag{G.84}$$

Hence, matrix $[r_{sz}]$ has the expression

$$[r_{sz}] = \tilde{s}^4 p_{rzz}^2 \, |J| \, \frac{1}{24} \begin{bmatrix} 2 & 1 & 1 & 0 & 0 & 0 \\ 1 & 2 & 1 & 0 & 0 & 0 \\ 1 & 1 & 2 & 0 & 0 & 0 \\ 0 & 0 & 0 & 0 & 0 & 0 \\ 0 & 0 & 0 & 0 & 0 & 0 \\ 0 & 0 & 0 & 0 & 0 & 0 \end{bmatrix} \tag{G.85}$$

Once the eigenvalue $\tilde{s}^2$ has been obtained and the vector of the degrees of freedom $\{d^{eg}\}$ of element Ω^e has been obtained, the integral $I_{z\Omega^e}$ corresponding to the surface residue due to the longitudinal term can be computed according to

$$I_{z\Omega^e} = \{d^{eg}\}^T \, [r_{sz}^l] \, \{d^{eg}\} \tag{G.86}$$

where $[r_{sz}^l] = [r_{sz}]$ as the sign of the first three degrees of freedom is equal to unity.

Hence, I_{Ω^e} can be written as

$$I_{\Omega^e} = I_{0\Omega^e} + \tilde{s}^2 I_{1\Omega^e} + \tilde{s}^4 I_{2\Omega^e} \tag{G.87a}$$

where

$$I_{0\Omega^e} = \{d^{eg}\}^T \left[r_{st0}\right] \{d^{eg}\} \tag{G.87b}$$

$$I_{1\Omega^e} = \{d^{eg}\}^T \, [r_{st1}] \, \{d^{eg}\} \tag{G.87c}$$

$$I_{2\Omega^e} = \{d^{eg}\}^T \left([r_{st2}] + [r_{sz}]\right) \{d^{eg}\} \tag{G.87d}$$

Now the computation of I_{Γ_k} for the first-order straight curl-conforming/Lagrange triangular element is carried out. Expressions of J_{Γ_k} and $\bar{a}_{nk}$ are given in Table G.3. Also

$$\bar{\nabla}_t \times \bar{N}_i = \frac{1}{|J|} \breve{\nabla}_t \times \breve{N}_i = \frac{2}{|J|} \bar{a}_z, \qquad i = 4,5,6 \tag{G.88a}$$

Hence, for (G.74e) one obtains

$$\{\bar{r}_{ed\tau k}\}^T = 2 L_k^{1/2} f_{rzz}^{-1} \; |J|^{-1} \bar{a}_{nk} \times \bar{a}_z \;\; [0,\ 0,\ 0,\ 1,\ 1,\ 1] \tag{G.88b}$$

Given that

$$\bar{a}_{nk} \times \bar{a}_z = -\bar{a}_{\tau k} \tag{G.88c}$$

it results in

$$\{\bar{r}_{ed\tau k}\}^T = -2 L_k^{1/2} f_{rzz}^{-1} \; |J|^{-1} \bar{a}_{\tau k} \, [0,\ 0,\ 0,\ 1,\ 1,\ 1] \tag{G.88d}$$

where $\bar{a}_{\tau k}$ is the unity vector tangential to the kth edge in the trigonometric sense.

It can be observed that the coefficients of $\{\bar{r}_{ed\tau k}\}^T$ are constant for each side of the element. Thus, to evaluate the first part of expressions (G.74a,b) it is enough to use one integration point (n_p=1) with a weight, w_τ, equal to unity.

It is convenient to write expression (G.74c) as a function of the global degrees of freedom according to

$$\{\bar{r}_{ed\tau k}\}^T \{d^{\,e}\} = \{\bar{r}^{\,l}_{ed\tau k}\}^T \{d^{\,eg}\} \tag{G.89a}$$

where

$$\{\bar{r}^{\,l}_{ed\tau k}\} = \{\bar{r}_{ed\tau k}\}^T \, [msign] \tag{G.89b}$$

where the 6×6 matrix [*msign*] is diagonal. Its nonzero coefficients are equal to the variable *sign* corresponding to each local degree of freedom (see Appendix C.4.3).

It can be observed that in order to evaluate the first part of expression (G.74a) it is not necessary to store the components of the unity vector tangential to the kth edge $\bar{a}_{\tau k}$, since it involves a scalar product of a vector in that direction by itself.

The equality

$$\bar{a}_{\tau k} = -\bar{a}_{\tau j} \tag{G.90}$$

relates the unit vectors tangential to the interface common between element Ω^e and Ω^m depending on whether it is calculated in one or other element. Hence, again it is not necessary to store the components of these vectors for the evaluation of the first part of (G.74b). It is enough with the allocation of the sign depending on whether the calculations are for Ω^e or for Ω^m. Therefore, one needs to store the kth edge of element Ω^e in a matrix of dimensions 1×6, whose coefficients are scalars and given by

$$\{r^l_{ed\tau k}\}^T = 2\,signe(k)\,L_k^{1/2} f_{rzz}^{-1}\,|J|^{-1}\,[0,\ 0,\ 0,\ 1,\ 1,\ 1\,]\,[msign] \tag{G.91a}$$

where $signe(k)$ is a variable of value +1 or -1 according to the value of the variable *sign* corresponding to the kth edge of element Ω^e, for k = 1, 2, 3. Since this variable coincides with the value of the variable *sign* corresponding to the (3+k)th degrees of freedom, it can be written as

$$\{r^l_{ed\tau k}\}^T = 2\,sign(3+k)\,L_k^{1/2} f_{rzz}^{-1}\,|J|^{-1}\,[0,\ 0,\ 0,\ 1,\ 1,\ 1]\ [msign] \tag{G.91b}$$

The terms of the first parts of (G.74a,b) can therefore be expressed as

$$\{\check{r}_{ed\tau k}\}^T\{d^{\,e}\}\,\{\check{r}_{ed\tau k}\}^T\{d^{\,e}\} = \left(\{r^l_{ed\tau k}\}^T\{d^{\,eg}\}\right)^2 \tag{G.92a}$$

and

$$\left(\{\check{r}_{ed\tau k}\}^T\{d^{\,e}\} + \{\check{r}_{ed\tau j}\}^T\{d^{\,m}\}\right)\left(\{\check{r}_{ed\tau k}\}^T\{d^{\,e}\} + \{\check{r}_{ed\tau j}\}^T\{d^{\,m}\}\right)$$
$$= \left(\{r^l_{ed\tau k}\}^T\{d^{\,eg}\} + \{r^l_{ed\tau j}\}^T\{d^{\,mg}\}\right)^2 \tag{G.92b}$$

With respect to (G.74f), the ith vector coefficient of the column vector $\{r_{edzk}\}$ turns out to be

$$\check{r}_{edzki} = \bar{a}_z\,L_k^{1/2}\{a_{nk}\}^T \begin{bmatrix} 0 & 1 \\ -1 & 0 \end{bmatrix} [f_{rt}]^{-1} \begin{bmatrix} 0 & 1 \\ -1 & 0 \end{bmatrix} \left([J]^T\right)^{-1}\{\hat{\nabla}_t \hat{N}_i\}, \qquad i=1,2,3 \tag{G.93a}$$

$$\check{r}_{edzki} = -\,\bar{a}_z \beta_a\,L_k^{1/2}\{a_{nk}\} \begin{bmatrix} 0 & 1 \\ -1 & 0 \end{bmatrix} [f_{rt}]^{-1} \begin{bmatrix} 0 & -1 \\ 1 & 0 \end{bmatrix} \left([J]^T\right)^{-1}\{\check{N}_i\}, \qquad i=4,5,6 \tag{G.93b}$$

Let $[D]$ be the 2×2 matrix

$$[D] = \begin{bmatrix} D_{11} & D_{12} \\ D_{21} & D_{22} \end{bmatrix} = \begin{bmatrix} 0 & 1 \\ -1 & 0 \end{bmatrix} [f_{rt}]^{-1} \begin{bmatrix} 0 & 1 \\ -1 & 0 \end{bmatrix} \left([J]^T\right)^{-1} \tag{G.93c}$$

and $[E]$ be the matrix of dimensions 2×6

$$[E] = \left[[\hat{\nabla}_t \hat{N}] \begin{matrix} 0 & 0 & 0 \\ 0 & 0 & 0 \end{matrix} \right] + \beta_a \left[\begin{matrix} 0 & 0 & 0 \\ 0 & 0 & 0 \end{matrix} \quad [\hat{N}] \right]$$

$$= \begin{bmatrix} -1 & 1 & 0 & -\beta_a(1-\hat{y}) & -\beta_a \hat{y} & -\beta_a \hat{y} \\ -1 & 0 & 1 & \beta_a \hat{x} & \beta_a \hat{x} & \beta_a(\hat{x}-1) \end{bmatrix} \tag{G.93d}$$

then

$$\{\check{r}_{edzk}\}^T = L_k^{1/2} \{a_{nk}\}^T [D]\,[E]\,\bar{a}_z \tag{G.93e}$$

Expression (G.74d) must be written as a function of the global degrees of freedom according to

$$\{\bar{r}_{edzk}\}^T \{d^{\,e}\} = \{\bar{r}^{\,l}_{edzk}\}^T \{d^{\,eg}\} \tag{G.94a}$$

where

$$\{\bar{r}^{\,l}_{edzk}\}^T = \{\bar{r}_{edzk}\}^T [msign] \tag{G.94b}$$

Vector $\bar{a}_z$ in (G.93e) can be neglected. Thus, it is enough to store a matrix of dimension 1×6 for each side, whose coefficients are scalars according to

$$\{r^{\,l}_{edzk}\}^T = L_k^{1/2} \{a_{nk}\}^T [D]\,[E]\,[msign] \tag{G.95a}$$

As it can be observed, some of the coefficients of matrix $[E]$ are polynomials of first order; so to evaluate the second part of expressions (G.74a,b), two integration points ($n_{pz} = 2$) on each side of the triangle can be used with weighting, w_z, equal to 1/2. Since on the kth edge of the element Ω^e the rest of the variables are constant,

$$\{ r_{edzk}^{l}(\hat{x}_{pi}, \hat{y}_{pi}) \}^T = L_k^{1/2} \{ a_{nk} \}^T [D] [E(\hat{x}_{pi}, \hat{y}_{pi})] [msign] \tag{G.95b}$$

where $(\hat{x}_{pi}, \hat{y}_{pi})$ are the coordinates of the ith integration point on the kth side, according to Table G.6.

Summarizing, the calculation of $I_{\Gamma k}$ can be done according to the following expressions:

$$I_{\Gamma k} = I_{\Gamma kN} = \left(\{r_{ed\tau k}^{l}\}^T \{d^{eg}\} \right)^2 + \frac{1}{2} \sum_{i=1}^{2} \left(\{r_{edzk}^{l}(\hat{x}_{pi}, \hat{y}_{pi})\}^T \{d^{eg}\} \right)^2, \quad \text{for} \quad \Gamma_k \subset \Gamma_N \tag{G.96a}$$

$$\begin{aligned} I_{\Gamma k} = I_{\Gamma ki} &= \left(\{r_{ed\tau k}^{l}\}^T \{d^{eg}\} + \{r_{ed\tau j}^{l}\}^T \{d^{mg}\} \right)^2 \\ &+ \frac{1}{2} \left(\{r_{edzk}^{l}(\hat{x}_{p1}, \hat{y}_{p1})\}^T \{d^{eg}\} + \{r_{edzj}^{l}(\hat{x}_{p2}, \hat{y}_{p2})\}^T \{d^{mg}\} \right)^2 \\ &+ \frac{1}{2} \left(\{r_{edzk}^{l}(\hat{x}_{p2}, \hat{y}_{p2})\}^T \{d^{eg}\} + \{r_{edzj}^{l}(\hat{x}_{p1}, \hat{y}_{p1})\}^T \{d^{mg}\} \right)^2, \quad \text{for } \Gamma_k \not\subset \Gamma \end{aligned} \tag{G.96b}$$

$$I_{\Gamma_k} = I_{\Gamma_{kD}} = 0, \qquad \Gamma_k \subset \Gamma_D \tag{G.96c}$$

where $\{r_{ed\tau j}^{l}\}^T$ and $\{r_{edzj}^{l}\}^T$ refer to the expressions (G.91b) and (G.95b) evaluated for the jth edge of element Ω^m that coincides with the kth edge of element Ω^e. $\{d^{mg}\}$ is the vector of global degrees of freedom of element Ω^m.

Once the integrals $I_{\Omega e}$ and $I_{\Gamma k}$ have been evaluated, the calculation of the error indicator for element Ω^e can be performed according to (G.63b) by taking into account that $h_e = \max(L_1, L_2, L_3)$ and $p = 1$.

Bibliography

BOOKS[1]

Alder, H., J. C. Heinrich, S. Lavanchy, E. Oñate, and B. Suárez (eds.) (1992), *Numerical Methods in Engineering and Applied Sciences*, CIMNE, Barcelona.

Akin, J. E. (1982), *Application and Implementation of Finite Element Methods*, Academic Press, London.

Akin, J. E. (1994), *Finite Elements for Analysis and Design*, Academic Press, London.

Ames, W. F. (1969, 1977, 1992), *Numerical Methods for Partial Differential Equations*, Academic Press, San Diego, CA.

Argyris, J. H., and S. Kelsey (1960), *Energy Theorems and Structural Analysis*, Butterworth Co., London.

Arthurs, A. M. (1980), *Complementary Variational Principles*, Clarendon Press, Oxford.

Aziz, A. K. (ed.) (1972, 1980), *The Mathematical Foundations of the Finite Element Method with Applications to Partial Differential Equations*, Academic Press, New York.

Babuška, I., O. C. Zienkiewicz, J. Gago, and E. R. de A. Oliveira (eds.) (1986), *Accuracy Estimates and Adaptive Refinements in Finite Element Computations*, John Wiley & Sons, London.

Bahl, I. J., and P. Bhartia (1988), *Microwave Solid State Circuit Design*, John Wiley & Sons, New York.

Bathe, K. J., and E. L. Wilson (1976), *Numerical Methods in Finite Element Analysis*, Prentice-Hall, Englewood Cliffs, NJ.

Bathe, K. J., J. T. Oden, and W. Wunderlich (eds.) (1977), *Formulation and Computational Algorithms in Finite Element Analysis*, MIT Press, Cambridge, MA.

Bathe, K. J. (1982), *Finite Element Procedures in Engineering Analysis*, Prentice-Hall, Englewood Cliffs, NJ.

Bernadou, M., P. L. George, A. Hassim, P. Joly, P. Lang, A. Perronet, E. Saltel, D. Steer, G. Vanderborck, and M. Vidrascu (1985), *MODULEF. Une Bibliothéque Modulaire d'Éléments Finis*, INRIA, Paris.

Binns, K. J., P. J. Lawrenson and C. W. Trowbridge (1992), *The Analytical and Numerical Solution of Electric and Magnetic Fields*, John Wiley & Sons, Chichester, England.

Blanc-Castillo, F. (1994), *Elementos de Arista en el Método de los Elementos Finitos para el Análisis de Estructuras Electromagnéticas en 2D,* Proyecto Fin de Carrera, E. T. S. I. de Telecomunicación, Universidad Politécnica de Madrid.

Booton, R. C. (1992), *Computational Methods for Electromagnetics and Microwaves*, John Wiley & Sons, New York.

Brebbia, C. A., and H. Tottenham (eds.) (1973), *Variational Methods in Engineering*, Southampton University.

Brebbia, C. A. (1978), *The Boundary Element Method for Engineers*, Pentech Press, London.

Brebbia, C. A., and M. H. Aliabadi (eds.) (1993), *Adaptive Finite and Boundary Element Methods*,

1 Together with books on other topics cited throughout the present book, a number of books on general theory and applications in several fields of engineering of the finite element method (FEM) are listed. The reader is referred to http://ohio.ikp.liu.se/fe/index.html where an almost complete and updated list of FEM books and FEM conferences proceedings is available as well as their abstracts.

Computational Mechanics Publications, Southampton, England.

Burnett, D. S. (1987), *Finite Element Analysis*, Addison-Wesley, Reading.

Calamia, M., G. Pelosi (eds.) (1992), *Il Metodo degli Elementi Finiti (FEM) nelle Applicazioni dell'Elettromagnetismo*, Giornate di Studio, Università degli Studi di Firenze, Italy.

Calamia, M., G. Pelosi, and P. P. Silvester (eds.) (1994b), *Second Int. Workshop on Finite Element Methods for Electromagnetic Wave Problems*, *COMPEL*, vol. 13, supp. A.

Carey, G. F., and J. T. Oden (1984), *Finite Elements. Computational Aspects. Volume III*, Prentice-Hall, Englewood Cliffs, NJ.

Celia, M. A., and W. G. Gray (1992), *Numerical Methods for Differential Equations*, Prentice-Hall, Englewood Cliffs, NJ.

Chari, M. V. K., and P. P. Silvester (eds.) (1980), *Finite Elements in Electrical and Magnetic Field Problems*, John Wiley & Sons, New York.

Cheung, Y. K. (1976), *Finite Strip Method in Structural Analysis*, Pergamon Press, Oxford.

Chung, T. J. (1978), *Finite Element Analysis in Fluid Dynamics*, McGraw-Hill, New York.

Ciarlet, P. G. (1978), *The Finite Element Method for Elliptic Problems*, Elsevier North-Holland, New York.

Collin, R. E. (1960), *Field Theory of Guided Waves*, McGraw-Hill, New York.

Collin, R. E. (1966), *Foundations for Microwave Engineering*, McGraw-Hill, New York.

Connor, J. J., and C. A. Brebbia (1976), *Finite Element Technique for Fluid Flow*, Butterworth Co., London.

Cook, R. D. (1974, 1981), *Concepts and Applications of Finite Element Analysis*, John Wiley & Sons, New York.

Courant, R. L., and D. Hilbert (1989), *Methods of Mathematical Physics*, vols. I, and II, John Wiley & Sons, New York.

Crandall, S. H. (1956), *Engineering Analysis*, McGraw-Hill, New York.

Dahlquist, G., and A. Björck (1974), *Numerical Methods*, Prentice-Hall, Englewood Cliffs, NJ.

Debnath, L., and P. Mikusiński (1990), *Introduction to Hilbert Spaces with Applications*, Academic Press, London.

Desai, C. S., and J. F. Abel (1972), *Introduction to the Finite Element Method: A Numerical Approach for Engineering Analysis*, Van Nostrand Reinhold, New York.

Desai, C. S. (1979), *Elementary Finite Element Method*, Prentice-Hall, Englewood Cliffs, NJ.

Dhatt, G., and G. Touzot (1981), *Une Présentation de la Méthode des Eléments Finis*, Maloine, Paris.

Djordjević, A., R. F. Harrington, T. K. Sarkar, and M. Bazdar (1989), *Matrix Parameters for Multiconductor Transmission Lines: Software and User's Manual*, Artech House, Norwood, MA.

Djordjević, A., M. Bazdar, G. Vitosevic, T. K. Sarkar, and R. F. Harrington (1990), *Scattering Parameters of Microwave Networks with Multiconductor Transmission Lines: Software and Users Manual*, Artech House, Norwood, MA.

Djordjević, A., M. Bazdar, T. K. Sarkar, and R. F. Harrington (1996), *LINPAR for Windows: Matrix Parameters for Multiconductor Transmission Lines*, Artech House, Norwood, MA.

Dold, A., and B. Ectmann,(1977), *Mathematical Aspects of Finite Element Methods*, Lect. Notes on Mathematics, vol. 606, Springer-Verlag, New York.

Duff, I. S., A. Erisman, and J. Reid (1986), *Direct Methods for Sparse Matrices*, Clarendon Press, Oxford.

Dwoyer, D. L., M. Y. Hussain, and R. G. Voigt (eds.) (1988), *Finite Elements Theory and Applications*, Springer-Verlag, New York.

Elliot, R. S. (1993), *An Introduction to Guided Waves and Microwave Circuits*, Prentice-Hall, Englewood Cliffs, NJ.

Feynman, R. P. (1987), *The Feynman Lectures on Physics*, Addison Wesley, Reading, MA.

Finlayson, B. A. (1972), *The Method of Weighted Residuals and Variational Principles*, Academic Press, New York.

Galéev, E., and V. Tijomírov (1989), *Short Course on the Theory of Extreme Problems* (Translated from

Russian), MIR Publishers, Moscow.

Gallagher, R. H. (1975), *Finite Element Analysis Fundamentals*, Prentice-Hall, Englewood Cliffs, NJ.

García-Castillo, L. E. (1992), *Análisis en Onda Completa de Estructuras de Guiado de Microondas mediante el Método de los Elementos Finitos Empleando Elementos de Arista*, Proyecto Fin de Carrera, E. T. S. I. Telecomunicación, Universidad Politécnica de Madrid.

Gardiol, F. E. (1987), *Lossy Transmission Lines*, Artech House, Norwood, MA.

Garrido-Díaz, L. M. (1992), *Análisis Paramétrico de Estructuras de Microondas Mediante el Método de los Elementos Finitos. Aplicación al Diseño de Elementos Pasivos Empleados en Circuitos Monolíticos*, Proyecto Fin de Carrera, E. T. S. I. Telecomunicación, Universidad Politécnica de Madrid.

George, A., and J. Liu (1981), *Computer Solution of Large Sparse Positive Definite Systems*, Prentice-Hall, Englewood Cliffs, NJ.

Ghatak, A. K., I. C. Goyal, and S. J. Chua (1995), *Mathematical Physics. Differential Equations and Transform Theory*, Macmillan India Limited, New Delhi.

Glowinski, R., E. Y. Rodin, and O. C. Zienkiewicz (eds.) (1979), *Energy Methods in Finite Element Analysis*, John Wiley & Sons, New York.

Golub, G. H., and C. F. van Loan (1989), *Matrix Computations*, The Johns Hopkins University Press, Baltimore, MD.

Gupta, K. C., R. Garg, and I. J. Bahl (1979), *Microstrip Lines and Slotlines*, Artech House, Dedham, MA.

Gupta, K. C., R. Garg, and R. Chadha (1981), *Computer-Aided Design of Microwave Circuits*, Artech House, Dedham, MA.

Harrington, R. F. (1961), *Time-Harmonic Electromagnetic Fields*, McGraw-Hill, New York.

Harrington, R. F. (1968), *Field Computation by Moment Method*, Macmillan Company, New York.

Hecht, E., and A. Zajac (1990), *Optics*, Addison Wesley, New York.

Hildebrand, F. B. (1952, 1965), *Methods of Applied Mathematics*, Prentice-Hall, Englewood Cliffs, NJ.

Hildebrand, F. B. (1956), *Introduction to Numerical Analysis*, McGraw-Hill, New York.

Hinton, E., and D. R. J. Owen (1977), *Finite Element Programming*, Academic Press, London.

Hirsch, C., J. Périaux, and E. Oñate (eds.) (1992a), *Computational Methods in Applied Sciences*, Elsevier, Amsterdam.

Hirsch, C., O. C. Zienkiewicz, and E. Oñate (eds.) (1992b), *Numerical Methods in Engineering '92*, Elsevier, Amsterdam.

Hoffmann, R. K. (1987), *Handbook of Microwave Integrated Circuits*, Artech House, Norwood, MA.

Holland, I., and K. Bell (eds.) (1969), *Finite Element Methods in Stress Analysis*, Tapir, Trondheim, Norway.

Hoole, S. R. H. (1989), *Computer Aided Analysis and Design of Electromagnetic Devices*, Elsevier, New York.

Howe, H. (1974), *Stripline Circuit Design*, Artech House, Dedham, MA.

Huebner, K. (1975), *The Finite Element Method for Engineers*, John Wiley & Sons, New York.

Irons, B. M., and S. Ahmad (1980), *Techniques of Finite Elements*, Ellis Horwood, Chichester, England.

Isaacson, E., and H. B. Keller (1966), *Analysis of Numerical Methods*, John Wiley & Sons, New York.

Itoh, T. (ed.) (1989a), *Numerical Techniques for Microwave and Millimeter Wave Passive Structures*, John Wiley & Sons, New York.

Itoh, T., G. Pelosi, and P. P. Silvester (eds.) (1996), *Finite Element Software for Microwave Engineering*, John Wiley & Sons, New York.

Jin, J. (1993), *The Finite Element Method in Electromagnetics*, John Wiley & Sons, New York.

Johnson, C. (1992), *Numerical Solution of Partial Differential Equations by the Finite Element Method*, Cambridge University Press, Cambridge.

Kajfez, D., and P. Guillon (eds.) (1986), *Dielectric Resonators*, Artech House, Dedham, MA.

Kantorovich, L. V., and V. I. Krylov (1964), *Approximate Methods of Higher Analysis*, (Translated from Russian), John Wiley & Sons, New York.

Kardestuncer, H., and D. H. Norrie (eds.) (1987), *Finite Element Handbook*, McGraw-Hill, New York.

Koshiba, M. (1992), *Optical Waveguide Theory by the Finite Element Method*, K. T. K. Scientific Publishers, Tokyo.

Kraus, J. D. (1992), *Electromagnetics*, McGraw-Hill, New York.

Kron, G. (1939), *Tensor Analysis of Networks*, John Wiley & Sons, New York.

Marcuvitz, N. (1951, 1986), *Waveguide Handbook*, Peter Peregrinus, London.

Martin H. C., and G. F. Carey (1973), *Introduction to Finite Element Analysis*, McGraw-Hill, New York.

Mikhlin, S. G. (1964), *Variational Methods in Mathematical Physics*, Pergamon Press, Oxford, NY.

Mikhlin, S. G. (1966), *The Problem of the Minimum of a Quadratic Functional*, Holden Day, San Francisco.

Mikhlin, S. G., and K. I. Smolitsky (1967), *Approximate Methods for Solution of Differential and Integral Equations*, Elsevier, New York.

Mirshekar-Syahkal, D. (1990), *Spectral Domain Method for Microwave Integrated Circuits*, John Wiley & Sons, New York.

Mitchell, A. R., and R. Wait (1977), *The Finite Element Method in Partial Differential Equations*, John Wiley & Sons, New York.

Mittra, R., and S. W. Lee (1971), *Analytical Techniques in the Theory of Guided Waves*, The Macmillan Company, New York.

Morita, N., N. Kumagai, and J. R. Mautz (1990), *Integral Equation Methods for Electromagnetics*, Artech House, Norwood, MA.

Nash, S. G. (ed.) (1990), *A History of Scientific Computing*, ACM, New York.

Navarrina, F., and M. Casteleiro (eds.) (1993), *Métodos Numéricos en Ingeniería*, SEMNI, La Coruña, Spain.

Noble, B., and J. W. Daniel (1988), *Applied Linear Algebra*, Prentice-Hall, Englewood Cliffs, NJ.

Norrie, D. H., and G. de Vries (1973), *The Finite Element Method*, Academic Press, New York.

Norrie, D. H., and G. de Vries (1978), *An Introduction to Finite Element Analysis*, Academic Press, New York.

Oden, J. T. (1972), *Finite Elements of Nonlinear Continua*, McGraw-Hill, New York.

Oden, J. T., and J. N. Reddy (1976), *An Introduction to the Mathematical Theory of Finite Elements*, John Wiley & Sons, New York.

Ottosen, N. S., and H. Peterson (1992), *Introduction to the Finite Element Method*, Prentice-Hall, London.

Paris, D. T., and F. K. Hurd (1969), *Basic Electromagnetic Theory*, McGraw-Hill, New York.

Pearson, C. E. (1974), *Handbook of Applied Mathematics*, Van Nostrand, New York.

Pérez-Yuste, A. (1996), *Método de Elementos Finitos de Arista en 3D para el Análisis Dinámico de Discontinuidades en Estructuras de Guiado de Microondas*, Proyecto Fin de Carrera, E. T. S. I. Telecomunicación, Universidad Politécnica de Madrid.

Pinder, G. F., and W. G. Gray (1977), *Finite Element Simulation in Surface Hydrology*, Academic Press, New York.

Przemieniecki, J. S. (1968), *Theory of Matrix Structural Analysis*, McGraw-Hill, New York.

Proceedings (1965, 1968, 1971), *Proc. 1st, 2nd and 3rd Conferences on Matrix Methods in Structural Mechanics*, Wright-Patterson AFB, Ohio.

Ramo, S., J. R. Whinnery, and T. van Duzer (1965), *Fields and Waves in Communication Electronics*, John Wiley & Sons, New York.

Raviart, P. A., and J. M. Thomas (1983), *Introduction à l'Analyse Numérique des Equations aux Dérivées Partielles*, Masson, Paris.

Recio-Peláez, J. M. (1995), *Paquete Software de Análisis de Estructuras de Guiado de Microondas Mediante el Método de los Elementos Finitos para PC Compatible Orientado a Uso Educativo*, Proyecto Fin de Carrera, E. T. S. I. Telecomunicación, Universidad Politécnica de Madrid.

Reddy, J. N. (1984, 1993), *An Introduction to the Finite Element Method*, McGraw-Hill, New York.

Rektorys, K. (1977), *Variational Methods in Mathematics, Science and Engineering*, D. Reidel, Dordrecht, Boston.

Robinson, J. (1973) (ed.), *Integrated Theory of Finite Element Methods*, John Wiley & Sons, London.

Roy, T. (1996), *A Hybrid Method to Terminate Finite Element Meshes (Frequency and Time Domain Analysis)*, Ph. D. Dissertation, Syracuse University, NY.

Sabonnadière, J. C., and J. L. Coulomb (1987), *Finite Element Methods in CAD*, Springer-Verlag, New York.

Sadiku, M. N. O. (1992), *Numerical Techniques in Electromagnetics*, CRC Press, Boca Raton, FL.

Schelkunoff, S. A. (1945), *Electromagnetic Waves*, D. van Nostrand Co., New York.

Segerlind, L. J. (1976, 1984), *Applied Finite Element Analysis*, John Wiley & Sons, New York.

Silvester, P. P., and R. L. Ferrari (1983), *Finite Elements for Electrical Engineers*, Cambridge University Press, Cambridge.

Silvester, P. P., and G. Pelosi (eds.) (1994), *Finite Elements for Wave Electromagnetics. Methods and Techniques*, IEEE Press, New York.

Silvester, P. P. (ed.) (1996a), *Software for Electrical Engineering Analysis and Design*, Computational Mechanics Publications, Southampton, England.

Simmons, G. F. (1993), *Differential Equations with Applications and Historical Notes*, McGraw-Hill, New York.

Sobolev, L. (1963b), *Applications of Functional Analysis in Mathematical Physics*, Translated from 1950 Russian edition by F. Browder, in *Translations Mathematical Monographs*, vol. 7, American Mathematical Society, Providence, RI.

Southwell, R. V. (1940), *Relaxation Methods in Engineering Sciences*, Oxford University Press, London.

Southwell, R. V. (1946), *Relaxation Methods in Theoretical Physics*, Clarendon Press, Oxford.

Stakgold, I. (1979), *Green's Functions and Boundary Value Problems*, John Wiley & Sons, New York.

Steele, C. W. (1987), *Numerical Computation of Electric and Magnetic Fields*, Van Nostrand Rheinhold, New York.

Strang, G., and G. J. Fix (1973), *An Analysis of the Finite Element Method*, Prentice-Hall, Englewood Cliffs, NJ.

Strang, G. (1986), *Introduction to Applied Mathematics*, Wellesley-Cambridge Press, Wellesley, MA.

Stratton, J. A. (1941), *Electromagnetic Theory*, McGraw-Hill, New York.

Synge, J. L. (1957), *The Hypercircle Method in Mathematical Physics*, Cambridge University Press, Cambridge.

Szabó, B., and I. Babuška (1991), *Finite Element Analysis*, John Wiley & Sons, New York.

Tai, C. T. (1992), *Generalized Vector and Dyadic Analysis: Applied Mathematics in Field Theory*, IEEE Press, New York.

Thomas, J. M., and P. Joly (1981), *Méthodes d'Éléments Finis Mixtes et Hybrides*, Cours D. E. A. 1980–1981. U. Pierre et Marie Curie, Paris, France.

Tong, P., and J. N. Rossettos (1977), *Finite-Element Method: Basic Techniques and Implementation*, MIT Press, Cambridge, MA.

Umashankar, K., and A. Taflove (1993), *Computational Electromagnetics*, Artech House, Norwood, MA.

Ural, O. (1973), *Finite Element Method, Basic Concepts and Applications*, Intext, Scranton, PA.

Van Bladel, J. (1991), *Singular Electromagnetic Fields and Sources*, Clarendon Press, Oxford.

Vorobev, Y. U. (1965), *Method of Moments in Applied Mathematics*, (Translated from Russian), Gordon & Breach, New York.

Washizu, K. (1976), *Variational Methods in Elasticity and Plasticity*, Pergamon Press, Elmsford, NY.

Whiteman, J. R. (ed.) (1973, 1976, 1979, 1982, 1985, 1988, 1991), *The Mathematics of Finite Elements and Applications*, vols. I, II, III, IV, V, VI, and VII, Academic Press, London.

Whiteman, J. R. (ed.) (1994), *The Mathematics of Finite Elements and Applications*, vol. VIII, John Wiley & Sons, Chichester, England.

Whitney, H. (1957), *Geometric Integration Theory*, Princeton University Press, Princeton, NJ.

Zienkiewicz, O. C., and G. S. Holister (1965), *Stress Analysis*, John Wiley & Sons, New York.

Zienkiewicz, O. C., and Y. K. Cheung (1967, 1968, 1970), *The Finite Element Method in Structural and Continuum Mechanics*, McGraw-Hill, London.

Zienkiewicz, O. C. (1967, 1971), *The Finite Element Method in Engineering Science*, McGraw-Hill,

London.
Zienkiewicz, O. C. (1977), *The Finite Element Method*, McGraw-Hill, London.
Zienkiewicz, O. C., and R. L. Taylor (1989), *Finite Element Method - Basic Formulation and Linear Problems*, vol. 1, McGraw-Hill, New York.
Zienkiewicz, O. C., and R. L. Taylor (1991), *The Finite Element Method - Solid and Fluid Mechanics, Dynamics and Nonlinearity*, vol. 2, McGraw-Hill, New York.

ARTICLES[2]

Abe, N. M., J. R. Cardoso, D. R. F. Clavunde, and A. Passaro (1997), "LMAG-2D: A Software Package To Teach FEA Concepts," *IEEE Trans. Magnetics*, vol. MAG-33, no. pp. 1986–1989.
Adamiak, K. (1992), "Local Error Indicator in Finite Element Analysis of Laplacian Fields Based on the Green Integration Formula," *Int. J. Numer. Meth. Eng.*, vol. 33, pp. 1625–1642.
Ahagon, A., K. Fujiwara, and T. Nakata (1996), "Comparison of Various Kinds of Edge Elements for Electromagnetic Field Analysis," *IEEE Trans. Magnetics*, vol. MAG-32, pp. 898–901.
Ahmed, S. (1968), "Finite Element Method for Waveguide Problems," *Electronics Letters*, vol. 4, pp. 381–389.

---

2 Papers dealing with the general theory and mathematical aspects of FEM may be found in the following scientific journals, among others.

- *Communications on Pure and Applied Mathematics*
- *Computer Methods in Applied Mechanics and Engineering*
- *Computers and Structures*
- *International Journal for Numerical Methods in Engineering*
- *Journal of Sound and Vibration*
- *Mathematical and Computer Modelling*
- *Mathematics of Computation*
- *Numerische Mathematik*
- *RAIRO Mathematical Modelling and Numerical Analysis*
- *SIAM Journal of Applied Mathematics*
- *SIAM Journal on Numerical Analysis*

Papers dealing with FEM applications to electromagnetic problems may be found in the following scientific journals, among others.

- *Applied Optics*
- *COMPEL—The International Journal for Computation and Mathematics in Electrical and Electronic Engineering*
- *Electromagnetics*
- *Electronics Letters*
- *IEE Proceedings-A*
- *IEE Proceedings-H*
- *IEEE Microwave and Guided Waves Letters*
- *IEEE Transactions on Antennas and Propagation*
- *IEEE Transactions on Education*
- *IEEE Transactions on Electromagnetic Compatibility*
- *IEEE Transactions on Electron Devices*
- *IEEE Transactions on Magnetics*
- *IEEE Transactions on Microwave Theory and Techniques*
- *IEEE Transactions on Power Engineering*
- *International Journal of Microwave and Millimeter-Wave Computer-Aided Engineering*
- *International Journal of Numerical Modelling: Electronic Networks, Devices and Field*
- *Journal of Applied Physics*
- *Journal of Computational Physics*
- *Journal of Electromagnetic Waves and Applications*
- *Journal of Lightwave Technology*
- *Microwave and Optical Technology Letters*
- *Optical and Quantum Electronics*
- *Proceedings of the IEE*
- *Proceedings of the IEEE*
- *Radio Science*

Ahmed, S., and P. Daly (1969a), "Waveguide Solutions by Finite Element Method," *Radio Electron Eng.*, vol. 38, pp. 217–223.

Ahmed, S., and P. Daly (1969b), "Finite-Element Methods for Inhomogeneous Waveguides," *Proc. IEE*, vol. 116, pp. 1661–1664.

Aiello, G., S. Alfonzetti, and S. Coco (1992), "Charge Iteration for N-Dimensional Unbounded Electrical Field Computations", *IEEE Trans. Magnetics*, vol. MAG-28, pp. 1682–1685.

Aiello, G., S. Alfonzetti, S. Coco, and N. Salerno (1993), "Axisymmetric Unbounded Electrical Field Computation by Charge Iteration," *IEEE Trans. Magnetics*, vol. MAG-29, pp. 2043–2046.

Aiello, G., S. Alfonzetti, S. Coco, and N. Salerno (1994a), "Convergence Analysis of the Charge Iteration Procedure for Unbounded Electrical Fields," *IEEE Trans. Magnetics*, vol. MAG-30, pp. 2873–2876.

Aiello, G., S. Alfonzetti, and S. Coco (1994b), "Charge Iteration: A Procedure for the Finite-Element Computation of Unbounded Electrical Fields," *Int. J. Numer. Meth. Eng.*, vol. 37, pp. 4147–4166.

Aiello, G., S. Alfonzetti, S. Coco, and N. Salerno (1995), "Treatment of Unbounded Skin-Effect Problems in the Presence of Material Inhomogeneities," *IEEE Trans. Magnetics*, vol. MAG-31, pp. 1504–1507.

Aiello, G., S. Alfonzetti, S. Coco, and N. Salerno (1996a), "Overrelaxing the Charge Iteration Procedure," *IEEE Trans. Magnetics*, vol. MAG-32, pp. 694–697.

Aiello, G., S. Alfonzetti, S. Coco, and N. Salerno (1996b), "An FEM Code for Electrical Engineering Research: ELFIN," in *Software for Electrical Engineering Analysis and Design*, Ed. P. P. Silvester, Computational Mechanics Publications, Southampton, England, pp. 357–366.

Aiello, G., S. Alfonzetti, S. Coco, and N. Salerno (1997a), "A Generalization of the Charge Iteration Procedure," *IEEE Trans. Magnetics*, vol. MAG-33, pp. 1204–1207.

Aiello, G., S. Alfonzetti, S. Coco, and N. Salerno (1997b), "Combining Non-Linearity and Current Iterations for the Solution of Boundless Skin-Effect Problems," *IEEE Trans. Magnetics*, vol. MAG-33, pp. 1291-1294.

Ainsworth, M., J. Z. Zhu, A. W. Craig, and O. C. Zienkiewicz (1989), "Analysis of the Zienkiewicz-Zhu *A-Posteriori* Error Estimator in the Finite Element Method," *Int. J. Numer. Meth. Eng.*, vol. 28, pp. 2161–2174.

Ainsworth, M., and A. Craig (1992), "*A Posteriori* Error Estimators in the Finite Element Method," *Numer. Math.*, vol. 60, pp. 429–463.

Ainsworth, M., and J. T. Oden (1993), "A Unified Approach to *A Posteriori* Error Estimation Using Element Residual Methods," *Numer. Math.*, vol. 65, pp. 23–51.

Akin, J. E. (1976), "The Generation of Elements with Singularities," *Int. J. Numer. Meth. Eng.*, vol. 10, pp. 1249–1259.

Alam, M. S., K. Hirayama, Y. Hayashi, and M. Koshiba (1993), "A Vector Finite-Element Analysis of Complex Modes in Shielded Microstrip Lines," *Microwave Optical Tech. Letters*, vol. 6, pp. 873–875.

Alam, M. S., K. Hirayama, Y. Hayashi, and M. Koshiba (1994a), "Finite Element Analysis of Propagating, Evanescent, and Complex Modes in Finlines," *IEE Proc.*, vol. 141, pt. H, pp. 65–69.

Alam, M. S., K. Hirayama, Y. Hayashi, and M. Koshiba (1994b), "Analysis of Shielded Microstrip Lines with Arbitrary Metallization Cross Section Using a Vector Finite Element Method," *IEEE Trans. Microwave Theory Tech.*, vol. MTT-42, pp. 2112–2117.

Alam, M. S., M. Koshiba, K. Hirayama, and Y. Hayashi (1995), "Analysis of Lossy Planar Transmission Lines by Using a Vector Finite Element Method," *IEEE Trans. Microwave Theory Tech.*, vol. MTT-43, pp. 2466–2471.

Alam, M. S., M. Koshiba, K. Hirayama, and Y. Hayashi (1997), "Hybrid-Mode Analysis of Multilayered and Multiconductor Transmission Lines," *IEEE Trans. Microwave Theory Tech.*, vol. MTT-45, pp. 205–211.

Álvarez Melcón, A., R. Molina, and M. Guglielmi (1994), "Modal Spectrum of the Ridged Circular Waveguide Using the FE Method," *COMPEL*, vol. 13, supp. A, pp. 353–358.

Amari, S. (1993), "Capacitance and Inductance Matrices of Coupled Lines from Modal Powers," *IEEE Trans. Microwave Theory Tech.*, vol. MTT-41, pp. 147–150.

Andersen, J. B., and V. V. Solodukhov (1978), "Field Behavior Near a Dielectric Wedge," *IEEE Trans. Antennas Propag.*, vol. AP-26, pp. 598–602.

Angelini, J. J., C. Soize, and P. Soudais (1993), "Hybrid Numerical Method for Harmonic 3-D Maxwell Equations: Scattering by a Mixed Conducting and Inhomogeneous Anisotropic Dielectric Medium," *IEEE Trans. Antennas Propag.*, vol. AP-41, pp. 66–76.

Anger, A. (1990), "Software Computes Maxwell's Equations," *Microwave Journal*, vol. 33, pp. 170–178.

Angkaew, T., M. Matsuhara, and N. Kumagai (1987), "Finite-Element Analysis of Waveguide Modes: A Novel Approach that Eliminates Spurious Modes," *IEEE Trans. Microwave Theory Tech.*, vol. MTT-35, pp. 117–123. See "Comments", M. Mrozowski (1991), same journal, vol. MTT-39, p. 611.

Arabi, T. R., A. T. Murphy, T. K. Sarkar, R. F. Harrington, and A. R. Djordjević (1991a), "Analysis of Arbitrarily Oriented Microstrip Lines Utilizing a Quasi-Dynamic Approach," *IEEE Trans. Microwave Theory Tech.*, vol. MTT-39, pp. 75–82.

Arabi, T. R., A. T. Murphy, T. K. Sarkar, R. F. Harrington, and A. R. Djordjević (1991b), "On the Modeling of Conductor and Substrate Losses in Multiconductor, Multidielectric, Transmission Line Systems," *IEEE Trans. Microwave Theory Tech.*, vol. MTT-39, pp. 1090–1097.

Aregba, D., J. Gay, and G. Mazé-Merceur (1994), "Modeling Multiport Using a Three-Dimensional Coupled Analytical/Finite Element Method Application to Microwave Characterization," *IEEE Trans. Microwave Theory Tech.*, vol MTT-42, pp. 590–594.

Argyris, J. H. (1954), "Energy Theorems and Structural Analysis. Part 1," *Aircraft Eng.*, vol. 26, pp. 347–356, pp. 383–387, p. 394.

Argyris, J. H. (1955), "Energy Theorems and Structural Analysis. Part 2," *Aircraft Eng.*, vol. 27, pp. 42–58, pp. 80–94, pp. 125–134, pp. 145–158.

Arlett, P. L., A. K. Bahrani, and O. C. Zienkiewicz (1968), "Application of Finite Elements to the Solution of Helmholtz Equation," *Proc. IEE*, vol. 115, pp. 1762–1766.

Arnold, D. N. (1990), "Mixed Finite Element Methods for Elliptic Problems," *Comp. Meth. Appl. Mech. Eng.*, vol. 83, pp. 281–300.

Aubourg, M., J. P. Villote, F. Godon, and Y. Garault (1983), "Finite Element Analysis of Lossy Waveguides — Application to Microstrip Lines on Semiconductor Substrates," *IEEE Trans. Microwave Theory Tech.*, vol. MTT-31, pp. 326–330.

Aubourg, M., S. Verdeyme, and P. Guillon (1996), "Microwave Passive Devices," in *Finite Element Software for Microwave Engineering*, Eds. T. Itoh, G. Pelosi, and P. P. Silvester, John Wiley & Sons, New York, pp. 53–78.

Babuška, I. (1971), "Error-Bounds for the Finite Element Method," *Numer. Math.*, vol. 16, pp. 322–333.

Babuška, I., and M. B. Rosenzweig (1972), "A Finite Element Scheme for Domains with Corners," *Numer. Math.*, vol. 20, pp. 1–21.

Babuška, I. (1973), "The Finite Element Method with Lagrange Multipliers," *Numer. Math.*, vol. 20, pp. 179–192.

Babuška, I., and A. K. Aziz (1974), "On the Angle Condition in the Finite Element Method," *SIAM J. Numer. Anal.*, vol. 13, pp. 214–226.

Babuška, I. (1976), "The Selfadaptive Approach in the Finite Element Method," in *The Mathematics of Finite Elements and Applications II, MAFELAP 1975*, Ed. J. R Whiteman, Academic Press, New York, pp. 125–142.

Babuška, I. (1977), "Singularity Problems in the Finite Element Method," in *Formulation and Computational Algorithms in Finite Element Analysis*, Eds. K. J. Bathe, J. T. Oden, and W. Wunderlich, MIT Press, Cambridge, MA, pp. 748–792.

Babuška, I., and W. C. Rheinboldt (1978a), "Error Estimates for Adaptive Finite Element Computations," *SIAM J. Numer. Anal.*, vol. 15, pp. 736–754.

Babuška, I., and W. C. Rheinboldt (1978b), "*A-Posteriori* Error Estimates for the Finite Element Method," *Int. J. Numer. Meth. Eng.*, vol. 12, pp. 1597–1615.

Babuška, I., R. B. Kellogg, and J. Pitkäranta (1979), "Direct and Inverse Error Estimates for Finite Elements with Mesh Refinements," *Numer. Math.*, vol. 33, pp. 447–471.

Babuška, I., and W. C. Rheinboldt (1979a), "Analysis of Optimal Finite-Element Meshes in $\mathbb{R}^1$," *Math. Comp.*, vol. 33, pp. 435–464.

Babuška, I., and W. C. Rheinboldt (1979b), "Adaptive Approaches and Reliability Estimations in Finite Element Analysis," *Comp. Meth. Appl. Mech. Eng.*, vol. 17/18, pp. 519–540.

Babuška, I., and M. R. Dorr (1981), "Error Estimates for the Combined *h* and *p* Versions of the Finite Element Method," *Numer. Math.*, vol. 37, pp. 257–277.

Babuška, I., and W. C. Rheinboldt (1981), "*A Posteriori* Error Analysis of Finite Element Solutions for One-Dimensional Problems," *SIAM J. Numer. Anal.*, vol. 18, pp. 565–589.

Babuška, I., B. A. Szabó, and I. N. Katz (1981), "The *p*-Version of the Finite Element Method," *SIAM J. Numer. Anal.*, vol. 18, pp. 515–545.

Babuška, I., and B. Szabó (1982), "On the Rates of Convergence of the Finite Element Method," *Int. J. Numer. Meth. Eng.*, vol. 18, pp. 323–341.

Babuška, I., and A. Miller (1984), "The Post-Processing Approach in the Finite Element Method — Part 3: *A Posteriori* Error Estimates and Adaptive Mesh Selection," *Int. J. Numer. Meth. Eng.*, vol. 20, pp. 2311–2324.

Babuška, I., and M. Vogelius (1984), "Feedback and Adaptive Finite Element Solution of One-Dimensional Boundary Value Problems," *Numer. Math.*, vol. 44, pp. 75–102.

Babuška, I. (1986), "Feedback, Adaptivity, and *A Posteriori* Estimates in Finite Elements: Aims, Theory, and Experience," in *Accuracy Estimates and Adaptive Refinements in Finite Element Computations*, Eds. I. Babuška, O. C. Zienkiewicz, J. Gago, and E. R. de A. Oliveira, John Wiley & Sons, London, pp. 3–23.

Babuška, I., and M. Suri (1987a), "The *h-p* Version of the Finite Element Method with Quasiuniform Meshes," *RAIRO Math. Model. Numer. Anal.*, vol. 21, pp. 199–238.

Babuška, I., and M. Suri (1987b), "The Optimal Convergence Rate of the *p*-Version of the Finite Element Method," *SIAM J. Numer. Anal.*, vol. 24, pp. 750–776.

Babuška, I. (1988), "The *p*- and *hp*-Versions of the Finite Element Method: The State of the Art," in *Finite Elements Theory and Applications*, Eds. D. L. Dwoyer, M. Y. Hussain, and R. G. Voigt, Springer-Verlag, New York.

Babuška, I., and B. Q. Guo (1988), "The *h-p* Version of the Finite Element Method for Domains with Curved Boundaries," *SIAM J. Numer. Anal.*, vol. 25, pp. 837–861.

Babuška, I., R, Durán, R. Rodríguez (1992), "Analysis of the Efficiency of an a Posteriori Error Estimator for Linear Triangular Finite Element," *SIAM J. Numer. Anal.*, vol. 29, pp. 947–964.

Babuška, I., and R. Rodríguez (1993), "The Problem of the Selection of an *A Posteriori* Error Indicator Based on Smoothing Techniques," *Int. J. Numer. Meth. Eng.*, vol. 36, pp. 539–567.

Babuška, I., T. Strouboulis, C. S. Upadhyay, S. K. Gangaraj, and K. Copps (1994), "Validation of *A Posteriori* Error Estimators by Numerical Approach," *Int. J. Numer. Meth. Eng.*, vol. 37, pp. 1073–1123.

Babuška, I., T. Strouboulis, and C. S. Upadhyay (1997a), "A Model Study of the Quality of *A Posteriori* Error Estimators for Finite Element Solutions of Linear Elliptic Problems with Particular Reference to the Behavior Near the Boundary," *Int. J. Numer. Meth. Eng.*, vol. 40, pp. 2521–2577.

Babuška, I., T. Strouboulis, C. S. Upadhyay, and S. K. Gangaraj (1997b), "*A Posteriori* Estimation and Adaptivity Control of the Pollution Error in the *h*-Version of the FEM," *Int. J. Numer. Meth. Eng.*, vol. 40, pp. 4207–4236.

Baehmann, P., M. S. Shepard, and J. E. Flaherty (1992), "*A Posteriori* Error Estimation for Triangular and Tetrahedral Quadratic Elements Using Interior Residuals," *Int. J. Numer. Meth. Eng.*, vol. 34, pp. 979–996.

Baillargeat, D., S. Verdeyme, and P. Guillon (1996), "Rigorous Design of Dielectric Resonator Filters Applying the Finite Element Method," in *Software for Electrical Engineering Analysis and Design*, Ed. P. P. Silvester, Computational Mechanics Publications, Southampton, England, pp. 167–175.

Bank, R. E., and A. Weiser (1985), "Some A Posteriori Error Estimators for Elliptic Partial Differential Equations," *Math. Comp.*, vol. 44, pp. 283–301.

Bank, R. E. (1986), "Analysis of a Local *A Posteriori* Error Estimate for Elliptic Equations," in *Accuracy Estimates and Adaptive Refinements in Finite Element Computations*, Eds. I. Babuška, O C Zienkiewicz, J. Gago, and E. R. de A. Oliveira, John Wiley & Sons, London, pp. 119–128.

Bank, R. E., and R. K. Smith (1997), "Mesh Smoothing Using *A Posteriori* Error Estimates," *SIAM J. Numer. Anal.*, vol. 34, pp. 279–297.

Bárdi, I., and O. Biro (1991), "An Efficient Finite-Element Formulation Without Spurious Modes for Anisotropic Waveguides," *IEEE Trans. Microwave Theory Tech.*, vol. MTT-39, pp. 1133–1139.

Bárdi, I., O. Biro, and K. Preis (1991), "Finite Element Scheme for 3D Cavities Without Spurious Modes," *IEEE Trans. Magnetics*, vol. MAG-27, pp. 4036–4039.

Bárdi, I., O. Biro, K. Preis, G. Vrisk, and K. R. Richter (1993), "Nodal and Edge Element Analysis of Inhomogeneously Loaded Waveguides," *IEEE Trans. Magnetics*, vol. MAG-29, pp. 1466–1469.

Bárdi, I., O. Biro, R. Dyczij-Edlinger, K. Preis, and K. R. Richter (1994), "On the Treatment of Sharp Corners in the FEM Analysis of High Frequency Problems," *IEEE Trans. Magnetics*, vol. MAG-30, pp. 3108–3111.

Barnhill, R. E., and J. R. Whiteman (1973), "Error Analysis of Finite Element Methods with Triangles for Elliptic Boundary Value Problems," in *The Mathematics of Finite Elements and Applications*, Ed. J. R. Whiteman, Academic Press, London, pp. 83–112.

Barrett, R. M. (1984), "Microwave Printed Circuits — The Early Years," *IEEE Trans. Microwave Theory Tech.*, vol. MTT-32, pp. 983–990.

Barton, M. L. (1982), "Dual-Order Parallelogram Finite Elements for the Axisymmetric Vector Poisson Equation," *IEEE Trans. Magnetics*, vol. MAG-18, pp. 599–604.

Barton, M. L., and Z. J. Csendes (1987), "New Vector Finite Elements for Three-Dimensional Magnetic Field Computation," *J. Appl. Phys.*, vol. 61, pp. 3919–3921.

Bayliss A., and E. Turkel (1980), "Radiation Boundary Conditions for Wave-Like Equations," *Comm. Pure Appl. Math.*, vol. 33, pp. 107–725.

Bayliss, A., M. Gunzburger, and E. Turkel (1982), "Boundary Conditions for the Numerical Solution of Elliptic Equations in Exterior Regions," *J. Appl. Math.*, vol. 42, pp. 430–451.

Bayliss, A., C. I. Goldstein, and E. Turkel (1985), "On Accuracy Conditions for the Numerical Computation of Waves," *J. Comp. Physics*, vol. 59, pp. 396–404.

Beckers, P., and H. G. Zhong (1992), "Error Estimation and Mesh Adaptation in Engineering Analysis — A Review of the Methods and Future Trends," in *Numerical Methods in Engineering and Applied Sciences*, Eds. H. Alder, J. C. Heinrich, S. Lavanchy, E. Oñate, and B. Suárez, CIMNE, Barcelona, pp. 19–29.

Bedair, S. S., and I. Wolff (1992), "Fast, Accurate and Simple Approximate Analytic Formulas for Calculating the Parameters of Supported Coplanar Waveguides for (M)MICs," *IEEE Trans. Microwave Theory Tech.*, vol. MTT-40, pp. 41–48.

Beer, G., and J. L. Meek (1981), "'Infinite Domain' Elements," *Int. J. Numer. Meth. Eng.*, vol. 17, pp. 43–52.

Beker, B. (1991), "Electromagnetic Field Behavior Near Homogeneous Anisotropic Wedges," *IEEE Trans. Antennas Propag.*, vol. AP-39, pp. 1143–1151.

Belytschko, T., and M. Tabbara (1993), "*h*-Adaptive Finite Element Methods for Dynamic Problems, with Emphasis on Localization," *Int. J. Numer. Meth. Eng.*, vol. 36, pp. 4245–4265.

Berenger, J. P. (1994), "A Perfectly Matched Layer for the Absorption of Electromagnetic Waves," *J. Comp. Physics*, vol. 114, pp. 185–200.

Berenger, J. P. (1996), "Perfectly Matched Layer for the FDTD Solution of Wave-Structure Interaction Problems," *IEEE Trans. Antennas Propag.*, vol. AP-44, pp. 110–117.

Berk, A. D. (1956), "Variational Principles for Electromagnetic Resonators and Waveguides," *IRE Trans. Antennas Propag.*, vol. AP-4, pp. 104–111.

Bermúdez de Castro, A., and M. D. Gómez-Pedreira (1992), "Mathematical Analysis of a Finite Element Method Without Spurious Solutions for Computation of Dielectric Waveguides," *Numer. Math.*, vol. 61, pp. 39–57.

Bettess, P. (1977), "Infinite Elements," *Int. J. Numer. Meth. Eng.*, vol. 11, pp. 54–64.

Bettess, P., and O. C. Zienkiewicz (1977), "Diffraction and Refraction of Surface Waves Using Finite and Infinite Elements," *Int. J. Numer. Meth. Eng.*, vol. 11, pp. 1271–1290.

Bettess, P. (1980), "More on Infinite Elements," *Int. J. Numer. Meth. Eng.*, vol. 15, pp. 1613–1626.

Bettess, P., and J. A. Bettess (1984), "Infinite Elements for Static Problems," *Eng. Comp.*, vol. 1, pp. 4–15.

Bettess, P. (1988), "Finite Element Modelling of Exterior Electromagnetic Problems," *IEEE Trans. Magnetics*, vol. MAG-24, pp. 238–244.

Biddlecombe, C. S., J. Simkin, and C. W. Trowbridge (1986), "Error Analysis in Finite Element Models of Electromagnetic Fields," *IEEE Trans. Magnetics*, vol. MAG-22, pp. 811–814.

Bila, S., D. Baillergeat, S. Verdeyme, and P. Guillon (1997), "Automated Electromagnetic Optimization Method for Microwave Devices," *IEEE Microwave Guided Wave Letters*, vol. 7, pp. 242–244.

Bindiganavale, S. S., and J. L. Volakis (1997), "A Hybrid FE-FMM Technique for Electromagnetic Scattering," *IEEE Antennas Propag.*, vol. AP-45, pp. 180–181.

Biswas, R., J. E. Flaherty, and M. Benantar (1991), "Advances in Adaptive Parallel Processing for Field Applications," *IEEE Trans. Magnetics*, vol. MAG-27, pp. 3768–3773.

Blanc-Castillo, F., M. Salazar-Palma, and L. E. García-Castillo (1995), "Linear and Second Order Edge-Lagrange Finite Elements for Efficient Analysis of Waveguiding Structures with Curved Contours," *25th European Microwave Conf. Proc.*, Bologna, Italy, pp. 444–448.

Blaschak, J. G., and G. A. Kriegsmann (1988), "A Comparative Study of Absorbing Boundary Conditions," *J. Comp. Physics*, vol. 77, pp. 109–139.

Blondy, P., V. Madrangeas, D. Cros, and P. Guillon (1996), "Mode Coupling Prediction in Whispering Gallery Dielectric Resonators Modes," *IEEE Microwave Guided Wave Letters*, vol. 6, pp. 229–231.

Boag, A., A. Boag, R. Mittra, and Y. Leviatan (1994a), "A Numerical Absorbing Boundary Condition for Finite-Difference and Finite-Element Analysis of Open Structures," *Microwave Optical Tech. Letters*, vol. 7, pp. 395–398.

Boag, A., A. Boag, and R. Mittra (1994b), "A Numerical Absorbing Boundary Condition for Edge-Based Finite-Element Analysis," *Microwave Optical Tech. Letters*, vol. 7, pp. 733–737.

Boag, A., and R. Mittra (1995), "A Numerical Absorbing Boundary Condition for Finite-Difference and Finite-Element Analysis of Open Periodic Structures," *IEEE Trans. Microwave Theory Tech.*, vol. MTT-43, pp. 150–154.

Borovchaki, H., and P. L. George (1997), "Aspects of 2-D Delaunay Mesh Generation," *Int. J. Numer. Meth. Eng.*, vol. 40, pp. 1957–1977.

Bossavit, A., and J. C. Verité (1982), "A Mixed FEM-BIEM Method to Solve 3-D Eddy Current Problems," *IEEE Trans. Magnetics*, vol. 18, pp. 431–435.

Bossavit, A. (1988), "A Rationale for Edge Elements in 3-D Fields Computations," *IEEE Trans. Magnetics*, vol. 24, pp. 74–79.

Bossavit, A. (1989), "Simplicial Finite Elements for Scattering Problems in Electromagnetism," *Comp. Meth. Appl. Mech. Eng.*, vol. 76, pp. 299–316.

Bossavit, A., and I. Mayergoyz (1989), "Edge Elements for Scattering Problems," *IEEE Trans. Magnetics*, vol. 25, pp. 2816–2821.

Bossavit, A. (1990), "Solving Maxwell Equations in a Closed Cavity and the Question of Spurious Modes," *IEEE Trans. Magnetics*, vol. MAG-26, pp. 702–704.

Bossavit, A. (1991), "The Computation of Eddy-Currents, in Dimension 3, by Using Mixed Finite Elements and Boundary Elements in Association," *Math. Comp. Mod.*, vol. 15, pp. 33–42.

Boyse, W. E., and A. A. Seidl (1991), "A Hybrid Finite Element Method for Near Bodies of Revolution," *IEEE Trans. Magnetics*, vol. MAG-27, pp. 3833–3836.

Boyse, W. E., D. R. Lynch, K. D. Paulsen, and G. N. Minerbo (1992), "Nodal-Based Finite-Element Modeling of Maxwell's Equations," *IEEE Trans. Antennas Propag.*, vol. AP-40, pp. 642–651.

Boyse, W. E., G. N. Minerbo, K. D. Paulsen, and D. R. Lynch (1993), "Applications of Potentials to Finite Element Modeling of Maxwell's Equations," *IEEE Trans. Magnetics*, vol. MAG-29, pp.

1333–1336.

Boyse, W. E., and A. A. Seidl (1994), "A Hybrid Finite Element Method for 3-D Scattering Using Nodal and Edge Elements," *IEEE Trans. Antennas Propag.*, vol. AP-42, pp. 1436–1442.

Boyse, W. E., and K. D. Paulsen (1997), "Accurate Solutions of Maxwell's Equations Around PEC Corners and Highly Curved Surfaces Using Nodal Finite Elements," *IEEE Antennas Propag.* vol. AP-45, pp. 1758–1767.

Bramble, J. H., and J. E. Pasciak (1982), "A New Computational Approach for the Linearized Scalar Potential Formulation of the Magnetostatic Field Problem," *IEEE Trans. Magnetics*, vol. MAG-18, pp. 357–361.

Bramble, J. H., and M. Zlámal (1970), "Triangular Elements in the Finite Element Method," *Math. Comp.*, vol. 24, pp. 809–820.

Braess, D., and R. Verfürth (1996), "*A Posteriori* Error Estimators for the Raviart-Thomas Element," *SIAM J. Numer. Anal.*, vol. 33, pp. 2431–2444.

Brauer, J. R., L. A. Larkin, and B. E. McNeal (1991), "Higher Order 3D Isoparametric Finite Elements for Improved Magnetic Field Calculation Accuracy," *IEEE Trans. Magnetics*, vol. MAG-27, pp. 4185–4188.

Brauer, J. R., S. H. Lee, and Q. M. Chen (1997), "Adaptive Time-Stepping in Nonlinear Transient Electromagnetic Finite Element Analysis," *IEEE Trans. Magnetics*, vol. MAG-33, pp. 1784–1787.

Bressan, M., and P. Gamba (1994), "Analytical Expressions of Field Singularities at the Edge of Four Right Wedges," *Microwave Guided Wave Letters*, vol. 4, pp. 3–5.

Brezzi, F., J. Douglas, Jr., and L. D. Marini (1985), "Two Families of Mixed Finite Elements for Second Order Elliptic Problems," *Numer. Math.*, vol. 47, pp. 217–235.

Brezzi, F., and D. Marini (1994), "A Survey On Mixed Finite Element Approximations," *IEEE Trans. Magnetics*, vol. MAG-30, pp. 3547–3551.

Brunotte, X., G. Meunier, and J. F. Imhoff (1992), "Finite Element Modeling of Unbounded Problems Using Transformations: A Rigorous Powerful an Easy Solution," *IEEE Trans. Magnetics*, vol. MAG-28, pp. 1663–1666.

Bryant, C. F. (1985), "Two Dimensional Automatic Triangular Mesh Generation," *IEEE Trans. Magnetics*, vol. MAG-21, pp. 2547–2550.

Bryant, J. H. (1984), "Coaxial Transmission Lines, Related Two-Conductor Transmission Lines, Connectors, and Components: A U. S. Historical Perspective," *IEEE Trans. Microwave Theory Tech.*, vol. MTT-32, pp. 970–982.

Bugeda, G., and E. Oñate (1992), "Adaptive Mesh Refinement Techniques for Aerodynamic Problems," in *Numerical Methods in Engineering and Applied Sciences*, Eds. H. Alder, J. C. Heinrich, S. Lavanchy, E. Oñate, and B. Suárez, CIMNE, Barcelona, pp. 513–522.

Bulutay, C., and S. Prasad (1993), "Analysis of Millimeter Waveguides on Anisotropic Substrates Using the Three-Dimensional Transmission-Line Matrix Method," *IEEE Trans. Microwave Theory Tech*, vol. MTT-41, pp. 1119–1125.

Burais, N., H. Manser, and A. Nicolas (1985), "Physical and Geometric Singularities Modelling Techniques for Electrostatic and Electromagnetic Problems," *IEEE Trans. Magnetics*, vol. MAG-21, pp. 2157–2160.

Burkley, V. J., and J. C. Bruch, Jr. (1991), "Adaptive Error Analysis in Seepage Problems," *Int. J. Numer. Meth. Eng.*, vol. 31, pp. 1333–1356.

Calamia, M., R. Coccioli, G. Pelosi, and G. Manara (1994a), "A Hybrid FEM/UTD Analysis of the Scattering from a Cavity Backed Aperture in the Face of a Perfectly Conducting Wedge," *COMPEL*, vol. 13, supp. A, pp. 229–236.

Cangellaris, A. C., and R. Lee (1990), "The Bymoment Method for Two-Dimensional Electromagnetic Scattering," *IEEE Trans. Antennas Propag.*, vol. AP-38, pp. 1429–1437.

Cangellaris, A. C., and R. Lee (1991), "Finite Element Analysis of Electromagnetic Scattering from Inhomogeneous Cylinders at Oblique Incidence," *IEEE Trans. Antennas Propag.*, vol. AP-31, pp. 645–650.

Cangellaris, A. C. (1996), "Numerical Error in Finite Element Solutions of Electromagnetic Boundary Value Problems," in *Finite Element Software for Microwave Engineering*, Eds. T. Itoh, G. Pelosi, and P. P. Silvester, John Wiley & Sons, New York, pp. 347–384.

Caorsi, S., P. Fernandes, and M. Raffetto (1994a), "On the Effect of the Spurious Modes in Deterministic Problems," *COMPEL*, vol. 13, supp. A, pp. 317–322.

Caorsi, S., A. Massa, and M. Raffetto (1994b), "Hybrid Finite-Element Numerical Approach to the Solution of Electromagnetic Scattering Problem," *COMPEL*, vol. 13, supp. A, pp. 371–376.

Caorsi, S., P. Fernandes, and M. Raffetto (1995), "Edge Elements and the Inclusion Condition," *IEEE Microwave Guided Wave Letters*, vol. 5, pp. 222–223.

Caorsi, S., P. Fernández, P. Fernandes, and M. Raffetto (1997), "Do Covariant Projection Elements Really Satisfy the Inclusion Condition?," *IEEE Trans. Microwave Theory Tech.*, vol. MTT-45, pp. 1643–1644.

Cano, G., F. Medina, and M. Horno (1989), "Frequency Dependent Characteristic Impedance of Microstrip and Finline on Layered Biaxial Substrates," *Microwave Optical Tech. Letters*, vol. 2, pp. 210–214.

Cano, G.,F. Medina, and M. Horno (1992), "Efficient Spectral Domain Analysis of Generalized Multistrip Lines in Stratified Media Including Thin, Anisotropic, and Lossy Substrates," *IEEE Trans. Microwave Theory Tech.*, vol. MTT-40, pp. 217–227.

Carey, G. F., and D. L. Humphrey (1981), "Mesh Refinement and Iterative Solution Methods for Finite Element Computations," *Int. J. Numer. Meth. Eng.*, vol. 17, pp. 1717–1734.

Carey, G. F., and M. Seager (1985), "Projection and Iteration in Adaptive Finite Element Refinement," *Int. J. Numer. Meth. Eng.*, vol. 21, pp. 1681–1695.

Carson, C. T., and G. K. Cambrell (1966), "Upper and Lower Bounds on the Characteristic Impedance of TEM Mode Transmission Lines," *IEEE Trans. Microwave Theory Tech.*, vol. MTT-14, pp. 497–498.

Casas, E. B. de las, and A. M. G. Figueiredo (1992), "An *r-h* Adaptive Multigrid Program for Finite Elements," in *Numerical Methods in Engineering and Applied Sciences*, Eds. H. Alder, J. C. Heinrich, S. Lavanchy, E. Oñate, and B. Suárez, CIMNE, Barcelona, pp. 483–492.

Cermark, I. A., and P. Silvester (1968), "Solution of Two-Dimensional Field Problems by Boundary Relaxation," *IEEE Proc.*, vol. 115, pp. 1341–1347.

Chan, C. H., Z. Pantic-Tanner, and R. Mittra (1988), "Field Behavior Near a Conducting Edge Embedded in an Inhomogeneous Anisotropic Medium," *Electronics Letters*, vol. 24, pp. 355–356.

Chang, C. N., Y. C. Wong, and C. H. Chen (1990), "Full-Wave Analysis of Coplanar Waveguides by Variational Conformal Mapping Technique," *IEEE Trans. Microwave Theory Tech.*, vol. MTT-38, pp. 1339–1344.

Chang, C. N., W. C. Chang, and C. H. Chen (1991), "Full-Wave Analysis of Multilayer Coplanar Lines," *IEEE Trans. Microwave Theory Tech.*, vol. MTT-39, pp. 747–750.

Chang, S. K., and K. K. Mei (1976), "Application of the Unimoment Method to Electromagnetic Scattering of Dielectric Cylinders," *IEEE Trans. Antennas Propag.*, vol. AP-24, pp. 35–42.

Chang, W. S. C. (1984), "The Birth of Lightwave Technology and Its Implications to Microwaves," *IEEE Trans. Microwave Theory Tech.*, vol. MTT-32, pp. 1140–1143.

Chari, M. V. K., and P. P. Silvester (1971), "Finite-Element Analysis of Magnetically Saturated D.C. Machines," *IEEE Trans. Power Appl. Syst.*, vol. PAS-90, pp. 2362–2371.

Chatterjee, A., J. M. Jin, and J. L. Volakis (1992), "Computation of Cavity Resonances Using Edge-Based Finite Elements," *IEEE Trans. Microwave Theory Tech.*, vol. MTT-40, pp. 2106–2108.

Chatterjee, A., J. M. Jin, and J. L. Volakis (1993), "Edge-Based Finite Elements and Vector ABCs Applied to 3-D Scattering," *IEEE Trans. Antennas Propag.*. vol. AP-41, pp. 221–226.

Chatterjee, A., and J. L. Volakis (1995), "Conformal Absorbing Boundary Conditions for 3-D Problems: Derivation and Applications," *IEEE Trans. Antennas Propag.*, vol. AP-43, pp. 860–866.

Chaves, M. B. F., C. G. Migliora, and H. J. C. Barbossa (1994), "Elimination of Spurious FE Solutions of Dielectric Waveguides," *COMPEL*, vol. 13, supp. A, pp. 323–328.

Chellamuthu, K. C., and N. Ida (1994), "*A Posteriori* Element by Element Local Error Estimation Technique and 2D & 3D Adaptive Finite Element Mesh Refinement," *IEEE Trans. Magnetics*, vol. MAG-30, pp. 3527–3530.

Chellamuthu, K. C., and N. Ida (1995), "Reliability Assessment of an *A Posteriori* Error Estimate for Adaptive Computation of Electromagnetic Field Problems," *IEEE Trans. Magnetics*, vol. MAG-31, pp. 1761–1764.

Chen, A. T., and J. R. Rice (1992), "On Grid Refinement at Point Singularities for *h-p* Methods," *Int. J. Numer. Meth. Eng.*, vol. 33, pp. 39–57.

Chen, J., W. Hong, and J. M. Jin (1998), "An Iterative Measured Equation Technique for Electromagnetic Problems," *IEEE Trans. Microwave Theory Tech.*, vol. MTT-46, pp. 25–30.

Chen, Q., and A. Konrad (1997), "A Review of Finite Element Open Boundary Techniques for Static and Quasi-Static Electromagnetic Field Problems," *IEEE Trans. Magnetics*, vol. MAG-33, pp. 663–676.

Cheng, N., F. Ci-Zhang, and N. Guang-Zheng (1987), "Determination of Electromagnetic Energy and Parameters by Complementary-Dual Energy Method," *IEEE Trans. Magnetics*, vol. MAG-23, pp. 2671–2673.

Chew, W. C., and M. A. Nasir (1989), "A Variational Analysis of Anisotropic, Inhomogeneous Dielectric Waveguides," *IEEE Trans. Microwave Theory Tech.*, vol. MTT-37, pp. 661–668.

Chew, W. C., and W. H. Weedon (1994), "A 3-D Perfectly Matched Medium from Modified Maxwell's Equations with Stretched Coordinates," *Microwave Optical Tech. Letters*, vol. 8, pp. 599–604.

Chiang, K. S. (1985), "Finite-Element Analysis of Optical Fibers with Iterative Treatment of the Infinite 2-D Space," *Optical Quantum Electronics*, vol. 17, pp. 381–391.

Choi, C. T. M., and J. P Webb (1996), "Wave-Envelope and Electromagnetic Wave Problems," *IEEE Trans. Magnetics*, vol. MAG-32, pp. 886–889.

Choi, C. T. M., and J. P. Webb (1997), "The Wave-Envelope Method and Absorbing Boundary Conditions," *IEEE Trans. Magnetics*, vol. MAG-33, pp. 1420–1423.

Chung, S. J., and C. H. Chen (1988), "Partial Variational Principle for Electromagnetic Field Problems: Theory and Applications," *IEEE Trans. Microwave Theory Tech.*, vol. MTT-36, pp. 473–479.

Ciarlet, P. G., and P. A. Raviart (1972a), "General Lagrange and Hermite Interpolation in $\mathbb{R}^n$ with Application to the Finite Element Method," *Arch. Ration Mech. Anal.*, vol. 46, pp. 177–199.

Ciarlet, P. G., and P. A. Raviart (1972b), "Interpolation Theory over Curved Elements," *Comp. Meth. Appl. Mech. Eng.*, vol. 1, pp. 217–249.

Ciarlet, P. G., and P. A. Raviart (1972c), "The Combined Effect of Curved Boundaries and Numerical Integration in Isoparametric Finite Element Methods," in *The Mathematical Foundations of the Finite Element Method with Applications to Partial Differential Equations*, Ed. A. K. Aziz, Academic Press, New York, pp. 409–474.

Ciarlet, P. G. (1973), "Orders of Convergence in Finite Element Methods," in *The Mathematics of Finite Elements and Applications*, Ed. J. R. Whiteman, Academic Press, London, pp. 113–129.

Clough, R. W. (1960), "The Finite Element Method in Plane Stress Analysis" in *Proceedings of the Second ASCE Conference on Electronic Computation*, Pittsburgh, PA.

Clough, R. W., and E. W. Wilson (1962), "Stress Analysis of a Gravity Dam by Finite Element Method," *Proc. Symposium on the Use of Computer in Civil Engineering*, Lab. Nacional de Engenharia Civil, Lisbon, Portugal.

Clough, R. W. (1989), "Original Formulation of The Finite Element Method," *Proc. ASCE Structure Congress*, San Francisco, CA, pp. 1–10.

Coccioli, R., T. Itoh, G. Pelosi, and P. P. Silvester (1996a), "Finite Element Methods in Microwaves: A Selected Bibliography," *IEEE Antennas Propag. Magazine*, vol. 36, pp. 34–48.

Coccioli, R., M. Mongiardo, G. Pelosi, and R. Ravanelli (1996b), "Design of Matched Bends in Rectangular Waveguides by FEM," *Int. J. Microwave Millimeter-Wave Computer-Aided Eng.*, vol. 6.

Coccioli, R., G. Pelosi, and P. P. Silvester (1996c), "Finite Element Package for Inhomogeneous H- and E-Plane Junctions," in *Software for Electrical Engineering Analysis and Design*, Ed. P. P. Silvester,

Computational Mechanics Publications, Southampton, England, pp. 187–195.

Cohn, S. B., and R. Levy (1984), "History of Microwave Passive Components with Particular Attention to Directional Couplers," *IEEE Trans. Microwave Theory Tech.*, vol. MTT-32, pp. 1046–1054.

Collins, J. D., J. L. Volakis, and J. M. Jin (1990a), "A Combined Finite Element-Boundary Integral Formulation for Solution of Two-Dimensional Scattering Problems via CGFFT," *IEEE Trans. Antennas Propag.*, vol. AP-38, pp. 1852–1858.

Collins, J. D., J. M. Jin, and J. L. Volakis (1990b), "A Combined Finite Element-Boundary Element Formulation for Solution of Two-Dimensional Scattering Problems via CGFFT," *Electromagnetics*, vol. 10, pp. 423–427.

Collins, J. D., J. M. Jin, and J. L. Volakis (1992), "Eliminating Interior Resonances in Finite Element-Boundary Integral Method for Scattering," *IEEE Trans. Antennas Propag.*, vol. 40, pp. 1583–1585.

Corr, D. G., and J. B. Davies (1972), "Computer Analysis of the Fundamental and Higher Order Modes in Single and Coupled Microstrip," *IEEE Trans. Microwave Theory Tech.*, vol. MTT-20, pp. 669–678.

Costache, G. I. (1987), "Finite Element Method Applied to Skin-Effect Problems in Strip Transmission Lines," *IEEE Trans. Microwave Theory Tech.*, vol. MTT-35, pp. 1009–1013.

Coulomb, J. L., Y. du Terrail, and G. Meunier (1985), "FLUX3D. A Finite Element Package for Magnetic Computation," *IEEE Trans. Magnetics*, vol. MAG-21, pp. 2499–2502.

Courant, R. L. (1943), "Variational Methods for the Solution of Problems of Equilibrium and Vibration," *Bull. Amer. Math. Soc.*, vol. 49, pp. 1–23.

Cousty, J. P., S. Verdeyme, M. Aubourg, and P. Guillon (1992), "Finite Elements for Microwave Device Simulation: Application to Microwave Dielectric Resonator Filters," *IEEE Trans. Microwave Theory Tech.*, vol. MTT-40, pp. 925–932.

Cristina, S., and A. Di Napoli (1983), "Combination of Finite and Boundary Elements for Magnetic Field Analysis," *IEEE Trans. Magnetics*, vol. MAG-19, pp. 2337–2339.

Crowley, C. W., P. P. Silvester, and H. Hurwitz, Jr. (1988), "Covariant Projection Elements for 3D Vector Field Problems," *IEEE Trans. Magnetics*, vol. MAG-25, pp. 397–400.

Csendes, Z. J., and P. P. Silvester (1970), "Numerical Solution of Dielectric Loaded Waveguides: I — Finite Element Analysis," *IEEE Trans. Microwave Theory Tech.*, vol. MTT-18, pp. 1124–1131.

Csendes, Z. J., and P. P. Silvester (1971a), "Numerical Solution of Dielectric Loaded Waveguides: II — Modal Approximation Technique," *IEEE Trans. Microwave Theory Tech.*, vol. MTT-19, pp. 504–509.

Csendes, Z. J., and P. P. Silvester (1971b), "Dielectric Loaded Waveguide Analysis Program," *IEEE Trans. Microwave Theory Tech.*, vol. MTT-19, p. 789.

Csendes, Z. J. (1976), "A Note on the Finite-Element Solution of Exterior-Field Problems," *IEEE Trans. Microwave Theory Tech.*, vol. MTT-24, pp. 468–473.

Csendes, Z. J. (1980), "The High-Order Polynomial Finite Element Method in Electromagnetic Field Computation," in *Finite Elements in Electrical and Magnetic Field Problems*, Eds. M. V. K. Chari, and P. P. Silvester, John Wiley & Sons, New York, pp. 125–143.

Csendes, Z. J., D. Shenton, and H. Shahnasser (1983), "Magnetic Field Computation Using Delaunay Triangulation and Complementary Finite Element Methods," *IEEE Trans. Magnetics*, vol. MAG-19, pp. 2551–2554.

Csendes, Z. J., and D. N. Shenton (1985), "Adaptive Mesh Refinement in the Finite Element Computation of Magnetic Fields," *IEEE Trans. Magnetics*, vol. MAG-21, pp. 1811–1816.

Csendes Z. J., and S. H. Wong (1987), "C^1 Quadratic Interpolation over Arbitrary Point Sets," *IEEE Comp. Graphics Appl.*, vol. 7, pp. 8–16.

Csendes, Z. J., and J. F. Lee (1988), "The Transfinite Element for Modeling MMIC Devices," *IEEE Trans. Microwave Theory Tech.*, vol. MTT-36, pp. 1639–1649.

Csendes, Z. J. (1990), "EM Simulators = CAE Tools," *IEEE Spectrum*, Nov., pp. 73–93.

Csendes, Z. J. (1991), "Vector Finite Elements for Electromagnetic Field Computation," *IEEE Trans. Magnetics*, vol. MAG-27, pp. 3958–3966.

Csendes, Z. J. (1995), "Spurious Modes in Finite-Element Methods," *IEEE Antennas Propag. Magazine*, vol. 37, pp. 12–24.

Curnier, A. (1983), "A Static Infinite Element," *Int. J. Numer. Meth. Eng.*, vol. 19, pp. 1479–1488.

Cvetkovic, S. R., and J. B. Davies (1986), "Self-Adjoint Vector Variational Formulation for Lossy Anisotropic Dielectric Waveguide," *IEEE Trans. Microwave Theory Tech.*, vol. MTT-34, pp. 129–134.

Cwik, T. (1992), "Coupling Finite Element and Integral Equation Solution Using Decoupled Boundary Meshes," *IEEE Trans. Antennas Propag.*, vol. AP-40, pp. 1496–1504.

Cwik, T., C. Zuffada, and V. Jamnejad (1996a), "The Coupling of Finite Element and Integral Equation Representations for Efficient Three-Dimensional Modeling of Electromagnetic Scattering and Radiation," in *Finite Element Software for Microwave Engineering*, Eds. T. Itoh, G. Pelosi, and P. P. Silvester, John Wiley & Sons, New York, pp. 147–167.

Cwik, T., C. Zuffada, and V. Jamnejad (1996b), "Modeling Three-Dimensional Scatterers Using a Coupled Finite Element-Integral Equation Formulation," *IEEE Trans. Antennas Propag.*, vol. AP-44, pp. 453–459.

Daigang, W., and J. Kexun (1994), "*p*-Version Adaptive Computation of FEM," *IEEE Trans. Magnetics*, vol. MAG-30, pp. 3515–3518.

Daly, P. (1971), "Hybrid-Mode Analysis of Microstrip by Finite-Element Methods," *IEEE Trans. Microwave Theory Tech.*, vol. MTT-19, pp. 19–25.

Daly, P. (1973a), "Singularities in Transmission Lines," in *The Mathematics of Finite Elements and Applications*, Ed. J. R. Whiteman, Academic Press, London, pp. 337–350.

Daly, P. (1973b), "Finite Elements for Field Problems in Cylindrical Coordinates," *Int. J. Numer. Meth. Eng.*, vol. 6, pp. 169–178.

Daly, P. (1974), "Polar Geometry Waveguides by Finite-Element Methods," *IEEE Trans. Microwave Theory Tech.*, vol. MTT-22, pp. 202–209.

Daly, P. (1984), "Upper and Lower Bounds to the Characteristic Impedance of Transmission Lines Using the Finite Element Method," *COMPEL*, vol. 3, pp. 65–78.

Daly, P. (1985), "Dual-Potential Problems in Transmission Lines with Limited or No Symmetry," *IEE Proc.*, vol. 132, pt. H, pp. 351–359.

Damjanic, F., and D. R. J. Owen (1984), "Mapped Infinite Elements in Transient Thermal Analysis," *Computers & Structures*, vol. 19, pp. 673–687.

D'Angelo, J., and I. D. Mayergoyz (1989), "On the Use of Local Absorbing Boundary Conditions for RF Scattering Problems," *IEEE Trans. Magnetics*, vol. MAG-25, pp. 3040–3043.

D'Angelo, J., and I. D. Mayergoyz (1990), "Finite Element Methods for the Solution of RF Radiation and Scattering Problems," *Electromagnetics*, vol. 10, pp. 177–179.

D'Angelo, J., and I. D. Mayergoyz (1991), "Three-Dimensional RF Scattering by the Finite Element Method," *IEEE Trans. Magnetics*, vol. MAG-27, pp. 3827–3832.

D'Angelo, J., and I. Mayergoyz (1996), "Phased Array Antenna Analysis," in *Finite Element Software for Microwave Engineering*, Eds. T. Itoh, G. Pelosi, and P. P. Silvester, John Wiley & Sons, New York, pp. 169–191.

Dasgupta, G. (1984), "Computations of Exterior Potential Fields by Infinite Substructuring," *Comp. Meth. Appl. Mech. Eng.*, vol. 46, pp. 295–305.

Davies, J. B., F. A. Fernández, and G. Y. Philippou (1982), "Finite Element Analysis of all Modes in Cavities with Circular Symmetry," *IEEE Trans. Microwave Theory Tech.*, vol. MTT-30, pp. 1975–1980.

Davies, J. B. (1993), "Finite Element Analysis of Waveguides and Cavities — A Review," *IEEE Trans. Magnetics*, vol. MAG-29, pp. 1578–1583.

Davies, J. B. (1994), "Uniform and Non-Uniform Mesh in Finite Element Analysis," *COMPEL*, vol. 13, supp. A, pp. 305–310.

Decreton, M. C. (1974), "Analysis of Open Structures by Finite-Element Resolution of Equivalent Closed-Boundary Problem," *Electronics Letters*, vol. 10, pp. 43–44.

Delaunay, B. (1934), "Sur la Sphere Vide," *Bull. Acad. Sci. USSR*, vol. 7, pp. 193–200.

Dibben, D. C., and M. Metaxas (1997), "Frequency Domain vs. Time Domain Finite Element Methods for Calculation of Fields in Multimode Cavities," *IEEE Trans. Magnetics*, vol. MAG-33, pp. 1468–1471.

Dillon, B. M., P. T. S. Liu, and J. P. Webb (1994), "Spurious Modes in Quadrilateral and Triangular Edge Elements," *COMPEL*, vol. 13, supp. A, pp. 311–316.

Dillon, B. M., and J. P. Webb (1994), "A Comparison of Formulations for the Vector Finite Element Analysis of Waveguides," *IEEE Trans. Microwave Theory Tech.*, vol. MTT-42, pp. 308–316.

Di Nallo, C., F. Frezza, and A. Galli (1996), "Analysis of BEM and FEM Codes for the Modelling of Microwave Passive Dielectric Devices," in *Software for Electrical Engineering Analysis and Design*, Ed. P. P. Silvester, Computational Mechanics Publications, Southampton, England, pp. 377–384.

Djordjević, A. R., T. K. Sarkar, and R. F. Harrington (1986), "Analysis of Lossy Transmission Lines with Arbitrary Nonlinear Terminal Networks," *IEEE Trans. Microwave Theory Tech.*, vol. MTT-34, pp. 660–666.

Djordjević, A. R., and T. K. Sarkar (1987), "Analysis of Time Response of Lossy Multiconductor Transmission Line Networks," *IEEE Trans. Microwave Theory Tech.*, vol. MTT-35, pp. 898–907.

Djordjević, A. R., T. K. Sarkar, and R. F. Harrington (1987), "Time-Domain Response of Multiconductor Transmission Lines," *Proc. IEEE*, vol. 75, pp. 743–764.

Djordjević, A. R., and T. K. Sarkar (1994), "Closed-Form Formulas for Frequency-Dependent Resistance and Inductance per Unit Length of Microstrip and Strip Transmission Lines," *IEEE Trans. Microwave Theory Tech.*, vol. MTT-42, pp. 241–248.

Djordjević, A. R., T. K. Sarkar, T. Roy, S. M. Rao, and M. Salazar-Palma (1995), "An Exact Method for Simulating Boundary Conditions for Mesh Termination in Finite-Difference Techniques," *Microwave Optical Tech. Letters*, vol. 8, pp. 88–90.

Donzelli, P. S., R. L. Spilker, P. L. Baehmann, Q. Niu, and M. Shepard (1992), "Automated Adaptive Analysis of the Biphasic Equations for Soft Tissue Mechanics Using *A Posteriori* Error Indicators," *Int. J. Numer. Meth. Eng.*, vol. 34, pp. 1015–1033.

Dorr, M. R. (1984), "The Approximation Theory for the *p*-Version of the Finite Element Method," *SIAM J. Numer. Anal.*, vol. 21, pp. 1180–1207.

Dorr, M. R. (1986), "The Approximation of Solutions of Elliptic Boundary-Value Problems via the *p*-Version of the Finite Element Method," *SIAM J. Numer. Anal.*, vol. 23, pp. 58–77.

Drago, G., P. Molfino, M. Nervi, and M. Repetto (1992), "A 'Local Field Error Problem' Approach for Error Estimation in Field Element Analysis," *IEEE Trans. Magnetics*, vol. MAG-28, pp. 1743–1746.

Duff, I. S. (1977), "A Survey of Sparse Matrix Research," *Proc. IEEE*, vol. 65, pp. 500–535.

Duffin, R. J. (1959), "Distributed and Lumped Networks," *J. Math. Mech.*, vol. 8, pp. 793–826.

Duffin, R. J., and T. A. Porsching (1959), "Bounds for the Conductance of a Leaky Plate via Network Models," *Proc. Symposium Generalized Networks*, Brooklyn, New York, pp. 411–422.

Dunavant, D. A., B. A. Szabó (1983), "*A Posteriori* Error Indicators for the *p*-Version of the Finite Element Method," *Int. J. Numer. Meth. Eng.*, vol. 19, pp. 1851–1870.

Durán, R., M. A. Muschietti, and R. Rodríguez (1991), "On the Asymptotic Exactness of Error Estimators for Linear Triangular Finite Elements," *Numer. Math.*, vol. 59, pp. 107–127.

Durán, R., M. A. Muschietti, and R. Rodríguez (1992), "Asymptotically Exact Error Estimators for Rectangular Finite Elements," *SIAM J. Numer. Anal.*, vol. 29, pp. 78–88.

Dyczij-Edlinger, R., and O. Biro (1996), "A Joint Vector and Scalar Potential Formulation for Driven High Frequency Problems Using Hybrid Edge and Nodal Finite Elements," *IEEE Trans. Microwave Theory Tech.*, vol MTT-44, pp. 15–23.

Emson, C. R. I. (1988), "Methods for the Solution of Open-Boundary Electromagnetic-Field Problems," *IEE Proc.*, vol. 135, pt. A., pp. 151–158.

English, W. J. (1971), "Vector Variational Solutions of Inhomogeneously Loaded Cylindrical Waveguide Structures," *IEEE Trans. Microwave Theory Tech.*, vol. MTT-19, pp. 9–18.

English, W. J., and F. J. Young (1971), "An *E* Vector Variational Formulation of the Maxwell Equations for Cylindrical Waveguide Problems," *IEEE Trans. Microwave Theory Tech.*, vol. MTT-19, pp.

40–46.
Engquist, B., and A. Majda (1977), "Absorbing Boundary Conditions for the Numerical Simulation of Waves," *Math. Comp.*, vol. 31, pp. 629–651.
Ergatoudis, I., B. M. Irons, and O. C. Zienkiewicz (1968), "Curved Isoparametric 'Quadrilateral Elements' for Finite Element Analysis," *Int. J. Solids Struct.*, vol. 4, pp. 31–42.
Eriksson, K., and C. Johnson (1988), "An Adaptive Finite Element Method for Linear Elliptic Problems," *Math. Comp.*, vol. 50, pp. 361–383.
Eswarappa, C., G. Costache, and W. J. R. Hoefer (1989), "Finlines in Rectangular and Circular Waveguide Housings Including Substrate Mounting and Bending Effects — Finite Element Analysis," *IEEE Trans. Microwave Theory Tech.*, vol. MTT-37, pp. 299–306.
Evans, D. J. (1973), "The Analysis and Application of Sparse Matrix Algorithms in the Finite Element Method," in *The Mathematics of Finite Elements and Applications*, Ed. J. R. Whiteman, Academic Press, London, pp. 427–447.
Ewing, R. E. (1990), "*A Posteriori* Error Estimation," *Comp. Meth. Appl. Mech. Eng.*, vol. 82, pp. 59–72.
Felippa, C. A. (1994), "An Appreciation of R. Courant's 'Variational Methods for the Solution of Problems of Equilibrium and Vibration' 1943," *Int. J. Numer. Meth. Eng.*, vol. 37, pp. 2159–2187.
Felippa, C. A. (1966), "Refined Finite Element Analysis of Linear and Nonlinear Two Dimensional Structures," *Report UC SESM, 66–22*, Dept. of Civil Eng., Univ. of California, Berkeley.
Fernandes, P., P. Girdinio, P. Molfino, and M. Repetto (1988), "Local Error Estimates for Adaptive Mesh Refinement," *IEEE Trans. Magnetics*, vol. MAG-24, pp. 299–302.
Fernandes, P., P. Girdinio, P. Molfino, G. Molinari, and M. Repetto (1990a), "A Comparison of Adaptive Strategies for Mesh Refinement Based on *A Posteriori* Local Error Estimation Procedures," *IEEE Trans. Magnetics*, vol. MAG-26, pp. 795–798.
Fernandes, P., P. Girdinio, P. Molfino, and M. Repetto (1990b), "An Enhanced Error Estimator Procedures for Finite Element Field Computation with Adaptive Mesh Refinement," *IEEE Trans. Magnetics*, vol. MAG-26, pp. 2187–2189.
Fernandes, P., P. Girdinio, G. Molinari, and M. Repetto (1991), "Local Error Estimation Procedures as Refinement Indicators in Adaptive Meshing," *IEEE Trans. Magnetics*, vol. MAG-27, pp. 4189–4192.
Fernandes, P., P. Girdinio, M. Repetto, and G. Secondo (1992), "Refinement Strategies in Adaptive Meshing," *IEEE Trans. Magnetics* vol. MAG-28, pp. 1739–1742.
Fernandes, P., and G. L. Sabbi (1992), "On the Spurious Modes in Electromagnetic Field Problems Applications," *Proc. ICEF '92*, International Academic Publishers, Hangzhou, pp. 89–92.
Fernandes, P. P. Girdinio, and G. Molinari (1994), "Techniques of *h*-Refinement in Adaptive Meshing Algorithms," *COMPEL*, vol. 13, supp. A, pp. 329–334.
Fernández, F. A., and Y. Lu (1991), "A Variational Finite Element Formulation for Dielectric Waveguides in Terms of Transverse Magnetic Fields," *IEEE Trans. Magnetics*, vol. MAG-27, pp. 3864–3867.
Fernández, F. A., Y. Lu, J. B. Davies, and S. Zhu (1993a), "Finite Element Analysis of Complex Modes in Inhomogeneous Waveguides," *IEEE Trans. Magnetics*, vol. MAG-29, pp. 1601–1604.
Fernández, F. A., Y. C. Yong, and R. P. Ettinger (1993b), "A Simple Adaptive Mesh Generator for 2-D Finite Element Calculations," *IEEE Trans. Magnetics*, vol. MAG-29, pp. 1882–1885.
Ferragut, L., R. Montenegro, and G. Winter (1993), "Estimación *A Posteriori* del Error por un Método de Dualidad en Problemas Elípticos," in *Métodos Numéricos en Ingeniería*, Eds. F. Navarrina, and M. Casteleiro, SEMNI, La Coruña, Spain, vol. 1, pp. 631–640.
Field, D. A., and Y. Pressburger (1993), "An *h-p* Multigrid Method for Finite Element Analysis," *Int. J. Numer. Meth. Eng.*, vol. 36, pp. 893–908.
Finlayson, B. A. (1975), "Weighted Residual Methods and Their Relation to Finite Element Methods in Flow Problems," in *Finite Elements in Fluids*, vol. 2, John Wiley & Sons, New York, pp. 1–29.
Fish, J., and R. Guttal (1997), "Adaptive Solver for the *p*-Version of FEM," *Int. J. Numer. Meth. Eng.*, vol. 40, pp. 1767–1784.
Fix, G. (1972), "Effects of Quadrature Errors in Finite Element Approximation of Steady State,

Eigenvalue, and Parabolic Problems," in *The Mathematical Foundations of the Finite Element Method with Applications to Partial Differential Equations*, Ed. A. K. Aziz, Academic Press, New York, pp. 525–556.

Foo, S. L., and P. P. Silvester (1992) "Boundary-Marching Method for Discontinuity Analysis in Waveguides of Arbitrary Cross Section," *IEEE Trans. Microwave Theory Tech.*, vol. MTT-40, pp. 1889–1893.

Foo, S. L., and P. P. Silvester (1993) "Finite Element Analysis of Inductive Strips in Unilateral Finlines," *IEEE Trans. Microwave Theory Tech.*, vol. MTT-41, pp. 298–304.

Forsman, K., and L. Kettunen (1994), "Tetrahedral Mesh Generation in Convex Primitives by Maximizing Solid Angles," *IEEE Trans. Magnetics*, vol. MAG-30, pp. 3535–3538.

Freeman, E. M., and D. A. Lowther (1988), "A Novel Mapping Technique for Open Boundary Finite Element Solutions to Poisson's Equation," *IEEE Trans. Magnetics*, vol. MAG-24, pp. 2934–2937.

Fremond, M. (1973), "Dual Formulations for Potential and Complementary Energies. Unilateral Boundary Conditions. Applications to the Finite Element Method," in *The Mathematics of Finite Elements and Applications*, Ed. J. R. Whiteman, Academic Press, London, 175–188.

Fried, I. (1973), "The l_2 and l_∞ Condition Numbers of the Finite Element Stiffness and Mass Matrices, and the Pointwise Convergence of the Method," in *The Mathematics of Finite Elements and Applications*, Ed. J. R. Whiteman, Academic Press, London, pp. 163–174.

Frey, W. H. (1987), "Selective Refinement: A New Strategy for Automatic Node Placement in Graded Triangular Meshes," *Int. J. Numer. Meth. Eng.*, vol. 24, pp. 2183–2200.

Fuenmayor, F. J., and J. L. Oliver (1993), "Criterios de Distribución del Error de Discretización en Procesos *h*-Adaptivos del MEF," in *Métodos Numéricos en Ingeniería*, Eds. F. Navarrina, and M. Casteleiro, SEMNI, La Coruña, Spain, vol. 1, pp. 878–888.

Gago, J. P. S. R., D. W. Kelly, O. C. Zienkiewicz, and I. Babuška (1983), "*A Posteriori* Error Analysis and Adaptive Processes in the Finite Element Method: Part II — Adaptive Mesh Refinement," *Int. J. Numer. Meth. Eng.*, vol. 19, pp. 1621–1656.

Gakũrũ, M. K., and R. L. Ferrari (1992), "Finite-Element Analysis of High-Frequency Axisymmetric Device Problems," *IEEE Trans. Magnetics*, vol. MAG-28, pp. 1635–1638.

Galerkin, B. G. (1915), "Rods and Plates. Series on Some Problems of Elastic Equilibrium of Rods and Plates," *Vestn. Inzh. Tech.*, vol. 19, pp. 897–908 (In Russian).

Gao, D. S., A. T. Yang, and S. M. Kang (1990), "Modeling and Simulation of Interconnection Delays and Crosstalks in High-Speed Integrated Circuits," *IEEE Trans. Circuits and Systems*, vol. CAS-37, pp. 1–9.

García-Castillo, L. E., and M. Salazar-Palma (1992a), "On the Use of Different Formulations Based on Edge Elements for the Dynamic Analysis of General Waveguiding Structures by Means of the Finite Element Method," *Proc. URSI Int. Symposium Electromagnetic Theory*, Sydney, pp. 31–33.

García-Castillo, L. E., and M. Salazar-Palma (1992b), "A Non-Standard Finite Element Method for the Dynamic Analysis of Microwave Waveguiding Structures and Transmission Line Structures," *22nd European Microwave Conf. Proc.*, Helsinki, pp. 1012–1017,

García-Castillo, L. E., T. K. Sarkar, and M. Salazar-Palma (1994), "An Efficient Finite Element Method Employing Wavelet Type Basis Functions," *COMPEL*, vol. 13, supp. A, pp. 287–292.

García-Castillo, L. E., M. Salazar-Palma, T. K. Sarkar, and R. S. Adve (1995), "Efficient Solution of the Differential Form of Maxwell's Equations in Rectangular Regions," *IEEE Trans. Microwave Theory Tech.*, vol. MTT-43, pp. 647–654.

Gasiorski, A. K. (1985), "A Finite Element Method for Analyzing Skin Effect in Conductors with Unknown Values of Surface Boundary Conditions," *Int. J. Numer. Meth. Eng.*, vol. 21, pp. 1641–1657.

Gavrilovic, M. M., and J. P. Webb (1997), "An Error Indicator for the Calculation of Global Quantities by the *p*-Adaptive Finite Element Method," *IEEE Trans. Magnetics*, vol. MAG-33, pp. 4128–4130.

Gedney, S., and R. Mittra (1991), "Analysis of the Electromagnetic Scattering by Thick Gratings Using Combined FEM/MM Solution," *IEEE Trans. Antennas Propag.*, vol. AP-39, pp. 1605–1614.

Gedney, S. D., J. F. Lee, and R. Mittra (1992), "A Combined FEM/MoM Approach to the Analysis of Plane Wave Diffraction by Arbitrary Gratings," *IEEE Trans. Microwave Theory Tech.*, vol. MTT-40, pp. 363–370.

Geisel, J., K. H. Muth, and W. Heinrich (1992), "The Behavior of the Electromagnetic Field at Edges of Media with Finite Conductivity," *IEEE Trans. Microwave Theory Tech.*, vol. MTT-40, pp. 158–164.

Gentili, G. G., and M. Salazar-Palma (1995), "The Definition and Computation of Modal Characteristic Impedance in Quasi-TEM Coupled Transmission Lines," *IEEE Trans. Microwave Theory Tech.*, vol. MTT-45, pp. 398–343.

Georges, M. K., and M. S. Shepard (1991), "Automated Adaptive Two-Dimensional System for the *hp*-Version of the Finite Element Method," *Int. J. Numer. Meth. Eng.*, vol. 32, pp. 867–893.

Giannacopoulos, D., and S. McFee (1994), "Towards Optimal *h-p* Adaptation Near Singularities in Finite Element Electromagnetics," *IEEE Trans. Magnetics*, vol. MAG-30, pp. 3523–3526.

Giannacopoulos, D., and S. McFee (1997), "An Experimental Study of Superconvergence Phenomena in Finite Element Magnetics," *IEEE Trans. Magnetics*, vol. MAG-33, pp. 4132–4139.

Gil, J. M., and J. Zapata (1994), "Efficient Singular Element for Finite Element Analysis of Quasi-TEM Transmission Lines and Waveguides with Sharp Metal Edges," *IEEE Trans. Microwave Theory Tech.*, vol. MTT-42, pp. 92–98.

Gil, J. M., and J. Zapata (1995), "A New Scalar Transition Finite Element for Accurate Analysis of Waveguides with Field Singularities," *IEEE Trans. Microwave Theory Tech.*, vol. MTT-43, pp. 1978–1982.

Gil, J. M., and T. P. Webb (1997), "A New Edge Element for the Modeling of Field Singularities in Transmission Lines and Waveguides," *IEEE Trans. Microwave Theory Tech.*, vol. MTT-45. pp. 2125–2130.

Gil, J. M., and J. Zapata (1997), "A Comparison of Scalar Singular Finite Elements Applied to the Analysis of Homogeneous Waveguides with Sharp Edges," *Int. J. Numer. Mod.: Electronic Net. Devices Fields*, vol. 10, pp. 343–354.

Girdinio, P., P. Molfino, G. Molinari, L. Puglisi, and A. Viviani (1983), "Finite Difference and Finite Element Grid Optimization by the Grid Iteration Method," *IEEE Trans. Magnetics*, vol. MAG-19, pp. 2543–2546.

Girdinio, P., P. Molfino, M. Nervi, and A. Manella (1996), "Non-Linear Magnetostatic Adaption Using a 'Local Field Error' Approach," *IEEE Trans. Magnetics*, vol. MAG-32, pp. 1365–1364.

Givoli, D., and J. B. Keller (1994), "Non-Reflecting Finite Elements," in *The Mathematics of Finite Elements and Applications*, Ed. J. R. Whiteman, John Wiley & Sons, Chichester, England, pp. 307–314.

Gluckstern, R. L., and E. N. Opp (1985), "Calculation of Dispersion Curves in Periodic Structures," *IEEE Trans. Magnetics*, vol. MAG-21, pp. 2344–2346.

Golias, N. A., and T. D. Tsiboukis (1991), "Adaptive Refinement in 2-D Finite Element Applications," *Int. J. Numer. Mod.: Electronic Net. Devices Fields*, vol. 4., pp. 81–95.

Golias, N. A., and T. D. Tsiboukis (1992), "Three-Dimensional Automatic Adaptive Mesh Generation," *IEEE Trans. Magnetics*, vol. MAG-28, pp. 1700–1703.

Golias, N. A., and T. D. Tsiboukis (1993), "Adaptive Refinement Strategies in Three Dimensions," *IEEE Trans. Magnetics*, vol. MAG-29, pp. 1886–1889.

Golias, N. A., and A. G. Papagiannakis (1994), "Efficient Mode Analysis with Edge Elements and 3-D Adaptive Refinement," *IEEE Trans. Microwave Theory Tech.*, vol. MTT-42, pp. 99–106.

Golias, N. A., T. D. Tsiboukis, and A. Bossavit (1994), "Constitutive Inconsistency: Rigorous Solution of Maxwell Equations Based on Complementary Approach," *IEEE Trans. Magnetics*, vol. MAG-30, pp. 3586–3589.

Golias, N. A., and T. D. Tsiboukis (1996), "Adaptive Methods in Computational Magnetics," *Int. J. Numer. Mod.: Electronic Net. Devices Fields*, vol. 9, pp. 71–80.

Gong, J., J. L. Volakis, A. C. Wong, and H. T. G. Wang (1994), "A Hybrid Finite Element-Boundary Integral Method for the Analysis of Cavity-Backed Antennas of Arbitrary Shape," *IEEE Trans.*

Antennas Propag., vol. AP-42, pp. 1233–1241.

Gong, Z., and A. Glisson (1990), "A Hybrid Equation Approach for the Solution of Electromagnetic Scattering Problems Involving Two-Dimensional Inhomogeneous Dielectric Cylinders," *IEEE Trans. Antennas Propag.*, vol. AP-38, pp. 60–68.

Goosen, K. W., and R. B. Hammond (1989), "Modeling of Picosecond Pulse Propagation in Microstrip Interconnections on Integrated Circuits," *IEEE Trans. Microwave Theory Tech.*, vol. MTT-37, pp. 469–478.

Gordon, R., R. Mittra, A. Glisson, and E. Michielssen (1993), "Finite Element Analysis of Electromagnetic Scattering by Complex Bodies Using an Efficient Numerical Boundary Condition for Mesh Truncation," *Electronics Letters*, vol. 29, pp. 1102–1103.

Gothard, G. K., S. M. Rao, T. Roy, T. K. Sarkar, and M. Salazar-Palma (1995), "Finite Element Solution of Open Region Problems Incorporating the Measured Equation of Invariance," *Microwave and Guided Wave Letters*, vol. 5, pp. 252–254.

Graglia, R. D., A. Freni, and G. Pelosi (1993), "A Finite Element Approach to the Electromagnetic Interaction with Rotating Penetrable Cylinder of Arbitrary Cross Section," *IEEE Trans. Antennas Propag.*, vol. AP-41, pp. 635–650.

Graglia, R. D., D. R. Wilton, and A. F. Peterson (1997), "Higher Order Interpolatory Vector Bases for Computational Electromagnetics," *IEEE Antennas Propag.*, vol. AP-45, pp. 329–342.

Gratkowski, S., and M. Ziolkowski (1992), "A Three-Dimensional Infinite Element for Modeling Open Boundary Field Problems," *IEEE Trans. Magnetics*, vol. MAG-28, pp. 1675–1678.

Gratkowski, S., L. Pichon, and A. Razek (1996), "New Infinite Elements for a Finite Element Analysis of 2D Scattering Problems," *IEEE Trans. Magnetics*, vol. MAG-32, pp. 882–885.

Greene, R. E., R. E. Jones, R. W. McKay, and D. R. Strome (1969), "Generalized Variational Principles in the Finite Element Method," *AIAAJ*, vol. 7, pp. 1254–1261.

Gui, W., and I. Babuška (1986a), "The *h*, *p* and *h-p* Versions of the Finite Element Method in 1 Dimension. Part I. The Error Analysis of the *p*-Version," *Numer. Math.*, vol. 49, pp. 577–683.

Gui, W., and I. Babuška (1986b), "The *h*, *p* and *h-p* Versions of the Finite Element Method in 1 Dimension. Part II. The Error Analysis of the *h*- and *h-p* Versions," *Numer. Math.*, vol. 49, pp. 577–683.

Gui, W., and I. Babuška (1986c), "The *h*, *p* and *h-p* Versions of the Finite Element Method in 1 Dimension. Part III. The Adaptive *h-p* Version," *Numer. Math.*, vol. 49, pp. 577–683.

Guo, B. Q. (1988), "The *h-p* Version of the Finite Element Method for Elliptic Equations of Order 2m," *Numer. Math.*, vol. 53, pp. 199–224.

Gupta, K. K., and J. L. Meek (1996), "A Brief History of the Beginning of the Finite Element Method," *Int. J. Numer. Meth. Eng.*, vol. 39, pp. 3761–3774.

Hahn, S., C. Camels, G. Meunier, and J. L. Coulomb (1988), "*A Posteriori* Error Estimate for Adaptive Finite Element Mesh Generation," *IEEE Trans. Magnetics*, vol. MAG-24, pp. 315–317.

Hammond, P., and J. Penman (1976), "Calculation of Inductance and Capacitance by Means of Dual Energy Principles," *Proc. IEE*, vol. 123, pp. 554–559.

Hammond, P., and J. Penman (1978), "Calculation of Eddy Current by Dual Energy Methods," *Proc. IEE*, vol. 125, pp. 701–708.

Hammond, P. (1982), "Use of Potentials in Calculation of Electromagnetic Fields," *IEE Proc.*, vol. 129, pt. A, pp. 106–112.

Hammond, P., and T. D. Tsiboukis (1983), "Dual Finite-Element Calculations for Static Electric and Magnetic Fields," *IEE Proc.*, vol. 130, pt. A, pp. 105–111.

Hammond, P. (1988), "Some Thoughts on the Numerical Modelling of Electromagnetic Processes," *Int. J. Numer. Mod.: Electronic Net., Devices Fields*, vol. 1, pp. 3–6.

Hanna, V. F., and D. Thebault (1984), "Theoretical and Experimental Investigation of Asymmetric Coplanar Waveguides," *IEEE Trans. Microwave Theory Tech.*, vol. MTT-32, pp. 1649–1651.

Hano, M. (1984), "Finite-Element Analysis of Dielectric-Loaded Formulation in Terms of the Magnetic Field Vector for Dielectric Waveguides," *IEEE Trans. Microwave Theory Tech.*, vol. MTT-32, pp.

1275–1279.

Hara, M., T. Wada, T. Fukasawa, and F. Kikuchi (1983), "A Three-Dimensional Analysis of RF Electromagnetic Fields by the Finite Element Method," *IEEE Trans. Magnetics*, vol. MAG-19, pp. 2417–2420.

Harari, I., P. E. Barbone, J. M. Montgomery (1997), "Finite Element Formulations for Exterior Problems: Application to Hybrid Methods, Non-Reflecting Boundary Conditions and Infinite Elements," *Int. J. Numer. Meth. Eng.*, vol. 40, pp. 2791–2805.

Harrington, R. F. (1967), "Matrix Methods for Fields Problems," *Proc. IEEE*, vol. 55, pp. 136–149.

Harrington, R. F., and C. Wei (1984), "Losses on Multiconductor Transmission Lines in Multilayered Dielectric Media," *IEEE Trans. Microwave Theory Tech.*, vol. MTT-32, pp. 705–710.

Hasegawa, T., S. Banba, H. Ogawa, and H. Nakamoto (1991), "Characteristics of Valley Microstrip Lines for Use in Multilayer MMIC's," *IEEE Microwave Guided Wave Letters*, vol. 1, pp. 275–277.

Hasegawa, T., S. Banba, and H. Ogawa (1992), "A Branchline Hybrid Using Valley Microstrip Lines," *IEEE Microwave Guided Wave Letters*, vol. 2, pp. 76–78.

Hayata, K., M. Koshiba, M. Eguchi, and M. Suzuki (1986), "Vectorial Finite-Element Method Without Any Spurious Solutions for Dielectric Waveguiding Problems Using Transverse Magnetic-Field Component," *IEEE Trans. Microwave Theory Tech.*, vol. MTT-34, pp. 1120–1124.

Hayata, K., M. Eguchi, and M. Koshiba (1988), "Self-Consistent Finite/Infinite Element Scheme for Unbounded Guided Wave Problems," *IEEE Trans. Microwave Theory Tech.*, vol. MTT-36, pp. 614–616.

Hayata, K., M. Eguchi, and M. Koshiba (1989), "Finite Element Formulation for Guided-Wave Problems Using Transverse Electric Field Component," *IEEE Trans. Microwave Theory Tech.*, vol. MTT-37, pp. 256–258.

Hazel, T. G., and A. Wexler (1972), "Variational Formulation of the Dirichlet Boundary Condition," *IEEE Trans. Microwave Theory Tech.*, vol. MTT-20, pp. 385–390.

Heighway, E. A., and C. S. Biddlecombe (1982), "Two Dimensional Automatic Triangular Mesh Generation for the Finite Element Electromagnetic Package PE2D," *IEEE Trans. Magnetics*, vol. MAG-18, pp. 594–598.

Heighway, E. A. (1983), "A Mesh Generator for Automatically Subdividing Irregular Polygons into Quadrilaterals," *IEEE Trans. Magnetics*, vol. MAG-19, pp. 2535–2538.

Heise, U. (1973), "Combined Application of Finite Element Methods and Richardson Extrapolation to the Torsion Problem," in *The Mathematics of Finite Elements and Applications*, Ed. J. R. Whiteman, Academic Press, London, pp. 225–237.

Helal, M., J. F. Legier, P. Pribetich, and P. Kennis (1994), "Full-Wave Analysis Using a Tangential Vector Finite-Element Formulation of Arbitrary Cross-Section Transmission Lines for Millimeter and Microwave Applications," *Microwave Optical Tech. Letters*, vol. 7, pp. 401–404.

Henneberger, G., G. Meunier, J. C. Sabonnadière, Ph. K. Sattler, and D. Shen (1990), "Sensitivity Analysis of the Nodal Position in the Adaptive Refinement of Finite Element Meshes," *IEEE Trans. Magnetics*, vol. MAG-26, pp. 787–790.

Henshell, R. D. (1973), "On Hybrid Finite Elements," in *The Mathematics of Finite Elements and Applications*, Ed. J. R. Whiteman, Academic Press, London, pp. 299–311.

Hernández-Figueroa, H. E., and G. Pagiatakis (1993), "Shape Function Optimization in the Finite Element Analysis of Waveguides," *IEEE Trans. Microwave Theory Tech.*, vol. MTT-41, pp. 1235–1238.

Hernández-Gil, J. F., and J. Pérez-Martínez (1985), "Analysis of Dielectric Resonators with Tuning Screw and Supporting Structure," *IEEE Trans. Microwave Theory Tech.*, vol. MTT-33, pp. 1453–1457.

Hernández-Gil, J. F., and J. Pérez-Martínez (1986), "Analysis of the Coupling Coefficient between a Cylindrical Dielectric Resonator and a Fin-Line," *IEEE MTT-S Int. Symposium Digest*, Baltimore, MD, pp. 221–224.

Hernández-Gil, J. F., R. Pérez-Leal, and A. Gebauer (1987), "Resonant Frequency Stability Analysis of Dielectric Resonators with Tuning Mechanisms," *IEEE MTT-S Int. Symposium Digest*, pp. 345–348.

Hinton, E., M. Özakça, and N. V. R. Rao (1991), "Adaptive Analysis of Thin Shells Using Facet

Elements," *Int. J. Numer. Meth. Eng.*, vol. 32, pp. 1283–1301.

Hirayama, K., and M. Koshiba (1989), "Analysis of Discontinuities in an Open Dielectric Slab Waveguide by Combination of Finite and Boundary Elements," *IEEE Trans. Microwave Theory Tech.*, vol. MTT-37, pp. 761–768.

Hirayama, K., and M. Koshiba (1990), "Numerical Analysis of Arbitrarily Shaped Discontinuities Between Planar Dielectric Waveguides with Different Thicknesses," *IEEE Trans. Microwave Theory Tech.*, vol. MTT-38, pp. 260–264.

Hirayama, K., and M. Koshiba (1992), "Analysis of Discontinuities in an Asymmetric Dielectric Slab Waveguide by Combination of Finite and Boundary Elements," *IEEE Trans. Microwave Theory Tech.*, vol. MTT-40, pp. 686–691.

Hirayama, K., M. S. Alam, Y. Hayashi, and M. Koshiba (1994), "Vector Finite Element Method with Mixed-Interpolation-Type Triangular-Prism Element for Waveguide Discontinuities," *IEEE Trans. Microwave Theory Tech.*, vol. MTT-42, pp. 2311–2316.

Hirayama, K., Y. Hayashi, and M. Koshiba (1996), "Analysis of Electromagnetic Waveguide Bends by Three-Dimensional Finite Element Method with Edge Elements," *Int. J. Microwave Millimeter-Wave Computer-Aided Eng.* vol. 6, pp. 474–481.

Hoefer, W. J. R. (1985), "The Transmission-Line Matrix Method — Theory and Applications," *IEEE Trans. Microwave Theory Tech.*, vol. MTT-33, pp. 882–893.

Hoffman, R. K. (1984), "Variational Formulation of Electromagnetic Field Problems in Gyrotropic Media," *AEÜ*, vol. 38, pp. 271–277.

Hoole, S. R. H. (1987), "Nodal Perturbations in Adaptive Expert Finite Element Mesh Generation," *IEEE Trans. Magnetics*, vol. MAG-23, pp. 2635–2637.

Hoole, S. R. H. (1988), "Flux Density and Energy Perturbations in Adaptive Finite Element Mesh Generation," *IEEE Trans. Magnetics*, vol. MAG-24, pp. 322–325.

Hoole, S. R. H., S. Jayakumaran, A. W. Ananadaraj, and P. R. Hoole (1989), "Relevant, Purpose Based Error Criteria for Adaptive Finite Element Mesh Generation," *J. Electromagnetic Waves Appl.*, vol. 3, pp. 167–177.

Hoole, S. R. H. (1990), "Eigen Value and Eigen Vector Perturbation and Adaptive Mesh Generation in the Analysis of Waveguides," *IEEE Trans. Magnetics*, vol. MAG-26, pp. 791–794.

Hoppe, D. J., L. W. Epp, and J. F. Lee (1994), "Hybrid Symmetric FEM/MOM Formulation Applied to Scattering by Inhomogeneous Bodies of Revolution," *IEEE Trans. Antennas Propag.*, vol. AP-42, pp. 798–805.

Hoppe, V. (1986), "Finite Elements with Harmonic Interpolation Functions," in *The Mathematics of Finite Elements and Applications*, Ed. J. R. Whiteman, Academic Press, London, pp. 131–142.

Horno, M., F. L. Mesa, F. Medina, and R. Marqués (1990), "Quasi-TEM Analysis of Multilayered, Multiconductor Coplanar Structures with Dielectric and Magnetic Anisotropy Including Substrate Losses," *IEEE Trans. Microwave Theory Tech.*, vol. MTT-38, pp. 1059–1068.

How, H., R. G. Seed, C. Vittoria, D. B. Chrisey, J. S. Horwitz, C. Carosella, and V. Folen (1992), "Microwave Characteristics of High-Tc Superconducting Coplanar Waveguide Resonator," *IEEE Trans. Microwave Theory Tech.*, vol. MTT-40, pp. 1668–1673.

Howe, H. (1984), "Microwave Integrated Circuits — An Historical Perspective," *IEEE Trans. Microwave Theory Tech.*, vol. MTT-32, pp. 991–996.

Hrenikoff, A. (1941), "Solution of Problems in Elasticity by the Framework Method," *J. Appl. Mech.*, vol. 8, A 169–175.

Hurd, R. A. (1976), "The Edge Condition in Electromagnetics," *IEEE Trans. Antennas Propag.*, vol. AP-4, pp. 70–73.

Hurwitz, Jr., H., (1984) "Infinitesimal Scaling — A New Procedure for Modeling Exterior Field Problems," *IEEE Trans. Magnetics*, vol. MAG-20, pp. 1918–1923.

Hurwitz, Jr., H., (1988) "The Surface Projection Approach to Vector Field Calculations," *Math. Comp. Mod.*, vol. 11, pp. 313–316.

Ihlenburg, F., and I. Babuška (1995), "Dispersion Analysis and Error Estimation of Galerkin Finite

Element Methods," *Int. J. Numer. Meth. Eng.*, vol. 38, pp. 3745–3774.

Ikeuchi, M., H. Sawami, and H. Niki (1981), "Analysis of Open-Type Dielectric Waveguides by the Finite-Element Iterative Method," *IEEE Trans. Microwave Theory Tech.*, vol. MTT-29, pp. 234–239.

Imhoff, J. F., G. Meunier, X. Brunotte and J. C. Sabonnadière (1990), "An Original Solution for the Unbounded Electromagnetic 2D and 3D Problems throughout the Finite Element Method," *IEEE Trans. Magnetics*, vol. MAG-26, pp. 1659–1661.

Iribarren, G., and M. C. Rivara (1992), "Estudio de un Algoritmo de Refinamiento de Triángulos," in *Numerical Methods in Engineering and Applied Sciences*, Eds. H. Alder, J. C. Heinrich, S. Lavanchy, E. Oñate, and B. Suárez, CIMNE, Barcelona, pp. 524–532.

Irons, B. M., and O. C. Zienkiewicz (1968), "The Isoparametric Finite Element System — A New Concept in Finite Element Analysis," *Proc. Conf. Recent Advances in Stress Analysis*, Royal Aeronautical Society, London.

Ise, K., and M. Koshiba (1988), "Numerical Analysis of *H*-Plane Waveguide Junctions by Combination of Finite and Boundary Elements," *IEEE Trans. Microwave Theory Tech.*, vol. MTT-36, pp. 1343–1351.

Ise, K., and M. Koshiba (1989), "Equivalent Circuits for Dielectric Post in a Rectangular Waveguide," *IEEE Trans. Microwave Theory Tech.*, vol. MTT-37, pp. 1823–1825.

Ise, K., K. Inoue, and M. Koshiba (1990), "Three-Dimensional Finite-Element Solution of Dielectric Scattering Obstacles in a Rectangular Waveguide," *IEEE Trans. Microwave Theory Tech.*, vol. MTT-38, pp. 1352–1359.

Ise, K., K. Inoue, and M. Koshiba (1991), "Three-Dimensional Finite-Element Method with Edge Elements for Electromagnetic Waveguide Discontinuities," *IEEE Trans. Microwave Theory Tech.*, vol. MTT-39, pp. 1289–1295.

Iskander, M. F. (1993), "Computer-Based Electromagnetic Education," *IEEE Trans. Microwave Theory Tech.*, vol. MTT-41, pp. 920–931.

Israel, M., and R. Miniowitz (1987), "An Efficient Finite Element Method for Nonconvex Waveguide Based on Hermitian Polynomials," *IEEE Trans. Microwave Theory Tech.*, vol. MTT-35, pp. 1019–1026.

Israel, M., and R. Miniowitz (1990), "Hermitian Finite Element Method for Inhomogeneous Waveguides," *IEEE Trans. Microwave Theory Tech.*, vol. MTT-38, pp. 1319–1327.

Itoh, T. (1980), "Spectral Domain Admittance Approach for Dispersion Characteristics of Generalized Printed Transmission Lines," *IEEE Trans. Microwave Theory Tech.*, vol. MTT-28, pp. 733–736.

Itoh, T. (1989b), "Overview of Quasi-Planar Transmission Lines," *IEEE Trans. Microwave Theory Tech.*, vol. MTT-37, pp. 275–280.

Jänicke, L., and A. Kost (1996), "Error Estimation and Adaptive Mesh Generation in the 2D and 3D Finite Element Method," *IEEE Trans. Magnetics*, vol. MAG-32, pp. 1334–1337.

Janković, D., M. LaBelle, D. C. Chang, J. M. Dunn, and R. C. Booton (1994), "A Hybrid Method for the Solution of Scattering from Inhomogeneous Dielectric Cylinders of Arbitrary Shape," *IEEE Trans. Antennas Propag.*, vol. AP-42, pp. 1215–1221.

Jansen, R. H. (1985), "The Spectral-Domain Approach for Microwave Integrated Circuits," *IEEE Trans. Microwave Theory Tech.*, vol. MTT-33, pp. 1043–1056.

Jeng, S., and C. Chen (1984), "On Variational Electromagnetics: Theory and Application," *IEEE Trans. Antennas Propag.*, vol. AP-32, pp. 902–907.

Jetvić, J. O., and R. Lee (1994), "A Theoretical and Numerical Analysis of the Measured Equation of Invariance," *IEEE Trans. Antennas Propag.*, vol. AP-42, pp. 1097–1105. See "Comments" by K. K. Mei, and Y. W. Liu (1995), in *IEEE Trans. Antennas Propag.*, vol. AP-43, pp. 1168–1171.

Jetvić, J. O., and R. Lee (1995a), "How Invariant is the Measured Equation of Invariance?," *IEEE Microwave Guided Wave Letters*, vol. 5, pp. 45–47. See "Comments" by K. K. Mei, and Y. W. Liu (1995), in *IEEE Microwave Guided Wave Letters*, vol. 5, p. 417.

Jetvić, J. O., and R. Lee (1995b), "An Analytic Characterization of the Error in the Measured Equation of Invariance," *IEEE Trans. Antennas Propag.*, vol. AP-43, pp. 1109-1115.

Jin, H., and R. Vahldieck (1992), "The Frequency Domain Transmission Line Matrix Method — A New Concept," *IEEE Trans. Microwave Theory Tech.*, vol. MTT-40, pp. 2207–2218.

Jin, H., and R. Vahldieck (1993), "Full-Wave Analysis of Guiding Structures Using a 2-D Array of 3-D TLM Nodes," *IEEE Trans. Microwave Theory Tech.*, vol. MTT-41, pp. 472–477.

Jin, J. M., and V. V. Liepa (1988a), "Application of Hybrid Finite Element Method to Electromagnetic Scattering from Coated Cylinders," *IEEE Trans. Antennas Propag.*, vol. AP-36, pp. 50–54.

Jin, J. M., and V. V. Liepa (1988b), "A Note on Hybrid Finite Element Method for Solving Scattering Problems," *IEEE Trans. Magnetics*, vol. MAG-24, pp. 1486–1490.

Jin, J. M., and J. L. Volakis (1991a), "A Finite Element-Boundary Integral Formulation for Scattering by Three-Dimensional Cavity-Backed Apertures," *IEEE Trans. Antennas Propag.*, vol. AP-39, pp. 97–104.

Jin, J. M., and J. L. Volakis (1991b), "Electromagnetic Scattering by and Transmission Trough a Three-Dimensional Slot in a Thick Conducting Plane," *IEEE Trans. Antennas Propag.*, vol. AP-39, pp. 543–550.

Jin, J. M., and J. L. Volakis (1991c), "A Hybrid Finite Element Method for Scattering and Radiation by Microstrip Patch Antennas and Arrays Residing in a Cavity," *IEEE Trans. Antennas Propag.*, vol. AP-39, pp. 1598–1604.

Jin, J. M., J. L. Volakis, and V. V. Liepa (1992), "Fictitious Absorber for Truncating Finite Element Meshes in Scattering," *IEE Proc.*, vol. 139, pt. H, pp. 472–476.

Johnson, C. (1994), "A New Paradigm for Adaptive Finite Element Methods," in *The Mathematics of Finite Elements and Applications*, Ed. J. R. Whiteman, John Wiley & Sons, Chichester, pp. 105–120.

Judd, S. V., I. Whiteley, R. J. Clowes, and D. C. Richard (1970), "An Analytical Method for Calculating Microstrip Transmission Line Parameters," *IEEE Trans. Microwave Theory Tech.*, vol. MTT-18, pp. 78–87.

Jung, H., G. Lee, and S. Hahn (1985), "3-D Magnetic Field Computations Using Finite-Element Approach with Localized Functional," *IEEE Trans. Magnetics*, vol. MAG-21, pp. 2196–2198.

Kagawa, Y., T. Murai, S. Kitagami (1982), "On the Computability of Finite Element-Boundary Element Coupling in Field Problems," *COMPEL*, vol. 1, pp. 197–217.

Kameari, A. (1990), "Calculation of Transient 3-D Eddy Current Using Edge-Elements," *IEEE Trans. Magnetics*, vol. MAG-26, pp. 466–469.

Kanellopoulos, V. N., and J. P. Webb (1990), "A Complete *E*-Plane Analysis of Waveguide Junctions Using the Finite Element Method," *IEEE Trans. Microwave Theory Tech.*, vol. MTT-38, pp. 290–295.

Kanellopoulos, V. N., and J. P. Webb (1991), "A Numerical Study of Vector Absorbing Boundary Conditions for the Finite-Element Solution of Maxwell's Equations," *IEEE Microwave Guided Wave Letters*, vol. 1, pp. 325–327.

Kelly, D. W., S. R. J. P. Gago, O. C. Zienkiewicz, and I. Babuška (1983), "*A Posteriori* Error Analysis and Adaptive Processes in the Finite Element Method: Part I — Error Analysis," *Int. J. Numer. Meth. Eng.*, vol. 19, pp. 1593–1619.

Kelly, D. W. (1984), "The Self-Equilibration of Residuals and Complementary *A Posteriori* Error Estimates in the Finite Element Method," *Int. J. Numer. Meth. Eng.*, vol. 20, pp. 1491–1506.

Kelly, D. W (1986), "The Self-Equilibration of Residuals and 'Upper-Bound' Error Estimates for the Finite Element Method," in *Accuracy Estimates and Adaptive Refinements in Finite Element Computations*, Eds. I. Babuška, O. C. Zienkiewicz, J. Gago, and E. R. de A. Oliveira, John Wiley & Sons, London, pp. 129–146.

Kelly, D. W., R. J. Mills, J. A. Reizes, and A. D. Miller (1987), "*A Posteriori* Estimates of the Solution Error Caused by Discretization in the Finite Element, Finite Difference and Boundary Element Methods," *Int. J. Numer. Meth. Eng.*, vol. 24, pp. 1921–1939.

Khebir, A., and A. B. Kouki, and R. Mittra (1990a), "An Absorbing Boundary Condition for Quasi-TEM Analysis of Microwave Transmission Lines via the Finite Element Method," *J. Electromagnetic Waves Appl.*, vol. 4, pp. 145–157.

Khebir, A., A. B. Kouki, and R. Mittra (1990b), "Higher Order Asymptotic Boundary Condition for the Finite Element Modeling of Two-Dimensional Transmission Line Structures," *IEEE Trans. Microwave Theory Tech.*, vol. MTT-38, pp. 1433–1438.

Kikuchi, F. (1987), "Mixed and Penalty Formulations for Finite Element Analysis of an Eigenvalue Problem in Electromagnetism," *Comp. Meth. Appl. Mech. Eng.*, vol. 64, pp. 509–521.

Kim, H. S., S. P. Hong, K. Choi, H. K. Jung, and S. Y. Hahn (1991), "A Three Dimensional Adaptive Finite Element Method for Magnetostatic Problems," *IEEE Trans. Magnetics*, vol. MAG-27, pp. 4081–4084.

Kingsland, D. M., J. Gong, J. L. Volakis, and J. F. Lee (1996), "Performance of an Anisotropic Artificial Absorber for Truncating Finite-Element Meshes," *IEEE Antennas Propag.*, vol.AP-44, pp. 975–982.

Kinsner, W., and E. della Torre (1974), "An Iterative Approach to the Finite Element Method in Field Problems," *IEEE Trans. Microwave Theory Tech.*, vol. MTT-20, pp. 221–228.

Kitazawa, T., and Y. Hayashi (1981), "Coupled Slots on an Anisotropic Sapphire Substrate," *IEEE Trans. Microwave Theory Tech.*, vol. MTT-29, pp. 1035–1040.

Kitazawa, T., and Y. Hayashi (1987), "Variational Method for Coplanar Waveguide with Anisotropic Substrates," *IEE Proc.*, vol. 134, pt. H., pp. 7–10.

Kitazawa, T. (1989a), "Variational Method for Multiconductor Coupled Striplines with Stratified Anisotropic Media," *IEEE Trans. Microwave Theory Tech.*, vol. MTT-37, pp. 484–491.

Kitazawa, T. (1989b), "Variational Method for Planar Transmission Lines with Anisotropic Magnetic Media," *IEEE Trans. Microwave Theory Tech.*, vol. MTT-37, pp. 1749–1754.

Kitazawa, T. (1993), "Loss Calculation of Single and Coupled Strip Lines by Extended Spectral Domain Approach," *IEEE Microwave Guided Wave Letters*, vol. 3, pp. 211–213.

Kiyoshi, I., and K. Masanori (1988), "Numerical Analysis of H-Plane Waveguide Junctions by Combination of Finite and Boundary Elements," *IEEE Trans. Microwave Theory Tech.*, vol. MTT-36, pp. 1343–1351.

Kobayashi, M. (1978), "Analysis of the Microstrip and the Electrooptic Light Modulator," *IEEE Trans. Microwave Theory Tech.*, vol. MTT-26, pp. 119–126.

Kobelansky A. J., and J. P. Webb (1986), "Eliminating Spurious Modes in Finite-Element Waveguide Problems by Using Divergence-Free Fields," *Electronics Letters*, vol. 22, pp. 569–570.

Koch, T. B., J. B. Davies, and D. Wickramasinghe (1989), "Finite Element-Finite Difference Propagation Algorithm for Integrated Optical Device," *Electronics Letters*, vol. 25, pp. 514–516.

Koh, D., H. Lee, and T. Itoh (1997), "Hybrid Full-Wave Analysis of Via Hole Grounds Using Finite Difference and Finite-Element Time-Domain Methods," *IEEE Trans. Microwave Theory Tech.*, vol. MTT-45, pp. 2217–2223.

Kolbehdari, M. A., M. Srinivasan, N. S. Nakhla, Q. J. Zhang, and R. Achar (1996), "Simultaneous Time and Frequency Domain Solutions of EM Problems Using Finite Elements and CFH Techniques," *IEEE Trans. Microwave Theory Tech.*, vol. MTT-45, pp. 1526–1534.

Konrad, A., and P. P. Silvester (1971), "Scalar Finite-Element Program Package for Two-Dimensional Field Problems," *IEEE Trans. Microwave Theory Tech.*, vol. MTT-19, pp. 952–954.

Konrad, A. (1976), "Vector Variational Formulation of Electromagnetic Fields in Anisotropic Media," *IEEE Trans. Microwave Theory Tech.*, vol. MTT-24, pp. 553–559.

Konrad, A. (1977), "High-Order Triangular Finite Elements for Electromagnetic Waves in Anisotropic Media," *IEEE Trans. Microwave Theory Tech.*, vol. MTT-25, pp. 353–360.

Konrad, A. (1985), "A Direct Three-Dimensional Finite Element Method for the Solution of Electromagnetic Fields in Cavities," *IEEE Trans. Magnetics*, vol. MAG-21, pp. 2276–2279.

Konrad, A. (1986), "On the Reduction of the Number of Spurious Modes in the Vectorial Finite-Element Solution of Three-Dimensional Cavities and Waveguides," *IEEE Trans. Microwave Theory Tech.*, vol. MTT-34, pp. 224–226.

Konrad, A. (1989), "A Method for Rendering 3D Finite Element Vector Field Solutions Non-Divergent," *IEEE Trans. Magnetics*, vol. MAG-25, pp. 2822–2824.

Kooi, P. S. (1985), "Finite Element Analysis of Shielded Cylindrical Resonator," *IEE Proc.*, vol. 136, pt.

H, pp. 7–16.

Koshiba, M., and M. Suzuki (1982), "Numerical Analysis of Planar Arbitrarily Anisotropic Diffused Optical Waveguides Using Finite-Element Method," *Electronics Letters*, vol. 18, pp. 579–581.

Koshiba, M., K. Ooishi, T. Miki, and M. Suzuki (1982a), "Finite-Element Analysis of the Discontinuities in a Dielectric Slab Waveguide Bounded by Parallel Plates," *Electronics Letters*, vol. 18, pp. 33–34.

Koshiba, M., K. Hayata, and M. Suzuki (1982b), "Approximate Scalar Finite-Element Analysis of Anisotropic Optical Waveguides," *Electronics Letters*, vol. 18, pp. 411–413.

Koshiba, M., K. Hayata, and M. Suzuki (1984a), "Vectorial Finite-Element Method Without Spurious Solutions for Dielectric Waveguide Problems," *Electronics Letters*, vol. 20, pp. 409–410.

Koshiba, M., K. Hayata, and M. Suzuki (1984b), "Approximate Scalar Finite-Element Analysis of Anisotropic Optical Waveguides with Off-Diagonal Elements in a Permittivity Tensor," *IEEE Trans. Microwave Theory Tech.*, vol. MTT-32, pp. 587–593.

Koshiba, M., K. Hayata, and M. Suzuki (1985a), "Improved Finite Element Formulation in Terms of the Magnetic Field Vector for Dielectric Waveguides," *IEEE Trans. Microwave Theory Tech.*, vol. MTT-33, pp. 227–233.

Koshiba, M., K. Hayata, and M. Suzuki (1985b), "Finite-Element Formulation in Terms of the Electric-Field Vector for Electromagnetic Waveguide Problems," *IEEE Trans. Microwave Theory Tech.*, vol. MTT-33, pp. 900–905.

Koshiba, M., H. Kumagami, and M. Suzuki (1985c), "Finite-Element Solution of Planar Arbitrarily Anisotropic Diffused Optical Waveguides," *IEEE J. Lightwave Tech.* vol. LT-3, pp. 773–778.

Koshiba, M., K. Hayata, and M. Suzuki (1986), "Finite-Element Solution of Anisotropic Waveguides with Arbitrary Tensor Permittivity," *IEEE J. Lightwave Tech.*, vol. LT-4, pp. 121–126.

Koshiba, M., and M. Suzuki (1986), "Finite Element Analysis of *H*-Plane Waveguides Junction with Arbitrarily Shaped Ferrite Post," *IEEE Trans. Microwave Theory Tech.*, vol. MTT-34, pp. 103–109.

Koshiba, M., and K. Inoue (1992), "Simple and Efficient Finite-Element Analysis of Microwave and Optical Waveguides," *IEEE Trans. Microwave Theory Tech.*, vol. MTT-40, pp. 371–377.

Koshiba, M., S. Maruyama, and K. Hirayama (1994), "A Vector Finite Element Method with the Higher-Order Mixed-Interpolation-Type Triangular Elements for Optical Waveguiding Problems," *IEEE J. Lightwave Tech.*, vol. LT-12, pp. 495–502.

Koshiba, M. (1996), "Computer-Aided Design of Integrated Optical Waveguide Devices," in *Finite Element Software for Microwave Engineering*, Eds. T. Itoh, G. Pelosi, and P. P. Silvester, John Wiley & Sons, New York, pp. 79–99.

Kost, A., and L. Jänicke (1992), "Universal Generation of an Initial Mesh for Adaptive 3-D Finite Element Method," *IEEE Trans. Magnetics*, vol. 28, pp. 1735–1738.

Křížek, M., and P. Neittaanmäki (1984), "Finite Element Approximation for a Div-Rot System with Mixed Boundary Conditions in Non-Smooth Plane Domains," *Aplikace Matematiky*, vol. 29, pp. 272–285.

Křížek, M. (1992), "On the Maximum Angle Condition for Linear Tetrahedral Elements," *SIAM J. Numer. Anal.*, vol. 29, pp. 513–520.

Kron, G. (1953), "A Set of Principles to Interconnect the Solutions of Physical Systems," *J. Appl. Phys.*, vol. 24, pp. 965–980.

Kuzuoglu, M., and R. Mittra (1996a), "Mesh Truncation by Perfectly Matched Anisotropic Absorbers in the Finite Element Method," *Microwave Optical Tech. Letters*, vol. 12, pp. 136–140.

Kuzuoglu, M., and R. Mittra (1996b), "Frequency Dependence of the Constitutive Parameters of Causal Perfectly Matched Anisotropic Absorbers," *IEEE Microwave Guided Wave Letters*, vol. 6, pp. 447–449.

Ladeveze, P., G. Coffignal, and J. P. Pelle (1986), "Accuracy of Elastoplastic and Dynamic Analysis," in *Accuracy Estimates and Adaptive Refinements in Finite Element Computations*, Eds. I. Babuška, O. C. Zienkiewicz, J. Gago, and E. R. de A. Oliveira, John Wiley & Sons, London, pp. 181–203.

Ladeveze, P., J. P. Pelle, and P. Rougeot (1991), "Error Estimation and Mesh Optimization for Classical Finite Elements," *Eng. Comp.*, vol. 9, pp. 69–80.

Lagasse, P., and J. van Bladel (1972), "Square and Rectangular Waveguides with Rounded Corners," *IEEE Trans. Microwave Theory Tech.*, vol. MTT-20, pp. 331–337.

Lager, I. E., and G. Mur (1996), "The FEMAX Package for Static and Stationary Electric and Magnetic Fields," in *Software for Electrical Engineering Analysis and Design*, Ed. P. P. Silvester, Computational Mechanics Publications, Southampton, England, pp. 295–304.

Lakshmipathi, R., Y. N. Rao, and G. S. Rao (1989), "Computation of Electrostatic Field Distribution in a Cable by Least-Squares Smoothing Finite Element Method," *Comm. Appl. Numer. Meth.*, vol. 5, pp. 15–22.

Lang, K. C. (1973), "Edge Condition of a Perfectly Conducting Wedge with Its Exterior Region Divided by a Resistive Sheet," *IEEE Trans. Antennas Propag.*, vol. AP-21, pp. 237–238.

Laura, P. A. A., K. Nagaya, and G. Sánchez Sarmiento (1980), "Numerical Experiments on the Determination of Cutoff Frequencies of Waveguides of Arbitrary Cross Section," *IEEE Trans. Microwave Theory Tech.*, vol. MTT-28, pp. 568–572.

Lebaric, J. E., and D. Kajfez (1989), "Analysis of Dielectric Resonator Cavities Using the Finite Integration Technique," *IEEE Trans. Microwave Theory Tech.*, vol. MTT-37, pp. 1740–1747.

Lee, C. F., R. T. Shin, J. A. Kong, and B. J. McCartin (1990), "Absorbing Boundary Conditions on Circular and Elliptical Boundaries," *J. Electromagnetic Waves Appl.* vol. 4, pp. 945–962.

Lee, C. K., and S. H. Lo (1992), "An Automatic Adaptive Refinement Finite Element Procedure for 2D Elastostatic Analysis," *Int. J. Numer. Meth. Eng.*, vol. 35, pp. 1967–1989.

Lee, C. K., and S. H. Lo (1997), "Automatic Adaptive 3-D Finite Element Refinement Using Different-Order Tetrahedral Elements," *Int. J. Numer. Meth. Eng.*, vol. 40, pp. 2195–2226.

Lee, J. F., and Z. J. Csendes (1987), "Transfinite Elements: A Highly Efficient Procedure for Modeling Open Field Problems," *J. Appl. Phys.*, vol. 61, pp. 3914–3915.

Lee, J. F., and Z. J. Csendes (1988), "The Transfinite Element Method for Modeling MMIC Devices," *IEEE Trans. Microwave Theory Tech.*, vol. MTT-36, pp. 1639–1649.

Lee, J. F., D. K. Sun, and Z. J. Csendes (1991a), "Full-Wave Analysis of Dielectric Waveguides Using Tangential Vector Finite Elements," *IEEE Trans. Microwave Theory Tech.*, vol. MTT-39, pp. 1262–1271.

Lee, J. F., D. K. Sun, and Z. J. Csendes (1991b), "Tangential Vector Finite Elements for Electromagnetic Field Computation," *IEEE Trans. Magnetics*, vol. MAG-27, pp. 4032–4035.

Lee, J. F., and R. Mittra (1992), "A Note on the Application of Edge-Elements for Modeling Three-Dimensional Inhomogeneously Filled Cavities," *IEEE Trans. Microwave Theory Tech.*, vol. MTT-40, pp. 1767–1773.

Lee, J. F., G. M. Wilkins, and R. Mittra (1993), "Finite-Element Analysis of Axisymmetric Cavity Resonator Using a Hybrid Edge Element Technique," *IEEE Trans. Microwave Theory Tech.*, vol. MTT-41, pp. 1981–1987.

Lee, J. F. (1994), "Finite Element Analysis of Lossy Dielectric Waveguides," *IEEE Trans. Microwave Theory Tech.*, vol. MTT-42, pp. 1025–1031.

Lee, J. F., R. Lee, and A. Cangellaris (1997), "Time-Domain Finite-Element Methods," *IEEE Antennas Propag.*, vol. AP-45, pp. 430–432.

Lee, R., and A. C. Cangellaris (1992), "A Study of Discretization Error in the Finite Element Approximation of Wave Solution," *IEEE Trans. Antennas Propag.*, vol. AP-40. pp. 542–549.

Leong, M. S., P. S. Kooi, A. L. Satya-Prakash (1983), "Finite-Element Analysis of Shielded Microstrip Disc Resonator," *IEE Proc.*, vol. 131, pt. H, pp. 126–128.

Levy, R., and S. B. Cohn (1984), "A History of Microwave Filter Research, Design, and Development," *IEEE Trans. Microwave Theory Tech.*, vol. MTT-32, pp. 1055–1067.

Levy, S. (1953), "Structural Analysis and Influence Coefficients for Delta Wings," *J. Aeronaut. Soc.*, vol. 20, pp. 449–454.

Lewis, R. W., H. C. Huang, A. S. Usmani, and J. T. Cross (1991), "Finite Element Analysis of Heat Transfer and Flow Problems Using Adaptive Remeshing Including Application to Solidification Problems," *Int. J. Numer. Meth. Eng.*, vol. 32, pp. 767–781.

Li, C., Z. Ren, and A. Razek (1994), "An Approach to Adaptive Mesh Refinement for Three-Dimensional Eddy-Current Computations," *IEEE Trans. Magnetics*, vol. MAG-30, pp. 113–117.

Lindell, I. V. (1981), "On the Quasi-TEM Modes in Inhomogeneous Multiconductor Transmission Lines," *IEEE Trans. Microwave Theory Tech.*, vol. MTT-29, pp. 812–817.

Lindell, I. V. (1982), "Variational Methods for Nonstandard Eigenvalue Problems in Waveguide and Resonator Analysis," *IEEE Trans. Microwave Theory Tech.*, vol. MTT-30, pp. 1194–1204. See "Comments" in *IEEE Trans. Microwave Theory Tech.* vol. MTT-31, pp. 786–789, and *IEEE Trans. Microwave Theory Tech.*, vol. MTT-32, pp. 474–476.

Lindell, I. V., and A. H. Sihvola (1983), "Dielectrically Loaded Corrugated Waveguide: Variational Analysis of a Nonstandard Eigenproblem," *IEEE Trans. Microwave Theory Tech.*, vol. MTT-31, pp. 520–526.

Lindell, I. V. (1992), "Variational Method for the Analysis of Lossless Bi-Isotropic (Nonreciprocal Chiral) Waveguides," *IEEE Trans. Microwave Theory Tech.*, vol. MTT-40, pp. 403–406.

Linner, L. J. P. (1974), "A Method for the Computation of the Characteristic Admittance Matrix of Multiconductor Striplines with Arbitrary Widths," *IEEE Trans. Microwave Theory Tech.*, vol. MTT-22, pp. 930–937.

Liskovets, O. A. (1965), "The Method of Lines," *Review. Differ. Uravneniya*, vol. 1, pp. 1662–1678.

Liu, C. T., and C. H. Chen (1981), "A Variational Theory for Wave Propagation in Inhomogeneous Dielectric Slab Loaded Waveguides," *IEEE Trans. Microwave Theory Tech.*, vol. MTT-29, pp. 805–811.

Liu, Y. C., and H. A. Elmaraghy (1992), "Assessment of Discretized Errors and Adaptive Refinement with Quadrilateral Finite Elements," *Int. J. Numer. Meth. Eng.*, vol. 33, pp. 781–798.

Lo, S. H. (1991), "Automatic Mesh Generation and Adaptation by Using Contours," *Int. J. Numer. Meth. Eng.*, vol. 31, pp. 689–707.

Löhner, R. K. Morgan, and O. C. Zienkiewicz (1986), "Adaptive Grid Refinement for the Compressible Euler Equations," in *Accuracy Estimates and Adaptive Refinements in Finite Element Computations*, Eds. I. Babuška, O. C. Zienkiewicz, J. Gago, and E. R. de A. Oliveira, John Wiley & Sons, London, pp. 281–297.

Lowther, D. A., C. B. Rajanathan, and P. P. Silvester (1978), "A Finite Element Technique for Solving 2-D Open Boundary Problems," *IEEE Trans. Magnetics*, vol. MAG-14, pp. 467–469.

Lowther, D. A., and D. N. Dyck (1993), "A Density Driven Mesh Generator Guided by a Neural Network," *IEEE Trans. Magnetics*, vol. MAG-29, pp. 1927–1930.

Lowther, D. A., R. Rong, and B. Forghani (1993), "Field Smoothing and Adaptive Mesh Generation," *IEEE Trans. Magnetics*, vol. MAG-29, pp. 1890–1893.

Lu, Y., and F. A. Fernández (1993a), "An Efficient Formulation for Guided-Wave Problems Using Transverse Electric Field Component," *IEEE Trans. Microwave Theory Tech.*, vol. MTT-41, pp. 1215–1223.

Lu, Y., and F. A. Fernández (1993b), "Finite Element Analysis of Lossy Dielectric Waveguides," *IEEE Trans. Magnetics*, vol. MAG-29, pp. 1609–1612.

Lu, Y., and F. A. Fernández (1994), "Vector Finite Element Analysis of Integrated Optical Waveguides," *IEEE Trans. Magnetics*, vol. MAG-30, pp. 3116–3119.

Lucas, E. W., and T. P. Fontana (1995), "A 3D Hybrid Finite Element Boundary Element Method for the Unified Radiation and Scattering Analysis of General Periodic Array," *IEEE Trans. Antennas Propag.*, vol. AP-43, pp. 145–154.

Luk, K. M., E. K. N. Yung, K. W. Leung, and Y. W. Liu (1995), "On the Measured Equation of Invariance for an Electrically Large Cylinder," *IEEE Microwave Guided Wave Letters*, vol. 5, pp. 445–447.

Luomi, J., and H. Rouhiainen (1988), "Adaptive Mesh Refinement for Magnetic Field Problems Involving Saturable Ferromagnetic Parts," *IEEE Trans. Magnetics*, vol. MAG-24, pp. 311–314.

Lynn, P. P., and H. A. Hadid (1981), "Infinite Elements with $1/r^n$ Type Decay," *Int. J. Numer. Meth. Eng.*, vol. 17, pp. 347–355.

Lyon, R. W., and J. Helszajn (1982), "A Finite Element Analysis of Planar Circulators Using Arbitrarily Shaped Resonators," *IEEE Trans. Microwave Theory Tech.*, vol. MTT-30, pp. 1964–1974.

Lyons, M. R., J. P. K. Gilb, and C. A. Balanis (1993), "Enhanced Dominant Mode Operation of a Shielded Multilayer Coplanar Waveguide Via Substrate Compensation," *IEEE Trans. Microwave Theory Tech.*, vol. MTT-41, pp. 1564–1567.

Lynch, D. R., and K. D. Paulsen (1993), "Origin of Vector Parasites in Numerical Maxwell Solutions," *IEEE Trans. Microwave Theory Tech.*, vol. MTT-39, pp. 383–394.

Lynch, D. R., K. D. Paulsen, and W. E. Boyse (1993), "Synthesis of Vector Parasites in Finite Element Maxwell Solutions," *IEEE Trans. Microwave Theory Tech.*, vol. MTT-41, pp. 1439–1448.

Ma, Y. C. (1991), "A Note on the Radiation Boundary Conditions for the Helmholtz Equation," *IEEE Trans. Antennas Propag.*, vol. AP-39, pp. 1526–1530.

Mabaya, N., P. E. Lagasse, and P. Vandenbulcke (1981), "Finite Element Analysis of Optical Waveguides," *IEEE Trans. Microwave Theory Tech.*, vol. MTT-29, pp. 600–605.

Madrangeas, V., S. Verdeyme, M. Aubourg, and P. Guillon (1994), "Modelling Microwave Boxed Structures by 2D and 3D Finite Element Method," *COMPEL*, vol. 13, supp. A, pp. 335–340.

Maiti, S. K. (1992), "A Finite Element for Variable Order Singularities Based on the Displacement Formulation," *Int. J. Numer. Meth. Eng.*, vol. 33, pp. 1955–1974.

Manges, J. B., and Z. J. Csendes (1997), "Tree-Cotree Decomposition for First-Order Complete Tangential Vector Finite Elements," *Int. J. Numer. Meth. Eng.*, vol. 40, pp. 1687–1700.

Marchetti, S., and T. Rozzi (1991a), "Electric Field Singularities at Sharp Edges of Planar Conductors," *IEEE Trans. Antennas Propag.*, vol. AP-39, pp. 1312–1320.

Marchetti, S., and T. Rozzi (1991b), "*H*-Field and *J*-Current Singularities at Sharp Edges in Printed Circuits," *IEEE Trans. Antennas Propag.*, vol. AP-39, pp. 1321–1331.

Marqués, R., and M. Horno (1985), "Propagation of Quasi-Static Modes in Anisotropic Transmission Lines: Application to MIC Lines," *IEEE Trans. Microwave Theory Tech.*, vol. MTT-33, pp. 927–932.

Marx, K. D. (1973), "Propagation Modes, Equivalent Circuits, and Characteristic Terminations for Multiconductor Transmission Lines with Inhomogeneous Dielectrics," *IEEE Trans. Microwave Theory Tech.*, vol. MTT-21, pp. 450–457.

Mathew, T. P. (1993a), "Schwarz Alternating and Iterative Refinement Methods for Mixed Formulations of Elliptic Problems, Part I: Algorithms and Numerical Results," *Numer. Math.*, vol. 65, pp. 445–468.

Mathew, T. P. (1993b), "Schwarz Alternating and Iterative Refinement Methods for Mixed Formulations of Elliptic Problems, Part II: Convergence Theory," *Numer. Math.*, vol. 65, pp. 469–492.

Matsuhara, M., H. Yunoki, and A. Maruta (1991), "Analysis of Open-Type Waveguides by the Vector Finite Element Analysis," *IEEE Microwave Guided Waves Letters*, vol. 1, pp. 376-378.

Matthaei, G. L., K. Kiziloglu, N. Dagli, and S. I. Long (1990), "The Nature of the Charges, Currents, and Fields in and about Conductors Having Cross-Sectional Dimensions of the Order of a Skin Depth," *IEEE Trans. Microwave Theory Tech.*, vol. MTT-38, pp. 1031–1036.

Mautz, J. R., R. F. Harrington, and C. I. G. Hsu (1988), "The Inductance Matrix of a Multiconductor Transmission Line in Multiple Magnetic Media," *IEEE Trans. Microwave Theory Tech.*, vol. MTT-36, pp. 1293–1295.

Mayergoyz, I. D., M. V. K. Chari, and A. Konrad (1983), "Boundary Galerkin's Method for Three-Dimensional Finite Element Electromagnetic Field Computation," *IEEE Trans. Magnetics*, vol. MAG-19, pp. 2333–2336.

Mazé-Merceur, G., S. Tedjini, and J. L. Bonnefoy (1993), "Analysis of a CPW on Electric and Magnetic Biaxial Substrate," *IEEE Trans. Microwave Theory Tech.*, vol. MTT-41, pp. 457–461.

McAulay, A. D. (1977), "Variational Finite-Element Solution for Dissipative Waveguides and Transportation Application," *IEEE Trans. Microwave Theory Tech.*, vol. MTT-25, pp. 382–392.

McDonald, B. H., and A. Wexler (1972), "Finite-Element Solution of Unbounded Field Problems," *IEEE Trans. Microwave Theory Tech.*, vol. MTT-20, pp. 841–847.

McDougall, M. J., and J. P. Webb (1989), "Infinite Elements for the Analysis of Open Dielectric Waveguides," *IEEE Trans. Microwave Theory Tech.*, vol. MTT-37, pp. 1724–1731.

McFee, S., and J. P. Webb (1992), "Adaptive Finite Element Analysis of Microwave and Optical Devices Using Hierarchal Triangles," *IEEE Trans. Magnetics*, vol. MAG-28, pp. 1708–1711.

McFee, S., and J. P. Webb (1993), "Automatic Mesh Generation for *h-p* Adaption," *IEEE Trans. Magnetics*, vol. MAG-29, pp. 1894–1897.

McFee, S., and D. Giannacopoulos (1996), "Optimal Discretization Based Refinement Criteria for Finite Element Adaption," *IEEE Trans. Magnetics*, vol. MAG-32, pp. 1357–1360.

McFee, S. (1997), "*h-p* Adaptive Numerical Integration Techniques for Finite Element Analysis in Magnetics," *IEEE Trans. Magnetics*, vol. MAG-33, pp. 4116–4118.

McHenry, D. (1943), "A Lattice Analogy of the Solution of Plane Stress Problems," *J. Inst. Civil Eng.*, vol. 21, pp. 59–82.

McQuiddy, D. N., J. W. Wassel, J. B. Lagrange, and W. R. Wisseman (1984), "Monolithic Microwave Integrated Circuits: An Historical Perspective," *IEEE Trans. Microwave Theory Tech.*, vol. MTT-32, pp. 997–1007.

Medina, F., and M. Horno (1985), "Upper and Lower Bounds on Mode Capacitances for a Large Class of Anisotropic Multilayered Microstrip-Like Transmission Lines," *IEE. Proc.*, vol. 132, pt. H, pp. 157–163.

Medina, F., and M. Horno (1987), "Capacitance and Inductance Matrices for Multistrip Structures in Multilayered Anisotropic Dielectrics," *IEEE Trans. Microwave Theory Tech.*, vol. MTT-35, pp. 1002–1008.

Medina, J. L., F. J. Mendieta, and W. H. Ku (1994), "Analysis of Shielding Effects on the Microstrip Effective Dielectric Constant by the Finite-Element Method," *Microwave Optical Tech. Letters*, vol. 7, pp. 414–419.

Mei, K. K. (1974), "Unimoment Method of Solving Antenna and Scattering Problems," *IEEE Trans. Antennas Propag.*, vol. AP-22, pp. 760–766.

Mei, K. K. (1987), "Unimoment Method for Electromagnetic Wave Scattering," *J. Electromagnetic Waves Appl.*, vol. 1, pp. 201–222.

Mei, K. K., R. Pous, Z. Chen, and Y. W. Liu (1992), "Measured Equation of Invariance — Breaking Through the Barrier of Radiation Condition," in *Numerical Methods in Engineering '92*, Eds. C. Hirsch, O. C. Zienkiewicz, and E. Oñate, Elsevier, Amsterdam, pp. 95–102.

Mei, K. K., R. Pous, Z. Chen, Y. W. Liu, and M. D. Prouty (1994), "Measured Equation of Invariance: A New Concept in Field Computations," *IEEE Trans. Antennas Propag.*, vol. AP-42, pp. 320–328.

Meinke, H. H., K. P. Lange, and J. F. Ruger (1963), "TE- and TM-Waves in Waveguides of Very General Cross Section," *Proc. IEEE*, vol. 51, pp. 1436–1443.

Meixner, J. (1972), "The Behavior of Electromagnetic Fields at Edges," *IEEE Trans. Antennas Propag.*, vol. AP-20, pp. 442–446.

Mercader, E., and M. C. Rivara (1992), "Herramienta para Resolver EDP en 2-D Mediante Métodos Adaptativos y Multimallas," in *Numerical Methods in Engineering and Applied Sciences*, Eds. H. Alder, J. C. Heinrich, S. Lavanchy, E. Oñate, and B. Suárez, CIMNE, Barcelona, pp. 543–551.

Mesa, F. L., G. Cano, F. Medina, R. Marques, and M. Horno (1992), "On the Quasi-TEM and Full-Wave Approaches Applied to Coplanar Multistrip on Lossy Dielectric Layered Media," *IEEE Trans. Microwave Theory Tech.*, vol. MTT-40, pp. 524–531.

Meyer, F. J. C., and D. B. Davidson (1994), "Error Estimates and Adaptive Procedures for the Two-Dimensional Finite Element Method," *Electronics Letters*, vol. 30, pp. 936–938.

Miniowitz, R., and J. P. Webb (1991), "Covariant Projection Quadrilateral Elements for the Analysis of Waveguides with Sharp Edges," *IEEE Trans. Microwave Theory Tech.*, vol. MTT-39, pp. 501–505.

Mishra, P. K., A. Sharma, S. Labroo, and A. K. Ghatak (1985), "Scalar Variational Analysis of Single-Mode Waveguides with Rectangular Cross Section," *IEEE Trans. Microwave Theory Tech.*, vol. MTT-33, pp. 282–286.

Mitchell, A. R. (1973), "An Introduction to the Mathematics of the Finite Element Method," in *The Mathematics of Finite Elements and Applications*, Ed. J. R. Whiteman, Academic Press, London, pp. 37–58.

Mitchell, K. G., and J. Penman (1992), "Self Adaptive Mesh Generation for 3-D Finite Element Calculation," *IEEE Trans. Magnetics*, vol. MAG-28, pp. 1751–1754.

Mittra, R., and T. Itoh (1971), "A New Technique for the Analysis of the Dispersion Characteristics of Microstrip Lines," *IEEE Trans. Microwave Theory Tech.*, vol. MTT-19, pp. 47–56.

Mittra, R., O. Ramahi, A. Khebir, R. Gordon, and A. Kouki (1989), "A Review of Absorbing Boundary Conditions for Two- and Three-Dimensional Electromagnetic Scattering Problems," *IEEE Trans. Magnetics*, vol. MAG-25, pp. 3034-3040.

Mittra, R. A. Boag, and A. Boag (1994), "A New Look at the Absorbing Boundary Condition Issue for FEM Mesh Truncation in Open Region Problems," *COMPEL*, vol. 13, supp. A, pp. 211–216.

Mittra, R., and Ü. Pekel (1995), "A New Look at the Perfectly Matched Layer (PML) Concept for the Reflectionless Absorption of Electromagnetic Waves," *IEEE Microwave Guided Wave Letters*, vol. 1, pp. 84–86.

Miyata, T., T. Takeda, and T. Kuwahara (1989), "Electric Fields on Edges of Anisotropic Dielectric Materials," *Electronics Letters*, vol. 25, pp. 1185–1186.

Moerloose, J. D., and M. A. Stuchly (1995), "Behavior of Berenger's ABC for Evanescent Waves," *IEEE Microwave Guided Wave Letters*, vol. 5, pp. 344–346.

Monk, P. (1991a), "An Analysis of Nédélec's Method for the Spatial Discretization of Maxwell's Equations," *Technical Report*, University of Delaware, Newark, DE.

Monk, P. (1991b), "On the *p* and *h-p* Extension of Nédélec's Curl Conforming Elements," *Technical Report*, University of Delaware, Newark, DE.

Monk, P. (1992a), "A Finite Element Method for Approximating the Time-Harmonic Maxwell's Equations," *Numer. Math.*, vol. 63, pp. 243–261.

Monk, P. (1992b), "Analysis of a Finite Element Method for Maxwell's Equations," *SIAM J. Numer. Anal.*, vol. 29, pp. 714–729.

Monk, P., A. K. Parrott, and P. J. Wesson (1994), "A Dispersion Analysis of Finite Element Methods on Triangular Grids for Maxwell's Equations," in *The Mathematics of Finite Elements and Applications*, Ed. J. R. Whiteman, John Wiley & Sons, Chichester, England, pp. 314–321.

Morgan, M. A. (1980), "Finite Element Computation of Microwave Scattering by Raindrops," *Radio Science*, vol. 15, pp. 1109–1119.

Morgan, M. A. (1981), "Finite Element Calculation of Microwave Absorption by the Cranial Structure," *IEEE Trans. Biomedical Eng.*, vol. BME-28, pp. 687–695.

Morgan, M. A., C. H. Chen, S. C. Hill, and P. W. Barber (1984), "Finite Element-Boundary Integral Formulation of Electromagnetic Scattering," *Wave Motion*, vol. 6, pp. 91–103.

Morishita, K., and N. Kumagai (1977), "Unified Approach to the Derivation of Variational Expressions for Electromagnetic Fields," *IEEE Trans. Microwave Theory Tech.*, vol. MTT-25, pp. 34–40.

Moyer, Jr., E. T., and E. A. Schroeder (1991), "Finite Element Formulations of Maxwell's Equations — Advantages and Comparisons between Available Approaches," *IEEE Trans. Magnetics*, vol. MAG-27, pp. 4217–4220.

Mur, G. (1981a), "Absorbing Boundary Conditions for the Finite-Difference Approximation of the Time-Domain Electromagnetic-Field Equations," *IEEE Trans. Electromagnetic Compat.*, vol. EC-23, pp. 377–382.

Mur, G. (1981b), "The Modeling of Singularities in the Finite-Difference Approximation of the Time-Domain Electromagnetic-Field Equations," *IEEE Trans. Microwave Theory Tech.*, vol. MTT-29, pp. 1073–1077.

Mur, G., and A. T. de Hoop (1985), "A Finite-Element Method for Computing Three-Dimensional Electromagnetic Fields in Inhomogeneous Media," *IEEE Trans. Magnetics*, vol. MAG-21, pp. 2188–2191.

Mur, G. (1991), "Finite-Element Modeling of Three-Dimensional Electromagnetic Fields in Inhomogeneous Media," *Radio Science*, vol. 26, pp. 275–280.

Mur, G. (1993), "The Finite-Element Modeling of Three-Dimensional Electromagnetic Fields Using Edge and Nodal Elements," *IEEE Trans. Antennas Propag.*, vol. AP-41, pp. 948–953.

Mur, G. (1994a), "Compatibility Relations and the Finite-Element Formulation of Electromagnetic Field Problems," *IEEE Trans. Magnetics*, vol. MAG-30, pp. 2972–2975.

Mur, G. (1994b), "Edge Elements, Their Advantages and Their Disadvantages," *IEEE Trans. Magnetics*, vol. MAG-30, pp. 3552–3557.

Mur, G. (1996), "The Finite-Element Modelling of Electromagnetic Fields," in *Software for Electrical Engineering Analysis and Design*, Ed. P. P. Silvester, Computational Mechanics Publications, Southampton, England, pp. 177–186.

Murty, A. V. K., and G. V. Rao (1973), "Assessment of Accuracies of Finite Element Eigenvalues," in *The Mathematics of Finite Elements and Applications*, Ed. J. R. Whiteman, Academic Press, London, pp. 379–386.

Nakata, T., N. Takahashi, K. Fujiwara, and M. Sakaguchi (1990), "3-D Open Boundary Magnetic Field Analysis Using Infinite Elements Based on Hybrid Finite Element Method," *IEEE Trans. Magnetics*, vol. MAG-26, pp. 368-372.

Nambiar, R. V., R. S. Valera, K. L. Lawrence, R. B. Morgan, and D. Amil (1993), "An Algorithm for Adaptive Refinement of Triangular Meshes," *Int. J. Numer. Meth. Eng.*, vol. 39, pp. 499–509.

Nath, B., and J. Jamshidi (1980), "The *w*-Plane Finite Element Method for the Solution of Scalar Field Problems in Two Dimensions," *Int. J. Numer. Meth. Eng.*, vol. 15, pp. 361–379.

Navarro, E. A., J. M. Femenía, and V. Such (1991), "Adapting Available Finite Element Structural Programs to Solve Three-Dimensional High Frequency Electromagnetic Problems," *Int. J. Microwave Millimeter-Wave Computer-Aided Eng.*, vol. 1, pp. 386–394.

Nédélec, J. C. (1980), "Mixed Finite Elements in $\mathbb{R}^3$," *Numer. Math.*, vol. 35, pp. 315–341.

Nédélec, J. C. (1986), "A New Family of Mixed Finite Elements in $\mathbb{R}^3$," *Numer. Math.*, vol. 50, pp. 57–81.

Ng, F. L. (1974), "Tabulation of Methods for the Numerical Solution of the Hollow Waveguide Problem," *IEEE Trans. Microwave Theory Tech.*, vol. MTT-22, pp. 322–328.

Nortier, J. R., and D. A. McNamara (1984), "A General-Purpose Finite Element Code for Microwave Transmission-Line Problems," *Microwave Journal*, vol. 27, pp. 109–120.

Nuño, L., J. V. Balbastre, and H. Castañé (1997), "Analysis of General Lossy Inhomogeneous and Anisotropic Waveguides by the Finite-Element Method (FEM) Using Edge Elements," *IEEE Trans. MTT Microwave Theory Tech.*, vol. MTT-45, pp. 446–449.

Oden, J. T.(1973), "Finite Element Applications in Mathematical Physics," in *The Mathematics of Finite Elements and Applications*, Ed. J. R. Whiteman, Academic Press, London, pp. 239–282.

Oden, J. T., L. Demkowicz, T. Strouboulis, and P. Devloo (1986), "Adaptive Methods for Problems in Solid and Fluid Mechanics," in *Accuracy Estimates and Adaptive Refinements in Finite Element Computations*, Eds. I. Babuška, O. C. Zienkiewicz, J. Gago, and E. R. de A. Oliveira, John Wiley & Sons, London, pp. 249–280.

Oden, J. T. (1990), "Historical Comments on Finite Elements," in *A History of Scientific Computing*, Ed. S. G. Nash, ACM, New York.

Oden, J. T., L. Demkowicz, W. Rachowicz, and T. A. Westermann (1990), "*A Posteriori* Error Analysis in Finite Elements: The Element Residual Method for Symmetrizable Problems with Applications to Compressible Euler and Navier-Stokes Equations," *Comp. Meth. Appl. Mech. Eng.*, vol. 82, pp. 183–203.

Oden, J. T. (1994), "Error Estimation and Control in Computational Fluid Dynamics," in *The Mathematics of Finite Elements and Applications*, Ed. J. R. Whiteman, John Wiley & Sons, Chichester, England, pp. 1–23.

Ohtaka, M., M. Matsuhara, and N. Kumagai (1976), "Analysis of the Guided Modes in Slab-Coupled Waveguides Using a Variational Method," *IEEE Quantum Elec.*, vol. QE-12, pp. 378–382.

Ohtaka, M., and T. Kobayashi (1990), "A New Vector Variational Expression for Spurious-Free Finite Element Analysis of Waveguide Eigenmodes," *Electronics Comm. Japan*, Part 2, vol. 73, pp. 1–8.

Ohtsubo, H., and M. Kitamura (1992), "Numerical Investigation of Element-Wise *A Posteriori* Error Estimation in Two and Three Dimensional Elastic Problems," *Int. J. Numer. Meth. Eng.*, vol. 34, pp.

969–977.

Okamoto, K., and T. Okoshi (1978), "Vectorial Wave Analysis of Inhomogeneous Optical Fibers Using Finite Element Method," *IEEE Trans. Microwave Theory Tech.*, vol. MTT-26, pp. 109–114.

Okoshi, T., and K. Okamoto (1974), "Analysis of Wave Propagation in Inhomogeneous Optical Fibers Using a Variational Method," *IEEE Trans. Microwave Theory Tech.*, vol. MTT-22, pp. 938–945.

Oksanen, M. I., and I. V. Lindell (1989a), "Variational Analysis of Anisotropic Graded-Index Optical Fibers," *IEEE Trans. Microwave Theory Tech.*, vol. MTT-37, pp. 87–91.

Oksanen, M. I., and I. V. Lindell (1989b), "Complex-Valued Functionals in Variational Analysis of Waveguides with Impedance Boundaries," *IEE Proc.*, vol. 136, pt. H, pp. 281–288.

Oldfield, L. C., and J. P. Ide (1985), "Precision Calculation of Microwave Cavity Parameters Using Finite Difference and Finite Element Techniques," *IEEE Trans. Magnetics*, vol. MAG-21, pp. 2321–2323.

Oliner, A. A. (1984), "Historical Perspectives on Microwave Field Theory," *IEEE Trans. Microwave Theory Tech.*, vol. MTT-32, pp. 1022–1045.

Oliveira, E. R. D. A. (1968), "Theoretical Foundations of the Finite Element Method," *Int. J. Solids Struct.*, vol. 4, pp. 929–952.

Olyslager, F., D. D. Zutter, and K. Blomme (1993), "Rigorous Analysis of Propagation Characteristics of General Lossless and Lossy Multiconductor Transmission Lines in Multilayered Media," *IEEE Trans. Microwave Theory Tech.*, vol. MTT-41, pp. 79–88.

Oñate, E., and G. Bugeda (1994), "Mesh Optimality Criteria for Adaptive Finite Element Computations," in *The Mathematics of Finite Elements and Applications*, Ed. J. R. Whiteman, John Wiley & Sons, Chichester, England, pp. 121–135.

Onuki, T. (1990), "Hybrid Finite Element and Boundary Element Method Applied to Electromagnetic Problems," *IEEE Trans. Magnetics*, vol. MAG-26, pp. 582–587.

Packard, K. S. (1984), "The Origin of Waveguides: A Case of Multiple Rediscovery," *IEEE Trans. Microwave Theory Tech.*, vol. MTT-32, pp. 961–969.

Pagear, S.S., and S. B. Biggers (1997), "Enrichment of Finite Elements with Numerical Solutions for Singular Stress Fields," *Int. J. Numer. Meth. Eng.*, vol. 40, pp. 2693–2713.

Pantic, Z., and R. Mittra (1986), "Quasi-TEM Analysis of Microwave Transmission Lines by the Finite-Element Method," *IEEE Trans. Microwave Theory Tech.*, vol. MTT-34, pp. 1096–1103.

Pantic-Tanner, Z., and R. Mittra (1988), "Finite-Element Matrices for Loss Calculation in Quasi-TEM Analysis of Microwave Transmission Lines," *Microwave Optical Tech. Letters*, vol. 1, pp. 142–146.

Pantic-Tanner, Z., G. Mavronikolas, and R. Mittra (1994), "A Numerical Absorbing Boundary Condition for Quasi-TEM Analysis of Microwave Transmission Lines Using the Finite-Element Method," *Microwave Optical Tech. Letters*, vol. 9, pp. 134–136.

Parks, D. M. (1979) "On the Convergence of Finite Element Meshes in the Presence of Singularities," *Comp. Meth. Appl. Mech. Eng.*, vol. 10, pp. 780–783.

Patterson, C. (1973), "Sufficient Conditions for Convergence in the Finite Element Method for any Solution of Finite Energy," in *The Mathematics of Finite Elements and Applications*, Ed. J. R. Whiteman, Academic Press, London, pp. 213–224.

Paul, C. R. (1975), "Useful Matrix Chain Parameter Identities for the Analysis of Multiconductor Transmission Lines," *IEEE Trans. Microwave Theory Tech.*, vol. MTT-25, pp. 756–760.

Paulsen, K. D., D. R. Lynch, and J. W. Strohbehn (1988), "Three-Dimensional Finite, Boundary and, Hybrid Element Solutions of the Maxwell Equations for Lossy Dielectric Media," *IEEE Trans. Microwave Theory Tech.*, vol. MTT-36, pp. 682–693.

Paulsen, K. D., and D. R. Lynch (1991), "Elimination of Vector Parasites in Finite Element Maxwell Solutions," *IEEE Trans. Microwave Theory Tech.*, vol. MTT-39, pp. 395–404.

Paulsen, K. D., W. E. Boyse, and D. R. Lynch (1992), "Continuous Potential Maxwell Solutions on Nodal-Based Finite Elements," *IEEE Trans. Antennas Propag.*, vol. AP-40, pp. 1192–1200.

Pavlidis, D., and H. L. Hartnagel (1976), "The Design and Performance of Three Line Microstrip Couplers," *IEEE Trans. Microwave Theory Tech.*, vol. MTT-24, pp. 631–640.

Pekel, Ü., and R. Lee (1995), "An *A Posteriori* Error Reduction Scheme for the Three-Dimensional Finite

Element Solution of Maxwell's Equation," *IEEE Microwave Theory Tech.*, vol. MTT-43, pp. 421–427.

Pekel, Ü., and R. Mittra (1995), "A First-Order, Hybrid FEM/MoM Approach for the Analysis of Two-Dimensional Structures," *Microwave Optical Tech. Letters*, vol. 8, pp. 66–70.

Pekel, Ü., and R. Mittra (1996), "Absorbing Boundary Conditions for Finite Element Mesh Truncation," in *Finite Element Software for Microwave Engineering*, Eds. T. Itoh, G. Pelosi, and P. P. Silvester, John Wiley & Sons, New York, pp. 267–312.

Pelosi, G., A. Freni, and R. Coccioli (1993), "A Hybrid Technique for Analyzing the Scattering from Periodic Structures," *IEE Proc.*, vol. 140, pt. H, pp. 65–70.

Penman, J., and J. R. Fraser (1982), "Complementary and Dual Energy Finite Element Principles in Magnetostatics," *IEEE Trans. Magnetics*, vol. MAG-18, pp. 319–324.

Penman, J., and J. R. Fraser (1983), "Dual and Complementary Energy Methods in Electromagnetism," *IEEE Trans. Magnetics*, vol. MAG-19, pp. 2311–2316.

Penman, J., and J. R. Fraser (1984), "Unified Approach to Problems in Electromagnetism," *IEE Proc.*, vol. 131, pt. A, pp. 55–61.

Penman, J., and M. D. Grieve (1985), "An Approach to Self Adaptive Mesh Generation," *IEEE Trans. Magnetics*, vol. MAG-21, pp. 2567–2570.

Penman, J., and M. D. Grieve (1988), "Self-Adaptive Finite Element Techniques for the Computation of Inhomogeneous Poissonian Fields," *IEEE Trans. Microwave Theory Tech.*, vol. MTT-24, pp. 1042–1049.

Perugia, I. (1997), "A Field-Based Mixed Formulation for the Two-Dimensional Magnetostatic Problem," *SIAM J. Numer. Anal.*, vol. 34, pp. 2382–2391.

Peterson, A. F. (1988), "Absorbing Boundary Conditions for the Vector Wave Equation," *Microwave Optical Tech. Letters*, vol. 1, pp. 62–64.

Peterson, A. F., and S. P. Castillo (1989), "A Frequency-Domain Differential Equation Formulation for Electromagnetic Scattering from Inhomogeneous Cylinders," *IEEE Trans. Antennas Propag.*, vol. AP-37, pp. 601–607.

Peterson, A. F., and R. J. Baca (1991), "Error in the Finite Element Discretization of the Scalar Helmholtz Equation over Electrically Large Regions," *IEEE Microwave Guided Waves Letters*, vol. 1, pp. 219–222.

Peterson, A. F. (1994), "Vector Finite Element Formulation for Scattering from Two-Dimensional Heterogeneous Bodies," *IEEE Trans. Antennas Propag.*, vol. AP-42, pp. 357–365.

Peterson, A. F., and D. R. Wilton (1996), "Curl-Conforming Mixed-Order Edge Elements for Discretizing the 2D and 3D Vector Helmholtz Equation," in *Finite Element Software for Microwave Engineering*, Eds. T. Itoh, G. Pelosi, and P. P. Silvester, John Wiley & Sons, New York, pp. 101–125.

Petre, P. (1987), "The Weighted Residual Method for Wave Propagation in a One-Dimensional Inhomogeneous Medium," *Periodica Polytechnica*, Budapest, vol. 31, pp. 85–97.

Petre, P., and L. Zombory (1988), "Infinite Elements and Base Functions for Rotationally Symmetric Electromagnetic Waves," *IEEE Trans. Antennas Propag.*, vol. AP-36, pp. 1490–1491.

Philippou, G. Y., G. S. Gupta, and J. B. Davies (1981), "Finite-Element Analysis of Angularly Dependent Modes in a General RF Periodic Cavity," *Electronics Letters*, vol. 17, pp. 588–589.

Picon, O. (1988), "Three-Dimensional Finite-Element Formulation for Deterministic Waveguide Problems," *Microwave Optical Tech. Letters*, vol. 1, pp. 170–172.

Pichon, L., and A. Bossavit (1993),"A New Variational Formulation, Free of Spurious Modes, for the Problem of Loaded Cavities," *IEEE Trans. Magnetics*, vol. MAG-29, pp. 1595–1600.

Pinchuk, A. P., and P. P. Silvester (1985), "Error Estimation for Automatic Adaptive Finite Element Mesh Generation," *IEEE Trans. Magnetics*, vol. MAG-21, pp. 2551–2554.

Pinchuk, A. R., C. W. Crowley, and P. P. Silvester (1988), "Spurious Solutions to Vector Diffusion and Wave Field Problems," *IEEE Transaction on Magnetics*, vol. MAG-24, pp. 158–161.

Pintzos, S. G. (1991), "Full-Wave Spectral-Domain Analysis of Coplanar Strips," *IEEE Trans. Microwave Theory Tech.*, vol. MTT-39, pp. 239–246.

Pissanetzky, S. (1984), "A Simple Infinite Element," *COMPEL*, vol. 3, pp. 107–114.

Povinelli, M. J., and J. D'Angelo (1981), "Finite Element Analysis of Large Wavelength Antenna Radome Problems for Leading Edge and Radar Phased Array," *IEEE Trans. Magnetics*, vol. MAG-27, pp. 4299–4302.

Pregla, R., and W. Pascher (1989), "The Method of Lines," in *Numerical Techniques for Microwave and Millimeter-Wave Passive Structures*, Ed. T. Itoh, John Wiley & Sons, New York, pp. 381–446.

Prouty, M. D., K. K. Mei, S. E. Schwarz, and R. Pous (1993), "A New Approach to Quasi-Static Analysis with Application to Microstrip," *IEEE Microwave Guided Wave Letters*, vol. 3, pp. 302–304.

Rahman, B. M. A., and J. B. Davies (1984a), "Finite-Element Analysis of Optical and Microwave Waveguide Problems," *IEEE Trans. Microwave Theory Tech.*, vol. MTT-32, pp. 20–28.

Rahman, B. M. A., and J. B. Davies (1984b), "Penalty Function Improvement of Waveguide Solution by Finite Elements," *IEEE Trans. Microwave Theory Tech.*, vol. MTT-32, pp. 922–928.

Rahman, B. M. A., F. A. Fernández, and J. B. Davies (1991), "Review of Finite Element Methods for Microwave and Optical Waveguides," *Proc. IEEE*, vol. 79, pp. 1442–1448.

Ramahi, O., and R. Mittra (1989), "Finite-Element Analysis of Dielectric Scatterers Using the Absorbing Boundary Conditions," *IEEE Trans. Magnetics*, vol. MAG-25, pp. 3043–3045.

Ramahi, O., A. Khebir, and R. Mittra (1991), "Numerically Derived Absorbing Boundary Condition for the Solution of Open Region Scattering Problems," *IEEE Trans. Antennas Propag.*, vol. AP-39, pp. 350–353.

Ramahi, O. M., and R. Mittra (1991a), "Finite Element Solution for a Class of Unbounded Geometries," *IEEE Trans. Antennas Propag.*, vol. AP-39, pp. 244–250.

Ramahi, O. M., and R. Mittra (1991b), "A Surface Integral Equation Method for the Finite Element Solution of Waveguide Discontinuity Problems," *IEEE Trans. Microwave Theory Tech.*, vol. MTT-39, pp. 604–608.

Ramahi, O. M. (1995), "Application of the Complementary Operator Method to the Finite-Difference-Time-Domain Solution of the Three-Dimensional Radiation Problem," *Microwave Optical Tech. Letters*, vol. 9, pp. 147–149.

Ramahi, O. M. (1997), "The Complementary Operator Method in FDTD Simulations," *IEEE Antennas Propag. Magazine*, vol. 39, pp. 33–45.

Ramakrishnan, C. V., S. Ramakrishnan, A. Kumar, and M. Bhattacharya (1992), "An Integrated Approach for Automated Generation of Two/Three Dimensional Finite Element Grids Using Spatial Occupancy Enumeration and Delaunay Triangulation," *Int. J. Numer. Meth. Eng.*, vol. 34, pp. 1035–1050.

Rank, E. (1986), "Adaptivity and Accuracy Estimation for Finite Element and Boundary Integral Element Methods," in *Accuracy Estimates and Adaptive Refinements in Finite Element Computations*, Eds. I. Babuška, O. C. Zienkiewicz, J. Gago, and E. R. de A. Oliveira, John Wiley & Sons, London, pp. 79–94.

Rank, E. (1993), "Adaptivity and Accuracy Estimation for FEM and BIEM," in *Adaptive Finite and Boundary Element Methods*, Eds. C. A. Brebbia, and M. H. Aliabadi, Computational Mechanics Publications, Southampton, England, pp. 1–46.

Rao, S. M., D. R. Wilton, and A. W. Glisson (1982), "Electromagnetic Scattering by Surfaces of Arbitrary Shape," *IEEE Trans. Antennas Propag.*, vol. AP-30, pp. 409–418.

Rappaport, C. M. (1995), "Perfectly Matched Absorbing Boundary Condition Based on Anisotropic Lossy Mapping of Space," *IEEE Microwave Guided Wave Letters*, vol. 5, pp. 90–92.

Raviart, P. A., and J. M. Thomas (1977a), "A Mixed Finite Element Method for 2nd Order Elliptic Problems," in *Mathematical Aspects of the Finite Element Method*, Eds. I. Galligani, and E. Mayera, Lect. Notes on Mathematics, vol. 606, Springer-Verlag, New York, pp. 293–315.

Raviart, P. A., and J. M. Thomas (1977b), "Primal Hybrid Finite Element Methods for 2nd Order Elliptic Equations," *Math. Comp.*, vol. 31, pp. 391–413.

Rayleigh, Lord (1870), "On the Theory of Resonance," *Trans. Roy. Soc. (London)*, vol. A 161, pp. 77–118.

Rayleigh, Lord (1876), "On the Approximate Solution of Certain Problems Relating to the Potential,"

Proc. London Math. Soc., vol. 7, pp. 70–75.

Reiter, J. M., and F. Arndt (1994), "A Full Wave Boundary Contour Mode Matching Method (BCCM) for the Rigorous CAD of Single and Cascaded Optimized H-Plane and E-Plane Bends," *IEEE MTT-S Int. Microwave Symposium Digest*, vol. 2, pp. 1021–1024.

Remacle, J. F., P. Dular, F. Henrotte, A. Genon, and W. Legros (1995), "Error Estimation and Mesh Optimization Using Error in Constitutive Relation for Electromagnetic Field Computation," *IEEE Trans. Magnetics*, vol. MAG-31, pp. 3587–3789.

Remacle, J. F., P. Dular, A. Genon, and W. Legros (1996), "A Posteriori Error Estimation and Adaptive Meshing Using Error in Constitutive Relation," *IEEE Trans. Magnetics*, vol. MAG-32, pp. 1369–1372.

Rheinboldt, W. C. (1980), "On a Theory of Mesh-Refinement Processes," *SIAM J. Numer. Anal.*, vol. 17, pp. 766–778.

Rheinboldt, W. C. (1981), "Adaptive Mesh Refinement Processes for Finite Element Solutions," *Int. J. Numer. Meth. Eng.*, vol. 17, pp. 649–662.

Riblet, G. P. (1996), "Two-Dimensional Bandpass Filters," in *Finite Element Software for Microwave Engineering*, Eds. T. Itoh, G. Pelosi, and P. P. Silvester, John Wiley & Sons, New York, pp. 127–146.

Richards, D. J., and A. Wexler (1972), "Finite-Element Solutions within Curved Boundaries," *IEEE Trans. Microwave Theory Tech.*, vol. MTT-20, pp. 650–657.

Rikabi, J., C. F. Bryant, and E. M. Freeman (1988), "An Error-Based Approach to Complementary Formulations of Static Field Solutions," *Int. J. Numer. Meth. Eng.*, vol. 26, pp. 1963–1987.

Ritz, W. (1908), "Über eine Neue Methode zur Lösung gewissen Variations-Probleme der Mathematischen Physik," *J. Reine Angew. Math.*, vol. 135, pp. 1–61.

Rivara, M. C. (1984), "Algorithms for Refining Triangular Grids Suitable for Adaptive and Multigrid Techniques," *Int. J. Numer. Meth. Eng.*, vol. 20, pp. 745–756.

Rivara, M. C. (1986), "Adaptive Finite Element Refinement and Fully Irregular and Conforming Triangulations," in *Accuracy Estimates and Adaptive Refinements in Finite Elements Computations*, Eds. I. Babuška, O. C. Zienkiewicz, J. Gago, and E. R. de A. Oliveira, John Wiley & Sons, London, pp. 359–370.

Rivara, M. C. (1987), "A Grid Generator Based on 4–Triangles Conforming Mesh-Refinement Algorithms," *Int. J. Numer. Meth. Eng.*, vol. 24, pp. 1343–1354.

Rivara, M. C. (1989), "Selective Refinement/Derefinement Algorithms for Sequences of Nested Triangulations," *Int. J. Numer. Meth. Eng.*, vol. 28, pp. 2889–2906.

Rivara, M. C. (1992), "Generación de Mallas para Métodos de Elementos Finitos Adaptivos," in *Numerical Methods in Engineering and Applied Sciences*, Eds. H. Alder, J. C. Heinrich, S. Lavanchy, E. Oñate, and B. Suárez, CIMNE, Barcelona, pp. 181–190.

Rivara, M. C., and P. Inostroza (1997), "Using Longest-Side Bisection Techniques for the Automatic Refinement of Delaunay Triangulation," *Int. J. Numer. Meth. Eng.*, vol. 40, pp. 581–598.

Rizzoli, V. (1979), "Highly Efficient Calculation of Shielded Microstrip Structures in the Presence of Undercutting," *IEEE Trans. Microwave Theory Tech.*, vol. MTT-27, pp. 50–157.

Rossow, M. P., and I. N. Katz (1978), "Hierarchical Finite Elements and Precomputed Arrays," *Int. J. Numer. Meth. Eng.*, vol. 12, pp. 977–999.

Roy, T., T. K. Sarkar, A. R. Djordjević, and M. Salazar-Palma (1995), "A Hybrid Method for Terminating the Finite Element Mesh (Electrostatic Case)," *Microwave Optical Tech. Letters*, vol. 8, pp. 282–287.

Roy, T., T. K. Sarkar, A. R. Djordjević, and M. Salazar-Palma (1996), "A Hybrid Method for Solution of Scattering by Conducting Cylinders (TM Case)," *IEEE Trans. Microwave Theory Tech.*, vol. MTT-44, pp. 2145–2151.

Roy, T., T. K. Sarkar, A. R. Djordjević, and M. Salazar-Palma (1998), "Time-Domain Analysis of TM Scattering from Conducting Cylinders Using a Hybrid Method," *IEEE Trans. Microwave Theory Tech.*, vol. MTT-46.

Rozzi, T., G. Cerri, M. N. Husain, and L. Zapelli (1991), "Variational Analysis of the Dielectric Rib Waveguide Using the Concept of 'Transition Function' and Including Edge Singularities," *IEEE Trans. Microwave Theory Tech.*, vol. MTT-39, pp. 247–257.

Rumsey, V. H. (1954), "The Reaction Concept in Electromagnetic Theory," *Phys. Rev.*, ser. 2, vol. 94, pp. 1483–1491.

Saad, S. M. (1985), "Review of Numerical Methods for the Analysis of Arbitrarily-Shaped Microwave and Optical Dielectric Waveguides," *IEEE Trans. Microwave Theory Tech.*, vol. MTT-33, pp. 894–899.

Sabonnadière, J. C., G. Meunier, and B. Morel (1982), "FLUX: A General Interactive Finite Elements Package for 2D Electromagnetic Fields," *IEEE Trans. Magnetics*, vol. MAG-18, pp. 624–626.

Sabonnadière, J. C., and A. Konrad (1992), "Computing EM Fields," *IEEE Spectrum*, vol. 29, pp. 52–56.

Sachdev, S., and Z. J. Csendes (1988), "Combined Tangential-Normal Vector Elements for Computing Electric and Magnetic Fields," *IEEE Trans. Microwave Theory Tech.*, vol. MTT-36, pp. 153–155.

Sacks, Z. S., D. M. Kingsland, R. Lee, and J. F. Lee (1995), "A Perfectly Matched Anisotropic Absorber for Use as an Absorbing Boundary Condition," *IEEE Trans. Antennas Propag.*, vol. AP-43, pp. 1460–1463.

Sadiku, M. N. O. (1989), "A Simple Introduction to Finite Element Analysis of Electromagnetic Problems," *IEEE Trans. Education*, vol. 32, pp. 85–93.

Sakellaris, J., G. Meunier, A. Raizer, and A. Darcherif (1992), "The Impedance Boundary Condition Applied to the Finite Element Method Using the Magnetic Vector Potential as State Variable: A Rigorous Solution for High Frequency Axisymmetric Problems," *IEEE Trans. Magnetics*, vol. MAG-28, pp. 1643–1646.

Salazar-Palma, M., and J. F. Hernández-Gil (1989), "Self Adaptive Mesh Scheme for the Finite Element Analysis of Anisotropic Multiconductor Transmission-Lines," *IEEE MTT-S Int. Symposium Digest*, Long Beach, CA, pp. 507–510.

Salazar-Palma, M., L. Ferragut, F. J. Mustieles, and J. F. Hernández-Gil (1991), "Assessing the Error of Transmission-Line Quasi-Static Analyses by Means of a Mixed Finite Element Method," *IEEE AP-S Int. Symposium Digest*, vol. 2, London, Ontario, Canada, pp. 1232–1235.

Salazar-Palma, M., F. J. Mustieles, and J. F. Hernández-Gil (1992), "Adaptive Finite Element Method for the Analysis of the Electro-Optic Effect in Optical Integrated Devices," in *Numerical Methods in Engineering '92*, Eds. C. Hirsh, O. C. Zienkiewicz, and E. Oñate, Elsevier, Amsterdam, pp. 457–463.

Salazar-Palma, M., L. E. García-Castillo, and G. G. Gentili (1994), "A Software Package for Accurate Computation of Frequency Dependent Propagation and Circuital Parameters of Inhomogeneous Anisotropic Arbitrary Shaped Multiconductor Transmission Line," *24th European Microwave Conf. Proc.*, Cannes, France, pp. 1709–1714.

Salazar-Palma, M., and J. F. Hernández-Gil (1995), "An Infinite Element for the Finite Element Analysis of Open Transmission Lines," *25th European Microwave Conference Proc.*, Bologna, Italy, pp. 747–750.

Salazar-Palma, M., and J. M. Recio-Peláez (1996), "A Convergence Study of a Self Adaptive Mesh Algorithm," in *Software for Electrical Engineering Analysis and Design*, Ed. P. P. Silvester, Computational Mechanics Publications, Southampton, England, pp. 385–394.

Salazar-Palma, M., and L. E. García-Castillo (1996), "Self-Adaptive Procedures for Waveguiding Structure Analysis," in *Finite Element Software for Microwave Engineering*, Eds. T. Itoh, G. Pelosi, and P. P. Silvester, John Wiley & Sons, New York, pp. 401–434.

Sandy, F., and J. Sage (1971), "Use of Finite Difference Approximations to Partial Differential Equations for Problems Having Boundaries at Infinity," *IEEE Trans. Microwave Theory Tech.*, vol. MTT-19, pp. 484–486.

Santos, A. F., and J. P. Figanier (1975), "The Method of Series Expansion in the Frequency Domain Applied to Multidielectric Transmission Lines," *IEEE Trans. Microwave Theory Tech.*, vol. MTT-23, pp. 753–756.

Saranen, J. (1981), "The Finite Element Method for the Electric Field of Maxwell's Boundary Value Problem in Polygonal Domains of Plane," *Annales Academia Scientiarum Fennice*, Series A. I. Mathematica, vol. 6, pp. 63–75.

Sarkar, T. K. (1983), "A Note on the Variational Method (Rayleigh-Ritz), Galerkin's Method, and the Method of Least Squares," *Radio Science*, vol. 18, pp. 1207–1224.

Sarkar, T. K. (1985), "A Note on the Choice of Weighting Functions in the Method of Moments," *IEEE Trans. Antennas Propag.*, vol. AP-33, pp. 436–441.

Sarkar, T. K., A. R. Djordjević, and E. Arvas (1985), "On The Choice of Expansion and Weighting Functions in the Numerical Solution of Operator Equations," *IEEE Trans. Antennas Propag.*, vol. AP-33, pp. 988–996.

Sarkar, T. K., K. Athar, E. Arvas, M. Manela, and R. Lade (1989a), "Computation of the Propagation Characteristics of TE and TM Modes in Arbitrarily Shaped Hollow Waveguides Utilizing the Conjugate Gradient Method," *J. Electromagnetic Waves Appl.*, vol. 3, pp. 143–165.

Sarkar, T. K, E. Arvas, and S. Ponnapalli (1989b), "Electromagnetic Scattering from Dielectric Structures," *IEEE Trans. Antennas Propag.*, vol. AP-37, pp. 673–676.

Sarkar, T. K. (1991), "From Reaction Concept to Conjugate Gradient: Have We Made Any Progress?," in *Application of Conjugate Gradient Method to Electromagnetics and Signal Analysis*, Ed. T. Sarkar, Elsevier, New York, pp. 1–25.

Sarkar, T.K., Z. A. Maricevic, and M. Kahrizi (1992), "An Accurate De-Embedding Procedure for Characterizing Discontinuities," *Int. J. Microwave Millimeter-Wave Computer-Aided Eng.*, vol. 2, pp. 135–143.

Sarkar, T. K., and M. Salazar-Palma (1994), "An Alternate Interpretation of Complex Modes in Closed Perfectly Conducting (Lossless) Structures," *AEÜ*, vol. 48, pp. 123–129.

Sarkar, T. K., R. S. Adve, L. E. García-Castillo, and M. Salazar-Palma (1994), "Utilization of Wavelet Concepts in Finite Elements for an Efficient Solution of Maxwell's Equations," *Radio Science*, vol. 29, pp. 965–977.

Sarkar, T. K., T. Roy, M. Salazar-Palma, and A. R. Djordjević (1998), "Finite Element in the Time Domain," in *Time Domain Electromagnetics*, Ed. S. M. Rao, Academic Press, New York.

Schatz, A. H., and L. B. Wahlbin (1978), "Maximum Norm Estimates in the Finite Element Method on Plane Polygonal Domains. Part 1," *Math. Comp.*, vol. 32, pp. 73–109.

Schatz, A. H., and L. B. Wahlbin (1979), "Maximum Norm Estimates in the Finite Element Method on Plane Polygonal Domains. Part 2," *Math. Comp.*, vol. 33, pp. 465–492.

Schellbach, K. (1851), "Probleme der Variationsrechnung," *J. Reine Angew. Math.*, vol. 41, pp. 293–363.

Scott, W. R. (1994), "Errors Due to Spatial Discretization and Numerical Precision in the Finite Element Method," *IEEE Trans. Antennas Propag.*, vol. AP-42, pp. 1565-1570.

Sears, G. J. (1988), "Hierarchical Quadrilateral Elements in Field Analysis," *IEEE Trans. Magnetics*, vol. MAG-24, pp. 404–407.

Seitelman, L. R. (1973), "Some Practical Solution Techniques for Finite Element Analysis," in *The Mathematics of Finite Elements and Applications*, Ed. J. R. Whiteman, Academic Press, London, pp. 471–481.

Selleri, S., and M. Zoboli (1995), "An Improved Finite Element Method Formulation for the Analysis of Nonlinear Anisotropic Dielectric Waveguides," *IEEE Trans. Microwave Theory Tech.*, vol. MTT-43, pp. 887–892.

Selleri, S., and M. Zoboli (1996), "A Comparison of Vector Finite Element Formulations for Waveguide Analysis," in *Software for Electrical Engineering Analysis and Design*, Ed. P. P. Silvester, Computational Mechanics Publications, Southampton, England, pp. 23–31.

Seshadri, T. H., S. Mahapatra, and K. Rajaiah (1980), "Corner Function Analysis of Microstrip Transmission Lines," *IEEE Trans. Microwave Theory Tech.*, vol. MTT-28, pp. 376–380.

Shenton, D. N, and Z. J. Csendes (1985), "Three Dimensional Finite Element Mesh Generation Using Delaunay Tessellation," *IEEE Trans. Magnetics*, vol. MAG-21, pp. 2535–2538.

Shepard, M. S. (1985), "Automatic and Adaptive Mesh Generation," *IEEE Trans. Magnetics*, vol. MAG-

21, pp. 2484–2489.

Shepard, M. S. (1986), "Adaptive Finite Element Analysis and CAD," in *Accuracy Estimates and Adaptive Refinements in Finite Element Computations*, Ed. I. Babuška, O. C. Zienkiewicz, J. Gago, and E. R. de A. Oliveira, John Wiley & Sons, New York, pp. 205–225.

Shepherd, P. R., and P. Daly (1985), "Modelling and Measurement of Microstrip Transmission-Line Structures," *IEEE Trans. Microwave Theory Tech.*, vol. MTT-33, pp. 1501–1506.

Sherwin, S. J., and G. E. Karniadakis (1995), "A New Triangular and Tetrahedral Basis for High-Order (*hp*) Finite Element Methods," *Int. J. Numer. Meth. Eng.*, vol. 38, pp. 3775–3802.

Shih, C., R. B. Wu, S. K. Jeng, and C. H. Chen (1988), "A Full Wave Analysis of Microstrip Lines by Variational Conformal Mapping Technique," *IEEE Trans. Microwave Theory Tech.*, vol. MTT-36, pp. 576–581.

Silvester, P. P. (1969a), "Finite Element Solution of Homogeneous Waveguide Problems," *Alta Frequenza*, vol. 38, pp. 313–317.

Silvester, P. P. (1969b), "A General High-Order Finite Element Waveguide Analysis Program," *IEEE Trans. Microwave Theory Tech.*, vol. MTT-17, pp. 204–209.

Silvester, P. P. (1969c), "High-Order Polynomial Triangular Finite Elements for Potential Problems," *Int. J. Eng. Science*, vol. 7, pp. 849–861.

Silvester, P. P., and M. V. K. Chari (1970), "Finite Element Solution of Saturable Magnetic Field Problem," *IEEE Trans. Power Appl. Syst.*, vol. PAS-89, pp. 1642–1651.

Silvester, P. P., and M. S. Hsieh (1971), "Finite-Element Solution of 2-Dimensional Exterior-Field Problems," *IEE Proc.*, vol. 118, pt. H, pp. 1743–1747.

Silvester, P. P., and K. K. Chan (1972), "Bubnov-Galerkin Solutions to Wire-Antenna Problems," *Proc. IEE*, vol. 119, pp. 1095–1099.

Silvester, P. P. (1973), "Finite Element Analysis of Planar Microwave Networks," *IEEE Trans. Microwave Theory Tech.*, vol. MTT-21, pp. 104–108.

Silvester, P. P., and Z. J. Csendes (1974), "Numerical Modeling of Passive Microwave Devices," *IEEE Trans. Microwave Theory Tech.*, vol. MTT-22, pp. 190–201.

Silvester, P. P., and P. Rafinejad (1974), "Curvilinear Finite-Element for Two-Dimensional Saturable Magnetic Fields," *IEEE Trans. Power Appl. Syst.*, vol. PAS-93, pp. 1861–1870.

Silvester, P. P., and A. Konrad (1977), "Planar Networks Analysis Programs - PNAP1 and PNAP2," *IEEE Trans. Microwave Theory Tech.*, vol. MTT-25, p. 959.

Silvester, P. P, D. A. Lowther, C. J. Carpenter, and E. A. Wyatt (1977), "Exterior Finite Elements for 2-Dimensional Field Problems with Open Boundaries," *Proc. IEE*, vol. 124, pp. 1267–1270.

Silvester, P. P. (1978), "Construction of Triangular Finite Element Universal Matrices," *Int. J. Numer. Meth. Eng.*, vol. 12, pp. 237–244.

Silvester, P. P. (1982a), "Universal Finite Element Matrices for Tetrahedra," *Int. J. Numer. Meth. Eng.*, vol. 18, pp. 1055–1061.

Silvester, P. P. (1982b), "Permutation Rules for Simplex Finite Elements," *Int. J. Numer. Meth. Eng.*, vol. 18, pp. 1245–1248.

Silvester, P. P. (1996b), "Profiles of Current Finite Element Software," in *Finite Element Software for Microwave Engineering*, Eds. T. Itoh, G. Pelosi, and P. P. Silvester, John Wiley & Sons, New York, pp. 455–476.

Slade, G. W., and K. J. Webb (1991), "Convergence in Microwave Circuit Parameters Obtained from a Vector Finite Element Analysis," *IEEE Trans. Magnetics*, vol. MAG-27, pp. 4048–4049.

Slade, G. W., and K. J. Webb (1992), "Computation of Characteristic Impedance for Multiple Microstrip Transmission Lines Using a Vector Finite Element Method," *IEEE Trans. Microwave Theory Tech.*, vol. MTT-40, pp. 34–40.

Sobolev, L. (1963a), "On a Theorem of Functional Analysis," *Transl. Amer. Math. Soc.*, vol. 2, pp. 39–68

Sorrentino, R. (1988), "Numerical Methods for Passive Components," *IEEE Trans. Microwave Theory Tech.*, vol. MTT-36, pp. 619–622.

Stochniol, A. (1992), "A General Transformation for Open Boundary Finite Element for Electromagnetic Problems" *IEEE Trans. Magnetics*, vol. MAG-28, pp. 1679–1681.

Stoval, R. E., and K. K. Mei (1975), "Application of a Unimoment Technique to a Biconical Antenna with Inhomogeneous Dielectric Loading," *IEEE Trans. Antennas Propag.*, vol. AP-23, pp. 335–341.

Strang, G. (1972), "Variational Crimes in the Finite Element Method," in *Mathematical Foundations of the Finite Element Method with Applications to Partial Differential Equations*, Ed. A. K. Aziz, Academic Press, New York.

Stupfel, B., and R. Mittra (1995), "A Theoretical Study of Numerical Absorbing Boundary Conditions," *IEEE Trans. Antennas Propag.*, vol. AP-43, pp. 478–487.

Stupfel, B., and R. Mittra (1996), "Numerical Absorbing Boundary Conditions for the Scalar and Vector Wave Equations," *IEEE Trans. Antennas Propag.*, vol. AP-44, pp. 1015–1022.

Su, C. C. (1986), "A Combined Method for Dielectric Waveguides Using the Finite-Element Technique and the Surface Integral Equation Method," *IEEE Trans. Microwave Theory Tech.*, vol. MTT-34, pp. 1140–1146.

Su, C. C. (1989), "Electromagnetic Scattering by a Dielectric Body with Arbitrary Inhomogeneity and Anisotropy," *IEEE Trans. Antennas Propag.*, vol. AP-37, pp. 384–389.

Su, C. C. (1993), "An Efficient Numerical Procedure Using the Shifted Power Method for Analyzing Dielectric Waveguides Without Inverting Matrices," *IEEE Trans. Microwave Theory Tech.*, vol. MTT-41, pp. 539–542.

Sumbar, E., F. E. Vermeulen, and F. S. Chute (1991), "Implementation of Radiation Boundary Conditions in the Finite Element Analysis of Electromagnetic Wave Propagation," *IEEE Trans. Microwave Theory Tech.*, vol. MTT-39, pp. 267–273.

Sun, D., J. Manges, X. Yuan, and Z. Csendes (1995), "Spurious Modes in Finite Element Methods," *IEEE Antennas Propag. Magazine*, vol. 37, pp. 12–24.

Sun, W., and C. A. Balanis (1994), "Vector One-Way Wave Absorbing Boundary Conditions for FEM Applications," *IEEE Trans. Antennas Propag.*, vol. AP-42, pp. 872–888.

Suri, M. (1990), "The *p*-Version of the Finite Element Method for Elliptic Equations of Order 2*l*," *RAIRO Math. Model. Numer. Anal.*, vol. 24, pp. 265–304.

Sussman, T., and K. J. Bathe (1985), "Studies of Finite Element Procedures on Mesh Selection," *Comp. Struct.*, vol. 21, pp. 257–264.

Svedin, J. A. (1989), "A Numerically Efficient Finite-Element Formulation for the General Waveguide Problem Without Spurious Modes," *IEEE Trans. Microwave Theory Tech.*, vol. MTT-37, pp. 1708–1715.

Svedin, J. A. (1990), "Propagation Analysis of Chirowaveguides Using the Finite Element Method," *IEEE Trans. Microwave Theory Tech.*, vol. MTT-38, pp. 1488–1496.

Svedin, J. A. (1991), "A Modified Finite-Element Method for Dielectric Waveguides Using an Asymptotically Correct Approximation on Infinite Elements," *IEEE Trans. Microwave Theory Tech.*, vol. MTT-39, pp. 258–266.

Swaminathan, M., E. Arvas, T. K. Sarkar, and A. R. Djordjević (1990), "Computation of Cutoff Wavenumbers of TE and TM Modes in Waveguides of Arbitrary Cross Sections Using a Surface Integral Formulation," *IEEE Trans. Microwave Theory Tech.*, vol. MTT-39, pp. 154–159.

Szabó, B. A., and A. K. Metha (1978), "*p*-Convergent Finite Element Approximations in Fracture Mechanics," *Int. J. Numer. Meth. Eng.*, vol. 12, pp. 551–560.

Szabó, B. A. (1986), "Estimation and Control of Error Based on *p* Convergence," in *Accuracy Estimates and Adaptive Refinements in Finite Element Computations*, Eds. I. Babuška, O. C. Zienkiewicz, J. Gago, and E. R. de A. Oliveira, John Wiley & Sons, London, pp. 61–78.

Tan, J., and G. Pan (1995), "A New Edge Element Analysis of Dispersive Waveguiding Structures," *IEEE Trans. Microwave Theory Tech.*, vol. MTT-43, pp. 2600–2607. See "Comments" by F. A. Fernández and Y. Lu, and "Author's Reply" in *IEEE Trans. Microwave Theory Tech.*, vol. MTT-44, pp. 1618–1619.

Tang, Y. Q., Y. P. Liang, H. R. Wu, X. L. Meng, S. C. Xu, and G. X. Fan (1994), "An Adaptive Finite

Element Computation of *h-p* Version for Magnetic Field Problems," *IEEE Trans. Magnetics*, vol. MAG-30, pp. 3519–3522.

Tanner, D. R., and A. F. Peterson (1989), "Vector Expansion Functions for the Numerical Solution of Maxwell's Equations," *Microwave Optical Tech. Letters*, vol. 2, pp. 331–334.

Tärnhuvud, T., K. Reichert, and J. Skoczylas (1990), "Problem-Oriented Adaptive Mesh-Generator for Accurate Finite-Element Calculation," *IEEE Trans. Magnetics*, vol. MAG-26, pp. 779–782.

Terakado, R. (1976), "The Characteristic Impedance of Rectangular Coaxial Line with Ratio 2:1 of Outer-to-Inner Conductor Side Length," *IEEE Trans. Microwave Theory Tech.*, vol. MTT-24, pp. 124–125.

Tessler, A., H. H. Riggs, and S. C. Macy (1994), "Application of a Variational Method for Computing Smooth Stresses, Stress Gradients, and Error Estimation in Finite Element Analysis," in *The Mathematics of Finite Elements and Applications*, Ed. J. R. Whiteman, John Wiley & Sons, Chichester, England, pp. 189–198.

Thatcher, R. W. (1976), "The Use of Infinite Grid Refinements at Singularities in the Solution of Laplace's Equation," *Numer. Math.*, vol. 25, pp. 163–178.

Thatcher, R. W. (1982), "Assessing the Error in a Finite Element Solution," *IEEE Trans. Microwave Theory Tech.*, vol. MTT-30, pp. 911–915.

Thorburn, M., A. Agoston, and V. K. Tripathi (1990), "Computation of Frequency Dependent Propagation Characteristics of Microstriplike Propagation Structures with Discontinuous Layers," *IEEE Trans. Microwave Theory Tech.*, vol. MTT-38, pp. 148–153.

Tippet, J. C., and D. C. Chang (1976), "A New Approximation for the Capacitance of a Rectangular-Coaxial-Strip Transmission Line," *IEEE Trans. Microwave Theory Tech.*, vol. MTT-24, pp. 602–604.

Tong, P., and T. H. H. Pian (1973), "On the Convergence of the Finite Element Method for Problems with Singularity," *Int. J. Solids Structures*, vol. 9, pp. 313–321.

Tortschonoff, T. (1984), "Survey of Numerical Methods in Field Calculations," *IEEE Trans. Magnetics*, vol. MAG-20, pp. 1912–1917.

Towers, M. S., A. McCowen, and J. A. R. Macnab (1993), "Electromagnetic Scattering from an Arbitrary, Inhomogeneous 2-D Object — A Finite and Infinite Element Solution," *IEEE Trans. Antennas Propag.*, vol. AP-41, pp. 770–777.

Trefethen, L. N., and L. Haalpern (1986), "Well-Posedness of One-Way Wave Equations and Absorbing Boundary Conditions," vol. 47, pp. 421–453.

Tripathi, V. K., and R. J. Bucolo (1985), "A Simple Network Analog Approach for the Quasi-Static Characteristics of General Lossy, Anisotropic, Layered Structures," *IEEE Trans. Microwave Theory Tech.*, vol. MTT-33, pp. 1458–1464.

Tsitsos, S., A. A. P. Gibson, and A. H. I. McCormick (1994), "Higher Order Modes in Coupled Striplines: Prediction and Measurements," *IEEE Trans. Microwave Theory Tech.*, vol. MTT-42, pp. 2071–2077.

Turner, M. J., R. W. Clough, H. C. Martin, and L. J. Topp (1956), "Stiffness and Deflection Analysis of Complex Structures," *Journal of Aeronautical Science*, vol. 23, pp. 805–823.

Tzuang, C. K. C., and T. Itoh (1986), "Finite-Element Analysis of Slow-Wave Schottky Contact Printed Lines," *IEEE Trans. Microwave Theory Tech.*, vol. MTT-34, pp. 1483–1489.

Tzuang, C. K. C., and T. Itoh (1987), "High-Speed Pulse Transmission Along a Slow-Wave CPW for Monolithic Microwave Integrated Circuits," *IEEE Trans. Microwave Theory Tech.*, vol. MTT-35, pp. 697–704.

Valor, L., and J. Zapata (1995), "Efficient Finite Element Analysis of Waveguides with Lossy Inhomogeneous Anisotropic Materials Characterized by Arbitrary Permittivity and Permeability Tensors," *IEEE Trans. Microwave Theory Tech.*, vol. MTT-43, pp. 2452–2459.

Valor, L., and J. Zapata (1996), "An Efficient Finite Element Formulation to Analyze Waveguides with Lossy Inhomogeneous Bi-Anisotropic Materials," *IEEE Trans. Microwave Theory Tech.*, vol. MTT-44, pp. 291–296.

Vandenbulcke, P., and P. E. Lagasse (1976), "Eigenmode Analysis of Anisotropic Optical Fibers or Integrated Optical Waveguides," *Electronics Letters*, vol. 12, pp. 120–122.

Vanmaele, M., A. Ženíšek (1995), "The Combined Effect of Numerical Integration and Approximation of the Boundary in the Finite Element Method for Eigenvalue Problems," *Numer. Math.*, vol. 71, pp. 253–273.

Venkataraman, J., S. M. Rao, A. R. Djordjević, T. K. Sarkar, and Y. Naiheng (1985), "Analysis of Arbitrarily Oriented Microstrip Transmission Lines in Arbitrarily Shaped Dielectric Media over a Finite Ground Plane," *IEEE Trans. Microwave Theory Tech.*, vol. MTT-33, pp. 952–959.

Verfürth, R. (1991), "*A Posteriori* Error Estimators for the Stokes Equations II Non-Conforming Discretizations," *Numer. Math.*, vol. 60, pp. 235–249.

Volakis, J. L., A. Chatterjee, J. Gong, L. C. Kempel, and D.C. Ross (1994), "Progress on the Application of the Finite Element Method for 3D Electromagnetic Scattering and Radiation," *COMPEL*, vol. 13, supp. A, pp. 359–364.

Volakis, J. L., J. Gong, and T. Ozdemir (1996), "FEM Applications to Conformal Antennas," in *Finite Element Software for Microwave Engineering*, Eds. T. Itoh, G. Pelosi, and P. P. Silvester, John Wiley & Sons, New York, pp. 313–345.

Waldow, P. (1989), "Feldtheoretische Berechnung vielfach gekoppelter Leitungsanordnungen," *Mikrowellen HF Mcg.*, vol. 15, pp. 449–453.

Wang, J. S., and N. Ida (1991), "Eigenvalue Analysis in Electromagnetic Cavities Using Divergence Free Finite Elements," *IEEE Trans. Magnetics*, vol. MAG-27, pp. 3978–3981.

Wang, J. S., and N. Ida (1992), "Eigenvalue Analysis in Anisotropically Loaded Electromagnetic Cavities Using Edge Finite Elements," *IEEE Trans. Magnetics*, vol. MAG-28, pp. 1438–1441.

Wang, J. S., and N. Ida (1993), "Curvilinear and Higher Order Edge Finite Elements in Electromagnetic Field Computation," *IEEE Trans. Magnetics*, vol. MAG-29, pp. 1491–1494.

Wang, J. S., and R. Mittra (1994), "A Finite Element Cavity Resonance Method for Waveguide and Microstripline Discontinuities," *IEEE Trans. Microwave Theory Tech.*, vol. MTT-42, pp. 433–440.

Wang, Y., P. Monk, and B. Szabó (1996), "Computing Cavity Modes Using *p*-Version of the Finite Element Method," *IEEE Trans. Magnetics*, vol. 32, pp. 1934–1940.

Wang, J. S. (1997), "Hierarchic Edge Elements for High-Frequency Problems," *IEEE Trans. Magnetics*, vol. MAG-33, pp. 1536–1539.

Warren, G. S., and W. R. Scott (1994), "An Investigation of Numerical Dispersion in the Vector Finite Element Method Using Quadrilateral Elements," *IEEE Trans. Antennas Propag.*, vol. AP-42, pp. 1502–1508.

Watson, D. F. (1981), "Computing the *n*-Dimensional Delaunay Tessellation with Application to Voronoi Polytopes," *Comp. J.*, vol. 24, pp. 167–172.

Webb, J. P. (1985), "The Finite Element Method for Finding Modes of Dielectric-Loaded Cavities," *IEEE Trans. Microwave Theory Tech.*, vol. MTT-33, pp. 635–639.

Webb, J. P., D. A. Lowther, and P. P. Silvester (1985), "A Finite Element Database for Electromagnetic Field Computation," *IEEE Trans. Magnetics*, vol. MAG-21, pp. 2503–2506.

Webb, J. P., and S. Parihar (1986), "Finite-Element Analysis of *H*-Plane Rectangular Waveguide Problems," *Proc. IEE*, vol. 133, pt. H, pp. 91–94.

Webb, J. P. (1988a), "Efficient Generation of Divergence-Free Fields for the Finite Element Analysis of 3D Cavity Resonances," *IEEE Trans. Magnetics*, vol. MAG-24, pp. 162–165.

Webb, J. P. (1988b), "Finite Element Analysis of Dispersion in Waveguides with Sharp Metal Edges," *IEEE Trans. Microwave Theory Tech.*, vol. MTT-36, pp. 1819–1824.

Webb, J. P., and V. N. Kanellopoulos (1989), "Absorbing Boundary Conditions for the Finite Element Solution of the Vector Wave Equation," *Microwave Optical Tech. Letters*, vol. 2, pp. 370–372.

Webb, J. P. (1990), "Absorbing Boundary Conditions for the Finite Element Analysis of Planar Devices," *IEEE Trans. Microwave Theory Tech.*, vol. MTT-38, pp. 1328–1332.

Webb, J. P., and S. McFee (1991), "The Use of Hierarchical Triangles and Finite-Element Analysis of Microwave and Optical Devices," *IEEE Trans. Magnetics*, vol. MAG-27, pp. 4040–4043.

Webb, J. P. (1993), "Edge Elements and What They Can Do for You," *IEEE Trans. Magnetics*, vol. MAG-29, pp. 1460–1465.

Webb, J. P., and B. Forghani (1993), "Hierarchical Scalar and Vector Tetrahedra," *IEEE Trans. Magnetics*, vol. MAG-29, pp. 1495–1498.

Webb, J. P., and B. Forghani (1994), "Adaptive Improvement of Magnetic Fields Using Hierarchical Tetrahedral Finite Element Methods," *IEEE Trans. Magnetics*, vol. MAG-30, pp. 3511–3514.

Webb, J. P., and R. Abouchacra (1995), "Hierarchical Triangular Elements Using Orthogonal Polynomials," *Int. J. Numer. Meth. Eng.*, vol. 38, pp. 245–257.

Weeks, W. T. (1970), "Calculation of Coefficients of Capacitance of Multiconductor Transmission Lines in the Presence of a Dielectric Interface," *IEEE Trans. Microwave Tech.*, vol. MTT-18, pp. 35–43.

Wei, C., R. F. Harrington, J. R. Mautz, and T. K. Sarkar (1984), "Multiconductor Transmission Lines in Multilayered Dielectric Media," *IEEE Trans. Microwave Theory Tech.*, vol. MTT-32, pp. 439–450.

Welij, J. C. V. (1985), "Calculation of Eddy Current in Terms of $\bar{H}$ on Hexahedra," *IEEE Trans. Magnetics*, vol. MAG-21, pp. 2239–2241.

Welt, D., and J. Webb (1985), "Finite Element Analysis of Dielectric Waveguides with Curved Boundaries," *IEEE Trans. Microwave Tech.*, vol. MTT-33, pp. 576–585.

Wexler, A. (1969), "Computation of Electromagnetic Fields," *IEEE Trans. Microwave Theory Tech.*, vol. MTT-17, pp. 416–439.

Wiberg, N. E., and F. Abdulwahab (1992), "An Efficient Postprocessing Technique for Stress Problems Based on Superconvergent Derivatives and Equilibrium," in *Numerical Methods in Engineering '92*, Eds. C. Hirsch, O. C. Zienkiewicz, and E. Oñate, Elsevier, Amsterdam, pp. 25–32.

Wiemer, L., and R. H. Jansen (1988), "Reciprocity Related Definition of Strip Characteristic Impedance for Multiconductor Hybrid-Mode Transmission Lines," *Microwave Optical Tech. Letters*, vol. 1, pp. 22–25.

Wilkins, G. M., J. F. Lee, and R. Mittra (1991), "Numerical Modeling of Axisymmetric Coaxial Waveguide Discontinuities," *IEEE Microwave Theory Tech.*, vol MTT-39, pp. 1323–1328.

Williamson, F. (1980), "A Historical Note on the Finite Element Method," *Int. J. Numer. Meth. Eng.*, vol. 15, pp. 930–934.

Wiltse, J. C. (1984), "History of Millimeter and Submillimeter Waves," *IEEE Trans. Microwave Theory Tech.*, vol. MTT-32, pp. 1118–1127.

Wong, M. F., O. Picon, and V. F. Hanna (1993), "Three-Dimensional Finite Element Analysis of *n*-Port Waveguide Junctions Using Edge-Elements," *Int. J. Microwave Millimeter-Wave Computer-Aided Eng.*, vol. 3, pp. 442–451.

Wong, S. H., and Z. J. Csendes (1988), "Combined Finite Element-Modal Solution of Three-Dimensional Eddy Current Problems," *IEEE Trans. Magnetics*, vol. MAG-24, pp. 2685–2687.

Wong, S. H., and Z. J. Csendes (1989), "Numerically Stable Finite Element Methods for the Galerkin Solution of Eddy Current Problems," *IEEE Trans. Magnetics*, vol. 25, pp. 3019–3021.

Worm, S. B., and R. Pregla (1984), "Hybrid-Mode Analysis of Arbitrarily Shaped Planar Microwave Structures by the Method of Lines," *IEEE Trans. Microwave Theory Tech.*, vol. MTT-32, pp. 191–196.

Wu, J. Y., D. M. Kingsland, J. F. Lee, and R. Lee (1997), "A Comparison of Anisotropic PML to Berenger's PML and Its Applications to the Finite-Element Method for EM Scattering," *IEEE Trans. Antennas Propag.*, vol. AP-45, pp. 40–50.

Wu, K., and R. Vahldieck (1991), "Hybrid-Mode Analysis of Homogeneously and Inhomogeneously Doped Low-Loss Slow-Wave Coplanar Transmission Lines," *IEEE Trans. Microwave Theory Tech.*, vol. MTT-39, pp. 1348–1360.

Wu, K., Y. Xu, and R. G. Bosisio (1994), "Theoretical and Experimental Analysis of Channelized Coplanar Waveguides (CCPW) for Wideband Applications of Integrated Microwave and Millimeter-Wave Circuits," *IEEE Trans. Microwave Theory Tech.*, vol. MTT-42, pp. 1651–1659.

Wu, K. L., G. Y. Delisle, D. G. Fang and M. Lecours (1989), "Waveguide Discontinuity Analysis with a Coupled Finite-Boundary Element Method," *IEEE Trans. Microwave Theory Tech.*, vol. MTT-37, pp. 993–998.

Wu, L. K., and Y. C. Chang (1991), "Characterization of the Shielding Effects and the Frequency

Dependent Effective Dielectric Constant of a Waveguide-Shielded Microstrip Using the Finite Difference Time Domain Method," *IEEE Trans. Microwave Theory Tech.*, vol. MTT-39, pp. 1688–1693.

Wu, R. B., and C. H. Chen (1985a), "On the Variational Reaction Theory for Dielectric Waveguides," *IEEE Trans. Microwave Theory Tech.*, vol. MTT-33, pp. 477–483.

Wu, R. B., and C. H. Chen (1985b), "A Variational Analysis of Dielectric Waveguides by the Conformal Mapping Technique," *IEEE Trans. Microwave Theory Tech.*, vol. MTT-33, pp. 681–685.

Wu, R. B. (1996), "A Wideband Waveguide Transition Design with Modified Dielectric Transformer Using Edge-Based Tetrahedral Finite-Element Analysis," *IEEE Trans. Microwave Theory Tech.*, vol. MTT-44, pp. 1024–1031.

Wu, Z., and M. Davidowitz (1992), "Capacitance of Microstrip Lines on Anisotropic Substrate Using the Method of Lines and Finite Elements," *Int. J. Microwave Millimeter-Wave Computer-Aided Eng.*, vol. 2, pp. 40–48.

Yamashita, E. (1968), "Variational Method for the Analysis of Microstrip-Like Transmission Lines," *IEEE Trans. Microwave Theory Tech.*, vol. MTT-16, pp. 529–535.

Yamashita, E., and R. Mittra (1968), "Variational Method for the Analysis of Microstrip Lines," *IEEE Trans. Microwave Theory Tech.*, vol. MTT-16, pp. 251–256.

Yamashita, E., B. Y. Wang, K. Atsuki, and K. R. Li (1985), "Effects of Side-Wall Grooves on Transmission Characteristics of Suspended Striplines," *IEEE Trans. Microwave Theory Tech.*, vol. MTT-33, pp. 1323–1328.

Yamashita, E., M. Nakajima, and K. Atsuki (1986), "Analysis Method for Generalized Suspended Striplines," *IEEE Trans. Microwave Theory Tech.*, vol. MTT-34, pp. 1457–1463.

Yamashita, E., H. Okoshi, and K. Atsuki (1989), "Simple CAD Formulas of Edge-Compensated Microstrip Lines," *IEEE MTT-S Int. Symposium Digest*, Long Beach, CA, pp. 339–342.

Yang, J. D., S. W. Kelly, and J. D. Isles (1993), "*A Posteriori* Pointwise Upper Bound Error Estimates in the Finite Element Method," *Int. J. Numer. Meth. Eng.*, vol. 36, pp. 1279–1298.

Yang, X., T. K. Sarkar, and E. Arvas (1989), "A Survey of Conjugate Gradient Algorithms for Solution of Extreme Eigen Problems of a Symmetric Matrix," *IEEE Trans. Acoustic, Speech, Signal Processing*, vol. ASSP-37, pp. 1550–1556.

Young, T. P., and P. Smith (1987), "Finite Element Analysis of the Electro-Optic Effect in Optical Waveguides," *GEC Research Ltd.*, pp. 164–167.

Young, T. P. (1988), "Design of Integrated Optical Circuits Using Finite Elements," *Proc. IEE*, vol. 135, pt. A, pp. 135–144.

Yuan, J. S., and C. J. Fitzsimons (1993), "A Mesh Generator for Tetrahedral Elements Using Delaunay Triangulation," *IEEE Trans. Magnetics*, vol. MAG-29, pp. 1906–1909.

Yuan, X. (1990), "Three Dimensional Electromagnetic Scattering from Inhomogeneous Objects by the Hybrid Moment and Finite Element Method" *IEEE Trans. Microwave Theory Tech.*, vol. MTT-38, pp. 1053–1058.

Yuan, X., D. R. Lynch and J. W. Strobehn (1990), "Coupling of Finite Element and Moment Methods for Electromagnetic Scattering from Objects," *IEEE Trans. Antennas Propag.*, vol. AP-38, pp. 386–393.

Yuan, X., D. R. Lynch and K. Paulsen (1991), "Importance of Normal Field Continuity in Inhomogeneous Scattering Calculation," *IEEE Trans. Microwave Theory Tech.*, vol. MTT- 39, pp. 638–641.

Zeng, L. F., N. E. Wiberg, and L. Bernspång (1992), "An Adaptive Finite Element Procedure for 2D Dynamic Transient Analysis Using Direct Integration," *Int. J. Numer. Meth. Eng.*, vol. 34, pp. 997–1014.

Ženîšek, A. (1995), "The Maximum Angle Condition in the Finite Element Method for Monotone Problems with Applications in Magnetostatics," *Numer. Math.*, vol. 71, pp. 399–417.

Zhu, N. Y., and F. M. Landstorfer (1995), "An Efficient FEM Formulation for Rotationally Symmetric Coaxial Waveguides," *IEEE Trans. Microwave Theory Tech.*, vol. MTT-43, pp. 410–415.

Zienkiewicz, O. C., and Y. K. Cheung (1965), "Finite Elements in the Solution of Field Problems," *The*

Engineer, vol. 200, pp. 507–510.

Zienkiewicz, O. C., P. L. Arlett, and A. K. Bahrani (1967), "Solution of Three Dimensional Field Problems by the Finite Element Method," *The Engineer*, vol. 224, pp. 547–550.

Zienkiewicz, O. C. (1973), "Finite Elements - The Background Story," in *The Mathematics of Finite Elements and Applications*, Ed. J. R. Whiteman, Academic Press, London, pp. 1–35.

Zienkiewicz, O. C., D. Kelly, and P. Bettess (1977), "The Coupling of the Finite Element Method and Boundary Solution Procedures," *Int. J. Numer. Meth. Eng.*, vol. 11, pp. 355–375.

Zienkiewicz, O. C., C. Emson, and P. Bettess (1983), "A Novel Boundary Infinite Element," *Int. J. Numer. Meth. Eng.*, vol. 19, pp. 393–404.

Zienkiewicz, O. C., J. P. de S. R. Gago, and D. W. Kelly (1983), "The Hierarchical Concept in Finite Element Analysis," *Comp. Struct.*, vol. 16, pp. 53–65.

Zienkiewicz, O. C., K. Bando, P. Bettess, C. Emson, and T. C. Chiam (1985), "Mapped Infinite Elements for Exterior Wave Problems," *Int. J. Numer. Meth. Eng.*, vol. 21, pp. 1229–1251.

Zienkiewicz, O. C., and A. Craig (1986), "Adaptive Refinement, Error Estimates, Multigrid Solution, and Hierarchic Finite Element Method Concept," in *Accuracy Estimates and Adaptive Refinements in Finite Element Computations*, Eds. I. Babuška, O. C. Zienkiewicz, J. Gago, and E. R. de A. Oliveira, John Wiley & Sons, London, pp. 25–59.

Zienkiewicz, O. C., and J. Z. Zhu (1987), "A Simple Error Estimator and Adaptive Procedure for Practical Engineering Analysis," *Int. J. Numer. Meth. Eng.*, vol. 24, pp. 337–357.

Zienkiewicz, O. C., Y. C. Liu, and G. C. Huang (1988), "Error Estimation and Adaptivity in Flow Formulation for Forming Problems," *Int. J. Numer. Meth. Eng.*, vol. 25, pp. 23–42.

Zienkiewicz, O. C., and J. Z. Zhu (1988), "Adaptive Techniques in the Finite Element Method," *Comm. Appl. Numer. Meth.*, vol. 9, pp. 197–204.

Zienkiewicz, O. C., and J. Z. Zhu (1989), "Error Estimates and Adaptive Refinement for Plate Bending Problems," *Int. J. Numer. Meth. Eng.*, vol. 28, pp. 2839–8953.

Zienkiewicz, O. C., J. Z. Zhu, and N. G. Gong (1989), "Effective and Practical *h-p* Version Adaptive Analysis Procedures for the Finite Element Method," *Int. J. Numer. Meth. Eng.*, vol. 28, pp. 879–891.

Zienkiewicz, O. C., and J. Z. Zhu (1991), "Adaptivity and Mesh Generation," *Int. J. Numer. Meth. Eng.*, vol. 32, pp. 783–810.

Zienkiewicz, O. C., and J. Z. Zhu (1992a), "The Superconvergent Patch Recovery and *A Posteriori* Error Estimates. Part I: The Recovery Technique," *Int. J. Numer. Meth. Eng.*, vol. 33, pp. 1331–1364.

Zienkiewicz, O. C., and J. Z. Zhu (1992b), "The Superconvergent Patch Recovery and *A Posteriori* Error Estimates. Part 2: Error Estimates and Adaptivity," *Int. J. Numer. Meth. Eng.*, vol. 33, pp. 1365–1382.

Zienkiewicz, O. C., and J. Z. Zhu (1992c), "Advances in Adaptive Accuracy Control," in *Numerical Methods in Engineering and Applied Sciences*, Eds. H. Alder, J. C. Heinrich, S. Lavanchy, E. Oñate, B. Suárez, CIMNE, Barcelona, pp. 96–109.

Zienkiewicz, O. C., and J. Z. Zhu (1992d), "Mesh Regeneration and Automatic Adaptive Analysis," in *Computational Methods in Applied Sciences*, Eds. C. Hirsch, J. Périaux, and E. Oñate, Elsevier, Amsterdam, pp. 13–24.

Ziolkowski, R. W. (1997), "The Design of Maxwellian Absorbers for Numerical Boundary Conditions and for Practical Applications Using Engineered Artificial Materials," *IEEE Trans. Antennas Propag.*, vol. AP-45, pp. 656–671.

Zlámal, M. (1968), "On the Finite Element Method," *Numer. Math.*, vol. 12, pp. 394–409.

Zlámal, M. (1973), "Some Recent Advances in the Mathematics of Finite Elements," in *The Mathematics of Finite Elements and Applications*, Ed. J. R. Whiteman, Academic Press, London, pp. 59–81.

Zuffada, C., T. Cwik, and J. Jamnejad (1997), "Modeling Radiation with an Efficient Hybrid Finite Element-Integral Equation Waveguide Mode-Matching Technique," *IEEE Trans. Antennas Propag.*, vol. AP-45, pp. 34–39.

About the Authors

Magdalena Salazar-Palma was born in Granada, Spain. She received the degree of *Ingeniero de Telecomunicación* and the Ph.D. degree from the *Universidad Politécnica de Madrid* (Spain), where she is a *Profesor Titular* of the *Departamento de Señales, Sistemas y Radiocomunicaciones* (Signals, Systems and Radiocommunications Department) at the *Escuela Técnica Superior de Ingenieros de Telecomunicación* of the same university. She has taught courses on electromagnetic field theory, microwave and antenna theory, circuit networks and filter theory, analog and digital communication systems theory, numerical methods for electromagnetic field problems, as well as related laboratories.

She has developed her research within the *Grupo de Microondas y Radar* (Microwave and Radar Group) in the areas of:

- Electromagnetic field theory, and computational and numerical methods for microwave structures, passive components, and antenna analysis;
- Design, simulation, optimization, implementation, and measurements of hybrid and monolithic microwave integrated circuits;
- Network and filter theory and design.

She has authored a total of 10 contributions for chapters and articles in books published by international editorial companies, 15 papers in international journals, and 75 papers in international conferences, symposiums, and workshops, plus a number of national publications and reports. She has delivered a number of invited presentations, lectures, and seminars. She has lectured in several short courses, some of them in the frame of European Community Programs. She has participated in 19 projects and contracts, financed by international, european, and national institutions and companies. She has been a member of the Technical Programme Committee of several international symposiums and has acted as reviewer for different international scientific journals, symposiums, and editorial companies. She has assisted the *Comisión Interministerial de Ciencia y Tecnología* (National Board of Research) in the evaluation of projects. She has also served in several evaluation panels of the Commission of the European Communities. She has acted in the past and is currently acting as topical editor for the disk of references of the triennial Review of Radio Science. She is a member of the editorial board of two scientific journals. She has served as vicechairman and chairman of the IEEE MTT/AP spanish joint chapter and is currently serving as chairman of the Spain Section of IEEE. She has received two individual research awards from national institutions.

Tapan Kumar Sarkar received the B.Tech. degree from the Indian Institute of Technology, Kharagpur, India, the M.Sc.E. degree from the University of New Brunswick, Fredericton, Canada, and the M.S. and Ph.D. degrees in Electrical Engineering from Syracuse University, Syracuse, New York, in 1969, 1971, and 1975, respectively. He has also received the degree of Doctor Honoris Causa from the Université Blaise Pascal, Clermont-Ferrand, France, in 1998.

From 1975 to 1976, he was with the TACO Division of the General Instruments Corporation. From 1976 to 1985, he was with the Rochester Institute of Technology, Rochester, NY. From 1977 to 1978, he was a Research Fellow at the Gordon McKay Laboratory, Harvard University, Cambridge, MA. He is now a Professor in the Department of Electrical and Computer Engineering, Syracuse University; Syracuse, NY. He has authored and coauthored 9 books, more than 15 book chapters, more than 185 articles in scientific journals, and a large number of presentations in conferences, symposiums, and workshops. His current research interests deal with numerical solutions of operator equations arising in electromagnetics and signal processing with application to system design which includes analysis of signal integrity in high speed computer systems.

Dr. Sarkar is a registered professional engineer in the State of New York. He was an Associate Editor for feature articles of the IEEE Antennas and Propagation Society Newsletter, and he was the Technical Program Chairman for the 1988 IEEE Antennas and Propagation Society International Symposium and URSI Radio Science Meeting. He was the Chairman of the Intercommission Working Group of International URSI on Time Domain Metrology. He is a member of Sigma Xi and International Union of Radio Science Commissions A and B. He received one of the "best solution" awards in May 1977 at the Rome Air Development Center (RADC) Spectral Estimation Workshop. He received the Best Paper Award of the IEEE Transactions on Electromagnetic Compatibility in 1979 and in the 1997 National Radar Conference. Dr. Sarkar is a Fellow of the IEEE.

Luis-Emilio García-Castillo was born in 1967 in Madrid, Spain. In 1992 he received the degree of *Ingeniero de Telecomunicación* from the *Universidad Politécnica de Madrid* (Spain). In 1993 he joined as research assistant the *Departamento de Señales, Sistemas y Radiocomunicaciones* (Signals, Systems, and Radiocommunications Department) at the *Escuela Técnica Superior de Ingenieros de Telecomunicación* of the same university.Since 1997 he has been an associate professor of the *Departamento de Ingeniería Audiovisual y de Comunicaciones* (Audiovisual and Communications Engineering Department) at the *Escuela Univesitaria de Ingeniería Técnica de Telecomunicación* of the same university where he teaches microwave theory and the related laboratory.

His research activity and interests are focused in the application of numerical methods, mainly finite elements, to electromagnetic problems, including the research on curl-conforming finite elements, the characterization of multiconductor and waveguiding structures, the analysis of scattering and radiation problems, and the use of wavelet theory in computational electromagnetics. Other research areas of his interest are network theory and filter design.

He has authored 4 contributions for chapters and articles in books published by international editorial companies, 6 papers in international journals, and 29 papers in international conferences, symposiums, and workshops, plus a number of national publications and reports. He has participated in 7 projects and contracts, financed by international, european, and national institutions and companies.

Tanmoy Roy received his B.Tech. degree from the Indian Institute of Technology, Kharagpur, India, in 1990, the M.S. and Ph.D. degrees on Electrical Engineering from Syracuse University in 1995 and 1996, respectively. From 1990 to 1991 he was a software consultant for Tata Consultancy Services, India. He was a summer intern at Schlumberger-Doll Research in Ridgefield, CT, in 1994 and 1995. Currently, he is with Sun Microsystems. His research interests deal with finite element techniques.

Antonije Djordjević was born in Belgrade, Yugoslavia, on April 28, 1952. He received the B.Sc., M.Sc., and D.Sc. degrees from the School of Electrical Engineering, University of Belgrade, in 1975, 1977, and 1979, respectively.

In 1975 he joined the School of Electrical Engineering, University of Belgrade, as a teaching assistant. In 1982 he was promoted to an assistant professor, in 1988 to an associate professor, and since 1992 he has been a professor at the same school. In 1983 he was a visiting associate professor at Rochester Institute of Technology, NY. Since 1992, he also has been an Adjunct Associate Professor with Syracuse University, NY. His main field of interest is numerical electromagnetics, in particular applied to multiconductor transmission lines, wire and surface antennas, and electromagnetic-compatibility problems. He is an corresponding member of the Serbian Academy of Sciences and Arts.

Index

The Artech House Antenna Library

Helmut E. Schrank, *Series Editor*

Advanced Technology in Satellite Communication Antennas: Electrical and Mechanical Design, Takashi Kitsuregawa

Advances in Computational Electrodynamics: The Finite-Difference Time-Domain Method, Allen Taflove, *editor*

Analysis Methods for Electromagnetic Wave Problems, Volume 2, Eikichi Yamashita, *editor*

Analysis of Wire Antennas and Scatterers: Software and User's Manual, A. R. Djordjević, M. B. Bazdar, G. M. Bazdar, G. M. Vitosevic, T. K. Sarkar, and R. F. Harrington

Antenna-Based Signal Processing Techniques for Radar Systems, Alfonso Farina

Antenna Engineering Using Physical Optics: Practical CAD Techniques and Software, Leo Diaz and Thomas Milligan

Antenna Design With Fiber Optics, A. Kumar

Broadband Patch Antennas, Jean-François Zürcher and Fred E. Gardiol

CAD for Linear and Planar Antenna Arrays of Various Radiating Elements: Software and User's Manual, Miodrag Mikavica and Aleksandar Nešić

CAD of Aperture-fed Microstrip Transmission Lines and Antennas: Software and User's Manual, Naftali Herscovici

CAD of Microstrip Antennas for Wireless Applications, Robert A. Sainati

The CG-FFT Method: Application of Signal Processing Techniques to Electromagnetics, Manuel F. Cátedra, Rafael P. Torres, José Basterrechea, and Emilio Gago

Electromagnetic Waves in Chiral and Bi-Isotropic Media, I.V. Lindell, S.A. Tretyakov, A.H. Sihvola, A. J. Viitanen

Fixed and Mobile Terminal Antennas, A. Kumar

Four-Armed Spiral Antennas, Robert G. Corzine and Joseph A. Mosko

Handbook of Antennas for EMC, Thereza Macnamara

Introduction to Electromagnetic Wave Propagation, Paul Rohan

Iterative and Self-Adaptive Finite-Elements in Electromagnetic Modeling, Magdalena Salazar-Palma, et al.

Mobile Antenna Systems, K. Fujimoto and J. R. James

Modern Methods of Reflector Antenna Analysis and Design, Craig Scott

Moment Methods in Antennas and Scattering, Robert C. Hansen, *editor*

Monopole Elements on Circular Ground Planes, M. M. Weiner et al.

Near-Field Antenna Measurements, D. Slater

Passive Optical Components for Optical Fiber Transmission, Norio Kashima

Phased Array Antenna Handbook, Robert J. Mailloux

Polariztion in Electromagnetic Systems, Warren Stutzman

Practical Simulation of Radar Antennas and Radomes, Herbert L. Hirsch and Douglas C. Grove

Quick Finite Elements for Elctromagnetic Waves, Giuseppe Pelosi, Roberto Coccioli, and Stefano Selleri

Radiowave Propagation and Antennas for Personal Communications, Second Edition, Kazimierz Siwiak

Small-Aperture Radio Direction-Finding, Herndon Jenkins

Solid Dielectric Horn Antennas, Carlos Salema, Carlos Fernandes, and Rama Kant Jha

Spectral Domain Method in Electromagnetics, Craig Scott

Understanding Electromagnetic Scattering Using the Moment Method: A Practical Approach, Randy Bancroft

Waveguide Components for Antenna Feed Systems: Theory and CAD, J. Uher, J. Bornemann, and Uwe Rosenberg

For further information on these and other Artech House titles, including previously considered out-of-print books now available through our In-Print-Forever™ (IPF™) program, contact:

Artech House
685 Canton Street
Norwood, MA 02062
781-769-9750
Fax: 781-769-6334
Telex: 951-659
e-mail: artech@artech-house.com

Artech House
Portland House - Stag Place
London SW1E 5XA England
+44 (0) 171-973-8077
Fax: +44 (0) 171-630-0166
Telex: 951-659
e-mail: artech-uk@artech-house.com

Find us on the World Wide Web at:
www.artech-house.com